Climate Change and Multi-Dimensional Sustainability in African Agriculture

Rattan Lal · David Kraybill
David O. Hansen · Bal Ram Singh
Theodosy Mosogoya · Lars Olav Eik
Editors

Climate Change and Multi-Dimensional Sustainability in African Agriculture

Climate Change and Sustainability in Agriculture

Springer

Editors
Rattan Lal
SENR-CMASC
The Ohio State University
Columbus, OH
USA

David Kraybill
AEDE/CFAES
The Ohio State University
Columbus, OH
USA

David O. Hansen
College of Food, Agricultural, and Environmental Sciences
The Ohio State University
Columbus, OH
USA

Bal Ram Singh
Norwegian University of Life Sciences
Ås
Norway

Theodosy Mosogoya
Sokoine University of Agriculture
Morogoro
Tanzania

Lars Olav Eik
Norwegian University of Life Sciences
Ås
Norway

ISBN 978-3-319-41236-8 ISBN 978-3-319-41238-2 (eBook)
DOI 10.1007/978-3-319-41238-2

Library of Congress Control Number: 2016949560

Printed on acid-free paper

This Springer imprint is published by Springer Nature
The registered company is Springer International Publishing AG
The registered company address is: Gewerbestrasse 11, 6330 Cham, Switzerland

Foreword 1

USAID/Tanzania Efforts to Promote Economic, Environmental, Social, and Institutional Sustainability in Tanzania in the Context of Climate Change: Priorities and Investments

"Sustainability" is a critical issue that should be at the heart of all interventions to improve agricultural production, food access, and agriculture-based livelihoods around the world, especially in Tanzania. This focus becomes even more critical when considering the context of climate change and extreme weather events associated with it.

It is important to think about what "sustainability" means, in terms of the four key dimensions discussed in the context of climate change, namely environmental, economic, social, and institutional. I will briefly touch on each category and how USAID/Tanzania's programs under Feed the Future (FTF) are seeking to address these four aspects of the sustainability issue.

As the donor partner that is supporting the iAGRI project's partnership with Sokoine University of Agriculture (SUA) to strengthen the institutional sustainability, USAID is rightfully proud of the many creative initiatives that result from this partnership, including diversifying SUA's financial support, for example, by broadening the university's mandate to undertake practical research that meets the needs of Tanzania's private sector; fostering leadership at multiple levels of the university; improving the quality and environment for teaching, including upgrading Sokoine University's IT systems through direct funding from USAID to SUA; and, of course, training the next generation of Tanzanian agricultural scientists to identify real and applied solutions to the problems of agricultural sustainability in Tanzania.

The iAGRI project has been central in organizing and implementing two recent climate change conferences at the Sokoine University of Agriculture. However, the project is just one of the many activities in which USAID is investing in Tanzania under the Feed the Future program. From 2010 to 2015, USAID invested US $350M in Tanzania through the Feed the Future program, the goal of which is to "sustainably reduce poverty and hunger."

Briefly, the USAID/Tanzania Feed the Future program has a value-chain focus—targeting increased productivity, input supply, and market access for three commodities: rice, maize, and horticulture. The approach is to work with farmer associations to promote good agricultural practices, using sustainable intensification approaches, such as the System of Rice Intensification (SRI), low-till and labor-saving technologies, maize–legume intercropping, and drip irrigation and soil management for horticulture. USAID is helping to foster private sector approaches to input supply and extension services through village-based agents and organizing agricultural marketing cooperatives to enable farmers to get the best prices for their inputs and production, while pooling resources for post-harvest storage. Investments are being made in rehabilitation of farm-to-market roads that are helping to increase competition among grain traders, thus bringing higher wholesale prices at the farm gate while lowering food prices for urban consumers due to lower transport costs. On the agriculture policy front, the focus is on commodity trade issues, agricultural taxation, and land tenure—all policy priorities identified by the Government of Tanzania. Underlying all of these efforts is a focus on *nutrition*. In addition to reducing poverty rates by 20 % in target areas of program investment, Feed the Future Tanzania is targeting a 20 % reduction in childhood stunting rates (which range from 35 % to more than 50 % of the under-five population in various regions of Tanzania) and a 20 % reduction in maternal anemia. We are addressing these nutrition targets by working with millers on cereal fortification and working at the household level to improve awareness of good nutrition and access to nutritious foods for mothers and their children, especially during the child's first 1000 days. Of the 13 high-level Feed the Future impact targets in Tanzania, 10 indicators are focused on nutrition. Nutrition is our investment in the future productivity and sustainable development of Tanzania's human capital.

For many years, USAID did not invest in Tanzania's agriculture sector; as recently as 2009, USAID funding for agriculture programs in Tanzania was around $2M. By 2012, this had risen to $77M/year, including funding for nutrition, a core element of FTF. Essentially overnight, Tanzania became the recipient of one of the largest USAID agriculture budgets in the world. Tanzania was seen as a "new frontier" for the administration's signature Feed the Future program: a country with abundant land and water resources, untapped potential to dramatically improve agriculture-led growth, and a government committed to prioritizing its agricultural economy.

Yet one of the first issues that USAID and the Government of Tanzania confronted when we launched our agricultural investments was the question of natural resource abundance—an issue at the heart of environmental sustainability. Tanzania is blessed with abundant freshwater resources on its border and in its rivers, but access to, and distribution of, water for multiple needs remains a challenge.

Currently, irrigated agriculture accounts for approximately 70 % of water use[1] in the country, while according to the 2002 Tanzania National Irrigation Master Plan, the country has reached only 2 % of its irrigation potential. How then to accommodate for water for Tanzania's future growing population? How to meet the growing need for water to supply households, and industrial and energy needs, while still ensuring adequate environmental flows for the survival of rivers, water bodies, and the species that depend upon them?

Abundance of land is another issue—abundance, but for what purposes? With its wealth of biodiversity, Tanzania has committed more of its land to conservation than perhaps any other country in the world—27 % of Tanzania's land area is under some form of protected status. The remainder—more than 70 %—is under village land tenure, and 20 % is government land. The challenge for social and economic sustainability is to provide sufficient land for those who want to cultivate crops or graze livestock, while providing adequate land for forests, fallow, and soil regeneration. Already there is intense competition for land in many parts of the country—with growing competition between farmers and livestock keepers, large investors, and small farmers. The challenge for social and economic sustainability is to develop land tenure policies that provide security of access for small farmers, including women, youth, herders, and investors at the same time.

Tanzania is promoting outgrower models with large-scale agricultural investors. This is a model that, if undertaken responsibly, can provide smallholders with access to improved inputs, post-harvest storage, and a ready and easily accessible market. However, for these models to be sustainable and beneficial to Tanzania's small farmers, it will be necessary for farmer organizations, civil society, and government actors to be actively and positively engaged as regulators to ensure that the economic relations between large investors and outgrowers do not become exploitative.

Finally, I'd like to focus a bit on economic sustainability and specifically the balance that Tanzanian policymakers must find between supporting fair prices for agricultural producers, while keeping urban food supplies and prices affordable for their populations. At USAID, we are working closely with the Government of Tanzania to improve their data on food production and prices, to avoid the temptation to allow imports of duty-free commodities such as rice and sugar—the very products their farmers are working hard to produce—just at the point in the season when market prices are highest. Last year, despite bumper rice harvests, duties were waived on the import of foreign rice, out of fear that staple commodity prices had risen too high for the urban consumer. The negative impact on farm gate prices for rice was immediate and devastating for producers. Similarly, export permits for maize restricted the number of traders and, thus, access to important regional

[1]Government of Tanzania, "Tanzania Integrated Water Resource Management and Development Plans," 2012: Of 8282 million cubic meters (MCM) of water available in Tanzania, irrigation accounted for 5796 MCM in 2012.

markets for maize producers, just at a time when they had surplus maize while other countries in the region were facing deficits.

To ensure economic sustainability of Tanzania's agriculture sector, USAID is supporting commodity associations to develop their capability to inform and influence agricultural trade and taxation policies. Additionally, USAID is supporting the Tanzanian Ministry of Agriculture, Food Security and Cooperatives to access reliable production and market data to ensure evidence-based agriculture policy decisions are made in the future. Without a supportive policy environment, the economic sustainability of Tanzania's agriculture sector will be compromised.

This first International Conference on Climate Change and Multi-Dimensional Sustainability in African Agriculture, held here in Morogoro, Tanzania, and hosted by Sokoine University of Agriculture, promises to be an important milestone in engaging representatives from academia, government, and civil society to share information with one another on important new developments in sustainable environmental, social, economic, and institutional approaches for African agriculture. There are few more important topics for the future food security and prosperity of this continent.

Mary Hobbs, Ph.D.
Economic Growth Office Director
USAID/Tanzania

Foreword 2

The Ohio State University and Its Commitment to Global Food Security

The Ohio State University (OSU) has long been committed to global food production and sustainability, particularly through building institutional and country educational capacity in developing regions of the world. These efforts date back to the 1950/1960s era and have continued to the current day. The USAID has been a key partner in many of these initiatives and continues that partnership through its investment in iAGRI, which OSU leads and manages in Tanzania. As with most significant efforts that are able to move the needle on our knowledge base, a consortium of capacity builders are committed. As part of that base with the UN/FAO, NORAD and the Norwegian University of Life Sciences all contributed to the conference and its significant theme. I was impressed by the contributions and involvement of our special host, the Sokoine University of Agriculture, who made a memorable conference a memorable experience.

As I reflect on the topic of this conference and the unique blend of talent assembled to share expertise and perspectives across multiple disciplines, I found myself reflecting on the structure and function of my own institution. The OSU is called a land-grant university (LGU), one of which was established in each state of the USA through the Morrill Act of 1862. This act, while initiating a network of institutions with significant colleges of agriculture, was augmented by the passage of the Hatch Act in 1887. The Hatch Act created a network of agricultural experiment stations to generate research findings to complement teaching efforts and fill knowledge gaps. In 1914, the Smith/Lever Act created the Cooperative Extension System to deliver the knowledge discovered through research to society, essentially creating a lifelong learning network. This knowledge triangle has been critical not only to turn the academic wheel at our universities, but to also connect our communities and regions in a way that has been critical to the prosperity of our states.

Ohio is comparatively small in area compared to other USA states, ranking 38 out of 50 in size. However, it is seventh in population (11.5 million people) and has multiple urban centers linked together by a rural landscape as an interface. This interface is comprised of some 78,000 farms covering about 50 million land acres. The Ohio agriculture sector annually generates over $100 billion US in economic value to the state. It is the state's premier economic sector and generates one in every seven jobs. Ohio is strategically located south of Lake Erie (Great Lakes) and west and north of the Ohio River, a major feeder into the Mississippi River. It is blessed with rich soils and a favorable agricultural climate. This productive agricultural region has continued to evolve and prioritize its goals to match societal needs and changes. The OSU has been an important contributor to this process. Function is impacted by priorities; however, the integrated teaching–research–extension structure in our LGU's has provided an integrated system that continues to be essential.

In the past decade our College of Food, Agricultural, and Environmental Sciences strategically identified three major themes that crosscut our academic departmental structure: "Food Production, Security and Human Health," "Environmental Quality and Sustainability" and "Advanced Bioenergy and Biobased Products." We conducted an economic analysis that demonstrated the value or return on investment for our university, state, and federal partners. It demonstrated that the economic value of the research investment exceeded investments tenfold. Currently, OSU has embraced these same themes in an institutional framework called Discovery Themes in order to integrate outstanding talent that exists across its 14 colleges and to build strength-on-strength with new investments, which will ensure continued commitment to and leadership in these strategic areas in the future. This approach is important both to make an impact on society nationally and internationally and to accrue the resources critical to success. Resources come from multiple public and private donors and are a mix of competitive and capacity grants, contracts, and gifts. The diversity of resource type and donor is key to maintain and growing a resilient system that reflects society.

It was a pleasure and an honor to participate in this conference and to represent OSU. The quality of presentations and the broad scope of presentations, ranging from agricultural to environmental to economic to social sustainability were impressive and the dialogue even richer. My congratulations to all involved!

Steven A. Slack
Associate Vice President for Agricultural Administration and
Director, Agricultural Research and Development Center
College of Food, Agricultural and Environmental Sciences
The Ohio State University

Foreword 3

Multi-dimensional Sustainability and Climate Change in African Agriculture

The Ohio State University was pleased to support the conference on climate change which produced the papers found in this volume. Major support for it was provided by the Innovative Agricultural Research Initiative (iAGRI) which is centered at Sokoine University of Agriculture (SUA). We are grateful to the USAID Mission in Tanzania for the financial support which it has provided to iAGRI. Collaborative research on food security topics is a major dimension of iAGRI and topic of great relevance to Tanzania and the region. We are especially grateful to our partners for their important contributions to the conference. They include SUA which provided the venue for it and the enthusiastic and competent participation of its staff and scientists.

I was very gratified that the conference organizers were able to bring together scientists from our US Consortium of universities, SUA, and the Ministry of Agriculture, Livestock and Fisheries (MALF), and from our Norwegian partner, the Norwegian University of Life Sciences. Our institutions have maintained important programs with SUA, and it is appropriate that we find common venues in which to collaborate. I would also like to recognize and thank the Food and Agricultural Organization for its important contributions to the conference as well as the support provided by international research centers located in the region, in particular Africa RISING and scientists from the International Institute of Tropical Agriculture and the World Agroforestry Center.

Sustainability in the context of climate change is a particularly relevant topic for Tanzania and other sub-Saharan Africa nations. Identification of sustainable agricultural systems requires that their various dimensions be considered. They need to be addressed within the production context that currently prevails where the large majority of rural inhabitants practice low-input and low-output agriculture on a semi-subsistence basis. A major challenge will be to identify sustainably more productive and more profitable systems for them which will facilitate adaptation and build resilience within the context of climate change.

Mounting evidence, including empirical evidence found in some of the papers included in this book, suggests that climate change is already having an important impact on agriculture systems and rural communities in sub-Saharan Africa. A major contribution of this volume is its consideration of how to mitigate some of these impacts and how to ameliorate their effects on the lives and livelihoods of rural inhabitants, particularly those who depend on agriculture for their subsistence. The conference was organized specifically around the impact of climate change on the environmental, economic, social and institutional dimensions of rural life and agriculture. It identifies adaptive strategies and highlights how rural communities, rural families and farmers are adjusting to the impacts of climate change in order to sustain their social institutions, way of life, and the land and water resourceson which agriculture depends.

As the director of the OSU Office of International Programs in Agriculture, I wish to acknowledge that the conference, which produced the papers in this volume, represents a continuation of collaborative activities between OSU and SUA, most of which have been in the area of agriculture and rural development. It is a particularly important for us given that we highly value our partnership with SUA. For us, SUA represents an important window on issues of global food security and climate change. It has provided us with opportunities to engage our academicians and researchers in the study of these issues and how they can better be addressed in sub-Saharan Africa.

Columbus, Ohio
March 2016

Mark Erbaugh

Preface

Food and nutritional insecurity have been major issues in sub-Saharan Africa (SSA) since the 1970s. One in four inhabitants of SSA (240 million) is vulnerable to food insecurity, and 79 million of them are also undernourished. The proportion of undernourishment has decreased from 33 % in 1990–1992 to 23 % in 2014–2016. While crop yields and agricultural production have marked an upward trend in some countries, major challenges lie ahead. Population of 800 million in 2000 and 962.3 million in 2015 is projected to be 1.1 billion by 2020, 1.8 by 2050, and 2.3 billion by 2100 and stabilized at $\sim$2.1 billion in 2150. Hot spots of hunger and malnutrition of SSA are Sahel, from Senegal to Chad, and the Horn of Africa. Food and nutritional insecurity are aggravated by civil strife, political instability, soil degradation, and harsh and uncertain climate. The GDP of SSA at the current market prices is $1729 trillion, and GNI is $1638 trillion by the Atlas method. The life expectancy at birth is 58 years, and the primary education completion rate (for both sexes) is 69 %. In 2014, 37 % of the population lived in urban centers, and per capita CO_2 emission is 0.8 Mg per year compared with the world average of 4 Mg CO_2 per year.

Despite numerous debates about sustainability, there is a strong need of critically re-examining the basic concept and taking a multi-dimensional approach with specific attention to environmental, economic, soil, and institutional sustainability. The involvement of private sector, providing a credit system and market-driven programs, is critical to the success. In addition to sustainability, there is also a strong need to consider other concepts, such as resilience and stewardship.

There is a strong link between hunger, poverty, and substance agriculture. The poverty rate (population living on less than $1.25 per day) has decreased from 57 % in 1990 to 41 % in 2015, but the challenge remains to be effectively addressed trough improvements in agriculture. While the proportion has decreased, the number of poor, hungry, and malnourished population has increased in absolute terms between 1990 and 2016. For example, the number of undernourished people has increased by 44 million since 1990.

Environmental sustainability is intricately linked with agricultural sustainability through deforestation, soil degradation, water contamination and eutrophication, decline in biodiversity, and the increase in emission of greenhouse gases, soot, dust, and other air pollutants. The problem of water scarcity may be exacerbated with the projected climate change and the related uncertainty. Clean potable water is neither available nor easily accessible to a large segment of both rural and urban population. Women, young girls, and boys have to walk long distances or join long queues waiting for the community water taps to open. Providing improved sanitation facilities to rural and slum dweller urban population is a high priority.

Economy of SSA grew at an impressive rate of 4.5 % during 2015. However, agricultural productivity has stagnated in several regions closely linked to low agricultural productivity, and land and environmental degradation pose strong risks to political instability and domestic insecurity. Thus, improving agricultural productivity as an engine of economic development remains to be a high priority. Growth in agricultural sector, through judicious management of soil and water resources, is also important to achieve the Sustainable Development Goals (SDGs) of the United Nations.

That said, it is important to note that several regions of SSA have registered impressively positive trends in economic growth, agricultural production, poverty alleviation, primary education, and availability/access to basic amenities of life. The momentum generated thus far must be sustained in all relevant sectors including the environment, economic, institutional, and social and political arenas. Yet, sustaining the momentum will become more difficult in the future as population grows, climate warms, water resources dwindle and pollute, soils erode and become salinized, pests and pathogens become more pervasive, and weather patterns change and become uncertain. Sustainability will become a bigger challenge than ever before.

Thus, an international conference was organized at the Sokoine University of Agriculture, Morogoro, Tanzania, from June 1, 2015, to June 6, 2015. Major objectives of the conference were to deliberate the importance of sustainability in the context of environmental, economic, institutional, political, and soil issues to advancing SDGs, improving nature conservancy, and restoring land and water resources.

This 35-chapter volume represents core of several oral and poster presentations made at the conference. In addition to the Introduction and Conclusion chapters, the book is divided into 8 sections, namely (1) Environmental Sustainability, (2) Economic Sustainability, (3) Institutional Sustainability, (4) Social and Political Sustainability, (5) Technological Innovations, (6) Landscape Restoration and Management, (7) Integration with the Private Sector, and (8) Challenges to Implementations of SDGs of the U.N.

The conference was attended by more than 100 participants from SSA countries as well as the USA, Norway, and Italy. It was organized by a Steering Committee with representatives from SUA, the Ohio State University, and the Norwegian University of Animal and Life Sciences, and Food and Agriculture Organization (FAO) of the U.N. The conference was funded by NORAD, USAID, and SUA. Primary funding for the conference was channeled through several programs at SUA, namely the USAID-funded International Agricultural Research Initiative (iAGRI), the Climate Change Impacts, Adaptation and Mitigation (CCIAM) project, and the Enhancing Pro-poor Innovations in Natural Resources and Agricultural Value-Chains (EPINAV) project. In addition, the conference benefitted from contributions from Africa Rising, the International Food Policy Research Institute (IFPRI), the World Agroforestry Center (ICRAF), and the Carbon Management and Sequestration Center (C-MASC).

The editors thank all authors for their outstanding contributions to this volume. Thanks are also due to staff at Springer for their timely efforts in publishing this volume. Our special thanks are due to Laura Alexander (iAGRI), Laura Hughes (C-MASC), Anthony Sangeda (iAGRI), and Ambonisye (iAGRI).

Columbus, OH, USA — Rattan Lal
Columbus, OH, USA — David Kraybill
Columbus, OH, USA — David O. Hansen
Ås, Norway — Bal Ram Singh
Morogoro, Tanzania — Theodosy Mosogoya
Ås, Norway — Lars Olav Eik
March 2016

Contents

Part I Multi-dimensional Sustainability

1 **Environmental Sustainability** 3
Rattan Lal

2 **Economic Sustainability** 13
David Kraybill

3 **Institutional Sustainability in the Face of Climate Change: Empirical Insights from Irrigation Institutions in the Iringa Rural District, Tanzania** 23
George C. Kajembe, Pål Vedeld, Innocent H. Babili, Dos Santos Silayo and Devotha B. Mosha

4 **Climate Change and Social Sustainability: A Case for Polycentric Sustainabilities** 43
Louise Fortmann, Abraham Ndhlovu, Alice Ndlovu, Rosina Philippe, Crystlyn Rodrigue and Ken Wilson

Part II Sustainable Management of Natural Resources

5 **Land Degradation Neutrality: Will Africa Achieve It? Institutional Solutions to Land Degradation and Restoration in Africa** 61
Luc Gnacadja and Liesl Wiese

6 **Extent of Salt-Affected Soils and Their Effects in Irrigated and Lowland Rain-Fed Rice Growing Areas of Southwestern Tanzania** 97
S. Kashenge-Killenga, J. Meliyo, G. Urassa and V. Kongo

7 **Sustainability of Intensification of Smallholder Maize Production in Tanzania** 127
Frank Brentrup, Joachim Lammel, Katharina Plassmann and Dirk Schroeder

8 **Potentials for Rehabilitating Degraded Land in Tanzania** 149
G.Z. Nyamoga, H.K. Sjølie, R. Malimbwi, Y.M. Ngaga and B. Solberg

9 **Economic Aspects of Genetic Resources in Addressing Agricultural Productivity in the Context of Climate Change** 171
Charles Nhemachena, Greenwell Matchaya, Sibusiso Nhlengethwa and Charity R. Nhemachena

10 **Soil and Nutrient Losses and the Role of Gender in Land Degradation in Southwestern Uganda** 185
C.L. Kizza, J.G.M. Majaliwa, R. Kunihira, G. Gabiri, A. Zizinga, E. Sebuliba, J. Nampijja and M.M. Tenywa

11 **The Social Dimension of Water Management in an Era of Increasing Water Scarcity in Tanzania** 201
Zebedayo S.K. Mvena

Part III Integrated Management of Natural Resources and Value Chains

12 **Managing Landscapes for Environmental Sustainability** 215
Rattan Lal

13 **Economic Impact of Drip Irrigation Regimes on Sorghum Production in Semi-arid Areas of Tanzania** 227
A.J. Mahinda, C.K.K. Gachene and M. Kilasara

14 **Social Aspects of Water Governance in the Context of Climate Change and Agriculture** 241
Richard Asaba Bagonza

15 **Assessment of Climate Change Impact on Common Bean (*Phaseolus Vulgaris* Savi, L.) Production in Tanzania** 259
Sixbert Kajumula Mourice, Siza Donald Tumbo and Cornell Lawrence Rweyemamu

16 **Institutional Water Resources Management and Livelihood Adaptation: A Case from Kilombero Rural Areas, Tanzania** 281
Paul Vedeld, Edgar Liheluka and Gimbage E. Mbeyale

17 **Institutional Aspects of Genetic Resources in Respect to Climate Change in Sub-Saharan Africa** 307
Denis T. Kyetere and Kayode Abiola Sanni

18 Crop Adaptation to Climate Change in SSA: The Role of Genetic Resources and Seed Systems 327
Ola T. Westengen and Trygve Berg

19 Updating Legacy Soil Maps for Climate Resilient Agriculture: A Case of Kilombero Valley, Tanzania 345
Boniface H.J. Massawe, Brian K. Slater, Sakthi K. Subburayalu, Abel K. Kaaya and Leigh Winowiecki

20 Measuring Agricultural Sustainability in Agroforestry Systems . . . 365
P.K. Ramachandran Nair and Gregory G. Toth

21 Experiences with Adopting the Catchment Approach in Sustainable Land Management: The Case of Kagera TAMP Tanzania 395
Fidelis Kaihura and Stefan Schlingloff

22 The Role of Forages in Sustainable Intensification of Crop-Livestock Agro-ecosystems in the Face of Climate Change: The Case for Landscapes in Babati, Northern Tanzania 411
F. Kizito, B. Lukuyu, G. Sikumba, J. Kihara, M. Bekunda, D. Bossio, K.W. Nganga, A. Kimaro, H. Sseguya, B. Jumbo and P. Okori

23 Smart Strategies for Enhanced Agricultural Resilience and Food Security Under a Changing Climate in Sub-Saharan Africa 431
Kennedy Were, Aweke Mulualem Gelaw and Bal Ram Singh

Part IV Outreach and Extension

24 Globalizing Environmental Sustainability: "2015 International Year of Soil" Transitioning to "2015–2024 International Decade of Soil" 457
Rattan Lal

25 From Ujamaa to Big Results Now: Sustainable Transformation of Tanzanian Agriculture in the Frame of Climate Change 467
Ruth Haug

Part V Private Sector Solutions

26 Effect of Improved Plant Nutrition on Maize (*Zea mays*) and Rice (*Oriza sativa*) Grain Chemical Nutrient Content Under Smallholder Farming Systems in Tanzania 485
Ephraim J. Mtengeti, Eva Mtengeti, Frank Brentrup, Lars Olav Eik and Ramadhani Chambuya

27 **Public-Private Partnership for Sustainable Production and Marketing of Goat's Milk in Light of Climate Change** 505
G. Msalya, H. Lie, V. Mfinanga, A.S. Ringheim, R. Sandvik, M. Åsli, O.A. Christophersen, A. Haug, D.E. Mushi, D. Mwaseba, G.C. Kifaro and Lars Olav Eik

28 **Private Sector Actions to Enable Climate-Smart Agriculture in Small-Scale Farming in Tanzania** 525
Sheryl Quail, Leah Onyango, John Recha and James Kinyangi

29 **The Role of Selected Agroforestry Trees in Temperature Adaptation on *Coffea arabica*: A Case Study of the Moshi District, Tanzania** 553
Jacqueline Kajembe, Ignas Lupala, George Kajembe, Wilson Mugasha and Faraji Nuru

Part VI Agricultural Risk, Insurance and Policy

30 **Addressing Climate Change Through Risk Mitigation: Welfare Implications of Index Insurance in Northeastern Tanzania** 569
Jon Einar Flatnes and Michael R. Carter

31 **Assessing the Economic Value of *El Niño*-Based Seasonal Climate Forecasts for Smallholder Farmers in Zimbabwe** 591
Ephias M. Makaudze

Part VII Agricultural Research for Sustainability

32 **Kinds of Research: Relationship with Agricultural Research for Sustainability** 615
Isaac J. Minde and Stephen A. Nyaki

33 **Biological Inoculants for Sustainable Intensification of Agriculture in Sub-Saharan Africa Smallholder Farming Systems** 639
C. Masso, R.W. Mukhongo, M. Thuita, R. Abaidoo, J. Ulzen, G. Kariuki and M. Kalumuna

34 **The Economics of Conservation Agriculture in Africa: Implications of Climate Change** 659
Philip Grabowski and Steven Haggblade

35 **Research and Development Priorities** 679
Rattan Lal, David Kraybill, David O. Hansen, Bal Ram Singh and Lars Olav Eik

International Conference on Climate Change and Multi-dimensional Sustainability in African Agriculture: June 3–5, 2015 695

Index 715

About the Editors

Rattan Lal, Ph.D. is a distinguished university professor of Soil Science and Director of the Carbon Management and Sequestration Center, The Ohio State University, and an Adjunct Professor of University of Iceland. His current research focus is on climate-resilient agriculture, soil carbon sequestration, sustainable intensification, enhancing use efficiency of agroecosystems, and sustainable management of soil resources of the tropics. He received honorary degree of Doctor of Science from Punjab Agricultural University (2001), the Norwegian University of Life Sciences, Aas (2005), and Alecu Russo Balti State University, Moldova (2010). He was the president of the World Association of the Soil and Water Conservation (1987–1990), the International Soil Tillage Research Organization (1988–91), and the Soil Science Society of America (2005–2007), and is the president-elect of International Union of Soil Science. He was a member of the Federal Advisory Committee on US National Assessment of Climate Change-NCADAC (2010–2013), member of the SERDP Scientific Advisory Board of the US-DOE (2011–), senior science advisor to the Global Soil Forum of Institute for Advanced Sustainability Studies, Potsdam, Germany (2010–), member of the Advisory Board of Joint Program Initiative of Agriculture, Food Security and Climate Change (FACCE-JPI) of the European Union (2013–), and chair of the Advisory Board of Institute for Integrated Management of Material Fluxes and Resources of the United Nation University (UNU-FLORES), Dresden, Germany (2014–2017). Professor Lal was a lead author of IPCC (1998–2000). He has mentored 102 graduate students and 54 postdoctoral researchers, and hosted 140 visiting scholars. He has authored/coauthored 780 refereed journal articles and has written 19 and edited/coedited 63 books. In 2015, Reuter Thomson listed him among the world's most influential scientific minds and having citations of publications among top 1 % of scientists in agricultural sciences.

Dr. David Kraybill currently serves as project director for the USAID-funded Innovative Agricultural Research Project in Tanzania. His primary research interest is economic development. His work focuses on regional and spatial development processes that interact with markets and national government policies to determine the economic and social well-being of individuals, households, and communities. Dr. Kraybill's recent research includes studies of household poverty, household savings, governmental decentralization, primary and secondary education, and adaptation to climate change. Fluent in French and a Swahili speaker, he has lived in Africa a total of five years, including a sabbatical year as Fulbright scholar at Makerere University in Uganda. Kraybill has served as an associate editor or member of editorial boards of seven academic journals. He has received research and advising awards at Ohio State and received the International Award of Merit from the Gamma Sigma Delta honor society. He has been consultant advisor to World Bank, Rockefeller Foundation, Southeast Consortium for International Development, US Department of Agriculture, and numerous governments and other organizations. He came to Ohio State in 1992.

David O. Hansen is currently employed by The Ohio State University to oversee the International Agricultural Research Initiative (iAGRI) a major USAID-funded higher education partnership project in Tanzania. Prior to assuming his current responsibilities, he was a senior fellow with the Association for Public and Land-grant Universities (APLU), Washington, DC. His responsibilities there included the Africa-US Higher Education Initiative which promoted partnerships among US and African Higher Education Institutions to build capacity in Africa. Many of them focused on agriculture, climate change, and related areas, such as water, range management, and ecosystems services. He collaborates with USAID, World Bank, USDA, land-grant universities, and NGO's to help define the parameters of higher education capacity building within the context of the US Government Feed the Future Initiative. He has dedicated his career to international development while at Ohio State University, having previously served as an associate dean and director of International Programs in Agriculture for over two decades. In this capacity, he helped design and manage major institution building projects in Uganda, India, Swaziland, Mexico, and the Dominican Republic. Early in his career he worked on long-term overseas assignments in Brazil where he served as Deputy Chief-of-Party and Chief-of-Party for institution building projects.

Bal Ram Singh is a professor in the Department of Environmental Sciences at the Norwegian University of Life Sciences (UMB). He earned his M.Sc. degree from the Indian Agricultural Research Institute, New Delhi, and his Ph.D. degree from G.B. Pant University of Agriculture and Technology, Pantnagar, India. His program focuses on bioavailability and mobility of heavy metals in the soil and plant system, fertility management, and agricultural sustainability in soils of the tropics and on carbon sequestration in soils under different land uses. He has served as chair of the program board "Soils and Plants" of the Research Council of Norway and the Research and Higher Education Committee in the department. He was chair of the Cost Action FA0905 on "**Mineral Improved Cop Production for Healthy Food and Feed**." Dr. Singh is a fellow of ASA and SSSA and a recipient of International SSSA award.

Theodosy Mosogoya is an associate professor in Horticulture and Plant Propagation in the Department of Crop Science and Production at Sokoine University of Agriculture, Tanzania. He holds a Ph.D. in Applied Plant Biology from the University of Essex in England and a M.Sc. in Horticulture from the National Institute of Horticulture and Landscaping in France. He teaches ornamental crop production, controlled environment horticulture, and plant propagation to undergraduate and graduate students. He has supervised over 20 graduate students and has published over 40 papers in areas of plant propagation and post-harvest of perishable crops. Currently, he is the head of the department of Crop Science and Production at Sokoine University of Agriculture, chairman of the Technical Committee of Fresh Fruits and Vegetables of Tanzania Bureau of Standards, and chairman of its Technical Committee of Processed Fruits and Vegetables. Msogoya is married with three children and lives in Morogoro town in Tanzania.

Lars Olav Eik is a professor at Noragric, Department of International Environment and Development Studies, Norwegian University of Life Sciences (NMBU), Norway. After graduating from NMBU in 1982, he joined Sokoine University of Agriculture (SUA), Tanzania, working on dry season feeding of ruminants and introduction of dairy goat keeping in Tanzania. After this assignment, he returned to NMBU and completed his Ph.D. based on work with dairy goats in Norway. Since 2005, he has been the Norwegian coordinator for three major research programs in collaboration with SUA. He has also participated in research projects in Ethiopia, Malawi, and South Africa. Working together with farmers and private sector, his main interest is multi-functional and innovative production systems for small ruminants, both in tropical and in temperate regions. At Zanzibar, he is currently involved in a value-chain research program aiming to link stallholder farmers with upmarket restaurants on the island. In Norway, he works with sheep farmers and indeed together they have observed that Norwegian lambs slaughtered directly from the mountain pastures have a distinct flavor and fatty acid profile, features that may be used for marketing of Norwegian lamb both in Norway and in abroad. He has supervised a number of Ph.D. students, particularly from East Africa. At his home university, he teaches small ruminant nutrition and production systems in temperate and tropical regions.

Part I
Multi-dimensional Sustainability

Chapter 1
Environmental Sustainability

Rattan Lal

1.1 Introduction

Dwight D. Eisenhower, the 34th president of the United States who held office from 1954–1961, stated during an address at Bradley University in Peoria, Illinois, on September 26, 1956 that "farming looks mighty easy when your plow is a pencil, and you are a thousand miles from a cornfield." While Eisenhower might not have had the opportunity to witness firsthand the challenges facing small landholders and resource-poor farmers in sub-Saharan Africa (SSA), his remarks are even more relevant now than ever. Among the principal environmental-sustainability challenges in SSA are: (1) providing food security to 250 million hungry people (1 in 4 people) in Africa; (2) reducing soil degradation; (3) preserving forests and improving soil and ecosystem C pools; (4) adapting to and mitigating the causes of changing and uncertain climate events; (5) eliminating poverty; and (6) alleviating drought. Drought is the single most natural-disaster in SSA, with strong adverse impacts on crop yield, animal productivity and human wellbeing. It is aggravated by anthropogenic factors and constrains agricultural production in SSA. The risks of drought are likely to be further exacerbated by projected climate changes, including ever increasing temperatures. In SSA, 95 % of agriculture is rain-fed, leaving it highly susceptible to drought. There are 6 types of drought (Fig. 1.1). Pedologic and agronomic droughts, in particular, are strongly aggravated by soil degradation and desertification and adversely impact crop growth and agronomic productivity.

The challenges facing resource-poor farmers are aggravated by the surging population of SSA and its growing affluence. The SSA population of approximately 0.8 billion (bn) is expected to reach 1.1 bn by 2020, 1.4 bn by 2030, 1.7 bn by 2040, 2.1 bn by 2050, and 3.8 bn by 2100 (Fig. 1.2; United Nations 2015). The

R. Lal (✉)
Carbon Management and Sequestration Center, The Ohio State University, Columbus, OH 43210, USA
e-mail: lal.1@osu.edu

R. Lal et al. (eds.), *Climate Change and Multi-Dimensional Sustainability in African Agriculture*, DOI 10.1007/978-3-319-41238-2_1

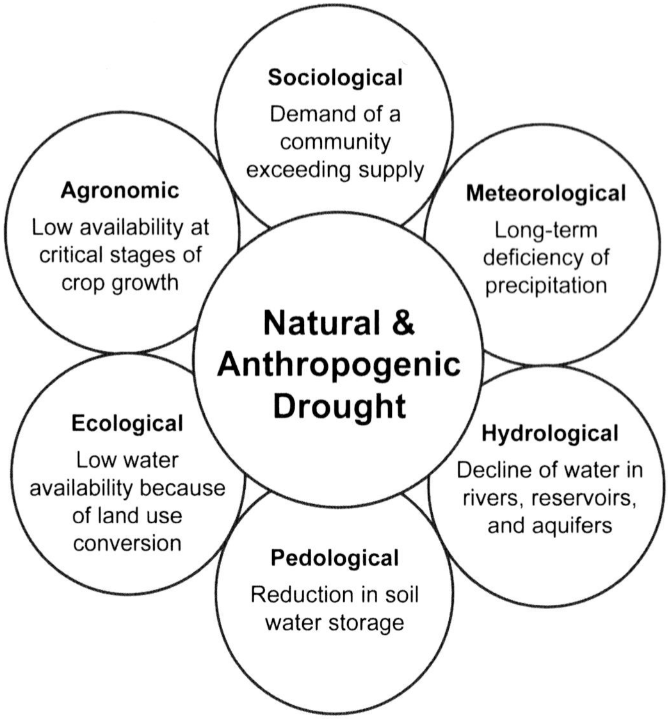

Fig. 1.1 Types of drought prevalent in sub-Saharan Africa

environmental impact (I = PAT) will depend on the population size (P), affluence (A) of the population's lifestyle, and technological (T) advances (Ehrlich and Holdren 1971). Thus, there exists a strong need for developing an environmental perspective in SSA (Sutton 2004), and it must be based on scientific principles (Webersik and Wilson 2009).

The objective of this chapter is to describe the principles, practices, factors and management strategies of environmental sustainability and management strategies in SSA. The goal is to outline a conceptual basis for environmental sustainability, explain the factors and processes that are the predominant controls of environmental sustainability, and outline relevant land use and management practices.

1.2 Environmental Sustainability

A natural environment encompasses the aggregate of the surroundings, milieu, context or simply the style of a place. It is comprised of physical, chemical, and biological components and the interactions among them (Fig. 1.3). The concept of sustainability, which has been discussed widely since the release of the Brundtland

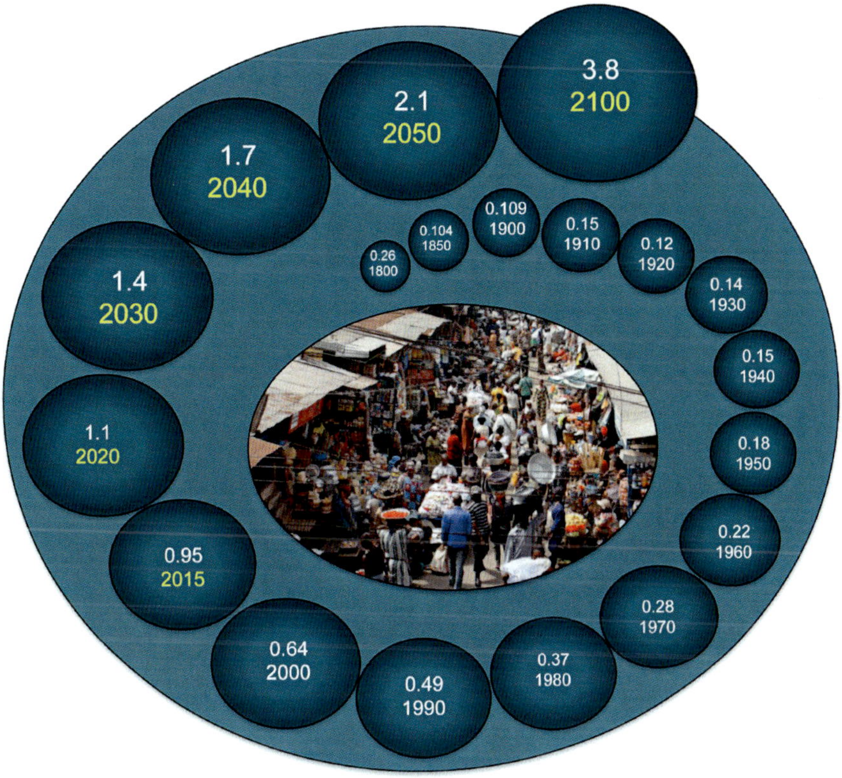

Fig. 1.2 Projected population increases in Africa (drawn from U.N. 2015 data)

Commission (1987) report, refers to the state in which demands placed on the environment can be met without reducing its capacity to allow all people to live well now and in the future. There are four pillars of sustainability, and all are discussed in this volume: (1) the environment (this chapter); (2) the economy (Chap. 2); (3) society (Chap. 3); and (4) institutions (Chap. 4). Sustainability refers to the integration or balancing by environmental, social, institutional, and economic issues. Morelli (2011) emphasizes the importance of the balancing act of explaining that environmental sustainability is a condition of balance, resilience and inter-connectedness that allows human societies to satisfy its needs while neither exceeding the capacity of its supporting ecosystems to continue to regenerate the services necessary to meet those needs nor by our action diminishing biological diversity.

This balancing act necessitates translating science into action to achieve environmental sustainability and sustainable intensification of agro-ecosystems. The latter involves conservation agriculture systems, improved germ plasm, including genetically modified organisms or GMO and measurement and monitoring of soil, vegetation, water, and other natural resources by remote sensing or information and communication technology (ICT). It also involves the adoption of innovative

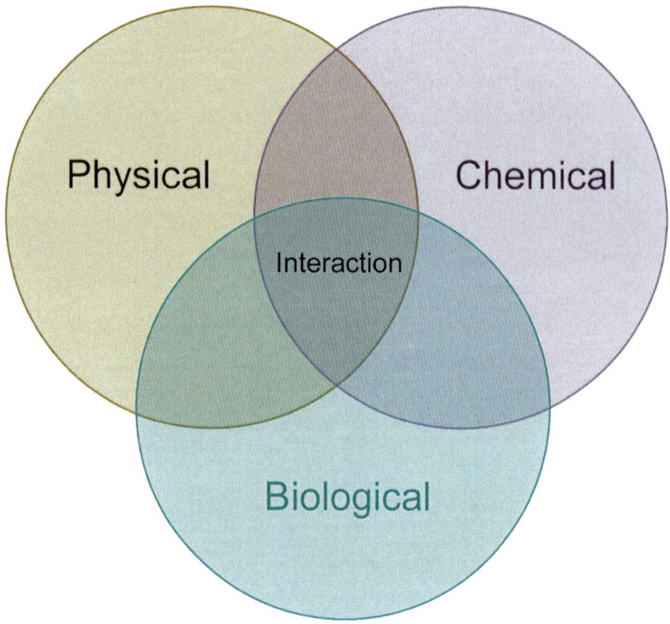

Fig. 1.3 Three interacting components of environment

modern technologies that integrate indigenous knowledge, improved organization, sound processes, judicious management, and effective governance. A strong link exists between environmental sustainability and political stability. Civil strife, social unrest, and political instability adversely affect economic returns, aggravate poverty, perpetuate hunger and malnutrition, exacerbate desperateness, and drastically reduce the cost of rebel soldiers. The latter is an important factor influencing the civil strife, political unrest, civil war, and migrant crises that affected Europe in 2015. Attributes of environmentally sustainable agroecosystems include physical, social, culture and economic (Fig. 1.4). On a broader scale, the attributes are also determined by components of the environment: hydrosphere, atmosphere, biosphere and the lithosphere (Fig. 1.4).

1.3 Agro-ecosystems and Their Resilience

Agro-ecosystems entail the management of energy transformation and the biochemical cycles of specific plant and animal communities within a landscape to generate essential ecosystem services, such as food, fiber, and fuel (Loucks 1977).

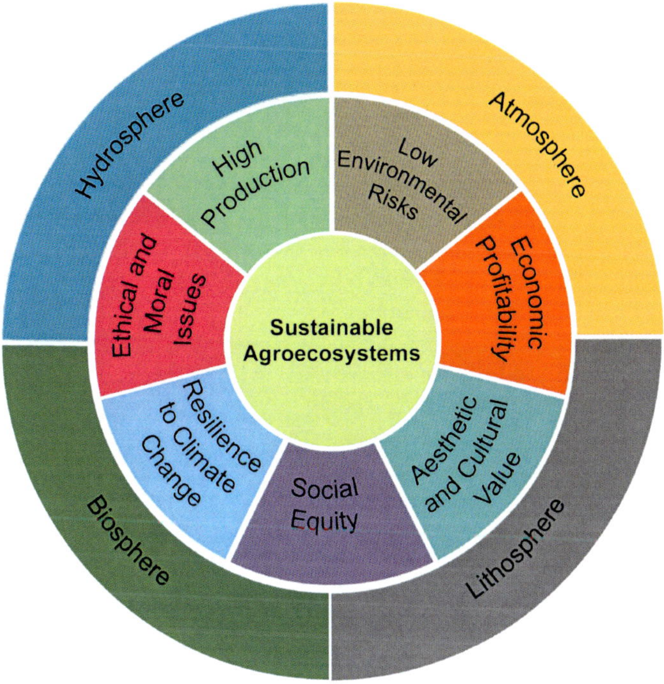

Fig. 1.4 Attributes of environmentally sustainable systems

Table 1.1 Properties of agro-ecosystems and their determinants

Property		Indication	Controls
1	Productivity	Total output	Soil quality, micro and meso-climate
2	Stability	Consistency	Management, availability of inputs
3	Equitability	Distribution and availability	Economic, social, and cultural factors
4	Autonomy	Independence	Political, economic, and social factors
5	Perpetuity	Forever	Prudent management, education, strong institutions
6	Efficiency	Input	Intensification, production of more from less

However, there can be tradeoffs or disservices, such as higher gaseous emissions, accelerated erosion, decreased biodiversity and eutrophication, and increased non-point source pollution. The key properties of agro-ecosystems depend on soil quality. Among the principal determinants of soil quality are the soil organic carbon (SOC) concentration and pool and their dynamics, which are affected by natural factors and management. The properties of agro-ecosystems outlined in Table 1.1 need to be managed in order to enhance provision of essential ecosystem services and to reduce the risks of tradeoffs or disservices.

1.4 Soil Quality and Agro-ecosystem Productivity

The soils in SSA are highly diverse, with strong spatial and temporal variability. However, they are prone to degradation due to harsh climate conditions and widespread use of extractive farming practices. Soil degradation is governed by biological processes, which are aggravated by social, economic, cultural, and political forces. Unless these human dimensions are addressed through effective governance and political willpower, the soil problems in SSA will persist. Strategies to alleviate these problems will be effective only if they are reinforced by relevant policies and their effective regulation.

A quantum jump in agronomic productivity (yield per unit area, time, energy-based input, and gaseous emissions) will be necessary given predicted population increases. However, it should occur while also mitigating climate change and improving the environment. Thus, any strategy for sustainable soil management in SSA must be based on the following basic principles: (1) replace what is removed (e.g., nutrients, SOC, topsoil); (2) respond wisely to what is changed (e.g., soil quality, topsoil depth, soil biodiversity, nutrient- and water-holding capacity); (3) predict the effects of anthropogenic and natural perturbations (e.g., climate change); and (4) increase soil/ecosystem resilience.

1.5 Sustainable Intensification of Agro-ecosystems

Three separate but related terms are used to describe how productivity from existing land resources can be improved with the judicious use of external inputs: agricultural, ecological, and sustainable intensification. The conceptual bases of these terms are outlined in Table 1.2. The principal goals of agricultural intensification

Table 1.2 Merits and challenges of sustainable intensification

Term	Merits	Challenges
Agricultural intensification	Low inputs High productivity Soil and resources conservation High biodiversity High resilience	Increasing productivity Reducing positive feedbacks Decreasing adverse impacts Minimizing trade-offs Increasing equity
Ecological intensification	Low byproducts Minimal material fluxes Low waste levels Low risks	Increasing productivity Increasing emissions Reducing drudgery Improving nutritional security
Sustainable intensification	High production High income Agro-industry Rural employment	Reducing inputs Decreasing environmental pollution Reducing soil degradation Decreasing emissions

are to (1) increase and sustain high productivity, (2) decrease positive feedbacks to climate change; (3) reduce off- and on-site impacts; (4) minimize trade-offs in ecosystem services; and (5) improve social and gender equity. Similarly, the principal objectives of ecological intensification are to (1) increase and sustain productivity; (2) reduce gaseous emissions; (3) alleviate human drudgery; and (4) improve nutritional security. Notable benefits of sustainable intensification are higher production and income levels, the development of agro-industries, and significant increases in rural employment. The principal challenges to sustainable intensification are high inputs and high risks of environmental pollution, soil degradation, and gaseous emissions (Table 1.2).

The principal challenge facing agricultural intensification is to increase productivity from lower inputs while reducing the risk of increased soil and environmental degradation. This will always remain a challenge.

1.6 Processes, Factors, and Causes of Environmental Sustainability

Processes refer to underpinning mechanisms which have impacts on ecosystem goods and services. Such important processes include the cycling of water and elements (C, N, P, S); weathering and new soil formation; changes in soil structure related to the formation and stabilization of soil aggregates; accumulation of SOC; changes in the activities and species diversity of soil flora and fauna; accelerated soil erosion caused by water and wind; salinization resulting from a salt imbalance in the soil; acidification or lower soil pH; leaching of plant nutrients and dissolved organic carbon; accumulation of some elements at toxic concentrations (Al, Fe, Mn); and deficiencies of essential plant nutrients (N, P, Zn).

Factors refer to causative agents which impact degradation processes, accelerated erosion, climate change, SOC depletion etc. Important factors are environmental parameters, including the climate, vegetation, terrain, and drainage density. The major causes are anthropogenic activities, such as deforestation, plowing, and drainage of wetlands. Strong interactions among these factors and causes lead to changes in the rates at which different processes occur. These interactions are manifest in such activities as land use and soil, water, and vegetation management.

Therefore, indicators of sustainability must be carefully selected. Indeed, indicators might be different for developed and developing countries. For example, Zhen and Routray (2003) outlined the following indicators for developing countries: (1) availability of data; (2) sensitivity to biotic and abiotic stresses; (3) information about threshold and critical values; (4) predictability; (5) scaling up or integratability; (6) knowledge of responses to perturbations (natural and anthropogenic); and (7) adaptability amid changes over time. The selection of indicators for specific issues (e.g., environmental, economic challenges) must be prioritized in the context of spatial and temporal variations. Therefore, identifying appropriate indicators is important in order to transfer technological and policy interventions for improving productivity across diverse settings in SSA.

1.7 Impacts of Land Use and Management

Decisions regarding land use and management (soil, water, vegetation, animals) have major effects on environmental sustainability. Any change in land use and vegetation cover, such as the conversion of a natural ecosystem into a managed, can deeply disturb the hydrological cycle, energy balance, ecosystem C pool, element reserves, and fluxes of material within and across ecosystems. Underlying these processes is the need to minimize the magnitude of perturbations and maintain a favorable elemental balance of C, N, P, and S. Soil, then, should always be covered by perennial vegetation, detritus material mulch, a gravel layer, or water in the case of rice paddies or wetlands. In addition, nutrients need to be recycled, rates of water infiltration increased, and water runoff and erosion decreased. The key is to increase nutrient and water reserves and minimize the drought-flood syndrome. The risks of pedologic and agronomic droughts can be minimized by conserving water in the root zone and decreasing losses from surface runoff and evaporation. The effects of droughts are exacerbated by the lower plant-available water capacity of soil. This results from the depletion of the SOC concentration below the threshold level, declines in soil structure, and reductions in effective rooting depths caused by the truncation of soil profiles from accelerated erosion.

Therefore, the choice and adoption of recommended management practices (RMP) are critical. Appropriate techniques can be identified from a wide range of RMPs and validated for site-specific situations determined by biophysical and socio-economic conditions. No silver bullet exists in this matter. Promising RMPs include conservation agriculture, agroforestry, integrated nutrient management, improved pastures, integration of crops with trees and livestock, and controlled grazing at low stocking rates (Lal 2015a, b, 2016).

1.8 Stewardship

More than 200 years ago, Chief Seattle (1780–1866, Chief of the Duwamish tribe) explained his conceptual understanding of the stewardship of natural resources to George Washington, the first president of the United States: “We are a part of the Earth, and it is part of us. The bear, the dear, the great eagle, these are our brothers. The earth does not belong to the man; man belongs to the Earth. How can you buy or sell the sky?” The key idea is that we cannot live apart from nature. Rather, we are part of it. This attitude reinforces the concept that one’s life is more important than one’s lifestyle. Yet, land managers must also be rewarded by the society, such as through payments for ecosystems services. An example of such a reward is for payments of soil C sequestration based on its societal value (Lal 2014). Simply put, the risks of technology adoption must be minimized (Graff and Lipper 2008).

1.9 Conclusion

For the nations of SSA to feed their present and future populations and for the world to feed its the populations of SSA or the globe, it will not be necessary to bring any new land under cultivation in SSA or elsewhere. Instead, the methods should be to restore depleted or desertified soils, improve productivity through sustainable intensification, and return large tracts of agriculturally marginal lands to nature conservation. Ultimately, the strategy should be to minimize the land area under managed ecosystems and to preserve pristine soil and environments for nature conservation. Re-carbonization of soil and the terrestrial biosphere, re-wilding of surplus land, and restoration of soil, water, and vegetation should be prioritized. Food should be produced using the best soil and best management practices. Other soil should be saved for nature conservation.

References

Brundtland Report (1987) Our common future: World commission on environment and development. Oxford University Press, Oxford, U.K.

Ehrlich PR, Holdren JP (1971) Impacts of population growth. Science 171:1243–1248

Graff-Zivin J, Lipper L (2008) Poverty, risk and the adoption of soil carbon sequestration. Environ Dev Econ 13(3):353–373

Lal R (2014) Societal value of soil carbon. J Soil Water Conserv 69: 186A–192A

Lal R (2015a) Sequestering carbon and increasing productivity by conservation agriculture. J Soil Water Conserv 70(3):55A–62A

Lal R (2015b) A system approach to conservation agriculture. J Soil Water Conserv 70(4):82A–88A

Lal R (2016) Beyond COP-21: Potential and challenges of the "4 per thousand" initiative. J Soil Water Conserv 71(1):20A–25A

Lemonick M (2009) Top 10 myths about sustainability. Sci Am. 1 Mar 2009. http://www.scientificamerican.com/article/top-10-myths-about-sustainability/

Louck O (1977) Emergence of research on agroecosystems. Annu Rev Ecol Syst 8:173–192

Morelli J (2011) Environmental sustainability: a definition for environmental professionals. J Environ Sustain 1(1):1–9

Sutton P (2004) A perspective on environmental sustainability? A paper for the Victorian commissioner for environmental sustainability. http://www.green-innovations.asn.au/A-Perspective-on-Environmental-Sustainability.pdf. Accessed 26 Feb 2016

U.N. (2015) World population prospects; key findings and advance tables. 2015 revision. U.N, New York

Webersik C, Wilson C (2009) Achieving environmental sustainability and growth in Africa: the role of science, technology, and innovation. Sustain Dev 17:400–413

Zhen L, Routray JK (2003) Operational indicators for measuring agriculture sustainability in developing countries. Environ Manag 32:34–46

Chapter 2
Economic Sustainability

David Kraybill

Abstract Lively debate has occurred among environmentalists over the past three decades about whether economic systems can be treated as independent of environmental systems. The mainstream economic view is that man-made inputs and natural resources are highly substitutable through technological innovation, and therefore economic analysis can proceed without reference to environmental stocks and flows. This assumption is increasingly untenable as climate change brings major changes in global and local economies. To meaningfully analyze the sustainability of economic systems, many analysts now use research frameworks that are conceptually rigorous with regard to both economic and natural systems. Understanding the nature and complexity of capital (assets) is an important step in analyzing economic sustainability. Emerging notions of capital toward the end of the 20th century included natural capital. A growing number of environmental analysts are attempting to incorporate these expanded notions of capital into theory and practice. Using the concept of natural capital, it is possible to analyze the sustainability of human and natural systems and to assess the impact of economic activity, including agriculture, on future generations as compared to the present generation. This chapter presents an overview of research approaches that attempt to incorporate both economic and environmental systems for the study of sustainability with a focus on relevance of these methods for the study of African agriculture.

D. Kraybill (✉)
The Ohio State University, Columbus, USA
e-mail: kraybill@iagri.org

D. Kraybill
Innovative Agricultural Research Initiative, Morogoro, Tanzania

R. Lal et al. (eds.), *Climate Change and Multi-Dimensional Sustainability in African Agriculture*, DOI 10.1007/978-3-319-41238-2_2

2.1 Introduction

A growing body of scientific evidence points to anthropomorphic causes of climate change (Oreskes 2004; Cook et al. 2013). As the evidence mounts that human activity contributes to climate change, it would foolish to continue to treat the economy as though it were independent of the biological and chemical processes that create and sustain the organismic life on planet Earth.

The objective of this chapter is to present alternative conceptual models of the economy with regards to its relationship with natural and social systems. The paper begins with the standard neoclassical model of the economy and proceeds to present alternative frameworks in which the economy is embedded within the natural environment. Three distinct research paradigms aiming to develop integrated models of natural and human systems are discussed.

Climate histories constructed by scientists for ancient and modern periods reveal that major shifts in climate can have calamitous effects that result in altered trajectories of life on Earth. Because of the potential path-altering effects of climate change, both scientists and laymen increasingly ask how current economic practices will affect the conditions for life in the future. The extent to which today's actions affect the world of tomorrow has come to be known as "sustainability".

What is sustainability? Solow (1991), a Nobel Prize-winning economist, defines *sustainability* broadly as a moral imperative: "An obligation to conduct ourselves so that we leave to the future the option or capacity to be as well off as we are." Costanza and Patten defined a *sustainable system* as a system that survives or persists over time. Anand and Sen (2000) defined *sustainable development* as "non-declining welfare," where welfare is measured either as utility (satisfaction) or income.

For the purpose of this chapter, I define *economic sustainability* as the preservation, replacement, or expansion of produced and natural capital so that the economy is able to generate non-declining welfare in the future. This definition is generally consistent with the definitions in the previous paragraph but it adds the notion of capital, which is a key concept for linking the actions of today with well-being in the future.

2.2 A One-Dimensional View of the Economy

Economic theories and studies can be categorized broadly into two types: whole-system analysis and partial-system analysis. For many decades and continuing today, students in most introductory economics classes are presented with a whole-system model of the economy known as the "circular flow diagram" (Fig. 2.1) in which factors of production (capital, labor, and land) are provided by households to firms that produce goods which are then sold to households. In the circular flow diagram, the natural environment makes only a minor appearance, showing up as "land". This is a catch-all term that includes soil, water, forests, and

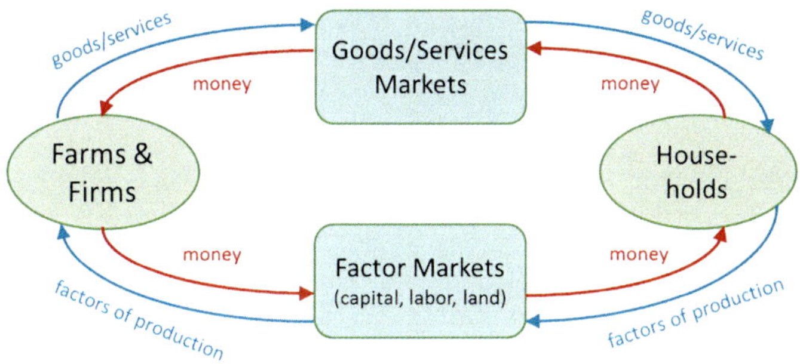

Fig. 2.1 The neoclassical circular flow diagram of the economy

other natural resources, but only those resources that are exchanged in markets. Omitted from this framework are natural resources that not traded in markets either because they are so abundant that they have no market value, as in the case of air, or because their use is not regulated by ownership-defining property rights. From an environmental sustainability standpoint, the absence of a mechanism for valuing nontraded natural resources is an important omission in the circular flow diagram, and one that is particularly relevant when considering issues of climate change.

A second important omission in the circular flow diagram from a sustainability standpoint is the absence of environmental "bads", which are generated as byproducts of human activity (Georgescu-Roegen 1971; Daly 1985). The diagram, which focuses on the economy in isolation from the natural environment, ignores the possibility that the economy may harm the natural environment and that this harm may, in turn, create harm to the economy. In essence, the circular flow diagram of the economy depicts the economic system as free from feedback from the natural environment.

While the complete absence or, at best, naïve representation of the natural environment that characterizes the circular flow diagram is evident in virtually all economy-wide analyses in mainstream economics, a separate branch of economics began to emerge more than 50 years ago to deal with problems of the environment. Environmental economics is a set of concepts and tools for analyzing environmental goods and services through partial-system analysis. Key among these concepts are market failure and time discounting.

The concept of market failure is relevant in situations where environmental goods, sometimes referred to as ecosystem services, or environmental bads are not traded in markets or where there are nonexistent or weak property rights to define ownership or responsibility for actions affecting the environment. Market failure also occurs where there are shared goods, such as a stable climate or community pastures, for which it is difficult or impossible to hold individuals accountable for how their actions affect the ecosystem services generated by these goods. A major focus of environmental economics is the imputation of the economic value of environmental goods and services where market failures occur so that gains and losses that are not formally recognized in market transactions are nevertheless taken into account by economic actors through governance processes. Market failures, if not addressed, threaten sustainability. Stern (2006), in his book on *The Economics of Climate Change*, asserts that "Climate change is the greatest market failure the world has ever seen."

Time discounting addresses future consequences of today's actions through the "social discount rate," which is the relative valuation that society places on the wellbeing of people today versus the wellbeing of people in the future. A zero discount rate implies future generations are treated the same as the current generation, while a positive discount rate implies the wellbeing of future generations matters less than the wellbeing of the current generation.

The concepts of market failure and time discounting have been used widely by mainstream economists to analyze the causes and consequences of economic activities that generate greenhouse gases. Based on these analyses, numerous solutions to curb greenhouse gases have been proposed, including emission trading schemes, carbon taxation, and transfer payments. Critics of environmental economics point out that analysts working in the mainstream tradition tend to treat the environment as a subsystem of the economy, using models and frameworks that fail to account for the complex dynamics of ecosystems. These dynamics are important for environmental sustainability but they are also important for economic sustainability if the environment and the economy are, in fact, interlinked.

Various economists have developed economic-environmental models, most of which are based on neoclassical economic theory with selected environmental flows appended. For example, the ENV-Growth model developed by OECD (2016) is a neoclassical growth model with three conventional drivers of change: exogenous technical change, changes in physical capital, and changes in the quantity and quality of labor. To this mainstream economic model, its creators have added two environmental drivers of growth: energy and natural resources, both modeled as economic flows (supply-demand and income-revenue); there is no attempt to model the biological and chemical systems from which energy and natural resources are derived.

2.3 Interlinked Human and Natural Systems

Representing interactions between human and natural systems in a formal model is difficult because the phenomena on which each of these systems focus operate at different spatial, temporal, and organizational scales. Consequently, the scholarly disciplines historically associated with these two systems, human and natural, use distinct concepts, methods, and tools of analysis that make conversation across disciplines a challenge. Despite these difficulties, over the past two decades, three groups of scholars have emerged with the aim of modeling human and natural system interactions using interdisciplinary and transdisciplinary frameworks that aim to embody fundamental principles of both ecosystems and human behavior. One of these scholarly groups is known as ecological economics, the second operates under the rubric of coupled human and natural systems, the third identifies its work as integrated assessment, and the fourth calls its approach sustainable livelihoods.

Sustainability is one of the central themes of the founders of ecological economics (Constanza et al. 1991) and the approach is generally transdisciplinary. Scholars affiliated with this group hold advanced degrees in a variety of disciplines, including ecology, economics, and philosophy. One of the major contributions of ecological economics is the development and application of tools for valuing natural capital, defined as bio-physical and chemical assets produced by nature. Among the tenets of ecological economics is the idea that produced capital has very limited capacity to substitute for natural capital, and therefore damage to the environment has major consequences for both natural and economic systems. In a paper published in *Nature*, Costanza et al. (1997) estimated that the annual value of global ecosystem services exceeds the annual value of global GDP as it is conventionally defined. Half of the value these authors impute for ecosystem services is related to nutrient recycling. Given the enormous value generated by nutrient flows, it follows that economic models that fail to account accurately for nutrient flows are inadequate tools for understanding and analyzing sustainability.

Created within the ecological economics tradition, an integrated human-ecosystem simulation model developed by Low et al. (1999) is based on the assumption that the stock of natural capital is jointly influenced by natural capital growth and depletion, ecological fluctuations, harvest rules, and biological transfers across ecosystems. Basic principles of both ecosystems and human systems are built into the model, which the authors use to analyze natural resource harvest patterns and the effect of resource movement across ecosystems. They conclude that inter-ecosystem movement reduces the threat of human-system collapse but can threaten the viability of ecosystems from which resources migrate.

According to Daly and Farley (2011), it is a tenet of ecological economics that the economy is "an open subsystem of a larger ecosystem that is finite, nongrowing and materially closed." Figure 2.2 is an environmental-economic model consistent with this view of the primacy of the ecosystem relative to the economic system. The stock of natural capital generates ecosystem services, some of which are used by producers and consumers. An important service of the ecosystem is to regenerate itself but this capacity is diminished by flows of pollution arising from economic activity. Natural capital is an important input into the making of goods and services but it is viewed as having a relatively low degree of substitutability with respect to physical capital or labor, a perspective known as "strong sustainability". The stock of natural capital is depreciated by the pollution generated by producers and consumers.

The extent to which ecological economics has adopted the methodological pluralism to which its founders aspire is debatable. Anderson and M'Gonigle (2012) contend that ecological economics, 20 years after its launch, has failed to deliver on its promise to move beyond the narrow confines of neoclassical economics and address seriously the ecological and political economy dimensions of the environment.

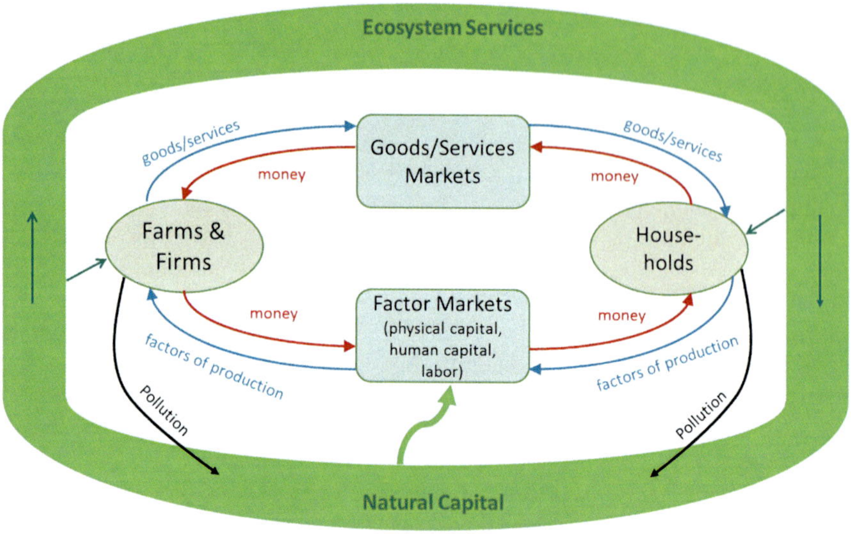

Fig. 2.2 The economy as a sub-system of the natural environment

Coupled human and natural systems (CHANS), a second major strand of international scholarship devoted to multidisciplinary analysis of sustainability, is arguably more equitably balanced than ecological economics in its treatment of ecological and human systems. According to McConnell et al. (2009), "what distinguishes the CHANS approach is an explicit acknowledgement that human and natural systems are coupled via reciprocal interactions, understood as flows (e.g., of material, energy, and information)." CHANS focuses on system complexity arising from feedback loops, nonlinearity and thresholds, legacy effects, and heterogeneity but also incorporates aspects of human behavior (Liu et al. 2007).

In 2007, the National Science Foundation (NSF) in the United States launched a formal program of funding for CHANS research. NSF restricts its funding to CHANS proposals that incorporate four elements: (1) dynamics of a natural system; (2) dynamics of a human system; (3) processes through which the natural system affects the human system; and (4) processes through which the human system affects the natural system.

The CHANS approach applied to African agriculture is illustrated by a research framework developed by Olson et al. (2008) to analyze two-way interactions of land use change and climate change. The framework consists of sub-models of land use change, agricultural productivity, land cover, and climate. Data were gathered from case studies, satellite, and secondary sources. A version of the framework was used to analyze the interaction of food production, land use and land cover, and climate change in East Africa (Moore et al. 2012). The model reveals that both climate change and land cover/land use are capable of bringing about significant changes in food production, though for a given exogenous change in climate or land cover/land use, the food production response varies enormously across the East African landscape.

A third interdisciplinary approach to analysis of the economy and the environment is known as integrated assessment. According to Center for International Earth Science Information Network (CIESIN 1995), the two defining characteristics of integrated assessment are "(1) that it seeks to provide information of use to some significant decision-maker rather than merely advancing understanding for its own sake; and (2) that it brings together a broader set of areas, methods, styles of study, or degrees of certainty, than would typically characterize a study of the same issue within the bounds of a single research discipline". Scholars of integrated assessment typically work in teams of experts from the disciplines of economics, political science, engineering, ecology, climatology and other disciplines to jointly construct integrated assessment models (IAMs) with physical and social science sub-models. Typically, within an IAM, the physical sub-model simulates climate while the social science sub-model simulates the economy.

IAM researchers are well organized and have been able to attract significant amounts of funding. An international organization, the Integrated Assessment Modeling Consortium (IAMC), was created in 2008 and holds annual meetings. Major funding for IAM research has been provided by governments of the United States, Netherlands, Japan, from foundations, and various other sources.

IAMs have been used to analyze the causes and consequence of climate change in Africa. For example, the ADAPTCost Project of the United Nations Environmental Program (UNEP) has been used to estimate the economic cost of climate change in Africa (United Nations Environment Program 2010). IAMs have been applied to the study of the interaction of climate change and agriculture (Valdivia et al. 2015). An international research initiative, the AgMIP Sub-Saharan Africa Regional Coordination Project, is developing agriculture-focused modules to be used with regional IAMs on Africa.

2.4 A Sustainability Framework for Extension and Outreach Practitioners

The integrated models described in the preceding sections are useful for research and policy-making on sustainability but they provide little specific guidance to practitioners working directly with farmers and farming communities. For the design of specific policies and programs, a less formal sustainability framework is needed to guide the practitioner. Such a framework is provided by the Sustainable Livelihoods (SL) framework.

For economies with poor, vulnerable populations engaged in informal employment, "livelihood" is a useful concept for analyzing multi-dimensional sustainability at the household level. According to Chambers and Conway (1991), "a livelihood comprises the capabilities, assets (stores, resources, claims and access) and activities required for a means of living: a livelihood is sustainable which can cope with and recover from stress and shocks, maintain or enhance its capabilities and assets, and provide sustainable livelihood opportunities for the next generation; and which contributes net benefits to other livelihoods at the local and global levels and in the short and long term."

The SL framework addresses environmental, economic, social, and institutional sustainability by analyzing expected or actual changes in the assets on which the household's livelihood is based. The livelihood assets on which SL analysts typically focus are physical, natural, human, social, and financial assets. The aim of the framework is to deepen understanding of the "vulnerability context" through which economic, political, social, and natural change can diminish the grip of households on their assets. Change is sustainable when it gives a household greater command over one or more of its livelihood assets without diminishing its other livelihood assets or the livelihood assets of other households. The household's asset portfolio determines the kinds of strategies available to the household for obtaining its livelihood. Figure 2.3 identifies various rural household strategies in developing countries. The particular strategies chosen by the household determine the livelihood outcomes it is able to reap. Outcomes are measured in terms of both the level and variability of well-being. The household's livelihood outcomes in the current period have implications, positive or negative, for its well-being in the next period. Good outcomes this year lead to greater assets next year. Bad outcomes this year lead to reduced assets next year, threatening the sustainability of the household.

The SL approach has been used widely by scholars and practitioners to undertake integrated, multidisciplinary analysis of the likely effects of climate change on

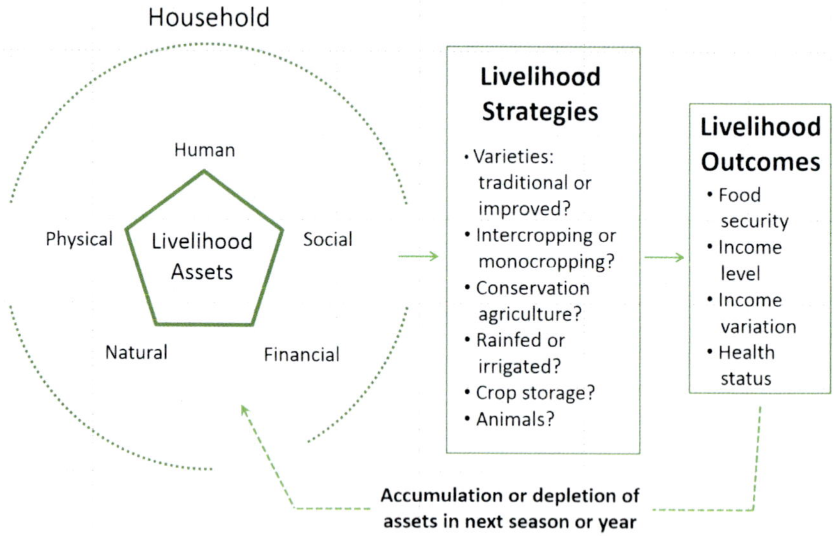

Fig. 2.3 Sustainable livelihoods framework

agricultural households. Connolly-Boutin and Smit (2015) reviews applications of the SL approach to the study of climate change in Africa.

2.5 Conclusion

In summary, analytical frameworks available for analyzing the economics of agriculture in Africa have expanded from economics-only theories and models to multidisciplinary and transdisciplinary models. In some of these frameworks, the economy is a sub-system of a larger, natural system while in others the economy is treated as the dominant system to which the natural system is appended. Ecological economics, coupled human and natural systems (CHANS), integrated assessment modeling (IAM), and the sustainable livelihoods (SL) approach all aim to take into account linkages between the economy and the environment, though the featured interactions among these systems vary across the approaches and even across studies within each of these approaches. All four approaches have been applied usefully to the study of sustainability of African agriculture.

References

Anand S, Sen A (2000) Human development and economic sustainability. World Development 28 (12): 2029–2049

Anderson B, M'Gonigle M (2012) Does ecological economics have a future? Contradiction and reinvention in the age of climate change. Ecol Econ 84:37–48. doi:10.1016/j.ecolecon.2012.06.009

Chambers R, Conway GR (1991) Sustainable rural livelihoods: practical concepts for the 21st century. IDS discussion paper 296. Institute of Development Studies, University of Sussex, Brighton

Connolly-Boutin L, Smit B (2015) Climate change, food security, and livelihoods in Sub-Saharan Africa. Regional environmental change. Springer, Berlin, pp 385–399

Consortium for International Earth Science Information Network (CIESIN) (1995) Thematic guide to integrated assessment modeling of climate change. CIESIN, Palisades. http://sedac.ciesin.org/mva/iamcc.tg/TGHP.html

Constanza R, Daly H, Bartholomew J (1991) Goals, agenda and policy recommendations for ecological economics. Environ Accoun Sustain. doi:10.1055/s-0029-1214156

Cook J, Nuncatelli D, Green SA, Richardson M, Winkler B, Painting R, Way R, Jacobs P, Shaw A (2013) Quantifying the consensus on anthropogenic global warming. Environ Res Lett 8:24–31

Costanza R, Arge R, De Groot R, Farberk S, Grasso M, Hannon B, Limburg K et al (1997) The value of the world's ecosystem services and natural capital. Nature 387:253–260. doi:10.1038/387253a0

Daly HE (1985) Matter-energy: a case of misplaced concreteness. Rev Soc Econ 43(3):279–297

Daly HE, Farley J (2011) Ecological economics: principles and applications. Island Press, Washington

Georgescu-Roegen N (1971) The entropy law and the economic process. Harvard University, Cambridge

Liu J, Dietz T, Carpenter SR, Alberti M, Folke C, Moran E, Pell AN et al (2007) Complexity of coupled human and natural systems. Science 317(5844):1513–1516. doi:10.1126/science.1144004

Low B, Costanza R, Ostrom E, Wilson J, Simon CP (1999) Human-ecosystem interactions: a dynamic integrated model. Ecol Econ 31(2):227–242. doi:10.1016/S0921-8009(99)00081-6

McConnell WJ, Millington JDA, Reo NJ, Alberti M, Asbjornsen H, Baker LA, Brozović N et al (2009) Research on coupled human and natural systems (CHANS): approach, challenges, and strategies. Bull Ecol Soc Am 92(2): 218–228

Moore N, Alagarswamy G, Pijanowski B, Thornton P, Lofgren B, Olson J, Andresen J, Yanda P, Qi J (2012) East African food security as influenced by future climate change and land use change at local to regional scales. Clim Change 110(3–4):823–844. doi:10.1007/s10584-011-0116-7

OECD (2016) Environmental-economic modelling. http://www.oecd.org/environment/indicators-modelling-outlooks/modelling.htm

Olson JM, Alagarswamy G, Andresen JA, Campbell DJ, Davis AY, Ge J, Huebner M et al (2008) Integrating diverse methods to understand climate-land interactions in East Africa. Geoforum 39(2):898–911. doi:10.1016/j.geoforum.2007.03.011

Oreskes N (2004) The scientific consensus on climate change. Science 305(5702):1686

Solow RM (1991) Sustainability: An economist's perspective. The Eighteenth J. Seward Johnson Lecture, Marine Policy Center. Woods Hole, MA: Woods Hole Ocenographic Institution, June 14, 1991

Stern N (2006) The economics of climate change. Stern Rev. doi:10.1257/aer.98.2.1

United Nations Environment Program (2010) ADAPTCost project : analysis of the economic costs of climate change adaptation in Africa

Valdivia RO, Antle JM, Rosenzweig C, Ruance AC, Vervoot J, Ashfaq M, Hathie I et al (2015) Representative agricultural pathways and scenarios for regional integrated assessment of climate change impact, vulnerability and adaptation. In: Hillel D, Rosenzweig C (eds) Handbook of climate change and agroecosystems: the agricultural model intercomparison and improvement project (AgMIP). American Society of Agronomy, Crop Science Society of America, and Soil Science Society of America. http://ntrs.nasa.gov/archive/nasa/casi.ntrs.nasa.gov/20150009315.pdf

Chapter 3
Institutional Sustainability in the Face of Climate Change: Empirical Insights from Irrigation Institutions in the Iringa Rural District, Tanzania

George C. Kajembe, Pål Vedeld, Innocent H. Babili, Dos Santos Silayo and Devotha B. Mosha

Abstract Adaptation and mitigation have been proposed as strategies for addressing climate change, which is a predominant challenge facing society today. It has been found that an increase of 1 °C leads to a 5 % reduction in grain yield. From 1901 to 2010, the average sea level rose by 19 cm, and global emission of CO_2 has increased by about 50 % since 1990. However, little attention has been paid to the link between institutional sustainability and climate change adaptation and mitigation. This paper argues that effective adaptation to and mitigation of climate change depends on the sustainability of both formal and informal institutions. Institutional sustainability is used in this paper to mean the ability of institutions, under particular conditions, to continue guiding actors to achieve desirable goals. Sustainable formal and informal institutions provide a framework that guide interactions among actors, including organizations and individuals, for mitigating and adapting to climatic change. Literature has pointed out a number of conditions that ensure institutional sustainability during climate change. Drawing from empirical insights from a case study in Tanzania, it is demonstrated that institutional sustainability is possible in the face of climate change and can be achieved through evolution and change, legitimacy, bricolage and performance and embracing practices of polycentric governance. Furthermore, a *multiple institutional logics of*

Paper presented by Innocent H. Babili at an International Conference on Climate Change and Multi Dimensional Sustainability in African Agriculture, June 3–5, 2015, Morogoro, Tanzania.

G.C. Kajembe · I.H. Babili (✉) · D.S. Silayo · D.B. Mosha
Sokoine University of Agriculture, Morogoro, Tanzania
e-mail: babhili@yahoo.com

G.C. Kajembe
e-mail: kajembegc@yahoo.com

P. Vedeld
Norwegian University of Life Sciences (NMBU), Ås, Norway

R. Lal et al. (eds.), *Climate Change and Multi-Dimensional Sustainability in African Agriculture*, DOI 10.1007/978-3-319-41238-2_3

action (MILA) theoretical framework was recommended for assessing institutional sustainability in the face of climate change.

Keywords Adaptation · Mitigation · Institutions · Institutional bricolage · Multiple institutional logics of action (MILA) · Conditions for institutional sustainability · Institutional performance · Institutional legitimacy · Good governance

3.1 Introduction

Adaptation and mitigation have been proposed as strategies for addressing climate change, which is a predominant challenge facing society today (IPCC/UNFCC 2014). On December 12th, 2015, during the Paris Conference on Climate Change (COP21), participating countries reached nine key agreements for adapting to and mitigating climate change (UNFCCC 2015). Climate change is defined as "a change of climate which is attributed directly or indirectly to human activities that alter the composition of the global atmosphere, and which is in addition to natural climate variability observed over considerable time periods (UNFCCC 2007). It has been found that an increase of 1 °C leads to a 5 % reduction ingrain yield. From 1901 to 2010, average sea levels rose by 19 cm, and global emissions of CO_2 have increased by about 50 % since 1990 (UN 2015). The precise predictions about the likelihood of global temperature changes and the time-scales for their occurrence vary and are subject to varying levels of confidence. However, there is much more agreement about the likely impacts on humans, with a general consensus that the poor in developing countries—given their dependence on climate sensitive sectors such as rain-fed agriculture—will suffer most (IPCC 2007; Stern 2006). Climate change directly threatens current progress towards alleviating poverty (IPCC 2007).

The reality of climate change and its effects is becoming more apparent, as exemplified by more frequent and severe draughts, hurricanes, floods and storms. These changes are increasingly threatening people's livelihoods, especially of those in the developing world. Tanzania, already stressed by other factors, is highly vulnerable to the effects of climate change and extreme weather variability. Therefore, the country will need to adapt at various levels in order to cope with the additional challenges. From a sociological point of view, adaptation to and mitigation of climate change depend on the institutional sustainability of both formal and informal institutions. While formal institutions provide constitutional frameworks in which organizations and individuals interact, informal institutions offer norms and informal sanctioning mechanisms to govern ways of doing things (Blomquist et al. 2004).

Institutional sustainability is used in this paper to mean the ability of institutions, under particular conditions, to continue guiding actors to achieve desirable goals. A number of studies have focused on the interaction between climate change and institutions. However, those studies pay little attention to the link between institutional sustainability and adaptation to and mitigation of climate change. This paper seeks to discover how bureaucratic and socially embedded institutions have been continuously successful in helping actors to adapt practices and implement effective measures to mitigate climate change. To achieve this objective, the paper draws on empirical insights from an irrigation case in the rural district of Iringa, Tanzania.

3.2 Understanding the Concepts of Climate Change, Adaptation and Mitigation

According to Katani and Babili (2012), climate change refers to long-term alterations in global weather patterns, including increases in temperature over time, rainfall fluctuations and storm activities, which are all caused by the greenhouse effect and aggravated by continuous deforestation. It is a change attributed directly or indirectly to human activities that alter the composition of the global atmosphere and which cannot be attributed to natural climate variations observed over a comparable period of time (UNFCCC 2007). Adaptation to climate change is defined as an adjustment in natural or human systems, in response to actual or expected climate stimuli or their effects, which moderates harm or exploits potential benefits (IPCC 2001). Adaptation is usually a long-term activity and is a continuous process that sustains results. It uses resources efficiently and sustainably, involving planning, combines new and old strategies and knowledge, and is focused on finding alternatives (IPCC 2001).

IPCC (2007) defines climate change mitigation as an anthropogenic intervention to reduce the causes or enhance the sinks of greenhouse gases. According to FEMA (2014), mitigation is an effort to reduce the loss of life and property by lessening the impact of disasters or any action taken to permanently eliminate or reduce the long-term risks and hazards of climate change to human life and property. Mitigation also involves taking action now, before the next disaster, to reduce human and financial consequences later. In this paper, climate change mitigation use as defined by IPCC (2007). anthropogenic intervention to reduce the amount of atmospheric greenhouse gases or enhance their sinks.

3.3 Institutions and Institutional Sustainability

3.3.1 Definitions of an Institution

Institutional scholars disagree on the definition of an institution (Msuya 2010). Institutional economists define institutions as the set of rules of the game in a society that defines and shapes human interactions (North 1990; Ostrom 1990,

2007). According to North (1990), institutions are the rules of the game for the society or, more formally, the humanly devised constraints that shape human interactions in economic, social and political life. Institutions can also be defined as the social cement (Davies 1997) which links stakeholders' different kinds of access capitals to the means of exercising power. Institutions define the gateways through which stakeholders contribute to positive or negative adaptations to climate change. In other words, institutions are normative factors that evolve in a society to regulate and standardize people's conduct (Smajgl and Larson 2006).

Ostrom (1990) defines institutions as the working rules or rules-in-use set by individuals to organize repetitive activities, the outcomes of which affect those individuals and potentially others. The working rules are those actually used, monitored and enforced when individuals choose to act in operational settings or when they make collective choices (Ostrom and Crawford 2005). These rules and their enforcement mechanisms, together with formal and informal sanctions, indicate what individuals must or must not do; what they may do without interference from other individuals; what they can do with the aid of collective power and what they cannot do (Ostrom and Crawford 2005). Encouragement to do something is an incentive, and the opposite is a sanction. In a very broad sense, institutions determine who is eligible to make decisions; what actions are allowed; which methods are used to reach consensus on communitywide rules, decisions, norms and choices; what information is available to individuals and what returns individuals obtain from collective actions (Ostrom 1990).

3.3.2 *Types of Institutions*

According to Cleaver (2002), institutions can be either "bureaucratic" or "socially-embedded." Bureaucratic institutions are formalized arrangements based on explicit organizational structures, contracts and legal rights, often introduced by governments or development agencies; while socially embedded institutions are those

Table 3.1 Differences between formal and informal institutions (Leach et al. 1997)

Aspects	Formal institutions	Informal institutions
Nature of evolution	Exogenous	Endogenous
Functional and structural arrangement	Commonly at district or national level	Site specific
Consideration of social and cultural aspects	Low	High
Embeddedness	Low	High
Ownership	Low	High
External input and material support	State	Local community
Enforcement and monitoring	Legal and state	Based on local agreements

based on culture, informal social organizations and daily practices (Cleaver 2002). Institutions are also classified as either formal or informal (Table 3.1). Formal institutions have written laws, regulations and procedures, while informal institutions have norms, mores, myths, practices and patterns of behavior (Kisoza 2007). Table 3.1 compares formal and informal institutions. Both formal and informal institutions mediate access to livelihood capitals and, in turn, affect the composition of climate change adaptation strategies.

3.3.3 *Institutions Versus Organizations*

Institutions and organizations are similar, which has led to a debate over how to distinguish the two. Some scholars and integrate organizations into the definition of institutions, use the two terms interchangeably and define them as overlapping (e.g., DfID 2003; Blomquist et al. 2004). Others make a clear distinction between the two (e.g., North 1990; Bandaragoda 2000). Although institutions are not identical to organizations, the two do influence each other (Giddens 1994).

While institutions are the rules of the game, organizations are groups of individuals or teams with definite roles, bound together by some common purpose and working within the framework of rules and procedures to achieve specific objectives and provide structures for human interaction (Bandaragoda 2000). Furthermore, the term organization can refer to players acting within an institutional framework, whereas institutions form the basis of organizations but are influenced by them at the same time (Diaz and Rojas 2006). Organizations evolve and come into existence through the influence of institutions, but they also influence how the institutional framework develops (Msuya 2010). Understanding the difference between institutions and organizations is important in designing human adaptations to climate change.

3.3.4 *Institutional Sustainability*

The literature presents various perspectives on institutional sustainability. One perspective defines institutional sustainability not only by continued existence but also in terms of sustainable development (Pfahl 2005). Another perspective views institutions as long-enduring by definition (Ostrom 1990). In this paper, institutional sustainability is used to mean the ability of institutions, under particular conditions, to guide actors to reach desirable goals. Sustainable formal and informal institutions provide a framework that guides interactions among actors, including organizations and individuals, for adapting to and mitigating climatic change.

3.4 Methodology

Data for the study were collected through a literature search and complemented by a case study on Farmer-Managed Irrigation Schemes (FMISs) in the Iringa rural district. Two villages were selected for this study. Selection criteria of the study villages included similarity in ecological characteristics and the presence of irrigation schemes with paddy as the main irrigated crop. Selected schemes and villages were the Mlenge semi-improved irrigation scheme in the village of Itunundu and the traditional Mkombozi scheme in the village of Mboliboli. Primary data collection was done through key informants, including farmers with experience in irrigation, District Natural Resources Officers and Rufiji Basin Water Officers, and one focus group discussion (FGD) in each study village. Participants in the FGD were chosen through informal gatherings, with assistance from village leaders and members of water user groups. The data was analyzed through content analysis.

3.5 Conditions Underlying Institutional Sustainability in the Face of Climate Change

3.5.1 Evolution and Change

Institutions are dynamic if they are liable to evolve and change with time (Msuya 2010). Evolution and change involve different mechanisms in formal and informal institutions (Greif and Laitin 2004).

3.5.1.1 Change of Formal Institutions

Formal institutions can be crafted, maintained and changed through deliberate reforms by the state or through political processes. Ostrom (1990) suggested eight design principles for crafting effective institutions for managing Common Pool Resources (CPRs). Table 3.2 presents the eight design principles suggested by Ostrom.

Some scholars have questioned the application of Ostrom design principles (e.g., Cleaver 2002; Babili et al. 2015). Babili et al. (2015) formulated an alternative theoretical framework, the *Multiple Institutional Logics of Action* (MILA), which combines the logic of discourse, the logic of appropriateness and the logic of rational choice to assess institutional change and institutional performance. Babili et al. (2015) used the logic of discourse in the MILA framework to explain the successful introduction of decentralized forest management (DFM) in the Babati District, Tanzania, and they used the MILA's logic of rational choice to explain compliance with agreed-upon rules. Babili et al. (ibid.) also indicated four areas of weakness in Ostrom's design principles and described Ostrom's response to the observed weaknesses.

Table 3.2 Ostrom's design principles for crafting effective CPR institutions (Ostrom 1990)

Design principle	Description
1	Clearly defined boundaries: Individuals or households who have rights to withdraw resource units from CPR must be clearly defined, as must the boundaries of the CPR itself
2	Congruence between appropriation and provision rules and local conditions: Appropriation rules restricting time, place, technology and/or quantity of resource units are related to local conditions and to provision rules regarding labor, material and/or money
3	Collective-choice arrangements: Most individuals affected by the operational rules can participate in modifying the operational rules
4	Monitoring: Monitors who actively audit CPR conditions and appropriator behavior are accountable to the appropriators or are appropriators
5	Graduated sanctions: Appropriators who violate operational rules are likely to be subjected to graduated sanctions (depending on the seriousness and context of the offence) by other appropriators, by officials accountable to these appropriators or by both
6	Conflict-resolution mechanism: Appropriators and their officials have rapid access to low-cost local arenas to resolve conflicts among appropriators or between appropriators and officials
7	Minimum recognition for rights to organize: The rights of appropriators to devise their own institutions are not challenged by external governmental authorities
8	Nested enterprises: Appropriators, provision, monitoring, enforcement, conflict resolution and governance activities are organized in multiple layers of nested enterprises

According to Babili et al. (ibid.), Ostrom's design principles have the following weaknesses: (1) an emphasis on the stability of institutions (2) non-applicability of Ostrom design principles in areas which have registered successes in managing CPRs (3) a disregard of the broader context that can affect management of CPRs and (4) a cost-benefit consideration in designing effective common pool institutions. Babili et al. (2015) also described Ostrom's response to the criticisms: Ostrom asserted that operational institutional change is possible, recommended the use of design principles depending on the situation and proposed new multi-level governance as an alternative.

In Tanzania, formal natural resource management institutions have undergone various changes, reflecting changes in socio-economic, political and ecological conditions as well as trends and pressures from international donors (Maganga et al. 2004; Van Koppem et al. 2007; Babili and Wiersum 2013, 2015). Petursson and Vedeld (2012) asserted that formal institutions evolve overtime through the entry and exit of particular actors and alterations in the interaction patterns of rules and power relations. Formal institutional change may take two distinct forms. Firstly, it may be a change in design, which may alter actors' incentives by changing the expected outcomes of formal rules (Helmke and Levitsky 2003). Secondly, it may be a change in institutional strength or effectiveness, which may change actors' expectations about whether formal rules will be enforced (Helmke and Levitsky 2003).

3.5.1.2 Evolution of Informal Institutions

In contrast to formal institutions, informal institutions cannot be changed through reforms, nor can they be crafted; they evolve through spontaneous socio-cultural processes (North 1990; Saleth and Dinar 2004). Babili et al. (2015) found that local actors in the Babati District, Tanzania applied the logic of appropriateness to match new DFM ideas with socially-embedded institutions before the adoption of new ideas. While the actors and rule-making involved in changes in formal institutions are relatively easy to identify, the evolutionary processes of informal institutions are less transparent, and the key actors and mechanisms are more difficult to identify (DfID 2003).

Existing explanations of the evolution of informal institutions treat these institutions as a historical given but also link their emergence to purported effects, failing to identify the actual mechanisms by which informal institutions are created (Greif and Laitin 2004). An analysis of the sustainability of informal institutions during climate change based on their evolution must go beyond historical contingency and functionalist views. Instead, such an analysis must consider the evolution of such institutions in the light of decentralization, bargaining and elite crafting (Cleaver 2002). In fact, analyzing the evolution of informal institutions is a difficult task (DfID 2003; Helmke and Levitsky 2003). This is because no descriptions of the original institutions were written down, and their evolution was not documented (Ostrom 2007). Notwithstanding these obstacles and since informal institutions evolve very slowly, scholars can take for granted such slow evolution process in studies of marginal changes in informal institutions (North 1990). Generally speaking, informal institutions develop in contexts in which power and resources are unevenly distributed, and, like their formal counterparts, they tend to produce winners and losers (Young 2005; Amanzi 2011).

Informal institutions may change via collapse, replacement by other informal institutions or replacement by formal institutions (formalization) (Ostrom 2007). According to Young (2005), within comparative politics, informal institutions are often characterized as highly resistant to change; hence, they are relatively sustainable as they possess "tenacious survival ability" and often persist even in the face of extensive formal institutional change. Informal institutions are deeply embedded in social practice. It is relatively simple to change the organizational forms of formal institutions, but it is much difficult to change accepted social practices and the ways in which people behave, which are the basis of informal institutions (Cleaver 2002).

In summary, formal institutions can be changed through deliberate reform programs, while the nature of social life within communities makes informal institutions difficult to change, though not all informal institutions fit this characterization of being robust and difficult to change (Ostrom 2003).

3.5.2 Institutional Bricolage

Institutional bricolage is the process of constructing and borrowing different existing institutional elements in order to create a different framework for decision-making and practice (Cleaver 2002). Mehta et al. (2001) point out that the concept of institutional bricolage assumes that actors' decision-making processes and related practices are guided by social norms, but actors can also analyze and react to situations. Of importance in the theory of institutional bricolage is the way informal institutions replace, modify or combine formal institutions while devising local adaptations to climate change. Other practices underlying the theory include cultural borrowing, the adaptation of institutions to multiple purposes and the prevalence of common social principles, which foster cooperation between groups in adaptations to climate change.

Institutional bricolage theory is based on the idea that institutions are constructed through analogies and styles of thought that are already part of existing institutions (De Koning and Cleaver 2012). This theory recognizes the agency of individual actors in negotiating, transforming and adapting newly introduced institutions for climate change interventions. Cleaver (2001) examined institutional bricolage, conflict and cooperation in the Usangu plains, Tanzania, and found that institutions are shaped by the prevailing cultural milieu implied by the concepts of design and crafting in Ostrom's theory (1990). Katani (2010) employed the theory of institutional bricolage in his study on the role of multiple institutions in micro-spring forest management in Ukerewe district, Tanzania, and found that farmers are capable of developing robust self-governing systems that may lead to sustainable forest and water resource management systems. In institutional bricolage, actors draw on existing mechanisms (social, cultural and symbolic resources and relationships as well as their indigenous knowledge systems) to ensure institutional sustainability under the stresses of climate change. Thus, institutions are shaped in the light of contemporary needs (i.e. adapting to climate change) by borrowing from different cultures, by incorporating rules and meanings from one area of life to another and by drawing on the repertoire of local forms of decision-making (Cleaver 2002). Institutional bricolage is an advanced paradigm which helps policy-makers to think beyond "getting the property rights right" to "getting institutions right" (Cleaver 2002). Cleaver (2002) points out that bricoleurs can form new institutions by looking at different institutions (institutional shopping) and selectively adapting norms of external organizations, gradually changing socially embedded institutions (reinventing).

3.5.3 Institutional Legitimacy

A simple definition of a legitimate institution is one that is accepted by those it concerns, not necessarily because they find it right or good, but because it was made in the right way (Vatn et al. in press). This links to people's beliefs about political authority. Legitimacy as acceptability is often called the "descriptive" understanding. There is, however, also a "normative" understanding, which emphasizes that a legitimate institution has to abide by some standards. While there can be no universal definition of these—due to historical and cultural specificities—the normative position demands a justification for the standards that is supported by reason and viewed favorably by society (Habermas 1996; Bernstein 2005). Hence, a distinction can be made between an internal (subjectivist) and external (objectivist) assessment of the legitimacy of an institution. The first concerns the judgment made by the members of a community itself. The second is an evaluation against the standards for a legitimate institution.

A distinction can also be made between the legitimacy of the institutional arrangement process, i.e. "input legitimacy"—and the legitimacy of the results—i.e. "output legitimacy." When considering input legitimacy, issues like participation, representation, accountability and transparency of institutional performance are typically emphasized. Effectiveness, efficiency and equity/justice are key aspects of institutional output legitimacy. Quack (2010) argued that the perception of legitimacy matters, since institutions only prosper if the public views them as legitimate. Legitimacy is granted through social consent, formally or informally given, and thus is an important source of rule compliance, as actors obey the institution once they perceive it as legitimate (Bäckstrand 2006).

Acceptance and justification of authority are core elements of legitimacy. Acceptance refers to whether institutions are accepted by a community as authoritative; justification relates to the reasons that justify an institution's authority (Biermann et al. 2010). Both elements are essential for ensuring institutional sustainability in the face of climate change.

3.5.4 Institutional Performance

Institutional performance is defined as the extent to which existing institutions succeed in achieving goals at the lowest cost (Ostrom 2003). The performance of a given institution is measured by its structure and composition, its decision-making processes, its effectiveness in augmenting integration, its conflict resolution procedure and its ability to make plans according to available resources and avenues for resource mobilization (Sakthivadivel et al. 2004). According to North (1990), institutional performance cannot simply be attributed to combination of formal or informal institutions perse. Rather, high-performing and combined formal and informal institutions operate in an environment at a low cost. This means that stakeholders need to adapt new institutions to the existing institutional setting and improve it by finding the most efficient low-cost equilibrium to ensure sustainability (Msuya 2010).

Lankina (2008) suggested a range of indicators for measuring institutional performance, such as output effectiveness in local policies and services; responsiveness, defined as the degree of congruence between policies, outputs and popular preferences; and process, which refers to, among other things, the transparency and fairness of local officials. Putnam (2001) argues that evaluation of institutional performance should be based on its responsiveness to the demands of the local people. This is because the experiences of local people are the best evaluator of institutional performance (Kajembe and Kessy 2000). The use of indicators to describe institutional performance can show if the defined indicators capture the most important dimensions of that institution's performance. Indicators can be qualitative or quantitative. The indicators chosen should be specific and relevant to the issue at hand, in this case adaptation to and mitigation of climate change. Efforts to find universal sets of indicators have proved futile because the choice of an indicator depends on the objective of the policy, program or project being assessed and the conditions under which the indicators are being used (Kitula 2012).

3.5.5 *Good Governance*

Some aspects discussed above that contribute to institutional sustainability (e.g., participation, transparency, accountability, democracy and legitimacy) highlight that good governance practices are important in ensuring institutional sustainability. The concept of governance means that multiple actors have a role to play in steering societal issues, including climate change. According to Rhodes (1997), governance is a shift from government to governance. Scholars show that the governance of public issue involves various actors, including the public, private actors, NGOs and local people (e.g., Kooiman et al. 2005; Katani and Babili 2012; Derkyi 2012).

The role of multiple actors in governing global issues such as the climate points to the importance of democratization and good governance in achieving institutional sustainability. However, the practice of democracy and good governance is inadequate in Africa and in most developing countries. According to the World Bank (2007), good governance practices are important for ensuring sustainable development. For example, scholars argue that transparency and active public engagement are important aspects of governance for sustainability; hence, good governance can be linked to sustainable development (Kemp and Parto 2005). The United Nations (UN) recently listed 17 issues that require good governance to ensure institutional sustainability and Sustainable Development Goals (SDGs) (see Table 3.3) in all continents, including Africa. The SDGs constitute a goal that requires countries to take urgent action to combat climate change and its impacts (UN 2015).

Table 3.3 Sustainable development goals (SDGs)

1.	No poverty
2.	Affordable and clean energy
3.	Climate action
4.	Zero hunger
5.	Decent work and economic growth
6.	Life below water
7.	Good health and well-being
8.	Industry, innovation and infrastructure
9.	Life on land
10.	Quality education
11.	Reduced inequalities
12.	Peace, justice and strong institutions
13.	Gender equality
14.	Sustainable cities and communities
15.	Partnerships for the goals
16.	Clean water and sanitation
17.	Responsible consumption and production

The key aspects of good governance include accountability, democracy, human rights and rule. Other elements include the rule of law, anti-corruption, transparency, empowerment and participation. Babili et al. (2015) observed that participatory introduction of, negotiation and deliberations on decentralized forest management in the Babati District, Tanzania, enabled institutional changes which were acceptable to a majority of local actors. Hence, adherence to good governance increases the likelihood of institutional sustainability in the face of climate change in Africa and other developing countries.

3.6 The Case Study

3.6.1 Overview

The Iringa rural district's Farmer Managed Irrigation Schemes (FMISs) provide food and income that is very important for a large number of ethnic groups distinguished by various abilities to utilize the schemes (Mosha et al. 2016). Traditional and semi-improved irrigation schemes are implemented along the Little Ruaha River. The Mlenge semi-improved irrigation scheme serves seven villages over a total of 4217 ha, while the traditional Mkombozi scheme servesan estimated 3000 ha, including communities in Itunundu and Mbolibili villages. Irrigators in Farmer-Managed Irrigation Schemes have developed institutions to manage water resources. Depending on the season, there are rules guiding decision-making, e.g., who can have access to water and who can be excluded.

3.6.2 *Empirical Evidence of Conditions Underlying Institutional Sustainability*

3.6.2.1 Evolution, Change and Bricolage

Evolution of Informal Institutions

Initial water management efforts during the pre-colonial period involved the use of traditional beliefs, norms and taboos that shaped people's behavior, encouraging them to use water in a sustainable way. Elders and clan heads were highly recognized and respected as cultural institutional actors. Clan leaders and chiefs were able to monitor and guide access to and use of water for irrigation. They were also responsible for sanctioning those who broke the rules. In the Iringa rural district, the study found that the process of controlling access to and use of irrigation water was crucial for the sustainability of the Little Ruaha River, which provides water downstream for domestic uses, energy production (Mtera plant) and ecological functions.

The informal institutions in the Iringa rural district have evolved over time in the face of climate change. In traditional schemes, there are still a few traditional practices and beliefs in operation. Young people are now questioning the validity of the belief that lack of sacrifices and respect for ancestors could lead to the drying up of water sources. Nevertheless, during the interviews, older people still asserted that violations of such beliefs had led to the current droughts and river water shortages. Generally, focus group discussions showed that "lack of will and social closeness among members led to weak adherence to rules and regulations." In contrast, Babili et al. (2015) reported that the evolution of socially embedded institutions, which included introducing a new rule banning thedirect grazing of calves in a forest that protected spring water in the village of Endanachan village, Babati District, Tanzania, resulted in increased flow and volume in the spring.

Changes in Formal Institutions

The most important radical change in formal institutions in the study area and in Tanzania in general occurred between 1972 and 1976, when the implementation of the Ujamaa (socialism and self-reliance) policy started reinforcing state control over natural resources. This saw compulsory re-allocation of rural dwellers into the newly established villages while abolishing clan settlements. Since then, all villages have been reorganized under a system of village government, giving formal power to elected village leaders. The villagization policy created more heterogeneous communities, potentially disrupting traditional practices as well as making it difficult to follow formal rules which could conserve water resources. It also significantly changed the ownership of water resources, responsibilities for water resources' management and the system of regulating and managing irrigation water in the study area, but it had little impact on how water is distributed in irrigation furrows.

A number of studies (Komakech et al. 2011; Liheluka 2014; Mosha et al. 2016) show the changes taking place in FMISs in Tanzania with regard to the sustainability of water institutions. Currently, the village governments in the study area own the irrigation systems and work closely with formal irrigator associations- the lowest entities in the hierarchy of irrigation water management and use. It was found that irrigator associations had formulated formal constitutions for both traditional and improved irrigation systems. Only an improved irrigation scheme had registered its 'Water Users Association' (WUA). However, in practice, these so-called WUAs are actually irrigator associations, as they did not include other water users. By definition, the WUAs should comprise water users (appropriators) of an irrigation system whose membership is based on ownership of land in the irrigation schemes. All appropriators of a system have to contribute labor and cash towards system upkeep. The irrigators' associations in the studied case had formulated written rules and regulations regarding the allocation and distribution of water, resource mobilization for repair and maintenance, fines for absence during repair and maintenance and default penalties.

Institutional Bricolage

During colonial and post-colonial times, control over and responsibility for resource management was increasingly taken out of the hands of local user groups by the state. The rights to access water and land in the schemes were guaranteed only for clan members, while outsiders who asked permission for access would either receive it or be denied (Mosha et al. 2016). In the Iringa rural district, the post-independency government established institutions to finance construct and manage irrigation canals and other infrastructure. For example, the Rufiji Basin Water Board (RBWB) was established to sustainably allocate and manage water. The introduction of formal institutions to existing informal institutions in the FMISs created a hybrid of the two kinds of institutions, demonstrating the process of bricolage. Although a mix of institutions was observed starting at the basin level, formal arrangements predominate. Informal institutions and practices are often silently embedded in local daily realities, while formal water rights are embodied in official water certificates that permit irrigators to access water. *Msaragambo*, a local term, originating in the Pare Mountains, for collective labor, is another example of institutional bricolage in the study area.

3.6.2.2 Institutional Legitimacy

Locally developed institutions were found to be more legitimate than externally introduced institutions. For example, a locally developed collective was popularly trusted to guide access to water and to address irrigation scheme management challenges. The rules requiring collective action were observed in Mlenge semi-improved FMISs when dealing with stream flooding and the cleaning of intake weirs, which was done at the beginning of a cropping season and at other times as needed (Mosha et al. 2016). The rules of legitimacy also operated in the Mboliboli traditional scheme when villagers collectively constructed temporary intake structures. Farmers adhered to calls to work together due to the legitimacy of locally developed rules. In each irrigator's group, a locally appointed person communicates information by going around the village announcing the need for *Msaragambo*. According to Msuya (2010, cited in Mosha et al. 2016), *Msarangambo* is an old institution associated with cultural integrity and social relations. Both males and females participate in *Msaragambo* work, but women are assigned less physical work such as collecting light wood materials and packing sandbags, while men do more physically challenging duties such as blocking the intake, digging the canals and carrying stones (Mosha et al. 2016).

In contrast, externally imposed institutions lacked legitimacy. For example, the compulsory establishment of formal institutions after the relocation of rural dwellers into villages was unpopular and abolished traditional institutions. Since these institutions were coercive, local people found it difficult to follow them, and they were not effective in guiding villagers to conserve water resources.

3.6.2.3 Institutional Performance

The literature shows that climate change has been associated with reductions in natural resources, including water (Nombo et al. 2013). However, there are other reasons for the decrease in water resources. Mosha et al. (2016) reported that the commercialization of irrigated paddy crops and institutional rule changes—done by the state or influenced by powerful local and national actors with capital—have led to the scarcity of and competition and conflict over water resources in FMISs.

This study found a number of adaptations to climate change that were linked to traditional institutions. Among these were digging small and narrow irrigation furrows to speed up water flow and reduce the water percolation into the soil in order to manage water in an irrigation system. This was done especially while constructing tertiary irrigation canals. In addition, the traditional use of locally available materials to construct intake structures was clearly observed in traditional FMISs. In the study area, it was observed that irrigation ditches are constructed according to indigenous practices that ensure significant water flows by gravity. These observations are a clear testimony to traditional institutional performance (Mosha et al. 2016).

The study found massive conflicts between water users and uses. These conflicts developed because new users from the cities were gaining access to land and water. This is possible due to their financial and political capital, which helps them gain exclusive access rights, which are often guaranteed by the state. As a consequence, resources are overused due to the new comers' use of better farming technologies. This indicates that introduction of formal institutions has decreased performance in terms of successful management, optimal use of water resources and resolution of water use conflicts in the face of climate change.

3.7 Good Governance

The case study in Iringa shows a number of actors in Farmers Managed Irrigation Schemes. These actors include immigrants, youth and older people. Other actors were the state, powerful private actors, small scale farmers and the Rufiji Basini Water Body. Governance of irrigation resources in the face of climate change in the study case will require good governance practices, including the democratization of decision-making in the allocation of irrigation land, water distribution and resource management. However, not much democratization (as good governance practices) has been observed in decisions regarding small holder farmers; for example, village land has been privatized to outside private actors without the endorsement of villagers. Dominance of the state leads to decisions which are not acceptable to the local people. Adherence to good governance practices, including participation, accountability and transparency, are pertinent to ensuring institutional sustainability in Iringa and other similar cases in Africa.

3.8 Conclusion

Adaptation to climate change depends on the sustainability of both formal and informal institutions. Generally speaking, informal institutions are often characterized as highly resistant to change, making them rather sustainable as they posses "tenacious survival mechanisms." However, in this study, informal institutions were found to be less sustainable, as they were often replaced by formal ones. While formal institutions provide constitutional frameworks in which organizations and individuals interact, informal institutions offer norms and informal sanctioning mechanisms to govern ways of doing things. Formal institutions can be crafted, maintained and changed through deliberate reforms by the state or through political processes. On the other hand, informal institutions cannot be changed through reforms, nor can they be crafted; rather, they evolve through spontaneous socio-cultural processes. Moreover, while the actors and rule-making involved in formal institutional change are relatively easy to identify, the evolutionary processes in informal institutions are less transparent, and their key actors and mechanisms are more difficult to identify. Informal institutions may change via collapse and replacement by formal institutions (formalization).

Sustainability of informal institutions is challenged by the introduction of formal institutions, market pressure and a money economy. Furthermore, conditions for institutional sustainability were partially met in the Iringa rural district. Institutional bricolage seemed to be a promising condition for sustainability in the study area. However, neither formal nor informal institutions were completely legitimate, as both formal and informal rules were questioned. Youth question traditional institutions, while old people question formal institutions, which seem to have failed in terms of performance. Good governance practices were found to be inadequate in the case study; for example, local people were not engaged in decision-making regarding the privatization of irrigation land to immigrant private actors. It is recommended that good governance practices and "participatory" institutional bricolage be adopted as solutions for increasing the legitimacy of both formal and informal institutions in the face of climate change. Since sustainable institutions guide the interactions of various actors, it is necessary to apply the *multiple institutional logics of action* (MILA) framework (Babili et al. 2015) as an analytical guide for actors who seek to change institutions to adapt to and mitigate climate change. Application of MILA insights will increase institutional legitimacy and ensure institutional sustainability. It is also recommended to apply the MILA framework in assessing institutional sustainability in the face of climate change.

References

Amanzi NS (2011) Power relations between upstream and downstream common pool resource users: winners and losers in the Uluguru Mountains. Dissertation, Sokoine University of Agriculture, Morogoro (Unpublished)

Babili IH, Wiersum F (2013) Evolution and diversification of community forestry regimes in Babati District, Tanzania. Small Scale For 12(4):539–557. doi:10.1007/s11842-012-9228-9

Babili IH, Mtalo GE, Kajembe GC, der Wal H (2015) Institutional change and institutional performance under decentralized forest management in Babati District, Tanzania. Small Scale For 14(3):381–400. doi:10.1007/s11842-015-9294-x

Bäckstrand K (2006) Multi-stakeholder partnerships for sustainable development: rethinking legitimacy, accountability and effectiveness. Eur Environ 16:290–306

Bandaragoda DJ (2000) A framework for institutional analysis of water resources management in river basin context. International Water Management Institute (IWMI) working paper no. 5. Colombo, p 25

Bernstein S (2005) Legitimacy in global environmental governance. Int Lawa Int Relat 1–2:139–166

Biermann F, Betsill MM, Gupta J, Kanie N, Lebel L, Liverman D, Schroeder H, Siebenhuner B, Zondervan R (2010) Earth system governance: a research framework. Int Environ Agreem Polit Law Econ 10(4):277–288

Blomquist W, Hekkila T, Schlager E (2004) Building the agenda for institutional research in water resource management. J Am Water Resour Assoc 40(4):925–936

Cleaver F (2001) Institutional bricologe, conflict and cooperation in Usangu, Tanzania. IDS Bull 32(4):26–35

Cleaver F (2002) Reinventing institutions: bricolage and social embeddedness of natural resources management. Eur J Dev Res 14(2):11–30

Davies S (1997) Livelihood adaptation: paper for the workshop 'making livelihoods work – women and children in Rajasthan'. Unpublished (Mimeo), Institute of Developement Studies (IDS) Brighton

DFID (2003) Promoting institutional and organizational development. http://www.dfid.gov.uk/document/publications. Accessed 22 Aug 2008

De Koning J, Cleaver F (2012) Institutional bricolage in community forestry—an agenda for future research. In: Arts B, Van Bommel S, Ros-Tonen M, Verschoor G (eds) Forest people interfaces. Wageningen Academic Publishers, the Netherlands, pp 277–290

Derkyi MAA (2012) Fighting over forest: interactive governance of conflicts over forests and tree resources in Ghana's high forest zone. Ph.D. thesis, University of Amsterdam, p 317

Diaz H, Rojas A (2006) Methodological framework for the assessment of governance institutions. http://www.academia.edu/9528931/Methodological_Framework_for_the_Assessment_of_Governance_Institutions. Accessed 13 Mar 2016

Ezekiel AA, Olawuyi SO, Ganiyu MO, Ojedokun IK, Adeyemo SA (2012) Effects of climate change on cassava productivity. Ilesa—East Local Government Area, Osun State, Nigeria. BJASS 10(2):157–158

FEMA (Federal Emergency Management Agency) (2014) What is mitigation? www.fema.gov/what-mitgation. Accessed 13 Mar 2016

Giddens A (1994) The constitution of society: an outline of the theory of structuration. University of California Press, Berkeley

Greif A, Laitin D (2004) A theory of endogenous institutional change. Am Polit Sci Rev 98(4):633–652

Habermas J (1996) Between facts and norms: contributions to a discourse. Theory of law and democracy. MIT, Cambridge

Helmke G, Levitsky S (2003) Informal institutions and comparative politics: a research agenda. University of Notre Dame working paper no. 307. Notre, p 33

Intergovernmental Panel on Climate Change (IPCC) (2001) Climate change 2001: impacts, vulnerability and adaptation. Contribution of working group III to the third assessment report. Intergovernmental panel on climate change. Cambridge University Press, Cambridge, p 256

IPCC (2007) Mitigation of climate change, Working group III. Fourth assessment report, Intergovernmental panel on climate change

IPCC (2014) Climate change 2014. Synthesis report, p 151

Kajembe GC, Kessy JF (2000) Joint forest management in Urumwa Forest Reserve, Tabora, Tanzania: a process in the making. In: Virten P, Nummelin N, Carelica S (eds) Forests, chiefs and peasants in Africa: local management of natural resources in Tanzania, Zimbabwe and Mozambique. University of Joensuu, Finland, Silva Carelica, pp 141–158

Katani JZ (2010) The role of multiple institutions in the management of micro spring forests in Ukerewe, Tanzania. Ph.D. thesis, Wageningen University, Wageningen, p 202

Katani JZ, Babili IH (2012) Exploring forest governance in Tanzania. In: Arts B, van Bommel S, Ros-Tonen M (eds) Forest-people interfaces. Wageningen Academic Publishers, Wageningen, pp 259–275. doi:10.3920/978-90-8686-749-3

Kemp R, Parto S (2005) Governance for sustainable development. Moving from theory to practice. Int J Sustain Dev 8(1,2):12–30

Kisoza LJ (2007) The role of local institutions in the management of pastoral and agro-pastoral systems: a case study of Kilosa and Ngorongoro districts, Tanzania. Ph.D. thesis, Sokoine University of Agriculture, Morogoro

Kitula RA (2012) Stakeholders' analysis and institutional performance in the use and management of mangrove resources in the Rufiji Delta, Tanzania. Ph.D. thesis, Sokoine University of Agriculture, Morogoro, p 314

Knight J, James J (2007) The priority of democracy: a pragmatist approach to political—economic institutions and the burden of justification. Am Polit Sci Rev 101(1):47–61

Komakech HC, Mul ML, Rwehumbiza F, Van der Zaag P (2011) Water allocation and management in an emerging spate irrigation system in makanya catchment, Tanzania. Agric Water Manag 98(11):1719–1726

Kooiman J, Bavink M, Jentoft S, Pullin R (2005) Fish for life. Interactive governance for fisheries. MARE publication series no. 3. Amsterdam University Press, Amsterdam

Lankina T (2008) Cross-cutting literature review on the drivers of local council accountability and performance. Social Dev Depart, Washington, DC, p 60

Leach M, Mearns R, Scooner I (1997) Environmental entitlement: a framework for understanding the institutional dynamics of environmental change. Discussion paper no. 359. Institute of Development Studies, Brighton, p 38

Liheluka EA (2014) Institutional water resources management and livelihood adaptation in Kilombero rural areas, Tanzania. Dissertation, Norwegian University of Life Sciences, Noragric, p 204

Maganga FP, Kiwasila HL, Juma I, Butterworth JA (2004) Implications of customary norms and laws for implementing IWRM: findings from Pangani and Rufiji basins, Tanzania. Phys Chem Earth 29:1335–1342
Mehta L, Leach M, Scones I (2001) Environmental governance in an uncertain world. IDS Bull 32(4):1–9
Mosha DB, Vedeld P, Kajembe GC, Tarimo AKPR, Katani JZ (2016) Reflections on evolving water management institutions and institutional bricolage: acase of irrigation schemes in Iringa Rural and Kilombero Districts, Tanzania. Water Policy 18:143–160
Msuya TS (2010) Developing integrated institutional framework for sustainable watershed management in Pangani River Basin, Tanzania. Ph.D. thesis, Sokoine University of Agriculture, Morogoro
Nombo CI, Urasa JK, Mbwambo JS, Mattee AZ, Mamiro DP, Kabote SJ, Matata L, Synnevåg G (2013) A gendered analysis of climate change impacts and adaptations in semi-arid area farming systems of Tanzania. J Contin Educ Ext 4(1):212–227
North D (1990) Institutions, institutional change and economic performance. Cambridge University Press, Cambridge, p 1000
Ostrom E (1990) Governing the commons: the evolution of institutions for collective action. Cambridge University Press, Cambridge, p 280
Ostrom E (2003) How types of goods and property rights jointly affect collective action. J Theor Polit 15(3):239–270
Ostrom E (2007) Institutional rational choice. An assessment of the IAD framework. In: Sabatier P (ed) Theories of the policy process. Westview Press, Boulder, pp 35–72
Ostrom E, Crawford S (2005) Classifying rules. In: Ostrom E (ed) Understanding institutional diversity. Princeton University Press, Princeton, pp 187–215
Petursson JG, Vedeld P (2012) The "nine lives" of protected areas. A historical—institutional analysis from trans-boundary Mt. Elgon, Uganda and Kenya. Land Use Policy 42:251–263
Pfahl S (2005) Institutional sustainability. Int J Sust Dev 8(1,2):80–95
Putnam R (2001) Social capital: measurement and consequences. J Can Policy Res 2(1):41–51
Quack S (2010) Law, expertise and legitimacy in trans-national economic governance. An Introduction. Socio Econ Rev 8:3–16
Rhodes R (1997) Understanding governance: policy networks, governance, reflexivity and accountability. Open University Press, Buckingham
Sakthivadivel R, Gomathanayagan P, Shah T (2004) Rejuvenating irrigation tanks through local institutions. Econ Polit Wkly 6:3521–3526
Saleth RM, Dinar A (2004) The institutional economics of water. A cross-country analysis of institutions and performance. The World Bank, Washington, DC, p 416
Smajgl A, Larson S (2006) Adapting rules for sustainable resource use. Townsville, CSRO Sustainable Ecosystems, p 289
Stern N (2006) The economics of climate change. Her Majesty's Treasury, London
United Nations (2015) Sustainable development goals: 17 goals to transform our world. www.un.org/sustainabledevelopment/#. Accessed 10 Feb 2016
United Nations Framework Convention on Climate Change (UNFCCC) (2015) COP21 summary, p 31
UNFCCC (2007) Outcomes of completed activities under the Nairobi work programme on impacts, vulnerability and adaptation to climate change. Bangkok, United Nations framework convention on climate change, p 25
Van Koppem B, Sokile CS, Lankford B, Hatibu N, Mahoo H, Yanda PZ (2007) Water rights and water fees in rural Tanzania. In: Molle F, Berkoff J (eds) Irrigation water pricing. CAB International, Wallingford, pp 143–163
Vatn A, Kajembe GC, Mosi E, Nantongo M, Silayo D (2013) Instituting REDD+: An analysis of the processes and outcomes of two pilot projects in Brazil and Tanzania. IIED, London, p 60
World Bank (2007) Worldwide governance indicators. pp 1996–2005. http://www.worldbank.org. Accessed 13 Mar 2016
Young O (2005) Vertical interplay among scale-dependent environmental and resource regimes. Ecol Soc 11(1):27

Chapter 4
Climate Change and Social Sustainability: A Case for Polycentric Sustainabilities

Louise Fortmann, Abraham Ndhlovu, Alice Ndlovu, Rosina Philippe, Crystlyn Rodrigue and Ken Wilson

> *Chikari chinopfumba kunobva chimwe.* (This means, in brief, "We learn from others." More deeply, this proverb suggests that pots are connected by paths made by people going back and forth—in other words, it celebrates a fair sharing of life and the exchange of knowledge.) Shona Proverb.

Abstract Neither climate change nor human societies are homogeneous. Thus, in addressing the issues of climate change and social sustainability, we must consider "*which climate change?*" and "*whose social sustainability?*" Recognition of this difference and complexity calls for diverse locally-based innovation to solve problems rather than top-down universal solutions. This suggests the need to

---

The helpful comments of David O. Hansen and Emery Roe and the assistance of Kristina Peterson are gratefully acknowledged.

Muonde Trust Team contributed significantly to the content of the article along with Abraham Ndhlovu, Alice Ndlovu.

L. Fortmann (✉)
UC Berkeley, Berkeley, CA, USA
e-mail: louisef@berkeley.edu

A. Ndhlovu · A. Ndlovu · K. Wilson
Muonde Trust Team, Zvishavane, Zimbabwe
e-mail: mawere61@gmail.com

A. Ndlovu
e-mail: ndlovuallyc@gmail.com

K. Wilson
e-mail: Ken@muonde.org

R. Philippe
Atakapa-Ishak Tribal Elder, Grand Bayou, LA, USA
e-mail: rpatakapa@yahoo.com

C. Rodrigue
Grand Caillou/Dulac Band of Biloxi-Chitimacha-Choctaw Tribe, Dulac, LA, USA
e-mail: c.rodrigue@ymail.com

R. Lal et al. (eds.), *Climate Change and Multi-Dimensional Sustainability in African Agriculture*, DOI 10.1007/978-3-319-41238-2_4

identify polycentric sustainabilities. For the purposes of this article, the concept of polycentric sustainabilities refers to the differential effects of the biophysical and social characteristics of a socio-ecological system in a particular place and time on what is defined as "sustainable" and what can be sustained by what mechanism. In effect, this multiscalar phenomenon means that policies that assume a single homogeneous sustainability will fail. The argument for taking polycentric sustainabilities into account is illustrated by case studies of two communities (one in Zimbabwe, one in the US) each of which refuses to wait for, much less accept, externally imposed solutions to the problems it faces. These case studies illustrate how these communities have taken the initiative to identify innovative, locally appropriate ways to address their problems.

4.1 Introduction

I begin with the story of Alice Ndlovu. She grew up in a small rural village in Zimbabwe but was able to overcome numerous obstacles to attend college. Her first job after graduating was with a rural development organization implementing programs in villages other than her own in the region where she grew up. Her awakening came when she was asked to implement these programs in her own village. She relates this as follows (Ndlovu 2015):

> When I got employed after graduating, there were many programs that I felt free to implement in other areas. But they were difficult for me to employ in my own community because I understood it better and knew they were never going to work. Then I discovered the Muonde Trust and decided to join my community in solving our problems ourselves.

I tell this story because I suspect that many readers know numerous people, perhaps including themselves, who with the best of intentions are in the business of devising "solutions" for people in "other villages." But these people would almost never be comfortable seeing their solutions applied in "their own villages," in their own places. The implication here is that these professionals almost never have to live with the adverse consequences of the "solutions" they propose because they are not part of the communities where the solutions are implemented.

A major lesson of experiences like Alice's is that it may be best to avoid imposing solutions devised outside of an affected community, particularly from a central place, if one hopes for them to be sustainable. This conclusion is equally applicable when broaching the issues of climate change and social sustainability and leads directly to two questions: *which climate change* and *whose social sustainability*?

Climate change is not a homogeneous biophysical phenomenon. It differs by location. Its effects on social sustainability vary and responses to those effects similarly vary. Likewise, we should not make the mistake of thinking that these are simply biophysical variations. Multiscalar human activities and governance can magnify the effects of climate change. This has obvious implications for action.

As Nobel Laureate Ostrom (2012) wrote in regard to climate change policy on the last day of her life:

> We cannot rely on singular global policies to solve the problem of managing our common resources…We have never had to deal with problems of the scale facing today's globally interconnected society. No one knows for sure what will work, so it is important to build a system that can evolve and adapt rapidly.

Hence, if we are going to talk sensibly about climate change and social sustainability, we need to get "down and dirty" in specific places and the socio-ecosystems in which they are enmeshed.

Clearly, we must also ask: are there social systems that should NOT be sustained? And, if so, who makes these decisions, and how are they made? Many might argue that we would be much better off without sustaining patriarchy, systemic poverty, epistemic injustice, global land grabbing, environmental injustice, violence and militarization. They would further argue that all of these are actually threats to broader social sustainability. How do we rid ourselves of these phenomena? Finally, we must grapple with the fact of climate injustice and how it is intertwined in some of these institutions (see Mearns and Norton 2009). Richard Krajeski, Lowlander Center Board member, (2015, pers. comm. 19 October) argues the following:

> The roots of climate disasters are found in the products of unsustainable folks trying to make their world of consuming appear sustainable—there is a strange sense in which sustainability has become the problem with and of the sustainable.

In this statement, Krajeski is pointing out the social paradox of climate disasters. Climate change is largely fueled by the consumption patterns of the better off. However, the most vulnerable people, many living as sustainably as possible, bear both the brunt of the effects of climate change and, usually, the burden of measures needed to deal with climate change as has been revealed in the struggles around the program for Reducing Emissions from Deforestation and Forest Degradation in Developing Countries (REDD) and indigenous peoples. Meanwhile, there is a discursive cottage industry based on criticizing those same vulnerable people for their ostensible contributions to climate change; these contributions supposedly include excessive birth rates, deforestation, desertification and so on. The fact that these allegations are mostly untrue has not undermined the power of this discourse, and it is important to note the pernicious nature of this discursive phenomenon.

The remainder of this paper is devoted to two topics. First, the argument is made for the need to consider polycentric sustainabilities. Second, the activities of two communities designed to address adverse climate-change-induced problems are discussed. These communities are working proactively. The first, located in a semi-arid area of Zimbabwe, suffers from increasingly erratic and declining rainfall. The second, located in the US, suffers from sea rise aggravated by that country's industrial practices. These communities have refused to wait for external solutions to the results of climate change or to even accept solutions generated outside their respective communities—unless those solutions are implemented on the communities' own terms. They have taken the initiative to seek innovative answers to the problems they face. Both serve as models of polycentric sustainabilities and of how working proactively helps to build and sustain social sustainability.

4.2 Polycentric Social Sustainabilities

The concept of polycentric social sustainabilities is rooted in the work on polycentric governance (most recently climate governance) stemming from the research of what is called the Bloomington School of Political Economy (Cole and McGinnis 2015; Ostrom 2009). Put simply, polycentric governance differs from systems in which decisions are made by a monocentric high-level authority. Polycentric systems are, instead, multiscalar, with decisions made at the most appropriate level of governance (Cole 2011, 2015). Sunderlin et al. (2015) augment this definition in their discussion of climate change governance (REDD+ in particular) by including multiple mechanisms and multiple actors.

The lens of polycentricity reveals new things when focused on sustainability. The concept of polycentric sustainabilities is a heuristic device that highlights the complex and multifaceted performance of sustainability from the ground up. It goes beyond the focus on institutions of governance to the realities of sustainability embedded and embodied in socio-ecological systems. The concept of polycentric sustainability brings our attention to the multilocal and multiscalar character of sustainability; it demands that we be concerned not with social sustainability but with social sustainabilities. The social structures, values and institutions that people want to sustain will, of course, vary by place and over time, and there will be many different sustainabilities. For the purposes of this article, polycentric sustainabilities is taken to mean that the biophysical and social characteristics of a socio-ecological system in a particular place and time affect what is defined as "sustainable" and what can be sustained by what mechanism.

Centrally held control, centrally mandated policies and centrally defined sustainabilities may be irrelevant at best and unjust and destructive at worst. But polycentric sustainabilities that are linked to and enmeshed with locally controlled interactive networks encompassing multiple locations and scales can identify, develop and promote flexible micro-approaches that can respond quickly to local particularities. Furthermore, they can utilize outside resources and insights as appropriate.

Polycentric sustainabilities inform us that the actions necessary for localized sustainabilities will, by this logic, be locally implemented as well. They will be based on local decision-making rather than on the top-down, external decision-making of so-called "subject matter specialists." Local decision-makers may be actively engaged with outside allies, or they may rely almost entirely on their own resources. This suggests that efforts to identify social sustainabilities in the face of climate change cannot be focused on centralized and top-down organizations. In contrast, solutions depend on searching for local remediation based on the experiences of individuals "living the problem" in local communities. Solutions will emanate from learning how these people are affected by climate change and what they propose as solutions.

While polycentricity implies many centers of ideas and actions on multiple scales, the sources of climate change are also to be found in different locations and at different scales than those of the communities that bear the injustices of the results. Local adaptations to climate change may mitigate the proximate cause of local effects while failing to address the external causes. This suggests the need to consider multiscalar horizontal and vertical exchanges of information and analyses and collaborative networks of action.

4.3 Case Studies: The Muonde Trust and the Lowlander Center

The Muonde Trust was founded by the people of Mazvihwa village, which is located in the driest part of Zimbabwe. Much of the information provided below comes from the people of the Muonde Trust themselves, particularly from Alice Ndlovu, the Director of Operations for the Muonde Trust (Ndlovu 2015).

The Muonde Trust describes itself as a "community-based organisation dedicated to fomenting locally-driven creativity and development in the Mazvihwa and neighbouring areas of south central Zimbabwe (Zvishavane District). Through locally driven educational, agricultural and community extension programs, and a healthy dose of action research, it supports indigenous development efforts that maintain the connections between spirit, community and ecology" (Muonde Trust n.d.). Like other Africans across the continent, the people of Mazvihwa have grown weary of externally imposed solutions and so-called "experts" who fail to listen to them or to take into account the specifics of their land and ecosystem. Past "solutions" to their problems failed to work out as promised. For the most part, such "solutions" had their origins in the colonial encounter, when communities faced the twin pressures of dispossession of their land and culture on the one hand and, on the other, "encouragement" to accept development as modernization. Modernization, colonial administrators argued, would result if people accepted external (i.e., colonial) ways of thinking and doing and abandoned their own knowledge and problem solving to rely on passively received external solutions. Since Independence, these same underlying approaches have too often continued through a variety of development initiatives; now, however, they are usually framed in new language. As the people put it, they wanted "no more waiting for someone to target or participate us" (Ndlovu 2015).

The people of Muonde are involved in an ongoing, 30-year, community-owned cultural movement. This movement has been informed by processes reflected in the writings of Freire (1970), a renowned Brazilian educator and philosopher. Building on an initial engagement in participatory research in which maverick researchers facilitated by Ken Wilson sought to de-colonize their doctoral research programs, they have focused on deploying their own knowledge and thinking to nurture, identify and disseminate innovative local solutions to environmental, development and human welfare problems (Ndlovu 2015). In undertaking "indigenous innovation," they have become organic intellectuals (Gramsci et al. 1971) and undertaken civil science action research (Fortmann 2008) (described below) to develop innovations that address the enduring aridity, the loss of trees and soil and the local consequences of climate change. The approaches they have developed are more effective than generic solutions that have been proposed from the outside. And they can be implemented and sustained with local resources. They have developed farmer knowledge and seed exchanges, as well as networks for action.

The Muonde Project Team met in July 10, 2015 to discuss the issue of climate change and how it is being addressed by Muonde village. Among the results of this meeting were the following reflections on climate change and specific examples of activities being undertaken to address its negative impacts.

4.4 A Conversation on Climate Change

A university student from the village explained about greenhouse gases generated by exhaust from automobiles and industrial consumption of fossil fuels. The following quotations are from this July 10, 2015 meeting.

- Vonai Ngwenya: "We hear that the main issues that are causing the problems that we have here (climate change) are caused by people. We wonder how we people from Mazvihwa have caused that. We would love to rectify that and stop this suffering, which is going to affect us and the next generations."
- Britain Madzoke and Sarah Tobaiwa: "We do not have any industries here. Can you explain how having many children is causing ozone layer depletion?"
- Sarah: "We can't waste our energy talking about industries that we don't have. It would have been better if we had the industries because our husbands and children and even [we] would have been employed. Now we have double sufferings: lack of employment and erratic rain patterns caused by 'developed' worlds. These double sufferings have developed because of the industries which are causing us problems."
- Anonymous: "We are not the causers of climate changes. We rarely use cars. We don't have industries; therefore, we have no employment. We are suffering twice: from lack of employment and from buying goods at inflated prices because they have been produced by others. And then we suffer the effects of climate change caused by those big industries which are not helping us."

In short, the people of Mazviwha are suffering from the effects of climate change without being able to affect its causes. Therefore, they have focused their efforts through the Muonde Trust on mitigation and adaptation.

4.5 Muonde Trust Activities

The response of the Muonde Trust to the consequences of climate injustice is one of developing their knowledge and creativity to regain the initiative. They were motivated by the crises in Zimbabwe's formal economy and institutions during the 2000s. Rather than wither away and depend on relief aid, they made a virtue of "indigenous innovation." As one of them put it, "We do not want to be part of the 'donor syndrome.' Everyone who likes to get things for free is like a goalkeeper.[1] You just have to take every ball, even the one hit by a defender. You have to take it" (Anonymous. pers. comm. 10 July 2015) Instead of being kicked at, they went out to score their own goals.

Seed sources provide an example. When Zimbabwe's hybrid maize production capacity and extension services collapsed in the 2000s, communities ostensibly had no resilience. Frantic, heartfelt, but inadequate programs have tried for years to keep them supplied with maize. But Mazviwha and other communities across the semi-arid south undertook a remarkable shift back to ancestral drought-tolerant small grains (i.e.), sorghums and millet).[2] The capacity of people in Mazvihwa to do this drew on indigenous knowledge and ancestral varieties, but it had also been strengthened by the right kinds of relationships with outsiders—relationships that left the people of Mazvihwa in control of the knowledge. A participatory study in this community in the 1980s had identified 55 landraces of local finger millet as well as several hundred other crop varieties (Wilson 1990). This raised local awareness of their heritage value. From 1989 to 1993, Abraham Ndhlovu, now Director of the Muonde Trust, undertook field trials of those and other indigenous varieties, working towards sharing seed-saving approaches. A cluster of like-minded NGO research efforts, especially those of Andrew Mushita and the Community Technology Development Trust, had similar enabling impacts elsewhere.

Muonde is now helping to diversify further in the face of drought through celebratory seed fairs and seed exchanges. During the 2014–2015 season, Mr.

[1]This reference is to the sport of football (i.e., soccer in American English).

[2]Interestingly, the exception proved the rule: they also resuscitated "mascia," a rare example of a sorghum variety that has now become a landrace in Zimbabwe. Mascia was developed in deep collaboration with village women in Mozambique under an International Crops Research Institute for the Semi-arid Tropics (ICRISAT) program funded by the International Development Research Centre (IDRC) in the early 1990s. As a result of experience on the ground in these years, a number of donors also shifted their policies towards open-pollinated varieties rather than hybrids so farmers could keep their own seed.

Magwisanye, a local farmer, did his own scientific trials on four varieties of finger millet, testing for responses to variable rainfall and to soil differences to provide guidance for planting decisions. The Muonde farmers have invited a student to work with them in the spring of 2016 to record observations of community members about traditional varieties of millet and other traditional crops. She will also teach people how to identify, photograph and make botanical illustrations of the crops. These locally designed and implemented research projects will help the community identify the local resources at their disposal to help sustain their community in the face of climate change. Note that national and international research organizations do not know about these resources or, if they do, they consider them unimportant.[3] Their focus is still on breeding varieties that focus on optimization, nowadays often emphasizing short season drought tolerance. Such varieties might mature in, say, 90 days regardless of if or when the rains come. Farmers facing increased variability and unpredictability need varieties with opportunistic, adaptive responses; varieties must also meet criteria regarding suitability for storage, food preparation, taste and vulnerability to birds. Farmers have subtly built these traits into their landraces through the years.

Crucial to Muonde's impact has been the decision to learn from maverick local innovators such as Zephaniah Phiri,[4] who used a watershed approach to make his land capable of harvesting and storing water from bursts of heavy rain. Many years of practical experimentation, carefully evaluated by these farmers and Muonde action researchers, are now being rolled out on a massive scale in the form of swales (i.e., dead level contours), infiltration pits, ponds, gulley reclamations and countless other structures that are carefully integrated into the precise topography and soil ecology of the residents' homes and fields. These structures enable crop survival through seepage and micro-irrigation; they use water supplies for other productive and welfare activities, as well, while rebuilding heavily eroded catchments. Cash-strapped communities with no access to capital or credit are using their ingenuity and labor to develop major farm-level assets (what Blaikie and Brookfield [1987] call "landesque capital") that are making food production more resilient to whatever patterns of rainfall future climates may deliver. These activities are coupled with reforestation and the replacement of brushwood fences with drystone walls (similar to those at Great Zimbabwe from the pre-colonial period) and live hedges.

Clearly, the term "polycentricity" is not one that will readily enter daily use in Mazvihwa, but their approach to sustainability is an exemplar of the concept. Indeed, their emphasis, drawing on culture, on the particularity of their indigenous knowledge and the need for their own solutions is precisely what is needed to

---

[3]The ICRISAT program described in Footnote 4 is, again, the exception that proves the rule.

[4]The Muonde Trust participated in the launch of the Phiri Award for indigenous innovation in food and agricultural systems, named for the great Zephaniah Phiri (1927–2015) (Muonde Trust 2016). This award is given in an effort to encourage the kind of local initiatives that fuel polycentric sustainability.

dynamize polycentric approaches. For example, when asked what kind of approach Muonde takes, members tell you things like the following:

> The Trust actively promotes indigenous environmental preservation methods rather than other modern exogenous methods and practices. For example, specific types of trees were not allowed to be used for firewood, which meant that people were not cutting them. Villagers noted that they had sacred trees and rivers in the past. By failing to cut them, they were actually managing their environment.

Valuing their traditions does not mean that Muonde people want to sustain all dimensions of those social systems. For example, as mentioned in the opening of this chapter, they have made the conscious decision to work towards making patriarchy less sustainable. Thus, the Muonde Trust empowers women with knowledge and information. This also facilitates full information dissemination in the community: the more people have knowledge and the right to take leadership, the more able they are to tackle complex and unpredictable changes. As noted by one of the women in the village, "If we base [our programs] on teaching more men than women, they will go to South Africa with all the information and skills" (Anonymous, pers. com., 13 March, 2015). Thus, residents are addressing social relations and institutions that affect social sustainability at the same time as they are undertaking biophysical experiments and innovation. Indeed, Muonde's approach can be best understood as a cultural movement. They have increased capacity for collective action and local governance through systematic strategic planning, local action research, and increasing women's leadership roles and capacities.

Polycentric sustainabilities may indicate the lack of relevance of centralized top-down science, but they offer the great opportunity of collaboration on new scales and on new terms. Most of Muonde's work is actually research-based, and they have learned much from visitors and volunteers. They say things like: We are doing development which mixes indigenous knowledges with appropriate external knowledges. What makes the mixing work is that different kinds of knowledge are assessed on equal terms and are measured not with someone's science but with the ruler of practicality, namely, whether they work.[5] Within Muonde, there is strong support for access to formal educational opportunities and continuing education for adults whose studies were interrupted by the liberation war of the 1970s and the economic crisis of the 2000s. But, at the same time, Muonde rarely uses formal texts or the written word to share knowledge. Instead, it has an active solar-powered digital video program, and it embeds its insights in the form of memorable songs and dances. This performance engages people physically while explaining such things as the purpose of water harvesting and how to construct a dead level contour to harvest water to irrigate crop fields through seepage. In doing so, they are creating a dialogue that involves everyone on an equal basis, expanding local networks working on social sustainability and taking responsibility for their landscape.

---

[5]As Scott (1998) has noted, "Unlike the research scientist or extension agent who does not have to take her own advice, the peasant is the immediate consumer of his own conclusions."

Muonde's people have allies elsewhere in Zimbabwe and around the globe, but they remain firmly in charge of developing their own sustainabilities. As Muonde Trust member Britain Hove says (pers. com. July 10, 2015), "*Muonde chisipiti mugwenga*," meaning "Muonde is a well in the desert." By this, he means that when things are dry and tough for people, Muonde continues to yield up solutions. That is the kind of sustainability needed for the coming decades of climate chaos.

4.6 The Lowlander Center

Native Americans have lived in the bayous of Louisiana in the US for centuries. Over the last century, the actions of outside oil companies, timber companies and commercial hunters and trappers have destroyed and/or polluted most of their traditional territory and the sources of their livelihoods. Forests have been clear-cut; fur-bearing animals have been hunted to near extinction; and fishing/shrimping and oyster grounds have been overused. Oil companies have destroyed and polluted the bayou ecosystem with a steady cascade of oil leaks and spills as well as the widespread and longstanding practice of digging canals for gas and oil exploration. In 2005, the impact of Hurricane Katrina included the destruction of many homes. Its long-term consequences were aggravated by indifference and incompetence on the part of state and federal agencies as well as NGOs. Five years later, in 2010, local livelihoods were destroyed when oyster/shrimping grounds were severely and often permanently polluted as a result of the infamous British Petroleum oil spill in the Gulf. Rising sea levels, aggravated by the actions of the oil industry, mean that many of these already-embattled communities will be submerged by 2050, perhaps earlier.

Local people are well aware that the natural phenomenon of climate change has been aggravated by actions of outsiders, particularly the oil companies. In words that echo the words of the members of the Muonde Trust Team above, tribal member Ruby Ancar (Oil Spill Threatens Native American "Water" Community 2010) says the following:

> Nature you can't control. We can't control hurricanes, and people can't control a tornado. But when you have things that man makes and that destroy a person's way of life or [an] entire village or [an] entire community—I mean, that's uncalled for.

4.7 Two Stories of Polycentric Sustainability in the Bayou

Asked to describe local issues of sustainability in the face of Oil Spill Threatens Native American "Water" Community. June 2010. *National Geographic* climate change and how her community was addressing them, Rosina Philippe, elder and tribal historian of the Atakapa-Ishak/Chawasha Tribe in Grand Bayou Village,

Louisiana, wrote the following for this chapter situating the specificities of her community historically, geographically, ecologically and socially.

> Grand Bayou Village has always depended on the land and waters surrounding the community for food and medicines. From the waters, we harvested the freshest fish, shrimp, crabs and other seafood. From the land, we harvested leafy plants, roots, fruits and berries. It was not so long ago that we grew many of these foods, along with other more common and domestic foods, in "homestead plots," each homestead having its own garden plot. In the true sense of community, what we did not use for our own family, we bartered in exchange for things that we did not grow. Good food and the pride of being self-sustaining are two of the components necessary for a resilient and viable community.In Grand Bayou Village today, we find ourselves in a battle to continue our subsistence lifeways because of environmental degradation and extreme climate change. Erosion and subsidence, two processes that are very familiar to coastal peoples, have accelerated. What used to take years to change, we see changed in months, due in large part to human activities which have induced changes and created negative impacts on the natural processes we had been able to predict and rely on. Canals cut through the coastlines, and some of the most aggressive extractive industries operating in our waters can be pointed to as major contributors to our dilemma. We also see "the big picture," [which] many who do not encourage nor respect our lifeways tend to use as one of the reasons—a.k.a., "excuses"—for their way of doing business; and, therefore, [they claim 'the big picture'] must trump (pun intended) our ways of life. The big picture we see is one of global environmental abuses that continue to promote a consumerism cycle…the more we use, the more we need: very destructive when the supply is finite. The very energy used to fuel that concept is a major contributor to the destruction of environmental protection. Life cycles and seasons are becoming less predictable, and the stressors of surviving these changes have multiplied exponentially.The big question is, "How does a subsistent community remain self-reliant and viable while maintaining the integrity of self as an environmental symbiont?" In Grand Bayou Village, we are drawing on our historic knowledge of adaptation and incorporating new elements to achieve the results that will help us retain our lifeways and maintain a presence in our sovereign lands. The ability to feed oneself is a basic necessity. Because of environmental challenges, we are no longer able to have our garden plots. Incoming tides bring salt water onto the land, weakening and/or eventually killing all but the salt-water-tolerant vegetation. We have brought in HESCO baskets[6] to use as raised-bed gardens; as the land continues to degrade around us, however, we must look to other means of growing. The homes in the village are constructed 12–15 feet above grade, thereby providing room for hanging gardens. (We have grown tomatoes in this manner.) We are only able to grow around the outside edges of the house, where the plants have sunlight. This year, we will be growing plants on "floating" gardens. We have tried growing in small containers. But when the tides of salt water rise to the land, those containers have to be moved, which has proven to be both time consuming and laborious as some of them are heavy. This past summer, containers that had been successfully moved to avoid the salt-water tides fell prey to predation from local wildlife (rabbits). We will be constructing raised-bed garden platforms with flotation devices ([comprised of] Styrofoam or small drums) attached to the undersides; as the incoming tides cover the land, our gardens will rise and settle again when it recedes.As we continue to mitigate how we live with the changes to our environment, we know that we must reach beyond our small community to find other lifeways of survival and being. One project currently underway is creating a "food forest," an idea that we are borrowing from our Pacific Islander family. We are

[6]A HESCO basket is a cloth-lined, welded wire basket (4′ × 4′) attached in a series of five, which, when filled with garden soil, serves as a sturdy raised-bed garden.

> planting food-producing trees ([which can] be grown in similar climates and soil) to create a forest that will not only be able to provide foods for people but will be a source of food for wildlife in our area, including migratory fowl.As coastal people, we cannot ignore our waters. We continue to advocate for clean waters, not only for our benefit but also to benefit the abundant marine populations they contain. We continue to depend on some of these marine species for our economic needs but, primarily, we use them as a fresh food source. With the modern advances in preserving and freezing foods, our people had weaned ourselves from the traditional ways of preserving foods. However, with the ever-increasing cost of energy bills to provide electricity to homes and the appliances for freezing and storing foods, we are turning more to our traditional ways of sun-drying foods and preserving with sea-salt; both [resources are] in abundant supply (sun and salt).Grand Bayou Village is not alone in the struggle to live more sustainably and [to be] environmentally conscious. We are joined by a global family of like-minded individuals who are not only working for their own benefit but are working to raise the awareness of others to benefit a global community. We are in this to win. The environmental challenges may differ in some places, overlap in others [and] even mirror in yet more places. What we know is that meaningful change to how we live with this planet will truly begin when people begin to use "we" in place of "them"; and the "we" will reflect all lifeforms. One planet…without meaningful change to sustainability, it's me today but you tomorrow. The life-healing message to our future lies in our past and with people who hold that information as a sacred trust.

Crystlyn Rodrigue, who, although still a young woman, is Deputy Chief of the Grand Caillou/Dulac Band of the Biloxi-Chitimacha-Choctaw Tribe. Asked the same question as Rosina Philippe, she wrote the following, reflecting on her growth as a young leader:

> The Grand Caillou/Dulac Band of [the] Biloxi-Chitimacha-Choctaw Tribe is located in Dulac, Louisiana, southwest of New Orleans on the Gulf Coast. Our existence is being threatened by land erosion, saltwater intrusion, sea level rise and big oil companies.Over the decades, we have experienced a tremendous amount of land loss. We are people who are very much connected to our surroundings. We have a vast amount of understanding when it comes to our environment because of centuries living so close with nature. We live off the richness of the land, having compassion for our plants, animals and our way of living. This is taught from generation to generation based upon our lived experience. I grew up on a boat shrimping with my father. My father was eight years old when my grandfather took him out of school to pursue shrimping as a way of life. Unfortunately, what once was a sustainable way of life is no longer. In my short lifetime, I have witnessed significant environmental challenges from land loss, declining fisheries, water quality problems and climate change. It is common where I live to have a river running behind your house. We call it a bayou. Unfortunately, this bayou causes flooding to the land, high tides and salt-water intrusion. The intrusion is a movement of saline water into fresh water causing an accelerated rate of vegetation death. This also leads to land loss. Water continues to get closer and closer to homes. In my backyard, you can actually see that [the area] where there once was land to play and plant is now gone. All of these factors make it hard to plant food, even to preserve our healing and medicine plants.I once read a quote: "When an elder dies, a library burns." We have always talked about our plants and how they work and what they were used for, but no one had ever written down the knowledge that was talked about until now. This opened the door to an ethnobotany project. I volunteered to work with two college students, which meant forming a relationship with each one by understanding and learning from each other. This is an ongoing project. Unfortunately, we are not the only coastal tribe dealing with all of these issues. We took it upon ourselves, in 2012, to form the First People's Conservation Council with four coastal tribes to establish and solve the

issues we face on a daily basis.I was 19 when I was specifically picked to be Deputy Chief of my tribe. I had no idea what I was getting myself into, but I figured there had to be a reason for why I was picked. The very first meeting I attended was a Costal Protection and Restoration meeting held by the State of Louisiana to form a state Master Plan. This meeting was about protection and restoration of the coast. The most at-risk coastal communities were cut out of the plan due to cost. I remember crying to my Chief, asking her, "How did we let it get to this point?" I partially blamed myself for being so blind for most of my life [to] what was happening all around me. My Chief let me know, in that very moment… my heart [was the reason] that she chose me to be her deputy. That was the moment I knew that this was my calling. I was not going to give up on my land or my people.

4.8 A Network of Polycentric Sustainabilities

Through the Lowlander Center, indigenous Bayou residents have undertaken actions to sustain their communities in the face of climate change as aggravated by big oil. This has involved working with other indigenous communities (including Native Alaskans who experienced the Exxon Valdez Disaster) and collectively through indigenous coalitions and state, national and international organizations to ensure that indigenous residents, not outside "experts," will decide the terms under which they act as sea levels rise.

In this network, we see polycentric sustainabilities operationalized. While in global terms, these communities are located in relatively small areas, they are still very different places. The forms of and mechanisms for achieving sustainability differ among them and simultaneously echo what is happening in far-away Mazvihwa, where too little water is the problem rather than too much. Some communities are willing to relocate and are organizing to ensure that they can relocate together, thus keeping their communities intact. In those communities, the ethnobotany referred to by Crystlyn Rodrigue and the site-specific knowledge of soils mentioned by the people of Muonde are essential to knowing what plants they can take with them to the new site. Similarly, other communities, determined to stay in their tribal homelands, are actively exploring the social and technological strategies that will enable them to do this. Rosina Phillippe and the people of Muonde mention technical strategies of adaptation. Bayou communities are undertaking participatory mapping of settlements, territories and fishing grounds. They are systematically educating scientists, policy makers and legislators who, under ordinary circumstances, would determine their fate from positions of ignorance.

4.9 Conclusion

I have used the concept of polycentric sustainabilities to draw attention to the multilocal and multiscalar nature of social sustainability. This concept highlights the diverse pool of experiments with processes to create and/or sustain pathways to social sustainability.

The concept of polycentric sustainabilities does useful work here. It enables us to see the similarities between the communities, one in Zimbabwe and the other in the Louisiana bayou. Both are instances of polycentric sustainability. The case studies show the processes through which people in very different socio-ecosystems are dealing with the challenges of climate change. While they use some of the same processes and tools, among them traditional knowledge, indigenous innovation, participatory mapping and strategic networks with outsiders, their actions and outcomes vary with the biophysical and social (including historical) characteristics of their socio-ecological system. No remote central authority could begin to accomplish what the Muonde and bayou people have accomplished. Using the lens of polycentric sustainability, would-be development workers can see that they need to step back and understand what is happening locally before they take actions which may be colossally counterproductive.

In our own places, while we certainly can look upward for ideas for dealing with climate change and social sustainabilities, we do not need to do so. Rather, we can look within and around us for inputs that can be used to craft our own sustainabilities and our own pathways towards them. And, as we do this, we must resist the temptation to impose our solutions on other people's villages.

References

Blaikie P, Brookfield H (1987) Land degradation and society. Methuen, London

Cole DH (2011) From global to polycentric climate governance. In: European University Institute RSCAS working papers. http://papers.ssrn.com/sol3/papers.cfm?abstract_id=1858852 . Accessed 1 Feb 2016

Cole DH (2015) Advantages of a polycentric approach to climate change policy. Nat Clim Change 5:114–118

Cole D, McGinnis M (eds) (2015) Elinor Ostrom and the Bloomington School of Political Economy: polycentricity in public administration and political science, vol 1. Lexington Books, Lanham, Maryland

Fortmann L (2008) Introduction. In: Fortmann L (ed) Participatory research on conservation and rural livelihoods: doing science together. Wiley-Blackwell, Oxford, pp 1–17

Freire P (1970) Pedagogy of the oppressed. Continuum, New York

Gramsci A, Hoare Q, Nowell-Smith, G [translator] (1971) Selections from the prison notebooks. International Publishers, New York

Mearns R, Norton A (2009) The social dimensions of climate change: equity and vulnerability in a warming world. World Bank. https://openknowledge.worldbank.org/handle/10986/2689. Accessed 12 Sept 2012

Muonde Trust (2016) The Muonde Trust homepage. http://www.muonde.org. Accessed 13 Jan 2016
National Geographic (2010) Oil spill threatens native American "water" community. https://www.youtube.com/watch?v=nKtCpRfxaE8. Accessed 17 Apr 2015
Ndlovu A (2015) Muonde Trust: no more waiting for external solutions. Talk presented at the Department of Environmental Science, Policy and Management, University of California at Berkeley. 17 April, 2015
Ostrom E (2009) A polycentric approach for coping with climate change. Background paper to the World Bank's world development report 2010. www.10.iadb.org/intal/intalcdi/pe/2009/04268.pdf. Accessed 1 Feb 2016
Ostrom E (2012) Green from the grassroots. http://www.project-syndicate.org/commentary/green-from-the-grassroots. Accessed 30 June 2012
Scott JC (1998) Seeing like a state: how certain schemes to improve the human condition have failed. Yale, New Haven, CT
Sunderlin WD, Sills EO, Duchelle AE et al (2015) REDD+ at a critical juncture: assessing the limits of polycentric governance for achieving climate change mitigation. Int For Rev 17:400–414
Wilson K (1990) Ecological dynamics and human welfare: a case study of population, health and nutrition in southern Zimbabwe. Unpublished dissertation, University College London

Part II
Sustainable Management of Natural Resources

Chapter 5
Land Degradation Neutrality: Will Africa Achieve It? Institutional Solutions to Land Degradation and Restoration in Africa

Luc Gnacadja and Liesl Wiese

Abstract Land has gained stature as a natural resource to be protected since the 2012 Rio+20 Summit. The decision to "strive to achieve a land-degradation neutral world in the context of sustainable development" (United Nations 2012, p. 54), along with the understanding that "this should act to catalyze financial resources from a range of public and private sources," served as the foundation for further discussions about land. In 2015, land was included in the 2030 agenda for sustainable development as target 15.3 of Sustainable Development Goal 15 (SDG 15). Land has been defined as a global strategic resource that is under stress due to rapidly growing demand for biomass production to accommodate increased food and energy consumption and population dynamics. More than half of the additional two billion people who will live on Earth by 2050 will be born in Africa. The population of sub-Saharan Africa (SSA) is predicted to grow from 900 million to about 1.4 billion by 2030. However, the region leads the world in poverty, hunger and food insecurity, youth unemployment, agricultural vulnerability to climate change, land degradation, resource-based conflicts, and migrations leading to economic, social, and political instability. Africa has the resources to respond to the exploding global demand for food, energy, and water. The region accounts for 60 % of the world's uncultivated arable land and more than one third (715 million ha) of the world's degraded land with potential for restoration. It is already a global hotspot for successful land restoration projects due to innovations in technology and social engineering, which mostly occur at a local level. The institutional aspects

A contribution to the International Conference on Climate Change and Multi-Dimensional Sustainability in African Agriculture, Sokoine University of Agriculture, Morogoro, Tanzania, June 3–5 2015.

L. Gnacadja (✉)
Governance and Policies for Sustainable Development - GPS-Dev,
01 BP 308, Cotonou, Benin
e-mail: luc.gnacadja@gps-dev.org

L. Wiese
Land and Water Division, Food and Agriculture Organization of the United Nations,
Via delle Terme Di Caracalla, Rome 00153, Italy

R. Lal et al. (eds.), *Climate Change and Multi-Dimensional Sustainability in African Agriculture*, DOI 10.1007/978-3-319-41238-2_5

relating to the establishment, management and support of restoration projects, however, are one of the major obstacles to scaling up these projects. Achieving land degradation neutrality will be key for the economic, social, and environmental transformations SSA needs to accomplish, given the challenges and potentials of the region, such as climate change, population dynamics, and globalization.

Keywords Land degradation · Desertification · Land restoration · Land degradation neutrality (LDN) · Agricultural expansion · Adaptation · Poverty · Food insecurity · Sustainable development goals (SDG)

5.1 Introduction

5.1.1 The Global Importance of Land Degradation

In a broad sense, land degradation (LD) encompasses the degradation of soil resources, vegetation, water, and other biota (Le et al. 2014) as well as the interactions between these types of degradation within terrestrial ecosystems. Degradation causes negative changes in the capacity of affected ecosystems to provide biological, social, and economic services (FAO 2011). Of the earth's total land surface, about 60 % is managed by humans, and of this land, about 60 % is used for agriculture (ELD Initiative 2015b). LD is a pervasive and extensive problem occurring in these managed areas across all regions and ecosystems. However, it disproportionately affects rural, poor areas in developing countries. In drylands, it may transform productive landscapes into deserts—hence the term "desertification," which is given to LD in arid, semi-arid, and dry-subhumid areas (UNCCD 1994). Desertification may therefore be considered a subset or type of LD.

Despite being a widespread and intensifying phenomenon, LD is a perception-laden concept. Its extent is unclear due to the different methods that are used to assess the problem as well as the different types of environments that are included/excluded in definitions. Consequently, estimates of the area affected by LD in the global drylands range from 10 to 20 %, a percentage that is increasing at an annual global rate of 12 million ha of soil lost "from desertification and drought alone" (ELD Initiative 2015a, p. 8). A report on the state of the world's land and water resources (SOLAW) for food and agriculture states that globally, 25 % of land is highly degraded, 8 % is moderately degraded, and 36 % is stable or slightly degraded, while land improvement is seen in only 10 % of global land (FAO 2011). According to some more recent estimates, 52 % of agricultural land worldwide is moderately or severely affected by LD (ELD Initiative 2015b) and directly affects 1.5 billion people (Bai et al. 2008). Based on its severity, extent and relationship with conflict risk, LD was recently qualified as an underestimated "threat amplifier" (Van Schaik and Dinnissen 2014, p. 11). In Africa, "land issues have played a major role in 27 major conflicts" since 1990 (ELD Initiative 2015a, p. 9). Overall, as much as one-third of the world's arable farmland was lost due to erosion in the last 40 years (ELD Initiative and UNEP 2015).

Beyond its direct impact on the vitality of affected ecosystems and the livelihoods of affected populations, LD hampers the economic development of affected territories and countries. Using total economic value and valuation methods, the Economics of Land Degradation (ELD) Initiative concluded that, globally, "the lower estimate of lost Ecosystem Service Values of USD 6.3 trillion/year is more than five times larger than the entire value of agriculture in the market economy" (ELD Initiative 2015b, p. 61). The estimated economic loss due to LD is USD 434–720 per hectare and USD 870–1450 per capita per year (ELD Initiative 2015b).

A primary multilateral agreement to combat desertification was entered into force in 1996 as part of the United Nations Convention to Combat Desertification (UNCCD). The efficacy of the UNCCD in terms of reversing desertification in affected countries, especially in Africa, has been hampered by weak scientific evidence; a poorly defined interface between science and policy; a lack of inclusion of issues related to LD in national strategies for economic growth, poverty alleviation, and overall sustainable development; and a consequent lack of financial support and investment to reverse desertification (UNCCD 2007).

5.1.2 The Strategic Importance of Land for Sustainable Development in SSA

In SSA, land is the main resource supporting the population and is a key asset for poor individuals. Agriculture in Africa accounts for 30–40 % of the region's GDP and employs 65–70 % of the labor force (World Bank 2013). The agricultural outputs in SSA are mostly produced by smallholder farmers, about 70 % of whom are women (AGRA 2014; Bayene 2014). Smallholder farms (2 ha or less) account for about 80 % of all the farms in the region (AGRA 2014), and they often include degraded land (Nachtergaele et al. 2008). In addition, the majority (70 %) of poor individuals in Africa live in rural areas (World Bank 2013), and they mainly run family farms or smallholder agriculture systems to support themselves (World Bank 2013; FAO et al. 2015). Degradation often entrenches these farmers into a vicious cycle of loss of livelihood, extreme poverty, further degradation, food and water insecurity, and a lack of overall security (UNEP/GRID-Arendal 2005).

The population of SSA is predicted to grow from 900 million to about 1.4 billion by 2030 (United Nations Population Division 2011) and to 2 billion by 2050 (United Nations Population Division 2011). The region's rural population will grow by almost 50 % between 2015 and 2050 (United Nations Population Division 2011). This growth will substantially increase the demand for land due to increased living space and food production needs. Substantial increases in agricultural output will be essential to meet the rising demand for food in Africa, which will, in part, require increasing the yields of existing croplands and/or expanding the land that is available for cultivation (Chamberlin et al. 2014). In Africa, the growth in crop production over the past four decades has largely been a result of area expansion rather than increased yield, with an estimated 40 % increase in cultivated land area between 1990 and 2012 (Heady 2015). Although not discussed in this paper, it should be mentioned that increased demands for food should be addressed not only through increased production but also by ensuring that crop losses and food losses from fields to markets are minimized, food waste is reduced, and food is distributed to areas where it is needed.

Climate change is likely to further increase the need for improved food production in Africa, where growing conditions are often already harsh (UNDP 2012). If not addressed, climate change is expected to increase the long-term stress on land resources through increased desertification, increased incidence and intensity of extreme weather events, and more erratic rainfall events in terms of both timing and intensity (The Montpellier Panel 2014). Increasing farms' resilience to climate change is thus essential for securing long-term sustainability in agriculture.

With the combination of high population growth, high vulnerability to climatic shocks, poor yield, and increased land degradation, Africa is facing major sustainability challenges regarding land use for agriculture. As a result of these increasing pressures, there have been increased efforts during the past ten years to include agriculture in Africa's economic transformation. The continent's economic development has largely been—and will continue to be—driven by agriculture (IFAD 2001). Productivity levels in Africa are generally low and are exacerbated by natural resource degradation and insufficient investment in rural infrastructure and agricultural inputs (Rosenzweig and Parry 1994). In SSA alone, an estimated 180 million people are affected by land degradation and the resulting economic losses are USD 68 billion per year (The Montpellier Panel 2014).

The growing land restoration movement at grassroots level in some African countries such as Ethiopia, Niger, and Malawi are making Africa a global hotspot for land restoration. However, economic growth has not yet reduced food insecurity and rural poverty; increased investments in land restoration are needed in SSA.

5.1.3 The Cure for Land Degradation: Prevention, Rehabilitation, Restoration

Depending on the context and expected outcomes, the cure for LD is either referred to as land rehabilitation or land restoration. Restoration generally involves returning degraded land to its original or natural state, and rehabilitation involves taking degraded land and restoring some of its productive functions, even if they are different from its previous functions. Therefore, land restoration is often hampered by insufficient information about the initial state of the degraded land. Also, with climate change altering the climatic conditions in some areas, restoration may be difficult. In these cases, rehabilitation may be easier, more climate-appropriate, and better in terms of economic return. In recent years, the distinction between restoration and rehabilitation has been blurred in political debates and in efforts to address LD.

Therefore, given the purpose of this paper, we have chosen to use the term restoration to indicate the return of degraded land to at least some form of productivity. Nevertheless, it is worth mentioning that preventing LD is more cost-effective than restoration or rehabilitation. According to the SOLAW report (FAO 2011), preventive interventions are successful for 36 % of land that is stable or slightly degraded.

5.2 Land Degradation Neutrality: Emergence of a Goal to Effectively Address Land Degradation

The concept of land degradation neutrality (LDN) was developed by the UNCCD secretariat and first came to the fore at the United Nations (UN) General Assembly meeting on desertification, land degradation, and drought (DLDD), which was held in New York on September 20th, 2011. The President of the General Assembly recalled that land degradation corrodes the three pillars of sustainable development. He defined "the four priority actions needed to seal the gaps hindering progress" and stressed that "if the international community was serious in its commitment about reversing land degradation and desertification, the time had come to commit for building a land degradation neutral world, to set measurable sustainable development targets towards zero net land degradation" (Al-Nasser 2011, pp. 1–2).

The concept was initially called "zero net land degradation" (ZNLD) and was defined during discussions about sustainable development goals (SDGs). ZNLD is achieved by reducing the rate of land degradation on one hand and increasing the rate of restoration of degraded land on the other (Grainger 2014). The overall goals of a neutral state of land degradation are to restore more land than that which is degraded, to improve the productivity of land resources through sustainable management and restoration of soil, water, and biodiversity, and to contribute to poverty reduction, food and water security, and climate change adaptation and mitigation (Gnacadja 2014).

The concept was further discussed at events such as the 10th session of the UNCCD Conference of the Parties (COP) which was held in October 2011 in Changwon, Republic of Korea, and the preparatory process meetings of the United Nations Conference on Sustainable Development (UNCSD or Rio+20). LDN was mentioned in the outcome document of the Rio+20 conference as part of the decision to "strive to achieve a land-degradation neutral world in the context of sustainable development" with the understanding that "this should act to catalyse financial resources from a range of public and private sources" (United Nations 2012, p. 54).

LDN is also expected to advance the effective integration and/or mainstreaming of issues related to the degradation and restoration of terrestrial ecosystems into local, national, and regional strategies for sustainable development. Implementation will be measured and monitored in the context of set targets and indicators. LDN is considered to be achieved when "the amount and quality of land resources necessary to support ecosystem functions and services and enhance food security remains stable or increases within specified temporal and spatial scales and ecosystems" (WBCSD 2015, p. 4). The SOLAW 2011 report featured some intervention options for each type of degradation (FAO 2011). Achieving LDN was cemented into future sustainability and land restoration targets through the development of SDG 15, which proposes the "[p]rotection and promotion of sustainable use of terrestrial ecosystems, to halt desertification, land degradation and biodiversity loss." Specifically, SDG target 15.3 stipulates that, "[b]y 2030, [we will] combat desertification, restore degraded land and soil, including land affected by desertification, drought and floods, and strive to achieve a land degradation-neutral world."

The concept of LDN involves (1) preventing and/or reducing LD, (2) restoring partly degraded land, and (3) reclaiming desertified land (Chasek et al. 2014; World Business Council for Sustainable Development 2015). However, the envisaged balance between land degradation and restoration should not be seen as a license to degrade or a way to compensate for land degradation in one area by restoring the land in another. As stated by Chasek et al. (2014), what counts is the combined LDN achieved by local communities across the globe, in line with the saying "think globally, act locally."

Land is desired by several competing groups such as the agriculture industry, the forestry industry, pasture owners, advocates for urbanization, and companies interested in the extraction of raw materials. Globally, according to the medium estimate for a business as usual (BAU) scenario of food production, additional cropland equivalent to the size of South Africa will be needed by 2030 (UNEP 2014). BAU implies that more forests and natural ecosystems will be depleted unless crop yields are improved and degraded lands are restored to production, especially in Africa. Land has therefore become a strategic commodity to be secured as a resource for addressing future biomass production and consumption needs. For instance, the 2008 food crisis led to large-scale land acquisitions, especially in Africa. In 2009, the demand for land exceeded that of the previous 20 years (Deininger and Byerlee 2011).

Achieving LDN by 2030 calls for multilevel and multisectoral approaches to address LD. This includes addressing specific forms of LD that take soil out of biomass production, such as mining and soil sealing for building urban infrastructure. However, this paper mainly focuses on achieving LDN in agriculture, the sector that is considered to be the most promising in terms of its contribution to land restoration in SSA. Its implementation is envisaged at various levels (i.e., local, national and regional) and will be achieved by setting sustainable land management (SLM) and ecosystem or land restoration targets. This involves setting baselines for monitoring, evaluating trade-offs, and prioritizing ground-level actions at appropriate scales. In order to use the concept of LDN in practice, the UNCCD launched the LDN Project at an inception meeting in 2015. The project aims to help achieve the outcomes of Rio+20 by providing the UNCCD with empirical evidence on the feasibility and cost-effectiveness of LDN. The project will work in close collaboration with the UNCCD Intergovernmental Working Group (IWG) to determine how SLM can contribute to LDN (UNCCD Press Release 2015).

In the context of preparatory consultations with the UNCSD, Africa was the first region to adhere to the concept of LDN (Rio+20 et al. 2011), perhaps because of the challenges associated with its agricultural land use and LDN's potential to serve as a framework for promoting potential solutions.

The primary focus of this paper is whether Africa will achieve LDN by 2030. This is determined by analyzing the information and data that are available in literature or ongoing initiatives related to agriculture, and recommendations for institutional solutions to LD are provided.

5.3 Land Degradation, Agriculture, Poverty, and Food Insecurity in SSA

5.3.1 Land Degradation in SSA

Africa is considered the world's most vulnerable region to LD and desertification, but estimates of its extent vary, with some estimates only focusing on certain areas. In 2001, it was estimated that more than 45 % of Africa was affected by desertification (Reich et al. 2001). According to a more recent publication, "more than one third of the land in Africa is under threat of desertification as well as one third of the population" (ELD Initiative 2015c, p. 2). Overall, "65 % of arable land, 30 % of grazing land and 20 % of forests are already damaged" by LD in Africa (The Montpellier Panel 2014, p. 1).

Le et al. (2014) identified LD hotspots across the globe using remotely sensed vegetation index data, indicated as a decline in biomass production. Figure 5.1 shows the hotspots in SSA with significant declines in biomass productivity, considering various weights and corrections for local conditions. Although such an assessment does not provide information on biomass-related degradation, such as bush encroachment, it does provide an overview of biomass decline that can be used to identify sites that require more detailed investigations. It is assumed that where biomass is declining due to human intervention, soil is likely to be exposed to the elements, which increases the risk of various forms of soil degradation.

In agriculture, extending crop production in underperforming or poor-quality soil is common in SSA and is often accompanied by poor management practices with little or no external input. This leads to further reduction in soil quality and stagnation or decline in crop production (Diagana 2003). As soil in SSA is particularly vulnerable to degradation, especially in dryland areas (Nachtergaele et al. 2008), agricultural productivity and food security in the region are equally vulnerable. Soil fertility is steadily declining due to constant nutrient mining in underperforming soil, which decreases soil productivity (The Montpellier Panel 2014).

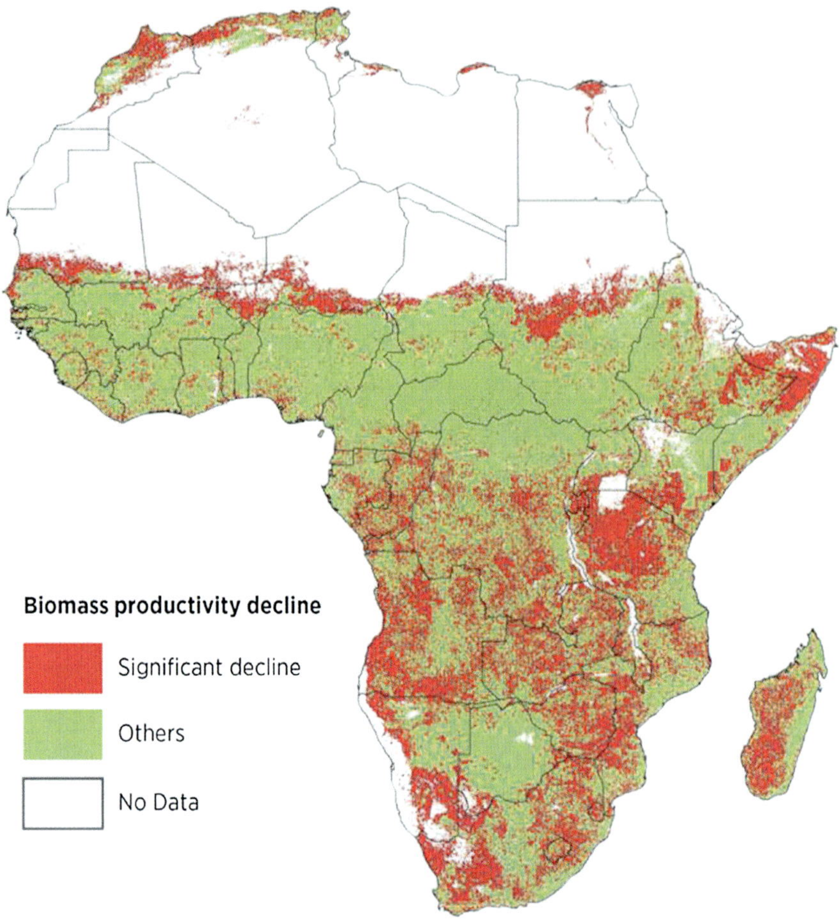

Fig. 5.1 Estimates of hotspots of biomass productivity decline in SSA, as determined by Le et al. (2014). Image taken from The Montpellier Panel (2014)

Soil degradation includes various forms of chemical, physical, and biological degradation such as loss of topsoil (erosion by wind or water), loss of organic matter, salinization/alkalization, acidification, pollution, compaction/crusting, and waterlogging (Diagana 2003). Such degradation is partially caused by a number of human actions like overgrazing, deforestation, cultivation of unsuitable soils (i.e. on steep slopes), and inappropriate agricultural activities and practices. Once soil becomes too degraded, most land users abandon the land, which often leads to the clearing of natural vegetation for cultivation purposes. This is especially true in poor rural areas, where land users do not have the financial means to invest in soil restoration.

Although estimates vary, it is clear that LD is a serious challenge in SSA and that concerted land restoration efforts are necessary, especially in the agricultural sector, if LDN is to be achieved by 2030.

5.3.2 Drivers of Land Degradation in Africa

Knowing and understanding the drivers of LD are key for developing appropriate institutional solutions to reduce degradation, increase land restoration, and, consequently, achieve LDN. The main drivers of LD are (Muchena et al. 2005; WMO 2005; Kiage 2013; The Montpellier Panel 2014; Tully et al. 2015):

- *Increasing population pressure*—The predicted population for 2030 (United Nations Population Division 2011) will need more land for food, ranges, shelter, and other uses, and the production pressure on arable land will continually increase. From an LDN perspective, a balance between sustainable intensified crop production and crop expansion is needed to meet these increasing needs while still conserving natural resources. If crop expansion is inevitable, it should not only focus on expanding into previously uncultivated areas but also include the restoration of suitable degraded areas.
- *Poor land management*—Poor, unsustainable land management practices are prevalent on most African farms, especially in small-scale rain-fed production systems. The main reason for poor land management decisions and implementation is farmers' lack of knowledge about and experience with alternative sustainable practices and technologies. This is worsened by the high cost of fertilizers and other external inputs as well as the lack of incentives to improve management practices.
- *Insecure land tenure*—Insecure land tenure occurs under both statutory and customary land tenure systems. Especially under customary systems, tenure may be loosely defined, often to the disadvantage of women who play a major role in farming. Furthermore, generally unclear tenure terms, small and fragmented landholdings, and a limited ability to mortgage or transfer land may disincentivize farmers from investing in improved and sustainable agricultural practices and technologies due to the risk of limited or no return.
- *Poor access to markets and services*—Farmers need markets as an incentive to produce excess goods to sell for economic benefit, which increases the resources they have available to improve land management. When markets are poorly developed, missing, or too far away from production sites, farmers are more likely to make subsistence-based management decisions and are less able to generate economic benefits from the land. Hence, fewer resources are available to improve land management practices and prevent degradation.
- *Climate change*– Climate directly affects vegetation production, especially in dryland areas, which in turn affects the availability of organic matter and cover to protect the soil surface. In this way, it influences various soil properties and

processes that are essential to the ecosystem. Rainfall is considered to be the most important factor affecting LD vulnerability, followed by temperature and wind (WMO 2005).

5.3.3 How the Cartographies of Rural Poverty, Agriculture, Hunger, and Food Insecurity Correlate with and Contribute to Land Degradation

Globally, an increase in poverty correlates with an increase in the proportion of highly degraded lands and a simultaneous decrease in land improvement (FAO 2011). The impact of LD is highest in the poorest and most vulnerable communities in developing countries because they are usually the most dependent on natural resources for their livelihoods. Hence, the previously mentioned correlation is especially important in Africa. Positive correlations were found between poverty and LD at a global level, with roughly 42 % of poor individuals depending on degraded or underperforming land to survive. For the moderately poor, this figure declines to 32 %, and for those who are not poor, it declines to 15 % (Nachtergaele et al. 2010).

LD is directly linked to declines in crop productivity (Obalum et al. 2012; IFAD 2013; The Montpellier Panel 2014). With decreased productivity, farmers grow less food, become less economically competitive, and receive lower incomes (FAO et al. 2015). SSA is the world's most hunger-ridden region, home to 9 of the 16 countries around the world in which food security conditions have deteriorated since 1990. SSA was further identified as "the region with the highest prevalence of undernourishment and where only modest progress has been made in recent years" to reduce hunger, despite its overall economic performance (FAO 2013). The distribution of rural poverty, hunger, and food insecurity therefore correlates with the distribution of degraded lands. Poverty, hunger and food insecurity reduces farmers' motivation to mitigate LD. Especially for the very poor, once degradation has occurred, a vicious cycle is started where an increase in poverty leads to an increase in land degradation which, in turn, leads to more poverty (Peprah 2014). It would be worth researching how LD as an "underestimated threat multiplier" contributes to the protracted food insecurity crisis in SSA.

5.4 Land Restoration in SSA

5.4.1 Potential for Restoring Degraded Land

Restoration of degraded land is an essential part of achieving LDN. It is estimated that about half of currently degraded areas are suitable for restoration (Global

Mechanism-UNCCD 2015). Restoring only 12 % (or 150 million ha) of the world's degraded lands would supply food for another 200 million people and boost smallholders' income by USD 35–40 billion per year (New Climate Economy 2014). With almost two thirds of its arable land degraded (FAO 2008), Africa is able to contribute a large portion of this 12 %, provided that the degraded areas can be restored to agricultural productivity.

The World Resources Institute (WRI) produced an updated global approximation map (WRI 2010) showing the extent and location of areas in which forest and landscape restoration opportunities exist and where more detailed analyses on a national scale are recommended. Figure 5.2 presents this map for SSA and indicates the three types of areas that are suitable for restoration:

- *Wide-scale restoration* areas with less than 10 people per km^2 and the potential to support closed forests,
- *Mosaic restoration* areas with between 10 and 100 people per km^2 (moderate human pressure), and
- *Remote restoration* areas with a density of less than one person per km^2 within a 500 km radius (very low human pressure).

A visual comparison between Figs. 5.1 and 5.2 shows that many of the hotspots of biomass productivity decline are also potential restoration areas.

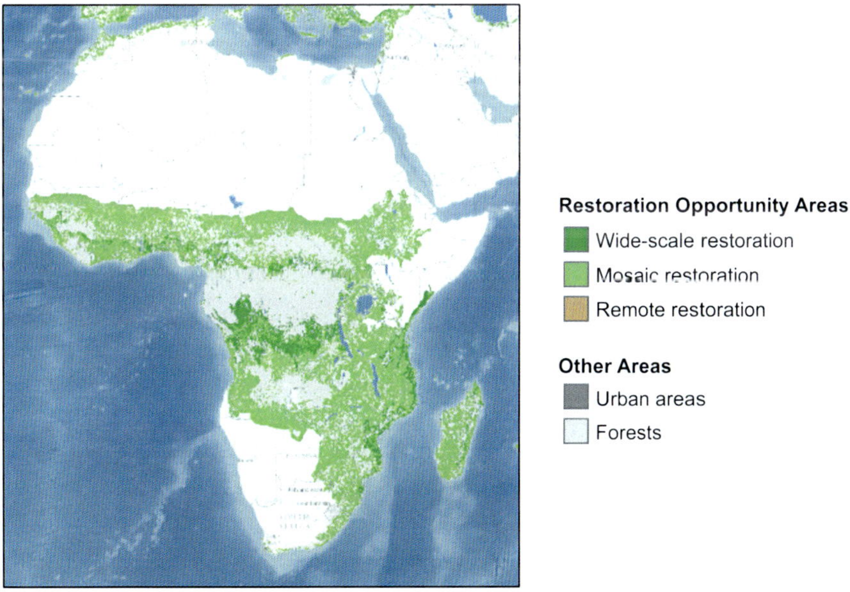

Fig. 5.2 Areas with the opportunity for restoration in Africa. Snapshot from the Interactive Atlas of Forest Landscape Restoration Opportunities, World Resources Institute (WRI 2010)

Following the development of the WRI restoration map, the International Union for Conservation of Nature (IUCN), in conjunction with the WRI, published a Restoration Opportunities Assessment Methodology (ROAM) handbook. Countries can use this handbook to assess how much of their land is suitable for restoration, map those areas, and determine which degraded areas offer the most value to society (IUCN and WRI 2014). The authors consider this publication to be a "road-test" edition, and a revised version incorporating experiences and lessons from countries' implementations of the guidelines will be published later. More detailed guidelines for the tools included in the methodology will be produced over time as part of a "ROAM technical series" of publications.

5.4.2 Degrade, Abandon, and Migrate: The Cost of Inaction in SSA

The cost of LD, also referred to as the cost of inaction regarding LD, can be measured in different ways. On-site losses in productivity are usually calculated as a percentage of agricultural GDP, while off-site costs resulting from consequences of degradation, such as dryland salinity, changes in stream flow, low-quality drinking water, and silting of rivers, are paid by parties that are not responsible for the degradation. Indirect costs can be incurred when, for example, a lower supply of agricultural products results in increased food prices, which in turn increases rural poverty, food insecurity, and malnutrition (UNCCD 2013).

Costs and benefits of land restoration

The long-term economic benefits of land restoration are crucial for motivating investment in land restoration activities. The impact of inaction on social cohesion, economic development, and environmental sustainability should be taken into account when deciding whether to act.

Decisions regarding land restoration and its potential economic benefits cannot be made without data and comparisons of all the costs and benefits of action (implementing alternative land use and restoration practices) and inaction (BAU). Such cost-benefit analyses link monetary values to non-market goods and services to allow comparison of the costs of marketed goods (UNCCD 2013). There are many examples of the economic benefits of reversing or halting LD (Olson and Berry 2003; Nkonya et al. 2011; Morales et al. 2012; ELD Initiative 2015a; ELD Initiative and UNEP 2015; Global Mechanism-UNCCD 2015). In 2003, it was estimated that on-site losses in Uganda due to environmental degradation ranged from 4 to 12 % of the continent's GDP. Of these losses, 85 % resulted from soil erosion, nutrient loss, and changes in crops (Olson and Berry 2003).

A recent cost–benefit analysis of the costs of erosion-induced depletion of soil nutrients in croplands in 42 African countries estimated that nutrient losses resulted in the loss of over 280 million tons of cereal per year. As a result, the cost of not controlling erosion and nutrient depletion amounts to about USD 4.6 trillion in purchasing power parity (PPP) at an annual value of USD 286 billion PPP. This is equivalent to 12.3 % of the combined GDP of the 42 countries. The cost of action, however, would only amount to USD 344 billion in PPP over the same period. Annually, this would be equivalent to only 1.15 % of the combined GDP for these countries. Such action would also enable the combined economy of these countries to grow at an average rate of 5.31 % per annum over the next 15 years. Overall, the benefits of SLM in Africa would outweigh the actual cost of by a factor of 7 (ELD Initiative 2015a, ELD Initiative and UNEP 2015). In another example, the IUCN reported that restoring 150 million ha of degraded land could reduce the CO_2 emissions gap by 11–17 % and increase additional crop yields to USD 6 billion (Global Mechanism UNCCD 2015).

The above evidence begs the question: Why are governments and institutions so hesitant to invest in LD prevention and active restoration of degraded lands? SDG 15 is key to promote increased action to reverse LD. Governments' adoption of this goal and subsequent actions would highlight the unacceptable costs of inaction, such as decreased social cohesion, economic development, and environmental sustainability. In many countries, such adoption and action would require a paradigm shift in how land use is viewed, from "degrade, abandon, migrate" to "protect, sustain, restore" (UNCCD 2014), which requires cooperation among various sectors as well as awareness of the delayed, long-term benefits obtained from land restoration (Tal 2014).

5.4.3 Land Restoration and Poverty Alleviation

Addressing LD is means for alleviating poverty (Grewel et al. 2001; Adams et al. 2004; Glasdottir and Stocking 2005; Halverson and McNeill 2006), as environmental protection is central to the eradication of poverty (Galizzi and Helklotz 2008). The UNCCD is the main global agreement linking land restoration to poverty alleviation, with special emphasis on Africa. During the 2002 World Summit on Sustainable Development held in Johannesburg, improved land and natural resource management and better agricultural practices and ecosystem conservation were listed as ways to contribute to poverty alleviation (United Nations 2002). Since then, land restoration has been seen as a crucial factor for eradicating poverty on a global scale (Glasdottir and Stocking 2005).

Because smallholder and family farms make up such a large portion of agricultural land in Africa and are so strongly linked to poverty and food production, they are considered to play a key role in reducing poverty and hunger through sustainable production methods (FAO et al. 2015). Hence, to achieve LDN, degradation in SSA must be addressed. These farms' link to poverty creates further incentives to achieve both environmental and economic benefits for land users.

Economic growth and investment lead to the alleviation of poverty and hunger by increasing household incomes and employment opportunities due to increased demand for labor. This, in turn, could help reduce rural poverty and thus increase the productivity of smallholder farms. For this to be successful, however, the benefits of economic growth need to reach poor individuals in rural areas. According to the latest FAO report on the state of food insecurity in the world, countries that have invested in agriculture have made significant progress towards achieving the millennium development goal (MDG) 1c hunger target (to halve, between 1990 and 2015, the proportion of people who suffer from hunger), especially if the investments benefited smallholder and family farms (FAO et al. 2015).

In Ghana, an economic growth rate of 3.3 % per year since 1990 has decreased the percentage of the population living in extreme poverty from 51 % in 1991 to 29 % in 2005 and reduced the prevalence of undernourishment. This was partially achieved through growth in the agricultural sector due to increases in cocoa production and significant increases in domestic food production, which were promoted by government policies, institutional reforms, and investments (FAO et al. 2015). Since the mid-1990s, the agricultural sector has accounted for about 40 % of Ghana's GDP and 60 % of the country's active labor force. In 2003, about 40 % of Ghana's population lived below the poverty line, and the majority of these people lived in rural areas and relied on agriculture for their livelihoods (Asurning-Brempong 2003). Investing in the agricultural sector as part of Ghana's economic growth was therefore key to reducing both poverty and hunger in the country. However, it is worth noting that although good progress was made at a national level, the country still suffers from inequality. Northern Ghana (i.e., the drylands) still has the highest poverty rates and lowest levels of food security in the country.

Tanzania's annual economic growth of 2.3 % since the early 1990s was largely due to rapid expansion of industry and services. Agriculture expanded at a much slower rate; in fact, the agricultural sector's contribution to Tanzania's GDP decreased from 50 % in 1992 to 26 % in 2013. Undernourishment in the country increased from 24.2 to 34.6 % over the same period. Since 2004, however, the rate of undernourishment has started to decline. Poverty in the country is still high, with 44 % of the population living in extreme poverty (less than USD 1.25 per day) in 2012, down from 72 % in 1992 (FAO et al. 2015). The lack of a direct correlation between economic growth on one hand and poverty alleviation and increased food insecurity on the other is mainly attributed to a lack of effective policies for modernizing agriculture and ensuring that poor and food-insecure citizens benefit from the distribution of earnings resulting from economic growth. This was explained in part by Tanzania's low investment in its agricultural sector, which is dominated by family farmers that focus mainly on subsistence and have limited access to local and international markets (United Republic of Tanzania 2011).

Despite the general consensus that land restoration would contribute to poverty alleviation, there is little empirical evidence of this fact. Peer-reviewed studies examining ecological restoration and its intrinsic value or benefit to society often do not explore the link between restoration and ecosystem services. They also do not link ecosystem services to the beneficiaries of ecosystem restoration. One reason for this was that the researchers involved in creating pathways to economic development overlooked the value of restoring natural ecosystems and their related services (Aronson et al. 2010). Some research papers do provide evidence that ecosystem services support people's well-being, but they do not provide much evidence of their contribution to poverty alleviation. It is still unclear how ecosystem services are connected to poverty, how change occurs, and how poverty may be alleviated through SLM (Suich et al. 2015).

Ecosystem services provided through restoration projects and their impact on communities can be measured using the social return on investment (SROI) approach, which "places a monetary value on the social impact (the benefit) of an activity, and compares this with the cost incurred in creating that benefit" (Weston and Hong 2013, p. 6). This approach identifies which project outcomes are most valuable to the beneficiaries and applies proxy financial values to those outcomes in order to evaluate the project's monetary value.

In order to accurately assess the impact of land restoration on poverty alleviation, evaluations at the project level need to be systematic, work from baselines, and measure not only direct financial returns but also changes in ecosystem services and social returns on investments. Poverty alleviation should serve as an incentive for governments and institutions to invest in land restoration and achieve LDN.

5.4.4 The Potential for Agricultural Expansion

Production increases in Africa from 1970 to 2010 were largely achieved by increasing the cultivated land area from 132 to 184 million ha (AGRA 2014). During this period, Africa was the only developing region in which the percentage of area expansion exceeded the growth in yield. The World Bank (2013) attributes this trend mostly to the small percentage of irrigated land in Africa (95 % of cultivated land is rain-fed) and the very low use of modern inputs such as fertilizer, improved crop varieties, pesticides, and mechanization. This was worsened by the generally low uptake of improved, sustainable management techniques aimed at increasing soil productivity to increase yields. In the absence of these techniques, low yields are not sufficient or sustainable long term and soils are gradually depleted of nutrients, resulting in further yield decreases and subsequent expansion to new land. This adds to the degraded land resulting from agriculture and leads to the clearing of new areas for production, which are degraded in turn if they are not managed sustainably.

SSA is often seen as an abundant source of land (Deininger and Byerlee 2011), and many studies have estimated that there is a large amount of potentially available cropland (PAC) on the continent (Chamberlin et al. 2014). Experts estimate that 52 % of the world's remaining arable land is in Africa (Deininger and Byerlee 2011). However, no precise figures are available, and the information used for these estimates is still very inconsistent. One of the main reasons for this is that estimates of PAC depend on assumptions about what renders land "potentially available." Well-designed agricultural development strategies will depend on accurate estimates of the quantity and spatial distribution of underutilized land that is suitable for crop production, so realistic calculations are essential (Hertel 2011).

According to Chamberlin et al. (2014), estimates of the PAC in SSA have ranged from 400 to 800 million ha, although some have said these values do not conform with many of the realities of smallholder agriculture in the region. Such criticism further suggests the large likelihood of overestimating the cultivable land, underestimating the land that has already been cultivated, and/or underestimating competing non-agricultural land use.

Agricultural expansion: assumptions

Care should be taken when estimating the extent of PAC in Africa. One should not only focus on the amount of land that is available but also consider:

- The current population threshold
- Whether forest land should be included or excluded
- Previously degraded areas that can be restored to productivity
- The miminum threshold of economic viability
- Spatial variability in prices
- The actual potential to narrow the gap in yield
- The extent and location of protected areas

In 2011, Deininger and Byerlee used geospatial data in conjunction with population distributions and agro-ecological potential to estimate the potential for cropland expansion in Africa. This was based on agronomic suitability, existing rural population densities, and communities' proximity to sites. They concluded that 198–446 million ha of underutilized arable land are available in Africa, depending on the assumptions used in the estimate, and that there is a large potential for production expansion (Deininger and Byerlee 2011).

Chamberlin et al. (2014) revisited Deininger and Byerlee's study and other estimates of PAC expansion based on a combination of biophysical and economic factors. Geographically, they identified areas not currently used for agricultural production with a population density of less than 25 people per km^2. Estimates of the potential biophysical production and conservative assumptions about profitability were taken into account to characterize the economic attractiveness of expansion within these areas. PAC was defined as "the reserve of moderately to highly productive land that could be utilized for rain-fed farming, that is not currently under intensive use or legally protected," and in certain scenarios explicitly included land under mature forest cover. Based on production potential alone and excluding forest land, the countries with the largest amounts of underutilized land are Sudan, Madagascar, the Democratic Republic of Congo, Mozambique, Angola, the Republic of the Congo, the Central African Republic, Ethiopia, and Zambia. These countries account for 65 % of the estimated 247 million ha of land that are available for cropland expansion in SSA. When the criteria of profitability were added to the smallholder expansion scenario, the PAC decreased by roughly 70 % to 80 million ha. However, when assumptions about commercial farming production were used, the PAC estimate decreased less drastically—by 32 %—to a total of 167 million ha.

If forests are protected and there are no insurmountable costs, roughly 247 million ha could be converted to cropland. However, a large investment in infrastructure would be required to access these areas and allow post-production market access. As the vast majority of this land is located in only a few countries, local policies, investments, and institutions would be crucial to enable such a venture. As an alternative, previously cultivated areas that are now degraded need to be assessed to determine whether restoration is feasible.

5.4.5 *The Potential for Sustainable Intensified Agricultural Production*

Africa has an opportunity to improve and increase its agricultural productivity, mainly because the low productivity in the past was not due to the inherent lack of productive potential, but to poor management and LD (World Bank 2013). Sustainable production intensification is key for increasing agricultural production in the region (Nkonya et al. 2008; IFAD 2013; Jayne et al. 2014), which would increase food availability, food security, and nutrition (FAO et al. 2015). The principle of sustainable intensification is vital to protect natural resources against degradation. Since smallholder farmers make up such a large proportion of the agricultural community, sustainable intensification of production on smallholder farms is important. As current agricultural practices largely lead to the degradation of natural resources and reduce agricultural productivity, production intensification would only be feasible if it is done using sustainable land management techniques and practices (IFAD 2013). Such practices include soil conservation, improved water management, diversified agricultural systems, and agroforestry. In order to increase their adoption, using such practices and techniques should be incentivized through public policies.

For intensified production to be sustainable and successful, smallholder farmers need a supportive environment and various levels of targeted support, including improved market access, incentives for sustainable land management, secure land tenure, strong institutional support for the technical implementation of sustainable practices, and affordable agricultural inputs (IFAD 2013). Incentives for sustainable land management are especially important for poor smallholder farmers since they are often more concerned about short-term survival than about long-term benefits.

If done sustainably, intensified production would contribute to LDN by preserving natural resources and reducing the need for agricultural expansion while simultaneously contributing to food security in the region.

5.4.6 The Role of Carbon Sequestration in Achieving LDN in Africa

Land restoration usually involves revegetation of degraded areas such as croplands, grazing lands, and forests, which increases the carbon stock in biomass and soil. Carbon sequestration in biomass and soil has become a global priority, and there is a strong relationship between vegetation cover and the organic carbon content of soil. In Africa, the economic and environmental benefits of carbon sequestration are particularly relevant (Rohit et al. 2006) as about 65 % of Africa's arable land is being degraded (FAO 2008). Due to the large amount of potentially usable land, Africa has the potential to sequester carbon through land restoration. The prevalence and extent of LD in Africa, combined with its extreme climate and fragile soil (Lal 2015), makes soil carbon sequestration essential to increasing soil fertility and agricultural production (Tiessen et al. 1994). Organic carbon is a crucial component of soil and an important driver of agricultural sustainability (Lal 2015).

Carbon sequestration is a major benefit of forest restoration and outweighs many other benefits of forest ecosystems, whether the ecosystems are comprised of forests only or forests combined with croplands (thus forming agroforestry systems) (Poverty-Environment Partnership 2005; Rohit et al. 2006; Mbow et al. 2012). The Reducing Emissions from Deforestation and Forest Degradation (REDD+) initiative of the United Nations Framework Convention on Climate Change (UNFCCC) was established to capitalize on this fact, addressing mitigation of and adaption to climate change by reducing deforestation and LD while contributing to poverty alleviation in poor communities (Mbow et al. 2012). However, in SSA, the implementation of REDD+ has been met with several challenges, such as a lack of understanding of the underlying drivers of changes in land use, the need for locally adapted monitoring systems, and additional incentives outside of the forestry sector (Henry et al. 2011). Many communities depend on land and forests for subsistence (Rohit et al. 2006; Henry et al. 2011), and forests are often cleared for agricultural production (Mbow et al. 2012). Limiting forest losses will therefore need to be accompanied by increased food production on existing croplands to reduce the pressure of cropland expansion (Henry et al. 2011), which requires soil carbon sequestration.

The World Bank conducted a meta-analysis of carbon sequestration in Africa and the cost effectiveness of implementing a range of land management technologies for climate-smart agriculture (World Bank 2012). In theory, the capacity for soil carbon sequestration is equal to the cumulative carbon loss to date, but in practice, only 50–66 % of these losses can be replaced by implementing SLM practices. Even so, the potential total private profits resulting from carbon sequestration in Africa were estimated to reach USD 105 billion by 2030, and governments would only have to pay an estimated USD 20 billion to enable farmers to implement SLM practices. The potential economic benefits of soil carbon sequestration are therefore high and may provide an important incentive for soil restoration.

Carbon sequestration provides strong environmental and economic incentives for land restoration and should be applied to develop a pathway towards LDN.

5.5 Institutional Aspects of Land Degradation and Restoration Trends in SSA

5.5.1 What Triggers Land Improvement Processes? Lessons Learnt from SSA's Restoration Hotspots

Land use in SSA's agricultural sector is facing structural stresses and seemingly intractable challenges such as rapid population growth (human and livestock) and climate change. In this context, important agricultural land acquisitions by foreign investors in Africa since the 2008 food crisis threatened to take land away from poor farmers in rural areas and signaled that those investors understand that, despite the structural stresses, land investments could still be profitable.

Grassroots movements have made changes to overcome the challenges related to land use. Since the late 1980s, substantial land restoration has taken place in SSA. It would be worth documenting how land improvement processes at grassroots level correlate with the results achieved by the 18 SSA countries that have met the MDG 1c target of halving the percentage of hungry people by 2015, compared to their baseline hunger levels of 2000. Among these countries are Africa's hotspots for land restoration, including Niger, Mali, Ethiopia, and Malawi (FAO et al. 2015). A growing number of studies and literature provide insights into how those initiatives emerged and were sustained. Although outside the scope of this paper, documenting such correlation may inform to what extent grassroots initiatives could contribute to land improvement and subsequent increased food security.

By using a combination of low-cost techniques to protect and grow naturally regenerated trees (allowing regrowth of tree stumps), many smallholder farmers have created important agroforestry parklands on highly degraded land. This is done through regreening, which is defined as "a process in which farmers protect and manage trees that naturally regenerate on their land, rather than cut them down" (WRI 2015b). A grassroots-level movement called farmer-managed natural regeneration (FMNR) has emerged and is taking root in many Sahelian countries (Weston and Hong 2013; Reij 2014; Reij and Winterbottom 2015). Farmers have often taken the lead, advancing innovation, sharing knowledge, and assisting peers scale up and roll out these techniques. Until recently, the outcomes have been largely overlooked by most governments and development partners. To the communities and ecosystems of these regreened areas, restoration has had substantial local impacts, such as:

- Improved food security,
- Adaptation, leading to resilience to climate change-related shocks,
- Poverty reduction,
- Increased firewood and fodder, and
- Reduced stress on livelihood supporting resources (land and water), leading to reductions in conflicts and migration.

In Niger, 5 million ha of land have been regreened through FMNR (Reij and Winterbottom 2015), which has enabled farmers to protect and manage over 200 million trees. During the same period, tree planting projects resulted in an increase of 65 million trees. However, the survival rate of these trees was often below 20 %. The cost of planting trees usually exceeds USD 1000 per hectare, and external funding for promoting regreening through Nigerian farmers has not exceeded USD 100 million (equivalent to USD 20 per hectare) in social and ecological impacts (Reij 2014).

In an attempt to account for the social, economic and environmental outcomes of an FMNR project implemented in the dry northern part of Ghana, compared to the in-cash and in-kind cost of the project, World Vision reported results in terms of SROI ratios. "A SROI ratio of 1:1 means that for every dollar (or Ghanaian cedi) invested into a project, one dollar of benefit has been created for the project's stakeholders. 2:1 means that two dollars of value was created for every dollar invested" (Weston and Hong 2013, p. 6). At the end of the 3-year project the SROI ratio was reported as 7:1. Should implementation continue, the SROI ratios were projected as 19:1 by year seven and 46:1 after year thirteen (Weston and Hong 2013).

Reij and Winterbottom (2015, p. 5) reported that "a combination of factors, including the emergence of effective sustainable land management practices aimed at improving food security and increasing fodder and fuelwood" as well as demographic and land use pressures have triggered these land improvement processes. They stressed that in most cases, "innovative farmers have taken the lead in regreening efforts" and have built up Africa's restoration hotspots.

To ensure success when scaling up and rolling out, Reij and Winterbottom (2015) suggested the following 6 steps:

1. Identify and analyze existing regreening successes,
2. Build a grassroots movement for regreening,

3. Address policy and legal issues to enable conditions for regreening,
4. Develop and implement a communication strategy,
5. Develop or strengthen agroforestry value chains, and
6. Expand research activities.

However, in the atlas for forest and landscape restoration opportunities in Africa (Fig. 5.2), some of these regreened areas, like the Zinder region in southwest Niger, have not yet been acknowledged as areas with the potential for restoration. Africa's potential may therefore be higher than currently estimated. A more systematic mapping of the restoration potential in the region is key for scaling up and rolling out restoration efforts and investments and achieving LDN by 2030. Such mapping should focus on not only forest restoration but also agricultural and agroforestry restoration hotspots.

5.5.2 *Institutional Challenges to Furthering Climate Change Adaptation in the Agricultural Sector in SSA*

With large portions of SSA's agricultural lands already degraded (FAO 2008) and the increasing impact of climate on LD (WMO 2005; Kiage 2013; The Montpellier Panel 2014), climate change adaptation in the agricultural sector is an important factor for preventing future degradation and contributing to LDN. According to FAO, the total crop and livestock production losses caused by natural disasters from 2003 to 2013 in Africa amounted to USD 26 billion (FAO 2015). This makes the continent the world's most affected region, with a 3.9 % decrease in expected crop and livestock production, primarily due to drought. During the same period, 77 % of all production losses caused by drought worldwide occurred in 27 countries in SSA and affected nearly 150 million people, causing USD 23.5 billion in crop and livestock losses (USD 19.2 billion, or 82 %, for crop losses and USD 4.2 billion, or 18 %, for livestock losses). Most of these losses were borne by rural populations in the affected areas (FAO 2015).

In a systematic literature review, Antwi-Agyei et al. (2013, p. 15) reported that in reviewed publications, "financial and institutional barriers constituted the majority of barriers" to climate change adaptation in SSA, followed by information and technical barriers. For households, "the most commonly identified barrier was the lack of financial resources."

Despite the magnitude of climate change impacts, governments are very slow to shift from a crisis response mode to preparedness and risk management, to help populations adapt to future changes, and to reduce their vulnerabilities. One of the systemic challenges common to countries in SSA is a lack of and/or ineffective decentralization in the systems that govern the agricultural sector. For instance, the National Adaptation Programmes of Action (NAPA) focuses on priority adaptation activities in the least developed countries, most of which are in SSA. The NAPAs have mainly been designed to address adaptation issues related to agriculture and natural resource management at subnational and local levels. However, since "local institutions are incorporated as the focus of adaptation projects in just about 20 percent of the projects described in the NAPA documents" (Agrawal and Perrin 2008, p. 14), this has little impact on building resilience at a local level due to insufficient local incorporation.

To take advantage of Africa's potential in the agricultural sector and achieve LDN, African governments must speed up the effective decentralization of climate adaptation to their local governments and territories, as adaptation to climate change primarily occurs on a local level. This is also an opportunity to improve local governments' resilience, which implies that most governance systems remain in an adaptive management mode. To support decentralization, Local Climate Adaptive Living (LoCAL) (UNDP et al. 2010) has been introduced by the United Nations Capital Development Fund in some SSA countries (Benin, Ghana, Mali, Mozambique, and Niger) since 2013. LoCAL uses the decentralization of climate adaptation finance via performance-based climate resilience grants as a catalyst for the transformations required at a local governance level in order to build adaptation and resilience.

5.5.3 Emerging Initiatives for Scaling up and Rolling Out Land Restoration in SSA

After more than a decade of inaction and failure to capitalize on the grassroots restoration movements in Africa, there is growing political momentum for land restoration in SSA, with a number of initiatives germinating at different scales and levels. Two global political processes have contributed to building this momentum: the preparation and follow up for the Rio+20 Summit and the much-needed regional cooperation to combat climate change.

5.5.3.1 The African Response to Rio+20

The Programme on Sustainable Land Management, Desertification, Biodiversity and Ecosystems-Based Adaptation to Climate Change (LDBE) is one of the Regional Flagship Programmes created by the African Union (AU) as a response to the political commitments made at Rio+20. It was included under the umbrella of the African Ministerial Conference on the Environment (AMCEN) and will be operationalized by the New Partnership for Africa's Development (NEPAD). Intended to be a framework to facilitate synergy and cooperation between national and regional actors, African governments, and their development partners (AMCEN et al. 2013), the LDBE is poised to play a major role in advancing LDN at national and regional levels in Africa.

5.5.3.2 The African Forest Landscape Restoration Initiative

The African forest landscape restoration initiative (AFR100) was launched during the UNFCCC COP21, held in December 2015 in Paris, to support the African Resilient Landscapes Initiative (ARLI), which was initiated by NEPAD "to promote integrated landscape management with the goal of adapting to and mitigating climate change" (WRI 2015a). It is a "pan-African country-led effort to restore 100 million ha of degraded or deforested landscapes by 2030," and 10 countries have already committed "at least 31.7 million ha of land for forest landscape restoration" (WRI 2015c). More countries are likely to join, given that more than USD 1.5 billion in development, finance, and private sector investments have already been allocated to AFR100. It is still unknown whether restoring 100 million ha by 2030 will be enough to achieve LDN, given the inaccuracies and discrepancies of the available information and data on degradation trends versus restoration potential. However, if these investments are made available in a timely manner and are invested soundly, AFR100 is likely to catalyze more national and local investments and speed up progress towards LDN.

5.5.4 Creating a Dynamic Enabling Environment for SLM and LDN

5.5.4.1 Landscape as the Unit (or Scale) for Integrated or Holistic Management

A landscape is a local agroecological unit that manages people, biodiversity, and ecosystem services with biophysical, socioeconomic, or historic/cultural commonalities or identities. It is "a generic and politically neutral term for a socio-ecological system that consists of a mosaic of natural and/or human-modified ecosystems" (LPFN 2014, p. ii). Therefore, studying a given landscape is crucial for understanding the causes of its degradation and developing appropriate prevention and restoration solutions that consider what worked for successful restoration endeavors elsewhere.

Holistic management (HM), a decision-making framework initially developed by Allan Savory (Savory and Butterfield 1999) to reverse desertification in Zimbabwe, is a values-based and common purpose-driven approach to landscape management. It helps farmers and all relevant decision makers to plan and manage biological resources and ecosystem functions and services according to the available investments while monitoring results and progress toward goals.

Integrated landscape management (ILM) is designed "to promote an inclusive use and management of natural resources, especially land and water, which is centred on people's social, economic and environmental welfare" (LPFN 2014, p. ii). Both HM and ILM are key complementary concepts for advancing SLM in Africa and should be considered during the development of land restoration initiatives. ILM-related approaches are gaining momentum throughout Africa, with 87 integrated landscape initiatives documented in 33 African countries (LPFN 2014).

At the Landscapes for People, Food and Nature in Africa Conference held in Nairobi in July 2014, 200 expert practitioners and policy makers from around the world developed the African Landscape Action Plan, "an ambitious agenda to scale up landscape initiatives in Africa" (LPFN 2014, p. iii) with six action themes:

- Policy
- Governance
- Business
- Finance
- Research
- Capacity Development

The action plan was endorsed by NEPAD in line with its mandate and, "implemented through its current strategies such as TerrAfrica, the Comprehensive African Agriculture Development Programme (CAADP) and upcoming ones like the AMCEN Regional Flagship Programme on Sustainable Land Management, Desertification, Biodiversity and Adaptation to Climate Change (LDBA), will contribute to lifting over 40 % of the Sub-Saharan African population out of poverty" (LPFN 2014, p. iii).

To achieve LDN by 2030, the African Landscape Action Plan should be fully mainstreamed at the national and local level throughout Africa.

5.5.4.2 Marginal Land Versus Underperforming Assets

By taking the lead during land improvement on seemingly "marginal land," farmers have demonstrated that degraded lands are not "marginal" but rather "underperforming assets" (Gnacadja 2013) that can still thrive if they receive the appropriate investments. Against the odds, they have challenged BAU by designing and advancing feasible pathways towards sustainable management. Labeling land as "marginal" implies high investment risk. As a result the potential to transform such land may not be achieved due to a lack of investment. Referring to marginal land as an "underperforming asset" stimulates more positive thoughts regarding the potential of the land and the development of existing assets. Semantics in science communication and a term's reception by the non-scientific community, especially in the context of the science–policy interface, is therefore important in order to stimulate positive responses to current challenges.

5.5.4.3 Address What Causes Farmers (Smallholder Farmers/Small Family Farms) to Make Poor Land Management Investment Decisions

The adoption of SLM practices amongst farmers in Africa remains low despite the considerable potential gains due to a number of factors that influence farmers' land management decisions. Too often, more affordable, less labor-intensive management options are chosen rather than making an effort to apply better land management practices (The Montpellier Panel 2014). Increasing successful land restoration activities therefore requires addressing what causes farmers to make poor land management and investment decisions (UNCCD Secretariat 2013). This may include the use of stronger incentives and better information to enable farmers to make informed decisions. If the causes are not addressed, farmers will continue to make the same choices, even if the results are clearly undesirable and detrimental to natural resources, and the cycle of poor land management will continue (The Montpellier Panel 2014).

A key factor for increasing farmers' investment in SLM practices is the development of innovative approaches to financing for smallholder farmers to reverse degradation in croplands. Winterbottom et al. (2015) identified four major barriers that discourage smallholder farmers from adopting more sustainable farming practices: (1) a lack of knowledge about and technical support for the sustainable land and water management practices that are needed to increase resilience against climate change, (2) limited access to equipment and other inputs, (3) an aversion to risk, or reluctance among existing investment vehicles and commercial backers to provide capital to smallholder farmers, and (4) shortcomings in government policies and a lack of favorable conditions for innovative financing. In many places, cultural practices also present important barriers to change.

The concept of "restoration bonds" was born as an approach to address the root causes of LD and help smallholder farmers overcome these challenges. Restoration bonds are issued by governments or financial institutions and provide financial support to enable smallholder farmers to invest in sustainable land management practices. Farmers apply for a bond and agree to the repayment terms, and then the capital is mobilized through the bond and used to support the adoption of SLM practices in a predetermined area (Winterbottom et al. 2015). Farmers also have access to decision-support tools and other types of support for restoration bonds as a result of collaboration with investors and bond facilitators to help demonstrate the feasibility and value of SLM practices and ensure successful implementation.

According to Gnacadja (2012, p. 3), "Africa must invest strategically in agriculture and its value chain to boost productivity and competitiveness in a sustainable manner, to respond to domestic markets and to become the breadbasket of the world. If Africa fails in this regard, the market will still support actions such as land grabbing and the opportunity will become yet another resource curse." In Nigeria, for instance, agriculture accounts for 42 % of the GDP and more than 60 % of employment, but receives only 1.4 % of bank loans. Thus, it is unsurprising that the sector is underperforming. However, some central banks are rising to the challenge. Programs such as the Nigerian Incentive-Based Risk Sharing System for Agriculture Lending (NIRSAL), launched in November 2011, are steps in the right direction. Addressing further causes of LD requires increasing political support for SLM and land restoration, capitalizing on local knowledge to support SLM, making information about SLM more user friendly, improving land tenure security, and improving rural infrastructure.

5.5.4.4 Options for Engaging the Private Sector and Business Community

The private and business sectors play a key role in achieving LDN in Africa since both directly or indirectly occupy or use land as, for instance, a source of raw materials or building sites. LDN targets can be supported either through businesses' own operations that involve land use or through inclusion in a larger restoration value chain. Companies directly using land for their operations can contribute by adopting SLM practices to minimize current and future degradation, or by actively restoring degraded and abandoned production lands prior to utilization (World Business Council for Sustainable Development 2015).

A Land Degradation Neutrality Fund (LDN Fund) was created under the auspices of the UNCCD and its Global Mechanism (GM) as a collaborative initiative among several institutions involved in identifying sustainable solutions to achieving LDN. The Fund is run by a fund manager and operates as a coordination platform to leverage private sector and business investments in land restoration, identify relevant projects, and monitor impacts. The Fund will manage investments in the form of a public–private partnership to support the transition to LDN through land restoration. The LDN Fund Investment Model was designed to ensure land restoration while simultaneously generating revenues for investors through sustainable production on restored land (Global Mechanism-UNCCD 2015).

The Fund will mainly, but not exclusively, focus on direct investments in large-scale restoration. Resources will also be allocated for small- and medium-scale projects that produce local and global benefits. However, many stakeholders and civil society organizations have raised concerns about whether the LDN Fund leads to land grabbing, especially in countries with weak land governance. Such risks require effective management in order to ensure the initiative's success.

5.5.4.5 Monitoring Progress Towards LDN and Improving Institutions Accordingly

To achieve LDN, countries are expected to establish voluntary country-level targets to stop and reverse LD, as emphasized in the 12th session of the COP to the UNCCD. This approach is expected to cause countries to take appropriate actions and recognize the unacceptable socioeconomic and environmental costs of inaction or continued LD. Each government is responsible for setting its own national targets based on its national priorities and circumstances (UNCCD 2014). National LDN targets would make it easier to assess progress towards LDN at a regional level, but a well-developed monitoring system is still required. Such a monitoring system would be used to delineate areas and populations affected by LD, identify priority areas for intervention, measure progress towards LDN (and SDG target 15.3), and assess related policies and investments (Gnacadja 2014).

Developing options for such a monitoring system is one of the functions of the UNCCD Intergovernmental Working Group (UNCCD 2014). The system would include well-defined indicators and an evaluation framework to measure future changes in the rates of LD and restoration over large areas, referring to suitable baselines (Grainger 2014) and using a combination of advanced technologies (e.g., remote sensing) and in situ observations (Gnacadja 2014). Relevant indicators should encompass biophysical and environmental conditions as well as increases in livelihood stemming from restoration. This would require many companies to improve their ability to monitor changes in land resources or, in some cases, to create a framework for this from scratch (Grainger 2014). To be effective, the monitoring system would need to be flexible and cater to the needs of local, national, regional, and global policy makers, which would require a combination of top-down and bottom-up approaches (Gnacadja 2014; Grainger 2014).

Given that the focus and priority of LDN are national voluntary targets, each country needs to estimate its degraded area and the proportion of this area that is suitable for restoration. Areas deemed feasible for restoration should be prioritized for investments in remediation efforts based on appropriate criteria, including socio-economic, institutional, and biophysical factors (Chasek et al. 2014). The aggregation of national trajectories would be sufficient to determine the regional and global baseline area of land affected by LD.

5.6 Conclusions and Recommendations

Land degradation and restoration are clearly not simple concepts, and achieving LDN is not easy. However, we believe that, based on the information presented in this paper, achieving LDN over time is possible if all the necessary institutional and policy steps are taken and effectively implemented to create a suitable environment for SLM. For the time being, and despite ongoing success in land restoration projects, governments and citizens are motivated to achieve LDN in the context of sustainable development, although such a task seems daunting.

In order to work towards LDN, we need to know how much land is currently degraded (which serves as a degradation baseline), where degradation is still occurring, and the rate at which it is occurring. We also need to know where restoring degraded land would be economically feasible and where the current state of natural resources could be improved to achieve productivity. Such areas should be assessed at a national level and relevant targets should be set to prevent degradation and ensure restoration over time. Target 15.3 of SDG 15 is to begin ensuring political commitment to LDN, which requires political and institutional support.

It is clear that LD and agriculture are inextricably linked and that land restoration should go hand in hand with improved management of agricultural land and production systems. This is necessary to contribute to the restoration of degraded land, to prevent further degradation of agricultural land, to intensify production on existing agricultural land (and thus produce economic and social benefits), and to reduce the need for expansion of agricultural land into new areas. The belief that agriculture holds the key to Africa's development is not new. What is new is the emergence of systemic trends (population dynamics, climate change, and globalization) that challenge agricultural land use in Africa and the opportunities they provide. Faced with these challenges, SSA's smallholder farmers deserve more than charitable attention; they need secure access to land, energy, information on SLM and related technologies, improved seeds, water, markets, other factors that enable them to increase their capacity, and incentives to increase agricultural production. With the support of appropriate policies and institutions interested in sustainable transformation, these farmers can thrive, significantly increasing productivity and causing poverty alleviation, improved food security, and GDP growth.

SSA cannot sustain economic growth and decrease rural poverty and hunger without investing in the restoration of underperforming land assets in poor rural areas. SSA must therefore unleash that potential in order to generate more inclusive and sustained growth. Empirical evidence suggests that "[In] sub-Saharan Africa, agricultural growth can be 11 times more effective in reducing poverty than growth in non-agricultural sectors" and 3 times more effective in low-income countries in other regions (FAO et al. 2015, p. 28). Helping the rural poor invest in the improvement of their land assets should be part of the strategy for reducing rural poverty and hunger in SSA and ultimately achieving LDN. Yet, to date, few countries have met the 10 % national budget allocation target for agricultural development set in the AU 2003 Maputo Declaration on Agriculture and Food Security. SSA's public policies for competitiveness and sustainability in agriculture must change, and we need to consider family farmers as investors in their own businesses that need a conducive environment to secure their investments. Smart investments in agriculture and agribusiness involve land restoration. We must debunk the deeply rooted beliefs that land degradation costs nothing and that restoration is too costly. Governments should consider setting up suitable environments to issue restoration bonds under public and private partnerships.

The benefits of taking action against soil erosion and nutrient mining in SSA through capital and recurring expenditures on SLM will exceed the actual cost of such action by almost 700 % (ELD Initiative and UNEP 2015). The LDN goal must therefore be mainstreamed in countries' agriculture policies and governance systems, including in Africa's CAADP. We must continuously celebrate grassroots leaders who actively ensure the success of land restoration projects in Africa, turning the tables on land degradation and rendering presumably "lost" land productive again. To help these farmers, the stubbornly inefficient policies and institutions that continue to fail them should be changed because they prevent farmers from scaling up and rolling out innovations. Smallholder farmers invest much more in agriculture than governments, donors, and private enterprises combined. However, they need an environment that helps them make better decisions. In Tanzania, for instance, a farm commodities firm enlisted 30,000 farmers as suppliers via a mobile phone system, boosting profitability for all. Given the challenges and potentials of the region, we feel that the future of farming is in Africa. The size of farms is not an issue; innovation and the democratization of innovation are of much more interest.

Institutional transformation is essential for creating a dynamic and suitable environment for the implementation and adoption of SLM by land users in order to achieve LDN. Although it is still unclear whether SSA will achieve LDN by 2030, we believe the following nine recommendations are essential to increase the likelihood of this outcome:

(i) **View farmers as champions of change and agents of transformation and provide political support for SLM and restoration.** A paradigm shift is needed to think of agriculture as a business and smallholder farmers as investors in their own businesses. Empirical evidence shows that in SSA, farmers invest more in agriculture than governments, the private sector, and development partners combined. Farmers are the best agents of change in their own communities because they can serve as examples of successful SLM implementation, restoration of degraded land, and increased agricultural production. When farmers are convinced of the need for change, they will become the drivers of change, even if the economic and social environments are not yet conducive to such efforts. Political support is therefore needed to create an environment in which farmers can benefit from investing in their farms and implementing SLM in order to start a positive cycle of transformation.

(ii) **Improve and capitalize on local knowledge.** Practical solutions can and should be sourced from farmers' collective experience and intimate knowledge of local conditions. Local conditions do not always lend themselves to external management systems, so adaptations may be necessary. Farmers need to be consulted when new management practices and systems are introduced to ensure that practical issues are identified and addressed.

(iii) **Democratize innovations, make sure information and knowledge sharing is user-friendly, and monitor progress.** In order for farmers to fully understand, adopt, and implement innovations, information should be packaged in a user-friendly, easy-to-understand format to improve communication and knowledge transfer. The progress of innovation implementations need to be monitored to enable the implementations' effectiveness and to identify any adjustments that are needed to ensure success.

(iv) **Perform inclusive cost–benefit analyses and decision-making processes to ensure that transformations, incentives, and subsidies are pro-poor and pro-SLM.** Inclusive decision making processes are needed to design effective pathways to agricultural transformation. Both short- and long-term costs and the benefits of action and inaction (as well as an analysis of the impact of potential trade-offs in decision-making) regarding SLM should be considered to identify appropriate areas for investment and transformation. Incentive and subsidy schemes need to be revised according to the current priorities for agricultural production and environmental protection to ensure their effective application. Incentives and subsidies need to benefit poor individuals who need support to increase their agricultural inputs in accordance with appropriate SLM-related practices and systems.

(v) **Secure land tenure and use rights and improve transparency in land management.** Land tenure and use rights need to be secured to provide farmers with increased incentives to invest in their farms. However, securing land tenure alone is not sufficient and should be accompanied by increased access to markets and incentives for farmers to practice SLM and increase production.

(vi) **Improve rural infrastructure and urban–rural linkages.** Rural infrastructure such as roads, markets, and stores to supply inputs are necessary to serve as incentives for farmers to increase production and produce surplus food. Urban–rural linkages are needed to provide additional markets for rural produce and outputs.

Extension services in SSA

In the late 1980s, extension services were phased out in many SSA countries due to the emergence of macro-economic structural adjustment programmes. These services were sometimes reintroduced as agricultural input services to be provided by the private sector. However, such services should not be bound to any particular (private) interest but rather dedicated to sustainable agricultural transformation. To achieve this, policies and institutions should be regularly assessed and monitored to ensure that extension services contribute to the achievement of agricultural transformation and effectively support farmers.

(vii) **Improve the science–policy interface, extension services, and knowledge about soil science.** Scientific data and results need to inform policy development to enable optimal support for SLM implementation. Effective extension services, in turn, are needed to provide continuous support to farmers during SLM implementation, and extension personnel need to be regularly informed of scientific developments and improvements in SLM. Africa has a limited ability to conduct soil research, develop soil-related innovations, and provide support for SLM implementation. This needs to be addressed by participating in education exchange with countries offering degrees in soil science and by strengthening collaboration between African soil research centers and those in other regions.

(viii) **Map soil fertility, vulnerability to degradation, and potential for restoration.** An understanding of the status and spatial distribution of soils is needed to determine its fertility, vulnerability to degradation, and potential for restoration. This is important for making informed decisions about priority areas for SLM implementation and support. There are large gaps in national soil and degradation data in Africa that need to be filled and regularly made available to enable the effective use of inputs. Such mapping should address local conditions and provide local, pertinent, and farmer-friendly information and recommendations. To achieve LDN, Africa needs a soil data revolution.

(ix) **Prioritize adaptation to climate change.** Farmers need knowledge, resources, and support to adapt to the effects of climate change and increase their resilience to climate shocks. Finances for climate adaptation from both national and international sources must be decentralized to the local level and integrated into local governments' planning and investment processes in order to promote integrated landscape management and build local communities that are resilient to climate change.

References

Adams WM et al (2004) Biodiversity conservation and the eradication of poverty. Science 306:1146–1149

AGRA (2014) Africa agriculture status report: climate change and smallholder agriculture in sub-Saharan Africa. Nairobi, Kenya

Agrawal A, Perrin N (2008) Climate adaptation, local institutions, and rural livelihoods. IFRI working paper # W08I-6. Natural resources and environment, University of Michigan, Michigan, USA

Al-Nasser NA (2011) President's summary. High-level meeting of the United Nations General Assembly to address desertification, land degradation and drought in the context of sustainable development and poverty eradication. New York, 20 Sept 2011. http://www.un.org/en/ga/president/66/Letters/PDF/PGA%20summary%20on%20Desertification%20-%2023%20September%202011.pdf. Accessed 9 Feb 2016

AMCEN, AU, UNEP (2013) Progress on the development of the regional flagship programmes for the implementation of Rio+20 outcomes. Paper presented at the fifth special session of the African ministerial conference on the environment, Gaborone, 15–16 October 2013

Antwi-Agyei P, Dougill AJ, Stringer LC (2013) Barriers to climate change adaptation in sub-Saharan Africa: evidence from northeast Ghana and systematic literature review. Centre for climate change economics and policy working paper No. 154. Sustainability Research Institute Paper No. 52

Aronson J et al (2010) Are socioeconomic benefits of restoration adequately quantified? A meta-analysis of recent papers (2000–2008) in *Restoration Ecology* and 12 other scientific journals. Restor Ecol 18(2):143–154

Asuming-Brempong S (2003) Policy module Ghana: economic and agricultural policy reforms and their effects on the role of agriculture in Ghana. Paper presented at the roles of agriculture international conference, Rome, 20–22 Oct 2003

Bai ZG, Dent DL, Olsson L, Schaepman ME (2008) Global assessment of land degradation and improvement 1. Identification by remote sensing. Report 2008/01, ISRIC—World Soil Information, Wageningen

Bayene A (2014) Small farms under stress play a huge role for Africa: smallholder agriculture and emerging global challenges. Policy Note 5. Nordic Africa Institute, Uppsala, Sweden
Chamberlin J, Jayne TS, Heady D (2014) Scarcity amidst abundance? Reassessing the potential for cropland expansion in Africa. Food Policy 48:51–65
Chasek P et al (2014) Operationalizing zero net land degradation: the next stage in international efforts to combat desertification? J Arid Environ 112:5–13
Deininger K, Byerlee D (2011) Rising global interest in farmland: can it yield sustainable and equitable benefits? World Bank Publications, Washington, DC
Diagana B (2003) Land degradation in sub-Saharan Africa: what explains the widespread adoption of unsustainable farming practices? Draft working paper, Department of Agricultural Economics and Economics, Montana State University, Bozeman, MT
ELD Initiative (2015a) Report for policy and decision makers: reaping economic and environmental benefits from sustainable land management. The Economics of Land Degradation, Bonn, Germany
ELD Initiative (2015b) The value of land: prosperous lands and positive rewards through sustainable land management. The Economics of Land Degradation, Bonn, Germany
ELD Initiative (2015c) The value of land: quick guide to the report. Economics of Land Degradation. The Economics of Land Degradation, Bonn, Germany
ELD Initiative and UNEP (2015) The economics of land degradation in Africa: benefits of action outweigh the costs. The Economics of Land Degradation, Bonn, Germany
FAO (2008) The state of food insecurity in the world. Food and Agriculture Organization of the United Nations, Rome, Italy
FAO (2011) The state of the world's land and water resources for food and agriculture (SOLAW)—managing systems at risk. Food and Agriculture Organization of the United Nations, Rome and Earthscan, London
FAO (2013) Hunger map 2013. http://www.fao.org/docrep/019/i2975e/i2975e.pdf. Accessed 9 Feb 2016
FAO (2015) The impact of disasters on agriculture and food security. Food and Agriculture Organization of the United Nations, Rome, Italy
FAO, IFAD, WFP (2012) The state of food insecurity in the world. Economic growth is necessary but not sufficient to accelerate reduction of hunger and malnutrition. FAO, Rome
FAO, IFAD, WFP (2015) The state of food insecurity in the world. Meeting the 2015 hunger targets: taking stock of uneven progress. Food and Agriculture Organization of the United Nations, Rome
Galizzi P, Helklotz A (2008) The role of the environment in poverty alleviation. Fordham University Press, New York
Glasdottir G, Stocking M (2005) Land degradation control and its global environmental benefits. Land Degrad Dev 16:99–112
Global Mechanism-UNCCD (2015) White paper. Land Degradation Neutrality Fund, London
Gnacadja L (2012) Africa: the future global engine for sustainable development? Keynote speech of UNCCD Executive Secretary, Africa Day 2012, Berlin, 23 May 2012
Gnacadja L (2013) Inclusive and sustainable territorial governancce for food security: sharing lessons learned from around the world. Keynote speech of UNCCD Executive Secretary at the 6th International Land Coalition Global Assembly of Members, Antigua, Guatemala, 23–27 Apr 2013. Antigua, Guatemala
Gnacadja L (2014) New challenges in science and policies to combat desertification. J Arid Environ 112:1–4
Grainger A (2014) Is land degradation neutrality feasible in dry areas? J Arid Environ 112 (Part A):14–24
Grewel SS, Dogra AS, Jain TC (2001) Poverty alleviation and resource conservation through integrated watershed management in a fragile foot-hill ecosystem. In: Stott DE, Mohtar RH, Stenhardt GC (eds) Sustaining the global farm. Selected papers from the 10th International Soil Conservation Organization meeting held May 24–29, 1999 at Purdue University and the

USDA-ARS National Soil Erosion Research Laboratory. http://tucson.ars.ag.gov/isco/isco10/SustainingTheGlobalFarm/P197-Grewel.pdf. Accessed 9 Feb 2016
Halverson E, McNeill CI (2006) Advancing a new paradigm. Institutional and policy breakthroughs toward poverty reduction and sound environmental management. In: Galizzi P, Herklotz A (eds) The role of the environment in poverty alleviation. Fordham University Press, New York
Heady DD (2015) The evolution of global farming land: facts and interpretations. Poverty, Health and Nutrition Division of the International Food Policy Research Institute, Washington, DC
Henry M et al (2011) Implementation of REDD+ in Sub-Saharan Africa: state of knowledge, challenges and opportunities. Environ Dev Econ 16:381–404
Hertel TW (2011) The global supply and demand for agricultural land in 2050: a perfect storm in the making? Am J Agric Econ 93(2):259–275
IFAD (2001) Rural poverty report 2000: the challenge of ending rural poverty. Oxford University Press, New York
IFAD (2013) Smallholders, food security, and the environment. International Fund for Agricultural Development, Rome, Italy
IUCN, WRI (2014) A guide to the restoration opportunities assessment methodology (ROAM): assessing forest landscape restoration opportunities at the national or sub-national level. Working paper (road-test edition), IUCN, Gland, Switzerland
Jayne TS, Chamberlin J, Heady DD (2014) Land pressures, the evolution of farming systems, and development strategies in Africa: a synthesis. Food Policy 48:1–17
Kiage LM (2013) Perspectives on the assumed causes of land degradation in the rangelands of Sub-Saharan Africa. Prog Phys Geog 37(5):664–684
Lal R (2015) Restoring soil quality to mitigate soil degradation. Sustain 7:5875–5895
Le QB, Nkonya E, Mirzabaev A (2014) Biomass productivity-based mapping of global land degradation hotspots, ZEF—discussion papers on Development Policy No. 193, Center for Development Research, Bonn, July 2014, pp. 57.LPFN (2014) The African landscapes action plan. Landscapes for people, food and nature in Africa conference, Nairobi, 1–3 July 2014
LPFN (2014) The African landscapes action plan. Landscapes for people, food and nature in Africa conference, Nairobi, 1–3 July 2014
Mbow C, Skole D, Dieng M, Justice C, Kwesha D, Mane L, El Gamri M, Von Vordzogbe V, Virji H (2012) Challenges and Prospects for REDD+ in Africa: Desk Review Of REDD+ Implementation in Africa. GLP Report No. 5. GLP-IPO, Copenhagen
Morales C et al (2012) Estimating the economic cost of land degradation and desertification: case studies in South America. Secheresse 23(3):168–176
Muchena FN et al (2005) Turning the tides of soil degradation in Africa: capturing the reality and exploring opportunities. Land Use Policy 22:23–31
Nachtergaele F, Biancalani R, Petri M (2008) Land degradation. SOLAW Background Thematic Report 3. Rome
Nachtergaele F et al (2010) Global land degradation information system (GLADIS). Beta version. An information database for land degradation assessment at global level. Land Degradation Assessment in Drylands Technical Report, no. 17, FAO, Rome
New Climate Economy (2014) Better growth, better climate. The New Climate Economy Report, The Global Commission on the Economy and Climate
Nkonya E et al (2008) Linkages between land management, land degradation, and poverty in Sub-Saharan Africa. The case of Uganda. Research Report 159 Abstract. International Food Policy Research Institute
Nkonya E et al (2011) The economics of desertification, land degradation, and drought toward an integrated global assessment. ZEF—Discussion Papers on Development Policy No. 150, Centre for Development Research, Bonn
Obalum SE et al (2012) Soil degradation-induced decline in productivity of Sub-Saharan African soils: the prospects of looking downwards the lowlands with the *Sawah* ecotechnology. Appl Environ Soil Sci 2012:1–10

Olson J, Berry L (2003) Land degradation in Uganda: its extent and impact. Mimeo. Commissioned by Global Mechanism with support from the World Bank
Peprah K (2014) Poverty and land degradation nexus: the case of Asunafo, Ghana. Int J Sci Res 3 (4):2319–7064
Poverty-Environment Partnership (2005) Investing in environmental wealth for poverty reduction. United National Development Programme, New York
Reich PF et al (2001) Land resource stresses and desertification in Africa. In: Bridges EM, Hannam ID, Oldeman LR et al (eds) Responses to land degradation. 2nd International conference on land degradation and desertification, Khon Kaen, Thailand. Oxford Press, New Delhi, India
Reij C (2014) African re-greening initiatives. http://africa-regreening.blogspot.co.za/. Accessed 9 Oct 2015
Reij C, Winterbottom R (2015) Scaling up regreening: six steps to success. A practical approach to forest and landscape restoration, World Resources Institute, Washington, DC
Rio+20, UNEP, UNDP (2011) Africa Consensus Statement to Rio+20. Statement No. E/ECA/CFSSD/7/Min./3. Addis Ababa, Ethiopia
Rohit J, Swallow B, Kerr J (2006) Status of carbon sequestration projects in Africa: potential benefits and challenges to scaling up. Working Paper #26. World Agroforestry Centre, Nairobi, Kenya
Rosenzweig C, Parry ML (1994) Potential impact of climate change on world food supply. Nature 367:133–138
Savory A, Butterfield J (1999) Holistic management: a new framework for decision making. Island Press, Washington, DC
Suich H, Howe C, Mace G (2015) Ecosystem services and poverty alleviation: a review of the empirical links. Ecosys Serv 12:137–147
Tal A (2014) The implications of environmental trading mechanisms on a future Zero Net Land Degradation protocol. J Arid Environ 112:25–32
The Montpellier Panel (2014) No ordinary matter: conserving, restoring and enhancing Africa's soils. Dec 2014
Tiessen H, Cuevas E, Chacon P (1994) The role of soil organic-matter in sustaining soil fertility. Nature 371:783–785
Tully K et al (2015) The state of soil degradation in Sub-Saharan Africa: baselines, trajectories and solutions. Sustain 7:6523–6552
UNCCD (1994) Elaboration of an international convention to combat desertification in those countries experiencing serious drought and/or desertification, particularly in Africa. Document no. A/AC.241/27. http://www.unccd.int/Lists/SiteDocumentLibrary/conventionText/conv-eng.pdf. Accessed 9 Feb 2016
UNCCD (2007) Follow-up to the Joint Inspection Unit report and strategy development to foster the implementation of the Convention. Situational analysis. Document No. ICCD/COP(8)/INF.5. http://www.unccd.int/Lists/OfficialDocuments/cop8/inf5eng.pdf. Accessed 9 Feb 2016
UNCCD (2013) The economics of desertification, land degradation and drought: methodologies and analysis for decision-making. Conference background document. UNCCD 2nd Scientific Conference, Bonn, Germany, 9–12 April 2013
UNCCD (2014) Land degradation neutrality: resilience at local, national and regional levels. Bonn, Germany
UNCCD Press Release (2015) Inception meeting of the UNCCD land degradation neutrality project. http://www.unccd.int/en/media-center/MediaNews/Pages/highlightdetail.aspx?HighlightID=355. Accessed 9 Oct 2015
UNCCD Secretariat (2013) A stronger UNCCD for a land-degradation neutral world. Issue Brief. Secretariat of the United Nations Convention to Combat Desertification, Bonn, Germany
UNDP (2012) Africa human development report 2012: towards a food secure future. United Nations Development Programme, New YorK, NY, USA

UNDP, UNCDF, UNEP (2010) Local governance and climate change. A discussion note. December 2010. http://www.unpei.org/sites/default/files/publications/LocalGovernanceAndClimateChangeDiscussionNote.pdf. Accessed 9 Feb 2016
UNEP (2014) Assessing global land use: balancing consumption with sustainable supply. A report of the working group on land and soils of the International Resource Panel. United Nations Environment Programme, Nairobi, Kenya
UNEP/GRID-Arendal (2005) Mapping indicators of poverty in West Africa. UNEP/GRID-Arenda, Norway. http://www.grida.no/graphicslib/collection/mapping-indicators-of-poverty-in-west-africa. Accessed 9 Feb 2016
United Nations (2002) Report of the world summit on sustainable development. Johannesburg, South Africa, 26 Aug–4 Sept 2002
United Nations Population Division (2011) World population prospects: the 2010 revision. UN Population Division, New York
United Nations (2012) The future we want. Outcome document of the United Nations conference on sustainable development, Rio de Janeiro, Brazil, 20–22 June 2012
United Republic of Tanzania (2011) Tanzania agriculture and food security investment plan (TAFSIP) 2011–12 to 2020–21. https://www.grain.org/attachments/2688/download. Accessed 6 Feb 2016
Van Schaik L, Dinnissen R (2014) Terra incognita: land degradation as underestimated threat amplifier. Research Report, Netherlands Institute of International Relations Clingendael, The Hague, Netherlands
WBCSD (2015) Land degradation neutrality: a business perspective. World Business Council for Sustainable Development, Geneva, Switzerland
Weston P, Hong R (2013) Social return on investment report. Talensi farmer managed natural regeneration project in Ghana. World Vision Australia, Victoria, Australia
Winterbottom R et al (2015) Restoration bonds: a catalyst to restore ecosystem services in agricultural landscapes. In: Burke L, Ranganathan J, Winterbottom R (eds) Revaluing ecosystems: pathways for scaling up the inclusion of ecosystem value in decision making. Issue Brief. World Resources Institute, Washington, DC
WMO (2005) Climate and land degradation. WMO-No. 989. World Meteorological Organization, Geneva, Switzerland
World Bank (2012) Carbon sequestration in agricultural soils. Report Number: 67395-GLB. The World Bank, Washington, DC
World Bank (2013) Unlocking Africa's agricultural potential. An action agenda for transformation. Sustainable Development Series Nr. 76990. The World Bank, Washington, DC
World Business Council for Sustainable Development (2015) Land degradation neutrality: issue brief for business. World Business Council for Sustainable Development, Geneva, Switzerland
WRI (2010) Atlas of forest landscape restoration opportunities. http://www.wri.org/applications/maps/flr-atlas/#. Accessed 9 Oct 2015
WRI (2015a) African forest landscape restoration initiative (AFR100). http://www.wri.org/our-work/project/AFR100/about-afr100. Accessed 16 Feb 2016
WIR (2015b) Re-Greening. http://www.wri.org/our-work/project/re-greening. Accessed 14 March 2016
WRI (2015c) RELEASE: African countries launch AFR100 to restore 100 million hectares of land. http://www.wri.org/news/2015/12/release-african-countries-launch-afr100-restore-100-million-hectares-land%20. Accessed 16 Feb 2016

Chapter 6
Extent of Salt-Affected Soils and Their Effects in Irrigated and Lowland Rain-Fed Rice Growing Areas of Southwestern Tanzania

S. Kashenge-Killenga, J. Meliyo, G. Urassa and V. Kongo

Abstract Increasing salt-affected soils have become a major abiotic constraint on rice production in lowland areas (both irrigated and rainfed) in southwestern Tanzania. Meager information on salt-affected soils distribution is available. This study was aimed at (a) establishing the salt status of salt-affected soils by type in selected irrigation schemes in the southwestern rice-growing corridor of Tanzania and (b) establishing farmers' perceptions of the extent of the salt problem and associated crop losses in these rice irrigation schemes. Participatory diagnosis and an observation survey were conducted in the four major rice-producing regions of Katavi, Rukwa, Mbeya, and Iringa. Composite samples were collected from salt-affected hot-spot areas in 19 selected irrigation schemes and analyzed. Visual observation showed that 100 % of all surveyed irrigation schemes had symptoms of salt affected soils. However, laboratory results showed that 67 % of the schemes had salt problems. Three types of salt-affected soils (saline, sodic, and saline–sodic) with extreme salinity (4–15 dSm^{-1}), sodicity (10–34 Sodium adsorption ratio—SAR), and high soil pH (up to 10) values recorded. Saline–sodic soil was the most common problem, followed by sodic soils. About 90 % of the surveyed irrigation schemes had inadequate irrigation infrastructures, which seems to contribute to the problem. Land loss ranged from 5 to 25 % of schemes, and yield losses ranged from 5 to 100 %. Urgent measures, including the rehabilitation of irrigation infrastructure to improve

S. Kashenge-Killenga (✉)
Chollima Agro-scientific Research Centre, P.O. Box 1892,
Dakawa, Morogoro, Tanzania
e-mail: sophykashenge@yahoo.com

J. Meliyo · G. Urassa
Mlingano Agriculture Research Institute, P.O. Box 5088,
Mlingano, Tanga, Tanzania

V. Kongo
Dar-es-Salaam University, P.O. Box 35131,
Dar es Salaam, Tanzania

R. Lal et al. (eds.), *Climate Change and Multi-Dimensional Sustainability in African Agriculture*, DOI 10.1007/978-3-319-41238-2_6

drainage systems, the use of salt-tolerant cultivars, and implementation of salt soil management options, should be taken to prevent arable land losses.

Keywords Abiotic stresses · Salt-affected soils · Salinity · Sodicity · Salt tolerance · Southern corridor

6.1 Introduction

6.1.1 Rice Production Situation in Tanzania

In Tanzania, rice is a very important food and cash crop, ranks second in production and consumption after maize, and is grown almost in all regions of the country at various levels of production. Consequently, the crop is exposed to various environmental stresses across the country. According to Kashenge-Killenga et al. (2012a), a large portion of rice (80 % of the production) comes from lowland agro-ecology, which comprises both irrigated and rainfed agro-ecologies. Irrigated ecosystems produce only 10 % of rice production in the country, while 70 and 20 % of rice comes from lowland and upland rainfed agro-ecologies, respectively.

Rice is largely grown by smallholder farmers in irrigated, rainfed, lowland and upland ecosystems throughout the country. Only 5 % is grown by large-scale farmers. The total area under rice cultivation in Tanzania has increased from 557,991 ha in 2007 to 799,361 ha in 2012. During the same period, paddy production also increased by 22.8 %, from 1,217,302.2 to 1,633,432.3 Mg, while the average yield remained at 2.3 Mg/ha (FAOSTAT 2004). The increases in cultivated area and total production of rice suggest that the yield has not been growing; rather, rice yields have been increased by farm expansion, which accelerates land degradation.

6.1.2 Available Rice Production Opportunities

Demand for rice has also increased recently. Approximately 60 % of the country's population consumes rice, and annual per-capita consumption has risen from less than 15 kg in the 1970s (FAO 2004) and 25.4 kg in 2012 (Stryker and Amin 2012). This demand results in more cultivation for both consumption and marketing. About 1.7 million households were engaged in rice farming in Tanzania in 2012 (PHC 2012). The national average productivity of 2.3 Mg/ha is slightly higher than the Sub-Saharan Africa average rice productivity of 2.0 Mg/ha (AATF 2013) but much lower than the global average productivity of 4.0 Mg/ha (FAO 2014).

Under the *Kilimo Kwanza* (Agriculture First) resolution of 2009, the government of Tanzania is committed to transform its agricultural industry into a modern commercial sector, following strategies such as Tanzania's Green Revolution.

Enhancing food security and sustainable agricultural production by developing large-scale and improving small-scale irrigation schemes in the coastal plains and lowlands has been among the main initiative (Kashenge-Killenga et al. 2014). Small-scale irrigation schemes in mostly semiarid marginal areas dominate the production of rice and vegetables. However, according to Alam (2006) and FAO (2000), irrigated agriculture is among the major sectors seriously endangered by salt-affected soils, especially in arid and semi-arid environments.

The southern rice-growing corridor of Tanzania is comprised of the Rukwa, Mbeya, Katavi and Iringa regions and produces approximately 30 % of all rice grown in the country (Kilimo Trust Report 2011) (Fig. 6.1). This corridor is treated as a priority area in a number of strategic interventions, programs, and projects seeking to fulfil the *Kilimo Kwanza* declaration. These initiatives include the Southern Agriculture Growth Corridor of Tanzania (SAGCOT), a government effort to provide an important practical instrument for implementing the national *Kilimo Kwanza* initiative.

SAGCOT is the first of a series of phased initiatives to develop agricultural corridors that connect the highly productive southern highlands to the port of Dar es Salaam as part of the large Result Now Initiative to transform Tanzania from a low- to a middle-income economy. These government initiatives, among others, aimed at addressing the constraints on agricultural growth, which are largely related to poor irrigation infrastructure, environmental degradation, the low productivity of land, under-development of irrigation potential, infestations and outbreaks of crop pests and diseases, and erosion of the natural resource base. These initiatives will ultimately contribute to increasing food production and alleviating poverty in the country.

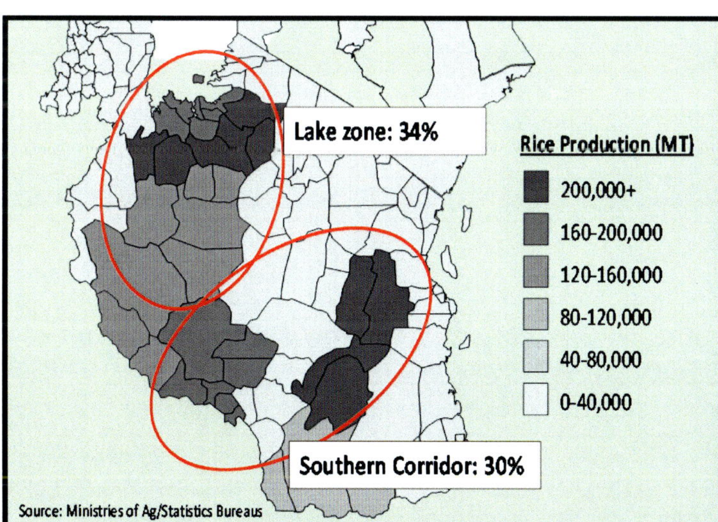

Fig. 6.1 Major rice-producing areas in Tanzania (*Source* Kilimo Trust 2011)

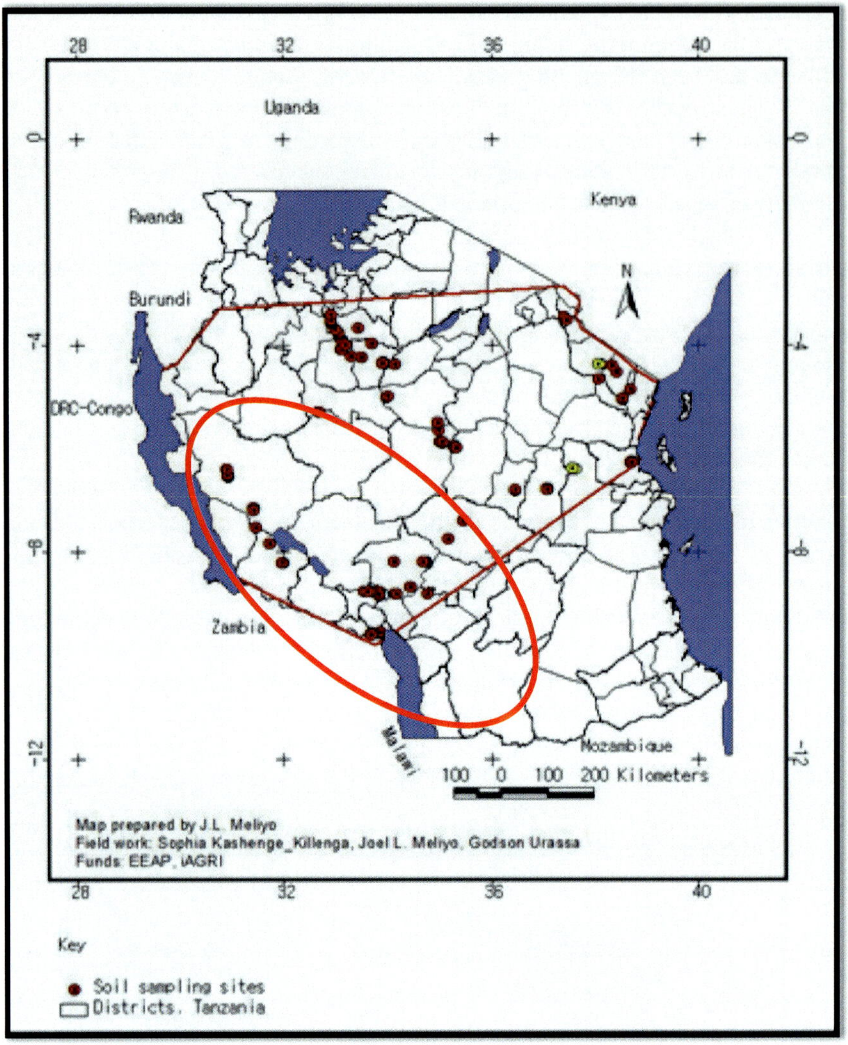

Fig. 6.2 Surveyed irrigation schemes of southwestern Tanzania

6.1.3 Factors Favoring Salt-Affected Soil Development and Associated Challenges

Given the prevailing climatic condition and geographic setting, the southern rice-growing corridor of Tanzania is characterized by a semi-arid and sub-humid climate suitable for crop production. Evapotranspiration is so high that it exceeds precipitation and favors salt accumulation (Kashenge-Killenga et al. 2012b). According to Chemura et al. (2014), salinization and sodification of soils are increasing, especially in arid and semi-arid regions of the world, and have negative impact on the production of cereals, such as rice. The loss of farmable land to salt-affected soil is in direct conflict with the needs of the world population, which is projected to increase by 1.5 billion over the next 20 years, and the challenge of maintaining the world's food supplies. The FAO (2014) predicted that, by 2020, Tanzania will have a deficit of 2.5 million Mg of rice. Therefore, a need to minimize production challenges and maximize productivity exists.

6.1.4 Salt-Affected Soils and Their Distribution

Soils are classified as saline or sodic (Fig. 6.3) based on the soluble salts in saturated extracts (EC) and the proportion of Na to Ca and Mg in saturated extracts (SAR = Na [Ca + Mg/2]½) or the proportion of exchangeable Na to the cation exchange capacity (ESP = Na/CEC/100) (Chemura et al. 2014). Saline soils have an EC of more than 4 dSm^{-1} and either a SAR of less than 13 or an ESP of less than 9 % (Chemura et al. 2014; Seilsepour et al. 2009). Sodic soils have an EC of less than 4 dSm^{-1} and either a SAR of more than 13 or an ESP of more than 9 % (Chemura et al. 2014; Kashenge-Killenga et al. 2012a; Seilsepour et al. 2009). Salt-affected soils have been extensively reported to be among the major problems in irrigated agriculture across the world. It is estimated that salinity and sodicity have impacts on more than 900 million ha of agricultural land, representing more than 6 % of all agricultural land and approximately 20 % of the world's irrigated land (Chemura et al. 2014). FAO (2012) reported that nearly 50 % of the irrigated lands in arid and semi-arid regions of the world have some degree of soil salinization problems, and TWAS (2006) reported that, by 2050, half of the arable land in the world will be salt affected.

FAO (2000) estimated that saline soils affect 1.7 million ha and sodic soils 300,000 ha in Tanzania. However, local estimations give an area of 2.9 million ha and 700,000 ha for saline and sodic soils, respectively. FAO (2003) reported that a total of 3.5 million ha in the country are affected by salt in the country, including 2.9 million ha affected by saline soils. The variations in these figures indicate that the extent of salt-affected soils has not been properly documented. Ten estates in the northern Tanzania have already encountered major salt-related problems (FAO 2003). Several smaller, community-managed irrigation schemes classified as traditional irrigation schemes are experiencing decreased rice yields due to salt (saline and sodic) problems (Kashenge-Killenga et al. 2012b). A study by Kashenge (2010) in the Northeastern Tanzania found that 7 of the 9 surveyed irrigation schemes were salt affected and that saline–sodic soil is a major problem. In regions such Kilimanjaro, some fields in these schemes have already been

Classification	Electrical Conductivity(ECe) (mmhos/cm)	Sodium Adsorption Ratio (SAR)	pH
Saline	> 4.0	< 13	< 8.5
Sodic	< 4.0	> 13	> 8.5
Saline - Sodic	> 4.0	> 13	< 8.5

Fig. 6.3 Salt-affected soils classification

abandoned due to the effects of hostile soil conditions. This phenomenon has also been reported by FAO (2003, 2005). Only minimal efforts have been made by the government to overcome this problem.

6.1.5 *Causes of Salts in Soils*

The factors leading to excessive accumulation of salts in soil can be natural or anthropogenic. According to SASO (2009), the environmental factors that result in salinization or sodification include geological events, which can increase the salt concentration in groundwater and consequently soils; natural factors, which can channel salt-rich groundwater to the surface, near the surface, or above the groundwater table; groundwater seepage into areas below sea level, that is, micro-depressions with little or no drainage; floodwaters from areas with geological substrates that release large amounts of salts; and wind action, which, in coastal areas, can blow moderate amounts of salts inland.

Human-induced factors that can lead to salinization or sodification include irrigation with salt-rich waters; a rising water table due to human activities; filtration from unlined canals and reservoirs; uneven distribution of irrigation water; poor irrigation practices; improper drainage; use of fertilizers and other inputs (especially where land under intensive agriculture has low permeability and limited leaching possibilities); salt-rich wastewater disposal in soils; and contamination of soils with industrial by-products. According to Kashenge-Killenga et al. (2013), salinization and sodification are often associated with irrigated areas where low rainfall, high evapotranspiration rates, or soil textural characteristics impede salts from washing out of the soil, so they build up in the surface layers. Irrigation with water with high salt content dramatically worsens the problem. Salinization also associated with the overexploitation of groundwater caused by the demands of growing urbanization, industry, and agriculture.

6.1.6 *Effects of Salts in Plants*

The value and yield of soils with high salt contents are significantly reduced, causing severe socio-economic and environmental problems. Strong salt-affected soils can affect plant growth, both physically (osmotic effect) and chemically (nutrition effect and/or toxicity). Consequently, plant growth and yield decline, and the quality or value of agricultural production decreases (Gonzalez et al. 2004). Moradi et al. (2003) reported that saline soils are characterized by a white surface crust, good soil tilth, high fertility, and poor yield. The white surfaces develop as water evaporates from saline soil, and salts which were in the water are left behind and accumulate on the soil surface. These excess salts keep the clay in a flocculated state so that these soils generally have good physical structure, tillage characteristics, and water permeability, even better than those of non-saline soils (Siyal et al. 2002). Salinity does not affect the physical properties of soils but is harmful because elevated soluble salts in the soil solution reduce the availability of soil water to plants (Kashenge 2010). The lower the salts content of the soil are, the lower the dSm^{-1} rating is, and the less effect on plant growth the salts have (FAO 2005).

Sodic soils have a high pH, and according to Kashenge (2010), high pH and excessive sodium are major characteristics which do not allow soil particles to attach to one another. As a result, the soil disperses and is not friable. These soils are very sticky, have a soapy feel when wet, and are very hard when dry. Cloudy water in puddles might form on the surface of this soil, and the surface crust is always black due to dispersion of organic matter (FAO 2005). Siyal et al. (2002) added that these soils have a poor structure that hinders smooth management. Seed germination in these soils is also poor due to the difficulty leveling seedbeds. This study, therefore, was aimed at (a) establishing the salt status of salt-affected soils by type in selected irrigation schemes in the southwestern rice-growing corridor of Tanzania and (b) establishing farmers' perceptions of the extent of the salt problem and associated crop losses in the respective rice-irrigation schemes.

6.2 Methodology

6.2.1 Description of Study Area

Soil characterization was conducted during July and August 2014 covered selected major rice-growing irrigation schemes in the southern rice-growing corridor of Tanzania. Four major rice-growing regions were involved: Katavi, Rukwa, Mbeya, and Iringa. Nineteen rice-irrigation schemes were randomly selected based on prior but unconfirmed knowledge of salt-affected soils: Karema, Mwamapuli, Ikaka, and Mwamkulu (Mpanda district, Katavi region); Mlele and Kamsis (Mlele district, Katavi region); Ng'ongo, Sakalilo, and Mpete (Sumbawanga rural district; Rukwa region); Ngana, Makweale, and Tenende (Kyela district, Mbeya region); Gwiri, Ruanda-majenje, Madibira, Bethania, and Mbuyuni (Mbarali district; Mbeya region); and Pawaga and Idodi (Iringa rural district, Iringa region) (Fig. 6.2).

The overall climate of the southern rice-growing corridor is semiarid to sub-humid, which is favorable for crop production. Rainfall in the region follows a typical monomodal pattern, with a single rainy season from November through May, and average annual rainfall of 700–2500 mm. This pattern only allows for only one large harvest during the year, and it is heavily dependent on rains. The altitude of the region ranges from 500 to 2981 m above sea level and is suitable for farming a wide variety of food and cash crops. The climate is generally tropical, with both seasonal and altitudinal temperature variations and distinctly defined dry and rainy seasons. The average temperature ranges from 16 °C in the highlands and to 25 °C in the lowland areas.

As long as the agricultural sector remains primarily within the hands of small-holder farmers, the available land area is not a constraint on production. The Rukwa region has been divided into the Katavi and Rukwa regions. The two regions account for 8 % of Tanzania's land area and are composed of 91 % land and 9 % water bodies. According to the Sumbawanga District Council Report (2013) this proportion of inland water is higher than the national average of 7 %, indicating the region's potential for superior water-management initiatives that could further boost productivity levels. Of the land area, 42 % is arable, but only 23 % of this is cultivated. Mbeya is the sixth-largest region in Tanzania in terms of total area. The region is not restricted by the amount of land available for cultivation, with 47 % of land arable. Of the total area of 63,429 km^2, 1757 km^2 is covered by water bodies, and 3314 km^2 serves as game and forest reserves. However, only approximately 28 % of the land is currently used for agricultural purposes, demonstrating high potential to increase agricultural activities. The Mbeya region is estimated to have 90,190 ha suitable for irrigation, but 36,449 ha are undeveloped, 25,456 ha are developed schemes, and 27,965 ha are under traditional irrigated schemes.

Iringa is the seventh largest region in Tanzania, covering an area of 58,936 km^2, of which 56,864 km^2 (96 %) is land. The remainder is covered by water bodies. The total land area that is arable is 41,973 km^2 (71 %), but only 7183 km^2 (17.5 %)

is under cultivation. The arable land available to smallholders that remains uncultivated is 34,790 km^2. Overall, land under irrigation increased by 58 % from 2005 to 25,784 ha in 2010, demonstrating the continued focus on water management as an important agricultural investment. The area of land under irrigation in Iringa region has increased from 10,382 ha in 2005 to 25,784 ha in 2010 (148 % increase) (Iringa District Council Report 2013).

6.2.2 Field Data Collection

Participatory diagnosis and observation were the main techniques used to generate field data collected from randomly selected areas in the irrigation schemes. Major rice-producing districts were selected from each region. In each district, a list of priority irrigation schemes was provided by district agricultural officials. A team comprised of researchers, a representative from the District Agriculture and Irrigation Cooperation Officer's (DAICO) Office, village or ward extension staffs, scheme leaders, progressive farmers, and village leaders visited selected irrigation schemes and randomly sampled top soils (0–20 cm) by auger or hand hoe. A focus group of 8–15 people—key informants, village leaders, influential village residents, progressive farmers, youth, and women—was used to gather information on perceptions and experiences of the salt-affected soil problem in the surveyed irrigation schemes. The focus group also explored the communities' responses and identified traditional adaptation measures to cope with the problem.

Information from the team, supplemented by various village reports from the district office, helped to qualitatively determine the extent of the salt problem in the irrigation schemes. Observation included randomly selected transect walks along farmers' fields and direct observations of water distribution systems, soil conditions, and rice plants. Based on farmers' experiences, extremely salt-affected fields were visited in each irrigation scheme. Five to 10 fields were selected depending on the size of the scheme and the intensity of the problem. Composite samples were taken from a depth of 0–20 cm. Every sampling site was geo-referenced using global positioning system (Fig. 6.2). Ten to 30 composite samples per scheme were collected. A total of 160 soil samples were collected and sent to a Mlingano soils laboratory for analysis.

6.2.3 Laboratory Analysis of Collected Soils Samples

The soils were analyzed following the standard procedure for soil analysis as outlined by Moberg (2000). Soil pH was measured potentiometrically in water at 1:2.5 soil–water ratio. For the assessment of salinity and sodicity, ECe was determined by paste extract, in which the leachate was used to take readings with an

EC meter. The cations exchange capacity and exchangeable bases were extracted by saturating soils with neutral 1 m NH_4OAc (Thomas 1982). The adsorbed NH_4^+ was displaced by K^+ using 1 M KCL and then determined via the Kjeldal distillation method for the estimation of the CEC of soil. The bases Ca^{2+}, Mg^{2+}, K^+, and Na^+ displaced by NH_4^+ were measured using an atomic absorption spectrophotometer. Assessment of soil-soluble and -insoluble salts was done by measuring electrical conductivity and calculating the sodium adsorption ratio (SAR) and exchangeable sodium percentage (ESP). Based on the pre-set criteria shown in Fig. 6.3, groups of soils were established.

6.3 Results

6.3.1 *Soil Characteristics of Rice Irrigation Schemes in the Southern Rice-Growing Corridor*

The southern rice-growing corridor is dominated by Cambisols soils, which are of moderate fertility and vary by climate, topography, depth, and stoniness (Fig. 6.3). Soils with Cambisols characteristics occupy 46.4 % (7,899,751.4 ha) of the land area of the southern rice-growing corridor, followed leptisols (14.7 %; 2,513,194 ha), which are shallow, stony, rocky, and vulnerable. Next, ferrasols, which have low natural fertility and tend to fix phosphates, occupy 1,252,603.8 ha (7.3 %). Lixisols, which are also naturally low in fertility and tend to slaking/crusting, compaction and erosion in sloping land, occupy 1,212,796.7 ha (7.1 %). Nitisols, which have low base status and low available phosphorus, cover 1,169,904 ha (7.0 %). Solonetz occupy 875,379.7 ha (5.4 %) and have strongly sodicity characteristics. Arenosols/Andosols, which have low available-water capacity and very low natural fertility and susceptible to erosion, occupy 819,060.4 ha (4.8 %). Fluvisols, which are susceptible to seasonal flooding, high groundwater levels, and salinity, cover 778,737.4 ha (4.6 %). Acrisols, which have low natural fertility, aluminum toxicity, strong phosphate fixation, and slaking/crusting and are highly susceptible to erosion, occupy 310,128.2 ha (2 %). Finally, very hard vertisols, which are difficult to manage, cover 205,328.6 ha (1.2 %) (Fig. 6.4).

6.3.2 *Categorizations of Soils Based on Their Salt Content*

Salt-affected soil symptoms were visually observed in all (100 %) of the surveyed irrigation schemes, with the greatest effects in Mbeya, Rukwa, Iringa, and Katavi in that order. Areas affected ranged from 2 to 20 %, while 2 to 15 % of irrigated land

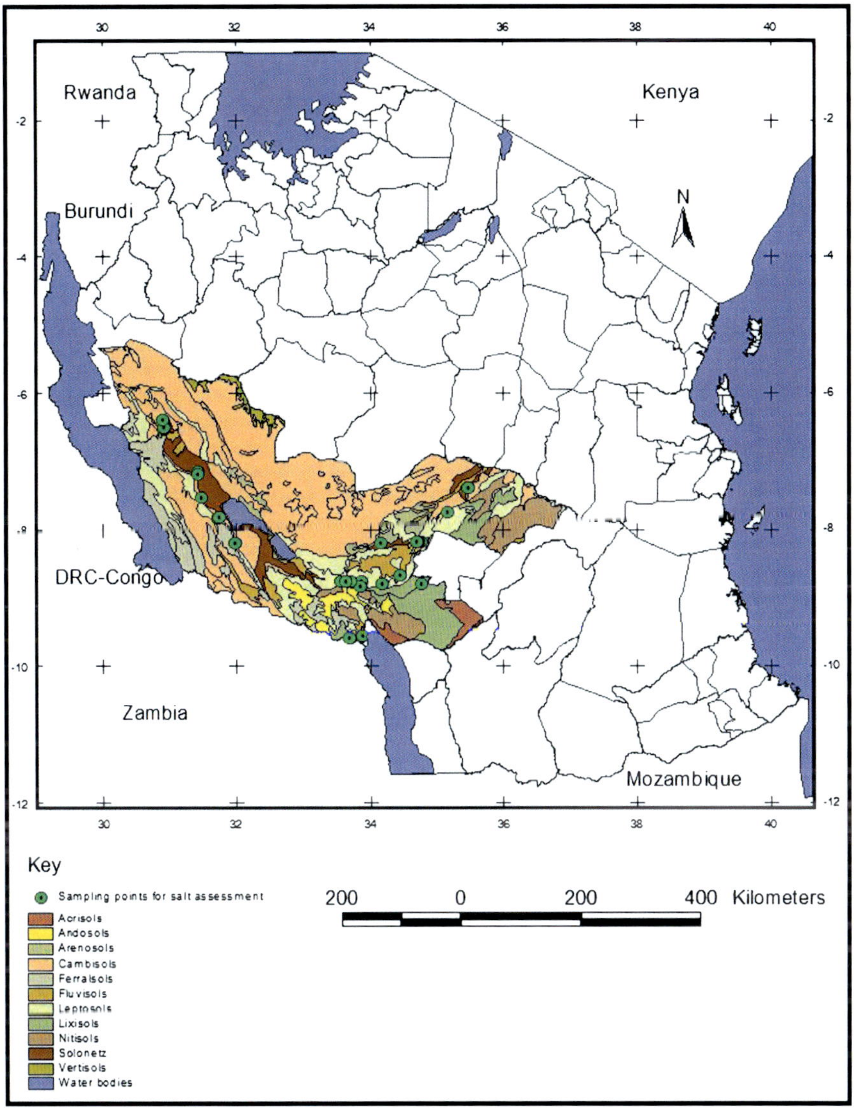

Fig. 6.4 Soils of the sampled rice irrigation schemes in the southern rice-growing corridor

had been completely abandoned. Yield loss of up to 100 % has been reported. However, laboratory analysis of the 160 samples collected from the 19 rice irrigation schemes in all four regions of the southern highland areas showed that 65 samples (41 %) had salt-affected soils characteristics.

Table 6.1 Salt-affected soil characteristics in rice irrigation schemes in Katavi region

Statistics	K cmol(+)/ kg soil	Ca cmol(+)/ kg soil	Mg cmol(+)/ kg soil	Na cmol(+)/ kg soil	SAR	pH (water)	ECe (dSm^{-1})
Mean	0.58	12.87	4.82	5.00	23.37	7.51	4.45
SE	0.06	0.96	0.58	2.09	9.85	0.24	0.94
Median	0.53	11.46	4.2	0.48	2.55	8.2	0.44
SD	0.28	4.79	2.92	10.43	49.25	1.20	4.70
Variance	0.08	22.98	8.52	108.87	2425.89	1.44	22.07
Range	1.32	20.54	12.3	40.69	203.64	3.4	15.17
Minimum	0.15	5.39	1.89	0.09	0.24	5.4	0.03
Maximum	1.47	25.93	14.19	40.78	203.88	8.8	15.2
Count (n)	25	25	25	25	25	25	25

6.3.2.1 Soils of Rice Irrigation Schemes in the Katavi Region

The results from the focus group discussion indicated that affected areas covered 2–12 % of the Katavi region, with no fields abandoned. However, depending on the intensity of salts in farmers' fields, yield losses of 5–70 % were reported. The laboratory results for samples collected from salt-affected hotspot areas indicated that 33 % (15 of 46) of samples had salt-affected soils. The mean SAR, pH, and ECe were 23.4, 7.5, and 4.45 dSm^{-1}, respectively, indicating that a combination of salinity and sodicity problems (saline–sodic soils characteristic) significantly affected the areas (Table 6.1). The Mwamapuli irrigation scheme (the largest irrigation scheme in the region with 13,000 ha) seemed to be more affected than the rest of the surveyed schemes in the region. Samples collected from this scheme had ECe values of 4.58–15.2 dSm^{-1}, indicating the presence of much soluble salt in the soil solution (Fig. 6.5). The SAR values of this soil were 13.2–204, which is greater than the standard for crop productivity (SAR of 13).

The higher SAR values indicated that the concentration of sodium in the soil solution is much higher than the concentration of calcium and magnesium. In this case, the Mwamapuli irrigation scheme has slightly to strong saline–sodic soil. However, soils from the Karema, Mlele, Kamsisi, and Ikaka irrigation schemes had lower SAR, pH, and ECe, indicating normal soils with no salt problem.

The results showed that the mean SAR was 23.37 for the samples collected from Katavi region with a maximum value of 203.9 indicating sodic to very strong sodic characteristics. Although high levels of SAR, pH, and ECe were recorded, the median values of these parameters were below the critical values for rice production (Table 6.1). These median values indicated that 50 % of the soils in the surveyed irrigation scheme have normal (neither saline nor sodic) characteristics. The results also showed that one site in the Mwalukulu irrigation scheme had an extremely high ECe level of 12.2 dSm^{-1}, indicating a high concentration of soluble salts and a

Fig. 6.5 Salt-affected field in Mwamapuli irrigation scheme, Katavi region

salinity problem. The rest of tested sites exhibited slightly alkaline characteristics (pH: 8.4–8.7), and other salt soil indicative parameters were very low (ECe: 0.3–0.6 dSm^{-1}; SAR: 0.24–2.14) (Annex Table 6.5).

6.3.2.2 Soils of Rice Irrigation Schemes in the Rukwa Region

In the Rukwa region, the affected area ranged from 7 to 15 % of total land area, 2–10 % of rice fields were abandoned, and yield loss was 5–70 %. The laboratory

Table 6.2 Salt-affected soil characteristics in rice irrigation schemes in Rukwa region

Statistics	Na cmol(+)/kg soil	K cmol(+)/kg soil	Ca cmol(+)/kg soil	Mg cmol(+)/kg soil	SAR	pH (water)	ECe (dSm^{-1})
Mean	11.55	1.61	14.25	4.31	52.48	8.14	6.10
SE	5.16	0.35	1.58	0.60	20.66	0.10	1.46
Median	1.46	0.69	13.32	3.15	8.19	8.2	4.65
SD	24.77	1.63	7.42	2.80	99.06	0.50	7.02
Variance	613.33	2.64	55.11	7.82	9813.58	0.25	49.27
Range	102.28	5.62	27.83	9.95	396.46	1.9	23.73
Minimum	0.01	0.1	2.71	2.35	0.02	6.9	0.27
Maximum	102.29	5.72	30.54	12.3	396.48	8.8	24
Count	23	23	23	23	23	23	23

results from 23 soil samples collected in selected irrigation schemes showed that 13 samples (57 %) had salt-affected soil characteristics. The mean SAR, pH, and ECe were 52.5, 8.1, and 6.1 dSm^{-1}, respectively, indicating a combination of salinity and sodicity soils characteristic. These values also indicated that 50 % of the soil samples collected from the surveyed irrigation scheme were slightly saline, but the rest of the collected samples had higher levels of SAR, pH, and ECe and, therefore, extreme levels of saline–sodic (Table 6.2).

Saline–sodic was shown to be a common problem in the affected irrigation scheme, mostly in the Sakalilo and Mpete irrigation schemes (Figs. 6.6 and 6.7). The ECe and SAR values of the soils collected from the Sakalilo and Mpete sites

Fig. 6.6 Abandoned rice field in Mpeta village, Sumbawanga, Rukwa region

Fig. 6.7 A farmer's field in the Sakalilo irrigation scheme, Sumbawanga, Rukwa region. The owner of the farm still grow rice in this field especially during the rainy season when salt levels are very low due to rain wash

were high to extremely high (ECe: 5.34–19 dSm^{-1}; SAR: 17.5–396.5) (Annex Table 6.6), indicating extreme levels of sodium concentration in the soil solution. Only 3 samples collected from three tested sites (1 in Ngongo, 2 in Mpete) showed purely saline characteristics with an ECe slightly greater than 4 dSm^{-1} and low pH (7.8–8.3) and SAR (2.12–9.07).

One sample collected from a salt-affected hotspot about 5 m from a water pond that served as an irrigation source had extremely high saline–sodic characteristic. A tremendous salinity value of more than 19 dSm^{-1} indicated the possibility of salt distribution after mixing with irrigation water (Annex Table 6.6).

6.3.2.3 Soils of Rice Irrigation Schemes in Mbeya Region

A similar situation was observed in the Mbeya region, where affected areas were 2–15 % of the total, abandoned land were 5–7 %, and yield losses of up to 100 % were reported (Fig. 6.8). Laboratory results indicated that 37 % (31 of 84) of the surveyed sites were salt-affected. Of the 31 affected samples, only 5 samples had sodicity soil characteristics, while the rest had a combination (saline–sodic characteristics). The mean SAR, pH, and ECe were 98, 8.4, and 5.3 dSm^{-1}, respectively, indicating a combination of salinity and sodicity soils characteristics. The median values for SAR, pH, and ECe were 31.4, 8.4, and 5.1, respectively, indicating that 50 % of the soil samples collected from the surveyed irrigation scheme had strongly saline–sodic characteristics (Table 6.3).

Fig. 6.8 Water drainage canal with strong salty symptom in Madibira Irrigation Scheme, Mbeya, 2014

Table 6.3 Salt-affected soil characteristics in rice irrigation schemes in Mbeya region

Statistic	Na cmol(+)/ kg soil	Ca cmol(+)/ kg soil	Mg cmol(+)/ kg soil	K cmol(+)/ kg soil	SAR	pH (water)	ECe (dSm^{-1})
Mean	25.43	12.31	3.39	3.16	97.58	8.39	5.29
SE	10.10	1.89	0.66	1.37	31.85	0.16	1.07
Median	10.33	8.81	2.40	0.89	31.37	8.50	3.12
SD	63.86	11.93	4.20	8.69	201.42	1.01	6.78
Variance	4077.61	142.34	17.64	75.47	40,569.42	1.02	45.94
Range	399.98	66.04	26.26	55.05	1156.31	5.1	23.96
Minimum	0.08	1.23	0.27	0.2	0.04	5.4	0.04
Maximum	400.06	67.27	26.53	55.25	1156.35	10.5	24
Count	40	40	40	40	40	40	40

Strong salt effects were also observed in the Bethania, Madibira, and Luanda Majenje irrigation schemes, where extreme ECe (>19 dSm^{-1}), pH (10), and SAR (1156.4) levels were recorded (Fig. 6.8 and Annex Table 6.7).

6.3.2.4 Soils of the Rice Irrigation Schemes in the Iringa Region

Two irrigation schemes (Pawaga and Idodi) in the Iringa region were involved the study. All 6 samples (100 %) collected from 6 hot-spot sites were salt affected, with extremely high levels of saline–sodic soil characteristics. The affected areas were

Fig. 6.9 Salt affected field in Pawaga irrigation scheme, Iringa region, 2014

2–20 % of the total land, the abandoned area was 5–15 %, and yield losses of up to 100 % were reported (Fig. 6.9). The mean SAR, pH, and ECe levels were 523, 8, and 18 dSm^{-1}, respectively, indicating a combination of extreme levels of salinity and sodicity soils characteristics. The median values for SAR, pH, and ECe were 136, 8.5, and 24, 18 dSm^{-1}, respectively, indicating that 50 % of the soil samples collected from the surveyed irrigation schemes had extremely saline–sodic characteristics (Table 6.4). Among all the samples collected from the southern rice-growing corridor, the highest values of saline (>19 dSm^{-1}) and sodic (SAR: 2472.97) were recorded in the Pawaga irrigation scheme in the Iringa region (Fig. 6.9 and Annex Table 6.8).

6.3.3 Farmers' Experiences, Perceptions and Management of Salt-Affected Soils for Irrigated Rice Production

Results from the interviewed farmers revealed that farmers associate the presence of salt with the prevailing weather situation, which they reported to have undergone

Table 6.4 Salt-affected soil characteristics in rice irrigation schemes in Iringa region

Statistic	K Mean	Ca cmol(+)/ kg soil	Mg cmol(+)/ kg soil	Na cmol(+)/ kg soil	SAR	pH (Water)	ECe (dSm^{-1})
	3.13	24.46	7.27	41.24	523.34	8.00	18.67
SE	1.66	6.18	2.77	21.79	393.78	0.55	3.39
Median	0.91	18.94	3.41	18.97	135.97	8.45	24.00
SD	4.07	15.14	6.78	53.38	964.55	1.34	8.30
Variance	16.59	229.24	45.97	2849.18	930,358.23	1.81	68.90
Range	9.69	40.25	16.39	138.01	2471.19	3.60	17.22
Minimum	0.35	11.88	2.91	0.48	1.78	5.30	6.78
Maximum	10.04	52.13	19.30	138.49	2472.97	8.90	24.00
Count (n)	6	6	6	6	6	6	6

drastic changes compared over the decades. Recent noted changes in the climate include rising temperatures, decreased precipitation, delay in rainfall by several weeks, and poor rainfall distribution over time. Farmers argue that the intensity of rain in the region has increased, from 3 to 4 days of steady precipitation per week. Dry spells are also on the rise, which increases salt accumulation on the soil surface and increasingly affects crop growth, resulting in patchy growth. When asked about their perceptions of the salt-affected soils problem, the majority of respondents rated the problem as either bad (39 %) or extremely bad (30 %), with the highest rates noted in the Iringa, Rukwa, and Mbeya regions (Fig. 6.10). As well, 20 % of respondents could not explain the situation and responded "don't know."

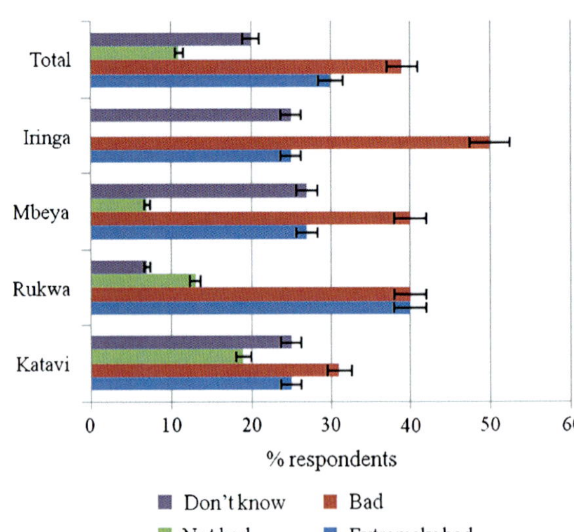

Fig. 6.10 Respondents' perceptions of the effects of salt-affected soils on rice production in the southern rice-growing corridor

Fig. 6.11 Respondents commonly methods used to manage salt-affected soils problems in rice fields

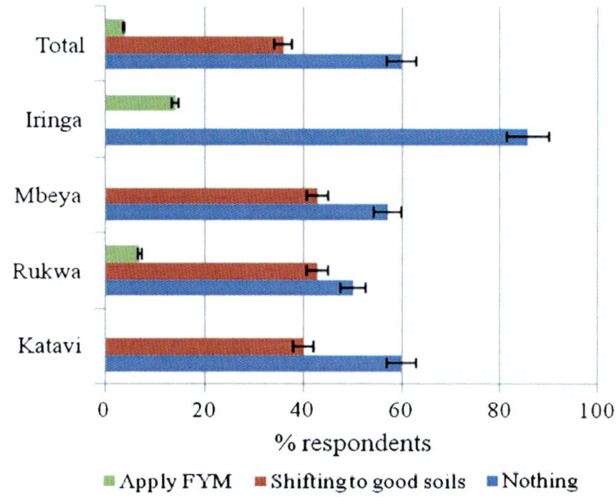

When asked about management practices to mitigate salt-affected soils problems in their areas, 60 % of focus-group members reported that they do nothing, while 36 % said they tend to shift to areas with good soil conditions (Fig. 6.11). Farmers sometimes described the problem as *toxicity of the soils* and shifted to better places or opened new land for cultivation. This practice implies increasing land degradation, water erosion, and water pollution as areas along river banks become farmers' main target for farm expansion due to the moisture availability and fertility.

6.4 Discussion

Both visual observations and the focus group discussion indicated the presence of salt-affected soils in the southern rice-growing corridor of Tanzania. This finding was supported by laboratory results which detected high to extremely high levels of salt indicators. Crop plants' salt stress responses throughout their growth cycle depend on several interacting variables, including the cultural environment, plants' developmental stage, salt concentration, and duration of the stress (Munns 2002).

Kashenge-Killenga et al. (2013), Studies have shown that crop yields are not significantly affected when the salt level is 0–2 dSm^{-1} (Kashenge-Killenga et al. 2013; FAO 2005; Chinnusamy et al. 2005). A level of 2–4 dSm^{-1} restricts some crops. Levels of 4–5 dSm^{-1} restrict many crops, and levels above 8 dSm^{-1} restrict all but very tolerant crops (Maas 1986). According to Chinnusamy et al. (2005), most grain crops and vegetables are glycophytes and are highly susceptible to soil salinity higher than an ECe of 2 dSm^{-1}. The estimated threshold above which rice yield decreases is 1.9–3.0 dSm^{-1} (Chinnusamy et al. 2005), indicating that the ECe in the southern rice-growing corridor is highly restrictive for rice cultivation. Furthermore, FAO (2005) highlighted yield losses for most salt-sensitive crops. If the ECe is less than 4 dSm^{-1}, yield loss will be less than 10 %; if the ECe is more than 6 dSm^{-1}, the yield loss will be 20–50 %, and if the EC(e is more than 10 dSm^{-1}, then yield loss will be more than 50 % (FAO 2005).

Generally, the ECe of all the affected areas was greater than 3 and 4 dSm^{-1}, the thresholds for rice cultivation and the value used to define saline soils, respectively, the thresholds for rice cultivation and the value used to define saline soils, respectively (Kashenge-Killenga et al. 2012a, b; Chinnusamy et al. 2005) (Fig. 6.1). Plant damage was obvious and supported by both data and farmers' statements. The higher recorded ECe indicates high concentrations of soluble salts (*salinity*) (>4 dSm^{-1}) (Buckland et al. 2002). These levels are detrimental to plant growth due to reduced water availability, direct toxicity from some ions, such as sodium and chloride, and ionic imbalances in the plants (Kashenge-Killenga et al. 2012a). All the surveyed areas are characterized by semi-arid and sub-humid conditions. Studies have found that salinity is a common problem in arid, semi-arid, and sub-humid areas, especially when there is not enough irrigation water to leach the soluble salts away from the root zone and a lack of proper irrigation infrastructures that allow easy movement of water in and out the irrigated fields (Kashenge 2010; FAO 2004; Tesfai et al. 2002; Buckland et al. 2002). Unfortunately, poor irrigation infrastructures were common in almost all the visited irrigation schemes.

Small-scale paddy farming in the lowlands regions of Tanzania depend, in most cases, on irrigation, which involves the collection of water in depressions or low-lying areas using mostly poorly managed traditional irrigation facilities (Kashenge 2010; Mnkeni 1996). Collection of water in these depressions and low-lying areas might be accomplished by purposeful blocking of drainage ways. In time, this increases salt movements from sub surface layers and accumualation of salts in plant rooting zone. Farmers might be unaware of the effects of doing so (Kashenge-Killenga et al. 2012b). Unfortunately, failure to apply principles of efficient water management in irrigated systems can result in wastage of water through seepage (Chemura et al. 2014; Tesfai et al. 2002; Buckland et al. 2002). According to Mnkeni (1996), seepage and runoff water which collect in inland depressions evaporate, leaving behind dissolved salts and causing salt build–up.

Maas (1994) added that high carbonates causes calcium and magnesium ions to form insoluble minerals, leaving sodium as the dominant cation in the solution; consequently, the soil becomes sodic. This process was well evidenced by low calcium and magnesium levels and extreme levels of sodium in most of the surveyed schemes where the soil pH values were as high as 10. In sodium-rich clays, HCO_3 reacts with water to form hydroxyl (OH-) ions, which cause high pH (Wong et al. 2010). These are characteristics of most Solonetzic soils (Fig. 6.4), which are very dense and high in clay and sodium. In some areas, the dominate soil is Solonetzic, which usually indicates that the soils developed from sodium-rich parent materials. Management of Solonetzic soils requires paying special attention to the timing of tillage and seeding operations with respect to moisture conditions. When the soil is too wet, tillage implements cannot be operated. Conversely, when the soil is too dry, tillage implements have difficulty penetrating the soil and create hard lumps. If serious measures are not taken, Tanzania will lose large areas of productive lands to sodium.

6.5 Conclusions and Recommendations

Small-scale paddy farming is normally done in lowlands using traditional irrigation facilities or depends entirely on rainfall. These areas are experiencing increasing levels of salt-affected soil due to soil mismanagement, global climatic change, poorly designed and managed irrigation and drainage principles and infrastructures, and excessive and irrational use of irrigation water. Adding to the problem are excessive irrigation, high temperatures, evapotranspiration greater than precipitation, and a lack of enough water, which increases the capillary rise of the saline groundwater table to the soil surface. The poor management of irrigation water, canals, and skills for water management worsens the situation in most of the local irrigation schemes. A severe problem is expected in the long term if corrective measures are not considered. Lack of knowledge about the salt-affected soil problem has contributed to the minimal efforts given to problem identification and management.

A combination of techniques, such as the use of soil amendment (sometimes expensive and unsustainable for resource-poor farmers) and salt-tolerant cultivars (a cheap, sustainable biological amendment) could help mitigate the problem. To achieve this, a clear understanding of the bio-physical characteristics of rice-growing environments is vital. Specifically, the existing type and extent of the salt problem needs to be known for efforts towards the development of salt-tolerant rice varieties. Chollima Agro-scientific Research Centre has developed a salt-tolerant rice population, and in 2016, a salt tolerant rice variety SATO1 were released. Among other achievement of the center is the release of another new high yielding variety SATO9 in the same year 2016, development of a widely growing TXD 306 rice variety which were released in 2001 as well as development of TXD 88 and TXD 85 which were released in 2000. The study also recommends, a massive awareness-creation program about the occurrence and effects of salt-affected soils is highly important. Any attempt to increase irrigated food production in the coming years must pay adequate attention to the improvement of the affected irrigation schemes and the prevention of further deterioration of productive soils through these degradation processes.

Acknowledgments We are grateful to iAGRI–USAID and the World Bank for their financial support through the EAAPP Program. Many thanks are due to the district and village officials throughout the surveyed areas for assisting in site identification and soil sample collection. Special thanks are owed to Jackson Shushu, the ARI–KATRIN driver who worked tirelessly in the collection of soil samples and to ARI—Mlingano Laboratory Technicians for their work in analysing the soil samples.

Annex

See Tables 6.5, 6.6, 6.7 and 6.8.

Table 6.5 Soluble salts, SAR, pH, and ECe values for soils from rice irrigation schemes in Katavi region

Scheme	K (meq/100 g)	Ca (meq/100 g)	Mg (meq/100 g)	Na (meq/100 g)	SAR	pH (1:2.5) (water)	ECe (dSm^{-1})	Remarks
Karema	0.57	25.93	13.53	0.54	6.76	6.5	0.12	Normal
Karema	0.40	20.24	6.32	0.67	6.02	6.0	0.17	Normal
Karema	0.81	23.04	14.19	0.94	9.53	5.6	0.41	Normal
Karema	0.64	14.46	5.50	1.00	5.77	5.4	0.42	Normal
Mlele	0.42	9.67	4.01	0.26	1.25	6.6	0.08	Normal
Mlele	0.26	6.77	2.42	0.26	1.06	6.4	0.04	Normal
Kamsisi	0.15	9.57	3.11	0.12	0.91	6.3	0.03	Normal
Kamsisi	0.28	12.86	4.36	0.35	2.55	5.7	0.04	Normal
Mwalukulu	0.40	9.78	4.15	0.14	0.70	8.5	0.44	Slightly alkaline
Mwalukulu	0.61	10.88	4.51	0.12	0.56	8.6	0.55	Slightly alkaline
Mwalukulu	0.82	14.07	5.41	0.20	1.05	8.7	0.46	Slightly alkaline
Mwalukulu	0.87	12.38	4.74	0.48	2.14	8.4	0.57	Slightly alkaline
Mwalukulu	0.47	10.08	4.00	0.17	0.79	8.5	0.34	Slightly alkaline
Mwalukulu	0.37	5.39	1.89	0.09	0.24	8.2	12.15	Strongly saline
Mwalukulu	0.49	13.77	4.20	0.09	0.50	8.2	0.33	Normal
Ikaka	0.31	9.78	3.58	0.32	1.69	8.5	0.39	Slightly alkaline
Ikaka	0.41	11.18	3.80	0.24	1.25	5.6	0.07	Normal
Mwamapuli	0.65	10.46	2.88	40.78	151.91	8.2	12.80	Strongly saline–sodic
Mwamapuli	0.79	11.46	3.08	13.02	50.04	8.4	5.50	Strongly saline–sodic
Mwamapuli	1.47	14.87	4.34	2.51	13.17	8.5	4.98	Slightly saline–sodic

(continued)

Table 6.5 (continued)

Scheme	K (meq/100 g)	Ca (meq/100 g)	Mg (meq/100 g)	Na (meq/100 g)	SAR	pH (1:2.5) (water)	ECe (dSm^{-1})	Remarks
Mwamapuli	0.40	7.97	2.48	8.45	29.84	7.8	6.75	Saline–sodic
Mwamapuli	0.74	11.36	3.19	12.55	49.22	8.5	9.43	Strongly saline–sodic
Mwamapuli	0.74	17.25	5.13	33.68	203.88	8.1	15.20	Strongly saline–sodic
Mwamapuli	0.53	16.37	5.18	4.24	27.70	7.8	6.43	Saline–sodic
Mwamapuli	0.98	12.08	4.62	3.77	15.80	8.8	0.73	Slightly sodic

Table 6.6 SAR, pH, and ECe values for soils from rice irrigation schemes in Rukwa region

Scheme	Na (meq/100 g)	K (meq/100 g)	Ca (meq)/ 100 g)	Mg (meq/100 g)	SAR	pH (1:2.5) (water)	ECe (mS/cm)	Remarks
Ngongo	0.42	0.55	15.87	4.51	2.12	7.8	4.72	Saline
Ngongo	0.07	0.38	10.58	3.01	0.25	8.0	0.77	Normal
Ngongo	0.33	0.58	10.78	2.67	1.12	8.0	4.6	Slightly saline
Ng'ongo	1.88	0.55	14.17	3.28	8.19	7.7	0.40	Normal
Ng'ongo	0.12	0.45	8.18	2.72	0.33	6.9	0.30	Normal
Sakalilo	0.07	0.35	7.09	2.45	0.18	8.0	0.78	Normal
Ng'ongo	0.01	0.34	7.49	2.37	0.02	8.2	0.46	Normal
Sakalilo	0.29	1.10	6.99	2.35	0.67	8.2	0.45	Normal
Sakalilo	4.72	4.23	2.71	12.30	17.71	8.6	5.32	Slightly saline–sodic
Sakalilo	12.86	3.82	26.63	5.05	101.88	8.8	7.81	Saline–sodic
Sakalilo	21.72	5.72	30.54	5.66	196.54	8.2	>19	Strongly saline–sodic
Sakalilo	28.16	3.14	22.34	6.29	201.54	8.5	10.45	Strongly saline–sodic
Sakalilo	7.08	0.79	7.09	2.80	17.51	8.7	8.43	Saline–sodic
Sakalilo[a]	4.25	2.45	12.46	3.04	17.70	8.7	>19	Strongly saline–sodic
Mpete	102.29	2.56	7.68	2.54	396.48	8.7	11.67	Strongly saline–sodic
Mpete	66.86	2.20	14.77	3.03	170.93	8.4	12.04	Strongly saline–sodic
Mpeta	10.00	0.11	22.26	12.21	44.49	8.7	11.71	Saline–sodic
Mpeta	0.84	0.10	21.16	4.21	7.24	7.9	0.50	Normal

(continued)

Table 6.6 (continued)

Scheme	Na (meq/100 g)	K (meq/100 g)	Ca (meq)/100 g)	Mg (meq/100 g)	SAR	pH (1:2.5) (water)	ECe (mS/cm)	Remarks
Mpeta	0.25	0.22	7.98	2.80	1.60	8.0	0.47	Normal
Mpeta	0.06	0.13	16.27	3.25	0.15	7.9	0.27	Normal
Mpeta	0.53	2.20	20.26	3.71	2.59	7.[illegible]	0.86	Normal
Mpeta	1.46	3.41	20.26	4.58	8.75	8.1	5.65	Saline
Mpeta	1.46				9.07	8.3	4.65	Saline

[a]Sample collected near a water source used for irrigation

Table 6.7 SAR, pH, and ECe values for soils from rice irrigation schemes in Mbeya region

Scheme	Na (meq/100 g)	Ca (meq/100 g)	Mg (meq/100 g)	K (meq/100 g)	SAR	pH (water)	ECe (mS/cm)	Remarks
Ngana	3.72	20.66	4.33	0.31	23.25	7.8	0.42	Normal
Ngana	0.08	3.39	0.76	0.70	0.08	8.0	0.28	Normal
Ngana	16.27	5.99	1.24	0.44	29.40	8.9	0.70	Sodic
Tenende	0.69	3.93	0.81	0.50	0.81	8.1	0.30	Normal
Tenende	0.58	3.80	1.03	0.72	0.70	5.4	0.04	Normal
Gwiri	90.15	7.22	2.04	7.23	208.82	10	>19	Strongly saline–sodic
Gwiri	39.54	9.62	2.00	5.63	114.77	8.3	7.61	Saline–sodic
Gwiri	14.23	14.91	5.07	2.61	71.05	8.8	0.44	Sodic
Gwiri	11.22	20.54	5.86	3.92	74.07	8.7	0.64	Sodic
Ruanda Majenje	7.65	21.06	5.92	4.11	51.59	8.5	5.45	Saline–sodic
Ruanda Majenje	37.60	37.13	9.06	4.31	434.15	8.9	8.22	Saline–sodic
Ruanda Majenje	1.43	9.59	2.27	9.14	14.33	10.5	6.73	Saline–sodic
Luanda-Majenje	2.93	7.79	2.53	0.94	7.54	8.9	8.78	Saline–sodic
Luanda-Majenje	0.22	2.43	0.45	0.29	14.50	8.5	0.26	Sodic
Bethania	35.53	5.99	1.11	3.44	63.11	10.3	0.17	Strongly sodic
Bethania	400.06	9.48	2.08	55.25	1156.35	8.1	>19	Extremely saline–sodic
Bethania	0.48	2.89	2.88	0.42	0.70	7.8	0.49	Normal
Bethania	20.19	5.49	1.11	1.80	33.33	8.8	15.8	Strongly saline–sodic
Bethania	37.58	9.88	2.57	3.20	117.03	8.1	8.92	Strongly saline–sodic
Mbuyuni	20.15	2.20	1.83	0.79	20.29	8.4	6.51	Saline–sodic

(continued)

Table 6.7 (continued)

Scheme	Na (meq/100 g)	Ca (meq/100 g)	Mg (meq/100 g)	K (meq/100 g)	SAR	pH (water)	ECe (mS/cm)	Remarks
Mbuyuni	35.36	11.68	2.46	2.95	125.04	8.3	7.76	Strongly saline–sodic
Mbuyuni	20.75	7.39	1.83	0.31	47.85	8.5	5.88	Saline–sodic
Mbuyuni	5.66	2.99	1.54	0.54	6.42	6.8	0.10	Normal
Mbuyuni	22.19	6.79	1.28	0.78	44.76	8.4	6.01	Saline–sodic
Mbuyuni	1.10	7.69	1.49	0.53	2.53	6.7	0.45	Normal
Manyanga	3.34	9.12	2.65	0.82	14.10	8.9	0.52	Slightly sodic
Manyanga	3.60	14.21	4.37	1.00	16.71	8.8	6.76	Slightly sodic
Manyanga	9.43	16.50	3.69	0.20	47.63	8.1	5.56	Saline–sodic
Manyanga	0.38	2.53	0.65	0.41	0.30	8.7	0.09	Slightly alkaline
Manyanga	0.11	1.23	0.27	0.34	0.04	8.6	0.18	Slightly alkaline
Manyenga	1.37	8.49	2.92	0.39	3.92	6.2	0.06	Normal
Manyenga	1.34	8.09	3.06	0.65	3.73	6.4	0.11	Normal
Madibira	26.23	19.40	3.92	1.00	152.91	9.8	0.76	Sodic
Madibira	4.76	27.54	6.25	1.40	40.21	8.7	5.64	Saline–sodic
Madibira	21.63	13.11	2.34	0.89	83.54	8.9	0.79	Sodic
Madibira	2.64	18.46	3.48	0.88	14.48	8.3	5.87	Slightly saline–sodic
Madibira	13.84	4.28	2.09	1.96	22.03	8.7	0.52	Sodic
Madibira	18.22	22.46	4.54	3.44	122.99	8.9	5.98	Saline–sodic
Madibira	73.56	19.26	5.41	0.38	453.76	8.3	14.88	Extremely saline sodic
Madibira	11.28	67.27	26.53	1.67	264.53	8.8	>19	Extremely saline sodic

Table 6.8 SAR, pH, and ECe values for soils from rice irrigation schemes in Iringa region

Scheme	K (meq/100 g)	Ca (meq/100)	Mg (meq/100 g)	Na (meq/100 g)	SAR	pH (Water)	ECe (mS/cm)	Remarks
Pawaga	1.31	52.13	19.30	138.49	2472.97	8.2	>19	Extremely saline–sodic
Pawaga	0.36	13.68	3.05	3.89	16.27	8.5	9.21	Saline–sodic
Pawaga	10.04	18.77	3.28	25.78	142.11	8.4	>19	Strongly saline–sodic
Pawaga	6.23	31.17	11.56	12.15	129.82	8.9	>19	Strongly saline–sodic
Idodi	0.51	19.10	3.53	66.67	377.07	8.7	>19	Strongly saline–sodic
Idodi	0.35	11.88	2.91	0.48	1.78	5.3	6.78	Saline

References

African Agricultural Technology Foundation (2013) Raising the productivity of rice in Africa. www.AATF-Africa.org/information/newsletters. Accessed Jan 2015
Alam SM (2006) Profiting from saline tolerant crops. http://www.dawn.com/2006/10/23/ebr4.htm. Accessed Feb 2015
Buckland G, Bennett D, Mikalson D, de Jong E, Chang C (2002) Soils salinization and sodification from alternate irrigation with saline–sodic water and simulated rainfall. Can J Soil Sci 82:297–309
Chemura A, Kutywayo D, Tapiwanashe M, Chidoko P (2014) An assessment of irrigation water quality and selected soil parameters at Mutema irrigation scheme, Zimbabwe. J Water Resour Prot 6:132–140
Chinnusamy V, Jagendorf A, Zhu J (2005) Understanding and improving salt tolerance in plants. Crop Sci 45:437–448
Food and Agriculture Organization (2000) Land resources information WAICENT FAO search. http://www.fao.org/WAICENT/FAOINFO/AGRICULT/AGL/swlwpnr/htm. Accessed Mar 2015
Food and Agriculture Organization (2003) Databases Food and Agricultural Organization of the united nation database. http://www.fao.org/statistics/databases/en/. Accessed May 2014
Food and Agriculture Organisation (2004) FAOSTAT annual report 2004–2005. FAO, Rome. Online publication. http://www.fao.org. Accessed March 2013
Food and Agriculture Organization (2005) Global network on integrated soil management for sustainable use of salt-affected soils. Food and Agriculture Land and Plant Nutrition Management Service, Rome. Handbook of plant and crop stress, 3rd edn. https://books.google.com/books?isbn=143981399X. Accessed May 2015
Food and Agriculture Organisation (2012) FAO statistic year book. http://www.fao.org/docrep/015/i2490e/i2490e00.htm. Accessed May 2014
Food and Agriculture Organization (2014) Statistics on crop production. FAO statistical yearbook 2014—Food and Agriculture Organization. www.fao.org/3/a-i3590e.pdf. Accessed Dec 2014
Gonzalez NLM, Toth T, Garcia D (2004) Integrated management for sustainable use of salt affected soils in Cuba, Universidad Y Ciencia. 20(40):86–102
Iringa District Council Report (2013)
Kashenge S (2010) Breeding investigations for salt tolerant rice genotypes, studies on the extent and farmers perceptions of salt affected soils in northern-eastern Tanzania. Ph.D. dissertation, University of KwaZulu-Natal, Pietermaritzburg, Republic of South Africa
Kashenge-Killenga S, Tongoona P, Derera J, Kanyeka Z, Meliyo J (2012a) Soil characterisation for salt problems in selected rice irrigation schemes: a key aspect towards improvement of salt tolerance rice genotypes in north eastern Tanzania. J Adv Dev Res 3(1):21–32
Kashenge-Killenga S, Tongoona P, Derera J, Kanyeka Z (2012b) Irrigated rice-based farm characteristics and production constraints in selected salt affected areas of north-eastern Tanzania. J Adv Dev Res 3(1):1–14
Kashenge-Killenga S, Tongoona P, Derera J (2013) Morphological and physiological responses of Tanzania rice genotypes under saline condition and evaluation of traits associated with stress tolerance. Int J Dev Sustain 2(2):1457–1475
Kashenge-Killenga S, Tongoona P, Derera J (2014) Farmers' perception of salt affected soils and rice varieties preferences in the northeastern Tanzania and their implications in breeding. Int J Dev Sustain 3(6):1257–1271
Kilimo Trust (2011) Making markets work to reduce poverty. Six years of Kilimo Trust in East Africa, 2005–2011. Plot 42, Princess Anne Drive, Kapala Uganda, p 101. www.kilimotrust.org
Land FAOSTAT—Food and Agriculture Organization (2004) FAOSTAT annual report 2004–2005. http://faostat.fao.org/site/377/default.aspx. Accessed Dec 2013
Maas EV (1994) Salt tolerance at different stages of growth. In: Proceedings of the dryland salinity workshop "Causes to Cures", Lethbridge, Alberta, Canada, 8–9 Feb 1994
Maas EV (1986) Salt tolerance of plants. Appl Agric Res 1:12–26

Mnkeni PN (1996) Salt affected soils: their origin, identification, reclamation and management. Inter Press Ltd., Dar es Salaam, p 250
Moberg GP (2000) Biological responce to stress: implication for animal welfare in the biology of animal stress. In: Moberg GP, Mench JA (eds) CAB International
Moradi F, Ismail A, Gregorio G, Egdane J (2003) Salinity tolerance of rice during reproductive development and association with tolerance at the seedling stage. Indian J Plant Physiol 8:105–116
Munns R (2002) Comparative physiology of salt and water stress. Plant Cell Environ 25:239–250
Population and Housing Census of Tanzania (2012) Population distribution by administrative areas. National Bureau of Statistics, Ministry of Finance, Dar es Salaam, Tanzania
Seilsepour M, Rashidi M, Khabbaz B (2009) Prediction of soil exchangeable sodium percentage based on soil sodium adsorption ratio. Am Eurasian J Agric Environ Sci 5(1):1–4
Siyal AA, Siyal AG, Abro ZA (2002) Salt affected soils, their identification and reclamation. Pak J Appl Sci 5:537–540
Stryker D, Amin M (2012) Study of policy options for increasing Tanzanian exports of maize and rice—improving food security to 2025. USAID/AIRD
Sumbawanga District Council Report (2013)
Sustainable Agriculture and Soil Conservation (2009) Soil degradation processes. Fact sheet no 4. http://esdac.jrc.ec.europa.eu/projects/SOCO/FactSheets/EN%20Fact%20Sheet.pdf. Accessed Feb 2014
Tesfai M, Dawod V, Abreha K (2002) Management of salt-affected soils in the NCEW "Shemshemia" irrigation scheme in the Upper Gash Valley of Eritrea. DCG Report No. 20
Thomas GW (1982) Exchangeable cations. In: Page AL, Miller RH, Keeney RD (eds) Methods of soil analysis, 2nd edn, ASA, SSSA Monograph no. 9. Madison, Wisconsin, pp 159–165
TWAS (2006) More drop per crop. TWAS Newsl 18(4):1–50
Wong V, Greene R, Dalal R, Murphy B (2014) Soil carbon dynamics in saline and sodic soils: a review. Soil Use Manag 26(1):2–11. doi:10.1111/j.1475-2743.2009.00251

Chapter 7
Sustainability of Intensification of Smallholder Maize Production in Tanzania

Frank Brentrup, Joachim Lammel, Katharina Plassmann and Dirk Schroeder

Abstract Tanzania's fast growing population will increase the demand for food crops, particularly maize. Sustainable intensification is needed to meet this demand while, at the same time, preserving the environment and climate. A joint project of Yara, Syngenta, Sokoine University of Agriculture, and the Norwegian University of Life Sciences demonstrated that a balanced supply of crop nutrients together with other improved practices has the potential to increase maize yields and farm profitability. A framework was developed and applied to assess the potential impacts of different cropping intensities on climate, soil, water, and biodiversity in order to evaluate the environmental sustainability of the measures. Maize yields increased by 49–163 % compared to prevailing farmer practice (FP). This in turn may reduce the need for arable land expansion and thus potentially avoid GHG emissions. If GHG emissions from potential arable land expansion into tropical scrubland are considered, GHG emissions from the low-yielding treatments would be 3.6–12 times greater than the CFP of the improved protocol. Low positive soil nutrient balances with the improved cropping protocol indicate sustainable fertilizer use, which can replenish the soil with sufficient nutrients. In contrast, FP often showed negative nutrient balances, signifying unsustainable nutrient mining even at low yield levels. The increased nitrogen rates and crop productivity lead to increased soil acidification, which needs compensation from liming. The improved maize protocol doubled maize stover biomass, which can be used to improve the organic matter and fertility of the soil either through direct incorporation into the soil or through feeding livestock and producing and applying manure. A water footprint calculation at one site revealed that the maize produced according to FP consumed 50 % more water per ton of grain compared to the improved protocol. Biodiversity was assessed in different ways. In-field biodiversity was reduced with the suggested protocol, while on-farm biodiversity was enhanced through the planting of additional trees instead of expanding cropland. About 50 % less land was needed to produce one ton of maize grain, reducing pressure for land expansion that would have potential negative effects on biodiversity on a larger scale.

F. Brentrup (✉) · J. Lammel · K. Plassmann · D. Schroeder
Yara International ASA, Research Centre Hanninghof, Dülmen, Germany
e-mail: frank.brentrup@yara.com

R. Lal et al. (eds.), *Climate Change and Multi-Dimensional Sustainability in African Agriculture*, DOI 10.1007/978-3-319-41238-2_7

Keywords Sustainable intensification · Mineral fertilizer · Balanced nutrition · Carbon footprint · Environmental footprint · Nutrient mining · Soil fertility · Public-private partnership

7.1 Introduction

Sub-Saharan Africa in general and Tanzania in particular have a fast-growing population, resulting in increased demand for food crops, including maize. The population in Tanzania is expected to grow from 45 million in 2010 to 137 million by 2050 (United Nations, Department of Economic and Social Affairs, Population Division 2015). Maize is currently the most prevalent food crop in the country (Mtengeti et al. 2015). Assuming that the demand for maize increases at a pace similar to population growth, the total maize production in Tanzania needs to roughly triple by 2050. This increased demand for maize can either be met by expansion of the maize cropping area, increasing crop yields per hectare, a combination of both, or by additional imports. Current yield trends indicate the extension of cropland into non-agricultural areas at the expense of other crops, as shown in Fig. 7.1.

The conversion of non-agricultural land, such as forests or savannahs, into cropland has significant consequences in terms of a loss of biodiversity and natural carbon stocks. Globally, the loss of carbon due to land-use change (LUC) is responsible for about 10 % of the total emissions of greenhouse gases (Tubiello et al. 2015). Therefore, the concept of sustainable intensification, i.e., the production of more food from the same area of land while reducing associated environmental impacts, is of particular importance (Beddington et al. 2012). If, for instance, the maize yield in Tanzania was increased from the 2010 level of 1.55 to 2 t/ha, it would theoretically save almost three million hectares of land for other purposes, e.g., nature conservation. For a yield level of five tons of maize grain per

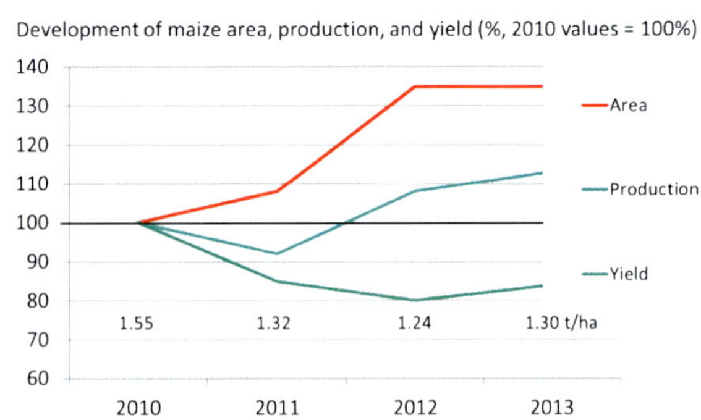

Fig. 7.1 Current trends in maize area, production, and yield in Tanzania (Food and Agriculture Organization of the United Nations, Statistics Division 2015)

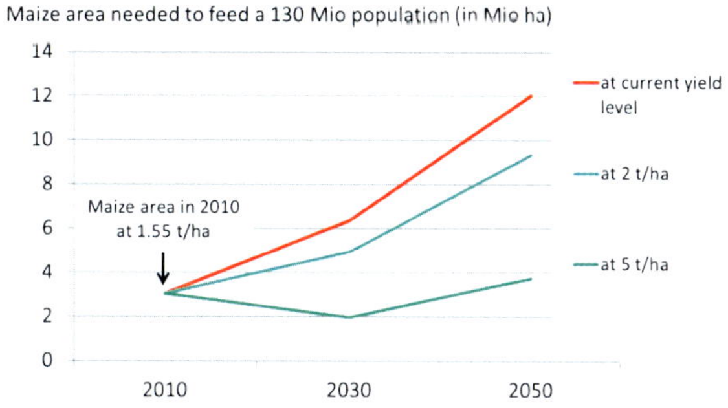

Fig. 7.2 Maize area needed to supply maize for a growing Tanzanian population at three different yield levels (based on data from Food and Agriculture Organization of the United Nations, Statistics Division 2015 and United Nations, Department of Economic and Social Affairs, Population Division 2015)

hectare, almost no additional land would be needed to provide sufficient quantities of maize for a population of about 130 million people in 2050 (Fig. 7.2).

An increased crop production per hectare can be achieved through more intensive and more efficient use of external and internal farm inputs, such as mineral and organic fertilizers, as well as appropriate plant protection in combination with other improved agricultural practices. Intensification in crop production has an impact on productivity per area and farm profitability as well as potentially negative effects on soil, water, air, climate, and the biodiversity of extensive agriculture.

A private-public partnership (PPP) project between two companies, Yara International (plant nutrition) and Syngenta (plant protection), and two universities, Sokoine University of Agriculture (Morogoro, Tanzania) and Norwegian University of Life Sciences (As, Norway), developed a maize cropping protocol, including the use of mineral multi-nutrient fertilizers and chemical crop protection with the aim to improve the productivity and profitability of smallholder farms. The suggested agronomic protocol was tested against a farmer practice (FP) treatment in a series of field experiments on smallholder maize farms over three seasons. The objective was to improve productivity and profitability without increasing the environmental impact per unit of output. Therefore, the project investigated selected environmental parameters alongside yield and profit in order to measure and compare the impact of the two maize production systems on the environment.

7.2 Materials and Methods

The Environment and Climate Compatible Agriculture (ECCAg) PPP project was launched in 2010 to investigate the impacts of an improved agronomic protocol for maize and rice production on productivity, profitability, and the environment compared to prevailing FP in the region. The study took place over four seasons (2011–2014) on smallholder farmer fields within Tanzania's Southern Agricultural

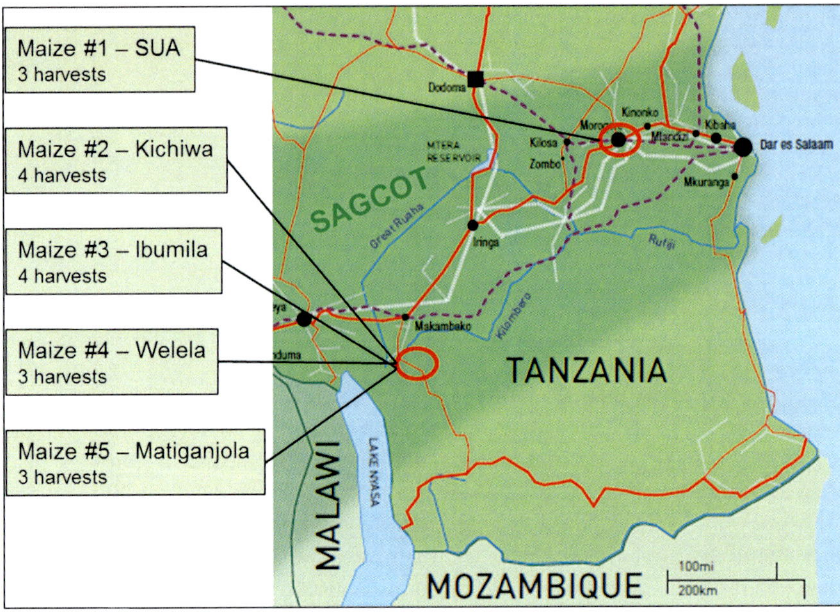

Fig. 7.3 Location of the five ECCAg maize experimental sites (map from Southern Agricultural Growth Corridor of Tanzania [SAGCOT] 2011)

Growth Corridor (Southern Agricultural Growth Corridor of Tanzania [SAGCOT] 2011) and in the fields of Sokoine University and Dakawa Rice Research Institute. These included five maize farms in the Morogoro and Njombe regions (Fig. 7.3), chosen with support from the project's core local partner, Sokoine University of Agriculture (SUA). This paper presents the results of the maize experiments.

Fig. 7.4 Example of a field trial comparing the intensified maize protocol (YSS) and prevailing FP at Kichiwa site in April 2012

At all trial sites, two test plots were established: (1) a Yara/Syngenta/SUA agronomy protocol (YSS) and (2) alongside the YSS plot. The farmers acted according to prevailing FP, which resulted in varying rates of inputs in the FP plots depending on the farmer and the season. The YSS treatments received 110–165 kg N ha^{-1} $year^{-1}$, 46–70 kg P_2O_5 ha^{-1} $year^{-1}$, 27–40 kg K_2O ha^{-1} $year^{-1}$, and micronutrient applications. The FP treatments received 0–79 kg N ha^{-1} $year^{-1}$, 0–57 kg P_2O_5 ha^{-1} $year^{-1}$, and no application of any other nutrients. This paper focuses on the role of crop nutrition as one possible measure of intensified crop production. Fertilizer application rates are shown in the Annex-Tables 7.1, 7.2, 7.3, 7.4 and 7.5. For three consecutive years, maize was planted at the beginning of the long rainy season in early December in the Njombe region and in early March in Morogoro. The trials were conducted on demo plots of about one acre on smallholder farms in Njombe (see example in Fig. 7.4) and as a replicated and fully randomized field trial at the Sokoine University campus area. At harvest the demo plots at the smallholder farms were separated into three sub-plots. Two sampling units 4 m long were then fitted in the middle of the three sub plots making a total of six sampling units per treatment. Two soil samples at 0–20 and 20–40 cm soil depth were collected in each sampling unit for physical and chemical analysis after each harvest. In order enable the calculation of nutrient removal and nutrient use efficiency, soils, grain, and biomass samples were taken and analyzed for N, P, K, and other nutrients. Several crop parameters were recorded separately for each sampling unit during each harvest: inter-row (between) and intra-row (within) spacing of plants, number of plants per sampling unit, plant height, cob weight, cob length, grain yield, grain-specific weight (1000 seed weight), stover biomass, and weed biomass (Mtengeti et al. 2015).

All field operations on the farms were carried out manually. The nutrient rates were derived from soil analysis data and the expected nutrient demand of the maize crops. Fertilizer rates were adjusted during the project depending on the actually realized maize yields and nutrient requirements at different sites. The fertilizer application rates on the FP plots varied between farmers and seasons. On two of the FP plots, no fertilizer was applied at all, whereas some nitrogen (11–79 kg N ha^{-1} $year^{-1}$) and phosphorus (28–57 kg P_2O_5 ha^{-1} $year^{-1}$) were applied on the other three FP plots. Weed control was mainly done manually on the FP plots, whereas plant protection agents were applied on the YSS plots when necessary. More details on the cropping systems and the applied sampling and harvesting protocols are available in Mtengeti et al. (2015).

Productivity was measured as marketable grain yield at 14 % moisture content, i.e., excluding rotten and non-marketable grain. Grain yields in the YSS treatments varied from 2.1 to 9.3 t/ha with an average of 5.9 t/ha. The average grain yield in the FP plots was 3.2 t/ha (1.1–5.5 t/ha). More detailed results on grain and stover yields are given in the results section.

Profitability was calculated as the difference between income (marketable yield * selling price) and costs for seeds, fertilizer, crop protection, and labor (e.g., planting, weeding, harvesting, shelling, winnowing).

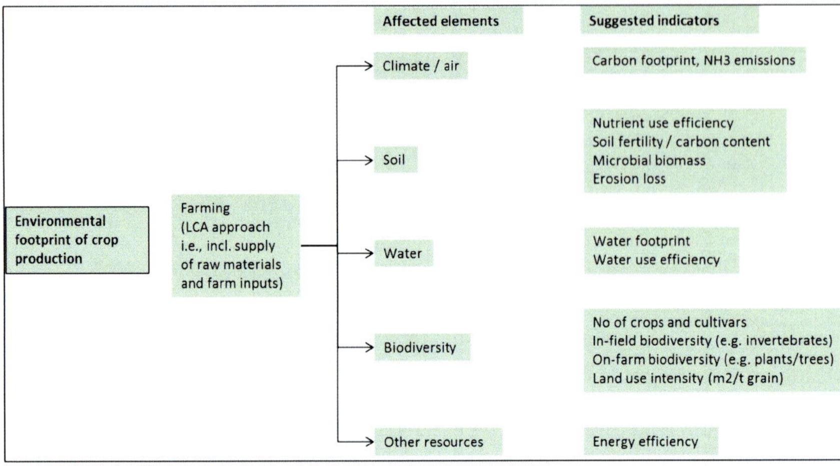

Fig. 7.5 Framework to measure and evaluate the environmental sustainability of crop production (own graph)

The environmental impact of arable farming was described by a structured and pragmatic framework, drawing on a thorough literature review with input from Yara, Syngenta, and external experts (Fig. 7.5). This paper presents data of the impact of maize cropping on greenhouse gas emissions ("carbon footprint"), N/P/K balances, soil pH, and land use intensity. For one site (Kichiwa), the water footprint was also calculated.

The carbon footprint (CFP) calculation included the production and transport of fertilizers, agrochemicals, and seeds, as well as all relevant on-farm activities up to the harvest for one project year. Primary data on the amounts of fertilizers, agrochemicals, yields, and crop residues were available for each treatment. Direct N_2O emissions from soils as a result of nitrogen inputs (mineral fertilizers, crop residues) were estimated using the Intergovernmental Panel on Climate Change (2006) Tier 1 method. Indirect N_2O emissions were calculated using the Bouwman et al. (2002) model for ammonia volatilization, and a nitrogen balance approach was used to estimate potential nitrate leaching losses. Mineral NPK fertilizers used in the YSS treatment were produced in Norway and transported to Tanzania by container ship. Mineral fertilizers applied on FP plots were assumed to be produced in China and Morocco and shipped in bulk. Emission factors for the production of inputs were obtained from Yara International, Brentrup and Palliere (2008), International Fertilizer Industry Association (2009), Saling and Kölsch (2008), and PE International's GaBi 4 LCA database.

In addition to the GHG emissions directly related to the maize production system, we also analyzed the consequences if low-yielding maize production systems, e.g., FP treatments, were expanded to match the higher yields of the YSS treatments. This expansion of cropland will cause GHG emissions outside the farms if it includes land use change (LUC) from land with higher carbon stocks into land containing less carbon. In this study, we assume that the FP production system expands within the same region and replaces tropical scrubland that is potentially available to support this expansion. First, we calculate the amount of land that needs to be converted to produce the same amount of maize harvested in YSS under the current FP management. This additional area of land is then assumed to be converted from tropical scrubland into cropland, leading to the emission of 12.4 t CO_2e ha^{-1} a year if emissions are allocated evenly across 20 years (Intergovernmental Panel on Climate Change 2006, Tier 1). In the following sections, this form of land use change is called 'potential LUC' (pLUC).

7.3 Results

In the following sections, the results are presented per experimental site aggregated over the whole trial period. Detailed results per site and year are available as Tables 7.1, 7.2, 7.3, 7.4 and 7.5 in the Appendix.

7.3.1 *Productivity*

Figure 7.6 summarizes the grain yield result for the five sites over 3 years. The three-year average grain yield of the YSS protocol is expressed as a percentage of the respective yield achieved with the FP.

The yield increases can be linked to the effective and balanced management of nutrients accompanied by crop protection application ensuring strong growth and safeguarding from pests and disease. The highest yield increases have been

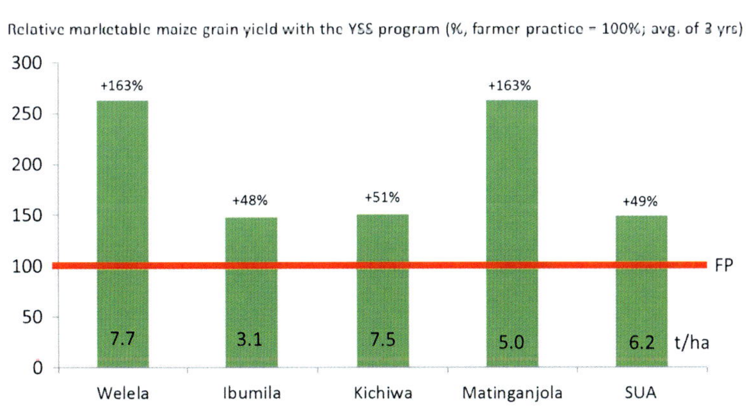

Fig. 7.6 Relative marketable grain yield at the five trial sites (FP = 100 %, average of three years)

achieved in Welela and Matingajola where the FP treatments received the lowest inputs of fertilizer. This reflects the actual situation in most of Tanzania's smallholder maize farms (average current maize yield: 1.3 t/ha). The yields achieved in these field trials show that it is possible to increase productivity considerably under the given soil and climatic conditions.

7.3.2 *Profitability*

Income and expense data were gathered following each season in order to calculate profitability (Fig. 7.7). In maize, the sharp increase in marketable grain is estimated to have increased household income by 70 on average in 2012 and 2013. However, a moderate application of nutrients in combination with good soil conditions, as found in Kichiwa and SUA, leads to comparable profitability.

We saw instances of income diversification by planting additional cash crops, such as fruit or timber trees, on the farms, or purchasing livestock, such as fowl. As part of a livelihood impact assessment, farmers responded that they spent part of their additional income on children's education and on improving their homes. At scale, such income creation would stimulate local economies to the benefit of wider rural communities.

7.3.3 *Environment*

To describe the environmental impact of arable farming, we developed a structured and pragmatic framework, which was derived from a thorough literature review with input from all project partners (Fig. 7.5). For this paper, we selected particular environmental indicators representing potential impacts on climate (carbon footprint), soil (N, P, K balance, soil pH), water (water footprint), and biodiversity (land use intensity).

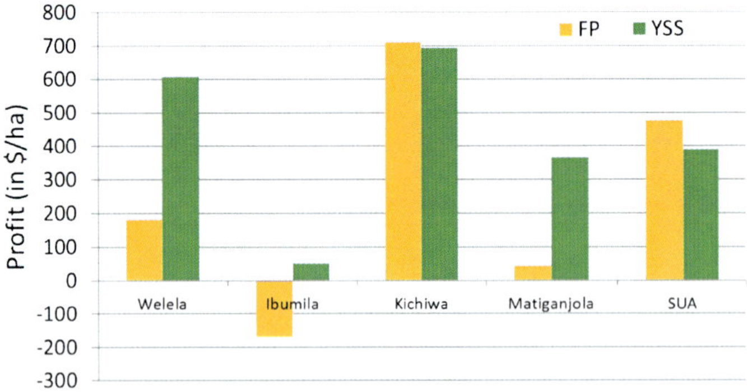

Fig. 7.7 Profitability of the YSS and FP protocols (average of two seasons)

7.3.4 Climate

GHG emissions were calculated for each protocol using Yara's carbon footprint calculation methodology, which includes emissions from the production and transportation of farm inputs (fertilizer, plant protection, seeds), on-farm energy use, nitrous oxide (N_2O) emissions from soil, and carbon loss from land use change.

The carbon footprint (CFP) per ton of maize grain at farm gate varied from 105–587 kg CO_2e. A comparison of the results for the paired YSS and FP systems reveals that FP systems had lower CFPs in all cases the (Fig. 7.8, left). The main emissions sources on all farms were related to fertilizer production and use and crop residues remaining in the field after harvest. The more mineral fertilizer was used, the more dominant became this source of emissions in relation to other sources. In cases with no or little fertilizer use (e.g., FP in Matingajola), N_2O emissions from N contained in crop residues were the main emissions source. In the YSS treatments, production and transport of mineral fertilizers to the farm and direct N_2O emissions from nitrification and denitrification of applied N were the main sources of GHG emissions.

However, taking into account the increasing demand for maize grain in Tanzania, we need to consider land use change across the wider agricultural system. In order to produce the additional YSS yield with the farmers' prevailing agricultural practice, a larger land area must be put in use. The carbon losses associated with such land use change (e.g., converting savannah or scrubland into arable land) would lead to a substantially higher carbon footprint per unit of grain versus the YSS protocol. For this analysis, we have assumed that available scrubland in the close proximity of the case study farms would be converted into maize fields. If GHG emissions from such potential land use change are considered, the

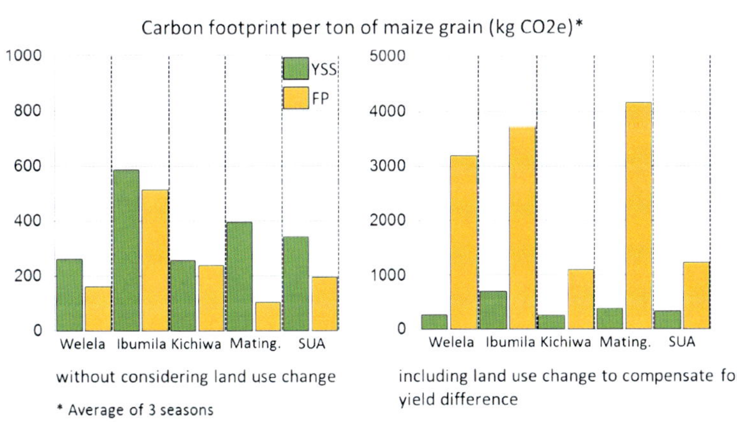

Fig. 7.8 *Left* Carbon footprint (CFP) per ton of maize grain up to the farm gate; *right* CFP including potential land use change needed to compensate for yield differences between the treatments (adapted from Plassmann et al. 2014)

conclusions drawn from considering the farm gate CFP change dramatically, with GHG emissions related to the extended FP 3.6–12 times greater than the CFP of the YSS protocol (Fig. 7.8, right).

7.3.5 Soil

Potential impacts of the two different cropping systems on soil fertility are presented in this paper in two ways: (1) as soil nutrient balances indicating the balance between nutrient inputs with fertilizer (or other available sources of nutrients) and the permanent export of nutrients with harvested crops and (2) as the development of soil pH in the top-soil layer of 20 cm.

Figure 7.9 shows the average nutrient balances (N, P, K) over three seasons of maize cropping. Negative nutrient balances indicate unsustainable nutrient mining of the soil even at low yield levels. Deficiencies and imbalance of nutrients are a limiting factor for yields and impair soil fertility over time. On the other hand, if more nutrients are applied than are being utilized by the crop, there is a risk of a loss of nutrients to the wider environment. It is difficult to determine the exact levels for acceptable and sustainable nutrient balances, but, for this study, we have set the upper boundary at 50 kg nutrient surplus per ha in accordance with levels for N discussed in Europe (EU Nitrogen Expert Panel 2015). Figure 7.9 reveals that for N and P, the YSS protocol in most cases complies with boundaries set; except for N in Ibumila and Matingajola, the surplus is classified as "unsustainably high." In these two cases, the yields achieved were on average below the expected level. In order to better balance N inputs and outputs, the N application rate was reduced by 20 % in Ibumila, while in Matingajola, the yield only dropped in the last season. For

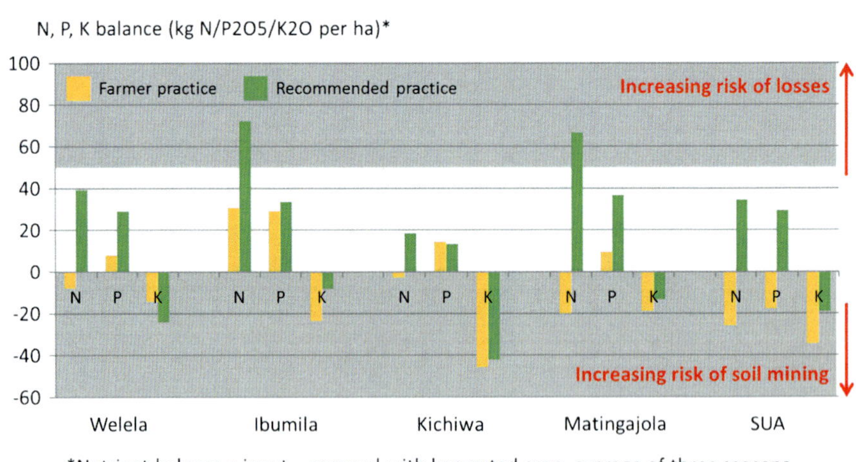

Fig. 7.9 Average nutrient balances (N, P, K) over three seasons

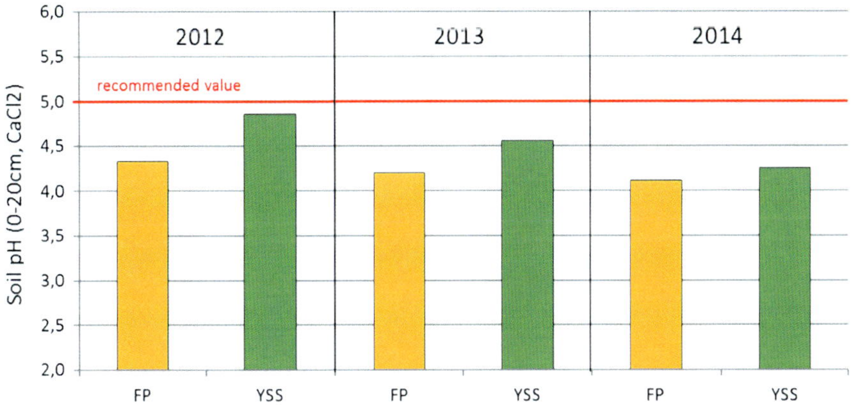

Fig. 7.10 Development of soil pH in the topsoil at the Welela site as an example for all other sites

potassium, the balance was negative at all sites and treatments. Although the YSS protocol included the application of K, this was never enough to balance the K removal. In most FP plots, no K was applied, leading to even more negative K balances in many cases in spite of the low yields. Whether mining of the soil for potassium should be regarded as a problem depends on the soil reserves available to the crop. Soil analysis after each harvest has shown moderate to low levels of available K in the top soil, particularly in Matingajola and Ibumila but also in the FP plots of Kichiwa and Welela, whereas the SUA soil was rich in K. However, in the long-term, K has to be applied in sufficient amounts to balance inputs and outputs and to ensure sustainable crop production on the same piece of land.

As a second important indicator for soil quality and fertility, we analyzed the soil pH over the course of the project. Figure 7.10 shows the development of soil pH in the upper 20 cm of the soil at the Welela site. This result is representative of all sites in the sense that the soil pH shows a consistent decreasing trend over the years. This trend is even more pronounced in the YSS treatment if the application of N fertilizer is not accompanied by the application of lime or other ameliorating measures. At the SUA and Kichiwa sites, the current status of soil pH is still at the recommended or only moderately low level and therefore does not currently affect productivity. However, to ensure the long-term productivity of the fields, liming is inevitable at these sites.

7.3.6 Water

Weather stations at two of the sites (Kichiwa and SUA, both since 2013) allowed the team to measure and assess water use across the protocols. All maize sites were rain-fed so yield increases were achieved with no additional water. According to the Global Water Footprint Assessment Standard (Hoekstra et al. 2011), one full water footprint analysis was performed for the last cropping season in Kichiwa.

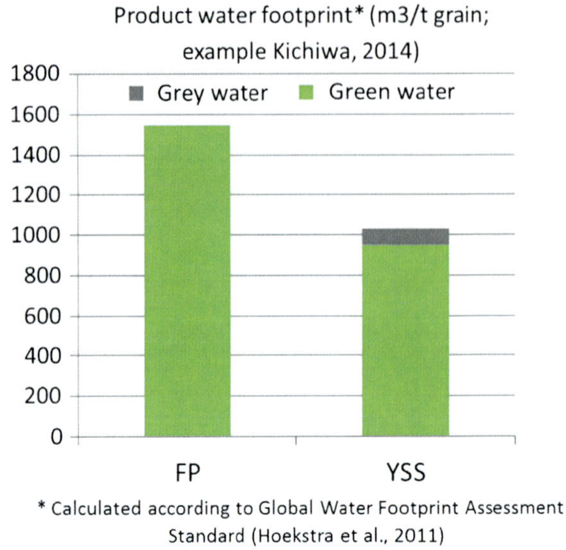

Fig. 7.11 Water footprint per ton of maize at site Kichiwa in 2014 (data from Holger Brück, personal communication)

Figure 7.11 shows the water footprint per ton of maize grain, which was 50 % higher in the FP treatment compared to YSS. In the YSS treatment, a small N surplus (11 kg N/ha) was responsible for the grey water component of the water footprint, while, overall, the available rain water was utilized more efficiently than in the FP, i.e., producing "more crop per drop."

7.3.7 Biodiversity

Biodiversity was assessed in different ways in the project: in-field, on-farm, and in terms of land-use efficiency.

In-field biodiversity was evaluated by tracking the number of invertebrates and earthworms in the plots, counting the number and variety of weeds, and analyzing the microbial biomass of soil samples from several sites. The biodiversity of the whole farm was recorded by creating biodiversity maps of each of the sites. In-field biodiversity assessments showed fewer species of weeds and insects in YSS plots due to the applied crop protection protocol; this was a critical enabler of improved yield as this allows plants to grow unhindered. In the areas immediately surrounding the growing area, biodiversity remained unchanged. Off-field biodiversity improved with YSS protocol use, as increased yields and additional income resulted in several farmers planting perennial trees.

In this paper, we focus on land-use efficiency as a measure of how much land is occupied by agriculture to provide a unit of agricultural output, in this case, one ton of maize grain. Figure 7.12 shows the average land-use efficiency for the five sites.

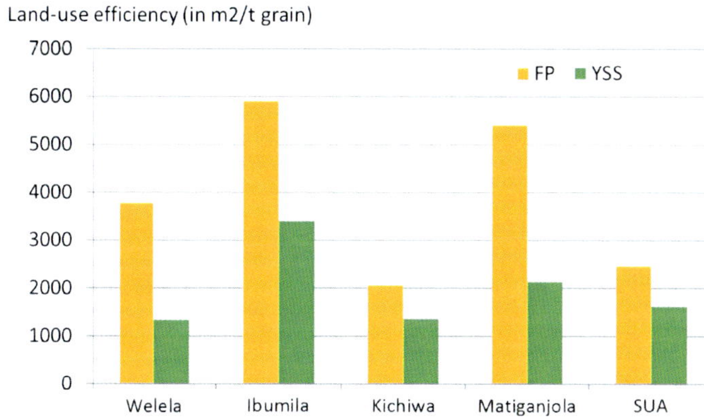

Fig. 7.12 Land-use efficiency of maize production at the trial sites (average of three seasons); the smaller the number the more efficient is the cropland used

On average, the FP required twice as much land area to produce a ton of maize grain as the YSS protocol. This is of course simply the reverse of the yield data, but it allows a different perspective on the data. The improved productivity not only provides surplus output for the farmer to sell and more food for the society, it also removes pressure for further land use change, thus enabling the preservation of biodiversity and carbon stocks at the landscape level. Therefore, land-use efficiency can be regarded as an additional indicator in the biodiversity context that informs about the efficiency the limited land area utilized for a certain purpose. This indicator does not include information about the biodiversity value of the area under use itself and whether differences in the biodiversity value of the cropping systems under investigation exist.

7.4 Conclusions

Tanzania has a fast growing population with an increasing demand for food crops, particularly maize. Sustainable intensification is needed to meet this demand while, at the same time, preserving the environment and climate. With a balanced supply of plant nutrients, it is possible to increase maize yields and farm profitability substantially. Accompanied good agricultural practices are important to ensure high productivity of the existing cropland in the long-term. Potential impacts on climate, soil, water, and biodiversity need to be assessed and monitored in a consistent and transparent way in order to evaluate the sustainability of intensification measures and their further improvement.

This case study showed:

- Significantly improved yields may in turn reduce the need for arable land expansion and can thus potentially avoid GHG emissions.

- In most cases, low positive soil nutrient balances for N and P with the improved cropping protocol indicate sustainable fertilizer use, replenishing the soil with sufficient nutrients. In contrast, FP often showed negative nutrient balances, signifying unsustainable nutrient mining even at low yield levels. K balances were negative in almost all cases.
- The maize produced according to the improved protocol at the Kichiwa site consumed less water per ton of grain compared to FP.
- About 50 % less land was needed to produce one ton of maize grain, reducing pressure for land expansion that has a potential negative effect on biodiversity on a larger scale.

Acknowledgments We thank our colleagues from the Sokoine University of Agriculture (Morogoro/Tanzania), the University of Life Sciences (As/Norway), Syngenta, and Yara Tanzania, who were all part of this public-private partnership and contributed substantially to the results and success of this project.

Appendix

Welela Site

See Table 7.1.

Nutrient management

- Increased fertilizer rate in YSS (+20 %) adjusted to high yield in 2013
- FP treatment with increasing fertilizer use during the project

Table 7.1 Nutrient application rates, productivity, profitability, and environmental parameters for the Welela site (profitability not available for 2014 due to lack of data)

Parameter	Site	Welela (GPS: S 09′01.218; E 034′ 48.236)							
	Treatment	YSS				FP			
	Unit/year	2012	2013	2014	Avg	2012	2013	2014	Avg
Nutrient management									
N	kg N/ha	138	138	166	147	0	40	56	32
P	kg P_2O_5/ha	58	58	70	62	0	29	24	18
K	kg K_2O/ha	34	34	40	36	0	0	12	4
Productivity									
Grain yield	t/ha (86 % dm)	6.1[a]	9.3[a]	7.6[a]	7.7	1.8[b]	3.2[b]	3.8[b]	2.9
Stover yield	t/ha (86 % dm)	13.1[a]	9.3[a]	8.3[a]	10.2	3.5[b]	2.2[b]	4.6[b]	3.4

(continued)

Table 7.1 (continued)

Parameter	Site	Welela (GPS: S 09′01.218; E 034′ 48.236)							
	Treatment	YSS				FP			
	Unit/year	2012	2013	2014	Avg	2012	2013	2014	Avg
Profitability									
Revenue	USD/ha	1159	1740	n.a.	1450	344	692	n.a.	518
Cost	USD/ha	567	1118	n.a.	843	205	467	n.a.	336
Profit	USD/ha	592	622	n.a.	607	139	225	n.a.	182
Environment									
Carbon footprint	kg CO_2e/t	324	188	275	262	98	190	200	163
CFP incl pLUC*	kg CO_2e/t	324	188	275	262	4944	2778	1863	3195
N balance	kg N/ha	52	8	58	39	−18	−8	2	−8
P balance	kg P_2O_5/ha	31	18	38	29	−4	17	11	8
K balance	kg K_2O/ha	−22	−30	−21	−24	−11	−20	−12	−14
Soil pH		4.9^{a}	4.6^{a}	4.3^{a}	4.6	4.3^{a}	4.2^{b}	4.1^{a}	4.2
Land use intensity	m^2/t	1647	1078	1311	1345	5556	3165	2653	3791

[a]*pLUC* potential land-use change; [b]*n.a.* not available
Means of YSS and FP in the same village and year with different superscripts are significantly different at $P < 0.05$ according to t test

Productivity

- High yield potential (9.3 t/ha in 2013)
- On average, 2.6 times higher grain yield and three times higher stover yield with YSS protocol
- FP yield increasing with increasing fertilizer use (1.8–3.8 t/ha)

Profitability

- On average, 3.3 times higher profit with YSS

Environment

- Average carbon footprint (CFP) with YSS 61 % higher without considering potential land use change (pLUC) needed to compensate for the yield difference between FP and YSS (for more information about the pLUC concept and definition, see Plassmann et al. 2014)
- Average CFP 12 times higher if pLUC is included
- Negative N and P balances with FP (i.e., reducing soil reserves)
- K balance negative in both treatments
- Decreasing pH in both treatments, indicating a need for lime application
- On average, FP requires almost three times more land to produce one ton of maize grain.

Ibumila Site

See Table 7.2.

Nutrient management

- Decreased fertilizer application rate (−20 %) in 2013 and 2014 due to low yield level
- Farmer applied relatively high nutrient rates from the beginning
- Farmer applied in addition some manure to FP plot

Productivity

- Low yields overall, low NUE, but still on average +48 % grain yield and +68 % stover yield with YSS protocol

Table 7.2 Nutrient application rates, productivity, profitability, and environmental parameters for the Ibumila site (profitability not available for 2014 due to lack of data)

Parameter	Site	Ibumila (GPS: S 09′ 06. 614; E034′50.978)							
	Treatment	YSS				FP			
	Unit/Year	2012	2013	2014	Avg	2012	2013	2014	Avg
Nutrient management									
N	kg N/ha	138	110	110	119	68	48	76	64
P	kg P_2O_5/ha	58	46	46	50	28	48	54	43
K	kg K_2O/ha	34	27	27	29	0	8	0	3
Productivity									
Grain yield	t/ha (86 % dm)	2.1[a]	4.3[a]	3.3[a]	3.2	1.8[a]	1.1[b]	3.6[a]	2.2
Stover yield	t/ha (86 % dm)	8.6[a]	4.4[a]	6.3[a]	6.4	7.0[a]	1.4[b]	3.1[a]	3.8
Profitability									
Revenue	USD/ha	395	930	n.a.	662	343	233	n.a.	288
Cost	USD/ha	495	729	n.a.	612	484	430	n.a.	457
Profit	USD/ha	−100	201	n.a.	51	−141	−197	n.a.	−169
Environment									
Carbon footprint	kg CO_2e/t	964	339	457	587	632	602	301	512
CFP incl pLUC	kg CO_2e/t	964	339	810	704	1502	9383	301	3728
N balance	kg N/ha	97	50	69	72	35	30	27	31
P balance	kg P_2O_5/ha	43	26	32	34	15	40	32	29
K balance	kg K_2O/ha	4	−15	−14	−8	−25	−17	−29	−24
Soil pH		4.2[a]	4.1[a]	4.0[a]	4.1	4.1[a]	4.1[a]	4.0[a]	4.1
Land use intensity	m^2/t	4854	2353	3040	3416	5556	9434	2755	5915

[a]*pLUC* potential land-use change; [b]*n.a.* not available

Means of YSS and FP in the same village and year with different superscripts are significantly different at $P < 0.05$ according to *t* test

- Profitability
- Small but still positive financial results with YSS compared to loss with FP
- Whether a profit or loss is realized depends on maize price; therefore, investing in inputs is high risk for the farmer

Environment

- High product carbon footprint in both treatments (>500 kg CO_2e/t grain)
- Five times higher CFP with FP when including potential land use change effects
- On average, unsustainably high N surplus in YSS treatment (+72 kg N/ha), negative K balances in both treatments
- Very low pH from the beginning, even decreasing during the project, Al/Mn toxicity is a possible reason for low yields; liming is required
- Additional liming experiment was established, but not presented here; some positive effects on soil pH, but not yet on grain yield
- On average, FP requires 1.7 times more area to produce one ton of maize grain.

Kichiwa Site

See Table 7.3.

Nutrient management

- Increased fertilizer rate in YSS (+20 %) adjusted to high yield in 2012 (not shown)
- Farmer adopted improved system partially (comparably high input rates and same products)

Productivity

- Very good crop response to fertilizer application
- High yield potential (8.7 t/ha in 2014), less fluctuation than in Welela (better soil)
- On average, 51 % higher yield with YSS protocol, only 14 % more stover yield
- FP yield much higher than TZ average (5 t/ha)

Profitability

- Similar profit with both strategies, but less risk with FP if investment in inputs fails, e.g., due to lack of rain

Environment

- Comparably low CFP in both treatments (about 250 kg CO_2e/t grain)
- On average, 4.3 times higher CFP with FP if potential land use change is considered
- Negative N and K balances with FP, negative K balance in YSS treatment

Table 7.3 Nutrient application rates, productivity, profitability, and environmental parameter for the Kichiwa site (profitability not available for 2011 and 2014; no nutrient balances in 2013 due to missing plant samples; no data for 2012 because FP plot was harvested without yield measurements)

Parameter	Site	Kichiwa (GPS: S 09′01.064; E 034′52.164)							
	Treatment	YSS				FP			
	Unit/Year	2011	2013	2014	Avg	2011	2013	2014	Avg
Nutrient management									
N	kg N/ha	138	165	138	147	79	110	57	82
P	kg P_2O_5/ha	58	70	58	62	57	86	25	56
K	kg K_2O/ha	34	40	33	36	0	12	12	8
Productivity									
Grain yield	t/ha (86 % dm)	6.1[a]	7.6[a]	8.7[a]	7.5	4.0[b]	5.5[b]	5.3[b]	5.0
Stover yield	t/ha (86 % dm)	8.3[a]	7.2[a]	9.3[a]	8.3	8.6[a]	9.2[a]	4.1[b]	7.3
Profitability									
Revenue	USD/ha	n.a.	1652	n.a.	1652	n.a.	1212	n.a.	1212
Cost	USD/ha	n.a.	957	n.a.	957	n.a.	501	n.a.	501
Profit	USD/ha	n.a.	695	n.a.	695	n.a.	711	n.a.	711
Environment									
Carbon footprint	kg CO_2e/t	292	277	203	257	313	281	124	239
CFP incl pLUC	kg CO_2e/t	292	277	203	257	1395	877	1027	1100
N balance	kg N/ha	26	n.a.	11	19	13	n.a.	−19	−3
P balance	kg P_2O_5/ha	9	n.a.	18	14	30	n.a.	−1	15
K balance	kg K_2O/ha	−38	n.a.	−46	−42	−49	n.a.	−42	−46
Soil pH		4.7[a]	4.7[a]	4.6[a]	4.7	4.4[a]	4.8[a]	4.7[a]	4.6
Water footprint	m3/t	n.a.		1029	1029	n.a.		1548	1548
Land use intensity	m^2/t	1639	1325	1148	1371	2513	1805	1876	2065

[a]*pLUC* potential land-use change; [b]*n.a.* not available
Means of YSS and FP in the same village and year with different superscripts are significantly different at $P < 0.05$ according to *t* test

- Higher soil pH compared to other Njombe sites; in acceptable range in both treatments
- 50 % higher water consumption per ton of grain in FP treatment, more efficient use of available water with YSS protocol
- On average, 51 % more area required with FP to produce one ton of grain.

Matingajola Site

See Table 7.4.

Nutrient management

- Very low input/output in FP; no adoption of improved practice

Productivity

- Good yield response in first year but declining yields in both treatments during the course of the project (lack of rain but also soil acidification may have contributed)
- On average, 2.6 times higher grain yield and 2.4 times higher stover yield with YSS protocol

Table 7.4 Nutrient application rates, productivity, profitability, and environmental parameters for the Matingajola site (profitability not available for 2014 due to lack of data)

Parameter	Site	Matingajola (GPS: S 09'13.881; E 034'53.470)							
	Treatment	YSS				FP			
	Unit/Year	2012	2013	2014	Avg	2012	2013	2014	Avg
Nutrient management									
N	kg N/ha	138	138	138	138	11	7	0	6
P	kg P_2O_5/ha	58	58	58	58	28	28	0	19
K	kg K_2O/ha	34	34	34	34	0	0	0	0
Productivity									
Grain yield	t/ha (86 % dm)	6.3[a]	5.4[a]	3.3[a]	5.0	2.3[b]	2.0[b]	1.5[a]	1.9
Stover yield	t/ha (86 % dm)	13.5[a]	6.3[a]	3.2[a]	7.7	5.7[b]	2.5[b]	1.3[b]	3.1
Profitability									
Revenue	USD/ha	1212	1174	n.a.	1193	433	436	n.a.	434
Cost	USD/ha	702	952	n.a.	827	293	488	n.a.	390
Profit	USD/ha	510	221	n.a.	366	140	−52	n.a.	44
Environment									
Carbon footprint	kg CO_2e/t	319	341	529	396	147	97	70	105
CFP incl pLUC	kg CO_2e/t	319	341	529	396	3654	4019	4793	4155
N balance	kg N/ha	51	66	84	67	−18	−20	−22	−20
P balance	kg P_2O_5/ha	32	36	42	37	18	19	−9	9
K balance	kg K_2O/ha	−15	−9	−16	−14	−18	−18	−21	−19
Soil pH		4.4[a]	4.3[a]	4.0[a]	4.2	4.4[a]	4.4[a]	4.2[a]	4.3
Land use intensity	m^2/t	1577	1862	2994	2144	4405	5025	6803	5411

[a] *pLUC* potential land-use change; [b] *n.a.* not available

Means of YSS and FP in the same village and year with different superscripts are significantly different at $P < 0.05$ according to t test

Profitability

- On average, more than eight times higher profit with YSS protocol
- Negative result in 2013 in FP, probably leading to the decision not to invest in fertilizer in 2013

Environment

- Very low CFP in FP treatment due to low or zero mineral N fertilizer use; 3.8 times higher CFP with YSS protocol on average
- Including potential land use change effects, the CFP of the FP treatment is about 10 times higher
- Consistently negative N and K balances in FP treatment, probably unsustainably low P balance
- Increasing N balances with YSS due to decreasing N use efficiency, need for N input adjustment
- Drop in soil pH in particular with YSS; liming is needed
- 2.5 times more land is needed to produce a ton of grain with FP.

Site Sokoine University of Agriculture (SUA)

See Table 7.5.

Nutrient management

- No fertilizer input in FP treatment in first season, which is typical for maize cultivation on newly re-established fields; low input in subsequent years

Productivity

- Comparably high yields with no or low input in FP due to high soil fertility of the re-established maize field (was left fallow for some years before)
- Good and consistent grain yield response to fertilizer input in YSS (+49 % compared to FP; +59 % more stover yield)

Profitability

- 23 % higher profitability with FP
- Profitability in FP treatment is based on natural soil fertility not sustainability, as negative nutrient balances indicate

Environment

- Very low CFP in FP treatment due relatively high yield with low N input, 75 % higher CFP with YSS
- Including potential land use change effects changes the result with 3.6 times higher CFP for FP

Table 7.5 Nutrient application rates, productivity, profitability, and environmental parameters for the Sokoine University of Agriculture (SUA) site (profitability not available for 2014 due to lack of data)

Parameter	Site	SUA (GPS: S 06″50.875; E 037″39.270)							
	Treatment	YSS				FP			
	Unit/Year	2012	2013	2014	Avg	2012	2013	2014	Avg
Nutrient management									
N	kg N/ha	138	138	138	138	0	57	57	38
P	kg P_2O_5/ha	58	58	58	58	0	0	0	0
K	kg K_2O/ha	34	34	34	34	0	0	0	0
Productivity									
Grain yield	t/ha (86 %dm)	6.2[a]	5.9[a]	6.4[a]	6.2	3.4[b]	4.8[a]	4.3[b]	4.1
Stover yield	t/ha (86 %dm)	13.3[a]	22.7[a]	13.4[a]	16.5	7.6[a]	15.8[a]	7.6[b]	10.4
Profitability									
Revenue	USD/ha	1177	1501	n.a.	1339	646	1223	n.a.	934
Cost	USD/ha	985	915	n.a.	950	451	461	n.a.	456
Profit	USD/ha	193	586	n.a.	390	195	762	n.a.	478
Environment									
Carbon footprint	kg CO_2e/t	334	402	294	343	83	263	241	196
CFP incl pLUC	kg CO_2e/t	334	402	294	343	1739	741	1221	1234
N balance	kg N/ha	30	36	38	35	−49	−21	−8	−26
P balance	kg P_2O_5/ha	27	31	31	29	−18	−19	−17	−18
K balance	kg K_2O/ha	−21	−32	−4	−19	−24	−49	−31	−35
Soil pH		5.0[a]	4.7[a]	4.3[a]	4.7	5.1[a]	5.1[a]	4.8[a]	5.0
Land use intensity	m^2/t	1623	1698	1563	1628	2959	2083	2353	2465

[a]*pLUC* potential land-use change; [b]*n.a.* not available

Means of YSS and FP in the same village and year with different superscripts are significantly different at $P < 0.05$ according to *t* test

- Consistently negative N, P, and K balances in FP treatment clearly indicate mining of natural soil fertility
- Positive but low N and P balances and negative K balance in YSS treatment
- Soil pH in recommended range but decreasing over the trial period; liming will become necessary
- On average, 51 % more land is needed with FP to produce one ton of grain.

References

Beddington J, Asaduzzaman M, Clark M, Fernandez A, Guillou M, Jahn M, Erda L, Mamo T, Van Bo N, Nobre CA, Scholes R, Sharma R, Wakhungu J (2012) Achieving food security in the face of climate change: final report from the Commission on Sustainable Agriculture and Climate Change. CGIAR Research Program on Climate Change, Agriculture and Food Security (CCAFS), Copenhagen

Bouwman AF, Boumans LJM, Batjes NH (2002) Estimation of global NH_3 volatilization loss from synthetic fertilizers and animal manure applied to arable lands and grasslands. Global Biogeochem Cycles 16(2):1–15

Brentrup F, Palliere C (2008) GHG emissions and energy efficiency in European nitrogen fertilizer production and use. In: Proceedings no. 639. International Fertilizer Society, New York

EU Nitrogen Expert Panel (2015) Nitrogen use efficiency (NUE)—an indicator for the utilization of nitrogen in agriculture and food systems. Wageningen University, Alterra. http://www.eunep.com. Accessed 28 Jan 2016

Food and Agriculture Organization of the United Nations, Statistics Division (2015). http://faostat3.fao.org/download/Q/QC/E. Accessed 05 Oct 2015

Hoekstra AY, Chapagain AK, Aldaya CM, Mekonnen MM (2011) The water footprint assessment manual. Setting the global standard. Earthscan, London

International Fertilizer Industry Association (2009) Fertilizers, climate change, and enhancing agricultural productivity sustainably. International Fertilizer Industry Association, Paris

Intergovernmental Panel on Climate Change (2006) IPCC 2006 Guidelines for national greenhouse gas inventories. In: Eggleston HS, Buendia L, Miwa K, Ngara T, Tanabe K (eds) Prepared by the National Greenhouse Gas Inventories Programme. IGES, Japan

Mtengeti EJ, Brentrup F, Mtengeti E, Eik LO, Chambuya R (2015) Sustainable intensification of maize and rice in smallholder farming systems under climate change in Tanzania. In: Lal R, Singh BR, Mwaseba DL, Kraybill D, Hansen DO, Eik LO (eds) Sustainable intensification to advance food security and enhance climate resilience in Africa. Springer, Cham, pp 441–465

Plassmann K, Brentrup F, Lammel J (2014) Trade-offs between agricultural product carbon footprints and land use: a case study from Tanzania. In: Schenck R, Huizen D (eds) Proceedings of the 9th international conference on life cycle assessment in the agri-food sector (LCA Food 2014), San Francisco, 8–10 Oct 2014, pp 1016–1021

Saling P, Kölsch D (2008) Ökobilanzierung: Energieverbräuche und CO_2-Emissionen von Pflanzenschutzmittlen. In: KTBL (eds) Energieeffiziente Landwirtschaft. KTBL-Schrift 463. Kuratorium für Technik und Bauwesen in der Landwirtschaft e.V., Darmstadt

Southern Agricultural Growth Corridor of Tanzania (SAGCOT) (2011) Southern agricultural growth corridor of Tanzania investment blueprint. http://www.sagcot.com/uploads/media/Invest-Blueprint-SAGCOT_High_res.pdf. Accessed 05 Oct 2015

Tubiello FN, Salvatore M, Ferrara AF, House J, Federici S, Rossi S, Biancalani R, Condor Golec RD, Jacobs H, Flammini A, Prosperi P, Cardenas-Galindo P, Schmidhuber J, Sanz Sanchez MJ, Srivastava N, Smith P (2015) The contribution of agriculture, forestry, and other land use activities to global warming, 1990–2012. Glob Change Biol 21:2655–2660. doi:10.1111/gcb.12865

United Nations, Department of Economic and Social Affairs, Population Division (2015) World population prospects: the 2015 revision, DVD edition. http://esa.un.org/unpd/wpp/Download/Standard/Population/. Accessed 28 Jan 2016

Chapter 8
Potentials for Rehabilitating Degraded Land in Tanzania

G.Z. Nyamoga, H.K. Sjølie, R. Malimbwi, Y.M. Ngaga and B. Solberg

Abstract In Tanzania, land rehabilitation seems promising for repairing damaged ecosystems and provide sustainable supply of forest and food products, thus securing vital environmental services including increased carbon sequestration for global climate change mitigation. Comprehensive estimates of how large areas Tanzania has of degraded land are however lacking. This study aimed to (i) assess the area of degraded land potentially available for rehabilitation in various regions of the country, and (ii) give a review of main experiences and economic results gained in previous land rehabilitation studies in the country. Based on new data from the National Forest Resource Monitoring and Assessment of Tanzania we found that about 49 % (43.3 mill ha) of the total land area in Mainland Tanzania is under either light (43 %, 37.7 mill ha), moderate (5 %, 4.4 mill ha) or heavy (1.3 %, 1.2 mill ha) erosion. These areas are substantial, and imply large opportunities for land rehabilitation. None economic studies were found which have calculated benefits and costs of land rehabilitation in Tanzania. Such studies are

---

G.Z. Nyamoga · Y.M. Ngaga
Department of Forest Economics, Faculty of Forestry and Nature Conservation, Sokoine University of Agriculture—SUA, P.O. Box 3011, Morogoro, Tanzania
e-mail: yngaga@yahoo.co.uk

R. Malimbwi
Department of Forest Mensuration and Management, Faculty of Forestry and Nature Conservation, Sokoine University of Agriculture—SUA, P.O. Box 3011, Morogoro, Tanzania
e-mail: remalimbwi@yahoo.com

G.Z. Nyamoga (✉) · H.K. Sjølie · B. Solberg
Department of Ecology and Natural Resources Management—INA, Norwegian University of Life Sciences (Norges Miljø-Og Biovitenskapelige Universitet)—NMBU, P.O. Box 5003, Ås, Norway
e-mail: nyamoga26@yahoo.co.uk; nyamoga@suanet.ac.tz

H.K. Sjølie
e-mail: hanne.sjolie@nmbu.no

B. Solberg
e-mail: birger.solberg@nmbu.no

R. Lal et al. (eds.), *Climate Change and Multi-Dimensional Sustainability in African Agriculture*, DOI 10.1007/978-3-319-41238-2_8

urgently needed in order to identify and prioritize among the most promising rehabilitation activities.

Keywords Deforestation · Reforestation · Afforestation · Agroforestry · Livelihoods · Land degradation · Ecosystem services · Community based, climate change · Income security · Sustainable land use · Economic benefits · Carbon sequestration · Soil erosion

8.1 Introduction

Tanzania experiences large land use changes. Between 2002 and 2012 the settlement and protected land increased by 26.7 and 8.5 % respectively, whereas wood and non-woody production land declined by 23.8 % and scattered settlement areas and agriculture land decreased by about 12.9 % (Malimbwi and Zahabu 2014). The area of degraded land is increasing, and land rehabilitation has been put forward as a promising path for repairing damaged ecosystems and secure ecosystem functions in order to enhance land productivity and provide essential goods and environmental services, including increased carbon sequestration for global climate change mitigation.

Forests supply about 92 % of the consumed energy in Tanzania through charcoal and firewood, as less than 15 % of the population, mainly in urban areas are connected to electricity (Mwampamba 2007). With a population increase of 2.7 % p.a. demand for wood and land is rapidly growing. About 55 % of the Tanzanian mainland is covered by forests (Malimbwi and Zahabu 2014; Tomppo et al. 2010, 2014), with an annual deforestation of about 3728 km^2, equivalent to 1.1 % of the total forest area (Bahamondez et al. 2010; Malimbwi and Zahabu 2014). Deforestation and land degradation due to over-exploitation and agricultural expansion leave the poor communities more vulnerable to poverty and causes multiple negative environmental effects (Appiah et al. 2009; Hartmana et al. 2014). Tanzania is among the 12 richest countries in the world with regard to biodiversity, in particular because of its forests (Myers et al. 2000). It has Africa's largest number of mammals, second largest number of plants, third largest number of birds, fourth largest number of reptiles and fourth largest number of amphibians (Burgess et al. 2002, 2007; Pettorelli et al. 2010).

Land degradation is a process of decline of natural resources due to improper practices and inability of the land to recover its natural state as results of disturbances of ecosystem functions (Bai et al. 2008; Bergsma et al. 1996; Rothman et al. 2007). Deforestation and land degradation is exacerbated by a range of factors like population growth, urbanization, rural-urban migration, overgrazing, types of land ownership, farming practices like shifting cultivation, slush and burn and monoculture practices and animal overstocking (Hartmana et al. 2014; Kajembe et al. 2005a; Mary and Majule 2009; Mbaga-Semgalawe and Folmer 2000). These agricultural activities result in reduced vegetation cover, decreased soil productivity, changes in species composition and severe soil erosion (Hartmana et al. 2014).

An important aspect here is the land ownership in Tanzania that is guided by the Land act and the Village land act of 1999 (Angelsen and Fjeldstad 1995; Shivji 1998, 1999). Under these two acts, land ownership can be under customary right of occupancy, granted right of occupancy, leasehold and residential occupancy rights. With poor land tenure systems and ownerships as well as deficient farming techniques, the agricultural land looses its fertility quickly forcing people to shift into new virgin and fertile lands.

Land rehabilitation is the process of repairing damaged ecosystems and ecosystem functions for the sake of raising ecosystem productivity to provide benefits to local people (Aronsod et al. 1993). These rehabilitation initiatives aims at restoring land to its original conditions by improving the soil and biodiversity conditions and forming a rational, effective and intensive land use pattern, increase effective cultivated land area and enhance land use efficiency (Angelsen and Fjeldstad 1995). If succeeding, rehabilitation can mitigate the need to shift to new areas hence reducing deforestation. Furthermore, rehabilitation activities are also accompanied by provision of multiple benefits such as sequestering carbon, improving food security and reducing poverty.

Despite the extensive deforestation and land degradation in Tanzania, to the best of our knowledge, very few if any studies exist on the potential to rehabilitate such lands in different land categories and regions of the country, and this study aims at filling parts of this void. The specific objectives of the study are to (i) assess the degraded land areas potentially available for rehabilitation in various regions in Tanzania, and (ii) give a review of main experiences and economic results gained in previous land rehabilitation studies in the country. Most efforts have been devoted to cover objective (i), where we provide information not published before from the newly established National Forest Resource Monitoring and Assessment of Tanzania (NAFORMA). We hypothesize that large areas are available for land rehabilitation in Tanzania, and that it is environmentally and economically viable to rehabilitate considerable parts of these areas.

The remaining part of the paper is organized as follows: In Chap. 2 the methods used for data collection are described, in Chap. 3 results and discussions are presented, and Chap. 4 provides conclusions and recommendations.

8.2 Methodology

The study is based on new data from NAFORMA and previous literature on socio-economic studies of rehabilitating degraded lands in Tanzania. Besides the review of articles, the websites of key organizations such as Tanzanian National Bureau of Statistics (NBS), Food and Agriculture Organization of the United Nations (FAO) and the World Bank (WB) were investigated. Information was likewise obtained from consultancy reports and personal communication with government officials and organization leaders to provide data about land rehabilitation projects undertaken by the government.

In the present study, NAFORMA has been essential for assessing potentials for rehabilitating degraded land in different land categories and regions in Tanzania, as the survey covers all regions and all main vegetation types of Mainland Tanzania (URT 2015; Tomppo et al. 2014; Vesa et al. 2010). The acquisition of NAFORMA data used a stratified systematic cluster sampling design (URT 2015; Tomppo et al. 2014; Vesa et al. 2010) considering estimation error and cost effectiveness. The sampling strata were located according to the distance between clusters and the number of plots within a cluster. Depending on the accessibility of the area, about 6–10 plots were established in each cluster. Data were collected from about 3419 clusters with a total of 32,660 plots. The distance between plots within a cluster was 250 m (URT 2015; Tomppo et al. 2014; Vesa et al. 2010).

We have used the data about erosion as indicator for land potentially available for rehabilitation. NAFORMA operates here with four erosion categories (URT 2015):

- *No* erosion—i.e. "No evidence of erosion".
- *Light* erosion—i.e. "Slight erosion where only surface erosion has taken place".
- *Moderate* erosion—i.e. "Erosion where mild gullies and rills are formed on the top surface of the soil".
- *Heavy* erosion—i.e. "Areas which have deep gullies, ravines, land slips etc.".

8.3 Results and Discussions

8.3.1 Degraded Land in Tanzania

8.3.1.1 NAFORMA Results

NAFORMA provides a lot of information, and we have just concentrated on the erosion data. Table 8.1 shows that in each region a significant size of the land area is affected by light erosion, followed by moderate and heavy erosion. Moderate erosion is more pronounced in Arusha (11 %), Iringa (10 %), Dodoma (9 %), Kilimanjaro (9 %), Kagera (9 %), Morogoro (8 %), Njombe (8 %), Tanga (7 %) and Ruvuma (7 %). Generally, about 49 % (43.3 mill ha) of the total land area in Mainland Tanzania is under either light (43 %, 37.7 mill ha), moderate (5 %, 4.4. mill ha) or heavy (1.3 %, 1.2 mill ha) erosion. These figures are substantial and imply large opportunities for land rehabilitation.

According to the national population census of 2012, Kagera and Arusha are among the regions with the highest annual population growth rates of 3.2 % and 2.7 % respectively, followed by Morogoro (2.4 %), Tanga (2.2 %), Dodoma (2.1 %) and Ruvuma (2.1 %) while Kilimanjaro, Iringa and Njombe have the lowest rates of 1.8, 1.1 and 0.8 % respectively (Tanzania 2012). Except for Dodoma, the rest of regions are found in more mountainous areas where other factors than population growth can be major causes for the experienced erosion—

Table 8.1 Extent of soil erosion in Tanzania distributed by regions, zones and erosion classification[a] (in ha and percentage of total land area)

Zones	Regions	Total land area (ha)	No erosion (ha, %)	Light erosion (ha, %)	Moderate erosion (ha, %)	Heavy erosion (ha, %)
Eastern	Dar es Salaam	150,745	90,451 (60)	53,717 (36)	5311 (4)	1248 (0.8)
	Morogoro	6,883,954	2,926,819 (43)	3,337,838 (48)	551,081 (8)	67,879 (1.0)
	Pwani	3,195,044	1,951,576 (61)	1,134,445 (36)	104,346 (3)	4677 (0.1)
Subtotal		10,229,742	4,968,846 (49)	4,526,000 (44)	660,738 (6)	73,803 (0.7)
Southern	Lindi	6,782,646	3,990,429 (59)	2,457,231 (36)	229,939 (3)	105,006 (1.5)
	Mtwara	1,794,089	933,321 (52)	771,830 (43)	74,532 (4)	14,364 (0.8)
	Ruvuma	6,335,334	3,450,650 (54)	2,370,111 (37)	417,215 (7)	97,265 (1.5)
Subtotal		14,912,070	8,374,400 (56)	5,599,172 (38)	721,686 (5)	216,635 (1.4)
Southern Highlands	Rukwa	2,166,572	1,091,288 (50)	964,313 (45)	88,910 (4)	22,042 (1.0)
	Njombe	2,197,169	1,043,671 (48)	899,995 (41)	175,259 (8)	77,845 (3.5)
	Iringa	3,453,407	1,692,785 (49)	1,306,373 (38)	328,241 (10)	125,914 (3.6)
	Mbeya	6,103,794	3,045,373 (50)	2,804,948 (46)	187,921 (3)	65,296 (1.1)
Subtotal		13,920,942	6,873,117 (49)	5,975,629 (43)	780,331 (6)	291,096 (2.1)
Central	Manyara	4,468,061	2,297,674 (51)	1,984,640 (44)	159,547 (4)	26,124 (0.6)
	Dodoma	4,131,412	1,681,043 (40)	1,987,238 (48)	383,986 (9)	129,146 (3.1)
	Singida	4,854,872	3,190,200 (66)	1,531,275 (32)	92,149 (2)	41,249 (0.8)
Subtotal		13,504,346	7,168,917 (53)	5,503,153 (41)	635,681 (5)	196,519 (1.4)
Lake	Mara	2,188,993	890,714 (41)	1,189,552 (54)	82,604 (4)	22,537 (1.0)
	Simiyu	2,344,077	1,499,860 (64)	752,691 (32)	57,632 (2)	33,894 (1.4)
	Mwanza	1,091,792	440,590 (40)	599,834 (55)	48,732 (4)	2636 (0.2)
	Kagera	2,526,237	683,880 (27)	1,575,018 (62)	237,196 (9)	30,124 (1.2)
	Geita	2,095,244	1,290,783 (62)	753,520 (36)	36,253 (2)	14,622 (0.7)
Subtotal		10,246,342	4,805,826 (47)	4,870,615 (48)	462,418 (5)	103,813 (1.0)

(continued)

Table 8.1 (continued)

Zones	Regions	Total land area (ha)	No erosion (ha, %)	Light erosion (ha, %)	Moderate erosion (ha, %)	Heavy erosion (ha, %)
Western	Tabora	7,592,764	4,109,529 (54)	3,370,599 (44)	89,632 (1)	23,003 (0.3)
	Shinyanga	1,853,143	1,227,797 (66)	574,694 (31)	36,292 (2)	14,360 (0.8)
	Kigoma	3,818,200	1,844,521 (48)	1,752,197 (46)	172,158 (5)	49,221 (1.3)
	Katavi	4,340,422	2,308,605 (53)	1,863,881 (43)	143,864 (3)	24,034 (0.6)
Subtotal		17,604,528	9,490,452 (54)	7,561,371 (43)	441,946 (3)	110,618 (0.6)
Northern	Kilimanjaro	1,249,964	621,355 (50)	494,883 (40)	114,404 (9)	19,228 (1.5)
	Arusha	3,821,292	1,386,870 (36)	1,876,091 (49)	425,106 (11)	133,094 (3.5)
	Tanga	2,809,417	1,295,337 (46)	1,318,183 (47)	185,180 (7)	10,567 (0.4)
Subtotal		7,880,673	3,303,562 (42)	3,689,157 (47)	724,690 (9)	162,889 (2.1)
Total		88,298,642	44,985 121 (51)	37,725,096 (43)	4,427,490 (5)	1,155,374 (1.3)

Source URT 2015: NAFORMA Biophysical Data and Report
[a]Soil erosion classification as defined in Chap. 2 of this article

like steepness, rainfall, soil and vegetation types. Although Dodoma region is located relatively in flat areas, the high erosion there may be due to severe droughts which have been common in the region for many years and the interactions between steep slopes, flatness and severe rainfalls. Dry conditions followed by heavy rainfall may also contribute to severe gully erosion in many places.

The high percentage of moderate erosion and heavy erosion in each region is also influenced by land use activities including logging and agriculture,—especially inappropriate farming practices like over cultivation and overgrazing (Cohen 2002). The high erosion rates experienced in Lindi may be due to emigration of pastoralists from other areas, especially the Sukuma and Masai. These people tend to settle in forested lands leading to severe deforestation and land degradation (Charnley 1997a, b). Furthermore, land use conflicts between pastoralists and farmers have been common in Morogoro and some parts of Tanga regions in the past 10 years. Poor land property regimes might have led to these conflicts leading to improper land use management hence the pronounced erosion in the areas.

Most of the regions experiencing highest erosion are also covered by miombo woodland. Other studies and empirical evidences from the field show that miombo woodland are subjected to severe harvesting for charcoal as well as clearing for agricultural activities (Hofstad 1997; Kajembe et al. 2005b; Luoga et al. 2000; Mwampamba 2007; Mbwambo et al. 2012).

Table 8.2 indicates that the two land-use categories Production forest (59 %) and Grazing land (60 %) are the main land-use categories having most eroded land relative to their land area. In the category of Production forests the distribution on light, moderate and heavy erosion classes is respectively 49, 8 and 2 % of the land area of the category, and about 1.98 mill ha is found to belong in the moderate and heavy erosion groups. The high rate of erosion in the grazing land category is most likely caused by the experienced uncontrolled movements of the pastoralists in the country, and is a strong indication of the need for land rehabilitation programs in this field.

Table 8.3 shows the erosion by vegetation types and we see that all vegetation types are affected by erosion although to varying degree. In forests, the Humid montane category has the highest erosion relative to other categorie's land area (61 % or about 530,000 ha having erosion), followed by Plantation (38 % or about 220 000 ha) and Lowland (37 % or about 660,000 ha). In Woodland, Scattered cropland has the highest relative erosion (76 % or about 1.9 mill ha). Light and moderate erosion is severe in the Humid montane forest (59 %), Woodlands (51 %), Grasslands (49 %), Cultivated agroforestry system (47 %) and in Other land uses (41 %), indicating significant land rehabilitation potentials. Delaying interventions and leaving these eroded areas unattended increase the economic losses in terms of crop yields, pasture quality, forest products and other woodlands (Misana et al. 2003). Changes in forest cover may also have strong impacts on biodiversity richness, water storage and supplies, carbon sequestration and climate regulation (Hansen et al. 2013).

The data obtained indicate rather strongly that there is a need to ensure that proper forest management practices are in place to safeguards the humid montane

Table 8.2 Extent of soil erosion[a] in Tanzania by land use[b] and erosion classification[a] (in ha and percentage of total land area)

Land use	Total land area (ha)	Types of erosion			
		No erosion (ha, %)	Light erosion (ha, %)	Moderate erosion (ha, %)	Heavy erosion (ha, %)
Production forest	19,807,566	8,200,141 (41)	9,724,176 (49)	1,493,086 (8)	388,402 (2.0)
Protection forest	9,384,775	5,639,694 (60)	3,143,094 (33)	504,690 (5)	97,296 (1.0)
Wildlife reserve	19,976,100	12,246,966 (61)	6,951,606 (35)	621,980 (3)	155,548 (0.8)
Shifting cultivation	5,844,356	2,640,180 (45)	2,950,033 (50)	202,498 (3)	51,645 (0.9)
Agriculture	20,219,956	10,425,878 (52)	8,896,612 (44)	726,792 (4)	170,674 (0.8)
Grazing land	9,161,425	3,715,878 (41)	4,565,264 (50)	633,026 (7)	243,456 (2.7)
Built-up areas	1,851,412	867,324 (47)	886,482 (48)	97,606 (5)	–
Other land uses	2,053,053	1,249,060 (61)	607,829 (30)	147,812 (7)	48,352 (2.4)
Total	88,298,642	44,985,121 (51)	37,725,096 (43)	4,427,490 (5)	1,155,374 (1.3)

Source URT 2015: NAFORMA Biophysical Data and Report
[a]Soil erosion classification as defined in Chap. 2 of this article
[b]Land use category as defined in NAFORMA Report—http://www.fao.org/forestry/17847/en/tza/

forests. Also in Woodlands, Bushlands, Grasslands, Bare lands and the Agroforestry systems land categories we see that tree planting programs could be of high interest for rehabilitating the already eroded and degraded lands. The Bare soils category on open lands might also include potential areas for rehabilitation through tree planting. In the Plantation forest category, tree gap-filling or replanting are examples of measures which can be undertaken to reduce and improve the area under light erosion. Tree planting has been suggested to be among the best techniques of increasing forest cover and may help in protecting and managing large areas of secondary forest or regrowth (Lamb et al. 2005).

Arusha, Kilimanjaro and Morogoro with high rates of moderate erosion are all regions with high opportunities for tourism businesses, having famous national parks and other types of tourist attractions. Continued erosion in these regions may cause significant damage to the existing infrastructures hence reduced income opportunities, implying negative impacts to the livelihood of people in those regions. Rehabilitation of degraded lands in those regions at early stages of the damage seems therefore of particular interest, both from economic and environmental point of view.

Table 8.3 Extent of soil erosion[a] in Tanzania distributed on vegetation types[b] and erosion classes (in ha and percentage of total land area)

Vegetation types	Total land (ha)	No erosion (ha, %)	Light erosion (ha, %)	Moderate erosion (ha, %)	Heavy erosion (ha, %)
Forest					
Humid montane	863,060	333,426 (39)	417,283 (48)	94,165 (11)	18,185 (2.1)
Lowland	1,740,987	1,084,587 (62)	562,103 (32)	73,760 (4)	20,537(1.2)
Mangrove	136,148	110,665 (81)	24,295 (18)	127 (0)	1061(0.8)
Plantation	573,382	352,694 (62)	186,416 (33)	30,832 (5)	3440 (0.6)
Sub-total	3,317,185	1,881,373 (57)	1,190,097 (36)	198,884 (6)	43,223 (1.4)
Woodland					
Closed (>40 %)	9,019,093	5,265,503 (58)	3,103,724 (34)	544,918 (6)	104,949 (1.2)
Open (10–40 %)	36,230,449	17,245,981 (48)	16,430,475 (45)	2,000,524 (6)	553,468 (1.5)
Scattered cropland	2,471,271	602,022 (24)	1,753,242 (71)	88,583 (4)	27,424 (1.1)
Sub-total	47,720,813	23,113,506 (48)	21,287,441 (45)	2,634,025 (6)	685,841 (1.4)
Bushland					
Thicket	938,847	734,939 (78)	181,248 (19)	15,194 (2)	7466 (0.8)
Dense	1,909,936	1,368,472 (72)	410,026 (21)	102,576 (5)	28,862 (1.5)
Scattered cultivated	1,183,258	576,410 (49)	525,444 (44)	66,711 (6)	14,693 (1.2)
Emergent trees	311,714	166,196 (53)	131,965 (42)	13,552 (4)	–
With emergent trees	316,734	237,201 (75)	61,644 (19)	15,980 (5)	1908 (0.6)
Open	2,682,269	1,390,003 (52)	1,087,750 (41)	167,026 (6)	33,690 (1.3)
Sub-total	7,342,757	4,473,221 (61)	2,398,077 (33)	381,040 (5)	86,619 (1.2)
Grassland					
Wooded	4,834,247	2,368,835 (49)	2,173,034 (45)	220,870 (5)	71,507 (1.5)
Bushed	438,000	253,602 (58)	139,293 (32)	34,940 (8)	10,164 (2.3)
Scattered cropland	559,625	287,194 (51)	224,501 (40)	24,423 (4)	23,507 (4.2)
Open	3,354,513	1,607,302 (48)	1,601,981 (48)	130,077 (4)	15,153 (0.5)
Sub-total	9,186,385	4,516,934 (49)	4,138,809 (45)	410,311 (4)	120,331 (1.3)
Cultivated land					
Agroforestry	1,300,338	353,878 (27)	869,196 (67)	77,264 (6)	–
Wooded crops	1,450,010	804,196 (55)	602,564 (42)	38,378 (3)	4872 (0.3)
Herbaceous crops	4,924,182	2,379,306 (48)	2,181,325 (44)	315,414 (6)	48,137 (1.0)
Mixed tree-cropping	134,658	48,817 (36)	55,316 (41)	27,910 (21)	2616 (1.9)

(continued)

Table 8.3 (continued)

Vegetation types	Total land (ha)	No erosion (ha, %)	Light erosion (ha, %)	Moderate erosion (ha, %)	Heavy erosion (ha, %)
Grain crops	9,670,874	5,560,203 (57)	3,847,191 (40)	186,997 (2)	76,483 (0.8)
Sub-total	17,480,063	9,146,400 (52)	7,555,592 (43)	645,963 (4)	132,108 (0.8)
Open land					
Bare soil	129,795	74,375 (57)	36,867 (28)	13,761 (11)	4792 (3.7)
Others					
Water, Swamp, Rock	3,125,253	1,779,313 (57)	1,118,212 (36)	143,505 (5)	80,614 (2.6)
Total	88,298,642	44,985,120 (51)	37,725,096 (43)	4,427,489 (5)	1155,373 (1.3)

Source URT (2015): NAFORMA Biophysical Data and Report
[a]Soil erosion classification as defined in Chap. 2 of this article
[b]Vegetation types as defined in NAFORMA Report—http://www.fao.org/forestry/17847/en/tza/

Although the economics losses due to erosion and land degradation are not quantified, the observed consequences to the communities and the nation as a whole are evident. Land rehabilitation projects should of course consider areas where erosion is a problem. However, in addition, thorough cost-benefit analyses are needed to prioritize between promising rehabilitation projects, as outlined some further in Sect. 8.3.2.

8.3.1.2 Other Data on Land Degradation

The NAFORMA data gives at present no information about changes over time as the survey has just had one "round" of registration. However, there are other studies from Tanzania which could indicate degree of land changes and deforestation rates. Hall et al. (2009) found that during the period 1955–2000 the rate of deforestation in the Eastern Arc Mountains increased as indicated in Tables 8.4 and 8.5. We see that the deforestation has varied rather much according to mountain block (Table 8.4) and according to ecological zones (Table 8.5), with highest deforestation in Lowland montane. Also, the data show that the deforestation rates in this area was higher during the period 1955–1975 than during 1975–2000.

Empirical evidence suggests that land use changes will continue in the coming decades because of the changes in causal factors such as population and demand for food and forest products (Swetnam et al. 2011). FAO (2010) reports that between 1990 and 2005 the category Forest in Tanzania decreased by about 15 %, the category Other wooded land by about 79 %, and that the two land categories together decreased by about 37 % (Table 8.6).

Table 8.4 The Eastern Arc Mountain blocks and rates of deforestation

Mountain block	Forest area (km^2)			1975–2000	
	1955	1975	2000	Change (km^2)	%
East Usambara	425	299	263	−36	−12.0
Mahenge	35	24	24	0	0.0
Malundwe	9	6	9	3	50.0
Nguru	–	313	297	−16	−5.1
Nguu	207	198	188	−10	−5.1
North Pare	36	27	26	−1	−3.7
Rubeho	652	532	477	−55	−10.3
South Pare	195	147	139	−8	−5.4
Udzungwa	1745	1402	1354	−48	−3.4
Ukaguru	200	181	167	−14	−7.7
Uluguru	338	321	279	−42	−13.1
West Usambara	579	348	323	−25	−7.2

Source Hall et al. (2009)

8.3.2 Experiences from Previous Land Rehabilitation Studies

8.3.2.1 More General Findings

In the past three decades, various projects have been established to combat land degradation problem especially in mountainous areas (Kajembe et al. 2005a). Initiatives being put in place to rehabilitate and conserve deforested and degraded land in different regions of Tanzania include HADO (Hifadhi Ardhi Dodoma), HASHI (Hifadhi Ardhi Shinyanga), HIMA (Hifadhi Mazingira), LAMP (Land Management Program), SECAP (Soil Erosion Control and Agroforestry Project) and HIAP (Handeni Integrated Agroforestry Project) projects implemented in Dodoma, Shinyanga, Iringa, Babati-Manyara, Lushoto and Handeni respectively (Table 8.7). The main goals of these projects were to help local communities

Table 8.5 Ecological zones and the rate of deforestation in the Eastern Arc Mountain

Zone	1955 (km^2)	1975 (km^2)	2000 (km^2)	Rate of change per year (%)	
				1955–1975	1975–2000
Lowland montane (200–800 m)	609	347	274	−2.84	−0.95
Sub montane (800–1200 m)	748	480	440	−2.25	−0.35
Montane (1200–1800 m)	1954	1649	1559	−0.85	−0.22
Upper montane (>1800 m)	1410	1309	1262	−0.37	−0.15

Source Hall et al. (2009)

Table 8.6 Forest and other wooded land changes in Mainland Tanzania

Forest land category	Area (1000 ha)			% Change
	1990	2000	2005	1990–2005
Forest	41,441	37,318	35,257	−14.9
Other wooded land	22,374	10,629	4756	−78.7
Forest and other wooded land	63,815	47,947	40,013	−37.3
Other land	24,544	40,412	48,346	−97.0
Total land area	88,359	88,359	88,359	
Inland water bodies	6150	6150	6150	
Total area of country	94,509	94,509	94,509	

Source FAO (2010)—http://faostat3.fao.org/download/F/FO/E (visited on 20/01/2016)

rehabilitate degraded land and ensure sustainable supply of woodfuel and fodder for livestock (Ghazi et al. 2005; Iddi 2002; Msuya et al. 2006) and ensuring sustainable environmental and land conservation.

These different projects resulted in mixed outcomes, for example the HADO project in Kondoa District rehabilitated only about 428 ha of land while the HASHI project in Shinyanga region rehabilitated about 350,000 ha of land using agro-forestry systems and participatory approaches involving local communities (Pye-Smith 2010). Experiences from these activities and other land use

Table 8.7 Soil and Land Conservation/Rehabilitation Initiatives and Projects in Tanzania

S/No.	Name of the project/initiative	Year
1.	Dodoma Region Soil Conservation Project (Hifadhi Ardhi Dodoma—HADO)	1973
2.	Rukwa Integrated Development Program	1985
3.	Shinyanga Soil Conservation and Afforestation Project (Hifadhi Ardhi Shinyanga—HASHI)	1986
4.	East Usambara Conservation and Agricultural Development Project	1987
5.	Kigoma Rural Integrated Development Program	1989
6.	East Usambara Catchment Forestry Project	1989
7.	Soil Erosion Control and Agroforestry Project (SECAP)	1989
8.	Environmental Conservation in Iringa (Hifadhi Mazingira Iringa—HIMA)	1990
9.	Land Management Program for Environmental Conservation (LAMP) in Babati District	1991
10.	Dodoma Village Afforestation Project (DOVAP)	1991
11.	Dodoma Land Use Management Project	1991
12.	Handeni Integrated Agroforestry Project (HIAP)	1992
13.	Tanga Coastal Zone Conservation and Development Program	1993
14.	Kilosa Environment Project	1994
15.	Kilimanjaro Environmental Conservation Management Trust Fund	1990
16.	Community Based Forest Management (Participatory Forest Management)	2000
17.	Tanzania Community Forest Network (MJUMITA)	2005

Source Schechambo et al. (1999), Personal Communication and Consultancy Reports interventions are reported in many studies (Abdallah and Monela 2007; Birch-Thomsen et al. 2001; Cleaver et al. 2010; Cleaver and Schreiber 1994; Dejene 1997; Dixon et al. 2001; Iddi 2002; Kajembe et al. 2005a, b; Lamb et al. 2005; Massao 1993: Mutuo et al. 2005; Msuya et al. 2006; Oba et al. 2008; Pye-Smith 2010; Reid et al. 2009; Schechambo et al. 1999). It is not possible in this article to cover all results reported in those studies, but the following points are in our opinion interesting findings from the studies regarding what are important factors to consider in land rehabilitation projects in Tanzania:

- The main causes of land degradation
- Land tenure system—property rights
- Rules and regulations for monitoring and governing land-use changes
- Local community involvement
- Education and awareness programs to enhance adaptive capacity of the local community
- Improved agricultural and forestry practices, including agro-forestry

In the following the latter two points—adaptive capacity and agroforestry—are elaborated some more. Adaptive capacity is an important aspect for local communities to cope with the effects of climate change at the local level (Cooper et al. 2008). Land rehabilitation can increase the adaptive capacity of local communities because it provides improved livelihood options through increased land productivity and income (Paavola 2008). However, the implementation and adoption of an effective land rehabilitation technique is affected by many factors including education level, perceptions of people of the problem, proper land tenure, tribe affiliation, gender, land location and size, labour availability and off-farm activities (Tenge et al. 2004). Also, expected increased utility and profit are the basis for adoptions of any innovation in the community (Mbaga-Semgalawe and Folmer 2000). Investing in education and awareness programs is therefore important for ensuring success on rehabilitating degraded lands, and the economic benefits associated with each rehabilitation technique should be studied and provided in order to motivate local communities and other stakeholders involved in the process.

Agroforestry, tree planting and reforestation practices as means of rehabilitating degraded lands have multiple benefits. First, agro-forestry can enhance agriculture profitability by increased crop yields due to fertilization and other effects of the trees. The trees provide supply of fodder, fibers and other forest related products demanded by the communities. These trees can provide alternative sources of energy and forest products hence reducing pressure on the existing plantations and natural forests. It can also contribute significantly to carbon sequestration and provide multiple benefits to farmers hence reducing their vulnerability and increase their adaptive capacity to climate change, as well as providing increased biodiversity conservation and economic benefits to the community (Daily 1995). By rehabilitating degraded lands, community members can also benefit from REDD+ initiatives, as the planted trees on the degraded lands will contribute to carbon sequestration and hence qualify for carbon payments according to the additional amount of carbon sequestered.

The integration of trees in farming systems is facing a number of constraints especially those related to economic and policy competition with the agricultural sector (Garrity et al. 2010, Linyunga et al. 2004). However, rehabilitating degraded croplands and pastures by converting it into a tree-based systems could increase the aboveground and belowground net carbon sequestration with about 10–70 Mg C ha^{-1} in the vegetation and 5–15 Mg C ha^{-1} in the topsoil within a period of about 25 years (Mutuo et al. 2005). The agroforestry tree-based systems is capable of sequestering carbon in vegetation up to more than 60 Mg C ha^{-1} compared to crop or pasture systems (Mutuo et al. 2005). In their opinion, rehabilitation of degraded land using agroforestry techniques is an important aspect with multiple benefits, including timber, wood fuel, soil nutrients, carbon sequestration and trade, reducing vulnerability, increasing adaptive capacity of people and reducing climate change impacts to local communities.

The multiple benefits obtained from the planted trees may therefore motivate local communities to participate in the rehabilitation process. With multiple tangible economic benefits it is possible to engage both private organizations and the government in rehabilitating degraded land. The government can promote rehabilitation activities by providing initial funding and awareness creation to local communities, profit and nonprofit organizations. Training, awareness creation and provision of startup funding can be potential motivation tools to local communities to increase the rate of adoption of the rehabilitation techniques.

However, to have a successful agroforestry system, it is important to understand land tenure and common practices of slash and burn agriculture which tend to affect tree planting and promotion of agroforestry practices. The complexity of causes behind deforestation and degradation and the importance of economic and policy frame conditions ask for combined efforts involving all relevant stakeholders such as individuals, private based organizations and the government.

8.3.2.2 Economic Results

To our surprise we were not able of finding any published economic cost-benefit analysis of land rehabilitation in Tanzania. However, several economic mechanisms, techniques and incentives for implementing effective rehabilitation programmes in tropical countries have been suggested. Paying the landowners for the ecological services and ensuring appropriate institutional, legal, and policy settings for providing defined land tenure systems and access to financial resources are among those mechanisms (Lamb et al. 2005). The growth of carbon markets for global climate change mitigation makes carbon sequestration a potential additional income to landowners (Montagnini and Nair 2004). They anticipate *op.cit.* that the extra income from the carbon trade could be an effective incentive to motivate local communities to undertake agroforestry practices and tree planting for land rehabilitation.

Forest goods and services support the economic livelihoods of more than 1 billion people, mainly in third world countries (Sunderlin et al. 2005). Both natural forests and plantations play a significant role in sustaining the livelihood of local communities. Products such as sawn wood, paper and fibre materials contribute directly to the economy, but also other goods derived from the forest ecosystems have significant economic value (De Groot et al. 2012). They claim that investments in afforestation and reforestation activities can potentially increase the value of forest related industry and carbon stored in forests notably. In many developing countries, wood is an important source of energy particularly in rural areas and at the same time providing raw materials for various forest related industries (Mwampamba 2007). Further, local communities collect different types of non-timber forest products (NTFP) from the forest for both domestic and commercial uses. These non-timber forest products have significant impacts to the livelihoods of households in some of the rural and peri-urban communities. In some communities especially in dry central parts of Tanzania, the NTFPs are the only source of food throughout the year. Forests also provide important services such as soil erosion control, biodiversity, catchment and watershed management, and protection of coastal areas. Forest produces wood fuels which can be used as an environmentally friendly alternative to fossil fuels in forms of biogas and bio-fuels (Ahlborg and Hammar 2014; Sheya and Mushi 2000; Van Eijck and Romijn 2008; Van der Werf et al. 2009). Other important forest benefits with potentially high economic values include tourism, biodiversity habitat protection, food sources, medicinal plants, forest products, regulation of the hydrologic cycle, protection of soil resources, recreational and spiritual benefits (Bonan 2008).

Carbon sequestration through afforestation projects and activities have proved to be a cost effective methods used for reducing carbon dioxide emissions (De Jong et al. 1997). More information on the availability of the potential lands for rehabilitation and carbon sequestration are however needed. According to De Jong et al. (1997), the estimation of costs for carbon sequestration may be simplified if proper information on land and land uses are available. Afforestation and reforestation activities may currently be cheap, but in the long run large-scale investment in these activities may encounter substantial cost increases because lands with higher productivity and opportunity costs must be used and transaction costs may increase (De Jong et al. 2000).

Lacking economic cost/benefit studies on land degradation/rehabilitation in Tanzania, we refer to Table 8.8 taken from Bojö (1996) to illustrate the economic losses caused by land degradation in some other African countries. The gross annual immediate loss and the discounted future loss reported there are the foregone benefits for not rehabilitating degraded lands. The analyses referred to there are based on many assumptions, but the results illustrate the high economic importance that land rehabilitation may have.

The literature indicates that the majority of the local communities practice shifting cultivation as an adaptive means of increasing and maintaining food security in their households (Dixon et al. 2001). Poor households consider deforestation rational because of the short-term benefits obtained. According to Gootee

Table 8.8 Economic loss due to land degradation in some African countries

Study	Country	Gross annual immediate loss (in million USD)	Gross discounted future loss (in million USD)	Gross discounted cumulative loss (in million USD)
FAO (1986)	Ethiopia	14.8	–	2993.0
Sutcliffe (1993)	Ethiopia	155.0	15.0	–
Bojö and Cassells (1995)	Ethiopia	130.0	22.0	2431.0
Convery and Tutu (1990)	Ghana	166.4	–	–
Bojö (1991)	Lesotho	0.3	3.2	31.2
World Bank (1988)	Madagascar	4.9–7.6	–	–
World Bank (1992)	Malawi	6.6–19.0	48.0–136.0	–
Bishop and Allen (1989)	Mali	2.9–11.6	19.3–76.6	–
MacKenzie (1994)	South Africa	18.0	173.0	503.0
Stocking (1986)	Zimbabwe	117.0	–	–
Norse and Saigal (1992)	Zimbabwe	95.5	–	–
Grohs (1994)	Zimbabwe	0.6	6.7	44.7

Source Bojö (1996)

et al. (2010) inadequate information on agricultural techniques and sustainable land uses and the increasing demand for forest products and agricultural land are the main reasons for the high rate of deforestation and land degradation by poor households. Thus, creating awareness and promoting adequate practices and systems are important for rehabilitating degraded lands in Tanzania. This will assist in mitigating climate change, enhancing adaptive capacity, providing tangible benefits to the communities and ensuring sustainable natural systems management (Alexander et al. 2011).

The majority of rural people in Tanzania rely on agriculture for their livelihoods. This is also common in many Sub-Saharan African countries where a large share of the world's poorest people are located. Without formulating and implementing proper measures, the available forests are likely to disappear even faster in the next few decades than experiencing now (Poore 2013). The limited available data on costs and income and lack of information on important benefits of forest conservation are among the serious problems facing people in making proper decisions on land rehabilitation methods (Angelsen and Rudel 2013).

The lack of analysis of the economic impacts of land rehabilitation in Tanzania is striking. Thorough cost-benefit analyses are strongly needed to prioritize between land rehabilitation projects. These analyses should emphasize to include all costs and benefits involved, and quantify them in economic terms as far as practically possible. Various techniques exist for that. However, some costs and benefits might be very difficult to quantify in economic terms, but in such cases one should at least try to quantify them in physical terms. An essential element in such analysis will be to quantify the distributional impacts—i.e. how costs, benefits and net surpluses—are distributed over time between different main stakeholders (e.g. poor and rich, local community, region, country at large).

8.4 Conclusions

Previously, only scant information was available on the extent and amount of land degradation in Tanzania, but now the NAFORMA data clearly show that the country is experiencing serious land degradation problems and where in the country they occur. The land degradations are exacerbated by significant increases in population, economic growth and demand of food and forest products.

These land use changes need proper attention to ensure sustainable supply of forest and food products and maintaining environmental benefits and services, including increased carbon sequestration. Appropriate measures to meet these changes may have significant implications to poor vulnerable households with weak adaptive capacities. Agroforestry and tree planting programs are potential techniques for rehabilitating degraded land in Tanzania because of the expected multiple economic and environmental benefits to the community and the country. Incorporating rehabilitation of degraded land through agroforestry, reforestation and other tree environmental protection activities is important also in order to benefit from the globally growing carbon markets.

Economic studies on the benefits and costs of land rehabilitation in Tanzania and their distribution on various stakeholders are urgently needed in order to identify and prioritize among the most promising rehabilitation activities.

References

Abdallah J, Monela G (2007) Overview of Miombo woodlands in Tanzania. MITMIOMBO–management of indigenous tree species for ecosystem restoration and wood production in semi-arid Miombo woodlands in eastern Africa. Working Papers of the Finnish Forest Research Institute, Helsinki, Finland, Number 50:9-23

Ahlborg H, Hammar L (2014) Drivers and barriers to rural electrification in Tanzania and Mozambique—Grid-extension, off-grid, and renewable energy technologies. Renew Energy 61:117–124

Alexander S, Nelson CR, Aronson J, Lamb D, Cliquet A, Erwin KL, Finlayson CM, De Groot RS, Harris JA, Higgs ES (2011) Opportunities and challenges for ecological restoration within REDD+. Restor Ecol 19:683–689

Angelsen A, Fjeldstad OH (1995) Land reforms and land degradation in Tanzania: alternative economic approaches. CHR. Michelsen Institute, Development Studies and Human Rights, Bergen, Norway. Working Paper WP 1995:3

Angelsen A, Rudel TK (2013) Designing and implementing effective REDD+ policies: a forest transition approach. Rev Environ Econ Policy 7:91–113

Appiah M, Blay D, Damnyag L, Dwomoh FK, Pappinen A, Luukkanen O (2009) Dependence on forest resources and tropical deforestation in Ghana. Environ Dev Sustain 11:471–487
Aronsod J, FledC Le, Floc'hl E, Ode C, Pontanier R (1993) Restoration and rehabilitation of degraded ecosystems in arid and semi-arid lands. I. A view from the south. Restor Ecol 1(1):8–17
Bahamondez C, Christophersen T, Csoka P, Drichi P, Filipchuk A, Gueye S, Johnson S, Kajarlainen T, Kapos V, Keenan R (2010) Global forest resources assessment 2010: main report. Food and Agriculture Organization of the United Nations (FAO). http://www.fao.org/forestry/fra/fra2010
Bai ZG, Dent DL, Olsson L, Schaepman ME (2008) Proxy global assessment of land degradation. Soil Use Manag 24:223–234
Bergsma E, Charman P, Gibbons F, Hurni H, Moldenhauer W, Panichapong S (1996) Terminology for soil erosion and conservation, ISSS/ITC/ISRIC, 6700 AJ Wageningen, The Netherlands
Birch-Thomsen T, Frederiksen P, Sano HO (2001) A livelihood perspective on natural resource management and environmental change in semiarid Tanzania. Econ Geogr 77:41–66
Bishop J, Allen J (1989) The on-site costs of soil erosion in Mali. World Bank Environment Working Paper no. 21. The World Bank, Washington, DC
Bojö J (1996) The costs of land degradation in Sub-Saharan Africa. Ecol Econ 16(2):161–173
Bojö J, Cassells D (1995) Land degradation and rehabilitation in Ethiopia: a re-assessment. The World Bank, Washington, DC
Bojö J (1991) The economics of land degradation: theory and applications to lesotho. The Stockholm School of Economics, Stockholm
Bonan GB (2008) Forests and climate change: forcings, feedbacks, and the climate benefits of forests. Science 320:1444–1449
Burgess N, Doggart N, Lovett JC (2002) The Uluguru Mountains of eastern Tanzania: the effect of forest loss on biodiversity. Oryx 36:140–152
Burgess ND, Butynski TM, Cordeiro NJ, Doggart NH, Fjeldså J, Howell KM, Kilahama FB, Loader SP, Lovett JC, Mbilinyi B, Menegon M, Moyer DC, Nashanda E, Perkin A, Rovero F, Stanley WT, Stuart SN (2007) The biological importance of the Eastern Arc Mountains of Tanzania and Kenya. Biol Conserv 134:209–231
Charnley S (1997a) Environmentally-displaced peoples and the cascade effect: lessons from Tanzania. Hum Ecol 25:593–618
Charnley S (1997b) Pastoralism and property rights: The evolution of communal property on the Usangu Plains, Tanzania. Afr Econ Hist 25:97–119
Cleaver J, Schram R, Wanga G (2010) Bioenergy in Tanzania: the country context. Bioenergy and food security: the Befs analysis for Tanzania, Rome, Italy: Food and Agriculture Organization of the United Nations Report, pp 23–45
Cleaver KM, Schreiber GA (1994) Reversing the spiral: the population, agriculture, and environment nexus in sub-Saharan Africa Report. World Bank, Washington, DC
Cohen E (2002) Land tenure and land degradation in eastern Africa. The context of the United Nations Convention to Combat Desertification. World Agroforestry Programs. Swedish International Development Cooperation Agency (SIDA), Regional Land Management Unit (RELMA), ICRAF House, Nairobi Kenya, Working Paper Number 15
Convery F, Tutu K (1990) Evaluating the costs of environmental degradation in Ghana. In: Applications of economics in the environmental action planning process in Africa. University College Dublin, Dublin
Cooper PJM, Dimes J, Rao KPC, Shapiro B, Shiferaw B, Twomlow S (2008) Coping better with current climatic variability in the rain-fed farming systems of sub-Saharan Africa: an essential first step in adapting to future climate change? Agric Ecosyst Environ 126:24–35
Daily GC (1995) Restoring value to the world's degraded lands. Science 269:350–353
De Groot R, Brander L, Van Der Ploeg S, Costanza R, Bernard F, Braat L, Christie M, Crossman N, Ghermandi A, Hein L (2012) Global estimates of the value of ecosystems and their services in monetary units. Ecosyst Serv 1:50–61

De Jong BH, Tipper R, Montoya-Gómez G (2000) An economic analysis of the potential for carbon sequestration by forests: evidence from southern Mexico. Ecol Econ 33:313–327

De Jong BH, Tipper R, Taylor J (1997) A framework for monitoring and evaluating carbon mitigation by farm forestry projects: example of a demonstration project in Chiapas, Mexico. Mitig Adapt Strat Glob Change 2:231–246

Dejene A (1997) Land degradation in Tanzania: perception from the village, Technical Paper Number 370. World Bank Publications, Washington, DC

Dixon JA, Gibbon DP, Gulliver A (2001) Farming systems and poverty: improving farmers' livelihoods in a changing world. Food and Agriculture Organization, Rome 412p

FAO (1986) Highlands reclamation study—Ethiopia. Final report, vols I and II. FAO, Rome

FAO (2010) Global forest resources assessment 2010. The main report: Food and Agriculture Organization of the United Nations, Rome, Italy

Garrity DP, Akinnifesi FK, Ajayi OC, Weldesemayat SG, Mowo JG, Kalinganire A, Larwanou M, Bayala J (2010) Evergreen Agriculture: a robust approach to sustainable food security in Africa. Food Secur 2:197–214

Ghazi P, Barrow E, Monela G, Mlenge W (2005) Regenerating woodlands: Tanzania's HASHI Project. In WRI et al.: The wealth of the poor: managing ecosystems to fight poverty. Report World Resources Institute with UNDP, UNEP, The World Bank, Washington, DC, pp 131–138

Gootee RS, Blatner KA, Baumgartner DM, Carroll MS, Weber EP (2010) Choosing what to believe about forests: differences between professional and non-professional evaluative criteria. Small Scale For 9:137–152

Grohs F (1994) Economics of soil degradation, erosion and conservation: A case study of Zimbabwe. Wissenschaftsverlag Vank Kiel KG, Kiel

Hall J, Burgess ND, Lovett J, Mbilinyi B, Gereau RE (2009) Conservation implications of deforestation across an elevational gradient in the Eastern Arc Mountains, Tanzania. Biol Conserv 142:2510–2521

Hansen MC, Potapov PV, Moore R, Hancher M, Turubanova S, Tyukavina A, Thau D, Stehman S, Goetz S, Loveland T (2013) High-resolution global maps of 21st-century forest cover change. Science 342:850–853

Hartmana BD, Chadwicka OA, D'antonio C, Bookhagen B (2014) Watershed Rehabilitation, Bofedal Restoration, and Increased Greenness in a Tropical Montane Area. https://scholar.google.no/scholar?q=Watershed+Rehabilitation%2C+Bofedal+Restoration%2C+and+Increased+Greenness+in+a+Tropical+Montane+Area.&btnG=&hl=en&as_sdt=0%2C5

Hofstad O (1997) Woodland deforestation by charcoal supply to Dar es Salaam. J Environ Econ Manag 33:17–32

Iddi S (2002) Community participation in forest management in the United Republic of Tanzania. In: Proceedings of the second international workshop on participatory forestry in Africa. Defining the way forward: sustainable livelihoods and sustainable forest management through participatory forestry, pp 59–67. https://scholar.google.no/scholar?q=Iddi%2C+S.+2002.+&btnG=&hl=en&as_sdt=0%2C5

Kajembe G, Julius F, Nduwamungu J, Mtakwa P, Nyange D (2005a) Impact of indigenous-based interventions on land conservation: a case study of a soil conservation and agroforestry project, Arumeru District, Tanzania. Land Degrad Dev 16:311–325

Kajembe G, Nduwamungu J, Luoga E (2005b) The impact of community-based forest management and joint forest management on the forest resource base and local people's livelihoods: case studies from Tanzania. CASS/PLASS Commons Southern Africa Occasional Paper Series 8

Lamb D, Erskine PD, Parrotta JA (2005) Restoration of degraded tropical forest landscapes. Science 310:1628–1632

Linyunga K, Matakala P, Chintu R (2004) Accelerating agroforestry adoption: a case of Mozambique. Centre international de recherche en agrosylviculture (ICRAF), projet d'agroforesterie, document présenté au Congrès de l'Union internationale des instituts de recherche forestière (IUFRO), FAO Report, Rome, pp 12–15

Luoga E, Witkowski E, Balkwill K (2000) Economics of charcoal production in miombo woodlands of eastern Tanzania: some hidden costs associated with commercialization of the resources. Ecol Econ 35:243–257

MacKenzie C (1994) Degradation of arable land resources: policy options and considerations within the context of rural restructuring in South Africa, vol 11. Land and Agriculture Policy Centre

Malimbwi R, Zahabu E (2014) NAFORMA process and biophysical results. In: Proceedings of the international conference on reducing climate change challenges through forestry and other land use practices. Report, Sokoine University of Agriculture, Morogoro, Tanzania

Mary A, Majule A (2009) Impacts of climate change, variability and adaptation strategies on agriculture in semi arid areas of Tanzania: the case of Manyoni District in Singida Region, Tanzania. Afr J Environ Sci Technol 3:206–218

Massao JF (1993) Soil conservation on the slopes of Mt. Meru: strategies developed on the basis of SCUAF model prediction. University of Wales (UCNW, Bangor: Environmental Forestry)

Mbaga-Semgalawe Z, Folmer H (2000) Household adoption behaviour of improved soil conservation: the case of the North Pare and West Usambara Mountains of Tanzania. Land Use Policy 17:321–336

Mbwambo L, Eid T, Malimbwi R, Zahabu E, Kajembe G, Luoga E (2012) Impact of decentralised forest management on forest resource conditions in Tanzania. For Trees Livelihoods 21:97–113

Misana SB, Majule AE, Lyaruu HV, Change LU (2003) Linkages between changes in land use, biodiversity and land degradation on the slopes of Mount Kilimanjaro, Tanzania, LUCID Project, International Livestock Research Institute. LUCID Working Paper Number 38

Montagnini F, Nair P (2004) Carbon sequestration: an underexploited environmental benefit of agroforestry systems. Agrofor Syst 61:281–295

Msuya T, Mndolwa M, Bakengesa S, Maduka S, Balama C (2006) Factors affecting adoption and dissemination of agroforestry technologies in Tanzania: a case of HASHI/ICRAF and Ruvu Fuelwood Pilot Project. In: Proceedings of the second national agroforestry and environmental workshop: partnerships and linkages for greater impact in agroforestry and environmental awareness, Mkapa Hall, Mbeya, Tanzania, pp 14–17

Mutuo PK, Cadisch G, Albrecht A, Palm C, Verchot L (2005) Potential of agroforestry for carbon sequestration and mitigation of greenhouse gas emissions from soils in the tropics. Nutr Cycl Agroecosyst 71:43–54

Mwampamba TH (2007) Has the woodfuel crisis returned? Urban charcoal consumption in Tanzania and its implications to present and future forest availability. Energy Policy 35:4221–4234

Myers N, Mittermeier RA, Mittermeier CG, Da Fonseca GA, Kent J (2000) Biodiversity hotspots for conservation priorities. Nature 403:853–858

Norse D, Saigal R (1992) National economic cost of soil erosion—the case of Zimbabwe. In: CIDIE workshop on environmental economics and natural resource management in developing countries, World Bank, Washington, DC, pp 22–24

Oba G, Weladji RB, Msangameno DJ, Kaitira LM, Stave J (2008) Scaling effects of proximate desertification drivers on soil nutrients in northeastern Tanzania. J Arid Environ 72(10):1820–1829

Paavola J (2008) Livelihoods, vulnerability and adaptation to climate change in Morogoro, Tanzania. Environ Sci Policy 11:642–654

Pettorelli N, Lobora A, Msuha M, Foley C, Durant S (2010) Carnivore biodiversity in Tanzania: revealing the distribution patterns of secretive mammals using camera traps. Anim Conserv 13:131–139

Poore D (2013) No timber without trees: sustainability in the tropical forest. Routledge, Earth Library Collection, Natural Resources Management, Earthscan Publishing for a Sustainable Future, London, Sterling

Pye-Smith C (2010) A rural revival in Tanzania: how agroforestry is helping farmers to restore the woodlands in Shinyanga Region. Report, The World Agroforestry Centre, Nairobi Kenya, Tree for Change No. 7
Reid H, Alam M, Berger R, Cannon T, Huq S, Milligan A (2009) Community-based adaptation to climate change: an overview. Particip Learn Action 60:11–33
Rothman D, Agard J, Alcamo J, Alder J, Al-Zubari W, Aus Der Beek T, Chenje M, Eickhout B, Flörke M, Galt M (2007) The future today. Report, United Nations Environment Programme, Global Environment Outlook GEO 4 Environment for Development (Section E-The Outlook-Towards 2015 and Beyond). United Nations Environment Programme-UNEP, Valeta, Malta
Schechambo F, Sosoveli H, Kisanga D (1999) Rethinking natural resource degradation in semi-arid sub-Saharan Africa: the case of semi-arid Tanzania. Report prepared for Overseas Development Institute, London, UK, pp 2–91
Sheya MS, Mushi SJ (2000) The state of renewable energy harnessing in Tanzania. Appl Energy 65:257–271
Shivji IG (1998) Not yet democracy: reforming land tenure in Tanzania. IIED/ HAKIARDHI/ Faculty of Law, University of Dar es Salaam, Dar es Salaam and London, Tanzania and UK, International Institute for Environment and Development (IIED), London, UK
Shivji IG (1999) The Land Acts 1999: a cause for celebration or a celebration of a cause?, Oxfam Report for Land Ownership in Tanzania, Dar es Salaam, Tanzania
Stocking M (1986) The cost of soil erosion in zimbabwe in terms of the loss of three major nutrients. Consultant's Working Paper No 3. FAO, Rome
Sunderlin WD, Angelsen A, Belcher B, Burgers P, Nasi R, Santoso L, Wunder S (2005) Livelihoods, forests, and conservation in developing countries: an overview. World Dev 33:1383–1402
Sutcliffe JP (1993) Economic assessment of land degradation in the Ethiopian highlands—A case study. National Conservation Strategy Secretariat, Ministry of Planning and Economic Development, Addis Ababa
Swetnam RD, Fisher B, Mbilinyi BP, Munishi PKT, Willcock S, Ricketts T, Mwakalila S, Balmford A, Burgess ND, Marshall AR, Lewis SL (2011) Mapping socio-economic scenarios of land cover change: a GIS method to enable ecosystem service modelling. J Environ Manag 92:563–574
Tanzania Bureau of Statisctics (TBS) (2012) Population and housing census: population distribution by administrative areas. Ministry of Finance, Dar es Salaam
Tenge A, De Graaff J, Hella J (2004) Social and economic factors affecting the adoption of soil and water conservation in West Usambara highlands, Tanzania. Land Degrad Dev 15:99–114
Tomppo E, Katila M, Mäkisara K, Peräsaari J, Malimbwi R, Chamuya N, Otieno J, Dalsgaard S, Leppänen M (2010) Report to the Food and Agriculture Organization of the United Nations (FAO) in Support of Sampling Study for National Forestry Resources Monitoring and Assessment (NAFORMA) in Tanzania. Rome, Italy
Tomppo E, Malimbwi R, Katila M, Mäkisara K, Henttonen HM, Chamuya N, Zahabu E, Otieno J (2014) A sampling design for a large area forest inventory: case Tanzania. Can J For Res 44 (8):931–948
URT (2015) National Forest Resources Monitoring and Assessment of Tanzania Mainland (NAFORMA), Main Results. Report, Ministry of Natural Resources & Tourism Tanzania Forest Services Agency in Collaboration with the Government of Finland and Food and Agriculture Organization (FAO) of the United Nations
Van Der Werf GR, Morton DC, Defries RS, Olivier JG, Kasibhatla PS, Jackson RB, Collatz GJ, Randerson J (2009) CO_2 emissions from forest loss. Nat Geosci 2:737–738
Van Eijck J, Romijn H (2008) Prospects for Jatropha biofuels in Tanzania: an analysis with strategic niche management. Energy Policy 36:311–325
Vesa L, Malimbwi R, Tomppo E, Zahabu E, Maliondo S, Chamuya N, Nssoko E, Otieno J, Miceli G, Kaaya A (2010) NAFORMA Field Manual: biophysical survey. Report, Dar es

Salaam, Tanzania: Forestry and Beekeeping Division, Ministry of Natural Resources and Tourism and FAO-Finland Forestry Programme, Forestry Department
World Bank (1988) Madagascar—environmental action plan. vol I. In cooperation with USAID, Coop. S., UNESCO, UNDP and WWF. The World Bank, Washington, DC
World Bank (1992) Malawi—economic report on environmental policy, vol I and II. Report No. 9888-MA1. The World Bank, Washington, DC

Chapter 9
Economic Aspects of Genetic Resources in Addressing Agricultural Productivity in the Context of Climate Change

Charles Nhemachena, Greenwell Matchaya, Sibusiso Nhlengethwa and Charity R. Nhemachena

Abstract The main objective of this paper is to discuss the economic aspects of genetic resources in addressing agricultural productivity in the context of climate change and variability in Africa. The paper synthesizes the published literature related to this topic, which has not been well integrated, especially with respect to economic improvements and the use of genetic resources in Africa. The focus is to understand the nexus between climate change, genetic resources, and agricultural productivity; the economic aspects involved in the conservation and improvement of genetic resources at farm-level use; and the adoption of these technologies to address agricultural productivity. The results show that climate change affects both genetic resources and agricultural productivity. The interaction of climate change and other stressors exacerbates the vulnerability of agricultural production systems and genetic resources. The conservation and improvement of genetic resources should address the urgent need to increase investments in conservation and the development of future adapted technologies. At the farm level, the focus should be on developing distribution and dissemination systems, including raising awareness and educating farmers on the role of genetic resource technologies in addressing agricultural productivity under climate change. Furthermore, it is critical to ensure that farmers have the means to purchase the improved genetic resource technologies to be able to use and adopt them. Efforts to conserve, improve, and promote the use of genetic resource technologies in addressing agricultural productivity should

C. Nhemachena (✉) · G. Matchaya · S. Nhlengethwa
International Water Management Institute, 141 Cresswell St., Weavind Park 0184, Pretoria, South Africa
e-mail: C.Nhemachena@cgiar.org; CNhemachena@gmail.com

G. Matchaya
e-mail: G.Matchaya@cgiar.org

S. Nhlengethwa
e-mail: S.Nhlengethwa@cgiar.orgg

C.R. Nhemachena
Department of Agricultural Economics, Extension and Rural Development, University of Pretoria, Lynnwood Rd., Hatfield Campus, Pretoria, South Africa
e-mail: cmudombi@yahoo.com

R. Lal et al. (eds.), *Climate Change and Multi-Dimensional Sustainability in African Agriculture*, DOI 10.1007/978-3-319-41238-2_9

integrate the distribution, accessibility, and use of the improved technologies at the farm level and be integrated in broader adaptation and development efforts.

Keywords Economic aspects · Genetic resources · Agricultural productivity · Climate change

9.1 Introduction

Climate change and variability is detrimental to most agricultural production systems, which are expected to experience increased warming, increased frequency and length of droughts, and floods that will compound stresses in these systems (IPCC 2014a; Fujisaka et al. 2010). Changes in climate will result in the shifting of agro-climatic systems, reducing the suitability of many areas to agricultural production systems. For example, shortened agricultural production seasons result in low yield potential, affecting agricultural productivity. In addition, the changes in climate will have both direct and indirect impacts on genetic resources. The direct impacts include detrimental changes in local climatic conditions that adversely affect genetic diversity in agricultural systems; and indirectly, changes in climate will affect priority actions regarding the conservation of genetic resources (Jarvis et al. 2010). The changes in local production conditions in many agricultural production systems due to climate change will outpace the adaptive capacity of a broad range of genetic resources and will thus require concerted efforts in agricultural research and technology development (especially for genetic resources) (FAO 2015; Pascual et al. 2011; Fujisaka et al. 2010; Jarvis et al. 2010).

The Intergovernmental Panel on Climate Change (IPCC) defines adaptation as the "adjustment in human and/or natural systems in response to actual or expected climate and its effects in order to moderate harm or exploit beneficial opportunities" (IPCC 2014b). Agricultural adaptation to climate change and variability occurs at different levels: (a) at the farm level, individual farmers make adjustments to their practices in response to changes in local climatic conditions and other factors; and (b) at the national and regional levels, governments invest in agricultural research and development, climate forecasting, infrastructure etc. (Scott et al. 2012; Hassan and Nhemachena 2008). Adaptation through genetic resources at both the farm and national levels needs to be incremental (maintaining the fundamental attributes of current genetic resources) and transformational (making fundamental changes to the attributes of current genetic resources in response to predicated changes in climate). Breeding programs in agriculture have to be proactive in developing future adapted genetic technologies from current investments that are relevant to predicted future changes in local production conditions due to climate change (Jarvis et al. 2010). These technologies should encompass different traits such as drought, heat, and salinity tolerance for specific crops, livestock, and regions.

Based on a synthesis of the published literature, this paper addresses the lack of a coherent discussion with regard to the economic aspects of genetic resource conservation, improvements, and use in addressing agricultural productivity in the context of climate change. The paper expands on the earlier efforts by Asfaw and Lipper (2011), who focused on the economics of Plant Genetic Resources for Food and Agriculture (PGRFA) management for adaptation to climate change, and by Pascual et al. (2011), who focused on the economics of agrobiodiversity conservation for food security under climate change. The current paper focuses on understanding the economic aspects involved in the conservation, improvement, development, and distribution of genetic resources and technologies to adapt agricultural systems within the framework of projected changes in climate with a particular focus on the African region. For example, it is useful to understand the economic incentive structures that determine the use of and access to technologies by farmers at the farm level. On the other hand, the availability of resources and systems that affect the development and distribution of improved agricultural technologies (including genetic resources) are of concern at the national, regional, and international levels.

9.2 The Nexus Between Climate Change, Genetic Resources, and Agricultural Productivity

Agricultural production systems in Africa are exceedingly vulnerable to climate change, because they rely mainly on dryland farming systems and experience high seasonal (inter and intra) climate variability and recurrent extreme weather events (droughts and floods), which have detrimental effects on agricultural productivity and have limited capacity to adapt (Boko et al. 2007). The projected changes in temperature and precipitation are expected to significantly affect agricultural production conditions in many parts of the region, which in turn will impact agricultural productivity. Climate change also results in an increase in pests, diseases, and weeds (IPCC 2014a), which affects both crop and livestock productivity. The interaction of climate change and non-climatic factors will amplify the vulnerability of agricultural systems, especially in semi-arid and arid agricultural areas. The IPCC (2014a) reports with high confidence that increases in warming and changes in precipitation patterns will have substantial impacts on agricultural productivity in Africa. The reports also predicts that yield losses for some crops could be as high as 50–80 %.

Developments in genetic resources can help farmers adapt their production systems against the negative impacts of climate change. Genetic resources underpin the ability of agricultural production systems to withstand a range of changed and harsh production conditions (FAO 2015). For example, the Leipzig Declaration, adopted by 150 countries in 1996, emphasizes the importance of genetic plant resources in sustaining agricultural production and in addressing challenges from changing environmental conditions such as climate change (FAO 1996). However, as indicated above, the strategic reservoirs of crop and genetic livestock resources required for the adaptation of production systems are also threatened by changes in climate (FAO 2015; Pascual et al. 2011; Nnadozie et al. 2002). For instance, changing production conditions due to climate change may force farmers to abandon some varieties and breeds. If no efforts are made to conserve them, the genetic resources that can help in the future development of adapted varieties and breeds will be lost. Furthermore, wild species that do not receive management interventions to help them adapt are vulnerable and face extinction due to the predicted changes in climate (Jarvis et al. 2008). The vulnerability of these species affects the genetic resources found in them, which that are critical in breeding programs aimed at developing adapted genetic technologies for the future. In addition, narrowly adapted species and endemics are vulnerable to the direct impacts of climate change exerting pressure on genetic diversity (FAO 2015; Jarvis et al. 2010).

The destruction of genetic resources (e.g., through climate change, land degradation, natural disasters, etc.) is a threat to farming systems and sustainable development in Africa (Nnadozie et al. 2002). Most crop and livestock adaptation strategies implemented by farmers to improve agricultural productivity have implications in regard to the development and management of genetic resources. For example, various adaptation strategies such as changing crop varieties and or animal breeds or diversifying crop and animal activities (varieties and breeds)

Table 9.1 Changes in climate, impacts on agricultural productivity, and use of genetic resources in adaptation

Changes in climate	Impacts on agricultural productivity	Use of genetic resources in adaptation
Changes in rainfall patterns	Projected increases in the variability of seasonal rainfall as well as increased drying in most parts of Africa affects agricultural potential and productivity	Strengthen the development, distribution, and use of moisture-stress-resistant genetic resources
Changes in temperature	Most parts of tropical and sub-tropical agricultural regions in Africa are expected to experience increased warming, which, together with increased dry conditions, affect agricultural productivity	Strengthen the development, distribution, and use of heat-tolerant genetic resource technologies
Changes in patterns and incidence of pests and diseases	Increased incidence of pests & diseases affect crop/livestock outputs	Strengthen the development, distribution, and use of pest- and disease-resistant genetic resource technologies
Changes in climate variability and incidence of extreme events	The projected increases in the frequency and intensity of extreme events such as droughts threaten agricultural productivity	Strengthen the development, distribution, and use of drought-tolerant genetic resource technologies

Source Scott et al. (2012)

require adapted genetic improvements that should be easily accessible for use by farmers. Asfaw and Lipper (2011) argue that broadening the genetic resource base for farmers is critical to enhancing adaptation to climate change and improving agricultural productivity in the future. Therefore, there is an urgent need to step up efforts to conserve genetic resources as a safeguard against the potential impacts of climate change on agricultural productivity in the future. The demand for both genetic crop and genetic animal resources adapted to changed production environments will increase with climate change (Jarvis et al. 2015; Pilling and Hoffmann 2015). Table 9.1 summarizes the projected changes in climate, its impacts on agricultural productivity, and the potential use of genetic resources to help farmers adapt.

9.3 Economic Aspects of Addressing Agricultural Productivity Through the Conservation, Improvement, and Use of Genetic Resources

9.3.1 Economic Aspects of the Conservation and Improvement of Genetic Resources

The discussion above shows how changes in rainfall patterns, increases in temperature and atmospheric carbon dioxide, changes in the patterns and incidence of pests and diseases and of extreme weather events (floods and droughts) affect agricultural productivity. This section focuses on the economic aspects of the conservation and improvement of genetic resources for adaptation to climate change.

Current conservation of genetic resources is critical for future adaptation of agricultural systems Genetic resources are central to the efforts by farmers and researchers in addressing agricultural productivity due to changing local production conditions arising from climate change. The conservation of diverse genetic resources that can survive and produce in future climates is critical for breeding programs and the development of adaptive technologies. For example, genetic plant and animal resources are required for the development of improved and adapted varieties and breeds, including the introduction of new species to help farmers adapt to climate change (FAO 2015, 2011; Jarvis et al. 2015; Pilling and Hoffmann 2015). Therefore, policies and programs that are focused on strengthening the current conservation of genetic resources are critical for the future adaptation of agricultural systems.

Systems should be put in place to ensure that genetic resources are not lost due to negligence. There is urgent need to improve and strengthen in situ and ex situ genetic resource conservation programs "for domesticated species, their wild relatives and other wild genetic resources" as well as policies to promote the sustainable use of genetic resources (FAO 2015; Jarvis et al. 2015; Scott et al. 2012). Rigorous efforts should be made to conserve genetic resources at the national and sub-regional levels by addressing critical challenges such as "limited financial resources, lack of institutional capacity, poor articulation of and support of conservation and sustainable use goals, limited technological infrastructure, and poor institutional linkages with development sector" (Nnadozie et al. 2002, p. 29). Despite the availability of gene banks around the world, further efforts are required to conserve more wild relatives and landraces to broaden the genetic diversity of crops and animal species that will be used for the development of future adapted genetic technologies.

Balancing between the ex situ and in situ conservation of genetic resources Genetic resources can be conserved ex situ (outside their natural setting) or in situ (within their natural setting). The challenge for policy and decision makers is to balance between the ex situ and in situ conservation of genetic resources. Rubenstein et al. (2005), Pascual et al. (2011) discussed some of the advantages and disadvantages of the ex situ versus in situ conservation of genetic resources; these are summarized in Table 9.2 below and should be taken into consideration when planning and implementing such programs and policies at different levels.

Strengthen regional and international collaboration in the conservation and development of genetic resources The extent of climate change in some regions that will push agricultural production systems beyond historical experiences means that the affected regions would mainly rely on "outside" genetic resources developed from either current species under production or completely new species (FAO 2015; Jarvis et al. 2015; Pilling and Hoffmann 2015; Fujisaka et al. 2010). Most of these would be derived from international sources, meaning that the global movement of genetic resources would be critical for farmers to maintain and/or improve their agricultural productivity. The demand for genetic resources is expected to increase internationally to support breeding programs focusing on developing varieties and breeds adapted to the harsh production conditions associated with climate change. The challenges posed by predicted changes in climate require the strengthening of collaboration efforts among plant breeders, seed systems, and conservation stakeholders. The conservation of genetic resources in Africa is currently well supported at the national and regional levels; however, there is limited material from the region in major international gene banks that support most public breeding programs (Burke et al. 2009). The efforts of the Consultative Group for International Agricultural Research (CGIAR) centers have been critical as sources of genetic material and as a host of some of the science for the development of genetic resource technologies in developing countries, such as in Africa. In addition, regional policies and programs such as the Southern Africa Development Community (SADC) Plant Genetic Resource Centre and SADC seed harmonization program led by the SADC seed project based in Lusaka, Zambia, are examples of regional efforts to bridge this gap. There is also a need for the increased active involvement of local stakeholders such as farmers and researchers in both the conservation and improvement of genetic resources for adaptation to climate change.

Table 9.2 Advantages and disadvantages of the ex situ and in situ conservation of genetic resources

Ex situ conservation		In situ conservation	
Advantages	Disadvantages	Advantages	Disadvantages
Costs generally centralized	Certain types of germplasm not readily conserved	Genetic resources used to produce valuable product	Costs borne by farmers (for landraces)
Can preserve large amounts of diverse germplasm	Regeneration can be costly, time-consuming	Evolutionary processes continue	May reduce on-farm productivity
Germplasm can be readily accessed by more breeders	Potential for genetic "drift" can reduce integrity of collection	May better meet the needs of certain farmers such as the livelihoods of many poor and marginalized farmers	Requires land
High-security storage impervious to most natural disasters	In practice, many collections lack the resources needed to organize, document, and maintain their samples	More efficient for some germplasm, e.g. animals, or crops that reproduce vegetatively	Farmer selections may not preserve targeted diversity
		Existing wild relatives can be preserved without collection	Loss of wild relatives when land use changes
Uses low costs conservation instruments		Genetic resources might be lost through changing environmental or economic conditions	
	Ex situ conservations refer to the static approach of maintaining information stock	In situ conservation is dynamic and integrates ecological and social relationships	
Enhances long-term adaptation to climate change	Fail to provide short-term insurance at the farm level		

Source Rubenstein et al. (2005), Pascual et al. (2011)

Timely development of improved genetic resources The changes in climate described above will make the current genetic resources suboptimal for future growing environments. Given the fact that developing new varieties and breeds takes a long time, the process of improving varieties and breeds for future climates should start many years in advance (FAO 2015). Timely investments in the development of adapted genetic technologies are crucial for farmers to improve and/or maintain their future productivity levels in the face of climate change.

Documentation of genetic resources There is also a critical need to document genetic resources, including the areas where they are found, their traits (e.g., resistance to drought, pests, or diseases), management options, how they can be used at the farm level, etc. (Jarvis et al. 2015). The diversity of these genetic resources is critical in the development of improved and adapted varieties and breeds that farmers can use in changed future conditions.

Adaptation through genetic resources should be part of broader development efforts Burke et al. (2009) argue that traditional breeding programs will be insufficient to adapt African agriculture. Genetic resources will be critical to sustaining the future of both traditional and formal breeding programs in the face of the expected changes in climate. However, it is important to note that the development and use of genetic resource technologies should be advanced as part of broader adaptation programs that address institutional, social, physical, and infrastructure needs; ecosystem services and environmental needs; and the financial and capacity needs of the affected communities (IPCC 2014a).

Estimating the benefits and costs of conservation and improvements in genetic resources to guide adaptation planning and investments The activities involved in the conservation and improvement of genetic resources are mainly public in nature (Rubenstein et al. 2005) and require more public investments. However, given the scarcity of public resources and an increasing need for accountability, it is critical to estimate the benefits and costs of conservation and improvements in genetic resources and technologies to inform planning and decision making on the alternative allocations of public funds and in setting priorities in agricultural adaptation.

9.3.2 Economic Aspects of the Use and Adoption of Improved Genetic Resources at the Farm Level

The interaction of various factors (institutional, political, social, cultural, biophysical, cognitive, and behavioral as well as gender-related) affects adaptation at national and local levels (IPCC 2014a). The barriers at the national level include, for example, the availability of resources and systems to develop and/or ensure that adapted technologies are readily available within the agricultural systems for farmers to use. At the local level, various factors constrain the capacity to respond to climate change and the use of available adapted technologies (such as improved genetic resource technologies). This section discusses some of the economic aspects that should be considered in efforts to promote the adoption and use of genetic resource technologies in addressing agricultural productivity.

Lack of institutional requirements for new technologies Ludi et al. (2012) argue that the lack of institutional requirements for new technological interventions contributes to low levels of adaptation at the local level. It is therefore critical that the development and promotion of genetic resource technologies incorporate institutional requirements across different agricultural systems to encourage farmer adoption and use. At the policy level, the focus areas include investments in national and regional genetic resource improvements, the development of input systems and other infrastructure to facilitate distribution, and access to these technologies by farmers.

Integration of formal and informal input distribution systems The informal structures and genetic systems among smallholder farmers in Sub-Saharan Africa remain a critical part of providing agricultural adaptation strategies (Westengen and Brysting 2014; Westengen et al. 2014; Louwaars and de Boef 2012). For example, smallholder farmers in Sub-Saharan Africa rely mainly on informal seed systems, with a small percentage of the farmers using formal seed systems that directly provide improved and hybrid seeds from plant breeding (Langyintuo et al. 2010). The availability of quality genetic resources adapted to prevailing climatic and other conditions and their utilization by farmers are important for improving agricultural productivity. Therefore, future adaptation strategies based on genetic resources should integrate the informal and formal seed structures and input systems that are inherent in smallholder farming systems in Africa to ensure easy accessibility. Policy efforts should focus on strengthening the distribution of and access to adapted varieties and breeds through both formal and local seed and input market systems.

Farmer awareness, capacity development, and effective extension services The low utilization and adoption of agricultural innovations and technologies in Africa is argued to be partly due to a lack of farmer awareness as well as adequate and effective extension services for farmers (IPCC 2014a; Hassan and Nhemachena 2008). In the conservation and development of genetic resources, systems of distribution, farmer awareness, and education are some of the critical areas that should be addressed to encourage the adoption and use of improved genetic resource technologies in adapting to climate change. At the farm level, the focus is on raising farmer awareness, building the capacity to use and facilitate the means to acquire the improved genetic resources for local use and adoption. In addition, sustainability is an important issue; basically, farmers shun some of the improved technologies because they are worried about losing local varieties/technologies that have been passed down through generations. There is need for awareness raising and capacity building for farmers to enable them to integrate local technologies and new improved genetic resource technologies.

Access to relevant information The IPCC (2014a) argues that a lack of reliable and accessible climate information remains a critical constraint to adaptation although it is not in itself sufficient to induce behavioral change and guarantee adaptation. The promotion of improved and adapted genetic resource technologies should critically address the reliability and accessibility of information on respective technologies and how they help address agricultural productivity in the context of climate change. This should be integrated within local distribution networks and local knowledge systems. The use of genetic plant resource technologies at the local level also requires the integration of new knowledge and local knowledge, including promotion through local channels (information and distribution) to facilitate change in farmer behavior and willingness to adopt.

Rights to ownership and access among women farmers One area that needs to be addressed in terms of access to genetic resource technologies is the rights to and access by women farmers. Despite their significant contribution to agricultural production in African systems, women face constrained rights and access to land and natural resources, credit, information and new ideas, etc. (IPCC 2014a). The success of conservation efforts at the farm level and use of genetically adapted technologies will be based on critical considerations of the rights to ownership and access among women farmers.

Addressing various adoption constraints in using agricultural technologies The various factors that affect the adoption and use of agricultural technologies at the farm level are discussed in the literature (e.g., IPCC 2014a; Scott et al. 2012; Below et al. 2010; Nhemachena 2009; Hassan and Nhemachena 2008). Examples of these factors include poor resource endowments, lack of readily available information on alternatives, poor input and output market access, poverty, etc. Efforts to promote the adoption and use of adapted genetic resource technologies should focus on addressing these constraints, particularly among smallholder farmers in Africa.

Other strategies discussed in Pascual et al. (2011) to facilitate farmer use of the improved genetic resources and technologies include participatory plant and animal improvement programs with the farmers to develop technologies that address the needs of the farmers; national research and extension services that can develop and widely distribute genetic resource technologies, including educating farmers; developing local registers of information about varieties and breeds; and strengthening local market and distribution channels to reduce transaction costs. It is also important to note that government expenditures on R&D are low across Africa, constraining the generation of Africa's own genetic resource technologies, and that some of the improved technologies developed from other regions are not adapted to local agro-climate systems. Increased investments are required to develop locally adapted genetic resource technologies in Africa.

9.4 Conclusion

The present paper reviewed and discussed the economic aspects of genetic resources in addressing agricultural productivity in the context of climate change. The primary focus was on understanding the economic aspects involved in the conservation, development, and distribution of genetic resources and technologies to adapt agricultural systems within the framework of projected changes in climate. The issues discussed include the nexus of climate change, genetic resources, and agricultural productivity; the economic aspects of conservation and improvement of genetic resources; and the economic aspects of the adoption and use of genetic resources and technologies at the farm level.

The vulnerability of agricultural production systems is worsened by the interactions of climate change and non-climatic stressors in most African farming systems. These changes cause shifts in the suitability of local agro-climatic systems for agricultural production, affecting both agricultural productivity and strategic reservoirs of genetic resources. Despite the threats posed by climate change, diverse genetic resources provide important inputs to help farmers and agricultural systems withstand the range of changed and harsh production conditions. Therefore, the destruction of the diversity of genetic resources is a serious threat to the agricultural productivity and sustainability of farming systems in Africa. Efforts to conserve and improve genetic resources should be stepped up to safeguard the diversity of genetic resources useful for the development of adapted agricultural technologies to support adaptation against changes in climate.

The economic aspects of the conservation and improvement of genetic resources and respective technologies include strengthening current efforts to conserve the diverse genetic resources that can survive and produce in future climates to support breeding programs and genetic resource technology development. The collaboration of diverse stakeholders at regional and international levels in the conservation and development of genetic resources and technologies for future adaptation, including the active involvement of farmers, policy makers, investors, researchers, etc., also needs to be stepped up. In addition, there is an urgent need for timely investments in the conservation and improvement of genetic resources and technologies (given the long time required to develop new varieties and breeds) to prepare farmers to adapt their future production systems to improve and/or maintain their productivity levels in the face of climate change. Efforts should also be made to strengthen and improve the documentation of genetic resources, including the areas where they are found, the traits they have (e.g., resistance to pests, diseases, drought), management options, how they can be used at the farm level, etc. The conservation and improvement of genetic resources for adaptation need to be part of a broader adaptation and development efforts in farming systems. It is also critical to estimate the benefits and costs of the conservation of and improvements in genetic resources to guide adaptation planning and investments.

At the farm level, the economic aspects for consideration entail addressing the various factors that constrain the ability of farmers to conserve as well as adopt and use agricultural genetic resource technologies. Examples of such efforts include ensuring that the conservation, improvement, and distribution of genetic resource technologies integrate institutional requirements to encourage active farmer engagement and use. There is also a need to strengthen the distribution of and access to adapted varieties and breeds through both the formal and local input market systems inherent in smallholder farming systems in Africa. Moreover, it is crucial to invest in the implementation of effective farmer training and extension services to raise awareness and to build and strengthen the capacity of farmers to use genetic resource technologies in addressing agricultural productivity in the context of climate change. Furthermore, measures should be put in place to ensure the reliability and accessibility of information (integrated within the local distribution networks and local knowledge systems) on the respective genetic resource technologies and how they can help address agricultural productivity in the context of climate change.

References

Asfaw S, Lipper L (2011) Economics of PGRFA management for adaptation to climate change: a review of selected literature. Commission on Genetic Resources for Food and Agriculture. FAO, Rome, Italy

Below T et al (2010) Micro-level practices to adapt to climate change for african small-scale farmers. IFPRI Discussion Paper 00953, International Food Policy Research Institute, Washington, D.C

Boko M et al (2007) Africa. Climate change 2007: impacts, adaptation and vulnerability. In: Parry ML et al (eds) Climate change 2007: impacts, adaptation and vulnerability. Contribution of Working Group II to the Fourth Assessment Report of the Intergovernmental Panel on Climate Change, Cambridge University Press, Cambridge UK, pp 433–467

Burke MB et al (2009) Shifts in African crop climates by 2050 and the implications for crop improvement and genetic resources conservation. Glob Environ Change 19(3):317–325

FAO (1996) Global plan of action for the conservation and sustainable utilization of plant genetic resources for food and agriculture and the Leipzig Declaration. Food and Agriculture Organization of the United Nations, Rome, Italy

FAO (2011) Genetic resources for food and agriculture and the international climate change policy agenda. Commission on Genetic Resources for Food and Agriculture. Thirteenth Regular Session, Food and Agriculture Organization of the United Nations, Rome, Italy

FAO (2015) Coping with climate change—the roles of genetic resources for food and agriculture. Food and Agriculture Organization of the United Nations, Rome, Italy

Fujisaka S et al (2010) The impact of climate change on countries' interdependence on genetic resources for food and agriculture. An executive summary. Bioversity International, Rome, Italy

Hassan R, Nhemachena C (2008) Adaptation to climate change in Africa: multinomial choice analysis of determinants of farm strategies. AfJARE 2(1):83–104

Intergovernmental Panel on Climate Change (2014a) Climate change 2014: impacts, adaptation, and vulnerability. Working Group II Contribution to the IPCC 5th Assessment Report. Cambridge University Press, Cambridge UK

Intergovernmental Panel on Climate Change (2014b) IPCC WGII AR5 glossary. Intergovernmental Panel on Climate, Cambridge University Press, Cambridge UK

Jarvis A et al (2008) The effect of climate change on crop wild relatives. Agric Ecosyst Environ 126:13–23

Jarvis A et al (2010) Climate change and its effect on conservation and use of plant genetic resources for food and agriculture and associated biodiversity for food security. Thematic background study. Food and Agriculture Organization of the United Nations, Rome, Italy

Jarvis A et al (2015) Plant genetic resources for food and agriculture and climate change. In: FAO (2015) Coping with climate change—the roles of genetic resources for food and agriculture. Food and Agriculture Organization of the United Nations, Rome, Italy

Langyintuo et al (2010) Challenges of the maize seed industry in eastern and southern Africa: a compelling case for private–public intervention to promote growth. Food Policy 35(4):323–331

Louwaars NP, de Boef WS (2012) Integrated seed sector development in Africa: a conceptual framework for creating coherence between practices, programs, and policies. J Crop Improv 26(1):39–59

Ludi E et al (2012) Changing focus? How to take adaptive capacity seriously. Evidence from Africa shows that development interventions could do more. ODI Briefing Paper 71, Overseas Development Institute, London, UK
Nhemachena C (2009) Agriculture and future climate dynamics in Africa: impacts and adaptation options. PhD Thesis, University of Pretoria
Nnadozie et al (2002) Plant genetic resources in Africa's renewal, policy, legal and programmatic issues under the New Partnership for Africa's Development. International Environmental Law Research Centre, Geneva, Switzerland
Pascual U et al (2011) The economics of agrobiodiversity conservation for food security under climate change. Economía Agraria y Recursos Naturales 11(1):191–220
Pilling D, Hoffmann I (2015) Animal genetic resources for food and agriculture and climate change. In: FAO (2015) Coping with climate change—the roles of genetic resources for food and agriculture. Food and Agriculture Organization of the United Nations, Rome, Italy
Rubenstein KD et al (2005) Crop genetic resources an economic appraisal. Economic Information Bulletin Number 2, United States Department of Agriculture
Scott M et al (2012) Agricultural adaptation to a changing climate: economic and environmental implications vary by U.S. region. ERR-136, U.S. Department of Agriculture, Economic Research Service
Westengen OT, Brysting AK (2014) Crop adaptation to climate change in the semi-arid zone in Tanzania: the role of genetic resources and seed systems. Agric Food Secur 3(3):1–12
Westengen OT et al (2014) Modern maize varieties going local in the semi-arid zone in Tanzania. BMC Evol Biol 14(1):1–12

Chapter 10
Soil and Nutrient Losses and the Role of Gender in Land Degradation in Southwestern Uganda

C.L. Kizza, J.G.M. Majaliwa, R. Kunihira, G. Gabiri, A. Zizinga, E. Sebuliba, J. Nampijja and M.M. Tenywa

Abstract Land degradation is rapidly increasing in Southwestern Uganda, leading to the loss of soil productivity, increased water-body pollution and reduced vegetation and wetland cover. Soil erosion is a leading observable indicator of land degradation. This study evaluated runoff, soil, and nutrient losses under the various land uses and landscape positions within the L. Bunyonyi catchment, which covers approximately 334 km^2 in the Kabale District in Southwestern Uganda. Erosion plots were established for the four major land uses: perennial crops, annual crops, woodlots, and grazing. Erosion trap plots measuring 15 m × 2 m were established for each land use and were replicated four times at each of the landscape positions of summit, shoulder, mid-slope, and foot slope. A pipe sampler was used to collect approximately 1 % of the plot runoff. Runoff was measured using a measuring cylinder the day after a rain event. All the runoff with its soil sediments was collected in plastic bottles and delivered to a laboratory for analysis of soil and nutrient losses. The role of women in land management and degradation was assessed using a social survey questionnaire administered to 120 adult female respondents. The most runoff was observed in the annuals (175 m^3 ha^{-1} $year^{-1}$), followed by woodlots and perennials (159 m^3 ha^{-1} $year^{-1}$) and (141 m^3 ha^{-1} $year^{-1}$), respectively, and, finally, grazing land (136 m^3 ha^{-1} $year^{-1}$). However, the most soil loss was observed in the woodlots (431 kg ha^{-1} $year^{-1}$), followed, in decreasing order, by grazing land (143 kg ha^{-1} $year^{-1}$), perennials (78 kg ha^{-1} $year^{-1}$), and annuals (72 kg ha^{-1} $year^{-1}$). The mid-slope had significantly more soil

C.L. Kizza · G. Gabiri · A. Zizinga · M.M. Tenywa (✉)
Department of Agricultural Production, Makerere University,
P.O. Box 7062, Kampala, Uganda
e-mail: tenywamakooma@yahoo.com

J.G.M. Majaliwa
Department of Geography, Geo-Informatics and Climatology,
Makerere University, P.O. Box 7062, Kampala, Uganda

R. Kunihira
Department of Geography, Kenyata University, P.O. Box 43844-00100, Nairobi, Kenya

E. Sebuliba · J. Nampijja
LesRams Consult Ltd., P.O. Box2 9832, Kampala, Uganda

R. Lal et al. (eds.), *Climate Change and Multi-Dimensional Sustainability in African Agriculture*, DOI 10.1007/978-3-319-41238-2_10

loss (274 kg ha^{-1} $year^{-1}$) than foot slopes (152 kg ha^{-1} $year^{-1}$). Soil loss was lowest at the summit shoulders (144 kg ha^{-1} $year^{-1}$). Runoff did not follow a similar pattern to that of soil loss. Although the variations were not statistically significant, more runoff was observed at the foot slopes (163 m^3 ha^{-1} $year^{-1}$) than the summit shoulder (157 m^3 ha^{-1} $year^{-1}$) and mid slope (138 m^3 ha^{-1} $year^{-1}$). The significantly high nutrient losses recorded in woodlots were more than for all other land uses. Women do not own land in Uganda but are primarily responsible for domestic food security and contribute more than 70 % of domestic labor. On average, women spend more than 8 h daily in the field providing manual labor. Occasionally, they engage in hired labor to supplement domestic agricultural activities and access other necessities. Most soil-conservation and soil-fertility improvement practices are too laborious and costly for women to undertake. This results in inevitable environmental and resource depletion problems and, subsequently, land degradation. The heavy burden of sustaining a rapidly increasing population on a diminishing resource base is accelerating land degradation, for which women share ample blame. If men fully participated in land management, especially contributing much-needed labor and resources, the state of land degradation in southwestern Uganda would be much better.

Keywords Catena · Soil erosion · Soil fertility · Hydraulic conductivity · Bulk density · Land management

10.1 Introduction

Traditionally in Southwestern Uganda, crops have been grown on terraces along steep hill slopes. However, a recent trend is to replace crops in most of these areas with tree woodlot plantations, primarily of Eucalyptus. Rapid diversification from cropland to tree plantation is driven by land degradation caused by loss of topsoil and, in some cases, exposure of the parent hard rock, resulting in reduced crop productivity. In other words, woodlot planting represents an alternative economic activity for severely degraded soils. Soil and nutrient losses occurs mostly through sheet and rill water erosion.

Globally, soil erosion is the most visible form of land degradation (Floor 2000; National Environment Management Authority (NEMA) 2001; Gobin et al. 2002) and a leading concern for hilly and mountainous farming systems (Price et al. 2011). Soil erosion, therefore, is a widely known cause of low productivity (Pender et al. 2004) through topsoil, water, and nutrient losses (Gorji et al. 2008). Water erosion in particular, is known to cause severely declining agricultural production, decreased food security, and sedimentation risk for water bodies (Meshesha et al. 2012).

Terracing should improve soil stabilization but is being destroyed by attempts to increase crop production (Nkonya 2002). Farmers started such attempts after realizing that crops close to the banks of the lower end of the terraces were much better than those higher up the terrace. The relative difference in crop performance along terraces is due partly to the nutrient gradient as a result of downward movement. Crop growing in southwestern Uganda is dominated by women (Puhalla 2009), whose capacity to implement soil and water conservation practices is severely constrained by limited labor and financial resources to hire the necessary labor. Attempts to increase food security for the growing population through the use of less arable land greatly hinder women's ability to achieve a decent living. Improvement of soil and water conservation practices in southwestern Uganda is imperative. Doing so will greatly minimize the rate of the conversion of marginal land, especially wetlands, for crop production, which has significantly contributed to climate change in this region. Hitherto, these wetlands used moderated and maintained the characteristic cool climatic environment of Uganda, which is steadily passing into history. However, to meaningfully and sustainably undertake soil and water conservation measures, it is a necessary to have a clear understanding of the current contribution of soil erosion to land degradation and the limitations faced by women, who are the prime users of land, in conserving soil.

The objectives of this study are to assess runoff, soil, and nutrient losses under the different land uses and landscape positions within the L. Bunyonyi catchment and to establish the role of women in land management and land degradation.

10.2 Materials and Methods

10.2.1 Study Area

The study was carried out in the L. Bunyonyi catchment (Fig. 10.1), Kabale District, which covers 332 km^2. Kabale District is located 1°20′42″ South, 29°51′1″ East. This highland area has hills with very steep slopes and deep valleys and an altitude of 1700–2200 m above sea level. The minimum and maximum temperatures are 12.8 and 21.4 °C, with a mean of 16.1 °C. The mean annual rainfall is 884 mm (Kristan et al. 2008). The area is densely cultivated from hilltops to the valleys, including along streambanks and lower terrace boundaries. The major crops grown in the area are potato, beans, maize, sorghum, barley, and wheat. Once exhausted and severely degraded, the soils of arable hillslopes are judged to be unproductive and then planted with Eucalyptus trees (FARA 2009).

10.2.2 Soil Erosion Measurements

Site characterization of all the land uses was carried out, examining the physio-chemical properties of the study area. A soil auger was used to pick composite soil samples for each land use at two depths: 0–15 and 15–30 cm. These samples were air-dried and analyzed for pH, organic matter (OM), percentage nitrogen (N), available phosphorus (P), and exchangeable bases (K, Ca, Mg, and

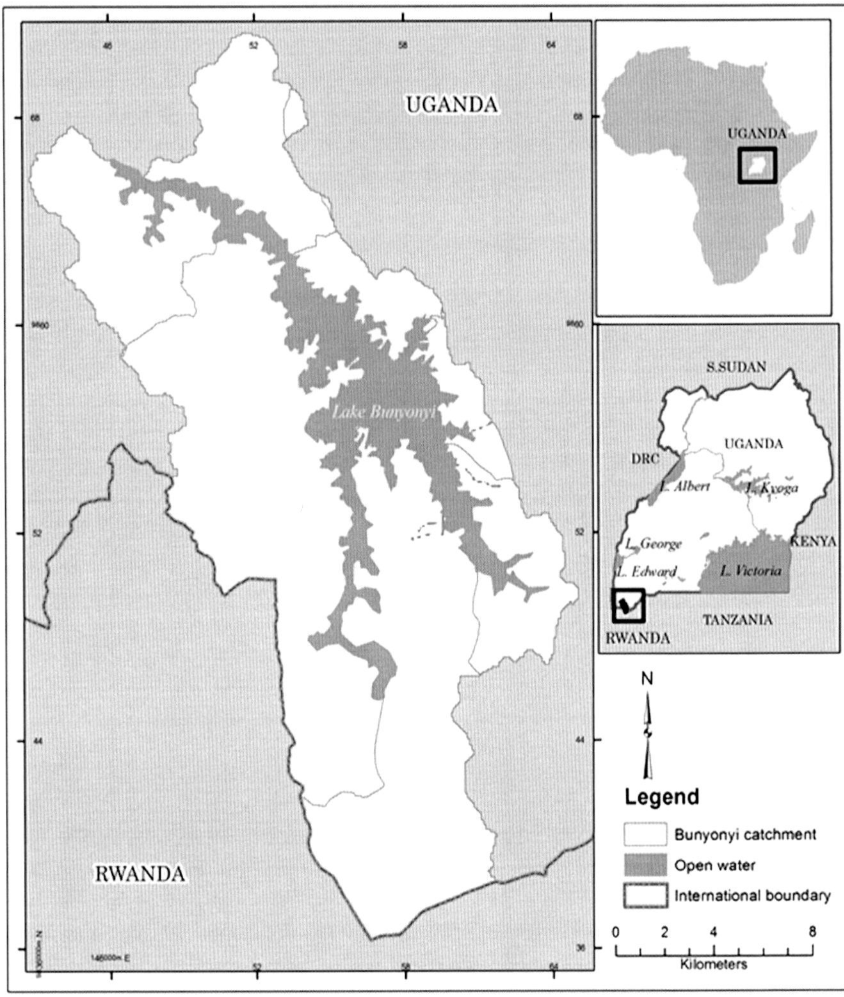

Fig. 10.1 Study area of L. Bunyonyi catchment in Kabale District, southwestern Uganda

Na). In addition, round soil cores were used to take soil samples to determine soil bulk density and saturated hydraulic conductivity (Ksat). Ksat was determined using the constant head method (Klute 1986).

Erosion study plots were established for the four major identified land uses: perennial crops, annual crops, woodlots, and grazing. The four land uses were studied at three landscape positions: hill summit shoulders, mid-slope, and foot slopes. At every landscape position for each land use, erosion trap plots measuring 15 m × 2 m were established (Fig. 10.2), and all treatments were replicated four times. After construction of the erosion trap plots, each plot, which was estimated to collect approximately 1 % of the runoff, was calibrated individually to establish the

Fig. 10.2 Erosion plot constructed in a woodlot in L. Bunyonyi catchment

plot runoff coefficient. This coefficient is the ratio of collected water to the total amount of water trapped. Rain-event runoff was measured using a measuring cylinder a day after a rain event. All the runoff with its soil sediments was collected into plastic bottles and delivered to a laboratory to test soil and nutrient losses. Event losses were summed to obtain seasonal data and then annual data. The runoff reciprocal of the coefficient was used to compute the total plot losses. The soil and nutrient losses were calculated next through filtration of the runoff samples using a Whatman No. 1 paper filter. The filtered soil was analyzed for soil loss by oven-drying half of the collection and air-drying the second half for nutrient loss analysis. The major nutrients—nitrogen, phosphorus, and potassium—were targeted and analyzed using standard laboratory operating procedures for each nutrient. Nutrient content was also determined in the runoff filtrate and later summed to assess total losses related to each land use.

Women's role in land management and degradation was assessed using a social survey questionnaire. The structured questionnaire was administered to 120 women respondents randomly selected from the sub-counties within the study catchment area. Erosion and gender data were analyzed through analysis of variance performed using Genstat 13th Version and the statistical package for social scientists (SPSS), respectively. Differences in erosion-study means were measured using Fisher's least significant difference (LSD = 0.05) test.

10.3 Results

10.3.1 Physio-Chemical Characteristics of Land Uses in the L. Bunyonyi Catchment

The variations of chemical parameter interactions between different land uses and landscape positions are found in Table 10.1. Soils in the L. Bunyonyi catchment were mostly acidic, but the level of acidity varied by catena and land use. In general, soils from the foot slopes were more acidic than those from the summit shoulders and mid-slope. At every landscape position, the annuals exhibited the lowest amounts of organic matter compared to the other two land uses (perennials and woodlots). With the exception of annuals on the mid-slopes, whose organic matter was below the critical value, all the other land uses had sufficient organic matter. The highest amount of organic matter was detected in the woodlots on the mid-slopes, followed by perennials at the foot slopes.

10.3.2 Saturated Hydraulic Conductivity (Ksat) and Bulk Density within Land Uses in the L. Bunyonyi Catchment

There was no clear trend in the Ksat among land uses, landscape positions, and soil depth (Table 10.2). However, Ksat ranged from moderate (21 cm h^{-1}) to very high conductivity (≥120 cm h^{-1}). Similarly, bulk density did not vary much between land uses, landscape positions, and soil depth. Except for grazing lands at the foot

Table 10.1 Chemical characterization of the different land uses within the L. Bunyonyi catchment

Landscape position	Land use	pH	OM	N	P	K	Ca	Na
			%		mg kg^{-1}	cmol kg^{-1}		
Foot slope	Annuals	5.3	4.32	0.36	9.90	0.51	2.71	0.05
	Woodlots	4.3	5.14	0.23	7.60	0.61	3.03	0.06
	Perennials	4.7	6.57	0.40	7.80	0.62	1.47	0.05
Mid-slope	Annuals	5.9	2.62	0.25	7.40	0.22	3.50	0.06
	Woodlots	5.0	7.31	0.15	4.40	0.58	3.71	0.07
	Perennials	5.2	4.40	0.38	6.37	0.46	1.23	0.06
Summit shoulder	Annuals	5.3	3.08	0.28	6.10	0.53	3.23	0.05
	Woodlots	4.7	3.72	0.43	7.40	0.22	1.79	0.05
	Perennials	6.0	4.80	0.20	4.40	0.95	1.28	0.06
	$LSD_{0.05}$	1.67	2.11	0.21	2.32	0.21	1.01	0.01

Table 10.2 Variation of Ksat (cm h^{-1}) in different land uses, landscape positions, and soil depth

Land use	Summit		Mid-slope		Foot-slope		$LSD_{0.05}$
	0–15 cm	15–30 cm	0–15 cm	15–30 cm	0–15 cm	15–30 cm	
Annuals	103	209	59	32	39	26	86.3
Perennials	21	83	70	130	169	116	
Woodlot	184	178	63	203	77	111	
Grazing	151	68	137	20	4	12	
$LSD_{0.05}$	74.8						

Table 10.3 Bulk density variation among land uses, landscape position, and soil depth within the L. Bunyonyi catchment

Land use	Summit		Mid-slope		Foot-slope		$LSD_{0.05}$
	0–15 cm	15–30 cm	0–15 cm	15–30 cm	0–15 cm	15–30 cm	
Annuals	0.818	0.883	1.032	0.902	0.902	1.145	0.067
Perennials	0.898	0.913	0.983	0.905	0.791	1.288	
Woodlot	0.938	0.848	0.993	1.077	0.791	1.008	
Grazing	0.883	1.057	1.187	0.760	1.621	1.510	
$LSD_{0.05}$	0.064						

slopes, all the other land uses at all landscape positions had very low to low soil bulk density (Table 10.3).

10.3.3 Runoff, Soil and Nutrient Losses by Different Land Use Type in the L. Bunyonyi Catchment

Runoff varied significantly ($p < 0.05$) according to land-use height and landscape position. The highest runoff came from perennials and annuals grown on foot slopes and mid-slopes, respectively, while the least amount of runoff came from perennials at the summit shoulders (Fig. 10.3). Except for perennials, other land uses on the summit shoulders experienced the same amount of runoff. At mid-slope, annuals were associated with significantly ($p < 0.05$) higher amounts of runoff than the other three land uses. No significant variations were recorded for the other land uses. On the foot slope, significantly ($p = 0.05$) higher rates of runoff came from perennials than woodlots, annuals, or grazing.

Figure 10.4 illustrates how soil loss varied by land use and landscape position. Generally, woodlots experienced significantly ($p < 0.05$) the highest soil losses, followed by grazing land at the foot slopes. For annual and perennial production, similar levels of soil loss were recorded. However, significant ($p = 0.05$) variations

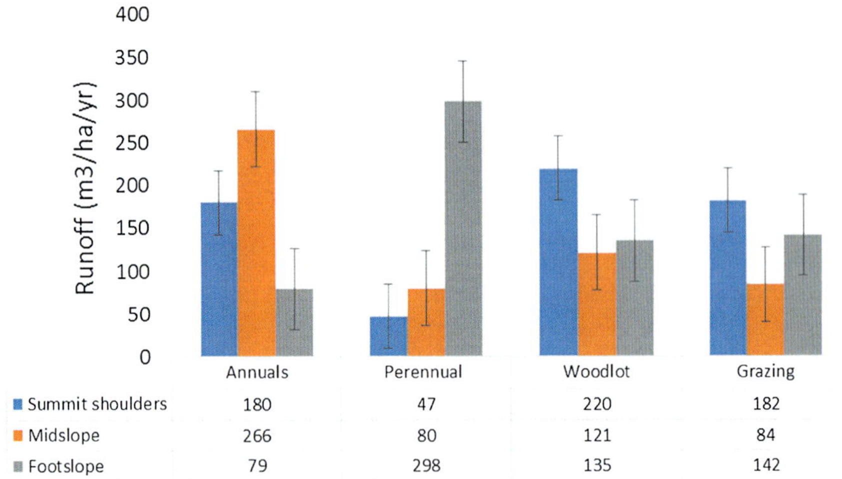

Fig. 10.3 Runoff within different land uses

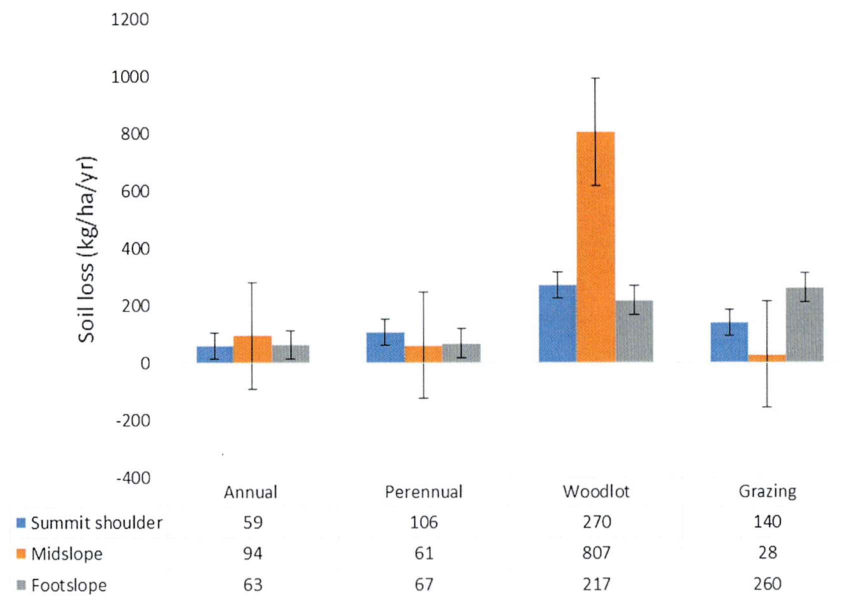

Fig. 10.4 Variation of soil loss from different land uses and landscape positions

in soil loss occurred in woodlots and grazing lands. Significantly more soil loss was observed in woodlots at mid-slope, while for grazing lands, the most soil loss occurred at the foot slope.

Table 10.4 Areas of women involvement in soil and water conservation activities

Land use	N	P	K	Mg	Ca	Total
	g ha^{-1} year^{-1}					g ha^{-1} year^{-1}
Annuals	715	139	828	277	628	2587
Perennials	389	213	1052	309	565	2529
Woodlot	1879	684	6833	1679	2528	13,603
Grazing	1005	292	1561	560	771	4189
$LSD_{0.05}$	209	114	602	256	493	

Nutrient losses for different land uses are presented in Table 10.4. Clearly, woodlots experienced the highest (significant at $p < 0.05$) levels of nutrient losses, followed by grazing lands. Nutrient losses related to the production of annuals and perennials did not vary significantly ($p < 0.05$). For all land uses, more potassium was lost than other nutrients (Table 10.4).

10.3.4 Gender in Land Management and Land Degradation

Figure 10.5 illustrates the extent to which women are involved in various land-use activities: crop farming, agroforestry, forestry, and animal rearing. Crop farming is clearly the leading activity, with 73 % of women involved in agriculture occupied in it. The next most prevalent form of activity is agroforestry (21 %). Women's involvement in tree plantation and animal production is minimal, with only 1 % involved in tree plantation and 5 % in livestock production. As indicated in Fig. 10.6, women are the most likely group to undertake soil and water conservation practices (73 %). Only a small percentage of women (27 %) indicated that they fail to practice soil and water conservation measures, while men contribute only 16 % of these activities.

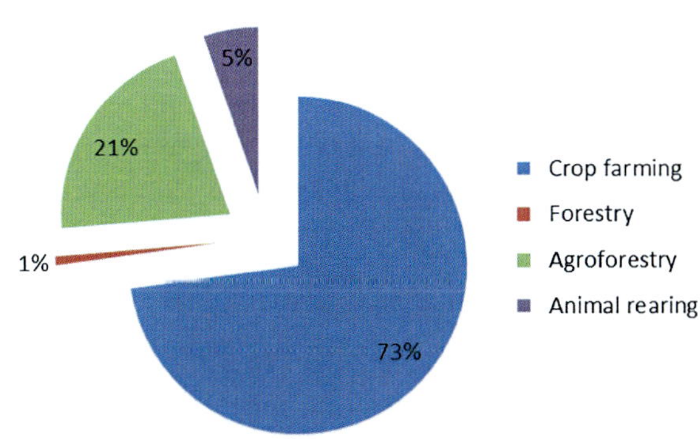

Fig. 10.5 Land use activities with highest involvement by women in southwestern Uganda

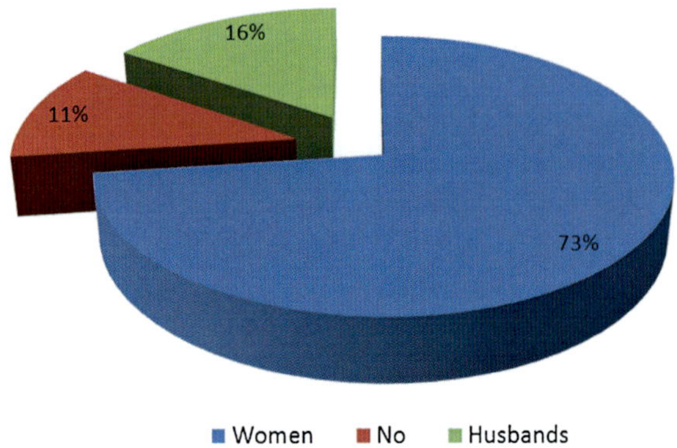

Fig. 10.6 Labor distribution of women and husbands in soil and water conservation

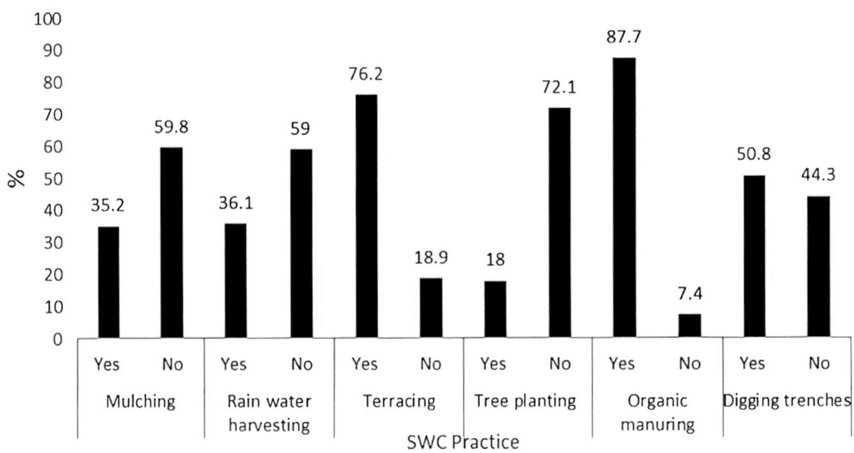

Fig. 10.7 Soil and water conservation practices and women's involvement

As shown in Fig. 10.7, the principal soil and water conservation activities carried out by women are mulching, harvesting rainwater, constructing terraces, planting trees, applying organic manure, and digging terraces. Women are more likely to apply organic manure (87.7 %) than to undertake other practices. The next most frequently practice among women is terrace construction (76.2 %). Tree planting is the conservation measure performed the least by women (7.4 %).

For the women involved in soil and water conservation, their roles differ widely. The majority (37.7 %) is actively involved as individuals. Only a small percentage (2.5 %) indicated that they participate by funding the process (Fig. 10.7). Finally, 21.3 % indicated that they undertake these activities but only with others (Fig. 10.8).

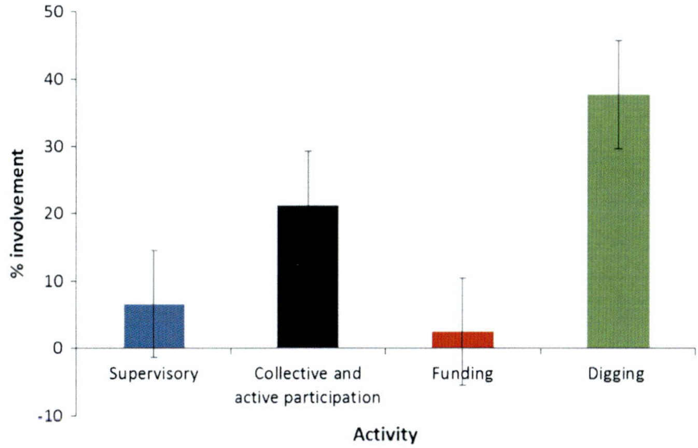

Fig. 10.8 Lake Bunyonyi catchment in Kabaale district, southwestern Uganda

10.4 Discussion

10.4.1 Physiochemical Characteristics

Soil pH was predominantly acidic, which can be attributed to the high levels of precipitation common to this region. The precipitation results in substantial leaching of basic cations, such as calcium, magnesium and sodium (McCauley et al. 2009). The acidic pH might also result from the release of a proton by carbonic acid atoms during the decomposition of organic matter when generated carbon dioxide dissolves in water (Nguyen and Ha 1984). Levels of soil organic matter (SOM) were high for all the land uses. For perennials and annuals, the high level of organic matter results from farming practices which increase its content. The addition of organic matter (Toor and Shober 2015) was the most commonly practiced soil and water conservation measure among women in Southwestern Uganda. In woodlots, the organic matter results from the decomposition of tree litter, though litter generation by eucalyptus trees is low compared to other trees. Decomposition levels are also lower compared to those reported in forest soils in general (Reversat 1993). The amount of nitrogen observed is relatively high, which might be the result of the application of organic manure by farmers. However, available phosphorus levels were low because the materials used for organic manure are crop based. These materials have low P levels and a very low N:P ratio ranging from 7:1 to 4:1, compared to a ratio of less than 4:1 for animal-manure-based sources. Similar soil chemical characteristics were reported by Majaliwa et al. (2015) and Nelson and Janke (2007).

Ksat was high for all land uses, likely because the soils included in the study are porous. The highest Ksat level, recorded in the woodlot, was due to larger soil-pore sizes caused by tree roots (Maynard et al. 2002). The bulky density of these soils is low probably due to lower levels of soil compaction.

10.4.2 Runoff, Soil and Nutrient Losses in Different Land Uses

The runoff trend is a somewhat complicated and, consequently, difficult to explain. Perennials and annuals experienced the highest levels of runoff in woodlots and grazing lands. In the case of the former, this can be explained in part by the lower ground vegetation cover on woodlots than grazing lands (Hubbard et al. 2004; Bolwig 2002). The least amount of runoff occurred for perennials on the summit shoulders. Generally, the runoff from all land uses was lower than that reported on similar hillslopes elsewhere (Bolwig 2002). The low runoff is attributed to the high infiltration rates of these soils. This observation was also reported by Siriri et al. (2006) and was further confirmed by the rapid soil-hydraulic conductivity and low soil bulk-density observed. Soils with low bulk density and high Ksat are known to have high infiltration and, subsequently, low runoff (Abel-Ahmed et al. 1987; Yadav et al. 2006).

The soil loss patterns were not consistent with the runoff patterns. Woodlots generally experienced more soil loss than other types of land uses. This is attributable to the low littering capacity of Eucalyptus (Reversat et al. 2008) which leaves the soil exposed. However, these findings contradict those findings of Siriri et al. (2006) where more runoff was experienced in the sole crops than under the tree-based systems though his was Calliandra instead of Eucalyptus. In addition, the general practice in Southwestern Uganda is to raise Eucalyptus plantations on highly degraded areas whose soils are already severely eroded. The recorded soil loss due to sheet/rill erosion in such steep hilly landscape is far too low compared to other reported findings for similar terrain (Yadav et al. 2006; Hicisalihoglu 2007; Bolwig 2002; Prasannakumar et al. 2011; Guo et al. 2013). This discrepancy is mainly because the soils in Kabale have low erodibility, which is influenced by their high porosity, infiltration rate (Bolwig 2002), and permeability. However, the low rate of soil loss has been cumulative, causing the observable widespread land degradation.

Nutrient loss followed a pattern similar to that of soil loss, suggesting that soil and nutrient losses are proportional to one another. This implies that more nutrients are lost in soil sediments than in liquid runoff. Like runoff and soil loss, the amount of nutrient loss is lower than that for similar hillslope types reported by Guo et al. (2013). The loss of soil fertility and the resulting low crop yield is due to the low soil fertility replenishment practices in southwestern Uganda (Okoboi and Barungi 2012; Okanya and Kroschel 2014).

10.4.3 Gender in Land Management and Degradation

Women are heavily involved in land-use activities in southwestern Uganda, particularly in agriculture due to their need to provide food security for household members. As reported by Okanya and Kroschel (2014) in a study of gender differences in access to and use of selected productive resources among sweet-potato farmers in Uganda, women in southwestern Uganda are responsible for food security. Byruhanga and Opedum (2011) reported a similar finding in a study on the impact of culture on food security in Uganda. Inevitably, women become fully involved in agricultural practices as their primary role is to ensure household food security. Tree plantations are generally for intended to generate income, so they are a preserve for men (Banana et al. 2012). A negligible amount of women are involved in woodlot plantations in Southwestern Uganda. Animal rearing is also a minor activity for women because of the limited land, which tends to be hilly in nature. As well, animal rearing, like eucalyptus growing, is a preserve of men due to traditional and cultural beliefs (Mwebaze 2006). The relatively high level of women's involvement in agroforestry is influenced by NGOs, which specifically promote this activity by women in recognition of their role and hardships in gathering wood for domestic fuel consumption (Williams 1992).

The high percentage of women's involvement in soil and water conservation practices (73 %), compared to 16 % performed by men, indicates that it is women's responsibility to undertake these practices. This finding accords well with that women's greater involvement in agricultural activities than men, as also observed by Okanya and Kroschel (2014). Women shoulder the burden for soil and water conservation practices. As they till the soil, they should be involved in conserving it. Tree planting is the conservation activity in which women in southwestern Uganda are least involved. Tree planting is a commercial activity in which men prefer to be involved while their wives focus on household food security (Okanya and Kroschel 2014). Men prefer tree production because it requires little management, especially weeding (Okorio 2006). Women are also less involved in mulching, primarily due to their limited access to mulching materials (Critchley et al. 1999). Most of the land has been cleared for agricultural intensification and has no fallows. The available materials for mulching are too distant to be accessible for this purpose, but women lack the financial resources to hire labor for mulching. Women's role in soil and water conservation in southwestern Uganda is mainly physical. The labor to undertake these activities is commonly provided only by women because they do not have funds to hire it. As in other regions of Uganda, women in southwestern Uganda are economically disadvantaged, and their limited incomes are reserved for meeting basic household necessities (Dunford and Watson 2008).

10.5 Conclusions and Recommendations

Soils in southwestern Uganda were characterized by rapid hydraulic conductivity and low bulk density. Runoff, soil, and nutrient losses varied with land use and landscape position. Runoff was highest and lowest under perennials at the foot slope and summit shoulders, respectively, while soil and nutrient losses were highest in woodlots, irrespective of landscape position. As a soil- and water-conservation woodlot, Eucalyptus plantations are not suitable because they do not accumulate the desirable organic matter. Overall, runoff and soil and nutrient losses are lower than those reported in other research on similar terrain. Women are more involved in land utilization activities than men. Agriculture is women's leading activity, and tree planting their least prominent. Women participate in soil and water conservation practices but lack adequate labor to implement effective practices and the necessary financial resources to hire labor. They rely solely on their own labor, which is not sufficient to conserve the resource base. Consequently, efforts by women to conserve soil and water have not yielded the desired goals.

Soil and nutrient losses are low, yet crop productivity has continuously declined, so it is important to conduct nutrient-leaching studies. The soils tested had low bulk density and high Ksat, which can promote the downward movement of water and dissolved nutrients. Traditional and cultural norms that place the burden of food security solely on women need to be revisited so that the much-desired male labor can be effectively utilized for crop production and soil and water conservation. This could help stem the high levels of land degradation experienced in the region.

References

Abel-Ahmed AH, Schuman GE, Hart RH (1987) Soil bulk density and water infiltration as affected by grazing systems. Range Manag 40(4):307–309

Banana AY, Bukenya M, Arinaitwe E, Birabwa B, Ssekindi S (2012) Gender, tenure and community forests in Uganda. A working paper 87. Centre for International Forestry Research (CIFOR), Bogor

Bolwig S (2002) Land use change and soil degradation in the Southwestern highlands of Uganda. A contribution to the strategic criteria for rural investments in productivity programme of USAID Uganda Mission, The International Food Policy Research Institute 2033 K Street, N. W. Washington, D.C. 20006

Byruhanga YB, Opedum PM (2011) The impact of culture on food security in Uganda. Academic essay on how a paternal society and traditions impact on food security in Uganda. Global Knowledge No. 1 (2008)

Critchley W et al (1999) Traditions and innovation in land husbandry: building on local knowledge in Kabale, Uganda. Regional land management unit, technical report no. 20, RELMA technical report series http://www.samsamwater.com/library/TR20_Traditions_&_innovat.pdf. Accessed 6 Nov 2014

Dunford C, Watson A (2008) Empowering women through integration of microfinance and health education. United Nations Population Fund (UNFP). http://www.unfpa.org/sites/default/files/pub-pdf/microfinancing_health_education.pdf. Accessed 30 Sept 2015

Floor JA (2000) Soil erosion and conservation. http://www.seafriends.org.nz/enviro/soil/erosion.htm. Accessed 4 Apr 2016
Forum for Agricultural Research in Africa (FARA) (2009) Status of social, financial, human and physical capital in Lake Kivu Pilot learning site. Sub-Saharan Challenge Programme, Accra
Gobin A, Govers G, Jones R, Kirkby M, Kosmas C (2002) Assessment and reporting on soil erosion. Background and workshop report. Technical report 94. European Environmental Agency, Copenhagen
Gorji M, Rafahi H, Shahoee S (2008) Effects of surface soil removal (simulated soil erosion) and fertilizer application on wheat yield. J Agric Sci Technol 10:317–323
Guo H, Sun G, Shi F, Lu T, Wang Q, Wu Y, Wu N (2013) Water, soil and nutrient losses caused by Wenchuan Earthquake: a case study in Pengzhou. Water Sci Technol 68(5):1055–1062
Hicisalihoglu S (2007) Determination of soil erosion in a steep hill slope with different land use types: a case study in Mertesdorf, Germany. Environ Biol 28(2):433–438
Hubbard RK, Newton GL, Hill GM (2004) Water quality and the grazing animal. J Anim Sci 82:255–263
Klute A (1986) Methods of soil analysis, part 1, physical and mineralogical properties, 2nd edn. American Society of Agronomy, Madison
Kristan M, Abeku TA, Beard J, Okia M, Rapuoda B, Sang J, Cox J (2008) Variations in entomological indices in relation to weather patterns and malaria incidence in East African highlands: implications for epidemic prevention and control. Malar J 7(231):1475–2875
Majaliwa MJG, Tenywa MM, Rao KPC, Musana B, Fungo B, Buruchara R (2015) Soil fertility in relation to landscape position and land use/cover types: a case study of the Lake Kivu pilot learning site. Adv Agric. doi:10.1155/2015/752936
Maynard JLJ, Graham RC, Wu L, Shouse PJ (2002) Modification of soil structural and hydraulic properties after 50 years of imposed chaparral and pine vegetation. Geoderma 110:227–240
McCauley A, Jones C, Jecobsen J (2009) Soil pH and organic matter. Nutr Manag Modul 8:1–12. Montana State University Extension 4449–8. http://landresources.montana.edu/nm/documents/NM8.pdf. Accessed 23 Aug 2016
Meshesha DT, Tsunekawa A, Tsubo M, Haregeweyn N (2012) Dynamics and hotspots of soil erosion and management scenarios of the Central Rift Valley of Ethiopia. Int J Sediment Res 27(1):84–99
Mwebaze SMN (2006) Country pasture/forage resources profile. http://www.fao.org/ag/agp/AGPC/doc/Counprof/Uganda/uganda.htm. Accessed 4 Apr 2016
Nelson NO, Janke RR (2007) Phosphorus sources and management in organic production systems. HortTechnology 17(4):442–454
National Environment Management Authority (2001) State of the environment report for Uganda, 2000/2001. NEMA, Kampala
Nguyen MT, Ha TKJ (1984) A theoretical study of the formation of carbonic acid from the hydration of carbon dioxide: a case study of active solvent catalysis. Am Chem Soc 106 (3):599–602
Nkonya E (2002) Soil conservation practices and non-agricultural land use in the south western highlands of Uganda. Strategic criteria for rural investment in productivity. SCRIP report. IFPRI, Washington
Okanya JS, Kroschel J (2014) Gender differences in access and use of selected productive resources among sweet potato farmers in Uganda. Agric Food Secur 3(1):1–10
Okorio J (2006) FAO Trans-boundary agro-ecosystem management program (tamp). Final report on status of forestry and agro-forestry in the lower Kagera, Uganda. www.fao.org/.../nr/.../kagera_brief_dec06.doc. Accessed 23 Aug 2016
Okuboi G, Barungi M (2012) Constraints to fertilizer use in Uganda: insights from Uganda Census of Agriculture 2008/9. Sustain Dev 5(10):99–113
Pender J, Jagger PA, Nkynya E, Sserunjuma D (2004) Development pathways and land management in Uganda. World Dev 32(5):767–792

Prasannakumar V, Vijith H, Abinod S, Geetha N (2011) Estimation of soil erosion risk within a small mountainous sub-watershed in Kerala, India, using revised universal soil loss equation (RUSLE) and geo-information technology. Geosci Front 3(2):209–215

Price MF et al (eds) (2011) Mountain forests in a changing world—realizing values, addressing challenges. FAO/MPS and SDC, Rome

Puhalla JM (2009) Land use and agricultural intensification in Mugandu wetland, Kabale District, Uganda (PhD). University of Florida, Gainesville

Reversat B (1993) Dynamics of litter and organic matter at the fast-growing tree plantations on the sandy ferrallitic soils (Congo). Acta Oecol 14(2):179–195

Reversat FB, Kimbatsa IB and Loumeto JJ (2008) Eucalyptus litter quality and sandy soils: addressing two cumulative effects on topsoil organic matter and soil activity in African plantations. In: Nambiar EKS (ed) Site management and productivity in tropical plantation forests: proceedings of workshops in Piracicaba (Brazil) 22–26 November 2004 and Bogor (Indonesia) 6–9 November 2006, Bogor, pp 275–281

Siriri D, Tenywa MM, Ong CK, Black CR, Bekunda MA (2006) Water infiltration, conductivity and runoff under fallow agroforestry on sloping terraces. Afr Crop Sci J 14(1):59–71

Toor GS, Shober AL (2015) Soils and fertilizers for master gardeners and organic amendments. University of Florida, IFS Extension. http://edis.ifas.ufl.edu/topic_series_soils_and_fertilizers_for_master_gardeners. Accessed 22 June 2015

Williams PJ (1992) NGOs, women and forestry activities in Africa. Unasylva 43(171):41–49

Yadav RK, Yadav DS, Rai N and Sanwal SK (2006) Soil and water conservation through horticultural intervension in hilly Areas. ENVIS Bull Himal Ecol 14(1):4–13.

Chapter 11
The Social Dimension of Water Management in an Era of Increasing Water Scarcity in Tanzania

Zebedayo S.K. Mvena

Abstract Tanzania has vast water bodies and rivers as well as substantial underground water reserves. Yet, the country has an inequitable distribution of this vital resource. This calls for a better understanding of the nature of water distribution and use by society both in the short and long term. The paper, therefore, begins by providing an assessment of the water situation in the country, including documentation of the different water sources, namely, rivers, lakes, and underground water. The paper highlights major factors contributing to a decline in water availability, such as inappropriate farming practices and misuse of irrigation waters, extensive or grazing livestock systems, and the domestic use of water. The open violation of government regulations in many areas in Tanzania inhibits the rectification of the water problem in the country. Human activities require a sustainable use of water, including behavioral changes, such as changes in cropping patterns and livestock options, as well as major shifts in eating habits. The paper concludes by highlighting some options for the future as humanity struggles to adapt to climate change. These options include enhanced water use efficiency and climate smart agricultural research to produce outcomes that are better adapted to climate change. Finally, that fact that little attention is given to climate change issues and their impacts on water supply is acknowledged. It is recommended that the government and the scientific community give more attention to these pertinent issues.

Keywords Water scarcity · Water management · Water harvesting · Drought resistant crops · Water withdrawals · Water use efficiency

---

Z.S.K. Mvena (✉)
Sokoine University of Agriculture, Morogoro, Tanzania
e-mail: zmvena2000@yahoo.com

R. Lal et al. (eds.), *Climate Change and Multi-Dimensional Sustainability in African Agriculture*, DOI 10.1007/978-3-319-41238-2_11

11.1 Introduction

Optimum management of global water resources presents a crucial challenge in the twenty-first century. The global population will increase by three billion or more over the next 50–75 years, and the number of people living in urban areas will more than double. Most of the world's population growth will occur in developing countries where water is already critically short and many of the residents are impoverished (Jury and Vaux 2005). In spite of this wake-up call, water as an issue has not featured that well on the priority list for much of Tanzania's development history since independence.

In a historical perspective, national issues come and go because they compete for attention from policy makers, scientists, and even the mass media. These avenues for attention have limited carrying capacities, that is, they can accommodate only a limited number of societal problems at any given time. Thus, in the history of Tanzania, we have witnessed a relay of national issues, such as poverty, illiteracy, the "politics is agriculture," and HIV/AIDS becoming national agendas and filtering in one at a time. Thus, the issue of water has not been defined as a national problem. There have been a number of policy pronouncements, but these have had limited or no impact on public attention. Water management is the responsibility of many different decision makers in both public and private sectors. The issue is how such shared responsibility can be turned into something constructive and elevated to a rallying point around which different stakeholders can gather and participate collectively to make informed decisions (World Water Assessment Programme (WWAP) 2015).

There has been mention of the need to protect water sources, but as this paper will show, little has been achieved toward this end. There have been government outcries about the need to protect water catchment areas, including resettling communities, but human activities, such as agricultural encroachment and timber harvesting, have continued unabated. This apparent public inaction on the fate of the water supply has dire consequences for humanity.

This paper includes a discussion of human activities that contribute to water-related problems and their subsequent impact on human populations. Several alternative ways to address the imminent water crisis in Tanzania and beyond are proposed. A major objective of this paper is to highlight the importance of water issues so that they are given priority status amongst conservation scientists, policy makers, and the public in Tanzania in order to better prepare for a future of water scarcity.

11.2 Water and Sustainable Socio-economic Development

The 2015 edition of the United Nations World Water Development Report (WWDR) highlights the role of water as a critical element of sustainable development (United Nations Educational, Scientific and Cultural Organization (UNESCO) 2015). In its 1987 report, *Our Common Future*, the United Nations'

Table 11.1 Water withdrawals in Tanzania *Source* UN Water (2013)

Sector	Percent use
Agriculture (total)	92.9
• Irrigated crops	88.7
• Livestock	4.1
• Aquaculture	1.0
Municipal	6.1
Industrial	1.0
Total	100.0

World Commission on Environment and Development (the Brundtland Commission) defined "sustainable development" as "development that meets the needs of the present without compromising the ability of future generations to meet their own needs. Water is the essential primary natural resource upon which nearly all social and economic activities and ecosystem functions depend. Sustainable development requires that we properly manage our freshwater resources and equitably share its benefits.

Globally, about 70 % of the total water withdrawals are used for agricultural activities, making it, by far, the largest water user. In Tanzania, it is 93 %. Table 11.1 presents water withdrawals in Tanzania. Water use or water withdrawals refer to the water removed from a water source and used for human needs, some of which may be returned to the original source and reused downstream with changes in water quantity and quality. Water depletion refers to the use or removal of water from a water basin that renders it unavailable for other uses, for example, through evaporation, flows to sinks, pollution, and incorporation into agricultural or industrial products.

In Tanzania, water withdrawals by sector totaled 4975 million m^3 in 2012. The distribution of these withdrawals is presented in Table 11.1. When agriculture is defined broadly to also include livestock, irrigated crops use nearly 89 % while livestock use 4 % and aquaculture uses 1 %.

Data in Table 11.1 informs policymakers about current uses of water in Tanzania. The data suggest where attention must be directed in order to ensure the sustainable use of water by the present and future generations.

11.3 Water Resource Endowment in Tanzania: So Much Water but Insufficient Water for Many

Tanzania enjoys vast water bodies and rivers as well as large deposits of underground water. The country shares eleven international water bodies with other riparian nations. They include Lake Victoria, Tanganyika and Nyasa, Natron, Chala, and Jipe as well as the rivers Kagera, Mara, Umba, Ruvuma, and Songwe (Water Sector Development Programme (WSDP) 2014). There are other major rivers within the country, including Pangani, Malagarasi, Great Ruaha, and Rufiji. Despite these vast water bodies, the country is facing major water problems of two types. First, many households are without adequate water, and the trend is getting worse. The average amount of available water per capita is projected to decrease by 30 % in the near future (WSDP 2014). The average per capita amount in 2015 is 1400 m^3, which is less than the defined water stress level of 1500 m^3. This condition is a result of the diminution of water resources and population increase (WSDP 2014). Projections indicate that population increases, which will result in increased consumption (World Bank 2006a), will be the major cause of increased water stress levels over the coming decade.

Worldwide, the decline in water availability is apparent in future water projections. "Approximately 80 % of the world's population already suffers serious threats to its water security, as measured by indicators including water availability, water demand and pollution" (Intergovernmental Panel on Climate Change (IPCC) 2015). By 2025, more than three billion people could be living in water stressed countries (defined as <1700 m^3/person/year) and 14 countries will drop from the water stress to water scarce category. This development will include intensifications of water stress across Sub-Saharan Africa, with the share of the region's population in water-stressed countries rising from just above 30–85 % by 2025 (United Nations Development Program (UNDP) 2006).

In Tanzania, many of the previously rich sources of water are now drying up. Once permanent rivers are now seasonal, while others are now permanent dry river beds. Mount Kilimanjaro, which previously had plenty of ice all year round, produced many small rivers flowing to the lowland areas. The route from Himo to Moshi crosses only seasonal rivers. Morogoro municipality once enjoyed plentiful supplies of water from the many permanent streams flowing down from the Uluguru Mountains, making it a swampland suitable for rice cultivation. What remains are a few seasonal rivers that cease to flow at the peak of the dry season between September and December.

Residents in the semi-arid areas of Tanzania, namely Dodoma and Singida, face serious water problems. They largely depend on underground water sources. Lake Haubi and Lake Bicha in Central Tanzania are now dry and have become grazing areas. The capital city of Tanzania, Dodoma, depends entirely on underground sources for its socio-economic development. The source consists of deep wells, with a water production capacity of about 40,000 m^3/day. This source is located in Mzakwe, Makutopora Basin, about 35 km from the center of Dodoma municipality. However, it is currently producing an average of 21 m^3/day (which meets only 53 % of the demand for water in the municipality).

As is true for other regions of the world, many underground water reserves are already running low. Rainfall patterns are predicted to become more erratic with climate change, thus reducing the recharge capacity of these reserves. As the world's population grows to an expected nine billion by 2050, more groundwater will be needed for farming, industry, and personal consumption (Deigle 2015). Yet, some reports indicate that management of groundwater is communal property which no one takes full responsibility for its conservation (Hardin 1968; UNESCO

2012). Countries, such as Ethiopia and Zimbabwe, which have recently become drought prone due to lack of rains, now have to rely on food handouts from the world community.

11.4 Causes of Water Scarcity

Although the major ultimate causes of water scarcity are climate change and population growth, certain human activities are reported to have a significant negative impact on water availability as well as causing climate change itself. Recurring short term droughts, such as those of California, South Africa, Zimbabwe, and Chile, cause considerable suffering. Human activities fall under three main categories: agriculture, domestic use and other land uses. The discussion below explains how these activities impact on the water resource.

11.4.1 Agriculture

As pointed out earlier, agriculture accounts for about 70 % of total water withdrawals globally, making it, by far, the largest water user sector (UNESCO 2015). In Tanzania, agriculture, defined broadly to include livestock, accounts for about 93 % of available water use, a total 4975 million m^3 in 2012 (United Nations Water (UN Water) 2013). Within the agricultural sector, irrigated crops account for 88.7 % of water usage while livestock production is responsible for 4 % and aquaculture the remaining 1 %. It is important to note that water wastage also occurs in rain-fed agriculture.

11.4.1.1 Water and Irrigated Crops

Water wastage related to irrigated crops occurs due to inefficient irrigation systems. Efficiency in water delivery to where it is needed very much depends on the design of the system as well as the quality of water conveyance systems. Locally-made earth canals permit large volumes of water to percolate into the soil, thus significantly reducing the amount of water reaching the intended destination. Many irrigation activities in rural Tanzania are of this type. Concrete canals can be poorly managed and maintained. Cracks can be found along these canals, which drain water from the irrigation system, hence reducing the amount of water actually applied to crops.

Faulty design of irrigation canals can also lead to large amounts of water being lost due to evaporation. This can happen in situations where canals are wide and the gradient is too small. Water thus travels slowly, thus leading to greater evaporation levels. Surface area is also positively correlated with water loss. The bigger the surface area the greater the loss of water through evaporation. Thus, where there are pools of shallow water covering large surfaces, the likelihood of water being lost through evaporation is greater than when a water body covers a small area but has greater depth.

11.4.1.2 Farming Practices

The disappearance of Lake Haubi and Lake Bicha as well as many other small lakes and ponds is ascribed to bad farming practices on the upper slopes of the water catchment areas. Due to bad farming practices, severe soil erosion has led to sedimentation in the lake zone. The depth of the lakes became shallower as more sediment is deposited year after year. Thus, instead of rainwater being stored within the lake basin, the water is left to drain away to the ocean. Today, both lakes have been turned into grazing areas all year round. Agricultural encroachment has led to the wanton clearance of vegetation in ecologically sensitive areas, such as water sources, leading to water decline.

The destruction of water sources by way of valley bottom farming or "vinyungu" in Iringa and Njombe regions significantly contributes to the water problem in many areas of Tanzania. The water in such areas, some of which may have been wetlands, is left to drain away and as the water table becomes shallow.

A study done by Mkavidanda and Kaswamila (2001) reveals that with current preparation and management practices "vinyungu" farming is detrimental to the environment in terms of water quality and quantity, soil erosion, and bush-fires. Because of relatively high yield returns from vinyungu farming, there is an over-cultivation of water source areas to the extent that even water conserving trees, such as *Syzygium cordatum*, S. and *Pappea capensis*, are cut to expand the "vinyungu" area. This has caused water sources to dry up, stream water flows to decrease, and lower slopes of valleys to erode. More and more hydrologically sensitive areas are cleared to give way to farming, especially during the dry season, despite the requirement that no human activity should be done within 60 m from water sources, such as rivers or dams, as prescribed by both the Environmental Management Act and the Water Resources Management Act (United Republic of Tanzania (URT) 2004, 2009).

11.4.1.3 Livestock Production

Extensive or grazing livestock production systems affect the sustainability of water in different ways. First, livestock affect water levels by denuding the vegetation cover that helps to trap rain water after which it percolates into the soil on its way to recharging aquifers. With no or little vegetation cover, rain water is allowed to flow to other areas as run-off. Second, livestock movements compact the soil, thus making it even harder for rainwater to percolate into the soil. This is reported to be the case, for example, with the Usangu wetlands where the presence of large herds of livestock had adverse impacts on the water flows of the Great Ruaha River (Msigwa and Mvena 2014). Third, livestock under the grazing livestock production systems has also been linked to soil erosion. Eroded soils are carried to water reservoirs, filling them and reducing their storage capacity.

11.4.2 Other Land Uses and Domestic Use of Water

Besides agriculture and industry, other human activities adversely impact water resources. Urban population growth has negatively impacted water resources due to deliberate efforts to dry up previous wetlands in order to use them for other activities, primarily residential, commercial, and industrial.

Household use of water has in many ways contributed to water scarcity albeit to a lesser extent. While some Tanzanian households have made attempts toward water recycling, as it is the case in some advanced countries, little has been invested towards this effort. Much of the water left after household use is left to drain away or lost. Such losses occur due to leaking water pipes and faucets, washing clothes and utensils, overuse in flower gardens, or taking showers.

It is reported, for example, that water waste is positively correlated to wealth. Richer water users waste much more water than poor water users (Winpenny et al. 2010). It is reported for example that the US is one of the world's biggest users of water: An American who takes a five-minute shower uses more water than the typical person living in a slum in the developing world uses all day. Other uses of water include: community sewage systems and toilets using water for the conveyance and disposal of both human and animal waste; household appliances, such as dishwashers and garbage grinders; domestic hot water devices that increase the use of water for bathing; leisure activities, such as golf courses and aquatic parks; urban greenery for local amenities; increased consumption of manufactured goods; dietary changes involving higher consumption of foodstuffs with greater water requirements; and recreational amenities, such as swimming pools.

11.4.3 Government Regulatory Failure

Agricultural encroachment into ecologically sensitive areas, the destruction of wetlands both in rural and urban areas, and uncontrolled grazing reflect in part the failure of government institutions to regulate these activities in Tanzania. Environmental laws are well articulated, but the ability to enforce them is largely lacking (Fig. 11.1).

Fig. 11.1 Violation of the 60 m minimum distance from a water source

11.5 Consequences and Options for the Future

11.5.1 Consequences

Reduced water availability can be catastrophic in many ways, including crop failures, ecosystem degradation, industry collapses, increases in disease and poverty, and increased conflicts over access to water (UNESCO 2015). Numerous changes may occur in the face of water scarcity.

The dietary shift from predominantly starch-based food to meat and dairy, which require more water, has had a significant impact on water consumption over the past 30 years. A dietary turnaround may result in a shift to vegetarianism, as it will be too expensive and water-consuming to grow crops only to be fed to livestock. It will be rational to consume the crops rather than the meat and other livestock products. For example, 1 kg of beef requires 15,000 L of water. Producing that 1 kg of meat requires as much water as an average domestic household uses over 10 months (50 L/person/day).

There may also be a shift in the type of crops to be grown for human consumption. High water demanding crops, such as rice, may have to be abandoned in favor of drought resistant crops, such as sorghum, millet, and cassava. Producing 1 kg of rice requires approximately 3500 L of water, while crops, such as sorghum and millet, require far less water to produce. This shift in eating habits was necessary for Isimani residents in the drier part of the Iringa region, an area that changed so much climatically that the once granary of Tanzania could not raise the maize crop due to recurrent spells of insufficient rainwater (Meena et al. 2008).

However, this shift has been resisted by some farmers in Ismani as well as much of the semi-arid areas of Tanzania where the planting of drought resistant crop varieties is recommended. A similar resistance has been observed in other countries. For example, although drought is blamed for this year's crop failure, Zimbabwean farmers are reluctant to grow small grains, such as sorghum and millet, which are more resistant to drought but are not widely eaten in Zimbabwe (Macdonald 2015). A shift will also be necessary in the consumption of certain crops that are known to require too much water to raise. Many of the fruits, for example, watermelons, and vegetables, such as cucumbers, fit this description. The scarcity of water has far reaching consequences, including the likely demise of aquaculture and the entire fish industry.

11.5.2 Options for the Future in Water Management

Addressing the problem of water scarcity will require a range of management strategies backed by collective responsibility towards enhancement of available water resources and collective restraint towards the prevailing wasteful propensities of using water. Society needs to harvest more rainwater rather than depleting existing ground water reserves. Harvesting of such water can be done in various ways. Rainwater can be harvested from rooftops, as is currently practiced, on a larger scale. It is also possible to harvest water by constructing dams and setting up percolation ponds to capture surface runoff, which is a much cheaper option. This water will eventually recharge underground water aquifers. This is particularly efficient with sandy soils. However, a major drawback is that water flows underground cannot be controlled.

Wasteful use of water occurs in many forms: in agricultural production, livestock production, industry, and the domestic use of water. Use of water will have to be managed better if water resources are to be sustained.

In the agriculture sector, several options exist. Feeding the growing population will likely require a number of different strategies from developing drought resistant and early maturing crop varieties to reducing food waste and diversifying food sources, including eating insects. Insects offer several advantages over other food sources: they have a high protein content, do not consume much water, can be easily bred, and do not require much land (Huis et al. 2013).

Breeding crops for specific traits that enable them to withstand biotic and abiotic stresses related to extreme weather conditions is another way to reduce losses due to climate change. Farmers will also need to improve crop management practices in order to deal with dry weather conditions. Research could capitalize on the genes that control hibernation, which will enable plants to hibernate when there is no water but continue to grow when adequate moisture is available.

Drought-prone regions will also have to shift crop production from crops that require substantial water inputs, such as rice, to less water demanding crops. Some indigenous crops, such as sorghum, millets, lablab beans, pigeon peas, and pulses are already widely grown in Africa and require less water inputs. They should be researched in order to increase their adaptability to drought conditions. Water-rich areas, such as Democratic Republic of Congo, can export crops that require substantial water inputs to produce, while drier areas, such as the Sahel, can continue to grow crops that require less water. This has obvious strategic implications for countries that are water-constrained, such as those found in the Southern African Development Community (SADC) region where water is a comparatively scarce resource. Nations in this region might need to focus on importing foods that require substantial water inputs to produce and export products or services that require less water.

These shifts can have important impacts on the inhabitants of poor countries or the poorer communities in them. Crop trading beyond borders may be too expensive for them to afford. Efforts will have to be directed towards changing eating habits to accommodate foods that can be easily raised locally. The resistance of Isimani farmers, who traditionally eat maize, to sorghum and millet consumption as a result of their environment becoming more drought-prone is a case in point. They will need to change their consumption habits. Agronomic practices will also need to change in response to environmental changes. Crop management systems will need to become more water use efficient. For example, farmers will need to learn how to cover crops with plastic or the much cheaper mulch to prevent evaporation and use drip irrigation to target water directly to plant roots. Irrigating in the evening rather than during the day can also significantly reduce water use in crop production. Options for better use of "gray water," the non-potable, moderately dirty wastewater produced by showers, bathroom sinks, laundry machines, etc., need to be pursued.

The production of livestock is likely change over time due to changes in water access. Livestock keepers will have to shift to livestock systems that use less water, such as extensive systems rather than industrial systems. Countries could exploit the virtues of comparative advantage and specialize in the production of commodities based on their respective resource endowment. Since beef requires roughly 15,000 L of water for every 1 kg produced, beef could be produced in the Amazon River Basin, the largest watershed in the world. In contrast, drier regions, such as the Middle East, could harvest fruits from drought-resistant crops, such as xerophyte plants, which are cactus-related plants that require less water (LiveScience Staff 2013).

Industrial production processes use a lot of water in the different stages of production. Research will have to be conducted with the sole objective of cutting down water use at every stage. Beverage industries are responsible for "water transfers" in the form of bottled water, juices, and water that is needed in the production of different industrial products. Industrial marketing managers will have to avoid “pushing” consumers to drink water beyond what they need physiologically.

Household use of water is a major source of wastage. Much needs to be done to get society to use the water resource more vigilantly. Water use for personal hygiene is wasted in many ways, including showers, washing clothes, and flush toilets. One Australian manufacturer, Caroma, is marketing innovative products such as the hybrid toilet sinks. These sinks allow one to wash hands and then direct the used water to a flush toilet tank (Smith 2015). Despite United Nations recognition that water is a human right, international financial institutions, such as the World Bank, have argued that water should be allocated through market mechanisms for effective rationing and to allow for full cost recovery from users (World Bank 2006b).

Deliberate drying of wetlands through planting Eucalyptus trees and digging deep furrows to drain away water is discouraged by environmentalists but is likely to be more legally enforced in the future. The protection of such wetlands is likely to become a priority at the national and local levels.

11.6 Conclusions

Although the water situation in Tanzania is precarious, little attention is being paid by the government and community to better managing water resources. Government regulations exist without the ability to enforce them. Thus, the water issue may be frequently mentioned by policymakers and politicians, but there is no visible commitment on the part of the government to make sure that water users adhere to these regulations for a better future.

It is also clear from the discussion above that if this situation is left unchanged, society is likely to face grave consequences, including a shift in food choices. This paper therefore recommends the following to the government: (1) enforce regulations in order to sustain the water supply, (2) invest more in climate-smart agricultural research to allow society adapt to climate change, and (3) invest more in community education related to water. Society must also generate a sense of preparedness for the impending predicament so that concrete measures are taken well before water supply shortages become critical. Integrated water resources management systems defined as a process which promotes the coordinated development and management of water, land and related resources in order to maximise the resultant economic and social welfare in an equitable manner without compromising the sustainability of vital ecosystems.” are necessarily due to the multi-dimensionality of water use and abuse.

References

Deigle K (2015) UN warns world could have 40 percent water shortfall by 2030. http://news.yahoo.com/un-warns-world-could-40-percent-water-shortfall-102838336.html. Accessed 21 Mar 2015

Global Water Partnership (GWP) (ed) (n.d.) Introduction to IWRM. Presentation by Thalmeinerova, based on GWP Toolbox Resources. Global Water Partnership, Stockholm

Hardin G (1968) The tragedy of the commons. Science 162:1243–1248

Huis VA, Itterbeeck JV, Klunder H, Mertens E, Halloran A, Muir G, Vantomme P (2013) Edible Insects: future prospects for food and feed security. Food and Agriculture Organization (FAO), Rome

Intergovernmental Panel on Climate Change (2015) Synthesis report. Contributions of working groups I, II and III to the 5th assessment report of the IPCC. World Meteorological Organization, Geneva, Switzerland

Jury WA, Vaux H Jr (2005) The role of science in solving the world's emerging water problems. PNAS 102(44):15715–15720

LiveScience Staff (2013) What 11 billion people mean for the planet. www.livescience.com/41316-11-billion-people-earth.html. Accessed 21 Mar 2015

Macdonald D (2015) Zimbabweans go hungry as drought hammers southern Africa. Thomson Reuters Foundation, New York

Meena HE, Lugenja MS, Ntikha OA, Hermes M (2008) Analysis of technological and policy options for adaptation to climate change. The Netherlands Climate Assistance Programme www.best-dialogue.org/wp.../TanzDairyScanAnnex-14Jan14-Final1.pdf?. Accessed 3 Apr 2015

Mkavidanda TAJ, Kaswamila AL (2001) The role of traditional irrigation systems (Vinyungu) in alleviating poverty in Iringa rural district, Tanzania. REPOA Research Report No. 01.2. Dar es Salaam

Msigwa GB, Mvena ZSK (2014) Change in livelihoods of evicted agro-pastoralists from Ihefu basin in Tanzania. Livest Res Rural Dev 26(1):1–11

Smith O (2015) California drought inspires these 5 creative ways people are conserving water. Good Morning America. https://gma.yahoo.com/california-drought-inspires-5-creative-ways-people-conserving-172438707–abc-news-topstories.html. Accessed 7 Apr 2015

UN Water (2013) Republic of Tanzania UN water country report. www.unwater.org/downloads/WCB_Final_Report_20131007.pdf. Accessed 27 Apr 2015

United Nations Development Programme (UNDP) (2006) Human development report, beyond scarcity: power, poverty and the global water crisis. UNDP, New York

United Nations Educational, Scientific and Cultural Organization (UNESCO) (2012) World's groundwater resources are suffering from poor governance. Natural Sciences Sector News. UNESCO, Paris

United Nations Educational, Scientific and Cultural Organization (UNESCO) (2015) United Nations world water development report (2015) water for a sustainable world. UNESCO, Paris

United Republic of Tanzania (URT) (2004) The environmental management act. National Environment Management Council, Dar es Salaam

United Republic of Tanzania (URT) (2009) The water resources management act. Government Printer, Dar es Salaam

Water Sector Development Programme (2014) The water sector status report. Ministry of Water, Dar Es Salaam, p. 42

Winpenny J, Heinz I, Koo-Oshima S (2010) The wealth of wastewater: the economics of wastewater use in agriculture. Food and Agriculture Water Report 35. Food and Agriculture Organization, Rome

World Bank (2006a) Water resources in Tanzania—factsheet. World Bank, Dar es Salaam

World Bank (2006b) Water resources in Tanzania, sustainable development of Tanzania's water resources. World Bank, Dar es Salaam

Part III
Integrated Management of Natural Resources and Value Chains

Chapter 12
Managing Landscapes for Environmental Sustainability

Rattan Lal

Abstract Landscape is a heterogeneous area consisting of a cluster of interactive ecosystems including the soil-scape, water, flora and fauna, micro- and meso-climate, and terrain or physiography. These components interact among one another, and are strongly influenced by anthropogenic activities. Misuse of landscape and mismanagement of its components can adversely impact biogeochemical cycles of water, carbon and other elements leading to degradation of natural resources and the environment. Thus, a high priority of restoring degraded landscapes includes enhancing soil and ecosystem carbon stock, conserving water and increasing the green water, reducing runoff, maintaining soil chemical fertility (nutrients), improving soil structure, enhancing soil biodiversity, and maintaining a favorable salt balance. Strategies of sustainable intensification must be adopted to restore the landscape. Sequestration of carbon in soil and vegetation within the landscape is important to adaptation and mitigation of climate change.

12.1 Introduction

A soil scape is a 4-dimensional body comprising of length, width, depth, and time. Therefore, soil scape is an integral component of a landscape. As for soil, a landscape is also a heterogeneous area comprising of a cluster of interactive ecosystems that are repeated in similar forms throughout (Fig. 12.1). Principal components of a landscape include: (1) soils varying over space and time, and affecting its immediate environments; (2) water, both above and below ground and continuously being transformed from blue to green and grey components; (3) flora and fauna, with a strong interdependence among them and with soil and water; (4) micro and meso climate, as determined by temperature and moisture regimes,

R. Lal (✉)
Carbon Management and Sequestration Center, The Ohio State University, Columbus, OH 43210, USA
e-mail: lal.1@osu.edu

R. Lal et al. (eds.), *Climate Change and Multi-Dimensional Sustainability in African Agriculture*, DOI 10.1007/978-3-319-41238-2_12

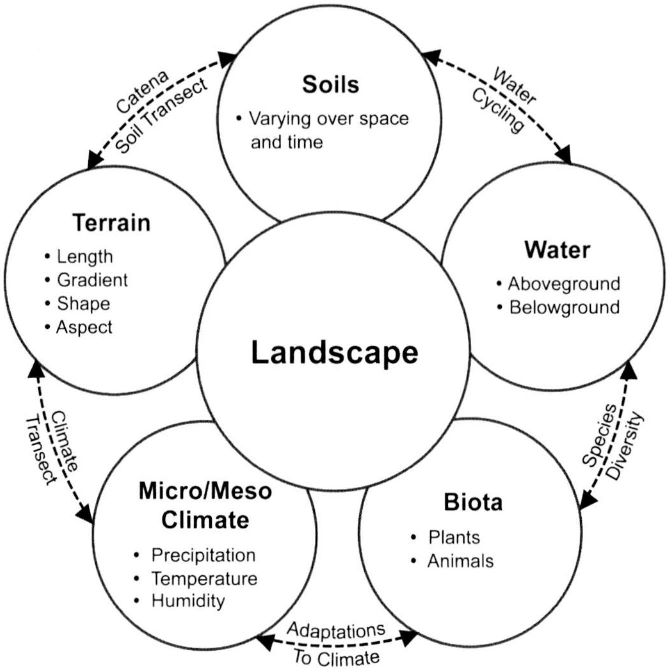

Fig. 12.1 Components of a landscape

annual precipitation and its distribution, and length of the growing season, and (5) terrain or the physiography, comprising of slope gradient, length, shape, and aspect along with the drainage density. These components strongly interact with one another, thus creating a unique landform with specific properties and processes which govern net ecosystem productivity (NPP).

It is the interaction among diverse yet interactive components with its biota and physiography that impacts unique attributes of a landscape. In addition, managing and monitoring human impacts on landscape (Hudson and LaFevor 2014) are important to sustaining landscapes.

Landscape ecology is the science of studying and managing relationships between ecological processes in the environment and particular ecosystems. Thus, a landscape may consist of two or three ecosystems in close proximity (Fig. 12.2). Misuse of a landscape and mismanagement of natural resources associated with it, such as soil, water and vegetation, can degrade a landscape through accelerated erosion, surface runoff, eutrophication (algal bloom), salinization and other negative processes. Managed ecosystems within a landscape are characterized by: (1) simplified food webs, (2) landscape horizonation, (3) high nutrient and energy input, and (4) low biodiversity (Western 2001). Therefore, landscape sustainability refers to the capacity of a landscape to consistently provide long-term landscape-specific ecosystem services essential to human wellbeing and nature conservancy (Wu 2013). In this context, human well-being is a journey, and not a well-defined destination.

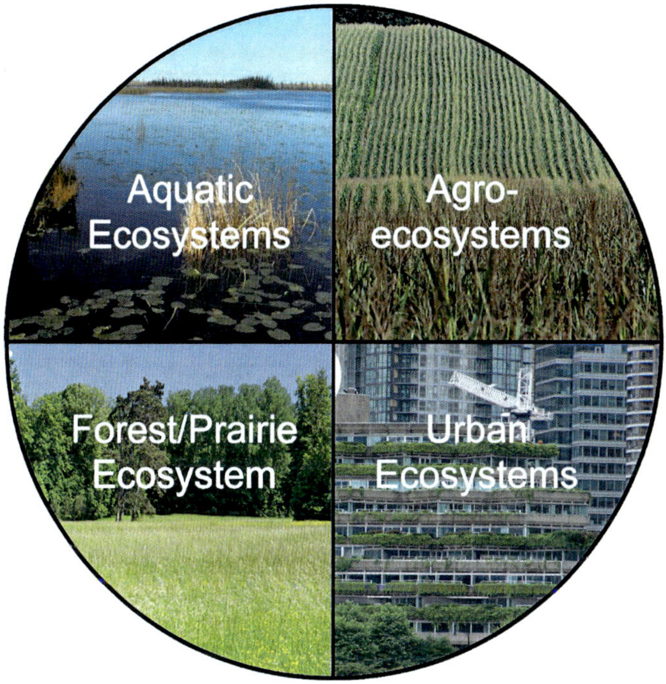

Fig. 12.2 Examples of ecosystems within a landscape

The importance of a landscape and its management on environmental sustainability, and the need for ecological restoration to adapt and mitigate climate change, advance food and nutritional security, improve biodiversity, and strengthen ecosystem services are discussed in this chapter.

12.2 Landscape Restoration

Restoring degraded landscapes is essential in order to be able to meet the demands of an ever increasing and affluent world population, including food and nutritional requirements, water amount and quality, aesthetic and livable landscape teaming with diverse flora and fauna, and amicable micro and meso-climate conditions. Landscape approaches provide tools and concepts for allocating and managing land to achieve specific objectives (Sayer et al. 2012). Landscape restoration tenets include: (1) creating positive budgets of C, N, and other essential elements, (2) conserving water and increasing the supply and renewability of green water, (3) controlling erosion and water runoff, (4) replacing plant nutrients harvested in crops and animals in soil, (5) improving soil structure and tilth, (6) enhancing activity and species diversity of soil biota, (7) alleviating constraints that limit root growth and proliferation, (8) maintaining a favorable salt balance in the soil solum/profile, (9) managing soil pH/reaction, and (10) creating disease suppressive soils. Implementing strategies include land use, and soil/water and nutrient management. They can enforce these tenets lead to sustainable landscapes and strengthened ecosystem services that are created in the process.

12.3 Principles of Ecological Restoration

Ecological restoration necessitates a holistic approach (Fig. 12.3) based on: (1) creating sustainable landscapes, (2) managing carbon (C) and nitrogen (N) pools, (3) improving ecosystem services and (4) managing water resources. The objective is to create and strengthen synergisms among these overlapping and strongly interactive strategies. For example, restoring an eroded landscape involves (1) providing a continuous vegetative cover with a strong root system that binds soil particles, (2) reducing runoff amount and velocity by enhancing water infiltration

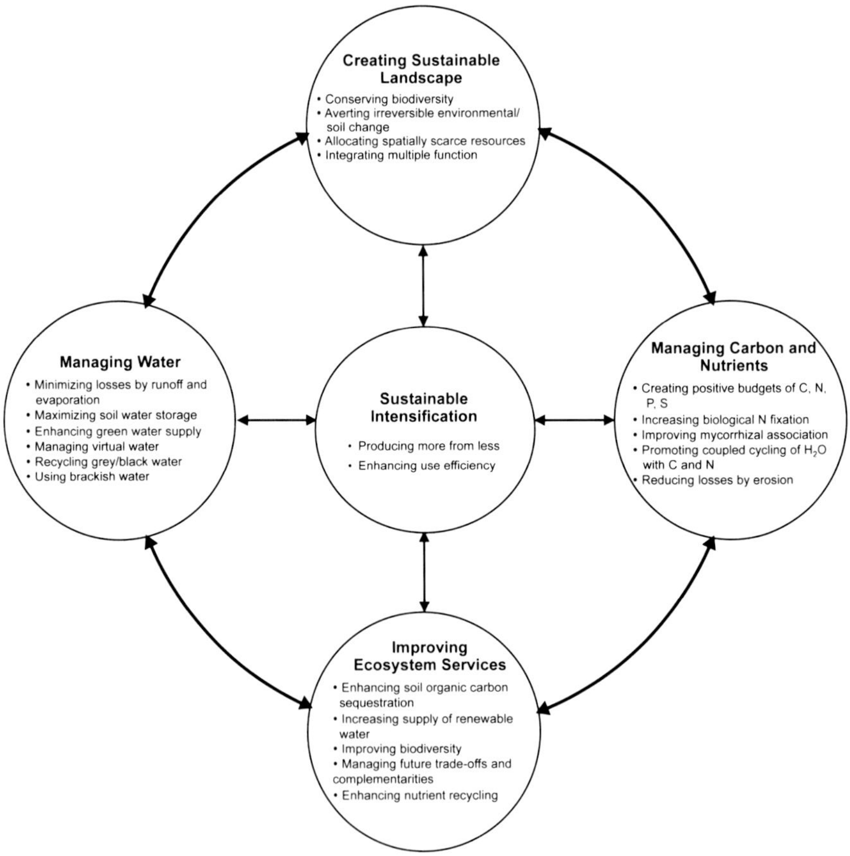

Fig. 12.3 Strategies to achieve sustainable intensification of landscapes in agroecosystems

rate and prolonging the time for water to infiltrate, (3) reducing velocity of water runoff, and (4) trapping the sediments.

In sum, basic principles of ecological restoration include, but are not limited to: (1) implementing innovative measures at a landscape level, (2) harmonizing the ecological effects in consideration of the current and future social and demographic changes, (3) developing and/or building traditional knowledge but with a strong emphasis on modern innovations, and (4) adopting principles of ecological engineering, such as managing and restoring wetlands. Indeed, sustainable development involves identification and implementation of an integrated approach to harness the landscape potential for environmental sustainability (Masnavi 2013). Attributes of a sustainable landscape include (Fig. 12.4): (1) strong biogeochemical cycles of nutrients (N, P, K, Ca, Mg) and C that accentuate transformations and recycling, (2) moderate rate of water cycling for conversion of blue water into green water with enhanced plant-available water capacity, (3) high above- and below-ground biodiversity, and (4) strong capacity to adapt to changing and uncertain climate while mitigating anthropogenic emissions of greenhouse gasses (GHGs) through soil/ecosystem C storage and reducing net emissions of CH_4 and N_2O (Fig. 12.4).

To create a sustainable landscape is to strengthen and enhance provisioning of essential ecosystem functions and services for human wellbeing and nature conservancy. Important among these ecosystem services are: (1) adaptation to and

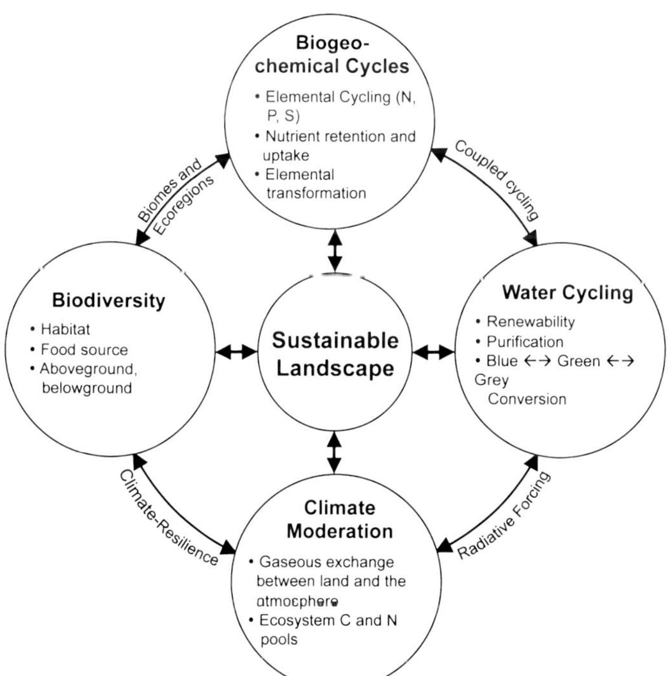

Fig. 12.4 Characteristics of a sustainable landscape

mitigation of climate changes, (2) conservation and enrichment of biodiversity, (3) management of gene flow and species migration, (4) enhancement and strengthening of ecological resilience, (5) moderation of natural and anthropogenic perturbations, (6) abatement of pollution of water, air, and soil, (7) management and purification of water quality, (8) enhancement availability of essential commodities to strengthen landscape functions, (9) preservation of cultural and artistic values, and (10) improvement of aesthetic and recreational functions.

12.4 Managing Water Over a Landscape

Drought or the lack of adequate water with accompanying supra-optimal (high or above optimal) temperatures, has strong adverse impacts on NPP and the wellbeing of human and animals. Many regions of Sub-Saharan Africa (SSA), especially the arid and semi-arid regions, suffer from perpetual droughts interrupted by flash floods. The drought-flood syndrome is symptomatic of an unsustainable landscape characterized by a perpetual imbalance in key components of the hydrological cycle.

Identification and implementation of strategies to alleviate droughts and to reduce their risks requires a thorough understanding of different types of drought. They may be: (1) *meteorological* or representing a long term decline in precipitation, (2) *hydrological* or representing reductions in stream flow and depletion of aquifers, (3) *pedological* or representing decline in soil water storage capacity caused by soil degradation (physical, chemical, biological), (4) *agronomic* or resulting in low available water reserves at critical stages of plant growth, and (5) *sociological* or resulting from significant increases in demand over the supply. Both pedological and agronomic droughts can be alleviated through creating sustainable landscapes and

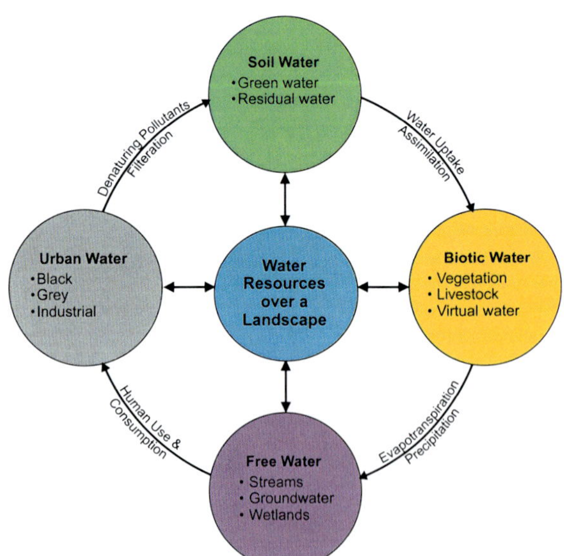

Fig. 12.5 The interconnectivity among different types of water over a landscape

Table 12.1 Population growth in African cities (UN Publication ST/ESA/SER. A/274 2008)

City	Population (10^6)			Growth (%/year)
	1975	2007	2025	
Accra	0.7	2.1	3.4	2.93
Addis	0.9	3.1	6.2	10.6
Dar	0.6	2.9	5.7	4.39
Kinishasa	1.5	7.8	16.8	3.89
Lagos	1.9	9.5	15.8	4.44
Nairobi	0.7	3	5.9	3.87

adopting recommended management practices (RMPs) for soil, water, crops, animals, vegetation, and terrain. Terrain management through the creation of micro-catchments and the installation of structures to harvest and recycle water (Fig. 12.5) are important. Converting gray water into green water is a very important component of the process of recycling water and nutrients among large and growing urban centers (Table 12.1), following the nexus approach (Lal 2013).

12.5 Sequestering Carbon Within a Landscape

Landscapes have a large C sequestration capacity and can be sinks for atmosphere CO_2. The latter can be sequestered in vegetation, soil, and wetlands (Fig. 12.6). The terrestrial C pool within a landscape is influenced by the vegetation cover. In general, the terrestrial C pool is more for landscapes containing woody perennials (trees) than annual grass coverage. However, the soil C pool may actually be more under grass-type terrains, such as prairies and steppes that under forest vegetation cover. The terrestrial C pool within a landscape can be enhanced by restorative land use and use of RMPs for soil, water, vegetation and animals (Fig. 12.7). Both create a positive ecosystem C budget. Important among RMPs are conservation agriculture (CA) mulch farming, cover cropping, integrated nutrient management (INM), complex rotation, and integrating crops with trees and livestock (Lal 2015). The overall objective is to reduce losses, recycle waste and biomass, and restore degraded lands of ecosystems.

12.6 Managing Urban Ecosystems

Urbanization is on the rise. More than 70 % of world population will live in urban centers by 2050. Urban ecosystems are strongly modified by human activities and are influenced by their side effects (Western 2001). Cities in SSA are expanding rapidly as data in Table 12.1 shows. These data include population increases for Accra, Addis, Dar Es Salaam, Kinshasa, Lagos and Nairobi.

Cities are becoming the predominant engines of socio-economic development. Unfortunately they are also the major sources of environmental problems (Wu 2013). Global sustainability will depend on urban ecosystems and their successful transition toward sustainability. Urban areas are ecological hot spots and principal drivers of environmental change at multiple scales (Grimm et al. 2008). Resilience and adaptability of urban landscapes are-essential to sustainable transform social/ecological systems (Walker et al. 2004). Improving the sustainability of urban landscapes is a

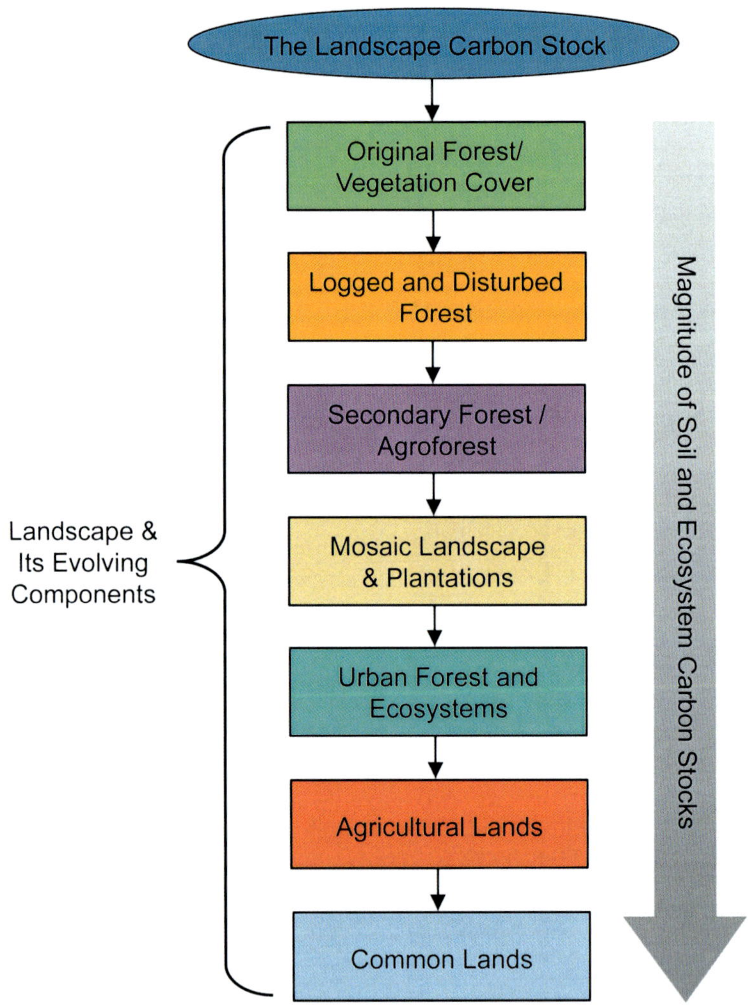

Fig. 12.6 Components of the landscape carbon stock

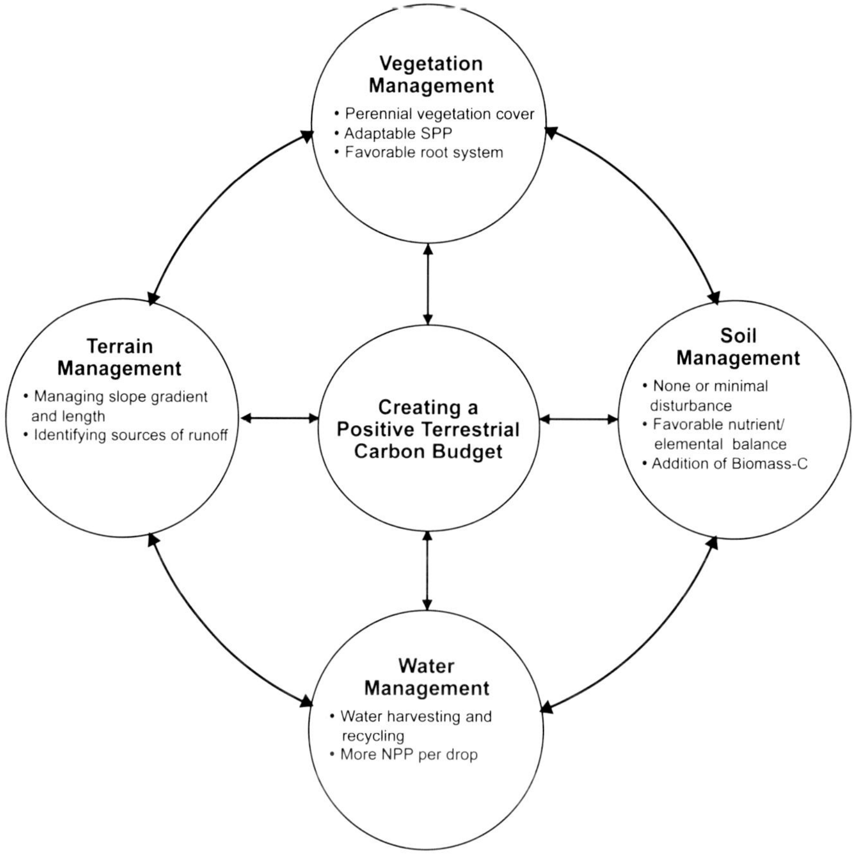

Fig. 12.7 The process of increasing the terrestrial carbon pool within a landscape

Table 12.2 Urban agriculture in Africa

City	Population in 2015 (10^6)	Total area (km^2)	Area cultivated		Average plot size (m^2)	Families practicing UA (%)
			(10^3 ha)	% of city area		
Harare	2.1	872	9.3	16	50–200	–
Nairobi	4	700	–	–	100	20
Dar Es Salaam	3.8	650	34	4	700–950	37

Modified from FAO (2008)

critical dimension of attempts to improve the environment broadly writ. Key activities that can help sustain urban landscapes are:

- Reducing storm water runoff (green roof),
- Enhancing water use efficiency (WUE) within landscapes,
- Bio-filtering of waste through wetlands,

- Using gray water responsible for irrigation purposes,
- Creating/enhancing wildlife habitat in urban environments (urban forestry),
- Developing energy efficient landscape design (shade trees),
- Providing permeable paving materials,
- Using composite wood products,
- Composting and recycling (kitchen/garden waste),
- Using renewable energy (solar panels),
- Promoting urban agriculture, and
- Promoting sustainable development goals of the U.N.

Implementation of these strategies involves: (1) creating green buildings with renewable energy, (2) growing food within the urban center to meet 10–20 % demand of fresh food (Table 12.2), (3) recycling gray water and nutrients in human and animal waste to produce food, and (4) creating green roots within urban center.

12.7 Greenhouse Farming and Use of Non-soil Culture

Creating synthetic soils for use in greenhouses (on roof top) to produce fresh vegetables is important to meeting the growing food demands of SSA. Similarly, recycling gray water (contaminated by human waste) is important for food production by using hydroponics, aquaponics, aeroponics, and aquaculture. These systems of food production, widely used in Asia, need to be adapted and used in SSA. Proper management of urban and suburban landscapes can also help substantially to mitigate climate change by enhancing biodiversity (Scyphers and Lerman 2014).

12.8 Research and Development Priorities

The traditional approach to ecological restoration has been criticized as fragmented and idealistic which does not relate to the real world situation (Choi et al. 2008). Therefore, appropriate and futuristic strategies must consider the dynamic nature of ecological communities with multiple trajectories, and connect landscape elements in order to strengthen ecosystem functions and services (Choi et al. 2008). Wu and Hobbs (2002) summarized key issues in landscape ecology. They highlighted the following: (1) using a multi-disciplinary approach, (2) combining basic and applied research, (3) developing conceptual/theoretical base to do so, (4) improving appropriate content in the curricula of school systems in order to impact education and training, (5) creating international cooperation, (6) strengthening communication with policy makers, (7) identifying causes, processes, and impacts land use/cover change, (8) understanding landscape complexity and non-linear dynamics, (9) developing procedures to scale up practices identified at the nanoscale or molecular level to kilometer scale at watershed level, (10) ensuring that relative ecological processes are applied to landscape ecologies, (11) integrating the human dimensions into landscape ecology, (12) optimizing landscape patterns, (13) promoting landscape sustainability, and (14) strengthening landscape databases, particularly their accuracy.

Rather than separately analyzing the impacts of humans and nature on landscapes, it is important to embed science in conservation and conservation in societal values (Western 2001). Thus, futuristic restoration strategies must consider: (1) establishing self-sustainable ecosystems, (2) identifying multiple alternative goals and trajectories for unpredictable endpoints, (3) focusing on rehabilitation of ecosystem functions and services, (4) applying ecological principles to restoring landscapes and (5) acknowledging that restoration ecology is a value-laden applied science.

References

Choi YD, Temperton VM, Allen B, Grootjans AP, Halassy M, Hobbs RJ, Naeth MA, Torok K (2008) Ecological restoration for future sustainability in a changing environment. Ecoscience 15:53–64

FAO (2008) Urban agriculture in Africa. Rome, Italy

Grimm NB, Faeth SH, Golubiewski NE, Redman CL, Wu J, Bai X, Briggs JM (2008) Global change and the ecology of cities. Science 319:756

Hudson PF, LaFevor MC (2014) Managing and monitoring human impacts on landscapes for environmental change and sustainability. J Environ Manag 138:1–3

Lal R (2013) The nexus approach to managing water, soil and waste under changing climate and growing demands on natural resources. In: UNU-FLORES "Advancing a nexus approach to the sustainable management of water, soil and waste: white book." UNU-FLORES

Lal R (2015) Sequestering carbon and increasing productivity by conservation agriculture. J Soil Water Conserv 70(3):55A–62A

Masnavi MR (2013) Environmental sustainability and ecological complexity developing an integrated approach to analyse the environment and landscape potentials to promote sustainable development. Int J Environ Res 7(4):995–1006

Sayer J, Sunderland T, Ghazoul J, Pfund J, Sheil D, Meijaard E, Venter M, Boedhihartono AK, Day M, Garcia C, van Oosten C, Buck LE (2012) Ten principles for a landscape approach to reconciling agriculture, conservation, and other competing land uses. PNAS 110:8349–8356

Scyphers SB, Lerman SB (2014) Residential landscapes, environmental sustainability and climate change. In: Research in urban sociology, sustainable cities: global concerns/urban efforts: vol 14. ISBN 978-1-7844-1-058-2

Walker B, Holling CS, Carpenter SR, Kinzig A (2004) Resilience, adaptability and transformability in social–ecological systems. Ecol Soc 9(2):5

Western D (2001) Human-modified ecosystems and future evolution. PNAS 98:5458–5465

Wu J (2013) Landscape sustainability science: ecosystem services and human well-being in changing landscapes. Landsc Ecol 28:999–1023

Wu J, Hobbs R (2002) Key issues and research priorities in landscape ecology. An idiosyncratic synthesis. Landsc Ecol 17:355–365

Chapter 13
Economic Impact of Drip Irrigation Regimes on Sorghum Production in Semi-arid Areas of Tanzania

A.J. Mahinda, C.K.K. Gachene and M. Kilasara

Abstract The need for increasing crops yield production to justify economic returns, improve food security, sustain scarce water resources, and safeguard the environment under the challenges of climate change and variability demands the development of promising drip-irrigation regimes. Field trials were conducted in the semi-arid area of central Tanzania with the aim of assessing the impact of three drip-watering regimes on the production and economic returns from sorghum (*Sorghum bicolor*). The irrigation treatments were—irrigating early in the morning (EM), late in the evening (EL), and both early in the morning and late in the evening (ELE). Each treatment was replicated three times in a randomized complete block design. The maximum yield of 13.12 t/ha with an economic return of Tanzania shilling 6,675,900/= was obtained when sorghum was irrigated twice a day during the dry season. Although irrigating twice a day in the dry season resulted in higher yield, net income was higher (7,607,780/=) in the dry–wet season. The results indicate that irrigating early in the morning or late in the evening resulted in more yield than the rainfed condition. However, it was economically viable to irrigate twice a day because this had the benefit of generating higher economic returns in the study area.

Keywords Drip irrigation · Semi-arid areas · Watering regimes · Sorghum production · Economic returns · Climate change

A.J. Mahinda (✉) · C.K.K. Gachene
Department of Land Resource Management and Agricultural Technology (LARMAT), University of Nairobi, P.O. Box 29053-00625, Nairobi, Kenya
e-mail: athumanmahinda@gmail.com

M. Kilasara
Department of Soil Science, Sokoine University of Agriculture, P.O. Box 3008, Chuo Kikuu, Morogoro, Tanzania

A.J. Mahinda
Agricultural Research Institute-Makutupora, P.O. Box 1676, Dodoma, Tanzania

A.J. Mahinda
Department of Agricultural Engineering, University of Dar es Salaam, P.O. Box 35091, Dar es Salaam, Tanzania

R. Lal et al. (eds.), *Climate Change and Multi-Dimensional Sustainability in African Agriculture*, DOI 10.1007/978-3-319-41238-2_13

13.1 Introduction

Climate change has had more significant impacts on livelihoods in developing countries than developed countries (Thorton and Jones 2003; Intergovernmental Panel on Climate Change (IPCC) 2014). Although the impacts vary across regions, farming and food systems, households, and individuals, climate change has been observed to affect more inhabitants in arid and semi-arid areas (ASALs) (Shemdoe 2011; Jerry et al. 2012; Kilasara et al. 2015). According to Thorton and Jones (2003), there will be more increasing crop failure in arid and semi-arid areas, which dominate most countries in Sub-Saharan Africa. Agricultural activities in this region relies on rainfed condition, so any adverse effects from climate change could have devastating effects on crop production and, consequently, the livelihood of the majority (Challinor et al. 2005, 2009, 2010; Rodima-Taylor 2012). The problem of water scarcity as a result of climate change and its variability is expected to worsen food security and economic status in semi-arid regions if farmers continue to rely on rainfed agriculture (Beddington 2010; Rodima-Taylor 2012; Gimbage et al. 2014).

In Tanzania, the central zone contains many arid and semi-arid areas. The zone receives a very low amount of rainfall with erratic and unpredictable trends (Food and Agriculture Organization (FAO) 1987; Ministry of Agriculture Food Security and Cooperatives (MAFC) 2013; Kilasara et al. 2015). In most cases, the amount received (300–600 mm) is insufficient and poorly distributed to meet crop water requirements (FAO 1987; International Institute for Sustainable Development (IISD) 2006; Famine Early Warning Systems Network (FEWSNET) 2014). Farmers have applied various soil and water conservation measures to improve crop production. Pitting, traditional terraces (Matuta), hillsides sheets, and runoff utilization have been used to collect water for growing various crops (Hatibu and Mahoo 1999). However, due to a high rate of evapotranspiration demand and great intra- and inter-annual rainfall fluctuations, these techniques have consistently been unpromising. For a number of years, there has been persistent crop failure.

As in many other arid and semi-arid areas of Sub-Saharan Africa, drought is the major cause of poverty in the central zone of Tanzania, and the most vulnerable groups are women, children, and people with disabilities (Thurlow and Wobster 2003; Ahmed et al. 2011). Although the area experiences water scarcity; underground water resources, intermittent rivers, and rainwater harvesting techniques can be used to increase both cash and food crop production through efficient means of irrigation during dry periods. Based on current technology, the area has the potential to utilize drip irrigation to meet crop evapotranspiration demands (Sijali 2001; Mvungi et al. 2005; Shemdoe et al. 2009).

Many farmers are willing to practice drip irrigation in growing sorghum to enhance resilience to climate change by increasing yield production, which justifies economic returns and improves food security while conserving scarce water resource and the environment. However, there are no research data available on the frequency of crops irrigation, the amount of irrigation needed, and the appropriate time for irrigation to be practiced to boost yield production. This research, therefore, is intended to fill some gaps in knowledge on appropriate drip-watering regimes and their economic returns as the driving tool in increasing the climate change resilience of sorghum production in arid and semi-arid areas of Tanzania and the tropics in general.

13.2 Materials and Methods

13.2.1 Study Area

The study was conducted in the periurban area of Dodoma Municipality at Makutupora Agricultural Research Institute (ARI-Makutupora), which is located at 05° 58′S and 35° 57′E. The area is classified as a semi-arid area, characterized by a mono-modal rainfall pattern. Rainfall commences from December to April, followed by a long dry season from May to November. The annual rainfall in the area ranges from 300 to 600 mm, with a mean of 450 mm, and is marked by high intra- and inter-annual fluctuations in amount and distribution. Monthly temperatures vary from 15 to 35.1 °C (Viticulture Research and Training Centre-Makutupora (VRTC-Makutupora) 2014; Tanzania Meteorological Agency (TMA) 2014).

Based on the United States Department of Agriculture (USDA) soil classification system, the soil temperature regime of the area is hyperthemic, and the soil moisture regime is aridic. The area has undifferentiated soil of alluvium, dominated by a sandy clay loam texture. The physiographic position of the area is foot slope (piedmont), with a flat surrounding landform with a slope of about 1–2 % (Food and Agriculture Organization-United Nations Education Scientific and Cultural Organization (FAO-UNESCO) 1989; Soil Survey Staff 1990; Msanya and Budotela 1994).

13.2.2 Methods

13.2.2.1 Experimental Design

A randomized complete block design was used. Treatments were assigned to blocks in a random manner to avoid bias. In this research study, the treatments were three irrigation regimes: irrigation in the morning (EM), irrigation in the evening (LE), and irrigation twice a day (early in the morning and late in the evening) (ELE). The three treatments were replicated three times in two seasons (dry season and dry–wet season).

The size of each plot was 4.5 m × 5 m, spaced by a 1-m path, creating a total experimental area of 214.5 m^2. Macia variety was sown at a spacing of 25 cm × 90 cm. Each experimental plot had a total of 100 plants, adding to a 900-plants population in an entire experimental area. Water was supplied to every plant on a daily basis (McWilliams 2003; Stichler and Fipp 2003; Assefa et al. 2010). Some adjustments were made depending on the sorghum growth stage, weather, and soil moisture condition (Seleshi et al. 2009; FAO 2013). The pan evaporation method was employed to compute crop evapotranspiration (ETcr), as follows:

$$ETcr = ETo * Kc \tag{13.1}$$

$$ETo = Epan * Kpan \tag{13.2}$$

where ETcr = crop evapotranspiration, ETo = reference crop evapotranspiration, Kc = crop coefficient, Epan = pan evaporation, and Kpan = pan coefficient.

Tensiometer and gravimetric water-content methods were used to monitor the amount of water in the sorghum root zone. Three tensiometers were inserted above, at, and below the active root zone at depths of 10, 25, and 40 cm respectively. Augering was performed to confirm the validity of the tensiometers installed at different depths. The augured soils were oven dried at 105 °C to a constant weight for determination of gravimetric and volumetric water content.

13.2.2.2 Soil and Water Sampling for Laboratory Analysis

Systematic soil sampling and soil testing were conducted to determine the physical and chemical properties of the soil. The soil chemical properties—nitrogen (N), phosphorus (P), potassium (K), potential hydrogen (pH), percentage organic carbon (%OC), and soil texture—were determined following standard laboratory procedures developed by Okalebo et al. (2002). The quality of water needed for drip irrigation was tested using appropriate laboratory procedures for its pH, electoral conductivity (EC), sodium adsorption ratio (SAR), chloride ion (Cl^-), sodium ion (Na^+), calcium ion (Ca^{2+}), magnesium ion (Mg^{2+}), and K^+ (Katerji et al. 2003).

13.2.2.3 Economic Returns Analysis

The economic impacts of sorghum production under different drip-irrigation regimes were determined. Justification of the net profit was obtained by subtracting the total production cost from total income (Olorunsanya and Akinyemi 2004; Kraybill and Michael 2009; Kuboja and Temu 2013). The total production cost in this study was incurred for drip irrigation facilities, irrigation water, seeds, fertilizer, pesticides, weeding, bird scaring, and winnowing. The fixed and variable costs for sorghum production were both based on the area market price. The price of grain sorghum used was that offered by East Africa Brewery Limited and Tanzania Brewery Limited.

$$ER = (TR - TC) / ha \tag{13.3}$$

where ER = economic returns, TR = total income from sales, TC − total cost of production, and ha = area of production in hectare.

The harvested sorghum grains from different irrigation regimes were sun dried, and their moisture content were determined using a digital grain-moisture meter. Adjustment of grain weight was made by interpolating all measured grain moisture to 12 % grain moisture content. The grains were then weighed to correlate the sorghum's yield response and economic returns under different irrigation watering regimes.

13.3 Statistical Analysis

Data were analyzed using the simple statistical method: the linear, additive model of two-way ANOVA, assuming that observations and errors were normally distributed and that all observations within and across the sample were independent, with the same variance. GenStat Discovery 20 Version was used as an additional software tool to analyzing the data at a 95 % level of significance. The means were separated by employing the New Duncan's Multiple Range (Wim et al. 2007)

13.4 Results and Discussion

13.4.1 Sorghum Yield Response and Its Economic Impacts Under Different Drip-Irrigation Regimes

13.4.1.1 Effect of Irrigation Regimes on Sorghum Grain Yield

The response of sorghum grain yield to drip irrigation under various watering regimes was significantly different ($p < 0.001$) for the sorghum irrigated twice a day (ELE) compared to that irrigated once a day (EM, EL). The highest yield (13.212 t/ha) was recorded in the sorghum irrigated twice a day, and the lowest yield (6.822 t/ha) from the sorghum irrigated late in the evening (LE) during the dry season, as indicated in Table 13.1.

The observed yield trends in the dry and dry–wet seasons showed that irrigating twice a day gives more yield than irrigating once a day. The results showed no significant differences ($p > 0.05$) among the plots irrigated once a day, although numerically, EM performed better than EL, as indicated in Fig. 13.1.

The results showed that irrigating twice a day (ELE) resulted into 50.5 % higher in grain yield than irrigating once a day. The increased sorghum yield was a function of water stored in the rhizosphere without percolating. The results also show that it is possible to double sorghum grain yield by irrigating twice a day to maintain a constant soil-moisture supply at the effective root zone. This finding

Table 13.1 Sorghum yield grown under different drip-irrigation regimes

Season	Irrigation regime	Yield (t/ha)
1	EM	6.833*
1	LE	6.822*
1	ELE	13.212**
2	EM	8.383*
2	LE	8.170*
2	ELE	12.703**
	Season	1.090
LSD, (p = 0.05)	Irrigation regime	1.335
	Season * Irrigation regime	1.887
%CV		11.1
	Season	ns
F-STAT	Irrigation regime	***
	Season * Irrigation regime	ns

Means with a column followed by the same superscript are not significantly different according to LSD at a probability level of 0.05

Key: Significance level: $^{ns}p > 5$ %, non-significant , $^{*}p$ = 5–1 %, significant; $^{**}p$ = 1–0.1 %, very significant; $^{***}p < 0.1$ %, very highly significant; [1]Dry season, [2]Dry–wet season

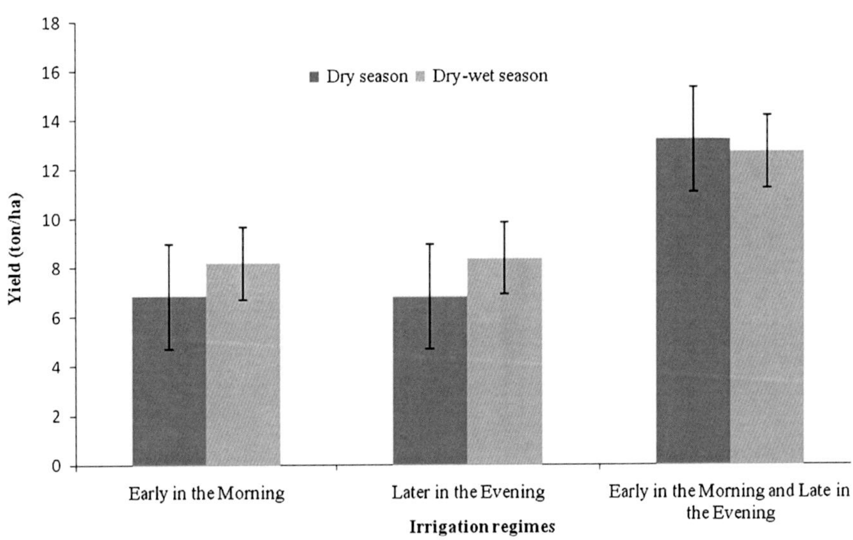

Fig. 13.1 Effect of different irrigation regimes on sorghum grain yield

proved that sorghum grain yield is more dependent on well-distributed soil moisture throughout the growing season based on crop water requirements than the total water amount. Irrigation twice a day kept additional soil moisture available for crop water requirement, which in turn created favorable microclimatic conditions for sorghum growth. The results suggested that irrigating once a day reduced the overall photosynthetic activities of the plant, as also reported by Zhang et al. (2004), Moroke et al. (2005), Stone and Schlegel (2006), Sepaskhah and Ghasemi (2008), and Dodd (2009).

The low yield from sorghum irrigated once in the morning was influenced by the higher evaporation demands and soil moisture deficit attributed in day time. The results also suggested that irrigating only late in the evening resulted in hidden water stress (water stress that decreases yield without crop leaves visibly wilting) which significantly reduced yield. Hidden water stress due to a moisture deficit reduces phosphoenolpyruvate carboxylase (PEPcase) activity, which physiologically decreases Ribulose-1 and 5-bisphosphate-carboxylase/oxygenase (Rubisco) regeneration and functionality. This process inhibits the functional activity of photosystem two (PSII) and consequently lowers the net photosynthetic rate and overall yield, as reported by Garrity et al. (1984) and Shangguan et al. (1999).

Compared to the average global sorghum production of 1.37 t/ha under rainfed conditions reported by Adzemi and Ibrahim (2014), the total yield obtained from this study was 5 and 6 times more for sorghum irrigated once a day; and 10 and 9 times more for sorghum irrigated twice a day in the dry and dry–wet seasons, respectively. The result showed that the lowest yield (8.38 t/ha) in the dry–wet season was 0.383 t/ha more than the maximum yield harvested under full-irrigated conditions in the United States, as reported by Blum (1996). This lowest yield is also far much more than the average yield of 0.9 t/ha produced in Tanzania under rainfed condition, as reported by Saadan et al. (2000), Mbwaga et al. (2007), and MAFC (2013).

The yields obtained from this study showed that a farmer could get 7.58, 7.59, and 14.68 t/ha more by irrigating LE, EM, and ELE, respectively, during dry season. It also showed that, if sorghum was supplemented with drip irrigation in LE, EM, or ELE during the dry–wet season, yields of 9.1, 9.3 and 14.1 t/ha more, respectively, would have been obtained.

13.4.1.2 Economic Returns of Sorghum Under Different Irrigation Regimes and Seasons

The highest net income (7,607,780/=) (1 TSH = USD 1500) was obtained from the treatment that was irrigated twice a day during the dry–wet season, while the lowest income (3,183,000/= TSH) was obtained from irrigation late in the evening during the dry season. It is important to note that all irrigation treatments generated better economic returns than solely rainfed farming (65,000/= TSH). However, the highest returns (7,607,780/= TSH) were obtained under the double irrigation regime (morning and evening) in the dry–wet season, as shown in Table 13.2.

Based on the cost–benefit ratio for sorghum production, the results showed that for every shilling invested, there was a return of 1403.5, 1399.6, and 1714.5/= TSH for EM, LE, and ELE, respectively, in the dry season. The corresponding values for the dry–wet season were 2752.4, 2614.9, and 2978.5/= TSH for the EM, EL and

Table 13.2 Economic return of sorghum grown under different irrigation regimes and seasons

Season	Irrigation regime	Yield (t/ha)	Market price (TSH/kg)	Other Cost + DI (000)	Irrigation water cost (000)	Total production cost (000)	Gross income (000)	Net income (000)	Cost–benefit ratio (000)
1	EM	6.833	800	655	1619.35	2274.35	5466.4	3192.05	1:1.4035
1	LE	6.822	800	655	1619.35	2274.35	5457.6	3183.25	1:1.3996
1	ELE	13.212	800	655	3238.70	3893.70	10,569.6	6675.9	1:1.7145
2	EM	8.383	800	655	949.81	1604.81	6706.4	4417.14	1:2.7524
2	LE	8.170	800	655	949.81	1604.81	6485.6	4196.34	1:2.6149
2	ELE	12.703	800	655	1899.62	2554.62	10,162.4	7607.78	1:2.9785

[1]Dry season, [2]Dry–wet season, *DI* Drip irrigation facilities

ELE treatments, respectively. Thus, despite the higher cost associated with the ELE treatment, it still had the best economic performance. It is also noted that, the dry–wet season outperformed the dry season. From an economic perspective, these results suggested that it is far more economically rewarding to do supplemental irrigation than pure irrigation.

The results indicated that, although ELE irrigation resulted in a higher yield in the dry season than the dry–wet season, net income from ELE was higher in the dry–wet season than the dry season. More water was bought for irrigation during the dry season (Table 13.3) than the dry–wet season (Table 13.4), which made the total production cost for the dry season higher by 931,880, 1,013,090, and 1,225,090/= TSH for ELE, LE, and EM, respectively.

In the dry season, sorghum was irrigated with gross amounts of water of 18, 93.9, 234.8, and 164 mm/crop in the initial development, middle and late growth stages, respectively, as indicated in Table 13.4. These were the largest gross amounts of water irrigated, compared to 18, 93.9, 101.8, and 100.3 mm/crop used for supplementary irrigation during the dry–wet season, as shown in Table 13.4. The total gross amount of water irrigated during the dry season was 510.7 mm/crop, while the total water amount used for the dry–wet season was 323.0 mm/crop.

Practicing supplementary drip irrigation saves approximately 187.0 mm/crop, generating the highest profitability comparable to pure drip irrigation. Thus, in the arid and semi-arid areas of Africa where more than 51 million ha are under sorghum production, ELE supplementary drip irrigation could save more than 423,866,666.7 m^3 of water in each cropping season.

13.5 Conclusion and Recommendations

Crop production is a business like any other. Therefore, quantifying crop yield against water use is of special importance in matching crop varieties with effective watering regimes. Guidelines on timing and levels of irrigation could maximize yield and economic returns under the challenges of climate change and variability.

The results showed that, under drip irrigation, the level of sorghum grain yield heavily depends on the irrigation regime practiced. When sorghum is irrigated both early in the morning and late in the evening, the yield is doubled compared to when it is irrigated either early in the morning or late in the evening.

The results also showed that harvesting a greater yield does not necessarily mean gaining more profit. The maximum average yield was 13.12 t/ha, with an economic return of 6,675,900/= TSH/ha, from sorghum irrigated ELE in the dry season, but the highest profit (7,607,780/= TSH/ha) was obtained from a 3.85 % less yield of sorghum irrigated ELE in the dry–wet season.

The result suggested that, in semi-arid areas, it is not only drip-irrigation watering that matters. The time and type of the irrigation regime also needs to compensate for the higher rate of evapotranspiration while replenishing the constant flow of water from soil to plant to atmosphere throughout the growth period without creating a deficit. Although irrigating early in the morning or late in the evening

Table 13.3 Consumptive water use and total cost of water incurred for pure irrigation during the dry season (August 18–December 19, 2013)

Crop stage	Crop kc	Growing days	Total rainfall (mm)	Gross water irrigation mm/crop/day	Supplemented water (mm/crop)	Cost of water Tsh/L	Gross water in L/ha	Cost of Irrigation water/ha
Initial growth	0.35	20	0	18.0	18	2.5	13,333.32	33,333.3
Development	0.75	30	0	93.9	93.9	2.5	104,333.229	260,833.073
Mid season	1.1	40	128.3	230.1	101.8	2.5	157,036.88	377,036.66
Late stages	0.65	30	58.7	159.0	100.3	2.5	116,999.883	278,610.833
Total		120	109.5	510.0	323.2	10	39,170.312	94,9813.866

Table 13.4 Ccnsumptive water use and total cost ɔf water supplemented during the dry–wet season (October 11, 2013–April 12, 2014)

Crop stage	Crop kc	Growing days	Total rainfall (mm)	Gross irrigation (mm/crop/day)	Supplemented water (mm/crop/day)	Water cost (TSH/L)	Gross water (L/ha)	Cost of irrigation (Water/ha)
Initial growth	0.35	20	0	18.0	18	2.5	13,333.32	33,333.3
Development	0.75	30	0	93.9	93.9	2.5	104,333.229	260,833.1
Mid-season	1.1	40	128.3	230.1	101.8	2.5	15,7036.88	377,036.7
Late stages	0.65	30	58.7	159.0	100.3	2.5	116,999.883	278,610.8
Total		120	187.0	510.0	323.0	10	39,170.312	949,813.9

produces more yield than rainfed conditions, it is more economically viable to irrigate twice a day to achieve higher yields and profits in both seasons. The results also suggested that, to get the most benefit from each shilling invested, it is more economically feasible to supplement water to the crop in dry–wet season than practicing pure irrigation. However, this might change depending on price fluctuations determined by market demand and supply.

References

Adzemi MA, Ibrahim W (2014) Effect of regulated deficit irrigation on photosynthesis, photosynthetic active radiation on yield of sorghum cultivar. J Biol Agric Healthc 4 (2):2224–3208

Ahmed SA et al (2011) Climate volatility and poverty vulnerability in Tanzania. Glob Environ Change 21(1):46–55

Assefa Y et al (2010) Grain sorghum water requirement and responses to drought stress: crop management. http://www.plantmanagementnetwork.org/pub/cm/reviw/2010. Accessed 24 May 2014

Beddington J (2010) Food security: contributions from science to a new and greener revolution. Philos Trans R Soc B 365:61–75

Blum A (1996) Crop responses to drought and the interpretation of adaptation. J Plant Growth Regul 20:135–148

Challinor AJ et al (2005) Simulation of the impact of high temperature stress on annual crop yields. Agric For Meteorol 135(1–4):180–189

Challinor AJ et al (2009) Crops and climate change: progress, trends, and challenges in simulating impacts and informing adaptation. J Exp Bot 60(10):2775–2789

Challinor AJ et al (2010) Increased crop failure due to climate change: assessing adaptation options using models and socio-economic data for wheat in China. Environ Res Lett 5 (3):034012

Dodd IC (2009) Rhizosphere manipulations to maximize 'crop per drop' during deficit irrigation. J Exp Bot 60:2454–2459

Famine Early Warning System Network (2014) Stable food prices across the country ensure food access for poor households. http://www.fews.net/east-africa/tanzania/remote-monitoring-report/wed-2014-04-30. Accessed 15 May 2014

Food and Agriculture Organization (2013) Yield response to water: the original FAO water production function. http://www.fao.org/docrep. Accessed 15 May 2014

Food and Agriculture Organization—United Nations (1987) Committee on agriculture (9th session). Improving productivity of dryland areas. http://www.fao.org/docrep/meeting/027/z1908e.pdf. Accessed 15 May 2015

Food and Agriculture Organization—United Nations Educational, Scientific and Cultural Organization (1989) Soil map of the world. Soil resources report 60. Food and Agriculture Organization of the United Nations, Rome

Garrity DP et al (1984) Change in grain sorghum stomata and photosynthetic response to moisture stress across growth stages. Crop Sci 24:441–446

Gimbage EM et al (2014) Livelihood option and food insecurity in marginal and semi-arid areas of Same district, Tanzania. Int J Agric Ext 2014:75–91

Hatibu N, Mahoo H (1999) Rainwater harvesting technologies for agricultural production. A case study for Dodoma, Tanzania. In: Kaumbutho PG, Simalenga TE (eds) Conservation tillage with animal traction. A resource book of the Animal Traction Network for Eastern and Southern Africa (ATNESA), Harare, Zimbabwe. ITDG publishing, pp 161–171

International Institute for Sustainable Development (2006) Arid and semi-arid lands: characteristics and importance. http://www.iisd.org/casl/asalprojectdetails/asal.htm. Accessed 15 May 2015
Intergovernmental Panel on Climate Change (2014) Climate change 2014: impacts, adaptation and *vulnerability: regional aspects*. Cambridge University Press. https://scholar.google.co.jp/scholar?q=Intergovernmental+Panel+on+Climate+Change+(2014)+Climate+change+2014:+impacts,+adaptation,+and+vulnerability&hl=en&as_sdt=0&as_vis=1&oi=scholart&sa=X&ved=0ahUKEwjY9b7MosjLAhVEI5QKHdx7CQsQgQMIGjAA. Accessed 5 April 2014
Jerry K et al (2012) Climate change impacts on crop productivity in Africa and South Asia. Environ Res Lett. doi:10.1088/1748-9326/7/3/034032
Katerji N et al (2003) Salinity effect on crop development and analysis of salt tolerance according to several classification methods. Agric Water Manag 62(1):37–66
Kilasara M et al (2015) Effect of in situ soil water harvesting techniques and local plant nutrient sources on grain yield of drought-resistant Sorghum varieties in semi-arid zone, Tanzania. In: Sustainable intensification to advance food security and enhance climate resilience in Africa. Springer, Berlin, pp 255–271
Kraybill D, Michael K (2009) Analysis of relative profitability of key Ugandan agricultural enterprises by agricultural production zone. Uganda Strategy Support Program (USSP) brief no. 7. International Food Policy Research Institute, Washington, DC
Kuboja NM, Temu AE (2013) comparative economic analysis of tobacco and groundnut farming in Urambo district, Tabora, Tanzania. J Econ Sustain Dev 4(19):2222–2855
Mbwaga AM et al (2007) Integrated striga management to meet sorghum market demands in Tanzania. In: Ejcte G, Giessel J (eds) Integrated new technologies for striga control towards ending the witch-hunt. Scientific Publishing Co Ltd., Singapore, pp 253–264
McWilliams D (2003) Drought strategies for corn and grain sorghum. Cooperative Extension Service, New Mexico State University, Las Cruces, New Mexico
Ministry of Agriculture and Food Security (2013) Irrigation water management, Technical handbook no. 6. Ministry of Agriculture and Food Security, Dar es Salaam, Tanzania, pp 112–118
Moroke TS et al (2005) Soil water and root distribution of three dry land crops. Soil Sci Soc Am J 69:197–205
Msanya BM, Budotela GMR (1994) Soil classification: Vyeyula—Makutupora; thereafter referred to as DU3. Makutupora—Dodoma Capital city. A part of thesis submitted in partial fulfillment of the requirement for the award of the degree of master of science in soil science and land management at Sokoine university of Agriculture, pp 94–17
Mvungi A et al (2005) Management of water for irrigation agriculture in semi-arid areas: problem and prospect. Phys Chem Earth Parts A/B/C 809–817(30):11–16
Okalebo JT et al (2002) Laboratory methods of soil and plant analysis; a working manual, 2nd edn. TSBF-CIAT and SACRED Africa, Nairobi
Olorunsanya EO, Akinyemi OO (2004) Gross margin analysis of maize based cropping system in Oyo State Nigeria. Int J Agric Dev 5:129–133
Rodima-Taylor D (2012) Social innovation and climate adaptation: local collective action in diversifying Tanzania. Appl Geogr 33(1):128–134
Saadan HM et al (2000) Performance of sorghum variety Macia under multiple environments in Tanzania. ISMN 41(10–12):41
Seleshi BA et al (2009) Soil-plant-water relationship; module 3. Improving Productivity and Market Success (IPMS) of Ethiopian Farmers Project and Livestock Research Institute (ILRI), Addis Ababa, Ethiopia
Sepaskhah AR, Ghasemi MM (2008) Every-other furrow irrigation with different irrigation intervals for sorghum. Pak J Biol Sci 11(9):1234–1239
Shangguan Z et al (1999) Interaction of osmotic adjustment and photosynthesis in winter wheat under soil drought. J Plant Physiol 154:753–758
Shemdoe RS (2011) Tracking effective indigenous adaptation strategies on impacts of climate variability on food security and health of subsistence farmers in Tanzania. In: African Technology Policy Studies Network Working Paper Series No. 51. http://www.atpsnet.org/Files/wps51.pdf. Accessed 14 June 2015

Shemdoe RS et al (2009) Increasing crop yield in water scarce environments using locally available materials: an experience from semi-arid areas in Mpwapwa District, Central Tanzania. Agric Water Manag 96(6):963–968

Sijali IV (2001) Drip irrigation; option for smallholder farmers in Eastern and Southern Africa. RELIMA, Technical Handbook, vol 24. Published by Sida's Regional Land Management Unit, pp 14–51

Soil Survey Staff (1990) Keys to soil taxonomy. United States Department of Agriculture. Soil Management Support Services. SMSS Technical Monograph No. 19. Virginia Polytechnic and State. United States Department of Agriculture. Natural Resources Conservation Service, p 644

Stichler C, Fipp G (2003) Irrigation in south and south central Texas. Texas Cooperative Extension Pub. L-54342-03. Texas A&M University, College Station, TX

Stone LR, Schlegel AJ (2006) Yield-water supply relationship of grain sorghum and winter wheat. Agron J 98:1359–1366

Tanzania Meteorological Agency (2014) Weather forecast; ISO 9001:2008 Certified. In: Aeronautical Meteorological Services. http://www.meteo.go.tz/index.php. Accessed 5 May 2014

Thorton PK, Jones PG (2003) The potential impacts of climate change in tropical agriculture: the case of maize in Africa and Latin America in 2055. Glob Environ Change 13:51–59

Thurlow J, Wobster P (2003) Poverty focused. Social accounting matrices for Tanzania. TMD discussion paper no. 112. International Food Policy Research Institute—Trade and Macroeconomics Division, Washington, DC

Viticulture Research and Training Centre-Makutupora (2014) Background information: Central zone is characterized by mono-modal rainfall pattern. http://www.erails.net/TZ/vrtc-makurupora/vrtc-makutupora/. Accessed 12 July 2014

Wim B et al (2007) GenStat discovery 4th edition for everyday use. ICRAF, Nairobi, p 117

Zhang Y et al (2004) Effect of soil water deficit on evapotranspiration, crop yield and water use efficiency in the North China Plain. Agric Water Manag 64:107–122

Chapter 14
Social Aspects of Water Governance in the Context of Climate Change and Agriculture

Richard Asaba Bagonza

Abstract Climate change is undoubtedly one of the greatest threats to humankind. Changes in temperature and precipitation will affect the supply, access and governance of water resources in many developing communities. The ecological, physical, technological and economic impacts of climate change on water have largely been studied, but there is little and sketchy evidence on the links between climate change, human relations, water governance and agriculture. The social dimensions of climate change, particularly in the governance of water in resource poor African agricultural-based contexts have not been given much attention, yet they are essential for adaptive and sustainable water management. This paper seeks to identify the key social issues affecting water governance in the face of climate change among small-holder farming communities in Sub-Saharan Africa. It is a sociological analysis of water governance, which encompasses the interrelated aspects of actors, resources, 'mechanisms' or arrangements of access to water, outcomes and water management processes. From the available evidence, the major social factors that are central to water governance in the context of climate change include: changes in water resources, which lead to reduced basic access to water for crops, livestock and humans; institutional and policy changes; changes in property and access rights; prohibitive water pricing or payments; conflicts resulting from climate change induced water scarcity and droughts; and gender norms and relations which cut across the water governance spectrum. These issues, which are not exhaustive, need to be integrated in adaptation planning and interventions so as to improve the adaptive capacity and sustainability of water governance among small-holder farmers in Sub-Saharan Africa.

Keywords Social · Water governance · Climate change · Vulnerability · Adaptation · Agriculture

R.A. Bagonza (✉)
Department of Environmental Management, Makerere University, P.O. Box 7062, Kampala, Uganda
e-mail: rasabab@yahoo.com

R. Lal et al. (eds.), *Climate Change and Multi-Dimensional Sustainability in African Agriculture*, DOI 10.1007/978-3-319-41238-2_14

14.1 Introduction

Climate change is regarded as one of the greatest threats to humankind. Projected changes in temperature and precipitation, as well as changes in the frequency and intensity of extreme events will have huge ecological, physical, economic and social impacts globally. The impacts of climate change will mostly affect developing countries due to their high vulnerability[1] (Adger 2006; Filho 2011; Godden et al. 2011; IPCC 2007; Parry et al. 2007; Toulmin 2009). Sub-Saharan Africa (SSA)[2] is said to be most vulnerable to climate change due to its geographical location in the tropics, dependence on climatic and natural resources, poverty and poor governance. All of these factors limit the capacity of crop and livestock farming communities to adapt to change (ADB 2011; Challinor et al. 2007; Chang et al. 2010; Dile et al. 2013; Downing et al. 1997; IPCC 2001; Kusangaya et al. 2014; Morton 2007; Ravindranath and Sathaye 2002). Climate change is already affecting agriculture, the major source of survival and livelihood in SSA (ADB 2011; Kurukulasuriya et al. 2006; Verhagen et al. 2004) as well as related areas such as water supplies and systems, food security, vulnerability and disaster risk reduction, and social protection (Birkmann et al. 2009; Brown et al. 2007; FAO and WWC 2015; Filho 2011; HLPE 2015; Lal et al. 2015; Lobell and Burke 2010).

The ecological (that is physical, chemical and biological processes) and technological impacts of climate change on water and other ecosystems, which are essential for sustainable adaptation[3] and resilience have largely been studied. For example, it has been reported that increases in temperature will increase the demand for water by plants and crops and may change the distribution and suitability of agro ecological zones (Challinor et al. 2007; Downing et al. 1997; Kusters and Wangdi 2013; Ravindranath and Sathaye 2002). A number of IPCC reports have highlighted the hydrological changes likely to result from global warming and how these are bound to reduce on the availability of water. They warn that by 2020, between 75 and 250 million people across 25 African nations will be at a greater risk of water stress (IPCC 2001, 2007). Others contend that the impact of climate change will be most strongly felt through changes in the distribution of water around the world and its seasonal and annual variability (e.g., Stern 2006; Mosello 2015). Research on the importance of water quality and quantity to smallholder farmers' and pastoralists' resilience to climate change in Sub-Saharan countries has

---

[1]In this paper, vulnerability is defined as the state of individuals, groups, and communities to cope with and adapt to any external stress placed on their livelihoods and well-being (see Adger 2006; Adger and Kelly 1999). Stresses here include extreme events related to climate variability, as pointed out by IPCC (2001). This definition defined vulnerability to be a socially constructed phenomenon that is key to the well-being of farmers and vulnerable groups with regard to their access to and governance of water, the core aspects of this analysis.

[2]This includes about 50 countries south of the Sahara Desert.

[3]Adaptation is used here to simply mean 'being prepared to cope with impacts' (Toulmin 2009) or to take actions to minimize any potential negative effects or enhance the ability of society and nature to adapt to changes (Chang et al. 2010).

been done in Uganda, Mali and Senegal (Cabell and Oelofse 2012; Dixon et al. 2014). Other studies on the economics of climate change have demonstrated how changes in temperature or rainfall lead to changes in demand for and supply of weather-sensitive goods and services (Burroughs 2005: 2), and how the changes or adaptation measures affect farmer's incomes through notable reductions in crop and livestock production (e.g., Chambwera and Stage 2010; Kurukulasuriya et al. 2006; Di Falco 2014).

Limited and sketchy evidence exists on social aspects of climate change. It has been documented that small but sustained changes in climate or weather anomalies can undermine social structures. It is also known that natural and social scientists need to work together to effectively address the complexities of climate change (Burroughs 2005: 8; Martens and Chang 2010: 4; Ravindranath and Sathaye 2002; Toulmin 2009). Various studies have indicated that social, cultural and governance issues determine vulnerability and can hamper adaptation, resilience and risk reduction (e.g., Adger 2006; Chang et al. 2010; McCornick et al. 2013; Miller and Belton 2014; Mosello 2015; Ribot et al. 1996). In this context, the links between climate change, human behaviour, water governance and agriculture are critical. Less attention has been given to studying the interactions between climate change and the dynamic socio-cultural processes that affect access to water and its governance among small-holder farming communities in SSA. Some studies on the social impacts of climate change have explored issues such as water technologies (e.g., Clements et al. 2011; Dile et al. 2013), extreme events (e.g., Morton 2007; Kusangaya et al. 2014) and conflicts and security (e.g., Brown et al. 2007). These studies have been general in nature with less attention to the overall dynamics of water governance. Understanding the social dimensions of climate change among resource poor farmers in SSA enables the recognition of the complex social, cultural and institutional factors that are essential for adaptive and sustainable water management (Burroughs 2005; Dietz et al. 2004; Godden et al. 2011).

Thus, this paper explores the key social issues underlying water governance in the face of climate change among farming communities in SSA. The impacts of climate change are multiple and diverse. For this reason, this paper will focus on the social impacts, or 'dislocations' (Burroughs 2005: 6) related to climate variability, mainly seasonal changes in precipitation and extreme events such as drought and floods, which are the most critical in SSA (Djalante et al. 2013; Downing et al. 1997; Dietz et al. 2004). Vulnerability, adaptation and resilience[4] are the core subjects of social scientific climate research (Chang et al. 2010). They are used interchangeably in the literature in regard to the access of farming communities to water and to water governance processes (e.g., Burroughs 2005: 3). This review also focuses on rural small-holder farming communities, whether rain-fed or irrigation-based rather than on analyses of management of large water basins.

---

[4]Resilience is the ability of a region, country, city, village or household to protect itself from adverse impacts and recover from damage (Toulmin 2009).

First, an understanding the importance of climate variability and extreme events to agriculture and water resources in Sub-Saharan Africa is given. It is followed by a sociological elaboration of the concept of water governance. This includes a discussion of the interrelated aspects of actors, resources, 'mechanisms' or arrangements of access to water (including water resources themselves, institutions, and water collection among others), outcomes and water management processes. Using evidence from the available literature, the key social issues in water governance in the context of climate change are then identified and discussed.

14.2 Impacts of Climate Change on Agriculture and Water: Climate Variability and Extreme Events

Climate change and its associated weather changes and extreme events affect agriculture and water, the sectors that are most vulnerable (ADB 2011; Birkmann et al. 2009; HLPE 2015; Ludwig et al. 2009) in various ways. In a review of climate change in Africa and its impacts on agriculture and water, Downing et al. (1997) note that much of the water in SSA is used for farming, an economic activity that is dependent on climate with minimal investment in irrigation. In 2002, only 6 % of the crop area in Africa was irrigated (Ravindranath and Sathaye 2002). Camilla Toulmin (2009: 34) provides more recent estimations, suggesting that 95 % of all land under cultivation for both food and industrial crops is rain-fed. The 5 % under irrigation is primarily for industrial crops, such as cotton, tobacco and sugar (Toulmin 2009: 34).

Small-holder agriculture is greatly affected by fluctuations in climate, which lead to low productivity, poor yields and food insecurity (Challinor et al. 2007; Dile et al. 2013; Glantz et al. 2009; Lal et al. 2015). Iglesias and Rosenzweig (2010) have analyzed the potential impacts of climate change on food production using detailed and up-to date data from IPCC emissions scenarios. They note that climate change will negatively impact agricultural yields of wheat, rice and maize from 1990 to 2080 in Africa, a fact that has been confirmed by other studies. The frequent droughts in the Horn of Africa for example led to a 45 % reduction in wheat yields in 2009 (Davies 2014, cited in HLPE 2015: 32) and more crop failures have been reported in the region, with countries such as Ethiopia and Somalia recording higher incidences of food insecurity and hunger among the poor rural farming communities.

Turning to water, a recent report by the UN High Level Panel of Experts on Food Security and Nutrition warns that climate change will put more strain on fresh water supplies, and that the situation will be worse in low rainfall areas (HLPE 2015). Emphasising the consequences previously highlighted by Downing et al. (1997) and Ludwig et al. (2009), the report states that climate change affects precipitation, runoff, hydrological flows, water quality, water temperature and groundwater recharge, and that it will impact rain-fed systems through precipitation patterns, and irrigated systems through availability of water at basin level. Perhaps a more pertinent observation made by the committee is that "climate change will modify crop and livestock water requirements, and impact water flows and water temperatures in water bodies, which will impact fisheries" (HLPE 2015: 12).

Brown et al. (2007), Toulmin (2009) and HLPE (2015) further state that reduced precipitation and increased evapotranspiration will result in more frequent extreme weather events, such as drought, floods and storms. These events will indisputably impact on the availability of water and food security, which will then have other social and economic consequences such as conflict, migration, crop failure and death or reduction in the value of livestock. Evidence exists that prolonged droughts and floods have led to water scarcity and stress which have disrupted pastoral livelihoods and markets in dry land areas in Eastern and Western Africa (e.g., Little and McPeak 2014; Nyong et al. 2006; Williams and Funk 2011).

14.3 Water, Water Governance, and Climate Change

Globally, there is abundant supply of water that can meet human needs. However, it is not evenly distributed (e.g., HLPE 2015; Rogers 2006; UNDP 2006). Much of SSA is endowed with rainfall and significant water resources. For example, Central Africa was estimated to have 1946 km^3 in the early 1990s (World Resources Institute 1994). This water is for household use, agricultural production, energy or hydropower and industry, and is of great importance to the economies of the countries in the region (e.g., Toulmin 2009: 32).

However, water systems in the region are vulnerable to climate change due to physical factors such as high seasonal and inter-annual variations, and river basins that span political boundaries. Vulnerability is also a result of poorly developed national and regional institutions and equity issues (Challinor et al. 2007; Cooper 2004; Downing et al. 1997; IPCC 2001; UNDP 2006). Various studies have shown how changes in temperature will affect most of the major river and lake basins in SSA, such as the Nile, Zambezi, Victoria and Tanganyika. Increases in temperature and precipitation are likely to result in decreases in river flow and significant reductions in water availability that may be reduced by investment in improved agricultural technology and irrigation (e.g., Strzepek et al. 1995; Downing et al. 1997; Toulmin 2009).

As already noted, social factors, or what Rogers (2006) calls 'human interference' in form of water governance, property rights and population, lead to changes in amounts of water used by different populations as well as the quality of water used by them. And since most of the water in SSA is used for agriculture, variations in climate and rainfall amounts further call for improved use and governance of water at farm, household, community and national levels.

Water governance is a relatively new concept which has been defined in various ways. Following their prominent work, *Effective Water Governance*, Rogers and Hall (2003: 16) define water governance as *the range of political, social, economic and administrative systems that are in place to develop and manage water resources, and the delivery of water services at different levels of society.* For Rogers and Hall, the 'systems' include policies, institutional frameworks and relationships that are important for water development and management, among socially accepted institutions. Rogers and Hall emphasize decentralisation, institutional reforms and gender, as important components of effective water governance (2003: 38). A less general definition of water governance that emphasizes decision-making and accountability is proposed by HLPE. They define water governance to be *a set of political, social, economic and administrative systems, rules and processes which determine the way decisions regarding the management and use of water resources, and the delivery of water services, are taken and implemented by the various actors; and through which decision-makers are held accountable* (HLPE 2015: 15).

A contextual definition of water governance is provided by Franks and Cleaver (2007). They propose a framework for analysing water governance among the poor. To do so they build on Rogers and Hall (2003), who studied empirical case studies from Africa, including the Kimani catchment and the Usangu wetland in South-Western Tanzania in particular. Franks and Cleaver also draw on Giddens and Long's social theories (e.g., Giddens 1984; Long 1992). They define water governance to be a *system of recursive actors, resources, mechanisms, outcomes and processes which mediate a society's access to water* (Franks and Cleaver 2007: 303). A*ctors* are 'non-state agents', such as the users of water, Non-governmental Organisations (NGOs) and the private sector, all of whom share an interest and role in addressing water issues in a socially acceptable manner. The *resources* are water, other water related infrastructure through which water is conveyed from the source to users and returned, and technology which refers to a 'human-made' resource or physical structure, such as hand pumps, irrigation canals and other irrigation technologies, needed to meet the demands of users. Related 'non-material' resources are rules and practices that determine how water is used or accessed which involve institutions, social resources (or structures); rights and entitlements, and human capacities. They add that *mechanisms*, or *arrangements that can be negotiated and which are likely to change over time* (Franks and Cleaver 2007: 295) include technology or water control structures for domestic or irrigation purposes, formal and informal water institutions, such as water user associations and local associations, customary and modern water rights (the 'ultimate result of water governance' according to Plummer and Slaymaker 2007: 2), payments and others such as queues and labour. They designate *outcomes*, in most cases gendered, as basic access to water of a particular quality, quantity and timing, livelihoods, and conflicts over access due to inclusion or exclusion. The *processes*, according to Franks and Cleaver include negotiation and decision-making and political voice, all of which are characterised by agency and power. In sum, water governance incorporates actors, water resources, mechanisms of access, positive or negative outcomes, and processes of management and practice.

A key question is how climate change affects access to and management of water resources. First, what social factors impact the ability of farmers, other individuals, communities or other vulnerable groups to access and effectively manage water resources in the face of climate change? The IPCC (2001) and Sen (2000) attempt to provide useful insights about the social determinants of adaptation. They argue that the ability of communities or socio-economic groups to adapt and cope with climate change is a function of governance (and national security strategies), wealth and economic development, technology, information, skills, infrastructure, institutions, and equity. In his earlier work, Sen (1990) suggests that capabilities motivate a person to act. Capabilities are sources of power and agency and enable individuals to realize their potential as human beings in the sense of their being healthy or adequately nourished and doing, such as developing skills, participating socially and making or exercising choices. Capabilities also represent assets or capital that people rely on to address unprecedented changes in climate which explains why the livelihoods framework (e.g., GDC 2003) has gained significance in climate change studies. This framework proposes five c's or 'the big capitals' as being key in determining individuals adaptive capacity, and these include: human (e.g., skills, knowledge, good health); natural (e.g., water resources, land types); financial (e.g., credit and saving); social (e.g., access to political power, social networks with neighbours, associations) and physical capital (basic infrastructure in this case for water management, transport and others) (GDC 2003: 3).

Chang et al. (2010: 1–10) discuss similar social and behavioural determinants of climate change. They conclude that technological options, economic resources, human and social capital and governance determine adaptive capacity. However, these factors are not specific to water governance among small-holder farming communities. Cooper (2004) notes that research on water management in rural areas in the context of climate variability and change requires, *inter-alia* understanding vulnerability-livelihood interactions and establishing related legal, policy and institutional frameworks. Following this line of inquiry identification of the key social factors in water governance in the face of climate change and agriculture was guided by three key questions. First, what is water governance and what social or human aspects of water governance could be critical under changing climatic conditions? Second, how does climate change impact access to and management of water among small-holder farming communities? And third, what social factors affect community vulnerability, adaptive capacity and resilience to access and management of water in the face of climate shocks? The social aspects of water governance, especially the 'realized' as stated in the literature are discussed in the next section.

14.4 Water Governance and Agriculture: Social Dimensions and Impacts

Several social aspects of water governance are central to understanding the changing climatic conditions in agricultural communities in SSA. Most salient among them are changing water resources and their associated technology or infrastructure; changing institutions (and policies); changing property rights; problematic water payments; conflicts; cultural norms and gender relations.

14.4.1 The Nature of Water Resources and Climate Change

Water resources, infrastructure or technologies can improve access and governance of water for domestic and irrigational or agricultural use (e.g., Franks and Cleaver 2007; Toulmin 2009; Kusangaya et al. 2014).

However, harsh conditions or climate-induced shocks may limit the ability of existing infrastructures to provide water to farmers and vulnerable communities. During droughts for example, some water sources dry up naturally due to reductions in water tables and other geogenic factors. This limits the ability of small-holder farmers or communities in general to physically access the water resources or technologies. The scarcity of water encourages communities to use the only few available water sources in their midst, leading to over-use that eventually limits basic access to water in terms of quality, quantity and timing. Even where there is irrigation, water resources become unreliable during long droughts, and may also be affected by salinization, further making them unsuitable for use by farmers (e.g., Downing et al. 1997; Fernald et al. 2015).

As to the resilience of pastoralism in SSA, particularly in the Horn of Africa, the Sahel in West Africa and parts of East Africa, Little and McPeak (2014) found that pastoralists, who also choose to migrate as an adaptation option are losing range lands due to socio-economic pressures such as population growth, encroachment by agriculturalists, irrigation, tourism, conservation programs and 'land grabbing' by outside investors. They further describe the loss of 'key resources', such as watering points, in range lands as "the greatest challenge to mobile pastoralism in the next 25 years" (Little and McPeak 2014: 3). In Kenya for example, climate-change induced drought limits physical access to water and pastures and forces pastoralists to go to overused areas, to encroach on conservation areas, such as forest reserves and national parks, and to go to cities such as Nairobi (Little and McPeak 2014: 3).

Climatic shocks also impact on water governance by changing or disrupting farmers' mechanisms of accessing water. For example, extreme events can destroy water sources, leading to inadequate access that also results in disease spread. The devastating floods followed by cyclone Eline in Southern Mozambique—particularly in Gaza, Sofala, Inhambane, Manica and Tete provinces in February and March 2000,[5] disrupted farmers' access to water, and many were forced to live in shelters (in temporary accommodation centres) with fluctuating water supplies, a situation that continues today. Due to inadequate access to safe water, farmers and 'flood survivors' as they were later called succumbed to dehydration, and diseases such as malaria and cholera (Africa Recovery 2000: 13).

[5]These floods killed more than 700 people, affected 1400 km of arable land and caused damage of over US$600 million (Martens and Chang 2010).

14.4.2 Institutional and Policy Changes

Governance structures or mechanisms of water governance consisting of formal and informal institutions with many 'actors', generally affect community vulnerability and adaptive capacity (Chang et al. 2010: 5). Water institutions, such as water user associations or irrigation associations, which are important in securing access to water and its management among smallholder farming communities may undergo positive or negative changes in the face of climate shocks (e.g., Toulmin 2009; Godden et al. 2011; Claruis et al. 2014).

Formal water user associations, irrigation associations or other formal, customary or indigenous institutions may sometimes change their organizational structure and operations during extreme events in ways that alter water governance among small holder farming communities. An example of such institutional 'bricolage' was reported in rural Iringa District in Tanzania, where formal and informal irrigation institutions interacted with other organisations to enable farmers to adapt to water scarcity and therefore mitigate the impacts of climate change (Kajembe et al. 2015). Another example is provided by Pavanello and Levine (in Little and McPeak 2014: 18). They describe how institutions, such as cross-border committees, blend formal and indigenous rules and mechanisms to create structures through which pastoral communities in parts of Western Africa negotiate or claim rights to access water resources during long droughts. This adaptation strategy has been hailed as a modern resilience strengthening approach to cross-border rangeland management.

14.4.3 Changing Property Rights

Property rights can be a water demand management tool (Downing et al. 1997; HLPE 2015) and an arrangement of access to water. They can change in the face of extreme events, to result in more efficient use of water.

Available evidence shows that agricultural communities respond to climatic changes through changing their arrangements or mechanisms of access to water by giving rights to particular groups of people or farmers to access water and also regulate the amounts to which they are entitled. In their attempt to measure the food security resilience of farming communities, Frankenberger and Nelson (2013) report that pastoral communities in West Africa and the Horn of Africa agree among themselves about how to use water and grazing lands during droughts. This suggests that climate change can induce changes in basic access to water, particularly the amounts that farmers can access. Property rights may be a good adaptive practice that ensures better distribution of water for all community members in the event of shocks or climatic hazards. However, they can also lead to conflicts, as discussed in Sect. 14.4.7.

14.4.4 Water Pricing and Payments

Water pricing is usually set by governmental institutions. They determine how water is governed under scarce conditions, including decisions about how it is used for domestic and irrigation purposes (FAO and WWC 2015). Water prices can change as a result of climatic changes as a consequence of changes in availability. Those most affected by these changes are communities and farmers as 'water users' who have to pay (Downing et al. 1997; UNDP 2006). Accessibility to water by poor farmers is conditioned on its affordability. Ideally, increases in water cost will not limit the capacity of users to pay for other basic services such as health and education (e.g., HLPE 2015: 102).

Extreme events such as droughts tend to create changes in the pricing of water, mainly in form of water user fees or irrigation fees for users or farmers. Fees generated are often intended to repair water sources that breakdown due to harsh conditions, to conserve available water, and to ensure optimum use. Unfortunately, fee changes often lead to access denial by vulnerable groups since they may not be able to afford the new prices. In a study of water collection among small-holder subsistence farmers in rural Uganda, Asaba (2014) verified that many households could not afford user fees for hand pumps which resulted in the malfunctioning and non-repair of many pumps for long periods especially during droughts, hence denying poor women, children, and other water fetchers access to safe water. Similar scenarios have also been reported in other parts of East and Central Africa (Coles and Wallace 2005; Tukai 2005; Osinde and Turner 2007). Failure of farmer households to pay user or maintenance fees, especially to the associations that regulate access to water at the local level, may also lead to conflicts, an issue that is explored further in the next section.

14.4.5 Conflicts

Climate change can cause conflicts by reducing access to resources, in terms of their quantity and quality, especially natural resources that sustain livelihoods (Barnett and Adger 2007). Similarly, conflicts, as outcomes of water governance (Franks and Cleaver 2007) in Sub-Saharan African communities affected by climate change can result from prolonged droughts that lead to water scarcity or inadequate access to water especially in dry land ecosystems, as well as the modification of local arrangements and rights for accessing and using water. As shown in Table 14.1, such conflicts have pitted pastoralists against various farming groups in many countries in SSA, leading to injuries and in extreme cases the loss of lives.

Little and McPeak (2014) describe land loss and endemic conflict and violence as key challenges to the resilience of pastoral systems in Africa to climate change. They argue that conflicts are driven by climate induced droughts that lead to the loss of key dry season grazing areas, such as 'patches' of high ecological

Table 14.1 Drought-related conflicts in selected countries in Africa

Country	Year (s)	Areas	Causes	Groups Involved
Kenya	2006, 2008, 2009	Northern region, Isiolo, Marsabit and Samburu districts	Access to water (and pasture)	Orma, Wardei, Somalis, and Samburu pastoralists, Turkana, Pokot
Somalia	2006, 2011	Southern region	Prolonged drought, limited pastures	Somali pastoralist groups and crop farmers
Ethiopia	2011, 2015	Southern region	Prolonged drought, inadequate water points	Somalis and indigenous Ethiopian herders
Sudan	1984–1986, 2011	Darfur (1984–1986)	Long drought that led to loss of cattle and cattle raids	Baggara, Janjawid militia, Darfurian pastoralists, Nubas

Adapted from Little and McPeak (2014), Nyong et al. (2006), Toulmin (2009), UNEP (2007), BBC News

value, and irrigation schemes in riverine areas. Little and McPeak further describe how "the loss of dry season water points and pasture crowds the herders onto less productive rangelands, undermining their economic welfare, and putting them into competition and conflict with other groups". Similar scenarios have been reported in parts of rural Uganda, where agro-pastoralists conflict with crop farmers over access to wetlands, the only sources of water during droughts (Mavuto 2015). Most of the conflicts in between herders and other farming groups therefore stem from competition over patches and the situation is not helped by armed conflicts in northern parts of countries such as Nigeria, Niger, Mali, and others such as the Central Africa, Republic, South Sudan and Somalia. These nations concentrate pastoralists on a few rangelands with limited access to water, hence making them even more vulnerable during droughts. The Darfur conflict in Sudan, in which Arab militias clashed with black rebels and farmers led to the death of an estimated 300,000 people. Some claim that it was driven by climate-induced water scarcity which led to tensions between farmers and herders over declining water-holes (UNEP 2007, in Brown et al. 2007: 1143). Other studies in West Africa, Northern and Eastern Kenya and Southern Ethiopia have also documented that conflicts between nomadic and semi-nomadic tribes over water points and cattle are more frequent during periods of harsh drought conditions (Nyong et al. 2006; Little and McPeak 2014; Staro 2014). The construction of private water points has reduced on this phenomenon in the latter case.

On the other hand, changes in irrigation water sharing arrangements as an adaptation measure to drought conditions have also increased conflicts between water users. Some of this conflict has resulted from subsequent illegal modes of access to water, such as diverting irrigation streams at night (Frankenberger and Nelson 2013).

14.4.6 Culture, Gender and Water Governance in the Face of Climate Change

Cultural[6] norms, values, ideas and symbols shape how people interpret and use resources, and they can influence their access and governance of water under varying weather conditions (FAO 2012; Miller and Belton 2014; UNDP 2006). Cultural norms also shape differences between groups that access and use water in Sub-Saharan Africa, as they are usually based on gender, ethnicity and other subjective criteria. Gender, defined as the socially constructed differences between men, women, boys and girls that differ geographically and change over time, determines access to and governance of water resources. Whereas inequality in governance and/or access to water result from prevailing gender related cultural and socio-political norms (CAP-NET and GWA 2006; Coles and Wallace 2005; Ray 2007), climate change induced events, such as droughts, can exacerbate these inequalities, increasing the vulnerability and lowering the adaptive capacity[7] of poor women and children in farming communities (Annecke 2002; Asaba 2014; Denton 2002; Figueiredo and Perkins 2013; HLPE 2015; UNDP 2006). Droughts also interfere with water collection, a mechanism of water governance by increasing the drudgery of women and children, who walk long distances and spend a lot of time at water points so as to be able to get for water for domestic or agricultural use (Arku 2010; Asaba et al. 2013; Coles and Wallace 2005; Geere et al. 2010; Kusangaya et al. 2014; Ray 2007; Sorenson et al. 2011; Toulmin 2009; Tukai 2005).

The unequal participation of men and women in decision-making in water governance is another gender issue that could have great social implications in the face of climate change. Women continue to be excluded from climate change decision-making processes and water governance in particular in most of Sub-Saharan Africa, despite the fact that they are largely responsible for water collection and management and farming at a household level. Studies in various countries in East, Central and Western Africa, most of which do not specifically focus on climate-induced events, show that most of the water actors are men. Women tend to be under-represented in domestic water and irrigation associations and have limited decision-making power (Asaba 2014; Coles and Wallace 2005; FAO 2015; Figueiredo and Perkins 2013; HLPE 2015: 98). This trend is likely to worsen in the face of changes in weather and extreme events.

[6]According to Llobera (2003), culture is the totality of a people's way of life, the whole complex of distinctive spiritual, material, intellectual and emotional features through which a society lives and reproduces itself.

[7]Adaptive capacity is defined as "the set of resources (natural, financial, institutional or human, and including access to information, expertise, and social networks) available for adaptation, as well as the ability or capacity of that system to use these resources effectively in the pursuit of adaptation," (Brooks and Adger 2004: 168).

Climate change often results in changes in gender relations and roles, both negatively and positively. Much of the literature on gender and climate change in agriculture in SSA has focussed on gender roles and vulnerability. Studies have demonstrated how women are burdened by household and agricultural tasks, and how their social position, poverty and cultural norms that limit their mobility make them and their children most vulnerable to hazards and extreme events (e.g., Asaba 2015; Denton 2002; Figueiredo and Perkins 2013; Fothergill 1996; Resurrección 2013; Wahlström 2012). Also, studies in Eastern Africa and the Horn of Africa indicate that changing climatic conditions such as droughts and floods have forced men to migrate to urban or other areas in search of alternative work, leaving women and children behind. This phenomenon has increased the domestic and productive tasks for women, who have taken on the traditionally male roles of not only being bread winners, but also undertaking water-related adaptation tasks. For example, seeking to provide options for water, agriculture and food security under climate uncertainty, McCornick et al. (2013) describe how "the exodus of men from rural areas has steered women into taking up their husbands' responsibilities, such as in the engineering aspects of irrigation or negotiating access to canal and groundwater" (McCornick et al. 2013:24).

14.4.7 Other Pertinent Factors

14.4.7.1 Perceptions of Climate Change and Water Availability

According to Chang et al. (2010:5), community perceptions of the risks of climate change can influence the acceptance of proposed adaptation measures.

Similarly, the way small holder farming communities comprehend and appreciate the potential impact of climate change on water availability may shape their access to and management of water for agriculture, as well as their adoption of adaptation measures. For example, devastating floods in Mozambique changed farmer perceptions of water scarcity for agriculture and drought risk in general, which in turn impacted their behaviour. Farmers in the five agricultural provinces affected by the floods had mixed feelings about adaptation actions, such as planting drought resistant varieties that did not require irrigation because they believed that varieties would result in reduced yields (Patt and Schröter 2008, in World Bank 2010: 325).

14.4.7.2 Access to Knowledge and Information About Water and Climate Change

Some scholars have argued that access to relevant knowledge and information, whether modern or indigenous, determines the ability of farmers and communities to adapt or adjust to climate variability and change in SSA (Brooks and Adger 2004; Challinor et al. 2007).

Community knowledge of climate information such as forecasts, changes in seasons and the likelihood of extreme events and their impacts on water resources or water scarcity can improve water governance by enhancing wise use and management of water and adoption of adaptation measures. Several studies indicate that many farmers lack information about weather and weather forecasts, and, therefore, are unable to adjust their growing seasons and plan their use of water accordingly. This sometimes leads to crop failures, food insecurity and increased scarcity of water (e.g., Downing et al. 1997; FAO and WWC 2015; McCornick et al. 2013).

14.5 Conclusion

From this expansive review, it is clear that whereas understanding the physical, ecological, and technological impacts of climate change on water governance is important, we cannot ignore the pertinent social issues that are critical to improving adaptation and enhancing resilience of poor agricultural communities in Sub-Saharan Africa. This review has shown that adjustments that address interrelated factors such as the nature of water technologies; formal and informal institutions; rights and entitlements; conflicts as an outcome of water governance; and cultural and gender relations, which change under varying climatic conditions are needed.

Given the multi-dimensional nature of climate change and its impacts, these social issues and their interconnections with the ecological, technological and economic impacts need to be given more attention in climate change adaptation interventions in the water and agriculture sectors. This is true, be they related to anticipatory actions, or related to institutional, research or development assistance activities at the local, regional and/or national levels. This will enhance resilience to climate change as well as promote more systemic and sustainable adaptation to it among farming communities in developing countries. In any case, sustainability can only be enhanced when the social, ecological, technological and economic aspects are interconnected; and in principle, social sustainability requires addressing issues of behaviour, justice, equality, resilience and capacity, many of which have been highlighted in the context of water governance.

Last but not least, more research needs to be done on other social determinants of access to water and its governance especially at household, farm, community and meso and macro levels in the face of changing climatic conditions. Important among them are the impacts of climate change on farmer water users and on water and adaptation service providers. For the farmers, issues pertaining to their capacities, such as their education, health and other social capital, and how these capacities influence water governance are very important. Other factors needing further research are indigenous knowledge; governance mechanisms such as water pricing or payments, customary and modern water rights, changing water institutions, water policies; and water management processes and practices and all their gender implications. In particular, undertaking more nuanced research on the gender dimensions of representation, decision-making and voice in water governance under varying farming systems in the context of climate change could be critical to successful attempts to adapt to climate change. Investigating these aspects will also inform climate and adaptation policy makers and service providers about future interventions in both domestic and agricultural water use and governance by communities in SSA.

References

ADB (2011) The cost of adaptation to climate change in Africa. African Development Bank, Tunis

Adger WN (2006) Vulnerability. Glob Environ Change 16:268–281

Africa Recovery (2000) Mozambique: growth with poverty. A difficult transition from prolonged war to peace and development, vol 14. United Nations, New York

Annecke W (2002) Climate change, energy-related activities and the likely social impacts on women in Africa. Int J Glob Environ Issues 2(3/4):207–222

Arku Frank S (2010) Time savings from easy access to clean water: implications for rural men's and women's well-being. Prog Dev Stud 10(3):233–246

Asaba RB (2014) Gender, power and local water governance in rural Uganda. PhD thesis, Department of Sociology, National University of Ireland, Maynooth, Maynooth

Asaba RB (2015) Gender and vulnerability to disasters and disaster/climate risk management in Uganda: a participatory characterisation. UNDP, Kampala

Asaba RB, Fagan GH, Kabonesa C, Mugumya F (2013) Beyond distance and time: gender and the burden of water collection in rural Uganda.wH2O: J of Gender and Water 2(1):31–38

Barnett J, Adger NW (2007) Climate change, human security and violent conflict. Political Geogr 1(26):639–655

Birkmann J, Tetzlaff G, Zentel KO (2009) Addressing the challenge: recommendations and quality criteria for linking disaster risk reduction and adaptation to climate change. German Committee for Disaster Reduction and DKKV Publication Series, Bonn

Brooks N, Adger N (2004) Technical paper 7: assessing and enhancing adaptive capacity. In: Lim B, Spanger-Siegfried E (eds) Adaptation policy framework for climate change. Cambridge University Press, Cambridge, pp 168–182

Brown O, Hammill AB, McLeman R (2007) Climate change as the 'new' security threat: implications for Africa. R Inst Int Aff 83(6):1141–1154

Burroughs WJ (2005) Does the weather really matter? The social implications of climate change. Cambridge University Press, New York

Cabell JF, Oelofse M (2012) An indicator framework for assessing agroecosystem resilience. Ecol Soc 17(1):18

CAP-NET, GWA (2006) Why gender matters: a tutorial for water managers. CAP-NET International network for Capacity Building in Integrated Water Resources Management, Delft

Challinor AJ, Wheeler TR, Garforth C, Craufurd P, Kassam A (2007) Assessing the vulnerability of food crop systems in africa to climate change. Clim Change 83(3):381–399

Chambwera M, Stage J (2010) Climate change adaptation in developing countries: issues and perspectives for economic analysis. International Institute for Environment and Development, London, p 39

Chang CT, Martens P, Amelung B (2010) The social and behavioural aspects of climate change: linking vulnerability, adaptation and mitigation. In: Martens P, Chang CT (eds) The social and behavioural aspects of climate change. Greenleaf Publishing Limited, Sheffield

Claruis MH, Fatichi S, Allan A, Fuhrer J, Stoffel M, Romerio F, Gaudard L, Burlando P, Beniston M, Xoplaki E, Toreti A (2014) Governing and managing water resources under changing hydro-climatic contexts: the case of the upper Rhone Basin. Environ Sci Policy 43:56–67. doi:10.1016/j.envsci.2013.11.005

Clements R, Haggar J, Quezada A, Torres J (2011) Technologies for climate change adaptation—agriculture sector. In: Zhu X (eds) UNEP Risø Centre, Roskilde

Coles A, Wallace T (2005) Gender, water and development. Berg, Oxford

Cooper P (2004) Coping with climatic variability and adapting to climate change: rural water management in dry-land areas. International Development Research Centre, Ottawa

Davies W (2014) Brazil drought: Sao Paulo sleepwalking into crisis. In: BBC News, BBC, 7 Nov 2014

Denton F (2002) Climate change vulnerability, impacts and adaptation: why does gender matter? In: Masika R (ed) Gender, development and climate change. Oxfam GB, Oxford

Di Falco S (2014) Adaptation to climate change in Sub-Saharan agriculture: assessing the evidence and rethinking the drivers. Eur Rev Agric Econ 41(3):405–430

Dietz AJ, Ruben R, Verhagen A (2004) The impact of climate change on drylands: with a focus on West Africa

Dile YT, Karlberg L, Temesgen M, Rockstrom J (2013) The role of water harvesting to achieve sustainable agricultural intensification and resilience against water related shocks in Sub-Saharan Africa. Agric Ecosyst Environ 181:69–79. doi:10.1016/j.agee.2013.09.014

Dixon JL, Stringer LC, Challinor AJ (2014) Farming system evolution and adaptive capacity: insights for adaptation support. Resources 3(1):182–214

Djalante R, Holley C, Thomalla F, Carnegie M (2013) Pathways for adaptive and integrated disaster resilience. Nat Hazards 69(3):2105–2135. doi:10.1007/s11069-013-0797-5

Downing T, Ringius L, Hulme M, Waughray D (1997) Adapting to climate change in Africa. Mitig Adapt Strat Glob Change 1(2):19–44

FAO (2012) Passport to mainstreaming gender in water programmes: key questions for interventions in the agricultural sector. Food and Agricultural Organisation, Rome

FAO, IFAD, WFP (2015) The state of food insecurity in the world 2015. Meeting the 2015 international hunger targets: taking stock of uneven progress. FAO, Rome

FAO, WWC (2015) Towards a water and food secure future: critical perspectives for policy makers. ood and Agriculture Organisation of the United Nations. World Water Council, Rome

Fernald A, Guldan S, Boykin K, Cibils A, Gonzales M, Hurd B, Lopez S, Ochoa C, Ortiz M, Rivera J, Rodriguez S, Steele C (2015) Linked hydrologic and social systems that support resilience of traditional irrigation communities. Hydrol Earth Syst Sci 19(1):293–307. doi:10.5194/hess-19-293-2015

Figueiredo P, Perkins PE (2013) Women and water management in times of climate change: participatory and inclusive processes. J Clean Prod 60:188–194. doi:10.1016/j.jclepro.2012.02.025

Fothergill A (1996) Gender, risk, and disasters. Int J Emerg Disasters 14(1):33–56

Frankenberger T, Nelson S (2013) Summary of the Expert Consultation on Resilience Measurement for Food Security. TANGO International, Tucson

Franks T, Cleaver F (2007) Water governance and poverty: a framework for analysis. Prog Dev Stud 7(4):291–306

GDC (2003) Manual: the livelihood system approach in project planning. Technical paper 28: integrated food security programme. German Development Cooperation, Trincomalee, pp 1–15

Geere JL, Hunter PR, Jagals P (2010) Domestic water carrying and its implications for health: a review and mixed methods pilot study in Limpopo Province, South Africa. Environmental Health 9:1–13

Giddens A (1984) The constitution of society: outline of the theory of structuration. Polity Press, London

Glantz M, Gommes R, Ramasamy S (2009) Coping with a changing climate: considerations for adaptation and mitigation in agriculture. Food and Agriculture Organisation, Rome

Godden L, Ison RL, Wallis PJ (2011) Water governance in a climate change world: appraising systemic and adaptive effectiveness. Water Resour Manag 25(1):3971–3976

HLPE (2015) Water for food security and nutrition. High Level Panel of Experts on Food Security and Nutrition of the UN Committee on World Food Security, Rome

Iglesias A, Rosenzweig C (2010) Effects of climate change on global food production: emissions and socioeconomic scenarios data from a crop modeling study. Socioeconomic data and applications center, Palisades, New York

IPCC (2001) Climate change 2001: impacts, adaptation, and vulnerability. Contribution of working group II to the third assessment report of the intergovernmental panel on climate change. Cambridge University Press, London

IPCC (2007) Fourth assessment report. Intergovernmental Panel on Climate Change, Cambridge University Press, London

Kajembe GC, Vedeld P, Silayo DS, Babili IH, Mosha DB (2015) Institutional sustainability in the face of climate change in Tanzania: empirical insights from irrigation institutions in Iringa rural district, Tanzania. Paper presented at the conference on climate change and multi-dimensional sustainability in African agriculture, Hilux Hotel, Morogoro, Tanzania, 3rd–5th June 2015

Kurukulasuriya P, Mendelsohn R, Hassan R, Benhin J, Deressa T, Mbaye D, Eid HM, Fosu YK, Gbetibouo G, Jain S, Mahamadou A, Mano R, Kabubo-Mariara J, El-Marsafawy S, Molua E, Ouda S, Ouedraogo M, Séne I, Maddison D, Seo NS, Dinar A (2006) Will African agriculture survive climate change? World Bank Econ Rev 20(3):367–388

Kusangaya S, Warburton ML, van Garderen EA, Jewitt GPW (2014) Impacts of climate change on water resources in Southern Africa: a review. Phys Chem Earth 67–69:47–54. doi:10.1016/j.pce.2013.09.014

Kusters K, Wangdi N (2013) The costs of adaptation: changes in water availability and farmers' responses in Punakha district, Bhutan. Int J Glob Warm 5(4):387–399

Lal R, Singh BR, Mwaseba DL, Kraybill D, Hansen D, Eik LO (2015) Sustainable intensification to advance food security and enhance climate resilience in Africa. Springer, Heidelberg

Leal Filho W (ed) (2011) Experiences of climate change adaptation in Africa. Springer, Berlin

Little PD, McPeak JG (2014) Resilience and pastoralism in Africa South of the Sahara, with a particular focus on the horn of Africa and the Sahel, West Africa. Paper presented at the 2020 conference paper 9, Washington, USA

Llobera J (2003) An invitation to anthropology: the structure, evolution, and cultural identity of human societies. Berghahn Books, London

Lobell D, Burke M (2010) Climate change and food security: adapting agriculture to a warmer world. Springer Science, New York

Long N (1992) From paradigm lost to paradigm regained? The case for an actor orientated sociology of development. In: Long N, Long A (eds) Battlefields of knowledge: the interlocking of theory and practice in social research and development. Routledge, London

Ludwig F, Kabat P, van Schaik H, van der Valk M (eds) (2009) Climate change adaptation in the water sector. Earthscan, London

Martens P, Chang CT (eds) (2010) The social and behavioural aspects of climate change. Greenleaf Publishing, Sheffield

Mavuto TD (2015) Understanding adaptive capacity on the ground: a case of agro-pastoralists in a Rural Parish in Uganda. In: Munck R, Asingwire N, Fagan H, Kabonesa C (eds) Water and development: good governance after neoliberalism. Comparative Research Programme on Poverty and Zed Books, London, pp 181–199

McCornick P, Smakhtin V, Bharati L, Johnston R, McCartney M, Sugden F, Clement F, McIntyre B (2013) Tackling change: future-proofing water, agriculture, and food security in an era of climate uncertainty. International Water Management Institute, Colombo, Sri Lanka

Miller KA, Belton V (2014) Water resource management and climate change adaptation: a holistic and multiple criteria perspective. Mitig Adapt Strat Glob Change 19(3):289–308. doi:10.1007/s11027-013-9537-0

Morton JF (2007) The impact of climate change on smallholder and subsistence agriculture. PNAS 104(50):19680–19685

Mosello B (2015) How to deal with climate change? Institutional adaptive capacity as a means to promote sustainable water governance. Springer International Publishing Switzerland, Heidelberg
Nyong A, Fiki C, McLeman R (2006) Drought-related conflicts, management and resolution in the West African Sahel: considerations for climate change research. Erde 137(1):223–248
Osinde R, Turner M (2007) Competition for water: are water riots greater threat than water wars? In: Cleaver F, Franks T (eds) New directions for water governance, Vol. 67. id21 insights
Parry ML, Canziani OF, Palutikof JP, van der Linden PJ, Hanson CE (2007) Climate change 2007: impacts, adaptation and vulnerability. contribution of working group II to the fourth assessment report of the intergovernmental panel on climate change. Cambridge University Press, Cambridge
Patt AG, Schröter D (2008) Climate risk perception and challenges for policy implementation: evidence from stakeholders in Mozambique. Glob Environ Change 18(1):458–467
Plummer J, Slaymaker T (2007). Rethinking governance in water services. Working paper 284. Overseas Development Institute, London
Ravindranath NH, Sathaye JA (2002) Climate change and developing countries. Kluwer Academic Publishers, New York
Ray I (2007) Women, water, and development. Annu Rev Environ Resour 32(1):421–449. doi:10.1146/annurev.energy.32.041806.143704
Resurrección BP (2013) Persistent women and environment linkages in climate change and sustainable development agendas. Women's Stud Int Forum 40(1):33–43
Ribot Jesse C, Magalhaes Rocha A, Panagides Stahis S (1996) Climate variability, climate change and social vulnerability in the semi-arid tropics. Cambridge University Press, Cambridge
Rogers P (2006) Water governance, water security and water sustainability. Taylor & Francis Ltd, London
Rogers P, Hall A (2003) Effective water governance. Background Paper No.7 . Global Water Partnership, Stockholm, p 44
Sen A (1990) Entitlements and the Chinese famine. Food Policy, Elsevier 15(3):261–263
Sen A (2000) Social justice and the distribution of income. In: Atkinson AB, Bourguignon F. (eds), Handbook of income distribution, Elsevier 1(1):59–85
Sorenson SB, Morssink C, Campos PA (2011) Safe access to safe water in low income countries: water fetching in current times. Soc Sci Med 72:1522–1526
Staro F (2014) Valuing "indigenous knowledge" related to water usage among garri pastoralists of Southern Ethiopia: which/whose knowledge? Afr Study Monogr 48(1):49–58
Stern N (2006) Stern review on the economics of climate change. HM Treasury, London
Strzepek KM, Onyeji SC, Saleh M, Yates DN (1995) An assessment of integrated climate change impacts on Egypt. In: Strzepek K, Smith JB (eds) As climate changes: international impacts and implications. Cambridge University Press, Cambridge, pp 180–200
Toulmin C (2009) Climate change in Africa. Zed Books, London
Tukai R (2005) Gender and access in pastoral communities: re-evaluating community participation and gender empowerment. Paper presented at the ESRC seminar on access, poverty and social exclusion. Overseas Development Institute, London. 1 Mar 2005
UNDP (2006) Beyond scarcity: Power, poverty and the global water crisis. human development report 2006. United Nations Development Programme, New York
UNEP (2007) Sudan: post-conflict environmental assessment. United Nations Environment Programme, Nairobi
Verhagen J, Put M, Zaal F, van Keulen H (2004) Climate change and drought risks for agriculture. In: Dietz A, Ruben R, Verhagen A (eds) The impact of climate change on dry lands: with a focus on West Africa. Kluwer, New York
Wahlström M (2012) Women, girls, and disasters. UNISDIR, Geneva
Williams AP, Funk C (2011) A westward extension of the warm pool leads to a westward extension of the walker circulation, drying eastern africa. Clim Dyn 1(37):2417–2424
World Bank (2010) World development report, 2010: development and climate change. World Bank, Washington
World Resources Institute (1994) World resources, 1994–95. World Resources Institute, New York

Chapter 15
Assessment of Climate Change Impact on Common Bean (*Phaseolus Vulgaris* Savi, L.) Production in Tanzania

Sixbert Kajumula Mourice, Siza Donald Tumbo and Cornell Lawrence Rweyemamu

Abstract Our understanding of how bean crops may respond to climate change is important in designing agronomic and breeding programs for the future. This study assessed the climate change impact on common bean yield in major producing regions of Tanzania. Five Coupled Mode Inter-Comparison Project Phase 5 (CMIP5) Global Circulation Models (GCMs) under two greenhouse gas emission scenarios [Representative Concentration Pathways (RCPs)] were evaluated against the baseline climate. The Decision Support System for Agrotechnology Transfer (DSSAT v4.5) model was used. Under RCP 4.5, the yield increased by 10–32 % for Bukoba, Manyara, Kigoma, and Mbeya, and it decreased by 3 % for Musoma in the near-term period. Under RCP 8.5, the yield increased by 5–30 % in the near-term, 15–40 % in the mid-century, and 20–48 % in the end-century periods for all bean growing areas. More on the climate change impact trajectory and further research recommendations for the common bean is discussed in this study.

Keywords AgMIP · C3 crops · DSSAT · RCP · Tanzania

15.1 Introduction

Agricultural production is greatly influenced by climate; therefore, any change in climate that results in an increase in temperature and a change in rainfall patterns as a result of increased greenhouse gases in the atmosphere will have a profound effect on agricultural productivity. Over the decades, atmospheric CO_2 has been increasing, from 276 ppm in the 1750s to 390 ppm in 2011 (IPCC 2013). Crop

S.K. Mourice (✉) · S.D. Tumbo · C.L. Rweyemamu
Sokoine University of Agriculture, Morogoro, Tanzania
e-mail: sixbert.mourice@gmail.com

S.D. Tumbo
e-mail: siza.tumbo@gmail.com

C.L. Rweyemamu
e-mail: crweyemamu60@gmail.com

R. Lal et al. (eds.), *Climate Change and Multi-Dimensional Sustainability in African Agriculture*, DOI 10.1007/978-3-319-41238-2_15

species respond differently to increases in CO_2, depending on the carbon fixation pathway (C3 vs. C4). The photosynthetic rate in C3 crops (wheat, common beans, rice, potatoes) is more responsive to increased CO_2 concentration than in C4 crops (maize, sugarcane, sorghum) (Porter et al. 2014). It is certain that increased atmospheric CO_2 concentration may cause a yield increase in C3 crops. However, increased warming trends are likely to offset the yield gains from elevated CO_2 if sound adaptation measures are not instituted to mitigate the high temperature consequences. In Tanzania, it has already been indicated that there will be a countrywide increase in mean temperature from 2 to 4 °C by 2100 (IPCC 2007). Higher temperatures can reduce crop duration in terms of increased growth rate, an increased vapor pressure deficit, causing a decreased crop water use efficiency that favors the development and spread of pests (Lobell and Gourdji 2012).

Like many other crop species, climate change is expected to have adverse effects on common bean production. For example, common beans are grown at mean air temperatures of 14–35 °C, whereas day temperatures or night temperatures above 30 or 20 °C, respectively, result in significant yield reduction (Beebe et al. 2011). The common bean is a very important food crop in Tanzania, contributing significantly to dietary protein for both urban and rural populations. It is also a source of livelihoods for growers. Therefore, climate change impacts on the agricultural sector will also affect the livelihoods of bean growers.

While previous climate change studies in Tanzania have focused mainly on cereals (Arndt et al. 2011; Kilembe et al. 2012), the common bean has not received due attention. A climate change impact assessment is pertinent to common bean production for two reasons. One, the common bean requires sufficient amounts of precipitation, between 300 and 600 mm, during a growth cycle (Katungi et al. 2009). As climate change progresses, it is speculated that areas, such as the northeast and the Lake Victoria basin might experience very wet conditions, while the Lake Tanganyika basin and the southern highlands might become drier toward the end of the century (Mwandosya et al. 1998). Excessive precipitation not only causes water logging conditions but also accelerates the development and spread of diseases. On the other hand, insufficient precipitation causes soil moisture stress which, in turn, reduces leaf expansion, halts photosynthesis, and consequently lowers grain yield (Egli and Bruening 2004). Second, high temperatures (particularly night temperatures) cause flower bud abortion and low pod filling (Konsens et al. 1991). With the anticipated changes in future climate, understanding the extent to which the climate change will affect common bean production is important. This will allow breeders to consider new frontiers for developing new bean cultivars that are tolerant of high or low moisture stress, high night temperatures, and high sink capacity, which is increasingly important because atmospheric CO_2 is expected to increase. Breeding to enhance sinks will be particularly important for target production zones where stresses, such as high temperatures or moisture stresses, can cause more damage to reproductive development (i.e. flowering, pod formation) than to the photosynthetic source (i.e. leaves) (Hall 2004).

Global circulation models (GCMs) are gaining popularity in projecting future climates and their impacts on ecosystems at coarse (continent) or fine resolution (river basin) spatial scales. Previous studies in Tanzania (Mwandosya et al. 1998; Matari et al. 2008) based their climate change projections on greenhouse gas (GHG) emission scenarios and doubling CO_2 concentration. Appropriate GCMs to use in the assessment were based on the lowest deviation between observed historical weather elements and the GCM data generated. Kilembe et al. used GCMs from the Coupled Model Intercomparison Project Phase 3 (CMIP3) climate change scenarios (A and B scenarios) and showed mixed results: maize yields would increase by 25 % in some areas and decline in other areas by 2050. CMIP5 scenarios emanating from the Fifth Assessment Report (AR5) (IPCC 2013) are more recent, providing a larger number of more complex models at higher resolution, more complete representations of external forcings, more scenarios, and more diagnostics stored (Knutti and Sedláček 2013). In this study, we employ a modeling approach using CMIP5 scenarios to evaluate the climate change effects on the productivity of the common bean (*Phaseolus vulgaris* Savi L.) in major growing regions of Tanzania. Specifically, we compare the average bean yield in a baseline climate against two climate change scenarios, and we determine the bean yield change as it is affected by general circulation model (GCM) projections with respect to baseline or historical yields.

15.2 Materials and Methods

15.2.1 Study Area

Bean growth is determined by average temperatures, which in turn affect duration to maturity and the incidence and severity of certain diseases and evapotranspiration rates. Precipitation determines the probability of soil moisture stress and the number of bean growing seasons per year (Wortmann et al. 1998). This study focused on six major producing regions in Tanzania: Kagera, Kigoma, Manyara, Mbeya, and Ruvuma (Fig. 15.1). Generally, these common bean producing areas are characterized by altitudes ranging from 1000 to 1500 m above sea level, with more than 400 mm of available soil moisture per growing season (Wortmann et al. 1998). Kagera, Mara, Arusha/Manyara, and the northern Kigoma regions experience bimodal rainfall patterns, i.e., short rains begin from October to December followed by long rains from March to May. Southern Kigoma, Mbeya, and Ruvuma regions experience unimodal rainfall patterns from December to May.

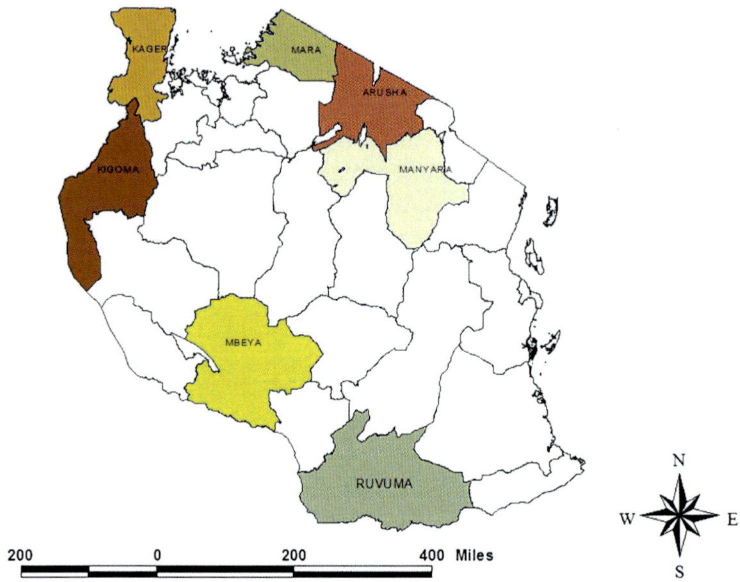

Fig. 15.1 Map of Tanzania indicating major common bean growing regions

15.2.2 The Decision Support System for Agrotechnology Transfer Model

The dry bean (BEANGRO) module of the Decision Support System for Agrotechnology Transfer (DSSAT v4.5) (Hoogenboom et al. 2010) was calibrated with observed data for the common bean (cf. Canadian Wonder) (Rweyemamu 1995). DSSAT was used because it simulates the effects of enhanced atmospheric CO_2 enrichment, which is in line with projected greenhouse gases emissions. Additionally, DSSAT simulates the effects of water and heat stresses, which are also anticipated in future climates due to increased or reduced rainfall and increased temperatures (Ritchie et al. 1998; Hoogenboom et al. 2010).

15.2.2.1 Model Inputs and Simulations

The BEANGRO model input files and simulations were organized using the Agricultural Model Inter-comparison and Improvement Project (AgMIP) modeling framework (Rosenzweig et al. 2013). Under this framework, three model input files were created. The survey import file consists of baseline weather stations, soil characteristics, and actual or observed bean yields from the National Bureau of Statistics (NBS) panel survey database (NBS 2012). The data overlay for multi-model export (DOME) files in terms of field overlay and seasonal strategy was created. Field overlay files consists of information, such as initial conditions (water, organic matter, and nitrogen) and all management practices and assumptions, which are not normally collected in surveys or were not measured in experiments. The seasonal overlay file specifies the span of years for which the assessment is required along with planting windows, which vary among regions.

15.2.3 Representative Concentration Pathways and Global Circulation Models

Representative concentration pathways (RCPs) describe the scenarios that result from the effects of climate change drivers in the future (van Vuuren et al. 2011). A detailed description of climate change scenarios has been given by Moss et al. (2010). There are four RCPs under the fifth phase of the Coupled Model Inter-comparison Project (CMIP5). RCP 2.6 is a mitigation scenario characterized by a very low forcing level. RCP 4.5/RCP 6 are medium stabilization scenarios, while RCP 8.5 is a high emission scenario (van Vuuren et al. 2011). In this study, two RCPs, namely RCP 4.5 and RCP 8.5, were selected for the assessment of climate change impacts on common bean production in Tanzania. RCP 4.5 represents a stabilization scenario in which total radiative forcing is stabilized shortly after 2100 without overshooting the long-run radiative forcing target level (van Vuuren et al. 2011). RCP 8.5 combines assumptions about high population and relatively slow income growth with modest rates of technological change and energy intensity improvements, leading, in the long term, to high energy demand and GHG emissions in the absence of climate change policies. Therefore, RCP 8.5 corresponds to the pathway with the highest greenhouse gas emissions (Riahi et al. 2011). Atmospheric CO_2 concentrations were included in the model simulations according to Meinshausen et al. (2011). CO_2 concentration for the baseline climate was 390 ppm, 423 ppm (RCP4.5 near term), 499 ppm (RCP4.5 mid-century), and 532 ppm (RCP 4.5 end-century). As for RCP 8.5, CO_2 concentrations were 432 ppm (near term), 571 ppm (mid-century), and 801 ppm (end-century). Five CMIP5 GCMs namely CCSM4, GFDL-ESM2M, HadGEM2-ES, MIROC5, and MPI-ESM-LR were used in this assessment to generate future climatic data. Historical weather data (1981–2010) from the synoptic stations in the study areas were obtained from the Tanzania Meteorological Agency (TMA).

15.2.3.1 Modeling Assumptions

In the model, planting was invoked automatically when a total of 25 mm or more of rain was received within five consecutive days, thus minimizing the possibility of a false start in the growing season. This also allowed for variations in thee start of rain from year to year and from one station to another. No inorganic fertilizer application was reported in the panel survey result; therefore, fertilizer application events were not activated in the model. As regards to other crop growth limitations that could not be explicitly explained by available data, e.g., insects and diseases, a soil fertility factor (SPLF) of 1.0 was assumed for all fields.

15.3 Results

15.3.1 Average Minimum Temperature

All GCMs under the RCP 4.5 emission scenario projected increased minimum temperatures (Tmin) with differing magnitudes at each location. The Musoma meteorological station showed a wide gap between historical and near-term climates, with GFDL-ESM2M indicating a higher Tmin increase from May to August. HADGEM-ES exhibited an increase in Tmin with respect to baseline in the December to April period, which is an important bean growing season (Fig. 15.2).

The average minimum temperature is expected to increase with respect to the baseline for all stations during the mid-century period, although the Bukoba station

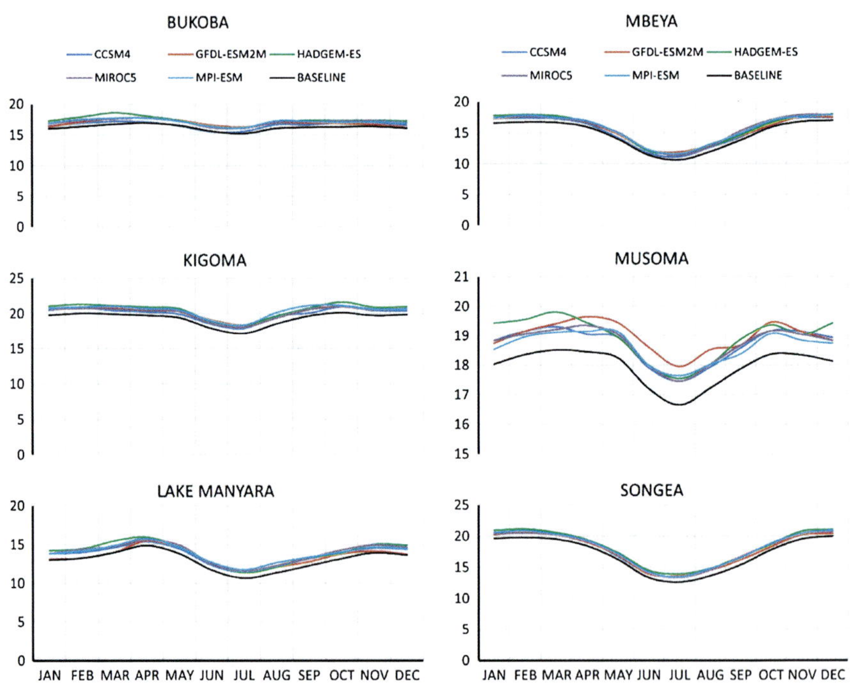

Fig. 15.2 Average minimum temperature for the near-term period under the RCP 4.5 scenario with respect to baseline mean minimum temperature

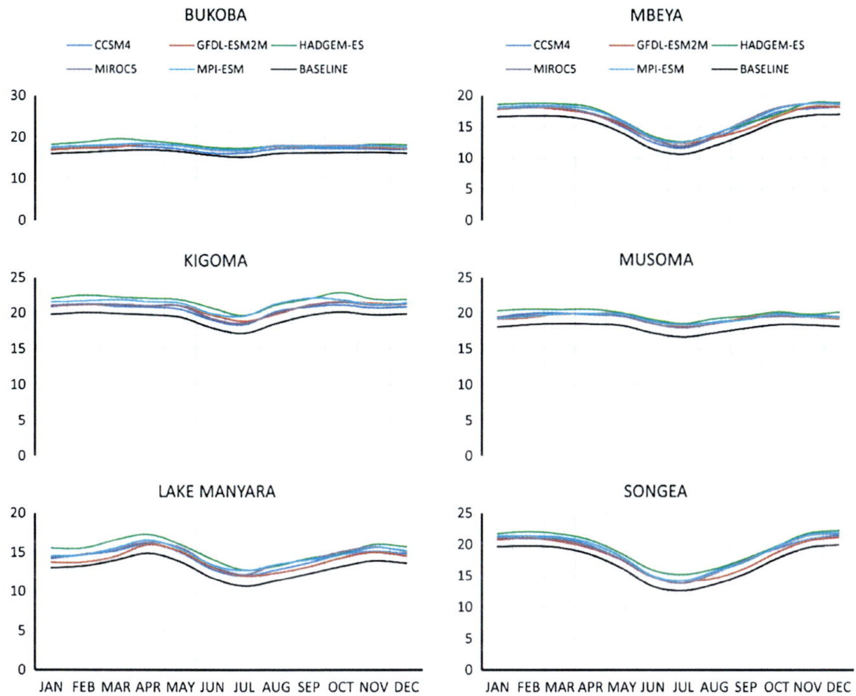

Fig. 15.3 Mean minimum temperature for the mid-century period under the RCP 4.5 scenario compared to baseline mean minimum temperature

exhibited a slight increase for all GCMs (Fig. 15.3). During the end-century period, minimum temperature is projected to increase from 1.5 to 2.5 °C under RCP 4.5 (Fig. 15.4). Under these scenarios, all study locations except Kigoma, Musoma, and Songea would still be suitable because minimum temperature requirement may still be within the suitability range of 15–22 °C.

Under the RCP 8.5 emission scenario, minimum temperature projections indicate an increase of 0.5–1 °C in the near-term (Fig. 15.5). During the mid-century period, all GCMs project an increase in daily minimum temperature for all stations (Fig. 15.6). For Bukoba, Lake Manyara, and Mbeya stations, the projected minimum temperature is less than 20 °C, whereas for Kigoma, Musoma, and Songea, the minimum temperature is projected to be above 21 °C (Fig. 15.7). Since a night temperature of above 20 °C leads to flower abortion and consequently a lower number of seeds per pod, it is likely that bean yields will be negatively affected in Kigoma, Musoma, and Songea stations during the mid-century period.

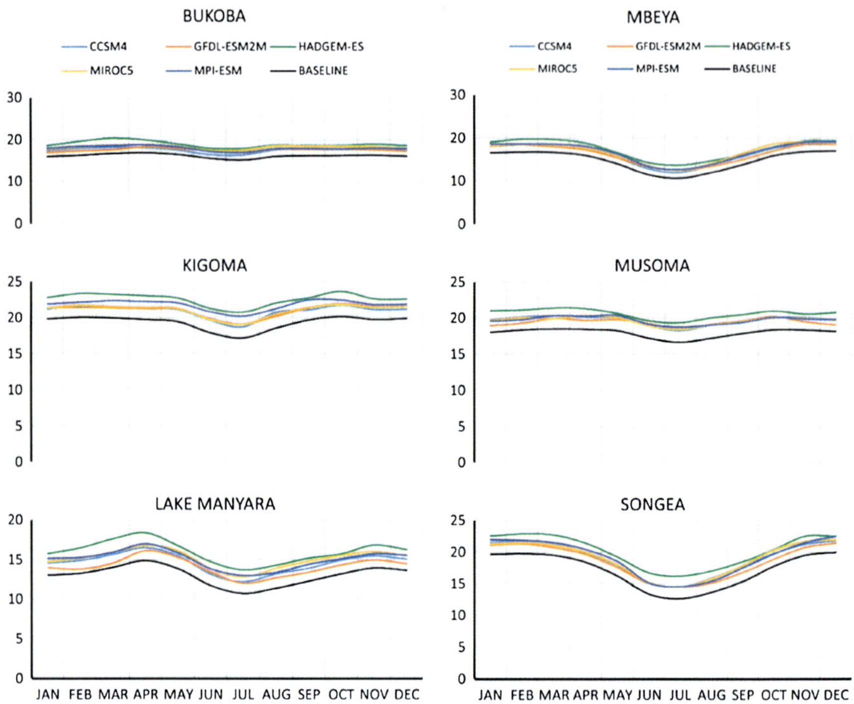

Fig. 15.4 Mean minimum temperature for the end-century period under the RCP 4.5 scenario compared to baseline mean minimum temperature

A further escalation of average minimum temperature is projected during the end-century period under the RCP 8.5 emission scenario. Projections for Bukoba and Mbeya stations indicate an increment of between 1 and 2 °C in the end-century period. For Lake Manyara and Kigoma stations, the minimum temperature is projected to rise from 2 to 3.5 °C by the end of the century. Moreover, 2.5–5 °C is projected for Musoma and Songea stations. This suggests that during the end-century period, the latter stations may become marginally suitable for bean production under the RCP 8.5 emission scenario.

15.3.2 Precipitation

Under the RCP 4.5 emission scenario, GCM projections in the near-term period exhibited no significant change in precipitation with respect to the baseline climate, although there were exceptions. GCM GFDL-ESM2M indicates high and low

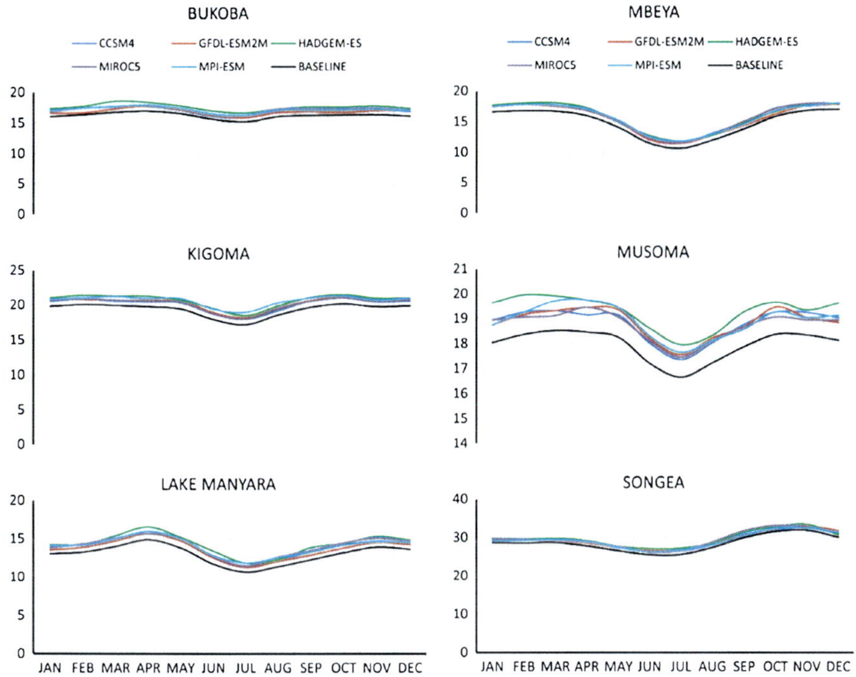

Fig. 15.5 Mean minimum temperature for the near-term period under the RCP 8.5 scenario compared to baseline mean minimum temperature

precipitation peaks at the Bukoba and Lake Manyara stations, respectively (Fig. 15.8). During the mid-century period, precipitation is projected to decline for some GCMs and increase for others across the country. GCM GFDL-ESM2M shows a unique pattern for the Lake Manyara and Musoma stations where precipitation is projected to drop during the month of March (Fig. 15.9). GCM projections for the end-century period indicate a decline in precipitation for the Bukoba station, except for MPI-ESM, which projects an increase during the September to October precipitation (Fig. 15.10).

Like the RCP 4.5 emission scenario, GCM precipitation projections under RCP 8.5 indicate a mixed pattern with respect to the baseline climate across all stations. As for the near-term period, GCMs exhibit rainfall patterns similar to the baseline climate. However, GCM MPI-ESM indicates an increase in precipitation for the Bukoba station during October and November and from June to August at the Musoma station (Fig. 15.11). During mid-century period, GCM GFDL-ESM2M exhibits an increase in precipitation during the peak months for the Bukoba, Lake

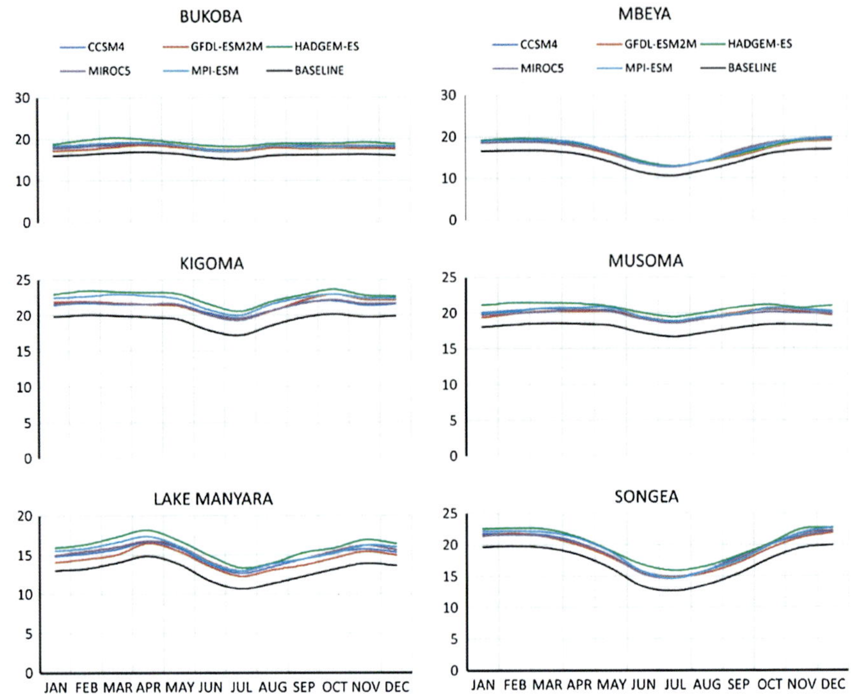

Fig. 15.6 Mean minimum temperature for the mid-century period under the RCP 8.5 scenario compared to baseline mean minimum temperature

Manyara, and Musoma stations (Fig. 15.12). Generally, GCM projections indicate declining precipitation for all stations. End-century period projections indicate that precipitation will increase for the Mbeya, Musoma, and Songea stations (Fig. 15.13).

15.3.3 Effects of Climate Change on Bean Yield

With respect to baseline yield, GCM projections under the RCP 4.5 emission scenario indicate that there will be yield increase for Bukoba (17 %), Kigoma (10 %), Lake Manyara (32 %), and Mbeya (10 %) in the near-term period. There will be a general decline in bean yield for the Musoma station but no change for the Songea station in the near-term period (Fig. 15.14a). GCM GFDL-ESM2M showed a 5 and 50 % yield decline for the Mbeya and Musoma stations, respectively. During the mid-century period, bean yield is generally projected to increase for all

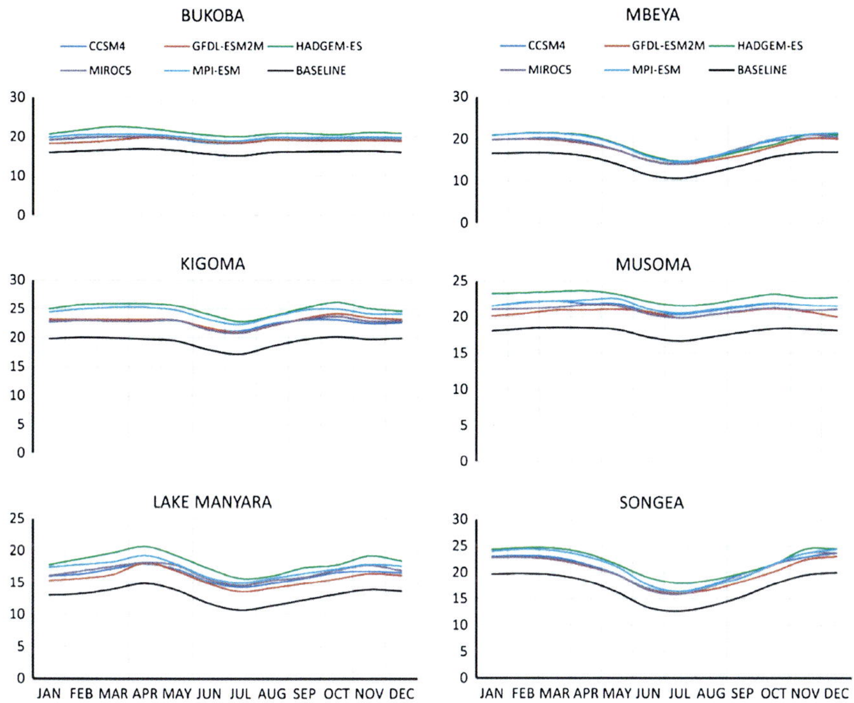

Fig. 15.7 Mean minimum temperature for the end-century period under the RCP 8.5 scenario compared to baseline mean minimum temperature

stations (Fig. 15.14b). Toward the end-century period, bean yield is projected to increase for all stations, 10 % for Musoma station, and up to 40 % for Mbeya station (Fig. 15.14c).

Under the RCP 8.5 emission scenario, bean yields are projected to increase in all periods in general. During the near-term period, bean yield is expected to increase most for the Mbeya station and the least for the Musoma and Songea stations (Fig. 15.15a). By the mid-century, bean yield increase will range from 14 to 37 % for the Musoma and Mbeya stations, respectively (Fig. 15.15b). Further yield increase will be expected in the end-century period for all stations. A 45 % bean yield increase is projected for the Mbeya station, whereas a 27 % bean yield increase is projected for the Musoma station (Fig. 15.15c).

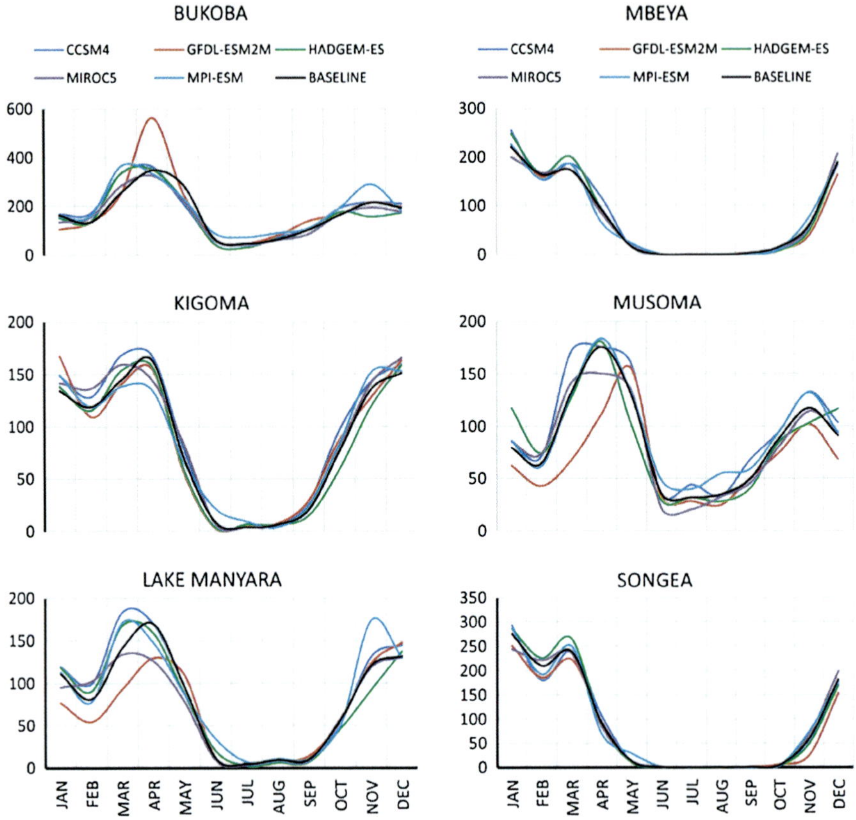

Fig. 15.8 Precipitation in the RCP 4.5 near-term period with respect to the baseline climate

15.4 Discussion and Conclusion

The common bean is significant because it is representative of most legumes with a C3 photosynthetic pathway; therefore, it is important to understand how increased greenhouse gases resulting from continued fossil fuel burning and other anthropogenic activities would affect the yield. It is indicated that the average minimum daily temperature will increase for RCPs and their respective periods compared to the baseline climate. Therefore, the bean producing areas of Musoma, Songea, and Kigoma may become marginally suitable for bean production. The common bean requires night temperatures of less than 20 °C (Beebe et al. 2011), suggesting that as the climate changes, most areas on average will get warmer with night

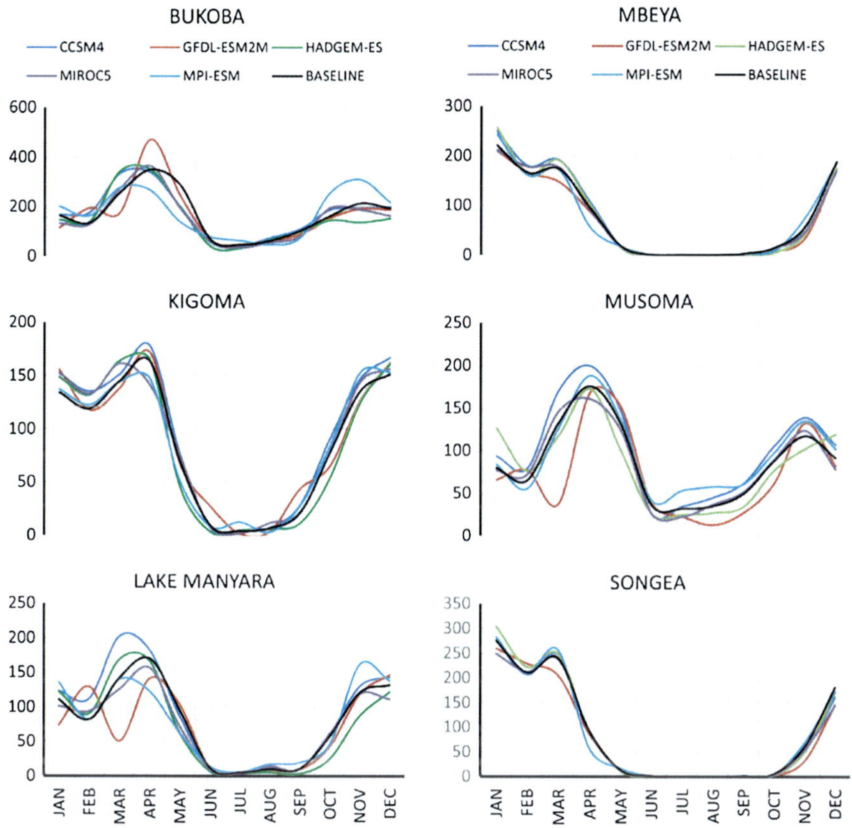

Fig. 15.9 RCP 4.5 mid-century precipitation with respect to the baseline climate

temperatures exceeding the common bean temperature requirement threshold. For locations where the common bean is currently cultivated at the margins of its temperature adaptation range, e.g., Kigoma and Songea, any increase in minimum temperature may render such areas unsuitable for cultivation. Warmer temperatures reduce the length of growing seasons and reduce pollen viability (Hall 2004). Warming leads to an increase in the saturation vapor pressure of the air, thus increasing the saturation vapor deficit (VPD) between the leaves and the air. Increased VPD has a bearing on the increased rate of evapotranspiration, which leads to the closure of stomatal pores and reduced photosynthesis (Lobell and Gourdji 2012). Moreover, high temperatures affect two important reproductive

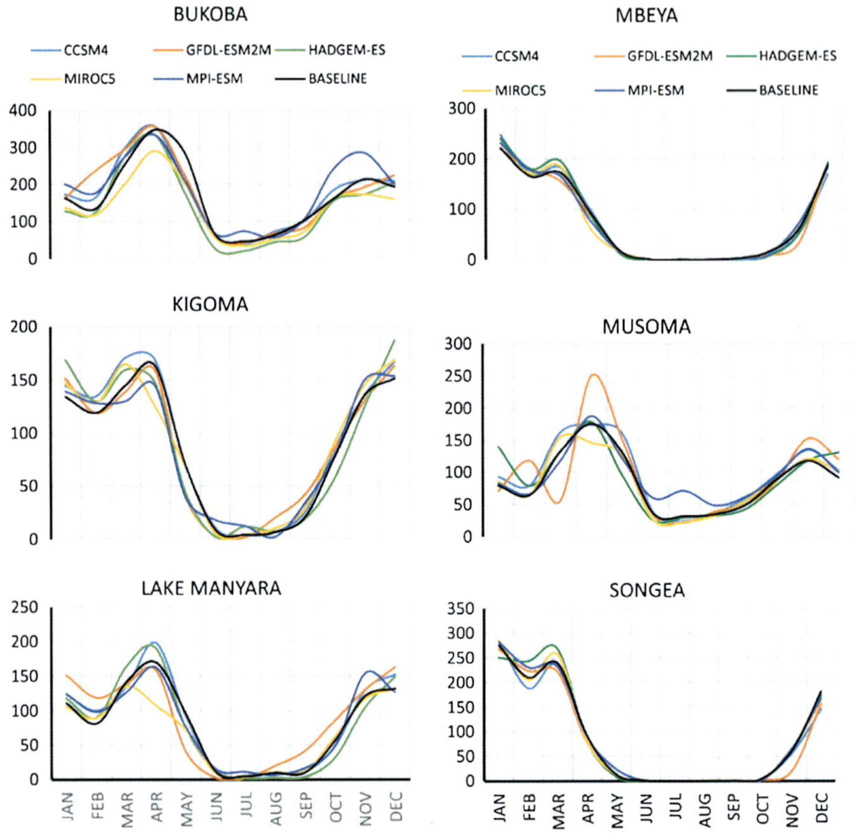

Fig. 15.10 RCP 4.5 end-century precipitation with respect to the baseline climate

stages, namely flower bud formation and the pod filling stages, in the common bean. Shonnard and Gepts (1994) point out that mean daily maximum and minimum temperatures of 35°/21 °C can considerably suppress flower formation in the susceptible genotypes of common beans. Indices for heat tolerance have been identified (Porch 2006) and can be applied for screening bean genotypes with good yield potential in future-like weather conditions.

Reduced or increased precipitation as indicated for some GCM emission scenarios and periods suggest that bean yields may be significantly affected by water stresses (in cases of reduced precipitation) or the development and spread of

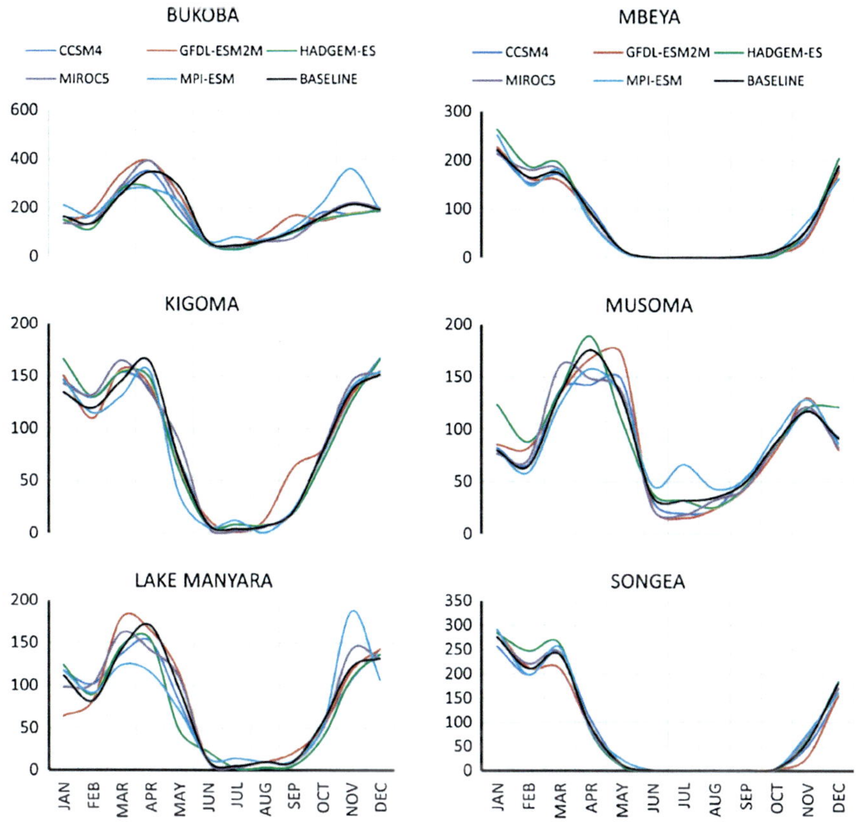

Fig. 15.11 RCP 8.5 near-term precipitation with respect to the baseline climate

diseases (in cases of increased precipitation). However, crop modeling results show that bean yields will increase despite the effects of climate change on night temperatures and precipitation. This increase may largely be due to enhanced atmospheric CO_2 concentrations projected in both RCP 4.5 and 8.5 (Meinshausen et al. 2011). In C3 plants, such as the common bean, increased CO_2 increases the photosynthesis rate, suppresses photorespiration, increases water use efficiency, and results in a yield increase (Kimball 1983; Lobell and Gourdji 2012). In this study, increased common bean yields suggest that high atmospheric CO_2 concentrations

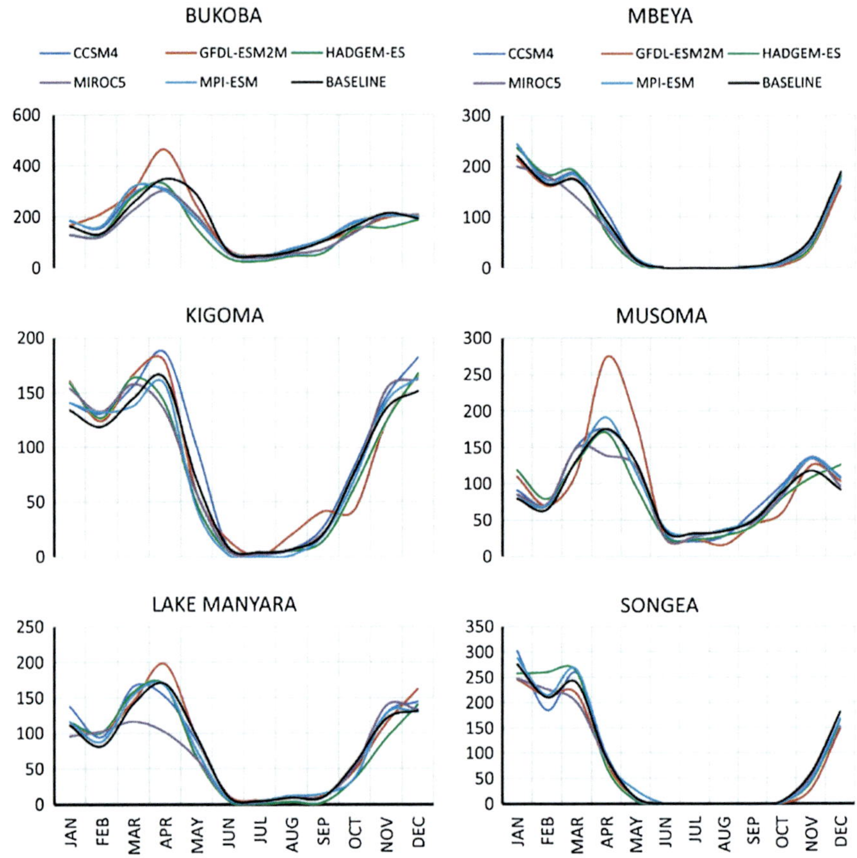

Fig. 15.12 RCP 8.5 mid-century precipitation with respect to the baseline climate

may counteract the otherwise negative effects of rising minimum temperatures and high or low precipitation. However, the yield increase could have been due to over-estimation in the model because yield is reduced under sub-optimal production conditions even if CO_2 is high (Rosenthal and Tomeo 2013). Overestimation may have been due to modelling assumptions in which pests and diseases were assumed non-existent.

This study assessed the impact of climate change on common bean production in major bean growing locations in Tanzania. It is concluded that the minimum temperatures that influence bean pollen viability are projected to increase and,

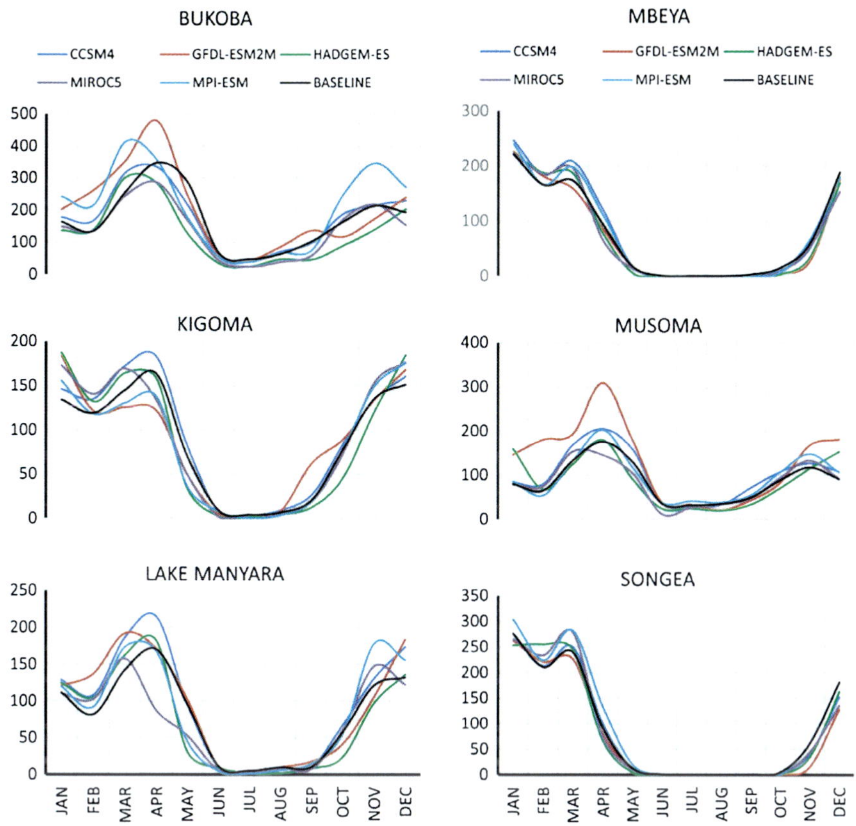

Fig. 15.13 RCP 8.5 end-century precipitation with respect to the baseline climate

consequently, in geographical areas where minimum temperatures are near the critical bean adaptation range, climate change may result in unsuitable for bean production. Since the common bean has C3 photosynthetic pathways in which carbon assimilation is normally limited by low atmospheric CO_2, increased GHG emissions will result in high photosynthetic rates and high yields. The outlook for future bean cultivars should be on those beans with high photosynthetic efficiency under elevated atmospheric CO_2 concentrations. When CO_2 efficiency in beans is coupled with elevated maximum and minimum daily temperature tolerance, climate change will likely have positive, rather than negative impacts on bean production. The CO_2 effects may offset the negative effects of the high minimum temperature

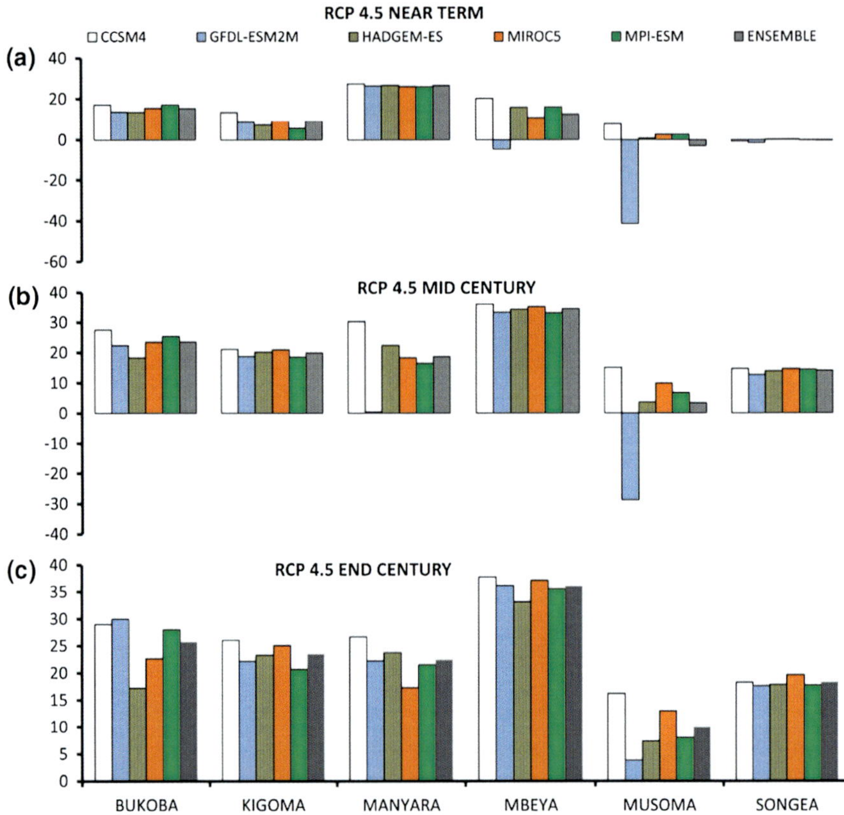

Fig. 15.14 Percent yield change due to climate change impacts under the RCP 4.5 emission scenario: **a** near-term, **b** mid-century, **c** end-century periods

and low or high precipitation exhibited by some of the GCMs. To develop an adaptation framework for climate change, a fine scale assessment is necessary in order to understand local growing conditions that should be incorporated for a comprehensive climate change impact and adaptation assessment.

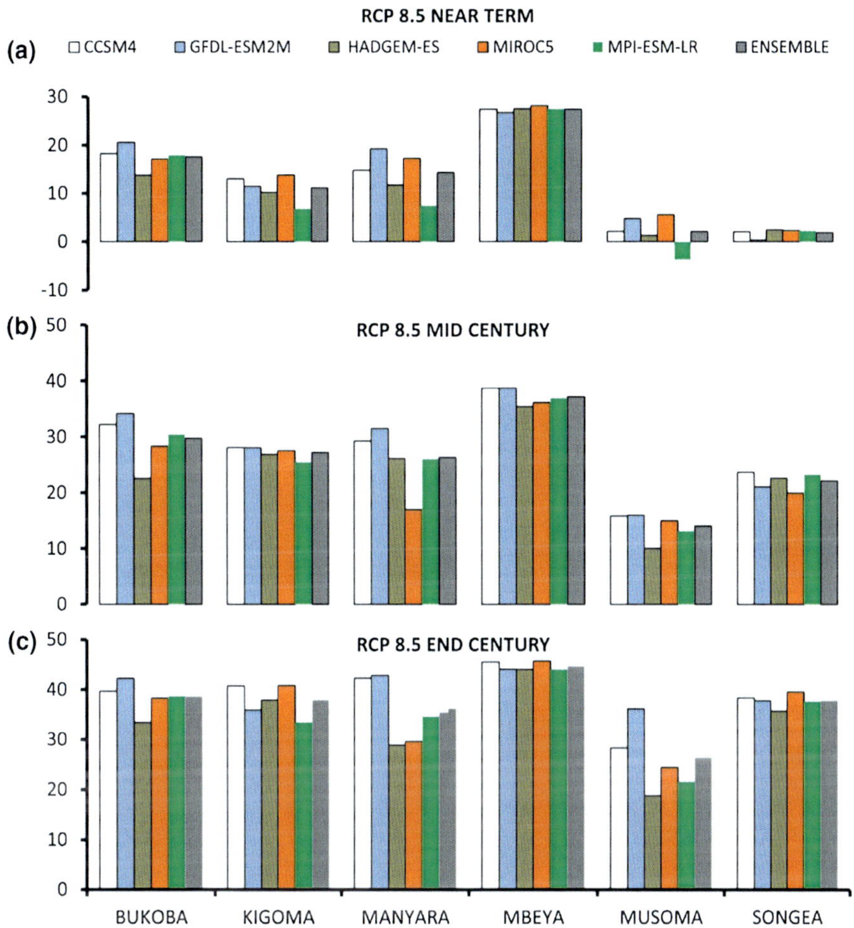

Fig. 15.15 Percent yield change due to climate change impacts under the RCP 8.5 emission scenario: **a** near-term, **b** mid-century, **c** end-century periods

Acknowledgments This study was supported by the Soil Water Research Group of the Sokoine University of Agriculture. Data were obtained from the Agricultural Model Inter-comparison and Improvement Project phase I (AgMIP1).

References

Arndt C, Farmer W, Strzepek K, Thurlow J (2011) Climate change, agriculture, and food security in Tanzania. Working paper no. 6188. United Nations University-World Institute for Economic Development Research (UNU-WIDER) 26

Beebe S, Ramirez J, Jarvis A, Rao IM, Mosquera G, Bueno JM, Blair MW (2011) Genetic improvement of common beans and the challenges of climate change. In: Yadav SS,

Redden RJ, Hatfield JL, Lotze-Campen H, Hall AE (eds) Crop adaptation to climate change. Wiley, Oxford, pp 356–369
Egli DB, Bruening WP (2004) Water stress, photosynthesis, seed sucrose levels and seed growth in soybean. J Agric Sci 142:1–8
Hall AE (2004) Comparative ecophysiology of cowpea, common bean and peanut. In: Nguyen HT, Blum A (eds) Physiology and biotechnology integration for plant breeding. Marcel Dekker Inc., New York, pp 271–325
Hoogenboom G, Jones JW, Wilkens PW, Porter CH, Boote KJ, Hunt LA (2010). Decision support system for agrotechnology transfer (DSSAT) version 4.5. University of Hawaii, Honolulu, HI. CD ROM
IPCC (2007) Climate change: the physical science basis. In: Solomon S, Qin D, Manning M, Chen Z, Marquis M, Averyt KB, Tignor M, Miller HL (eds) Contribution of working group I to the fourth assessment report of the intergovernmental panel on climate change. Cambridge University Press, Cambridge, pp 747–846
IPCC (2013) Annex II: climate system scenario tables. In: Prather M, Flato G, Friedlingstein P, Jones C, Lamarque J-F, Liao H, Rasch P (eds) Climate change 2013: the physical science basis. Contribution of working group I to the fifth assessment report of the intergovernmental panel on climate change. Cambridge University Press, Cambridge, pp 1395–1445
Katungi E, Farrow A, Chianu J, Sperling L, Beebe S (2009) Common bean in eastern and southern Africa: a situation and outlook analysis. International Centre for Tropical Agriculture Cali, Colombia
Kilembe C, Thomas TS, Waithaka M, Kyotalimye M, Tumbo S (2012) East African agriculture and climate change: a comprehensive analysis Tanzania. http://www.ifpri.org/sites/default/files/publications/aacccs_tanzania_note.pdf. Accessed 20 May 2015
Kimball BA (1983) Carbon dioxide and agricultural yield: an assemblage and analysis of 430 prior observations. Agron J 75:779–788
Knutti R, Sedláček J (2013) Robustness and uncertainties in the new CMIP5 climate model projections. Nat Clim Change 3:369–373. doi:10.1038/nclimate1716
Konsens I, Ofir M, Kigel J (1991) The effect of temperature on the production and abscission of flowers and pods in snap bean (*Phaseolous vulgaris* L.). Ann Bot Lond 67:391–399
Lobell DB, Gourdji SM (2012) Influence of climate change on global crop productivity. Plant Physiol 160:1686–1697
Matari ER, Chang'a LB, Chikojo GE, Hyera T (2008) Climate change scenario development for second national communication—Tanzania. TMA Res J 1:40–54
Meinshausen M, Smith SJ, Calvin K, Daniel JS, Kainuma MLT, Lamarque J-F, Matsumoto K, Montzka SA, Raper SCB, Riahi K, Thomson A, Velders GJM, van Vuuren DPP (2011) The RCP greenhouse gas concentrations and their extensions from 1765 to 2300. Clim Change 109:213–241
Moss RH, Edmonds JA, Hibbard KA, Manning MR, Rose SK, van Vuuren DP, Carter TR, Emori S, Kainuma M, Kram T, Meehl GA, Mitchell JFB, Nakicenovic N, Riahi K, Smith SJ, Stouffer RJ, Thomson AM, Weyant JP, Wilbank TJ (2010) The next generation of scenarios for climate change research and assessment. Nature 463:747–756
Mwandosya MJ, Nyenzi BS, Luhanga ML (1998) The assessment of vulnerability and adaptation to climate change impacts in Tanzania. Centre for Energy, Environmental Science and Technology Dar es salaam, Tanzania
Porch TG (2006) Application of stress indices for heat tolerance screening of common bean. J Agron Crop Sci 192:390–394
Porter JR, Xie L, Challinor AJ, Cochrane K, Howden SM, Iqbal MM, Lobell DB, Travasso MI (2014). Food security and food production systems. In: Field CB, Barros VR, Dokken DJ, Mach KJ, Mastrandrea MD (eds) Climate change 2014: impacts, adaptation, and vulnerability. Part A: global and sectoral aspects. Contribution of working group II to the fifth assessment report of the intergovernmental panel on climate change. Cambridge University Press, Cambridge, pp 485–533

Riahi K, Rao S, Krey V, Cho C, Chirkov V, Fischer G, Kindermann G, Nakicenovic N, Rafaj P (2011) RCP 8.5-A scenario of comparatively high greenhouse gas emissions. Clim Change 109:33–57

Ritchie JT, Singh U, Godwin DC, Bowen WT (1998) Cereal growth development and yield. In: Tsuji GY, Hoogenboom G, Thornton PK (eds) Understanding options for agricultural production. Kluwer, Dordrecht, pp 79–98

Rosenthal DM, Tomeo NJ (2013) Climate, crops and lacking data underlie regional disparities in the CO_2 fertilization effect. Environ Res Lett 8:031001

Rosenzweig C, Jones JW, Hatfield JL, Ruane AC, Boote KJ, Thorburn P, Antle JM, Nelson GC, Porter C, Janssen S, Asseng S (2013) The agricultural model intercomparison and improvement project (AgMIP). Protocols and pilot studies. Agric For Meteorol 170:166–182

Rweyemamu CL (1995). A systems assessment of phosphorus fertilizer for use in *Phaseolus vulgaris* L. production in Tanzania. PhD thesis, Michigan State University, p 218

Shonnard GC, Gepts P (1994) Genetics of heat tolerance during reproductive development in common bean. Crop Sci 34:1168–1175

The National Bureau of Statistics (2012) National panel survey 2009–2010. http://www.nbs.go.tz. Accessed 22 Apr 2015

van Vuuren DP, Edmonds JA, Kainuma M, Riahi K, Weyant J (2011) A special issue on the RCPs. Clim Change 109:1–4

Wortmann CS, Kirkby RA, Eledu CA, Allen DJ (1998) Atlas of common bean (*Phaseolus vulgaris* L.) production in Africa. CIAT publication no. 297: Centro Internacional de Agricultura Tropical, Cali, Colombia

Chapter 16
Institutional Water Resources Management and Livelihood Adaptation: A Case from Kilombero Rural Areas, Tanzania

Paul Vedeld, Edgar Liheluka and Gimbage E. Mbeyale

Abstract The impacts of irrigation schemes on poor people's livelihoods are studied in Kilombero, Tanzania. Total household income is 2 times higher for improved irrigation scheme farmers, and their farm income is 3 times higher than in traditional rainfed farmers. We further find that reported land productivity is 4–6 times higher in improved rice-irrigation fields. While the income of these farmers has gone up, so have their costs (3 times higher input costs). Looking at local people's dependence on water, households on average report to derive 43 % of their income from irrigation, and the dependence is even higher for poorer groups of households (57 %). Improved schemes come with formalized systems of rights and duties, monitoring, control, sanctions and water-user fee structures. This necessitates introducing new institutions on top of existing traditional systems for resource management. The new systems are bricolaged into existing systems, so in practice, traditional and modern irrigation schemes are not conducted very differently. Local people generally seem to manage these irrigation systems well within reasonable conflict levels. There is, however, concern that the new policy, advertised as the devolution of water rights to local communities, could lead to increased central control over rural water, especially when the hydropower sector's priorities (40 % of total water) sector's priorities constrain dry season irrigation. Within the agricultural sector large-scale commercial farmers may further access the majority of irrigation water at the expense of small-scale farmers.

P. Vedeld · E. Liheluka
Norwegian University of Life Sciences, Ås, Norway

G.E. Mbeyale (✉)
Sokoine University of Agriculture, Morogoro, Tanzania
e-mail: mbeyale@suanet.ac.tz

R. Lal et al. (eds.), *Climate Change and Multi-Dimensional Sustainability in African Agriculture*, DOI 10.1007/978-3-319-41238-2_16

16.1 Introduction

There is a 7-million-sq-km belt of arable land across the Great African Guinea Savannah. Presently, less than 10 % is permanently cultivated. The belt has a precipitation rate of 800–1500 mm and can support crops 150–220 days/year. According to the World Bank (Binswanger and Gautam 2010), this area is one of largest under-used agricultural land reserves in the world (Fig. 16.1).

Most of Tanzania's land falls into this category. Tanzania has been termed "an agricultural sleeping giant" with an estimated 44 million ha of arable land (Binswanger and Gautam 2010). Only 23 % of total land is presently under cultivation. Of the 29 million ha irrigable land, only 1 % of this land is presently under use. There is furthermore a substantial stock of livestock (20 million cattle) that is not much commercially exploited.

Tanzania's agricultural sector is the country's main productive sector, providing a livelihood for more than 70 % of the total population. Agricultural development thus indisputably remains key to the country's economic and social development, at least in the foreseeable future (Binswanger and Gautam 2010).

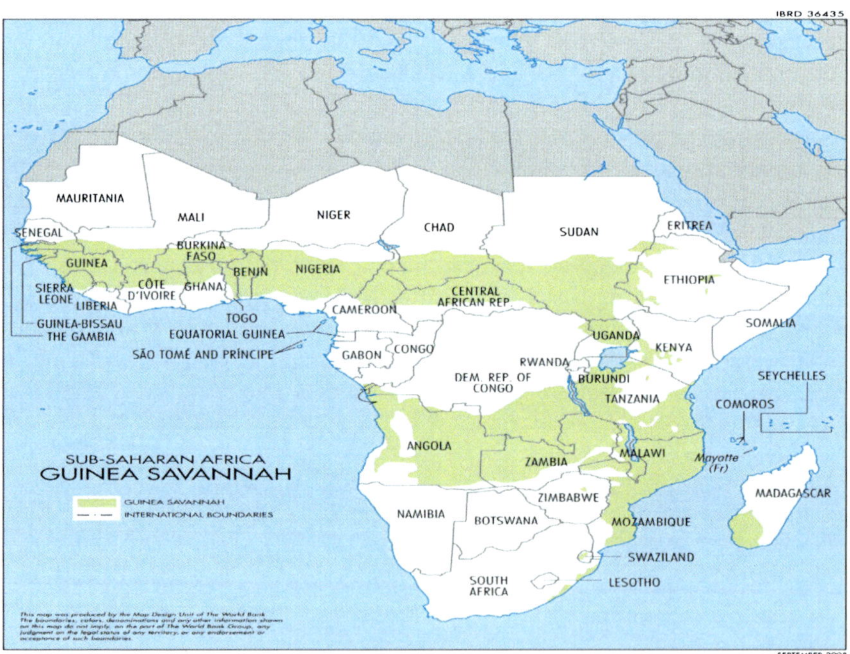

Fig. 16.1 Map of the Great African Guinea Savannah

Since Tanzania gained independence, there has been a striking lack of investment in agriculture. Tanzania annually invested some 5 % in agricultural infrastructure, whereas Asian countries spend 20–30 % of state investments In Tanzania, we further find that 80–90 % of the agricultural production increase has been achieved through the expansion into new agricultural land (land clearing)—not through increased productivity/ha (Binswanger and Gautam 2010). There are also other constraints, both economic and political, on productivity increases.

Irrigation is seen as an important measure to address several key challenges facing agricultural development. At the global level, irrigation is used on 20 % of all land, provides 40 % of all agricultural outputs, and consequently is crucial in agricultural intensification strategies (Food and Agriculture Organization 2007). The global movement of Integrated Water Resource management IWRM and the accompanying water policies have also been implemented in Tanzania. These water policies can be characterized as part of a neoliberal policy trend, rolling back the central state through devolution and decentralization, involving local communities in formalizing water and land rights, and introducing payments for environmental services, water fees, and other economic instruments. The World Bank and International Monetary Fund have been quite instrumental in national development of these policies, which have also found support among national elites and central state bureaucracies (Goldin and Kibassa 2009; ActionAid 2004).

Water use associations have been formed and linked to individual schemes and used to distribute sector-allocated water among members. The state still controls overall water-management planning and distribution of water rights among and within economic sectors. The government also controls the appropriation of funds through water fee allocations.

This paper investigates the substantial potential that decentralized, local water management and irrigation schemes offer for agricultural development and improved rural livelihoods in Tanzania. Poor people in rural areas in Tanzania depend on agriculture for survival and their livelihood. In areas where irrigation schemes are developed or improved, income levels and land productivity tend to increase. Household incomes often increase markedly relative to surrounding communities without this infrastructure. The scope for developing irrigation schemes throughout Tanzania is enormous, given that only 23 % of potential irrigation land is developed. There are several challenges related to the schemes introduced, including lower yield levels than expected and great input costs, debt, household vulnerability, local conflicts over water, and differences between rich and poor (Kissawike 2008).

This paper explores Tanzania's water policies in the context of rural livelihoods. How do people make a living? To what extent do they depend on water resources? How do they organize around the use of a common-pool resource, such as irrigation water? This paper uses the sustainable livelihood approach (SLA), economic indicators of irrigation-water dependence, and institutional theory on the challenges to introducing local-level, participatory, water-management organizational forms and institutions. We also offer policy recommendations.

16.2 Agricultural and Water Policies in Tanzania

Irrigation is not a new intervention in Tanzania. There are records of pre-colonial gravity-fed irrigation activities in certain high-potential areas conducive for irrigation. These activities were managed and controlled under customary rule systems (Chiza 2005; Kikula 1997; Kissawike 2008). Good examples include old irrigation systems in the northern highlands, where the Chagga people irrigated fields through canals accelerated by gravity. Certain areas in the Usangu Plains in the southern highlands also had permanent water supply. Individual German missionaries already introduced irrigated cash-crop systems in pre-colonial times, while Arabs introduced irrigated rice production in the Iringa, Mbeya and Tabora areas during the slave-trade period (Kissawike 2008; Pipping and Chale 1976). During the colonial era (1884–1960), irrigation policies mostly focused on supplying water to commercial farmers and settlers. As seen in Table 16.1, increasing formalization and organizational and management capacity developed over time throughout the colonial period.

Table 16.1 Timeline on evolution of water management institutions in Tanzania, 1885–2009

Year	Events
Before 1885	Customary systems used to manage water resources and irrigation systems, particularly in Kilimanjaro, Arusha and Mbeya regions; not in Kilomebro (as far as we know)
1914	Formation of first water law in Tanganyika under German rule
1923	First water law approved under British rule
1948	Water Rights Ordinance introduced. Recognized rights of native Africans to water for customary use
1959	Water Rights Ordinance introduced ownership of and right to use of water
	Mandated establishing institutions for water supplies in urban and rural areas
1950ies	Flood control and storage dam measures implemented throughout country
1961	Independence
1965–1973	Ujamaa villagization policies supporting irrigation scheme development
1974	Water Utilization (Control and Regulation) Act No. 42, 1974 introduced. Replaced 1948 Water Ordinance. Regulated river, streams, and internal lakes resources. Established institutions and organizations through principal water office and Central Water Board. Established the Regional Basin Water Board system. The Principal Water Act
1981	Water Utilization Act No. 42, 1974 amended to integrate the concept of river basin management
1981	Designation of Tanzania water resources into nine river basins
1989	Water (Misc.) Act. No. 17 of 1989 and General (Regulations) Act provided regulatory and institutional framework for water resources management. All water vested with the state. Principal Water Office authorized to be responsible for setting policy and allocating water rights at the national level. Basin Water Office given responsibility for designating water drainage basins

(continued)

Table 16.1 (continued)

Year	Events
1980ies	National Village Irrigation Development Programme (NVIP) supports farmer-managed village irrigation programs (DANIDA, CIDA, JICA)
1991	Water Policy of 1991 emphasizing the free provision of clean and safe water to all Tanzanians
1991	Establishment of Pangani Basin Water Board and Office
1992	Dublin principles established water as a right, economic good, and finite resource needing participatory, engendered management. No explicit trade-offs made between the principles
1993	Establishment of Rufiji Basin Water Board and Office
1994	National Irrigation Development Plan (1994–2014) and the National Irrigation Policy aimed at expanding irrigation activities
2000	Revision of national water policy to include aspects of integrated water resource management
2002	Development of new National Water Policy (NAWAPO), forming institutional basis for new policy regulating rights to irrigation and management at the national, basin, community and individual levels. New Ministry of Water and Irrigation. New policies on economic incentives and efficiency in water use
2003	Amendment of water legislation to reflect the new policy
2009	New Water Resource Management Act No. 11 of 2009

Sources Maganga et al. (2002), Sokile (2003), Van Koppen et al. (2007), URT (2009), Kissawike (2008), Patel et al. (2014) and Mosha et al. (2016)

In the post-independence era (1961–), managerial interventions and institutional evolution have continued. The Water Utilization (Control and Regulation) Act, 1974 was enacted to regulate the use of rivers, internal lakes, and streams. In 1981, adopted the river basin management concept and established nine basins (Sokile and van Koppen 2004). The Water Utilization Act (1974) and its Amendment Acts (1981 and 1989) provided the regulatory and institutional framework for water resource management. All water was vested with the state. The Principal Water Office was responsible for policy formation and allocation of water rights at the national level, while the Basin Water Office had charge of designating and administrating water-drainage basins. The 2002 National Water Policy and 2013 National Irrigation Act are the present cornerstones for water-sector policies and are aimed at increasing efficiency in water use through economic and legal incentives (Liheluka 2014).

The Rufiji Basin Water Office (RBWO) is the responsible body for the study area. The functions of RBWO includes administration of water utilization law, collection of various water-user fees, allocation of water rights, legalization of water use, and modification and controlling of water abstractions. Other functions include providing grants, monitoring water use, resolving conflicts, holding stakeholders meetings, and researching issues (Maganga 2003; RBWO 2007).

16.3 Theory

This paper employs an economic version of the sustainable livelihood approach (Scones 1998; Ellis 2000; Patel et al. 2014) to assess the socioeconomic effects of irrigation. We apply an institutional approach drawing on Ostrom's (1990) design principles and a critical institutional interpretation of them (Cleaver 2012) to investigate challenges to managing an irrigation scheme as a "long-enduring," local social institution (Ostrom 1990) (Table 16.2).

Table 16.2 Comparing institutional perspectives

Themes	Mainstream, rational-choice institutionalism	Emerging, critical institutionalism
Livelihood and natural resources management	Clear links between single resources and use	Multiple users, complex and diverse livelihood systems
Human agency	Rational, clear, consistent, consequential logic	Socially constructed, diffuse, interpretive, negotiable, logic of appropriateness
Community	Local, specific-user-group, homogenous, bounded rationality	Multiple locations, diffuse, heterogeneous, diverse, multiple social identities and groups
Institutions	Static, rules, managerial, functionalist, formal organizations and institutions emphasized	Institutions as socially constructed and embedded in practice, struggles over meaning, formal–informal, interlinked with knowledge and power
Property regimes	CPR as a set of rules based on collective action, determined, strategic outcomes, and clear universal boundaries	Determined by practice and not by formal rules, overlapping rights and responsibilities,ambiguity, inconsistency, flexibility interpretation, negotiation
Resources	Material, economic, direct use-values, clear sets of interests	Symbols, resources are locally and historically embedded and socially constructed, carriers of meaning and identity
Power and control	Transaction cost focus, elites, community leaders, common interests and perspectives	Differentiated actors, gender, conflict, central bargaining, negotiation and power relations

Sources Ostrom (1990), Mehta et al. (2001), Cleaver (2012), Vedeld (2002) and Patel et al. (2014)

16.4 Methods and Study Area

16.4.1 Methods, Data and Models

In a mixed-methods approach (Bryman 2008), quantitative and qualitative data were collected at the regional, village/community, and individual levels between September and December 2012. We used focus group discussions, key informants, and structured household surveys with open-ended and closed questions to collect and generate material (Liheluka 2014).

We selected 2 of 81 villages in Kilombero District where both rainfed agriculture and traditional and improved irrigation schemes were present. We interviewed 103 households, of whom 31 were members of improved irrigation schemes, 12 members of traditional schemes, and 60 without irrigation scheme memberships. We also held meetings with irrigation groups and key informants in the villages. The data were analyzed using descriptive and analytical models and SAS/JMP software packages.

16.4.2 Study Area

Mkula and Msolwa A are the two case study villages in the Kilombero Valley beside the Udzungwa Mountains in the Kilombero District in Morogoro Region, Tanzania (Fig. 16.2). The district covers 14,918 sq km, including the Selous Game Reserve, Udzungwa National Park, and Kilombero wetlands.

The mean annual air temperature is 26–32 °C, with precipitation of 1200–1600 mm spread across two rainy seasons. Most areas in the Kilombero Valley experience flooding during the rainy seasons (Region Commissioner's Office Morogoro 2008).

The Kilombero district has two main vegetation types: wooden grassland and Miombo woodland. These areas also contain wildlife populations in both national parks, game reserves, and outside national park areas (Haule et al. 2002). The district has a water surface area of 1341 sq km, with 38 perennial and seasonal rivers. The Mkula River is sourced at the Udzungwa Mountains and drains through Mkula village and passes several villages downstream before discharging into the Kilombero River. The dominant ethnic tribes in Kilombero district are the Wambunga, Wandamba, Wabena, and Wahehe. The district's population is 407,880, with a growth rate of 2.5 % (National Bureau of Statistics 2013).

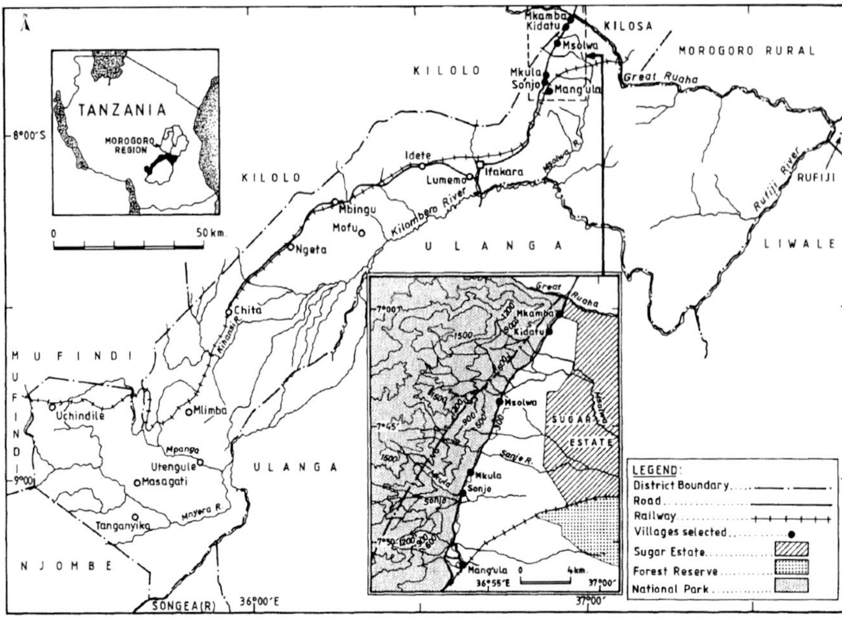

Fig. 16.2 Kilombero District and the case study area. *Source* Cartographic Unit, Geography Dept, UDSM

16.5 Results and Discussion

We present the results on people's reported livelihoods, dependence on water use, and capabilities to manage irrigation institutions, as well as the impacts from different irrigation-water-management institutions and water access on livelihoods.

16.5.1 Livelihood Adaptation and Irrigation Schemes

16.5.1.1 Wealth Groups, Location and Assets

The average household in the Kilombero district has 5.7 members and low education level (5 years of schooling) and owns very little land (2.1 ha). Of these households, 36 % have access to irrigation, only half have electricity, 68 % have bicycles, and 80 % have mobile phones (Table 16.3). They are hard-core poor, with an estimated per-capita income of USD 0.84 per day.

There are significant relationships among household assets, reported income, and various types of access. Looking at variation by income groups, we find substantial variations in land and capital access, labor, and education levels as the less poor groups generally have more capital from which to generate income. We also see

Table 16.3 Socio-economic assets by wealth groups, Kilombero district, Tanzania, 2012

Household assets	Very poor (N = 33)	Poor (N = 37)	Less poor (N = 39)	Sample (N = 109)
Sex (% male)	75 %	87 %	81 %	81 %
Household size (members)*	4.9[a]	5.3[a]	6.8[b]	5.7
Age (years)	50.4	52.5	52.25	51.7
Education level*	5[a]	6[ab]	7[b]	5.4
Land owned (ha)	1.7	1.7	3	2.2
Land used (ha)	1.7	1.8	3	2.2
Access to commons (%)*	86 %[a]	95 %[a]	67 %[b]	82.3 %
NGOs social network (%)	33 %	46 %	64 %	48 %
Hired labor (TZS)	86 %	97 %	92.3 %	91.7 %
Access to bicycles (%)	61 %	70 %	72 %	68 %
Electricity access (%)	33 %	43 %	52 %	43 %
Access to credit (yes)	9 (25 %)	7 (18 %)	18 (50 %)	11.3 (31.19 %)
Amount of credit	260,000	418,000	836,000	506,000
Location Msolwa Mkula	25 (35.7 %) 8 (20.5 %)	27 (38.6 %) 10 (25.6 %)	18 (25.7 %) 21 (53.9 %)	23.3 (100 %) 13 (100 %)
Main type of agriculture				
Improved irrigation	4 (12.5 %)	9 (28.1 %)	19 (59.4 %)	11 (100 %)
Traditional irrigation	4 (30.8 %)	4 (30.8 %)	5 (38.5 %)	4 (100 %)
Rain-fed agriculture	25 (38.5 %)	24 (36.9 %)	15 (23.4 %)	22 (100 %)
Total costs of production	497,915	1,229,757	2,842,349	1,585,172
Net income TZS (USD/cap and day)	498,165 (0.17)	2,485,410 (0.78)	5,851,736 (1.4)	2,893,732 (0.84)

USD 1 = TZS 1650 (2012). N = 109. *Significant differences in the absolute incomes from a source between income groups ($p < 0.05$); income groups with different superscript letters differ significantly in their level of dependence on a given source ($p < 0.05$) in a Tukey Kramer HSD test ($R^2 = 0.27$; DF24; Chi-square 63.74; Prob > ChiSq 0.0001)

that there is an overrepresentation of farmers with improved irrigation in the less poor groups, and there are more rainfed-dependent farmers in the very poor group. Land—and water during the dry season—are increasingly scarce in the area as the surrounding land has been incorporated into in Selous National Park and Udzungwa Mountain National Park on the western and northern side and in the Illovo Sugar Company plantation on the eastern side. This constrains agricultural land expansion and reduces access to various environmental resources.

In addition, the population growth rate of 2.5 % (doubling in less than 20 years) has precipitated a rapid, significant subdivision of land. The RBA has banned water access in the dry season, much to the dismay of farmers. Family labor access constrains production, and more than 90 % of study participants report hiring labor for certain tasks, such as land clearing, bunding, rice planting, land and soil preparation, and harvesting. Rice and sugarcane production are both labor-intensive activities, and the costs involved in commercial rice production are quite high, up to 50 % of net incomes in many cases.

There is substantial loaning activity in the area: 31 % of participants report borrowing money, and most are from the less-poor group. In the whole sample, households have an average household debt of 506,000 TSH, or some 17.5 % of the total household income. Among those who have loans, the average is TSH 1.4 million. Loans are given on a 1–3-year basis, and the interest rate varies between 2 and 20 %. Most of the loans are informal (70 %), while 30 % are through credit and saving institutions, national marketing Boards, and similar institutions. Farmers with improved schemes utilize much more credit (TSH 766,000) than traditional-scheme farmers (TSH 499,000) and rainfed farmers (TSH 384,000).

We selected two villages in Kilombero district, one (Mkula) where inhabitants have access to both a modern and a traditional irrigation scheme and one village (Msolwa) without such access. Msolwa had access to sugarcane production, an important alternative market option within agriculture. The two villages are both old Tanzanian Ujamaa villages from the 1970s, when people were centralized into villages and provided with infrastructure, schools, water supply, and social services. These public policies still affect the land layout and distribution in Msolwa. The government allocates 50 % of land, and only 21.7 % is inherited. In contrast, Mkula has a longer history with a more traditional land acquisition pattern, so 64 % of the land is inherited, and 25.6 % is allocated through the government. Approximately 10–12 % of the land in the two villages is reported to have been purchased.

16.5.1.2 Wealth Groups, Location, Activities and Outcomes

Households diversify their assets through variations in on-farm and livestock production, migration and remittances, and various off- and non-farm activities. Looking at Table 16.4, we see that agriculture is by far the main income source

Table 16.4 Income source by wealth group in the studied villages, Kilombero district, Tanzania, 2012

Income sources	(N = 36) Very poor		(N = 37) Poor		(N = 36) Less poor		(N = 109) Total	
	Income (TZS)	% Total	Income (TZS)	% Total	Income (TZS)	% Total	Income (TZS)	%Total
On-farm*	297,915	60	2,059,062	83	4,548,431	78	2,299,575	77
Off-farm	73,028	15	67,973	3	86,361	1	75,716	3
Non-farm*	102,500	20	327,297	13	1,108,889	19	554,404	18
Remittances	24,722	5	31,081	1	108,056	2	54,403	2
Total income*	498,165	100	2,485,410	100	5,851,737	100	2,984,098	100

N = 109. *Significant differences among households' wealth groups ($p < 0.0001$; Rsq 0.91; Prob > Chi-sq 1). Standard deviation in brackets. USD 1 = 1650 TZS

Table 16.5 Annual income source by type of production system in the studied villages, Kilombero district, Tanzania, 2012

Household variables	Rain-fed agriculture farmers (N = 64)	(%)	Traditional irrigation farmers (N = 13)	(%)	Improved irrigation farmers (N = 32)	(%)	Sample average (N = 109)	(%)
On-farm*	1,474,575[a]	72	2,988,615[b]	88	3,669,653[c]	81	2,299,575	77
Off-farm	95,453	5	46,076	2	48,281	1	75,716	3
Non-farm	460,000	22	317,692	9	692,188	15	554,404	18
Remittances	26,250	1	26,923	1	121,875	3	54,403	2
Total	2,056,278	100	3,379,307	100	4,531,997	100	2,984,098	100

USD 1 = 1650 TZS. N = 109. *Significant differences among types of agriculture ($p = 0.0001$; RSq 0.16; Prob > Chi-sq 0.97). Standard deviation in brackets. *Significant differences in the absolute incomes from a source between income groups ($p < 0.05$). Income groups with different superscript letters differ significantly in the extent of dependence on a given source ($p < 0.05$) in a Tukey Kramer HSD test

(some 80 % of all incomes) in the two villages, followed by non-farm income (18 %). We observe that the poor-income group has the highest share of income from activities outside agriculture, particularly piecework, environmental incomes, and various non-farm activities. There are in general low environmental incomes in the area, most likely due to the lack of available commons created by the use of land for protected areas and plantations. Agriculture accounts for 76 % of total income in Msolwa and 81 % of total income in Mkula. Within agriculture, rice accounts for 58 % of total income in Mkula and sugarcane 62 % of total income in Msolwa. Sugarcane and rice are the second-largest income sources in the two villages. Remittances are not important to these villages.

How does the type of farming system affect the overall diversification patterns among different groups of households (Table 16.5)? First, improved irrigation scheme farmers have more than twice the income of rainfed farmers, along with higher overall non-farm and remittances incomes. Rainfed farmers have lower total incomes and adapt by procuring nearly 30 % of their income from non-farm and off-farm activities. This difference likely reflects their lower ability to depend on agriculture due to less access to water, land, credit, and labor.

If we further dissect on-farm income, we see that farmers depend on either rice or sugarcane as their main livelihood (Table 16.6). Improved irrigation scheme farmers depend heavily on rice production, while traditional-irrigation and rainfed farmers depend more on sugarcane. Most of the sugarcane is produced outside irrigation areas. The poor and the less-poor income groups depend on sugarcane and rice production. Rice provides higher income in Mkula than Msolwa, where there is little irrigation.

Table 16.6 On-farm income by type of farmer, Kilombero district, Tanzania, 2012

Crop production income	Rainfed agriculture farmers (N = 109)		Traditional irrigation farmers (N = 13)		Improved irrigation farmers (N = 32)		Sample average (N = 109)	
Rice*	358,075[a]	24.3	818,000[b]	27.3	2,306,716[c]	62.7	985,006	42.8
Sugarcane	836,953	56.9	1,586,385	52.99	1,231,656	33.5	1,042,211	45.3
Maize	104,750	7.1	148,077	5.0	76,563	2.1	101,642	4.4
Vegetables	94,532	6.5 %	289,231	9.7	43,124	1.2	935,27	4.1
Livestock	77,797	5.3 %	151,923	5.1 %	20,000	0.5 %	69,690	3 %
Total on-farm income*	1,472,107[a]	100	2,993,615[b]	100	3,678,059[c]	100	2,299,575	100

N = 109. *Significant differences among types of agriculture ($p < 0.05$). Income types with different superscript letters differ significantly in the extent of dependence on a given source ($p < 0.05$) in a Tukey Kramer HSD test (RSq 0.30; DF 200; Prob > Chi-sq 0.9996)

16.5.1.3 Vulnerability Contexts

Important vulnerability challenges in the area arise from government policy failures, natural vagaries, a lack of efficient, fair markets for both inputs and outputs, and, of course, the recurrent themes of land and water shortages that have intensified over the past 5–10 years, according to respondents.

Risk Management

Farmers report engaging in various household risk-management strategies as they seek to earn a livelihood and survival. These strategies are related to, among others, the diversification choices between agriculture and other activities. These choices are more pronounced among the very poor farmers who tend to take on more off-farm activities (working for others). Many farmers are also involved in some agricultural production that does not depend on irrigated water. Many farmers also report storing part of their grains to avoid or reduce price and income fluctuations and maintain a store for unforeseen events. Farmers also attempt to grown more than one crop a year to build up a reserve of savings, both in kind and cash. They might also opt to plant some crops that are flood and drought resistant as a back-up to avoid the effects of recurrent crop failures.

Coping Strategies

This repertoire provides responses to crises of various types, including floods, drought, price variations, wildlife raiding crops and livestock, the illness or death of productive family members, and the loss of remittances and other non-farm activities that can bring in cash but are often accompanied by substantial uncertainties. Farmers respond to such crises by selling or renting assets such as bicycles, livestock, and land. They might migrate for periods of time (20 %). Poor people are more vulnerable to shocks than those who are less poor.

Irrigation introduces a new type of risk: a more capital-intensive form of production. Farmers report borrowing some 20 % of their total income every year for purchased inputs and heavily depend on secure outcomes. Irrigation farmers use more credit and capital than rainfed farmers. Although these practices are reflected in higher incomes and yields in good years, there certainly are also increased risks in years when irrigation water is scarce or crops fail for other reasons. Interest rates averaging 11.2 % add to the risk of irrigation farmers forfeiting loans and encountering problems, as is reported by other studies from Tanzania (Kissawike 2008).

16.5.2 *Irrigation Water Dependence and Sustainable Livelihoods*

In this livelihood assessment, we explore the level and type of household dependence on irrigation-water income in relation to livelihood outcomes and vulnerability. The share of total household income from rice irrigation was used to calculate an index for water household-livelihood dependence. It was found that 47 % of overall income from *households with irrigation* is derived from irrigation-dependent production. Traditional-irrigation farmers (24 %) but especially improved-scheme farmers (50 %) exhibit high income dependence on irrigation water.

If we look at the wealth groups of those involved in irrigation (Table 16.7), we find that the less poor have much higher total incomes from water but are also less dependent on water (43 %). The very poor have 8 times lower income from irrigation but still derive 50 % of their income from irrigation.

Table 16.7 Water income dependence of irrigators by household wealth group, Kilombero district, Tanzania, 2012

Irrigation and non-irrigation incomes for irrigation farmers	Very poor (N = 8)		Poor (N = 11)		Less poor (N = 20)		Total irrigation (Sample N = 39)	
	Average	%	Average	%	Average	%	Average	%
Irrigation income*	384,375[a]	50	1621,127[b]	62	2,656,125[c]	43	1,898,203	47
Non-irrigation income*	384,125[a]	50	997,000[b]	38	3,461,750[c]	57	2,135,256	53
Total household income*	768,500[a]	100	2618,127[b]	100	6,117,875[c]	100	4,033,459	100

N = 39. *Significant differences among wealth groups ($p < 0.05$). Income groups with different superscript letters are statistically significantly different ($p < 0.05$) in a Tukey Kramer HSD test (RSq 0.8321; DF 4; Prob > Chi-sq 0.0001)

Table 16.8 Water income dependence of Irrigators and None Irrigators by household wealth group, Kilombero district, Tanzania, 2012

Irrigation and non-irrigation incomes for all farmers	Very poor (N = 36)		Poor (N = 37)		Less poor (N = 36)		Total farmer sample (N = 109)	
	Average	%	Average	%	Average	%	Average	%
Irrigation income*	85,417[a]	17	481,956[b]	19	1,475,625[c]	25	679,173	23
Non-Irrigation income*	412,748[a]	83	2,003,457[b]	81	4,376,112[c]	75	2,304,925	77
Total household income*	498,165[a]	100	2,485,413[b]	100	5,851,737[c]	100	2,984,098	100

N = 109. *Significant differences among wealth groups ($p < 0.05$). Income groups with different superscript letters are statistically significantly different ($p < 0.05$) in a Tukey Kramer HSD test (RSq 0.8959; DF 4; Prob > Chi-sq 0.000)

It seems then that the very poor invest whatever land and labor they have into irrigation agriculture, whereas the less poor can diversify and expand their income into other agricultural and non-farm activities. Looking at averages for the total sample including those without irrigation income, we see that the average dependence is 23 %, somewhat higher among less-poor households. The average total income for the whole sample is more than 11 times higher than that of less-poor households, reflecting the limited access to both irrigated land and capital among the poorest households (Table 16.8).

16.5.3 Institutional Arrangements Around Water Use in Kilombero

We address to what extent people can manage water resources locally in the context of official policies, organizational structures, and institutions that frame local water management.

16.5.3.1 Brief Description of the Schemes

There are 14 schemes in the Kilombero district, covering 17,600 ha of irrigated land (Ngasonwa 2007). The Mkula village has two different irrigation systems: the Mkula improved irrigation scheme and the MAKI traditional-irrigation scheme. There were no irrigation schemes in the other village studied (Msolwa A).

The population in Kilombero has gradually increased after the founding of a sugar company in 1962, the construction of the Tazara railway, and the designation of Mkula as an Ujamaa village. There were no irrigation schemes before 1979, when Mkula was established with a traditional system and later upgraded to an improved scheme. Land was generally abundant before 1979, and the wetlands were not cultivated. The irrigation scheme launched in 1979 opened the wetlands to agricultural use. It was later transformed into an improved scheme, and MAKI, a traditional scheme, was established in 1994.

The Mkula scheme has 254 ha and 91 members, and the MAKI system 320 ha and 120 farmers. The two schemes are partly fed from the same river (Mkula River), while MAKI also draws water from the Msufini River. Both rivers have annual flows. Both schemes have written constitutions approved by the responsible authorities and the same administrative arrangements with an elected chairman, vice chairman, secretary bursar, board members, and various committees (maintenance, disciplinary, administration, and planning). The Rufiji Basin Office issues water permits, collects water fees, and provides planning and advisory services for the schemes.

The two types of schemes are still somewhat different. The traditional scheme has a poor physical infrastructure with limited water conveyance and hydraulic water-distribution mechanisms. Much work is spent repairing annual flood damage to the canal systems. The improved scheme has better water control, which reduces both water-management problems and workloads. The concrete intake and main canals facilitate control of the required water levels and reduce flooding costs and labor use. There is a general problem of leveling in both schemes because water flow is gravity fed, and even within a single field, amounts of water might differ around the plot.

16.5.3.2 Water Management Institutions

We investigated whether farmers perceive the institutions in place to manage water resources as water effective, cost efficient, and legitimate through the use of Likert-type statements (Table 16.9). We asked respondents to what extent they (strongly) agreed or disagreed to the statements and used a modified version of Ostrom's (1990) design principles to organize a discussion of the institutional arrangements. We discuss the findings related to Cleaver's (2012) more critical institutionalism.

(1) Clear Boundaries on Water Access and Rights

The clear boundaries principle posits that delimitations on physical resources, amount of water accessed, and land boundaries improves management and reduces conflicts. The scheme administrators at the local level identify users and allocates their water access from the main, secondary, and tertiary canals at given times and days. Water flows also vary over the year, causing seasonal shortages and excesses of water. Overall water access is regulated by RBWO through issuing water-user permits.

The formal boundaries might be clear as to how much water one can expect to get and who is eligible to get water, but practice conflicts with these principles, creating a sense of unclear boundaries. Examples given include farmers having to watch their fields at night to prevent others from diverting water to their own fields

Table 16.9 Farmers' statements on robust institutions, Kilombero, Tanzania, 2012

Institution and management	Improved	Traditional	All	Comparing schemes
Are boundaries clearly assigned? –Should others be restricted?	39 54	46 61	44 68	There are boundary issues in both schemes. There are somewhat clearer boundaries and a will to exclude in the traditional scheme. Rights also imply exclusions. An improved system would most likely be better for downstream users
Are memberships and rights clear and fairly distributed? –Do some have more water rights?	52 27	53 48	53 32	About half find the issues to be clear. The improved plan is clearer but has more exclusion (pastoral, in-migrants? Who plans? Who owns?) The scheme was clan–based before and is now more focused on individual rights. There is a more skewed rights distribution in the traditional system
Does everyone carry out their duties? –Are water charges paid regularly?	53 48	34 0	39 23	Many do not carry out duties. More work is done by poor farmers, and some avoid communal labor and get more water access in both schemes. More effective water use and higher outcomes in improved scheme
Is there a fair say in decision making?	84	64	70	Most report being involved. Village elders lead in the traditional scheme; in the improved scheme, the WUA and VC do. The improved scheme requires paying fees, but this has met with strong resistance. There are more meetings in the improved scheme
Is the monitoring system effective? –Is the water maintenance system satisfactory? –Do people take more water than allowed? –Do rich people have more water access?	41 34 83 66	30 53 76 62	35 42 84 65	Monitoring and control are generally low There seems to be overuse of water under both systems, especially by less poor people There is more control of water use in the improved scheme, but the control is less legitimate A majority seems to agree that less poor people have better water access
Does the sanction system work well?	56	61	56	There are many conflicts (between users, pastoral, in-migrants) in both systems, but somewhat more in the traditional schemes. Sanctions by village elders function somewhat better than in the improved WUA, where more formal (court) systems are involved

(continued)

Table 16.9 (continued)

Institution and management	Improved	Traditional	All	Comparing schemes
Are internal conflicts managed and resolved fairly?	60	53	57	Conflicts were seen as common by 80 % of the sample, a bit more in improved schemes. There is more competition over water, and 50 % complain about too little water
–Are conflicts over water use common?	83	76	81	The traditional scheme is cheaper and results in less water scarcity and fewer conflicts
–Are conflicts are resolved quickly?	72	76	74	
–Has water competition increased?	91	76	88	
–Do you get the right amount of water?	50	50	50	
Do external authorities interfere in local water arrangements?	17	41	23	There is a perception of local control of water and not much external interference in local water use. There is more formal and state control under the improved system
–Do local authorities have power over water?	76	88	81	The appropriation of fees, plus possible taxation basis and sector division or ban in dry season under the improved produce conflicts with the state
–Does the traditional water management system function?	51	83	79	

outside the schedule in the dry season or diverting water away onto other farmers' fields during the wet season. The major reported conflicts occur in the dry season. Interestingly, members of the improved irrigation scheme complain about blurred boundaries (61 %) more than traditional-scheme members (54 %). The very poor group (62 %) complains more than less poor groups (47 %) about blurred boundaries, limited water access, and weakened water rights. Clarifying boundaries is interpreted, negotiated, and molded into local institutions for resource management (Cleaver and Franks 2005). The physical boundaries of the resource typically are not clear because water is shared among people from different villages upstream and downstream, and these individual actors might be involved in various schemes. Additionally, people may develop or have kinship relations in other villages and engage in informal land exchange to secure water from within different social boundaries. Seasonality is an additional issue blurring boundaries because the dry season migration of pastoralists can generate blurred boundaries of both land and social memberships. As Cleaver and Franks (2005:9) claim one should expect boundaries to be "permeable and often fluctuating" and that boundaries are entrenched in the existing social institutions and networks where people access resources and make a living. Establishing formal rules and regulations concerning resource management boundaries on top of these existing social institutions can easily create more problem than they solve.

(2) Fair Distribution of Rights and Memberships

While some new schemes are established in pristine areas, the studied schemes have a land-use history, and those who have acquired water rights often have prior land-access rights. In the case of the traditional system, much of the land had already been distributed by 1994, when the scheme was developed. When water was made available, access rights were allocated to those who had land adjacent to the canals. The improved scheme was an upgrade of the existing traditional scheme, so to some extent, there already was an infrastructure for the physical layout and distribution system. In both cases, however, the placement of canals was decisive in distribution of water rights.

The prevailing rights system seems to be somewhat contested within both schemes, even if a slight majority of participants (53 %) states that the rights are reasonably fairly distributed. Complaints state that rights are not fulfilled (due to insufficient water supply or others taking water) and that they have to pay for rights, not access, while others get away with theft and force. There are no major, systematic differences among different scheme members' and wealth groups' views on rights and memberships. People renting irrigation land enjoy the same rights as others in the schemes. There were no reported tensions in the local community between people renting and owning irrigated land. Whereas rational-choice institutionalists emphasize rights as formally established, rational, consistent, and consequence-oriented devices, critical institutionalists interpret rights as more informal, social, negotiable, and interpretive and apply the logic of social appropriateness in their interpretation and management practices. Even if there is an ambition in the improved scheme to adhere to the rational-choice view of rights, it seems difficult to do so in practice. Coupled with Cleaver's (2012) institutional bricolage, it seems that the two types of schemes are more similar than expected in the management of rights.

(3) Duties, Rights, and Congruence

Ostrom's (1990:91) principles assume that a "congruence between appropriation and provision" is necessary for a robust institution to endure. In our case, there is a division of labor between the government and local people. The government is responsible for distributing water among sectors. Within agriculture, the government divide rights and duties between and within schemes. The government is also responsible for the establishment and major maintenance of intakes, main canals, and like infrastructure. For this service, the government charges an establishment fee and an annual user fee collected by the scheme administration. Members of the improved scheme arrange regular cleaning of the canals, report canal breakages, and carry out minor repairs. The traditional scheme does not have provisions for user fees or any formalized maintenance duties.

Neither local communities nor the government tend to carry out these duties and responsibilities. The government is rarely physically present at local level, so farmers are reluctant to pay fees and participate in maintenance work, even more so in the traditional scheme. The lack of joint action is a serious challenge.

Even if the government is not physically present, it still controls and restrains water access and rights, and there is a clear challenge to distributing water among small-scale farmers and commercial farmers within the watershed. While a rights-based clarification seems formally reasonable, what is seen as congruence in a local context is in practice an object for interpretation, negotiation, and social assessment. For example, upon deliberation, it might seem proper to relieve widows, the disabled, and the sick of communal duties.

(4) Collective Choice Arrangements

Scheme leaders have a substantial role in distributing water through main, secondary, and tertiary canals, but also members take part in these discussions. There are membership meetings at which issues concerning management and maintenance are raised. The level of participation in duties and responsibilities is reportedly low in both groups. In both improved (47 %) and traditional irrigation (66 %) schemes, farmers claim that there is little participation in carrying out duties and responsibilities. Scheme officials report regular confrontations with farmers who avoid participating in maintenance.

Tanzania has a dual-rights system to land and water, and clans still operate alongside formal rights systems in a number of resource-use issues in rural Tanzania. In many ways, clan leaders are still accepted as important authority persons, and clan elders can facilitate collaboration, involvement, and help in times of conflict. The clan system, however, is under pressure from the imposed official, legal, formal management system but also from within. Approximately 30 % of residents in Mkula and 70 % in Msolwa are (now) outsiders, which reduces social consistency and cohesion and might lead to more conflicts. However, the clans and traditional system still operate in most of the design principles discussed.

(5) Monitoring at the Local Level

Water use should be monitored at the basin level down to watersheds and individual rivers and locally down to the initial intake and through main, secondary, and tertiary canals and then farmers' fields until the remaining water drains back into the river. Daily water allocation and distribution falls under the responsibility of the Infrastructure Committee in improved schemes and the Canal Committee in traditional irrigation. These committees under the supervision of the scheme secretary are responsible for monitoring water schedules and distribution.

Monitoring water distribution is a source of numerous conflicts in irrigation schemes, especially in tertiary canals. Around 84 % of farmers report water overuse by others. The situation is reported to be severe during water shortage periods. As well, 59 % in the improved irrigation scheme and 70 % in the traditional system claim that monitoring is inefficient. More people in lower-wealth groups complain about a lack of monitoring.

Water use for domestic purposes, irrigation, and livestock occur at different times and in different places, making monitoring complex and challenging. Even in the irrigation canals themselves, little monitoring takes place; the central water authorities themselves often fail to do monitoring.

There are also conflicts between villages, and the Mkula and Magombela village leaders report holding meetings over joint monitoring because both villages depend on the Mkula River. The complexity of monitoring creates a need for legitimate, competent local institutions that are both efficient at monitoring and accepted by a broad range of local people.

(6) Legitimate Systems for Graduated Sanctions and Conflict Resolution

Farmers who violate operational rules face graduated sanctions. RBWO has the ultimate authority over issuing water rights and is formally obliged to impose graduated sanctions. The Disciplinary Committee is responsible for dealing with sanctions up to a certain level. Complex cases can be referred to higher-level authorities. Both formal and informal arrangements are used to resolve water-related conflicts, and many observed conflicts were resolved through informal arrangements. Elders and scheme officials are often called to resolve conflicts together. Farmers clearly prefer low-cost informal arrangements and less formal rule-based and more interpretive, negotiable solutions kept within the community. There were no reported court cases over water in the villages.

Guilty parties are to be penalized according to locally approved by-laws but are not so in practice. Most farmers choose to yield formal rights or duties and settle matters informally. Most sanctions need reshaping to work effectively, even if 56 % of farmers reported that the system works reasonably well (Table 16.9). There are no major differences in these views by type of scheme or wealth group. It seems as if there is a fine balance between farmers' desire to avoid conflict and maintain peace and harmony but also to make these systems water efficient and, not the least, fair.

Concerning conflict resolution, many farmers (57 %) report that, while conflicts over water use are common, so is conflict resolution. Conflict resolution mechanisms are claimed to be in place by farmers in both improved irrigation (60 %) and traditional irrigation schemes (53 %). At the nexus between formal and customary institutions which handle conflicts, we see, as Cleaver and Franks (2005:11) describe in the neighboring Usangu area, "a deeply held preference for conflict avoidance … and the desire for reconciliatory rather than adversarial solutions (fines and punishments)." This situation presents the dilemma of "public confrontation versus negotiated reconciliation" (Franks and Cleaver 2005:14). Traditional and modern systems of authority to issue sanctions co-exist. An important point is that social identity and individual context matter when sanctions are considered. Traditional systems might, of course, reproduce existing social structures but at the same time maintain social capital and cooperative relations.

(7) Right to Organize

Farmers are the ultimate owners and implementers of irrigation schemes and the direct beneficiaries of local irrigation policy. Internally, farmers manage the schemes and their rules and regulations through representatives, and practices molded in local institutions and agencies emerge as reasonably well functioning at the local level. The Zone Irrigation Officer and District Irrigation Engineer are external authorities who directly collaborate with small-scale irrigation farmers and their organizations. Approximately 77 % of respondents claim that local authorities control the local scheme and that there is little interference from external authorities once water is allocated. More people in the improved scheme (87 %) than the traditional scheme (59 %) claim no external interference in decisions made locally.

Cleaver and Franks (2005) stress the challenge of local actors' tendency to not see aggregated or nested issues, and indeed, many researchers also focus on local-level issues, conducting case studies at the expense of the larger picture. Looking at scale of vertical power, there are at least two concerns that constrain local institutions of water management. The first relates to the allocation of water among users. Agriculture competes with other water uses, and the hydropower producer TANESCO (40 %) and the wildlife sector takes much water, leaving less for irrigation at large. The RBA, for instance, does not allow water use by local people during the dry season. Moreover, local people have little say in these important water-distribution fora.

The second concern arises from the distribution of irrigation water between agricultural schemes and users of very different sizes. Research in the Wami/Ruvu basin reveals that 89 % of the water is consumed by 3 % of users, leaving 11 % to the remaining 97 % who are small-scale farmers. This situation might be similar in Kilombero, where large landowners, such as the Kilomebero sugar company, Kilombero plantation, Rubada, and Chilimo Cha Yesu, hold as much as 40,000–50,000 ha, of which some is and more might be irrigated. In contrast, the two small-scale farmer schemes we studied in Kilombero involve approximately 600 ha and 200 farmers.

Upstream and downstream water-use concerns are also relevant here. The Mkula River is sourced at Udzungwa Mountains (within the Mkula boundary). It drains across the Mkula village, discharging downstream in the Mkula River, and is used for both irrigation and domestic purposes in the villages. The RBWO has ruled that water used in the irrigation schemes should be channeled back to the river for the benefit of downstream users. During PRA sessions, most small-holder farmers in the village seemed to have little concern for downstream water users. Water shortages are mostly experienced during the dry season (June–November), and smallholders expend much effort and time to maintain water access to their farms during this period. This is also the peak period for water-use conflicts.

16.6 Conclusions and Recommendations

The Kilombero district has experienced significant changes over the past 30–40 years, with increasing infrastructure development and high population growth and rural–rural in-migration from other parts of the country. Migrants include not only farmers looking for land but also business owners, merchants, and pastoralists displaced from pastoral land in other places. Land and water scarcity have increased, and conflicts have emerged over competing uses, both within agriculture and between agriculture and other sectors.

Within agriculture, land use has increasingly shifted to irrigation agriculture and rice production over the past 10–20 years. Despite increasing production through irrigation, farmers in area remain poor. They have income levels below the poverty line (per-capita average USD 0.84/day). Even the least-poor group has an average daily income of only USD 1.4, while the poorest group reports a daily income average of USD 0.17. Households that access irrigation water have more than twice the income of rainfed farmers.

Households depend heavily on agriculture (77 % of total income). There are some differences: the poorest group and the rainfed farmer groups depend less on agriculture and more on off-farm (working for others) and non-farm activities. Within agriculture, irrigation farmers produce rice, while rainfed farmers grow more sugarcane.

Farmers close to the poverty line approach risks and uncertainties through risk management strategies and through coping strategies when crises occur. We see that farmers in this area diversify in agriculture and other activities but less than is common in other rural areas in Tanzania. Asset-poor farmers diversify more into off- and non-farm activities, especially working for others. Many farmers also diversify into rainfed production and flood- and drought-resistant cropping in case irrigation fails. Farmers also store crops over the year to avoid food-price fluctuations.

Irrigation introduces a new type of risk: capital-intensive production. Farmers report borrowing some 20 % of their total income on average and depending on high income to serve these loans. Interest rates of 10–20 % increase the consequences of crop failure from irrigation water failure or pests. In these circumstances, farmers could default on their loans and lose assets, such as land.

Among households with irrigation, we find that 47 % of the income is water dependent. The poorest households derive higher shares of their income from irrigation water, while less-poor households diversify into non-farm activities.

The two types of schemes are not very different as assessed by local people. They face similar issues of a lack of clear boundaries, membership, and access rights and neglect of performing duties, carrying out monitoring, paying water fees, and executing sanctions. People, though, do feel that, to some extent, they are participating in the water management system and that conflicts are reasonably well handled.

We have applied Ostrom's (1990) design principles, which clearly have a distinct neo-institutional flavor. She leads us to think that the irrigation schemes we have studied will function better the more they are intentional, formalized, and functional and that they will be effective when purposes, rights, duties, and rules are clear and transparent, and the operation legitimate and open.

Cleaver and Franks (2005:16) offer the contrasting critical institutionalist and ethnographic approach which focuses not only narrowly on purpose and outcomes but also more broadly on the complex relations between the "natural and social worlds." Rather than an instrumental view of institutions, Cleaver and Franks

(2005:16) view institutions as "formally and socially embedded, often multipurpose, intermittent and semi-opaque in operation." An institution should not be seen as a constraint or as a thing but as constitutive of humans as social beings. Practice is not formal and rule determined; instead, there are overlapping rights and responsibilities. There is ambiguity, inconsistency, flexibility, interpretation, negotiations, and appropriateness in contrast to rule-based clarity, intentionality, consistency, and consequentiality. In practice, we see that local people resort to elders and clan members in times of conflict and seek to legitimatize monitoring and sanctions based on traditional systems and clan institutions.

What does this study contribute to water policy formulation in Tanzania? There is at present some devolution of water rights and duties to village and scheme levels in Tanzania. These rights and duties, however, are contingent because priority is given first to hydropower (40 %), protected areas, and domestic and industrial water needs. Within agriculture, there is further division between commercial and local farmers, with some 70–80 % of water going to large-scale commercial users. This situation is documented for the Wami/Ruvu basin, but the figures are most likely similar for Kilombero (IDS 2014). A scenario is drawn in which improved schemes enable increased control of local water and imposition fees for water on small-scale farmers. Water is transformed from a common-pool resource accessed by local people into an increasingly state-controlled and -influenced resource regime. Taxation and user fees alienate people from water resources, and few services are offered in return for the user fees, leading to substantial levels of mistrust and perceived illegitimacy of the state by citizens. The TANESCO fees paid are not allocated to the basin or local governments but to the central state authorities. The priority of external actors increases. There has also been an attempt to establish new, parallel catchment councils and WUA outside the existing district councils and local governments. There is need for a revised water use policy, which allows local people more access to and control of resources and which returns fees to local use. At the same time, there is also a need for better control over the development of small- and large-scale commercial schemes. At present, it seems as if new and mostly commercial schemes are entered into without consultation with local- and district-level authorities.

References

ActionAid (2004) Turning off the taps; donor conditionality and water privatization in Dar es Salaam. Report Action Aid International, Dar es Salaam

Binswanger H, Gautam M (2010) Towards an internationally competitive Tanzanian agriculture. Draft report, World Bank, Washington

Bryman A (2008) Social research methods, 3rd edn. Oxford University Press, Oxford

Chiza CK (2005) The role of irrigation in agriculture, food security and poverty reduction. Presented to the 3rd annual engineer's day, 2005—learned discourse, 14-15-3-2005. Dar es Salaam, Tanzania

Cleaver F (2012) Development through bricolage: rethinking institutions for natural resource management. Routledge, London
Cleaver F, Franks T (2005) How institutions elude design: river basin management and sustainable livelihoods. BCID research paper No. 12. University of Bradford, Bradford, UK
Ellis F (2000) Rural livelihoods and diversity in developing countries. Oxford University Press, Oxford
Food and Agriculture Organization (2007) Irrigation management transfer: worldwide efforts and results. Food and Agriculture Organization of the United Nations, Rome
Goldin J, Kibassa D (2009) Past, present and future landscapes of water policy in Tanzania. In: Huitema BD, Meijerink S (eds) Water policy entrepreneurs—a research companion to water transitions around the globe. Edward Elgar, Massachusetts, pp 215–232
Haule KS, Johnsen FM & Maganga SLS (2002) Striving for sustainable wildlife management: the case of Kilombero Game Controlled Area, Tanzania. J Environ Manag 66(1):31–42
IDS (2014) Ensuring integrated water resource management benefits all. Policy briefing issue 76. IDS, Brighton, UK, October 2014
Kikula IS (1997) Policy implications on environment: the case of villagisation in Tanzania. The Nordic Africa Institute, Uppsala
Kissawike K (2008) Irrigation-based livelihood challenges and opportunities: a gendered technography of irrigation development intervention in the Lower Moshi irrigation scheme in Tanzania. PhD thesis, Wageningen University, Wageningen, Holland
Liheluka EA (2014) Institutional water resources management and livelihood adaptation in Kilombero rural areas Tanzania. Norwegian University of Life Sciences, Ås
Maganga FP (2003) Incorporating customary laws in implementation of IWRM: some insights from Rufiji River Basin, Tanzania. Phys Chem Earth 28(20–27):995–1000
Maganga FP, Butterworth JA & Moriarty P (2002) Domestic water supply, competition for water resources and IWRM in Tanzania: a review and discussion paper. Phys Chem Earth Parts A/B/C 27(11–22):919–926
Mosha D, Vedeld P, Kajembe GC, Tarimo AKPR, Katani JZ (2016) Reflections on evolving water management institutions and institutional bricolage: a case of irrigation schemes in Iringa Rural and Kilombero districts, Tanzania. Water Policy 18(2016):143–160
Mehta L, Leach M & Scoones I (2001) Environmental governance in an uncertain world. IDS Bull 32(4). IDS, Brighton, UK
National Bureau of Statistics (2013) Morogoro region population: 2012 Tanzania population and housing census. National Bureau of Statistics, Dar es Salaam
Ostrom E (1990) Governing the commons: the evolution of institutions for collective action. Cambridge University Press, Cambridge
Patel SP, Vedeld P & Tarimo A (2014) Irrigation management, institutions and local livelihood adaptation on Usangu Plains, Tanzania. Noragric working paper 49(26). Department of International Environment and Development Studies, Noragric Norwegian University of Life Sciences (NMBU) ISSN: 0809-4934
Pipping K, Chale HF (1976) Land holding in the Usangu plain: a survey of two villages in the southern highlands of Tanzania. Scandinavian Institute of African Studies, Uppsala, Sweden
Region Commissioner's Office Morogoro (2008) Morogoro region. TAMISEMI: Region Commissioner's Office, Morogoro
Rufiji Basin Water Office (2007) About RBWO. Rufiji Basin Water Office, Iringa
Scones I (1998) Sustainable rural livelihoods: a framework for analysis, vol 72. IDS Publications Office, Brighton
Sokile CS (2003) Towards an integrated water resource management in Tanzania: the role of appropriate institutional framework in Rufiji Basin. Phys Chem Earth 28(20–27):1015–1023
Sokile CS, van Koppen B (2004) Local water rights and local water user entities: the unsung heroines of water resource management in Tanzania. Phys Chem Earth 29(15–18):1349–1356

URT (2009) The water resources management act, 2009. Tanzania No. 20:90. Government Printers, Dar es Salaam

Van Koppen B, Sokile CS, Lankford BA, Hatibu N, Mahoo H & Yanda YZ (2007) Water rights and water fees in rural Tanzania. In: Molle F, Berkoff J (eds) Irrigation water pricing. CAB International, London, pp 143–163

Vedeld P (2002) The process of institutional building to facilitate local biodiversity management. Noragric working paper. Norwegian University of Life Sciences. Ås, Norway

Chapter 17
Institutional Aspects of Genetic Resources in Respect to Climate Change in Sub-Saharan Africa

Denis T. Kyetere and Kayode Abiola Sanni

Abstract Genetic resources are central to strengthening food security and building a more-resilient agricultural system in the face of climate change. They underpin today's production and provide the raw material needed for the challenges of tomorrow. The speed and complexity of climate change pose new constraints that are expected to make the task of achieving food security more challenging in sub-Saharan Africa (SSA). By 2080, viable arable land for production is predicted to decline, with 9–20 % of arable land becoming much less suitable for agriculture, and SSA is predicted to become the most food insecure region, with 40–50 % of undernourished people globally inhabiting the region, compared with 24 % today. Hence, there is the need to develop varieties that are well adapted to a new and unstable environment. Developing these new crop varieties with traits adaptive to present and future climatic stresses envisaged in SSA will increase the demand for genetic resources. Unfortunately, climate change also threatens the sustainable existence of agricultural biodiversity, increasing genetic erosion of landraces and threatening crop wild relatives. Adequate attention should, therefore, be given to collection, conservation, sharing, and use of genetic resources to mitigate climate change and be aimed at developing timely interventions across national borders. Institutional aspects toward sustainable conservation and use of this reservoir of genetic resources should be enhanced and clear strategies put in place. This presentation will review and focus on the national and regional capacity framework to provide an understanding of the institutional aspects of the adaptive capacity toward the conservation and use of genetic resources in relation to climate change in SSA.

Keywords Adaptive capacity · Developing countries · Food security · Sustainable agriculture · Varieties

D.T. Kyetere (✉) · K.A. Sanni
AATF, Nairobi, Kenya
e-mail: d.kyetere@aatf-africa.org

K.A. Sanni
e-mail: k.sanni@aatf-africa.org

R. Lal et al. (eds.), *Climate Change and Multi-Dimensional Sustainability in African Agriculture*, DOI 10.1007/978-3-319-41238-2_17

17.1 Introduction

The strength of Africa lies in its natural resources. Of particular importance are genetic resources that are the foundations for the growth and stability of agriculture, forestry, and the environment. They are essential for sustainable agriculture, food security, livelihoods, and the provision of environmental services in the continent. Africa's economies, cultures, and political systems are primarily dependent on how plant genetic resources are conserved and utilized. Considering this, the economic transformation and the ability of Africa to integrate itself into the evolving global system is largely dependent on its agricultural transformation, which is based on plant genetic resources (Nnadozie et al. 2003).

Genetic diversity, or genetic resources, for food and agriculture provides the building blocks for farmers, breeders, and biotechnologists to develop new plant varieties necessary to cope with unpredictable human needs and changing environmental conditions, including those due to climate change.

The effects of climate change are expected to reduce agricultural productivity, stability, and income in many areas of the world, some of which already face high levels of food insecurity. As a result, it is likely to become increasingly difficult to meet the goal of increasing food and agricultural production to keep pace with the projected growth of the human population. Food-insecure people in the developing world, especially women and indigenous peoples, are among the most vulnerable to climate change and are likely to be hit hardest (Food and Agriculture Organization [FAO] 2015).

The use of crop genetic resources to develop varieties more tolerant to rapidly changing environ-mental conditions will be an important part of agricultural adaptation to climate change. Unfortunately, climate change also threatens the survival of the strategic reservoir of crop genetic resources needed to adapt production systems to future challenges. As conditions change, varieties and breeds may be abandoned by farmers and may be lost forever if steps are not taken to ensure their conservation.

This paper will review and focus on the national and regional capacity framework to provide an understanding of the institutional aspects of the adaptive capacity toward the conservation and use of genetic resources in relation to climate change in Africa.

17.2 Impact of Climate Change on Agriculture

Sustainable productivity of agriculture and incomes in many part of the world, some of which are already facing high levels of food insecurity, is expected to be reduced as a result of the effects of climate change.

Agriculture development in Africa contends with a major hurdle in climate change. Climate change is a complex biophysical process, thus it is not possible to predict precise future climate conditions. But the scientific consensus is that global land and sea temperatures are warming under the influence of greenhouse gases and will continue to warm regardless of human intervention for at least the next two decades (Intergovernmental Panel on Climate Change [IPCC] 2007). The increasingly unpredictable and erratic nature of weather systems on the continent has placed an extra burden on food security and rural livelihoods.

Agricultural production and the biophysical, political, and social systems that determine food security in Africa are expected to be placed under considerable additional stress by climate change (FAO 2007). Agriculture, therefore, is expected to pay a significant cost of the damage caused by climate change.

Christensen et al. (2007) gave a summary of the possible trend of climate change in Africa and predicted that warming is very likely to be greater in Africa than the global annual mean throughout the continent and in all seasons, with drier subtropical regions warming more than the moister tropics. Annual rainfall is likely to decrease in much of Mediterranean Africa and the northern Sahara, with a greater likelihood of decreasing rainfall as the Mediterranean coast is approached. Rainfall in southern Africa is likely to decrease in much of the winter rainfall region and western margins. There is likely to be an increase in annual mean rainfall in East Africa, and thereby the possibility of increased flooding. Also, it is unclear how rainfall in the Sahel, the Guinean Coast, and the southern Sahara will evolve.

Some of the induced changes are expected to be abrupt, while others involve gradual shifts in temperature, vegetation cover, and species distributions. Climate change is expected to, and in parts of Africa has already begun to, alter the dynamics of drought, rainfall, and waves and trigger secondary stresses such as the spread of pests, increased competition for resources, the collapse of financial institutions, wildfire occurrence, and attendant biodiversity losses.

In many parts of Africa, warmer climates and changes in precipitation are likely to destabilize agricultural production. This is expected to undermine the systems that provide food security (Gregory et al. 2005). The general consequences of these changes on African farmers, as reported in Christensen et al. (2007), are expected to be adverse, particularly for the poor and the marginalized, who do not have the means to withstand shocks and changes.

A number of countries in Africa already face semi-arid conditions that make agriculture challenging, and climate change is likely to reduce the length of the growing season and force large regions of marginal agriculture out of production. Projected reductions in yield in some countries could be as much as 50 % by 2020; crop net revenues could fall by as much as 90 % by 2100, with small-scale farmers being the most affected (Kurukulasuriya et al. 2006; Benhin 2008).

Without extensive adaptation, the effects of climate change on agriculture is expected to exacerbate Africa's deepening food crisis, narrowing channels of food access and slowing efforts to expand food productivity. To ensure a food-secure future in Africa, climate change adaptation measures must be mainstreamed rapidly in national and regional agricultural policies and programs. This is an important social responsibility, which the region ought not to ignore, as the costs of inaction could be very high.

17.3 Importance of Genetic Resources in Africa

The abundance of plant genetic resources in Africa is associated largely with its ecological variability and diversity. Most of the region enjoys a tropical climate favorable to the evolution of unique plant genetic resources. The role played by these resources is of great importance to the well-being of Africa. Nnadozie et al. 2003 elucidated the importance of in africa as follow:

17.3.1 Agriculture and Food Security

The largest numbers of food-insecure people are in South Asia, while the largest proportion of food-insecure people are in Sub-Saharan Africa (SSA), where 27 % of people were undernourished in 2010–2012 (Vermeulen 2014). However, World Bank (2008) forecasts show that SSA will surpass Asia as the most food-insecure region, with 40–50 % of undernourished people globally inhabiting the area in 2080, compared with 24 % today.

Increasing food production to help meet the region's growing demand presents a major challenge in Africa. Even though the majority of Africans depend on agriculture for their livelihood, the levels of viable arable land for production are predicted to decline by 2080, with 9–20 % of arable land becoming much less suitable for agriculture (FAO 2009a). Therefore, using plant genetic resources for crop improvement, particularly if accompanied by improvements in agronomic practices, could bring great benefits in the short, medium, and long term, mainly as a result of increased yields and nutritional value throughout the year, enhanced crop resilience to pests, diseases, drought, and flooding, and reduced food imports in the region. Such improvements, if combined with the development of new options for value addition and direct marketing, could greatly enhance food security in Africa.

17.3.2 Source for Health Care and Nutrition

Plant and herbs have been used for a long time in Africa to cure diseases and heal injuries. There are numerous underutilized and neglected species with great potential for addressing nutrition and health that have not been given sufficient attention by research and development sectors. Promoting food biodiversity at the local, national, and regional levels is a key to improving nutrition through ensuring the availability of a broad range of nutrients, micronutrients, and bioactive compounds. Food systems that promote the consumption of a wide diversity of food plants at both the species and subspecies levels will help populations to better balance their diets. Combinations of energy-rich crops, animal and/or fish as major sources of protein, and vitamin-, mineral- and phytonutrient-rich fruit and vegetables could constitute the types of diet that are needed for a balanced nutrition and a healthy and productive population.

17.3.3 Environment Protection and Ecosystem Balance

Africa's biodiversity, including its flora and fauna and rainforests, is an important global resource in combating the environmental degradation posed by the depletion of the ozone layer and climate change, as well as the pollution of the air and water by industrial emissions and toxic effluents. African resources include the rainforests, the virtually carbon-dioxide-free atmosphere above the continent, and the minimal presence of toxic effluents in the rivers and soils that interact with the Atlantic and Indian oceans and Mediterranean red seas. The conservation of these genetic resources is necessary to face the unpredictable environmental changes and future human needs.

17.3.4 Poverty Alleviation

The IPCC's 2007 report estimates that Africa will be the most vulnerable globally due to the multiple stresses of poor infrastructure, poverty, and governance. Temperatures are likely to increase by 1.5–4 °C in this century. Projections on yield reduction show a drop of up to 50 %, and crop revenue is forecast to fall by as much as 90 % by 2100, thereby increasing the poverty level. The agriculture sector also is likely to experience periods of prolonged droughts and/or floods during El-Niño events. Agriculture losses of 2–7 % of the Gross Domestic Product (GDP) are expected by 2100 in parts of the Sahara, 2–4 % in Western and Central Africa, and 0.4–1.3 % in northern and southern Africa, thus increasing the poverty level.

The conservation and sustainable use of plant genetic resources are essential to improving agricultural productivity and, therefore, reducing poverty in Africa. Improvement of agricultural performance is a prerequisite of economic development in Africa. The resulting increase in the purchasing power of the rural people also will lead to higher demand for African industrial goods. The induced dynamics would constitute a significant source of economic growth and, thus, poverty alleviation.

17.3.5 Scientific and Technical Research and Training

Genetic resources are the foundation for all crop improvement research. Research institutes and universities across the continent make use of the available genetic resources and the knowledge available on them to create new innovations. Technology innovation and diffusion hold enormous potential to accelerate agricultural output and productivity, and they could serve as incubating grounds for innovation, pioneering practical solutions to both imminent and emerging problems, particularly in the area of agricultural production, medicinal plants, and plant genetic resource issues in general.

17.4 Impact of Climate Change on Genetic Resources

Agricultural diversity and genetic resources that support food crops efficiently need to maintain current levels of food production and to confront future challenges (FAO 2007). Increasing yields of major food crops or even maintaining them in the face of climate change will depend on combining genetic traits found in materials of a wide range of origins (National Research Council 1993; Petit 2001), including wild species. Unfortunately, these important resources also are vulnerable to climate change because they do not receive management interventions that help them adapt to changing conditions. Narrowly adapted species and endemics are especially vulnerable to the direct effects of climate change. Indirect effects also may have important impacts through changes in biotic interactions, including changes in pest and disease pressure (Newton et al. 2008), competition and successional dynamics, and changes in symbiotic compositional interactions.

Climate change will cause shifts in areas suitable for cultivation of a wide range of crops. Using current and projected climate data to about 2055, Lane and Jarvis (2007) predicted the impact of climate change on areas suitable for several staple and cash crops. The study revealed a general trend of loss in suitable areas in sub-Saharan Africa, the Caribbean, India, and northern Australia, and gains in the northern USA, Canada, and most of Europe. Lobell et al. (2008) further examined the likely impacts of climate change on food security. They found a number of hot spots where the yields of key crops are predicted to fall markedly.

The IPCC (2014) reported that there is a medium level of confidence that, if temperatures rise by 2 °C or more above late twentieth-century levels, without developing varieties that adaptable to such rise in temperature, production of the major staple crops (wheat, rice, and maize) will be affected negatively. There is evidence that climate change already negatively has affected wheat and maize yields in many regions (Lobell et al. 2011). At the level of individual species, a study of 43 crops predicted that 23 would gain suitable area for cropping as a result of climate change, while 20 would lose (Lane and Jarvis 2007). It is predicted that there will be substantial falls in the yields of key crops in a number of food-insecure regions, with serious implications for food security (Lobell et al. 2008). The areas in question include southern Africa, where land suitable for growing maize, a major staple crop in the region, is predicted to disappear almost completely by 2050.

In addition to its impact on domesticated crops, climate change will affect the ability of many wild relatives of crop species to survive in their current locations. Species that are unable to migrate quickly will be particularly vulnerable to extinction. It has been estimated that 16–22 % of wild relatives of crop species may be in danger of extinction within the next 50 years. This includes 61 % of peanut species, 12 % of potato species, and 8 % of cowpea species (Jarvis et al. 2008).

Although farmers always have adapted their cropping systems to adverse climatic and environmental conditions, the speed and complexity of climate change poses a new magnitude of problems. New within-crop diversity will be needed to adapt to future conditions, and under extreme conditions, new crops will be required.

Areas that are currently most food-insecure will be affected most by climate change and, thus, have the greatest need for new crop varieties tolerant to extreme climate conditions, such as drought, heat, submergence, and salinity. Adapting crop varieties to local ecological conditions can reduce risk due to climate change, but the need for adapted germplasm is urgent and requires characterization, evaluation, and the availability of materials now housed in genebanks. Comprehensive assessments are needed of both adaptation needs and suitable available genetic resources (Ainsworth et al. 2008). Crop wild relatives will play a crucial role in providing the genes and traits to help confront these challenges.

Climate change, therefore, will create the following challenges for genetic resources. First, climate change will accelerate genetic erosion and create a critical need to collect and conserve endangered plant genetic resources and their wild relatives before it is too late. Second, greater use of plant genetic resources will become vital to the development of varieties able to adapt to new and unstable environmental conditions and able to buffer and eventually overcome the negative effect of climate change in agriculture development and food production. Because of the interdependency of countries on matters related to plant genetic resources, international cooperation will become crucial. Also, important institutional and legal challenges will arise.

17.5 Adaptation to Climate Change Using Crop Genetic Resources

Adapting agriculture to climate change is one of the most urgent challenges of our time. One of the most important steps we can take to prepare for climate change is to ensure that the crops that feed humanity are able to thrive in the new climates that are developing all over the world. The need for new crop varieties that can be productive in the new climates of the future is becoming more widely recognized. But our ability to breed these new varieties should not be taken for granted, as it so often is. To breed new varieties, we need genetic diversity.

Since the development of plant varieties that can adapt to the changing climate is a major scientific and technical challenge, the following needs to be understood:

- The magnitude of change will require significant adaptation.
- New genetic diversity, within and between species, is likely to be needed. This will increase the potential of underutilized crops and other promising species.
- Novel and unstable production environments would require different breeding approaches.
- There is an increasing need for adaptability and resilience, properties that to date have not been embedded in traditional breeding.
- Institutional commitment and framework will create an enabling environment for the strategic implementation of the required approaches.

All of these will require research, not only on the diversity itself but also on how it can be deployed most effectively to maintain productivity. There also will be research needed on how genetic resources can be used to support mitigation strategies. The way in which diversity functions in different kinds of production systems, including organic agriculture, conservation agriculture, and the like, is also a relevant entry point. However, the starting point is the conservation and accessibility of genetic resources for use in crop improvement, which requires legal instruments and rules for its fair use.

17.6 International Legal Instruments on the Use of Genetic Resources

Interdependence among countries with regard to crop genetic resources is increasing as a result of climate change. It also implies an increased need for international cooperation as to conservation and use, and suggested that international exchange of genetic resources will be crucial for adaptation to climate change (Burke et al. 2009). There are, therefore, legal and institutional challenges, as well as a need to promote international cooperation, to ensure conservation and continuous access to plant genetic resource, in a way that they are used in adapting to climate change across the region. These factors all point to a growing role for the different legal instruments, such as the Convention on Biological Diversity (CBD), with its Nagoya Protocol, and the Plant Treaty, among others. However, a review of relevant provisions of international legal agreements may be needed to render operational the potential of plant genetic resources to feed human beings in a changing and challenging socio-economic environment.

17.7 Convention on Biological Diversity (CBD)

The CBD covers all biodiversity and underlines that states have sovereign rights over their natural resources, and Article 15 specifies that national governments have the authority to determine access to genetic resources. Articles 15.4 and 15.5 establish that access, when granted, should be on mutually agreed terms and subject to the prior informed consent of the contracting party in question. Article 15 also stipulates that the contracting parties must "endeavor to create conditions to facilitate access to genetic resources for environmentally sound uses by other Contracting Parties" (United Nations 1992) and that restrictions running counter to the convention's objectives are not to be imposed.

"The fair and equitable sharing of the benefits arising out of the utilization of genetic resources" (United Nations 1992) is one of the three stated objectives of the CBD. This gives benefit-sharing the same importance as the conservation of biological diversity and the sustainable use of its components, the two other main objectives of the convention. Further, "by appropriate access to genetic resources and by appropriate transfer of relevant technologies, taking into account all rights over those resources and to technologies, and by appropriate funding" (United Nations 1992) is included in fair and equitable sharing of benefits. Access to and sharing of genetic resources and technologies also are referred to in the Preamble as being "essential"' with regard to meeting the "food, health and other needs of the growing world population" (United Nations 1992). Access to genetic resources and the fair and equitable sharing of benefits arising from their utilization usually are referred to as "access and benefit-sharing".

The CBD has defined genetic resources as "genetic material of actual or potential value" (United Nations 1992). Since the convention covers all biological diversity, this also includes crop genetic resources. It further defined "genetic material" as "any material of plant, animal, microbial or other origin containing functional units of heredity" (United Nations 1992). The emphasis on "material" suggests that the resources in question are tangible, although, as Marrero-Girona and Vogel (2012) pointed out, it makes more sense scientifically to acknowledge their immaterial and intangible nature by using the term "information."

The concept of country of origin is central, and the "country of origin of genetic resources" is defined in Article 2 as "the country which possesses those genetic resources in in situ conditions" (United Nations 1992). In situ conditions are defined further as "conditions where genetic resources exist within ecosystems and natural habitats, and, in the case of domesticated or cultivated species, in the surroundings where they have developed their distinctive properties" (Article 2).

The definition of domesticated or cultivated species, a "species in which the evolutionary process has been influenced by humans to meet their needs" (United Nations 1992), is fairly broad, and crop species easily fall within this definition. However, the concept of country of origin, its definition, and related definitions are not, as pointed out by Fowler (2001), particularly well suited, especially with regard to crop genetic resources. Fowler (2001) emphasized that a crop species or a farmers' variety may contain many different properties and, therefore, also have many countries of origin. In addition, also a specific trait may have more than one country of origin (Fowler 2001). Many of the most-used modern crop varieties consist of genes from hundreds of varieties from different countries and regions. Identifying one particular country of origin for a crop variety, a seed sample, or the genetic information contained within, therefore, will be complicated and often impossible. There is a need to revisit negotiations of its International Regime on Access and Benefit Sharing.

17.8 The International Treaty on Plant Genetic Resources for Food and Agriculture

The objectives of the International Treaty on Plant Genetic Resources for Food and Agriculture (the Plant Treaty) (FAO 2009b) are "the conservation and sustainable use of plant genetic resources for food and agriculture and the fair and equitable sharing of benefits arising out of their use, in harmony with the CBD, for sustainable agriculture and food security" (FAO 2009b). They are based on the objectives of and "in harmony with" the CBD, but with some minor differences, such as the reference to food security and sustainable agriculture. Most importantly, unlike the CBD, the Plant Treaty does not cover biological diversity in general, but only plant genetic resources for food and agriculture (here the term "crop genetic resources" is used).

These resources are defined as "any genetic material of plant origin of actual or potential value for food and agriculture" (FAO 2009b), and "genetic material" is defined as "any material of plant origin, including reproductive and vegetative propagating material, containing functional units of heredity" (FAO 2009b). Again, it is clear that the language of the CBD has been taken as the point of departure. The Plant Treaty is clear and specified that the definitions "are not intended to cover trade in commodities" (FAO 2009b), whereas no such clarification is offered directly in the CBD. Such a specification is required due to the focus on "material" rather than "information," and the reference to "propagating material" perhaps makes it even more necessary to draw up the distinction to trade in the context of the Plant Treaty because seed and other propagating material are traded regularly as commodities.

A main component of the Plant Treaty is its Multilateral System for Access and Benefit Sharing (the Multilateral System, for short). This system is expected to be "efficient, effective and transparent" (FAO 2009b) and was set up to facilitate access to crop genetic resources and enable the benefits arising from their utilization to be shared fairly and equitably. Whereas the Plant Treaty itself covers all crop genetic resources, the Multilateral System covers only a subset of these resources: those listed in Annex 1 to the Plant Treaty and "under the management and control of the Contracting Parties and in the public domain" (FAO 2009b, Article 11.2).

As specified in Article 11.1 of the Plant Treaty, food security and interdependence were set as the criteria to be used for deciding which crops to include on the list. However, the selection process was quite political, and important crops were left out. A total of 64 crops made up of 35 food crops and 29 forage crops were included in the list. In addition to the mandatory inclusion of resources that are under the management and control of the contracting parties and in the public domain, the Plant Treaty encourages voluntary contributions from "all other holders" (FAO 2009b, Article 11.2) of crop genetic resources listed in its Annex 1. Contracting parties also agree to take "appropriate measures" (FAO 2009b, Article 11.3) to encourage natural and legal persons holding Annex 1 resources to include these in the Multilateral System. The Multilateral System provides facilitated access to the included crop genetic resources through a Standard Material Transfer Agreement adopted by the governing body (FAO 2009b, Article 12.4). According to the Plant Treaty, facilitated access is to be provided solely for "utilization and conservation for research, breeding and training for food and agriculture, provided that such purpose does not include chemical, pharmaceutical and/or other non-food/feed industrial uses" (FAO 2009b, Article 12.3a). It also is specified that access should be given expeditiously, that a fee should be charged only to cover the minimal costs involved, and that available plant passport data, and "subject to applicable law, any other associated available non-confidential descriptive information" (FAO 2009b, Article 12.3c), also shall be made available.

Recipients cannot claim intellectual property rights to the crop genetic resources, their genetic parts, or components in the form received by the Multilateral System. Any subsequent transfers of crop genetic resources accessed through the Multilateral System also are to be performed based on the provisions of the Standard Material Transfer Agreement. In addition, crop genetic resources accessed under the Multilateral System and conserved "shall continue to be made available to the Multilateral System by the recipients" (FAO 2009b, Article 12.3g). Contracting parties are obligated to provide access both to other contracting parties and to natural and legal persons under the jurisdiction of other contracting parties. However, the governing body has the power to strip legal and natural persons that have not included crop genetic resources in the Multilateral System of this privilege (FAO 2009b, Article 11.4). At the fifth meeting of the governing body in September 2013, the lack of inclusions by such entities was discussed, but no strong measures were agreed upon.

The Plant Treaty recognizes facilitated access itself as one of the benefits of the Multilateral System. In line with the CBD, benefits accruing from such access are to be "shared fairly and equitably" (FAO 2009b, Article 13.1). The Plant Treaty lists the following mechanisms for the sharing of benefits arising from the use of crop genetic resources accessed through the Multilateral System: exchange of information, access to and transfer of technology, capacity building, and sharing of monetary and other benefits of commercialization. As to the sharing of monetary benefits arising from commercialization of crop genetic resources from the Multilateral System, the Plant Treaty (and the Standard Material Transfer Agreement) distinguishes between mandatory benefit-sharing, when the commercialized product is not available without restriction to others for further research and breeding, and voluntary benefit-sharing, when the commercialized product is available without restriction to others for future research and breeding. However, the Plant Treaty provides the governing body with the opportunity also to extend mandatory payment to the latter cases.

In line with Article 13.2d and Article 19.3f of the Plant Treaty, a fund that could receive such payments, generally referred to as the Benefit-sharing Fund of the Plant Treaty, was established by the governing body at its first session. According to Article 13.3 of the Plant Treaty, the benefits arising from the use of crop genetic resources in the Multilateral System "should flow primarily, directly and indirectly, to farmers in all countries, especially in developing countries, and countries with economies in transition, who conserve and sustainably utilize such resources" (FAO 2009b).

Benefit-sharing is mentioned also in Article 9 on farmers' rights, where the right to participate equitably in the sharing of benefits arising from the utilization of crop genetic resources is listed among the elements that contracting parties should take measures to protect and promote as part of the realization of these rights.

Climate change is not mentioned specifically in either the CBD or the Plant Treaty, but in the Plant Treaty, the subject of climate change adaptation is addressed indirectly in its Preamble, paragraph 6: crop genetic resources are acknowledged as "essential in adapting to unpredictable environmental changes and future human needs" (FAO 2009b). As the section linking access and fair and equitable benefit-sharing to climate change adaptation shows, climate change increasingly has been accorded greater prominence in the implementation of the Plant Treaty since it was adopted in 2001.

There is a need to review the list of crops in the Annex 1 list in such a way that the species coverage of the Multilateral System for access and benefit-sharing should be broadened to include the diversity necessary to cope with climate change.

17.9 Agreements Dealing with Intellectual Property Rights, Such as the Union for the Protection of New Varieties of Plants (UPOV) and the World Trade Organization (WTO) (Especially the Agreement on Trade-Related Aspects of Intellectual Property Rights TRIPS)

It is also important to look at the possibility of reviewing agreements dealing with intellectual property rights, such as the Union for the UPOV and the WTO (especially the Agreement on [TRIPS]) due to their implications on the access to genetic resources and their use.

17.9.1 WTO-TRIPS Agreement

The WTO's TRIPS Agreement (WTO 1994) obliges its signatories to provide for such intellectual property protection in their laws and sets out the minimum standards that must be contained therein. Plant variety protection often is excluded from lists of intellectual property categories. However, the adoption of the TRIPS Agreement has done more to encourage the legal protection of plant varieties than any other international agreement. As the plant variety protection (PVP) debate has continued, a school of thought has evolved that considers it a form of industrial property right. The TRIPS Agreement does not require that a specific system be put into place to secure intellectual property rights for plant varieties. The agreement mandates its signatories to provide patent protection for any invention in all fields of technology, provided that the inventions are "new, involve an inventive step and are capable of industrial application" (WTO 1994). Such protection for plant varieties is covered by Article 27.3(b) of the agreement, which partly states the protection is to be provided "either by patents, or by an effective sui generis system or by any combination thereof" (WTO 1994). This means that the range of options is unlimited, provided some requirements are met. This also means that WTO member states, including Tanzania, are allowed to develop legislation that takes into account its unique features. Furthermore, Article 27.3(b) of the agreement states that members may exclude from patentability "plants and animals other than micro-organisms, and essentially biological processes for the production of plants or animals other than non-biological processes" (WTO 1994). Thus, as presently written, the TRIPS Agreement would permit WTO members to decline to protect plant varieties using the patent method, provided they protect such varieties with an effective *sui generis* plant variety protection system. There is no guidance or agreed formulation as to what constitutes "effective," nor what a sui generis system should entail at minimum. The International Seed Federation (ISF) recommends that countries that envisage the development of such sui generis systems ensure that, as a minimum, they conform to the requirements of the 1991 Act of the UPOV Convention, Least Developing Countries (LDCs), which were given an extension until July 1, 2013, to set up the appropriate protection framework under the TRIPS Agreement. The TRIPS Agreement provides for a review of Article 27.3(b), which began in 1999.

Most recently discussed are proposals on disclosing the source of biological material and associated traditional knowledge. The African group at the World Intellectual Property Organization (WIPO) has made a specific proposal on disclosure in patent applications of information on the origin of genetic resources and traditional knowledge on which invention is based.

17.9.2 UPOV 1961, 1972, 1978, 1991—Plant Breeders' Rights

Through a succession of international laws, the UPOV aims to harmonize national laws for protecting plant varieties. The UPOV was established "to provide and promote an effective system of plant variety protection, with the aim of encouraging the development of new varieties of plants, for the benefit of society" (UPOV 1992). In the UPOV community the benefits of PVP and UPOV are cited for encouraging the development of improved varieties and giving farmers in all UPOV member countries access to new, improved varieties for farmers. This is said to result in increased levels of agricultural produce after a country joins UPOV and also to increase the diversity of seeds available worldwide.

PVP is one type of intellectual property right, alongside others like patents, copyright, and trademarks. PVP is designed specifically for plant varieties and grants breeders exclusive rights on propagating material (such as seeds) of new plant varieties that they have developed. PVP is intended as an incentive for research and development by enabling breeders to recoup the costs of researching and developing improvements to pre-existing biological resources. In the absence of such exclusive rights, third parties could use breeders' innovations freely because plant genetic material is naturally self-replicating and easily susceptible to unauthorized exploitation. PVP differs from patents, for instance, by allowing more expansive public interest flexibilities, such as allowing access to PVP-protected materials for research, for further breeding, and for and for noncommercial use by farmers.

In addition to intellectual property rights, breeders also use technology or contract law to protect their knowledge and ensure that they can derive revenue from plant varieties that they have developed. Contract law commonly is applied through the use of licenses that purchasers must agree to and which may be more restrictive than PVP rules. An ideal plant variety intellectual property regime needs to provide incentives and attract research investment in at least two directions. First, and most importantly, it should support breeding targeted to the nutritional and other needs of the whole populace without unduly disrupting existing traditions, farming systems, and diversity. Secondly, such a system should support the development of nonfood, premium, or other food crops that can be sold to generate wealth that, to the greatest extent possible, is captured at local and national levels. In any event, the PVP regime should be for the benefit of society.

International protection of plant varieties facilitates access to new varieties created in other states. When breeders are assured that their rights will be protected in other countries, they will be more willing to make their new varieties available there. The UPOV is the only international PVP system.

The UPOV and the WTO need to avoid legal obstacles to the development and trade of plant varieties with needed adaptability and resilience to cope with climate change. This would imply reducing the degree of uniformity and stability currently required for the commercialization of new varieties. For instance, the current UPOV provisions on Distinctness, Uniformity, and Stability (DUS) might not be adequate.

17.10 Institutional Aspect of Genetic Resources Conservation and Utilization in Africa

There are a lot of regional plant genetic resource conservation initiatives in Africa. Under its Comprehensive African Agriculture Development Plan (CAADP) in the New Partnership for Africa's Development (NEPAD), the African Union sets out a number of clear goals for Africa's agricultural research, which formed the four pillars of its operation. Out of the four pillars of CAADP, two are dependent on Africa's ability to manage sustainably its plant genetic resources both in situ and ex situ (defined as the preservation of components of biological diversity outside their natural habitats (Bretting and Duvick 1997)):

- increasing food supply and reducing hunger across the region by increasing smallholder productivity and improving responses to food emergencies; and

- improving agricultural research and systems to disseminate appropriate new technologies, and increasing the support given to help farmers to adopt them (NEPAD 2005).

In addition, individual countries and subregional organizations such as Southern African Development Community (SADC), Economic Community of West African States (ECOWAS), East African Community (EAC), Common Market for Eastern and Southern Africa (COMESA), West and Central African Council for Agricultural Research and Development (CORAF/WECARD) and Association for Strengthening Agricultural Research in Eastern and Central Africa (ASERECA) in Africa also are pursuing development strategies that clearly consider agriculture as being central to national economic development and rural transformation.

The high dependence of the population on agriculture means that, by creating a dynamic agriculture sector, each of these countries is able to reach a big percentage of the population.

Several countries in Africa have established gene banks, but most lack the necessary resources and infrastructures, or the capacity to effectively carryout their functions, as a result of which a number of the gene banks actually are deteriorating. Currently, several international organizations, notably Bioversity International and the Global Crop Diversity Trust (the Trust), already have provided support to national gene banks within the region.

Bioversity International (formerly International Plant Genetic Resources Institute [IPGRI]) has been a longtime supporter and ally in the development of national and regional plant genetic resources programs in Africa. For decades, Bioversity has supported projects geared toward strengthening the capacity of partners in Africa to conserve ex situ plant genetic resources and sustainably manage in situ agro-biodiversity. Bioversity has been a partner in projects targeting the collection of plant genetic resources; providing technical backup to gene banks; working with national partners and farmers to identify biodiversity hot spots and to support sustainable management on the farm, in gene banks, and in the wild; developing markets for neglected and underutilized species; and strengthening the capacity of stakeholders at local, national, and regional levels to analyze and develop policy options in support of the sustainable use, conservation, and exchange of plant genetic resources.

Recently, the Trust has supported both national and international gene banks in the safety backup of their collection in the global seed vault in Norway. Also, the Trust has issued six grants for the collection of plant genetic material from priority areas and populations likely to harbor traits of use in adapting crops to climate change. These projects target landraces and wild relatives of cowpea, pearl millet, finger millet, pigeon pea, and sorghum in Ghana, Kenya, Malawi, Nigeria, Tanzania, and Uganda. In all cases, collection is being undertaken with national partners. This has allowed the Trust to provide material support for Article 5.1b of the International Treaty that calls on parties to "promote the collection of plant genetic resources for food and agriculture and relevant information on those plant genetic resources that are under threat or are of potential use" (FAO 2009b).

This support notwithstanding, more technical and material support is needed to sustain the national genebanks. As with most other areas, a lack of coherent policies exist with respect to ex situ conservation of plant genetic resources, and a well-articulated policy framework and appropriate legislative instruments could greatly enhance conservation and sustainable use of plant genetic resources in the individual countries and the region as a whole.

There is a need to maintain national gene banks based on national priorities and policies, especially for the conservation of genetic resources of particular national importance. They are necessary for building national capacity and competence in activities related to plant genetic resources and also are essential for academic national breeding programs and encouraging domestic research and development designed to respond to national and local conditions and needs.

Regional efforts include the establishment of regional gene banks. The most notable is the SADC's Plant Genetic Resources Center (SPRGC) based in Zambia, with financial and technical assistance from Nordic countries through a special technical assistance program under the Nordic gene bank. SPRGC is a network that services the ten SADC countries and provides long-term storage of the countries' plant accessions.

In addition to this, other institutions and organizations operating in the region also maintain gene banks. The most active and sustained gene banks are those maintained by the Consultative Group on International Agricultural Research (CGIAR) centers. The CGIAR was established in 1971 and consists of 16 international agriculture research centers. One of the CGIAR's principal research objectives is to contribute to the preservation of biodiversity by establishing ex situ collection and conservation of plant genetic resources. The agricultural research centers of the CGIAR in Africa that maintain gene banks in Africa are the International Center for Research in Agroforestry (ICRAF), the International Institute of Tropical Agriculture (IITA), Bioversity International, the Africa Rice Center (AfricaRice) formerly West Africa Rice Development Association (WARDA), and the International Livestock Research Institute (ILRI). Other CGIAR centers, such as the International Rice Research Institute (IRRI), the International Crops Research Institute for the Semi-Arid Tropics (ICRISAT), and the International Center for Agriculture Research in the Dry Areas (ICARDA), have programs in Africa and have conserved germplasm originating from the continent in their gene banks, which are located in other regions of the world. Because of the large number of accessions they maintain, their technical capacity, and other resources, the CGIAR centers are key players in the region and must be taken into account one way or another in the regional program and policies relating to the conservation, management, and utilization of plant genetic resources. They, however, have served as a source of support to the national gene banks in the region in the conservation of their plant genetic resources.

17.11 Institutional Capacity of Africa to Conserve Sustainable Use of Genetic Resources

The capacity of African nations to conserve and use their plant genetic resources effectively, to obtain maximum benefits from plant genetic resources endowment, and to participate effectively in international negotiations is limited by:

- inadequate expertise in the science of plant genetic resources;
- insufficient conservation infrastructures and facilities;
- disjointed efforts for collective bargaining abilities in international negotiations;
- genetic resources that are not adequately institutionalized into education curricula; and
- inadequate expertise in domestication of international treaties and development of national policies and legal frameworks.

17.12 Conclusion and Recommendation

Progress in delivering food security for Africa will be compromised significantly by the negative effects of climate change. Measures to adapt to climate change must be entrenched rapidly in national and regional mainstream agricultural policies and programs to ensure a food-secure Africa. Therefore, climate change is expected to place large demands for agricultural adaptation, of which genetic resources have great potential for strengthening.

Using genetic resources for crop improvement can increase yields and nutritional value and make crops more resilient in the face of pests, diseases, drought, and flooding, thus reducing the region's dependence on food imports.

Plant genetic resources that are essential for food security, either at the national or regional level, should receive high priority in the development of climate change adaptation strategies in Africa.

Institutional capacity should be enhanced to maximize the potentials of genetic resources in coping with the impact of climate change on agriculture and food security.

Information about the important characteristics of individual genetic resources and interactions between the public and private sectors and academia should be better articulated on the role of genetic resources in climate change in Africa.

It is imperative that greater efforts be made to ensure that the vital plant genetic resources are collected and conserved for the benefit of current and future generations.

References

Ainsworth EA, Leakey ABD, Ort DO, Long SP (2008) Letters: FACE-ing the facts: inconsistencies and interdependence among field, chamber and modeling studies of elevated [CO_2] impacts on crop yield and food supply. New Phytol 179(1):5–9

Benhin JKA (2008) South African crop farming and climate change: an economic assessment of impacts. Global Environ Chang 18:666–678

Bretting PK, Duvick DN (1997) Dynamic conservation of plant genetic resources. Adv Agron 61:1–51

Burke MB, Lobell DB, Guarino L (2009) Shifts in African crop climates by 2050, and the implications for crop improvements and genetic resources conservation. Glob Environ Change 19:317–325

Christensen J, Hewitson BC, Busuioc A et al (2007) Regional climate projections. In: IPCC fourth assessment report "Climate Change 2007: the scientific basis". Cambridge University Press, Cambridge

Food and Agricultural Organization [FAO] (2007) Climate change and food security: a framework for action. FAO, Rome

Food and Agricultural Organization [FAO] (2009a) Climate change in Africa: the threat to agriculture. FAO ftp://ftp.fao.org/docrep/fao/012/ak915e/ak915e00.pdf. Accessed 22 May 2015

Food and Agricultural Organization [FAO] (2009b) Text of the international treaty on plant genetic resources for food and agriculture. FAO ftp://ftp.fao.org/docrep/fao/011/i0510e/i0510e.pdf. Accessed 22 May 2015

Food and Agricultural Organization [FAO] (2015) Coping with climate change—the roles of genetic resources for food and agriculture. FAO, Rome

Fowler C (2001) Protecting farmer innovation: the convention on biological diversity and the question of origin. Jurimetrics 41:477–488

Gregory PJ, Ingram JSI, Brklacich M (2005) Climate change and food security. Philos Trans R Soc B 36:2139–2148

Intergovernmental Panel on Climate Change [IPCC] (2007) Summary for policymakers. In: Parry ML, Canziani OF, Palutikof JP, van der Linden PJ, Hanson CE (eds) Climate change 2007: impacts, adaptation and vulnerability. Contribution of working group II to the fourth assessment report of the IPCC. Cambridge University Press, Cambridge, pp 7–22

Intergovernmental Panel on Climate Change [IPCC] (2014) IPCC fifth assessment report: climate change 2014. IPCC, Geneva. http://www.ipcc.ch/report/ar5/index.shtml. Accessed 22 May 2015

Jarvis A, Lane A, Hijmans R (2008) The effect of climate change on crop wild relatives. Agr Ecosyst Environ 126:13–23

Kurukulasuriya P, Benhin JKA, Deressa T (2006) Will African agriculture survive climate change? World Bank Econ Rev 20(3):367–388

Lane A, Jarvis A (2007) Changes in climate will modify the geography of crop suitability: agricultural biodiversity can help with adaptation. Paper presented at the 35th anniversary symposium of ICRISAT/CGIAR, Climate-proofing innovation for poverty reduction and food security, Patancheru, India, 22–24 Nov 2007. http://www.icrisat.org/Journal/SpecialProject/sp2.pdf. Accessed 26 May 2015

Lobell DB, Burke MB, Tebaldi C et al (2008) Prioritizing climate change adaptation needs for food security in 2030. Science 319:607–610

Lobell DB, Schlenker W, Costa-Roberts J (2011) Climate trends and global crop production since 1980). Science 333(6042): 616–620. http://www.sciencemag.org/content/333/6042/616.full.pdf Accessed 22 May 2015

Marrero-Girona G, Vogel JH (2012) Can "monkey business" resolve the most contentious issue in the convention on biological diversity? Int J Psychol Stud 4:55–65

National Research Council (1993) Managing global genetic resources: agricultural crop issues and policies. National Academy Press, Washington DC

[NEPAD] (2005) Introducing the Comprehensive Africa Agriculture Development Programme (CAADP) NEPAD. http://www.nepad.org/system/files/CAADP_brochure1_visual.pdf. Accessed 22 May 2015

Newton AC, Johnson SN, Lyon GD, Hopkins DW, Gregory PJ (2008) Impacts of climate change on arable crops—adaptation challenges. In: Proceedings of the crop protection in Northern Britain Conference. http://www.scri.ac.uk/scri/file/climatechange/Climatechangearablecrops.pdf. Accessed 26 May 2015

Nnadozie K, Lettington R, Bruch C, Bass S, King S (2003) African perspective on genetic resources. In: A handbook on laws, policies, and institutions governing access and benefit-sharing. Environmental Law Institute, Washington DC

Petit M (2001) Why governments can't make policy: the case of plant genetic resources in the international arena. Commission on Intellectual Property, Lima

Union for the Protection of New Varieties of Plants UPOV (1992) Act of 1961/1972. International convention for the protection of new varieties of plants. http://www.upov.int/export/sites/upov/upovlex/en/conventions/1961/pdf/act1961.pdf. Accessed 22 May 2015

United Nations (1992) The text of convention on biological diversity. https://www.cbd.int/doc/legal/cbd-en.pdf. Accessed 22 May 2015

Vermeulen SJ (2014) Climate change, food security and small-scale producers. CCAFS Info Brief. CGIAR Research Program on Climate Change, Agriculture and Food Security CCAFS, Copenhagen, Denmark. https://cgspace.cgiar.org/rest/bitstreams/30971/retrieve. Accessed 25 May 2015

World Bank (2008) Climate change response strategies for agriculture: challenges and opportunities for the 21st century. Agriculture and rural development discussion paper 42

World Trade Organization [WTO] (1994) Agreement on trade-related aspects of intellectual property rights. https://www.wto.org/english/docs_e/legal_e/27-trips.pdf. Accessed 22 May 2015

Chapter 18
Crop Adaptation to Climate Change in SSA: The Role of Genetic Resources and Seed Systems

Ola T. Westengen and Trygve Berg

Abstract Crop adaptation plays a key role in enabling farmers to adapt to the impacts of climate change. According to the latest report from the Intergovernmental Panel on Climate Change (IPCC), *cultivar adjustment* is the single most effective on-farm adaptation strategy, but the report is largely silent on the modalities of cultivar adjustment; what are the assumptions with regard to the cultivar types used and the institutional context in which the adjustments will take place? The objective of the current paper is to explore these modalities and enhance our understanding of the potential for crop adaptation in Sub Saharan Africas's agriculture. We identify the key environmental impacts and the adaptation options vis-à-vis these impacts. Drawing on insights and perspectives from the international scholarly literature on genetic resources and seed systems, we report on a local case study from the semi-arid zone in Tanzania. Farmers use a range of varieties and seed systems to cope with current climatic stress and our findings from Tanzania illustrates that crop adaptation is not only a question of switching from one modern variety to another as commonly assumed in the Climate Change impact and adaptation literature. In our case study, only 24 % of the maize seeds and 8 % of the sorghum seeds were sourced through seed supply channels classified as formal. However, in the case of maize, we found that at least 11 % of the seeds sourced from informal seed supply channels were farm-saved modern varieties. Open Pollinated Varieties of maize developed in public breeding programs in collaboration with international programs in the 1980s are still an important part of farmers' adaptive capacity. Our results further indicate that crop adaption can happen through *creolization* between modern and local varieties in the local seed

The empirical basis and parts of this chapter is adopted from the Ph.D. thesis of the first author.

O.T. Westengen (✉) · T. Berg
Noragric, Department of International Environment and Development Studies, Norwegian University of Life Sciences, Ås, Norway
e-mail: ola.westengen@nmbu.no

T. Berg
e-mail: trygve.berg@nmbu.no

R. Lal et al. (eds.), *Climate Change and Multi-Dimensional Sustainability in African Agriculture*, DOI 10.1007/978-3-319-41238-2_18

system. We argue that seed security in the face of climatic change depends on *adaptive seed systems*, which integrate formal and informal seed system approaches to the development, release and distribution of varieties.

Keywords Crop adaptation · Genetic resources · Crop varieties · Seed systems

18.1 Introduction

The negative impacts of anthropogenic climate change are already evident in food production in Sub-Saharan Africa (SSA) and will most likely grow stronger over the course of the next decades. According to the majority of the climate change models, our crops will face a gradual increase in mean temperatures and changes in rainfall patterns, as well as increased frequencies of extreme weather events. Today's extremes are set to become tomorrow's normal and many areas will experience conditions outside the historical ranges of the Holocene climate—the geological period in which our crops have evolved and thus become adapted. There are basically four ways in which plant populations can react to changes in growing conditions: adaptation, plasticity, migration or extinction. Which of the four outcomes we get depends on our management of the available genetic diversity. Crops are domesticated and their future is, by definition, in our hands. This chapter is about the basic components of successful crop adaptation: genetic resources, which are the raw material for adaptation, and the seed systems, which are the institutions farmers and breeders rely on when they deploy these resources to adapt to novel climates. The chapter proceeds as follows: first, we provide a review of the impact projections for SSA and single out the key environmental factors to which our crops will have to adapt. Secondly, we review the adaptation concept in the climate change literature and explore what the literature on genetic resources and seed systems can contribute to broaden the perspective on crop adaptation. Thirdly, we present a local case study of the role of genetic resources and seed systems for adaptation from Tanzania, drawing on the expanded toolbox for crop adaptation studies introduced in the preceding section. In the final section, we discuss some key features that we believe are needed to create adaptive seed systems.

18.2 Impact Projections and the Factors of Change

The latest assessment report from the Intergovernmental Panel on Climate Change (IPCC) concludes that the negative impacts of climate change on global food production already are evident (Porter et al. 2014). The global aggregate impacts vary across crops, but the overall global production of maize and wheat has seen losses of 3.8 and 5.5 %, respectively, between 1980 and 2008 compared to a counterfactual without climate change (Lobell et al. 2011b). The *current* impacts are the backdrop for the large body of studies considering the likely *future* impacts of climate change on crop production and food security. By combining crop simulation models or historical crop production statistical analysis with future climate projections from general circulation models (GCMs), these studies tell us how and where the future impacts (and thus adaptation needs) will be largest.

SSA is one of the regions predicted to experience large negative production impacts for many crops. Schlenker and Lobell (2010) coupled national production data reported by the Food and Agriculture Organization (FAO) with predictions for temperature and precipitation changes from 16 GCMs and projected aggregate production losses of 17 % for sorghum and 22 % for maize by mid-century (Schlenker and Lobell 2010). A meta-level study by Knox et al. (2012) projected mean yield changes of 17 % (wheat), 5 % (maize), 15 % (sorghum) and 10 % (millet) across Africa by mid-century (Knox et al. 2012). While global and regional analyses of crop production impacts are important for assessing overall impacts on food production and availability, it is necessary to study impacts on a more local scale in order to arrive at relevant information for farmers within the region. In recognition of the need for finer scale projections for Africa's highly heterogeneous agriculture, Thornton et al. (2009) used biophysical crop models and assessed ~ 18 km resolution grid cells across East Africa. The study projected that up to 33 % of the maize area and 56 % of the bean area were likely to see yield losses of 20 % or more by 2050 (Thornton et al. 2009).

A study by Lobell et al. (2011a) used historical crop-trial data and daily weather data to study the effects of increased heat and moisture stress and projected that a 1° C warming will lead to yield losses in 65–100 % of the maize growing areas in East and West Africa. The study is particularly interesting for our subject matter as it is one of the few impact studies that considers intra-specific variation in crops' vulnerability to temperature and drought stress. The study found significant varietal differences depending on the maturity period of the varieties and on whether the variety was hybrid or open pollinated (OPV). The intermediate to late maturing varieties produced better than early maturing varieties under optimal management conditions while the early maturing varieties performed better under drought conditions. Likewise, the hybrid varieties consistently produced better yields than the OPVs under optimal moisture management conditions, while the OPVs were consistently higher yielding under drought conditions (Lobell et al. 2011a). The finding that the varieties with the highest potential yields are those that produce least under drought conditions is important to understand the impact study results. For example, Schlenker and Lobell (2010) found that countries with the current highest yields in the region (such as South Africa) are likely to see the largest yield losses with climate change. The explanation for this is probably that these countries have the highest use of hybrid varieties, which give high yields under good conditions, but also high losses under stress conditions.

In order to find out how the negative impacts can be avoided and how crops can adapt to climate change, we need to consider what specific ecological changes our crops will be facing. The major factors are arguably: (1) increased CO_2 concentration in the atmosphere, (2) higher temperatures, (3) altered rainfall patterns and water availability, and (4) indirect effects through changes in distribution and severity of weeds, pests and diseases. Adapting to these changes will require strategies that are both genetic (relying on crop evolution) and non-genetic (relying on management of the crops' environment).

The main driver of climate change, CO_2 itself, affects crop production through its role in photosynthesis. Crops, especially species with C_3 photosynthesis, tend to yield more in an elevated CO_2-atmosphere, but less so than reported in the earliest studies (Long et al. 2006). However, in most cases, such studies are done with a single variety. A rare CO_2 response study by Moya et al. (1998) included four different varieties of rice. The researchers found clear differences in relative response to increased CO_2. While screening of relevant genetic resources, including modern varieties, farmers' varieties and wild crop relatives, may not be practically feasible in such experiments, the finding of genetic diversity for CO_2-response in this experiment indicates the existence of a crop evolutionary potential with respect to this trait.

Climate change brings rising temperatures globally. Where temperatures are already high, such as in the low altitude areas of Sub-Saharan Africa, normal distribution of temperature during the growing season is projected to move towards a situation with more days with above optimal temperatures for many crops. The mean temperature increase comes out as the single most important yield reducing factor in impact projections (Challinor et al. 2014; Lobell and Burke 2008). Shortened crop growing seasons and heat stress during the crop's reproductive period are strongly associated with yield losses (Cairns et al. 2013). The physiological effects of abiotic stress are complex, but heat above critical thresholds during vulnerable stages of the growing cycle (e.g., flowering time) is critical. Cereals are more vulnerable to heat stress during the days before anthesis when microsporogenesis takes place as well as during anthesis (Martínez-Eixarch and Ellis 2015). Some studies have shown that increased night temperature reduces yield more than increased daytime temperature. Peng et al. (2004) found that yields of the same rice varieties declined by about 10 % for every 1 °C rise in nighttime temperature. Farmers can switch to the more heat-tolerant crop species and ensure ample availability of moisture. Crops suffer less from excessive heat if enough moisture is available. Otherwise, farmers cannot do anything in terms of the management of heat stress. Thus, crop evolution is going to be critical for our ability to adapt to this factor of change.

The reason why rainfall change typically comes out as a less important driver of yield loss under climate change relative to temperature increase is that rainfall patterns already show considerable year-to-year variability and are less predictable than temperature trends. However, it is clear that the combined effect of heat and moisture stress will have worse effects compared to heat stress alone. Obtaining acceptable crop yield under changed rainfall patterns requires soil and water management, and often supplementary irrigation to bridge the dry spells. While such management may help to maintain high yields, genotypic adaptation may be about the ability to withstand stress and thereby avoid crop failure. We do assume that improved management will be the most important response to this factor of climate change.

Also, yield-reducing organisms, pests, diseases and weeds are exposed to climate change. Unlike modern crop varieties that are uniform and supposed not to change, weeds are present with evolving populations. Ziska (2003) selected six invasive weed species that are rated as noxious in the US and studied their growth responses under past, present and future CO_2 concentrations. They all responded and that response is discussed as a factor in the invasiveness of those weeds. New strains evolve, become more aggressive and invade areas where they did not occur before (Ziska 2003). Farmers' management of diseases and pests could include broad integrated pest management approaches as well as the use of pesticides. However, genetic resistance is also important, and in some diseases, critically important. While traditional farmer-breeding facilitates and depends on co-evolution of crops and diseases, modern varieties depend on continuous resistance-breeding. The breeders need access to resistance genes and those are normally sourced in farmers' varieties or in wild crop relatives. When, however, the diseases respond to climate change, needed resistance genes are less likely to be found in old gene bank accessions and more likely to be found in currently evolving populations where plants and diseases coexist at locations where locally important diseases are present.

In addition to the changes in means, there is mounting evidence for the increased likelihood of extreme weather events under continued climate change. Such extreme weather events will doubtlessly aggravate the negative mean-based trends. Past and current day association between climate extremes and reduced crop production demonstrates this point. The latest IPCC report partially attributes the global production dips preceding the global food price crisis in 2007 to extreme weather events caused by anthropogenic climate change (Porter et al. 2014).

Thus, needs for crop evolutionary response to climate change include simultaneous adaptation to CO_2-enriched air, higher temperatures, altered rainfall patterns, new diseases and pests, as well as new strains of familiar diseases. All of these factors will change far beyond what the plant species have historically been exposed to and therefore require evolutionary change beyond the range of existing phenotypic variation. We know from evolutionary history and long-term experiments that quantitative traits can change far beyond the current range. However, this is only possible if the necessary genetic variation is available in the populations at hand.

18.3 Crop Adaptation

The latest IPCC report (AR5) defines *adaptation* as the process of adjustment to actual or expected climate and its effects. In this chapter, we use the term *crop adaptation* to refer to all kinds of crop changes that take place as a response to current climatic stress or anticipated climate change. Crop adaptation can happen at the interspecific level (e.g., switching from maize to sorghum) or at the intraspecific level (e.g., switching from one variety of maize to another).

There are several typologies of adaptation to climate change of relevance to the concept of crop adaptation, but to stick to the IPCC terminology, we distinguish between *incremental* adaptation and *systemic* and *transformative* adaptation. Incremental adaptation is defined as adaptation actions where the central aim is to maintain the essence and integrity of a system or process at a given scale. Transformational adaptation is defined as adaptation actions that change the fundamental attributes of a system in response to climate and its effects. According to Vermeulen et al. (2013), breeding of new and "climate ready" varieties represents systemic adaptation, while the use of existing varieties represents incremental adaptation. In the typology of Kates et al. (2012), the efforts to develop and disseminate "Water Efficient Maize for Africa"—a transgenic maize developed and promoted by a public/private partnership—is classified as transformative adaptation because it represents a novel approach to crop development and dissemination in Africa (Kates et al. 2012). It is also important to remember that crop adaptation on the farm takes place in an institutional and political context. In the words of the IPCC,

> There is an increasing recognition in the literature that while many adaptation actions are local and build on past climate risk management experience, effective adaptation will often require changes in institutional arrangements and policies to strengthen the conditions favorable for effective adaptation including investment in new technologies, infrastructure, information, and engagement processes. (Porter et al. 2014).

The AR5 presents a table with simulated median benefits (difference between the yield change from the baseline for the adapted and non-adapted cases) for a range of common on-farm adaptation strategies commonly assessed in the literature. Figure 18.1 shows that *cultivar adjustment*, which is the same as what we call crop adaptation, comes out as the adaptation with the largest potential for reducing yield loss.

The common assumption in the 33 studies from which the data to generate Fig. 18.1 originates is that farmers will switch to a better adapted modern variety developed by a plant breeding program. While this is the typical situation for a farmer in Europe, North America and other modernized farmlands, this is a problematic assumption for agriculture in developing countries. In order to explain what we mean, we provide a brief review of the terminology on genetic resources and seed systems.

The need to adapt crops to changing environmental conditions is not new. In traditional agriculture, adaptation is an interactive process between natural selection and farmers' selection. With the birth of commercial seed supply in the 19th century, breeders started to take over the farmers' role in crop improvement in Europe (Murphy 2007). Today, nearly all crops grown in industrialized countries are products of professional breeding. Since the Green Revolution began in the 1960s, varieties grown in Asia and Latin America are also, to a large extent, from professional breeding (Evenson and Gollin 2003). Simply put, all crop improvement can be described as crossing of plants with desired characteristics and selection of offspring

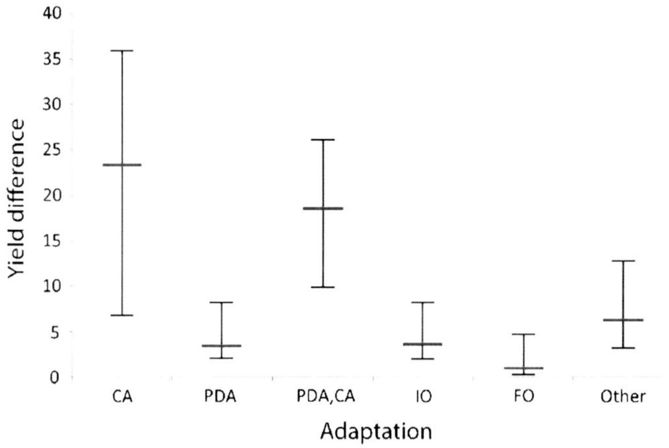

Fig. 18.1 The percentage benefit (yield difference between cases with and without the adaptation) for different crop management adaptations: cultivar adjustment (*CA*); planting date adjustment (*PDA*); adjusting planting date in combination with cultivar adjustment (*PDA*, *CA*); irrigation optimization (*IO*); fertilizer optimization (*FO*); and other management adaptations (*Other*). The simulated median benefit is marked with a *red line* and the *whiskers* indicate the 25th and 75th percentile. The figure is based on data from Challinor et al. (2014) and Porter et al. (2014)

combining those desirable characteristics. Thus, both on-farm crop improvement and professional plant breeding rely on access to genetic diversity as raw-material for the development of new varieties. This role of crop diversity is reflected in the term *genetic resources*—seeds, plants and plant parts useful in crop breeding, research or conservation for their genetic attributes (Fowler and Hodgkin 2004).

To distinguish between the varieties produced by farmers and breeders, it is common to refer to the former as *traditional* varieties or *landraces* (Zeven 1998) and the latter as *improved* or *modern* varieties. Unlike for modern varieties, which are released by a formal seed supply system as distinct, uniform and stable (DUS) varieties, there is no standard definition of the term landrace (Berg 2009; Zeven 1998), but Harlan (1975) offered the following definition:

> Landraces have a certain genetic integrity. They are recognizable morphologically; farmers have names for them and different landraces are understood to differ in adaptation to soil type, time of seeding, date of maturity, height, nutritive value, use, and other properties. Most important, they are genetically diverse.

While both landrace improvement and professional plant breeding ultimately rely on genetic resources, the way in which these resources are accessed and the new varieties are disseminated is quite different. In traditional agriculture, genetic resources for adaptation are sourced from the farmers' own field, through gifts and trade with other farmers, and sometimes through gene flow from wild relatives of the crop. In modernized agriculture, the plant breeder acts as an intermediary between the genetic resources and the farmer; the genetic resources are sourced from gene banks and genetic stocks and the modern varieties are distributed by private and public formal institutions. The terms *informal* and *formal* seed systems are used to distinguish between the two different systems of variety development and dissemination (Almekinders et al. 1994; Sperling et al. 2008). The formal seed system refers to the chain of public and private sector activities and institutions that produce and release certified seeds of officially registered varieties (Louwaars and de Boef 2012; Sperling et al. 2008). Informal seed systems include saving from one's own harvest, farmer-to-farmer seed exchange and purchase from local markets (Almekinders et al. 1994). In SSA's smallholder agriculture, most seeds are sourced through the informal seed system (DeVries and Toenniessen 2002; Langyintuo et al. 2010). Not only traditional varieties are sourced through informal channels, but also farm-saved and recycled improved varieties (Gibson et al. 2005; Lunduka et al. 2012; Mortimore and Adams 2001). For a farmer to be *seed secure*, three elements need to be in place: seeds must be *available* in the area at the right time; seeds must be *accessible* for the farmers; and the seeds must be of the desired quality with regard to genetic and health properties (Sperling et al. 2008). All three dimensions are potentially affected by climate change and the different seed system models offer different adaptation pathways. In Table 18.1, we outline the

Table 18.1 Dimensions of seed security and associated constraints and adaptation options

	Informal system	Formal system	Integrated system
Seed availability	Seeds sourced from production stock. Major supply system in developing countries	Seeds sourced from breeding stock separated from production stock. Supply limited by policy and institutional bottlenecks in formal system	Coexistence of formal and informal models. Seed flow between informal and formal system. Draws on strengths of informal system in seed production and distribution (e.g., Quality Declared Seeds schemes)
Seed accessibility	Accessible from own harvest, social networks or local markets. Socio-economic constraints (related to purchasing power and social network access)	Access from private agrodealers, NGOs or government programs. IP-rights and seed regulations can constrain access. Economic and infrastructure related constraints	Balance between variety protection and farmers' right to save seeds. Access to seeds is enhanced by coexistence of plural seed sources
Seed quality	Local adaptations to low-input, high-stress environments. Often low yields. Seed health a challenge	High yielding, specific resistance and biofortification traits. Long breeding, delivery and adoption time. Hybrids have large yield penalty for farm-saved seeds. Problem of counterfeit seeds in market	Combination and complementarity of quality from improved varieties and landraces. OPVs more appropriate for integrated system than hybrids. Seed quality can be strengthened by QDS

relationship between the three dimensions of seed security and the constraints/adaptation options farmers have.

Having introduced the terminology for genetic resources and seed systems, we will now turn to a case study of the role of different forms of genetic resources and seed systems in crop adaptation. We use Table 18.1 as a conceptual framework to understand the local seed security. The case study was carried out in two village areas in the semi-arid agroecological zone in Tanzania and our findings indicate that crop adaptation is indeed not only a question of switching from one modern variety to another.

18.4 Case Study from Morogoro

Equipped with the fine geographic level impact projections from Thornton et al. (2009) and in consultation with local crop experts at Sokoine University of Agriculture (SUA) and in the National Agricultural Research Organization (NARO), we chose two study sites where agricultural livelihoods are climatically stressed today and are likely to experience increased stress in the future (see Fig. 18.2).

The two most important grain crops in the study area were maize and sorghum. Sorghum is a drought tolerant grain crop of African origin, while maize was introduced on the African continent in the so-called Colombian exchange in the 16–17th centuries. The two crops differ in respect to commercialization of their seed supply. While the formal seed system only supplies a minor share of the sorghum seeds planted by smallholders in SSA, the maize seed supply is the most formalized and commercialized of all major crops in the region. Public maize breeding in Tanzania has a history back to the establishment of the National Maize Research Program with the assistance of International Maize and Wheat Improvement Center (CIMMYT) and the International Institute of Tropical Agriculture (IITA) in 1973 (Moshi and Marandu 1985). Today, maize is the most important staple crop in the country and was produced on 58 % of the total cereal area in 2010 (FAO 2014). Over the years, a number of modern maize varieties have been released and distributed through the formal seed system. However, the informal seed system continues to dominate the maize seed supply. The reason is partly the country's heterogeneous agricultural conditions with few suitable modern varieties available for many areas, but mainly the low purchasing power of smallholders, which hinders access to the varieties that exist (Ngwediagi et al. 2009).

Our survey showed that only 24 % of the maize seeds and 8 % of the sorghum seeds were sourced through seed sources classified as formal (Westengen and Brysting 2014). The majority of the seeds were sourced from farmers' own harvest and the local seed market was the second most important channel. However, a substantial part of the seeds sourced through informal channels were seeds of modern varieties that at one point had entered the local seed system from formal channels and were reused as farm saved seeds. We found that 11 % of the maize

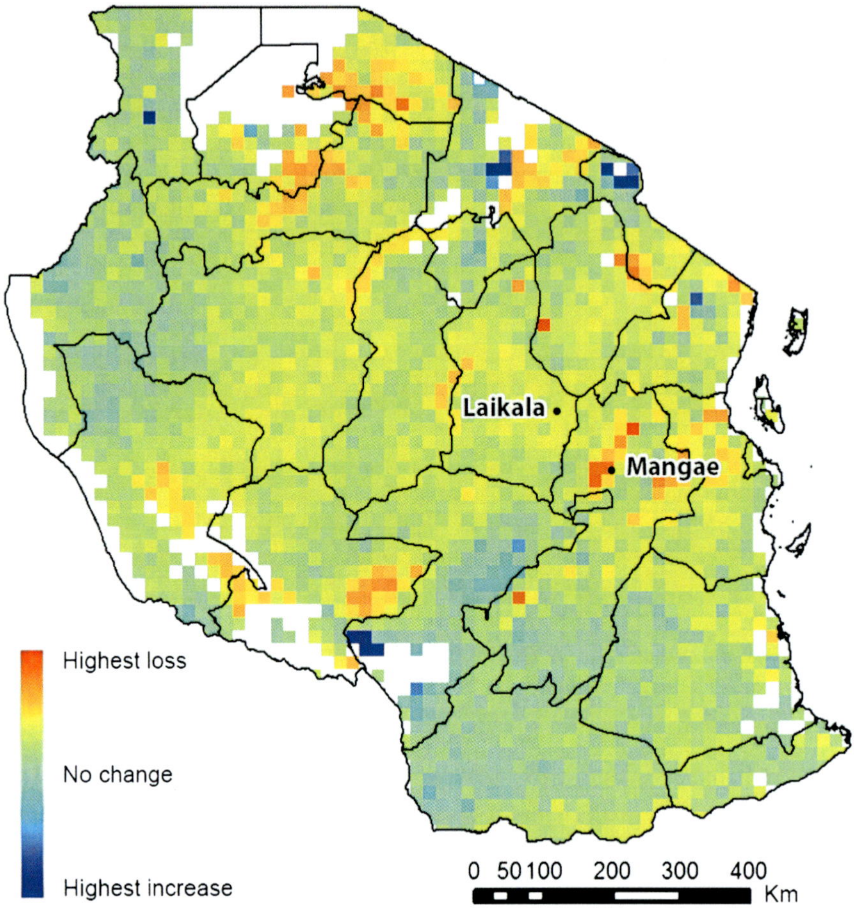

Fig. 18.2 The study sites in Morogoro and Dodoma, Tanzania. The map is a reproduction of the analyses done by Thornton et al. (2009) and shows that the study sites are located in areas that are projected to see large maize production losses in 2050 unless agriculture adapts. Figure from Westengen and Brysting (2014)

growing households in our study area reported growing recycled modern varieties. A similar figure (14 %) was reported in a recent study from Malawi (Lunduka et al. 2012). However, since the distinction between local varieties and recycled varieties has become blurred after decades of introduction of modern varieties in the region, the proportion of farm-saved seeds is likely to be higher than we could determine based on variety names.

In order to assess what role the different types of varieties played in adaptation, we classified all reported varieties as local, improved or farmer-recycled, and asked a range of questions about their cultivation and consumption qualities. Several findings from this assessment shows that farmer's in the area navigate climatic stress by drawing on different sorts of crop adaptation strategies. First, we found that a higher proportion of the households reported drought tolerance as a reason for cultivating local sorghum than for cultivating any of the maize genetic resource types, reflecting that sorghum is a considerably more drought-tolerant crop species than maize. Second, drought-tolerance was more frequently reported as a reason for growing local varieties than for growing improved varieties of both maize and sorghum. Third, local varieties of both sorghum and maize were reported to be significantly later maturing than improved ones. That local varieties were ranked as more drought-tolerant, despite their significantly later maturity than improved varieties, indicates that people distinguish between *drought tolerance* and *drought avoidance* traits. Also, studies of farmers' perceptions of drought tolerance in maize in Mexico have found that farmers rank drought tolerance traits as more important than short maturation period (Bellon and Risopoulos 2001; Bellon and Taylor 1993).

The high share ($\sim$3/4) of households reporting to practice selection suggested that incremental evolutionary change of current varieties might play a role in crop adaptation in the study area. In the village from which we report details about farmers' selection practices (Westengen et al. 2014), the two most frequently selected characteristics in maize were cob filling and drought tolerance. In order to investigate the genetic effects of the mixing of formal and informal seed system elements, we conducted a genetic analysis of the local maize seed system. We collected seed lots from different stages in the formal seed system (breeder's seeds, foundation seeds and commercial seeds) as well as "informal seeds" from the farmers' fields and granaries. We analyzed seeds from the two most common modern maize varieties in the area, the two OPVs Staha and TMV1, some local varieties and some reportedly recycled seed lots of the two OPVs (Fig. 18.3). The purpose was to see if we could detect creolization and signals of directional change as the OPVs entered the informal seed system. In Westengen et al. (2014), we presented what, to our knowledge, is the first genetic assessment of creolization. We used population genetic tools to detect genetic admixture between modern OPVs and presumably local varieties. Analysis of the genetic structure and differentiation between seedlots from different formal seed system stages, from breeder's seeds to commercial seeds, indicated that the formal system supplied seeds that were true to type, meaning that the seeds sold in the shop were genuine. Thus, in our case, we did not find that seed security is compromised by counterfeit certified seeds, which is commonly reported as a problem in formal seed systems across Eastern Africa (Langyintuo et al. 2010, Warburton et al. 2010) (see Table 18.1). When the OPVs entered the local seed system through farm-saving and recycling, a genetic effect was visible. Seedlots of the OPVs sampled on-farm after one or more growing seasons were significantly differentiated from those sampled from the formal system. Using association analysis, we detected signals of directional selection on three single nucleotide polymorphisms (SNPs) (see Fig. 18.4). Two of these SNPs were located in putatively protein coding genes and can potentially code for adaptation to the local agroecology (Westengen et al. 2014).

Fig. 18.3 The open pollinated variety "Staha", developed and released in the 1980s, is still a popular improved variety. *Left* A Staha exhibition plot in the 2010 Nane Nane Agricultural Exhibition in Morogoro. *Right* A farmer with creolized Staha. *Photos* Ola T. Westengen

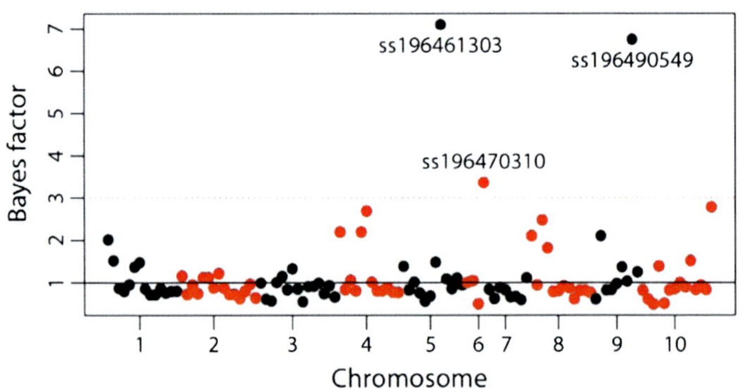

Fig. 18.4 Evidence for directional selection. Manhattan plot of SNP association with seed system stage. The SNPs are plotted according to chromosome and position at chromosome along the X-axis. Chromosomes 1–10 alter between *black* and *red color*. The *red dotted line* indicates the level of significance. Figure from Westengen et al. (2014)

In conclusion, we found that farmers draw on both formal and informal seed systems to adapt to climate change. Crop adaptation involves the adoption of improved maize varieties combined with continued use of local varieties of both maize and sorghum. We, therefore, suggest that access to drought tolerant OPVs in combination with farmer seed selection is likely to enhance seed system security and farmers' adaptive capacity in the face of climate change (see Table 18.1).

18.5 Adaptive Seed Systems

When crop varieties are highly adapted to target environments, they are likely to possess less adaptability under change (Cooper et al. 2001). Considering this inverse relation between adaptability and adaptation, we must carefully consider the genetic preconditions for the needed adaptation. For if varieties are typically adapted to a narrow ecological niche, how will they be able to adapt to new climates? Again, we believe the key to answer this is by considering the interaction between the technology (the genetic resources) and the institutions (the seed systems).

Considering first the technology, it is clear that significant potential for adaptation exists in standing diversity in open pollinated populations and landraces. Both general evolutionary biology and results from long-term selection experiments show us that populations can change far beyond the range of a source population. The most remarkable example is provided by the long-term selection experiment for oil and protein in maize started at the University of Illinois in 1896. The population responded steadily and did not reach the limit when stock was taken after 100 generations (Dudley and Lambert 2004). Landraces are, as noted by Harlan, commonly genetically diverse and therefore harbor a significant potential for adaptation. Studies of maize landraces in Mexico indicate that some display both plasticity and adaptive genetic potential, while others, notably the highland maize landraces in southern Mexico, are threatened by extinction unless population size and management allows for gene flow and adaptive evolution (Mercer and Perales 2010). A study of pearl millet landraces from Niger showed evidence of increased frequency of an allele associated with drought avoidance between 1976 and 2003—a period with a marked shift to drier conditions in the study area (Vigouroux et al. 2011). Such adaptation based on standing variation in landraces is probably the outcome of both natural selection and farmer selection and represents the most fundamental level of incremental adaptation to climate change. The same phenomenon is apparent also in breeding programs taking place year after year in the same environments, even when the breeding is targeting other traits than those relevant for climate change. The rice study from International Rice Research Institute (IRRI) showing declining yields with increased night temperatures also found that newer varieties showed less decline than those developed in the 1960s. The breeder, Robert Zeigler, takes this as an indication of an unintended, but adaptive, gradual change (Zeigler et al. 2013). However, Zeigler is also quick to point out that this is a rather unique example and that we cannot count on a *laissez faire* approach to climate change adaptation in modern rice breeding. Indeed, in predominantly inbred crops like rice, the variation in modern varieties is low and adaptation will require human mediation of a targeted sort.

Technological adaptation can only succeed if the institutions involved in the different stages of innovation and development are conducive. For crop adaptation, the most basic level is availability of genetic diversity. It is estimated that by 2050, the majority of African countries will experience "novel" climates over at least half their current crop area that lie outside the range currently experienced within the country (Burke et al. 2009). The authors of this study argue that international movement of genetic resources will be important for adaptation. The logic is that landraces from climatic areas that are "climate analogues" to the novel climates projected represent a valuable source of genetic variation. This points to an overarching question about institutional preparedness for climate change adaptation. It is clear that no country is self-reliant in genetic resources. All countries depend on crops that originated outside their present borders and their future evolution will also depend on availability and access to genetic resources from other environments. The institutions responsible for conservation and distribution of genetic resources are commonly called gene banks. The major international legal and policy framework for conservation and use of genetic resources is the International Treaty on Plant Genetic Resources for Food and Agriculture (ITPGRFA), which entered into force in 2004. This treaty lays out the principles and rules for access and benefit sharing for a list of crops that the parties have agreed to share in a multilateral system. There are still some challenges in the implementation of the treaty. Not all countries in the world, not even all those that are parties to the treaty, share genetic resources held in their gene banks according to the agreed terms (Bjørnstad et al. 2013). There can be many reasons why a gene bank fails to distribute seed samples to bona fide users, from limited staff capacity to national legal and political obstacles, but there are several international initiatives underway under the framework of the treaty that bide hope for a gradual improvement in the collaborative regime (Moore et al. 2013). The FAO, the Global Crop Diversity Trust and the Consultative Group on International Agricultural Research (CGIAR) work together to establish a well-functioning and securely funded global system for ex situ conservation of PGRFA. Since it's opening in 2008, the Svalbard Global Seed Vault, the back-up storage site for this conservation system, has become the world's largest repository of genetic diversity (Westengen et al. 2013). Another example of an international project that relies on the treaty is the joint project of the Global Crop Diversity Trust and the Kew Botanical Garden entitled "Adapting agriculture to climate change: collecting, protecting and preparing crop wild relatives". Under this project, wild botanical relatives of crop species are mapped and collected and will be utilized in pre-breeding programs. In pre-breeding programs, the genetic base is broadened in order to overcome the narrow genetic foundation for adaptation in many of our important food crops.

Speaking about base-broadening and international policies and programs to develop climate ready varieties, it is easy to forget the local realities of farmers in developing countries like we described them in the case study from Tanzania. Even if breeding programs are able to come up with impressive modern varieties with traits that seemingly make them ideal for the new climatic conditions, this does not necessarily lead to widespread adoption and use of the new varieties. Adoption of modern varieties is low in developing countries because of poor adaptation to local agroecological conditions and/or failure to meet the various end use preferences. These factors are likely to remain important as the climate ready varieties are released from formal seed systems. Successful adaptation through plant breeding is therefore contingent on the functioning of the seed systems they are introduced in.

Current development initiatives aiming at replacing informal seed systems with formal seed systems modeled after those found in industrialized countries are contested in recent seed system literature. A number of scholars in this field argue that a more suitable way forward for most developing countries is to enable the development of *integrated* seed systems (Louwaars and de Boef 2012; McGuire and Sperling 2013; Mercer et al. 2012; Scoones and Thompson 2011; Sperling and McGuire 2012). Integrated seed systems allow for coexistence of formal and informal seed system elements by, for example, relaxing the DUS requirements for variety approval and setting up less costly and more flexible certification schemes such as the quality declared seed approach of the FAO (see Table 18.1). Thus, it is important to strike a balance between breeders' need for variety protection, the market need for affordable seeds and farmers' right to save seeds. Furthermore, breeding programs should, to a much larger degree than what is common today, include farmers in the variety development process (forms of participatory plant breeding) to ensure that the crop's cultural, socioeconomic and local environmental context is heeded (Ceccarelli and Grando 2007). There is also a potential in breeding for local adaptation using locally preferred landraces as a basis and introducing genetic resources from the broader genepool (Hellin et al. 2014). The central lesson from the literature on genetic resources and seed systems is that adaptation to climate change cannot be considered a technological challenge alone. There are no silver bullets, but with accessible diversity and flexible and enabling institutional frameworks from the international to the local level, there is great potential for developing adaptive seed systems.

References

Almekinders CJM, Louwaars NP, Debruijn GH (1994) Local seed systems and their importance for an improved seed supply in developing countries. Euphytica 78:207–216

Bellon M, Risopoulos J (2001) Small-scale farmers expand the benefits of improved maize germplasm: a case study from Chiapas, Mexico. World Dev 29:799–811

Bellon MR, Taylor JE (1993) "Folk" soil taxonomy and the partial adoption of new seed varieties. Econ Dev Cult Change 41:763–786

Berg T (2009) Landraces and folk varieties: a conceptual reappraisal of terminology. Euphytica 166:423–430

Bjørnstad Å, Tekle S, Göransson M (2013) "Facilitated access" to plant genetic resources: does it work? Genet Resour Crop Evol 60:1959–1965

Burke MB, Lobell DB, Guarino L (2009) Shifts in African crop climates by 2050, and the implications for crop improvement and genetic resources conservation. Global Environ Change Hum Policy Dimens 19:317–325

Cairns JE, Hellin J, Sonder K, Araus JL, MacRobert JF, Thierfelder C, Prasanna B (2013) Adapting maize production to climate change in sub-Saharan Africa. Food Secur 5:1–16

Ceccarelli S, Grando S (2007) Decentralized-participatory plant breeding: an example of demand driven research. Euphytica 155:349–360

Challinor A, Watson J, Lobell D, Howden S, Smith D, Chhetri N (2014) A meta-analysis of crop yield under climate change and adaptation. Nat Clim Change 4:287–291

Cooper H, Spillane C, Hodgkin T (2001) Broadening the genetic base of crops: an overview. Broadening the Genetic Base of Crop Production Wallingford, New York. CABI Publishing, Wallingford, pp 1–23

DeVries J, Toenniessen G (2002) Securing the harvest: biotechnology, breeding and seed systems for African crops. CABI, New York

Dudley J, Lambert R (2004) 100 generations of selection for oil and protein in corn. Plant Breed Rev 24:79–110

Evenson RE, Gollin D (2003) Assessing the impact of the green revolution, 1960 to 2000. Science 300:758–762

FAO (2014) FAOSTAT. http://faostat3.fao.org/home/E

Fowler C, Hodgkin T (2004) Plant genetic resources for food and agriculture: assessing global availability. Annu Rev Environ Resour 29:143–179

Gibson R, Lyimo N, Temu A, Stathers T, Page W, Nsemwa L, Acola G, Lamboll R (2005) Maize seed selection by East African smallholder farmers and resistance to Maize streak virus*. Ann Appl Biol 147:153–159

Harlan JR (1975) Our vanishing genetic resources. Science 188:617–621

Hellin J, Bellon MR, Hearne SJ (2014) Maize landraces and adaptation to climate change in Mexico. J Crop Improv 28:484–501

Kates RW, Travis WR, Wilbanks TJ (2012) Transformational adaptation when incremental adaptations to climate change are insufficient. Proc Natl Acad Sci 109:7156–7161

Knox J, Hess T, Daccache A, Wheeler T (2012) Climate change impacts on crop productivity in Africa and South Asia. Environ Res Lett 7:034032

Langyintuo AS, Mwangi W, Diallo AO, MacRobert J, Dixon J, Banziger M (2010) Challenges of the maize seed industry in eastern and southern Africa: a compelling case for private-public intervention to promote growth. Food Policy 35:323–331

Lobell DB, Burke MB (2008) Why are agricultural impacts of climate change so uncertain? The importance of temperature relative to precipitation. Environ Res Lett 3 http://dx.doi.org/10.1088/1748-9326/3/3/034007

Lobell DB, Bänziger M, Magorokosho C, Vivek B (2011a) Nonlinear heat effects on African maize as evidenced by historical yield trials. Nat Clim Change 1:42–45

Lobell DB, Schlenker W, Costa-Roberts J (2011b) Climate trends and global crop production since 1980. Science 333:616–620

Long SP, Ainsworth EA, Leakey AD, Nösberger J, Ort DR (2006) Food for thought: lower-than-expected crop yield stimulation with rising CO_2 concentrations. Science 312:1918–1921

Louwaars NP, de Boef WS (2012) Integrated seed sector development in Africa: a conceptual framework for creating coherence between practices, programs, and policies. J Crop Improv 26:39–59

Lunduka R, Fisher M, Snapp S (2012) Could farmer interest in a diversity of seed attributes explain adoption plateaus for modern maize varieties in Malawi? Food Policy 37:504–510

Martínez-Eixarch M, Ellis RH (2015) Temporal sensitivities of rice seed development from spikelet fertility to viable mature seed to extreme-temperature. Crop Sci 55:354–364

McGuire S, Sperling L (2013) Making seed systems more resilient to stress. Global Environ Change Hum Policy Dimens 23:644–653

Mercer KL, Perales HR (2010) Evolutionary response of landraces to climate change in centers of crop diversity. Evol Appl 3:480–493

Mercer KL, Perales HR, Wainwright JD (2012) Climate change and the transgenic adaptation strategy: smallholder livelihoods, climate justice, and maize landraces in Mexico. Global Environ Change Hum Policy Dimens 22:495–504

Moore G, Hawtin G, Jackson M, Ford-Lloyd B, Parry M (2013) International mechanisms for conservation and use of genetic resources. Plant Genet Resour Clim Change 4:98

Mortimore M, Adams W (2001) Farmer adaptation, change and 'crisis' in the Sahel. Global Environ Change Hum Policy Dimens 11:49–57

Moshi AJ, Marandu W (1985) Maize research in Tanzania. In: To feed ourselves: a proceedings of the first eastern, central and southern Africa Regional Maize workshop, Lusaka, Zambia, March 10–17, 1985 CAB Direct, Lusaka

Moya TB, Ziska LH, Namuco OS, Olszyk D (1998) Growth dynamics and genotypic variation in tropical, field-grown paddy rice (Oryza sativa L.) in response to increasing carbon dioxide and temperature. Glob Change Biol 4:645–656

Murphy DJ (2007) People plants and genes: the story of crops and humanity. Oxford University Press, Oxford

Ngwediagi P, Maeda E, Kimomwe H, Kamara R, Massawe S, Akonaay HB, Mapunda LND (2009) Tanzania country report on the state of plant genetic resources for food and agriculture, Dar es Salaam

Peng S, Huang J, Sheehy JE, Laza RC, Visperas RM, Zhong X, Centeno GS, Khush GS, Cassman KG (2004) Rice yields decline with higher night temperature from global warming. Proc Natl Acad Sci USA 101:9971–9975

Porter JR, Xie L, Challinor AJ, Cochrane K, Howden SM, Iqbal MM, Lobell DB, Travasso MI (2014) Food security and food production systems. In: Field CB, Barros VR, Dokken DJ, Mach KJ, Mastrandrea MD, Bilir TE, Chatterjee M, Ebi KL, Estrada YO, Genova RC, Girma B, Kissel ES, Levy AN, MacCracken S, Mastrandrea PR, White LL (eds) Climate change 2014: impacts, adaptation, and vulnerability part a: global and sectoral aspects contribution of working group II to the fifth assessment report of the intergovernmental panel on climate change. Cambridge University Press, Cambridge, pp 485–533

Schlenker W, Lobell DB (2010) Robust negative impacts of climate change on African agriculture. Environ Res Lett 5:014010

Scoones I, Thompson J (2011) The politics of seed in Africa's green revolution: alternative narratives and competing pathways. IDS Bull 42:1–23

Sperling L, McGuire S (2012) Fatal gaps in seed security strategy. Food Secur 4:569. doi:10.1007/s12571-012-0205-0

Sperling L, Cooper HD, Remington T (2008) Moving towards more effective seed aid. J Dev Stud 44:586–612

Thornton PK, Jones PG, Alagarswamy G, Andresen J (2009) Spatial variation of crop yield response to climate change in East Africa. Global Environ Change Hum Policy Dimens 19:54–65

Vermeulen SJ, Challinor AJ, Thornton PK, Campbell BM, Eriyagama N, Vervoort JM, Kinyangi J, Jarvis A, Läderach P, Ramirez-Villegas J (2013) Addressing uncertainty in adaptation planning for agriculture. Proc Natl Acad Sci USA 110:8357–8362

Vigouroux Y, Mariac C, De Mita S, Pham J-L, Gérard B, Kapran I, Sagnard F, Deu M, Chantereau J, Ali A (2011) Selection for earlier flowering crop associated with climatic variations in the Sahel. PLoS One 6:e19563

Warburton ML, Setimela P, Franco J, Cordova H, Pixley K, Banziger M, Dreisigacker S, Bedoya C, MacRobert J (2010) Toward a cost-effective fingerprinting methodology to distinguish maize open-pollinated varieties. Crop Sci 50:467–477

Westengen OT, Brysting AK (2014) Crop adaptation to climate change in the semi-arid zone in Tanzania: the role of genetic resources and seed systems. Agric Food Secur 3:3

Westengen OT, Jeppson S, Guarino L (2013) Global ex-situ crop diversity conservation and the svalbard global seed vault: assessing the current status. PLoS One 8:e64146

Westengen OT, Ring KH, Berg PR, Brysting AK (2014) Modern maize varieties going local in the semi-arid zone in Tanzania. BMC Evol Biol 14:1

Zeigler RS, Jackson M, Ford-Lloyd B, Parry M (2013) Food security, climate change and genetic resources. Plant genetic resources and climate change UK. CAB International, Wallingford, pp 1–15

Zeven AC (1998) Landraces: a review of definitions and classifications. Euphytica 104:127–139

Ziska LH (2003) Evaluation of the growth response of six invasive species to past, present and future atmospheric carbon dioxide. J Exp Bot 54:395–404

Chapter 19
Updating Legacy Soil Maps for Climate Resilient Agriculture: A Case of Kilombero Valley, Tanzania

Boniface H.J. Massawe, Brian K. Slater, Sakthi K. Subburayalu, Abel K. Kaaya and Leigh Winowiecki

Abstract Since the first documented soil survey in Tanzania by Milne (J Ecol 35:192–265, 1936), a number of other soil inventory exercises at different scales have been made. The main challenge has been the fragmented nature of the often outdated detailed soil maps and small-scale less-informative country-wide soil maps. Recent advances in information and computational technology have created vast potential to collect, map, harness, communicate and update soil information. These advances present favorable conditions to support the already popular shift from qualitative (conventional) to quantitative (digital) soil mapping (DSM). In this study, two decision tree machine learning algorithms, J48 and Random Forest (RF), were applied to digitally predict *k*-means numerically classified soil clusters to update a soil map produced in 1959. Predictors were derived from 1 arc SRTM digital elevation data and a 5 m RapidEye satellite image. J48 and RF predicted the soil units of the legacy maps with greater detail. However, RF showed superiority for predicting clusters J48 could not predict and for showing higher pixel contiguity. No significant difference ($P = 0.05$) was observed between the soil properties

B.H.J. Massawe (✉) · A.K. Kaaya
Department of Soil and Geological Sciences,
Sokoine University of Agriculture, P.O. Box 3008, Morogoro, Tanzania
e-mail: bonmass@yahoo.com

A.K. Kaaya
e-mail: abelkaaya@yahoo.co.uk

B.K. Slater · S.K. Subburayalu
School of Environment and Natural Resources, The Ohio State University,
210 Kottman Hall, 2021 Coffey Road, Columbus, OH 43210, USA
e-mail: slater.37@osu.edu

S.K. Subburayalu
e-mail: subburayalu.1@osu.edu

L. Winowiecki
World Agroforestry Centre, United Nations Avenue, Gigiri, Nairobi, Kenya
e-mail: L.A.WINOWIECKI@CGIAR.ORG

R. Lal et al. (eds.), *Climate Change and Multi-Dimensional Sustainability in African Agriculture*, DOI 10.1007/978-3-319-41238-2_19

of the predicted soil clusters and the actual field validation points. Young soils (Entisols and Inceptisols) were found to occupy about 56 % of the study site's 30,000 ha followed by Alfisols, Mollisols and Vertisols at 31, 9 and 4 %, respectively. This study demonstrates the usefulness of DSM techniques to update conventionally prepared legacy maps to offer soil information at improved detail to agricultural land use planners and decision makers of Tanzania to make evidence-based decisions for climate-resilient agriculture and other land uses.

Keywords Kilombero valley · Digital soil mapping · Machine learning · Legacy soil maps

19.1 Introduction

19.1.1 Climate Change Verses Agriculture

Climate change and its associated results such as rainfall shifts and rises in temperature pose great challenges to the sustainability of agriculture worldwide (Anwar et al. 2015). However, the global climate models indicate that Sub-Saharan Africa will be one of the most affected regions because of its high dependency on rain-fed agriculture and the low level of investments and technology applied in farming (Zinyengere et al. 2013; Arslan et al. 2015). Yields for major food crops are projected to suffer a decrease of up to 20 % (Cline 2008).

Although the current ability to contain the climate change pace is limited (Rogelj et al. 2011; Intergovernmental Panel on Climate Change [IPCC] 2014), scientists suggest that agriculture might adapt to some new climatic conditions (Anwar et al. 2015; Arslan et al. 2015). This is possible if sufficient technology and capital are available to facilitate management modifications. In order to help farmers and decision makers choose climate change adaptation strategies, a strong scientific evidence base is needed (Anwar et al. 2013; Challinor et al. 2014). The evidence must be derived from all factors that contribute to climate change, are affected by climate change and directly reduce climate change or can help agriculture adapt to climate change effects.

Soil has been identified as an important factor in managing climate change. Soil acts as a source and sink of greenhouse gases and provides a medium on which food and fiber are grown (Palm et al. 2007). It is therefore very crucial to have an in-depth understanding of soils and how they can be managed to better address the issues of food security and climate change. It is projected that by 2050 the world will have a population of 9 billion people; most of this increase will take place in Sub-Saharan Africa (You et al. 2014) as the population of Africa will more than double during this period compared to the current level. The main challenge is how to feed this anticipated population while agricultural production is projected to decrease due to climate change and soil degradation.

Spatial soil information provides a basis for allocation of land into appropriate land use type. Soil properties associated with different soil types provide information about management requirements for a particular crop for its sustainable productivity. Although crops can be bred for climate change resilience such as drought resistance, prior information on the ability of the soil to provide ecosystem services such as water, nutrients and air required for the crop should not be sidelined (McBratney et al. 2014). The spatial soil information is obtained in the process known as a soil survey.

Traditionally, spatial soil information has been represented as soil maps in which polygons are considered to enclose homogeneous soils differing abruptly from the adjacent polygon (Fridland 1974; McBratney and Odeh 1997; Heuvelink and Webster 2001). These types of maps have been a result of conventional soil mapping (CSM) techniques that are generally qualitative. Recently, digital soil mapping (DSM) techniques have been used to produce continuous soil maps that are developed based on soil-forming factors (Jenny 1941). DSM approaches are quantitative and have been supported by recent advances in information technology together with the availability of new and more reliable soil-related data sources such as digital elevation models and satellite imagery. The DSM approaches use relationships between soil properties or soil classes and the soil-related environmental correlates at sample points to predict the soil properties or classes over a study area (Scull et al. 2003; McBratney et al. 2003).

Many developing countries, including Tanzania, experience inadequacies in spatial soil information (Msanya et al. 2002; Cook et al. 2008). The inadequacy can be categorized as a lack of desired details due to coarse mapping scale, use of outdated standards or under-coverage and the presence of disjointed maps prepared at different scales. Most of the soil information in Tanzania is stored in legacy maps that have these limitations. The maps' usability for land use planning and management is therefore also limited. However, these legacy soil maps provide useful soil data that can be easily updated using DSM techniques to get more detailed spatial digital soil information that is in high demand for land management and research for climate change adaptation.

Several studies have employed legacy soil surveys to produce digital soil maps. Such studies include those by Bui and Morgan (2001), Baxter and Crawford (2008), Kempen et al. (2009), Sulaeman et al. (2013), Malone et al. (2014), Vaysse and Lagacherie (2015) and Cambule et al. (2015). Some of these studies have also tried to develop methods through which the legacy maps can be updated and provide more useful information for land management and crop production.

19.1.2 Spatial Soil Information in Tanzania

The first documented soil survey in Tanzania was conducted by Milne (1936), who made an exploratory soil survey of some regions of the country. He classified

Tanzanian soils into nine groups: volcanic soils, plain soils, saline soils, podzolized soils, desert soils, non-laterized red earths, laterized red earths, plateau soils and loose sands. The usefulness of his classification was limited by it being based on a very general reconnaissance and limited analytical data. In 1954, Calton distinguished eluvial from illuvial soil types and several catenary associations to produce a 1:4,000,000 soil map of Tanganyika—now Mainland Tanzania (Calton 1954). In Scott (1962) developed a regional soil map of East Africa based on the landscape approach with a resultant scale of 1:4,000,000. In D'Hoore (1964) produced a soil map of the African continent at a scale of 1:5,000,000. This scale is too small for practical planning purposes at the national and regional level. Baker (1970) classified the soil of Tanzania into 31 units by studying only 40 soil profile pits countrywide, throwing doubt on its representativeness. Generally, these works lacked reliability as they covered only accessible areas and had very coarse scales (Wickama 1997).

Several other soil mapping activities that can be considered legacy were performed in later years, including work by Samki (1977, 1982) and Hathout (1983). No systematic soil mapping has been carried out to cover the entire country for a soil map of a scale of larger than 1:1,000,000 since 1970 by Baker. De Pauw (1984) produced soil, physiography and agroecological zones maps of the country at a scale of 1:2,000,000, and in 1998, the United Nations Food and Agriculture Organization (FAO) through the Soil and Terrain (SOTER) project produced a soil and terrain database for Tanzania at a scale of 1:2,000,000 (Eschweiler 1998).

This brief review shows that Tanzania has a long history of collecting soil information, but the data can mostly be classified as legacy soil data. Therefore, there is an opportunity and need to use these legacy soil data as base maps or data sources to produce more refined and informative soil maps using current soil mapping techniques. This study applies digital soil mapping techniques to update a map of Kilombero Valley prepared in 1959 (FAO 1961).

19.2 Materials and Methods

19.2.1 The Study Area

The study was conducted in Kilombero Valley, Tanzania. The valley is about 300 km east of the Indian Ocean and covers about 11,000 km^2. The study site is located in zone 37 south of the universal transverse Mercator (UTM) coordinate system, occupying the area lying between 9064697 and 9089031 m northing and 175422–197033 m easting. The map in Fig. 19.1 indicates the location of the study area in Tanzania.

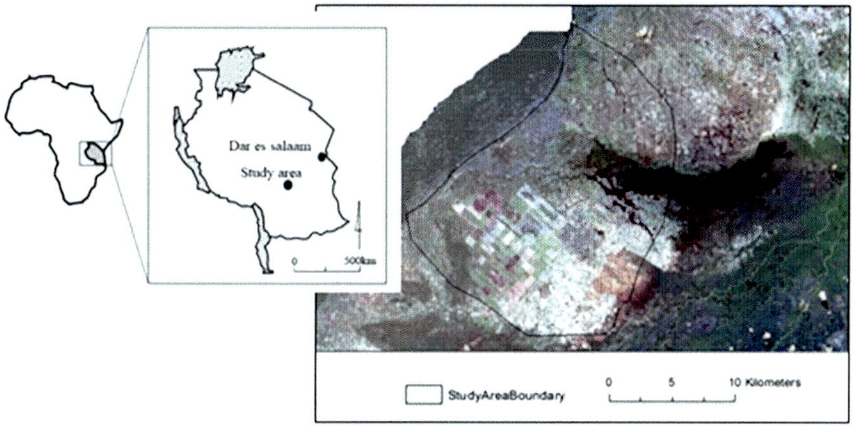

Fig. 19.1 Study area location (modified from Kato 2007)

19.2.2 Methods

19.2.2.1 Soil Data Collection

A field survey was performed to collect soil data. A legacy map (FAO 1961) with 10 soil units was used as the base map to guide the soil sampling. A stratified elevation map from the space shuttle radar topography mission (SRTM) digital elevation model (DEM; U.S. Geological Society [USGS] 2000) was used to decide where else to make observations in addition to the ten representative for the base map soil units. A total of 33 soil profiles (Fig. 19.2) were excavated and described according to FAO soil profile description guidelines (FAO 2006). Bulk soil samples were taken from designated soil horizons and were submitted to a soil laboratory for analysis of attributes as discussed below.

The soil pH was determined using a pH meter in water and in $CaCl_2$ at the ratio of 1:2.5 soil:water and soil:$CaCl_2$ as described by McLean (1986) while electrical conductivity (EC) was determined with a conductivity meter in a 1:2.5 soil:water suspension following a method by Rhoades (1982). Soil texture was determined with the hydrometer method using Calgon (5 %) as the dispersing agent (Gee and Bauder 1986). Organic carbon was determined with the Walkley and Black wet oxidation method as outlined by Nelson and Sommers (1982). The total nitrogen in the soil samples was determined with the Kjeldahl method (Bremner and Mulvaney 1982). Available phosphorus was extracted with the Bray and Kurtz-1 method (Bray and Kurtz 1945) for soils with pH_{water} less than 7 and the Olsen method for soils with pH_{water} above 7 (Watanabe and Olsen 1965). The cation exchange capacity of the soil (CEC) and exchangeable bases were determined by saturating the soil with neutral 1 M NH_4OAc (ammonium acetate) and the adsorbed NH_4^+ were displaced using 1 M KCl. The exchangeable bases (Ca^{2+}, Mg^{2+}, Na^+, K^+)

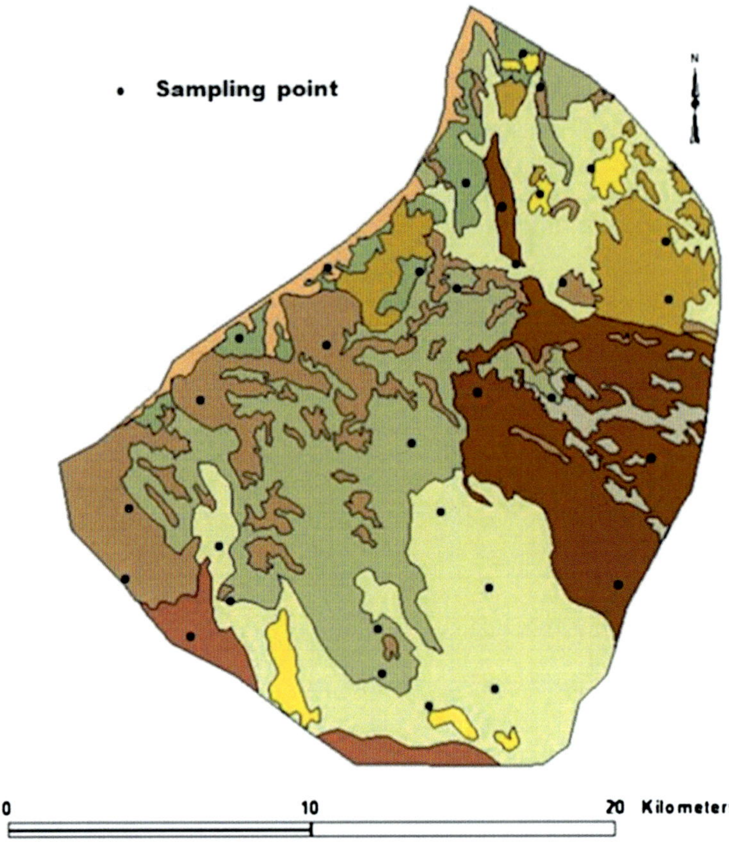

Fig. 19.2 Sampling points on the legacy soil map. *Note* The units of the legacy soil map are described in Fig. 19.3

were determined with an atomic absorption spectrophotometer (Thomas 1982) while the CEC was determined with the Kjeldahl distillation method (Schollenberger and Simon 1945). The diethylenetriaminepentaacetic acid (DTPA) method (Lindsay and Norvell 1978) was used to extract four micronutrients: iron, manganese, copper and zinc.

19.2.2.2 Clustering of Similar Soil Profiles

The studied soil profiles were clustered using distance metrics numerical classification (Carré and Jacobson 2009). The numerical clustering was done on the Outil Statistique d'Aide à la Cartogénèse Automatique (OSACA) application (Jacobson and Carré 2006) that is based on the *k*-means clustering algorithm (Diday 1971). In this method, similar soil profiles have shorter calculated distance metrics between them while dissimilar soil profiles have larger distances between them. A total of 13 soil clusters were generated. The soil cluster attributes values and modeled continuous vertical variability using equal area spline functions (Bishop et al. 1999; Malone et al. 2009) showed that these clusters were well separated and represented soils of the study area.

19.2.2.3 Prediction of the Soil Clusters Using DSM

The attributes for predicting the soil classes were chosen based on *scorpan* formulation (McBratney et al. 2003). The *scorpan* framework for digital soil mapping is based on factors of soil formation by Jenny (1941) but with *s* and *n* added. In the framework, *s* stands for soil properties, *c* for climate factors, *o* for organisms, *r* for relief (topography), *p* for parent material, *a* for age and *n* for spatial location.

In this study, *s* was derived from a 1959 legacy soil map of the area by Andeson (FAO 1961), *c* was not used, *o* was from a 5 m resolution RapidEye satellite image, *r* was from 30 m resolution SRTM and advanced spaceborne thermal emission and reflection radiometer (ASTER) DEMs, *p* was not used, *a* was not used, and *n* was recorded for each attribute used in the prediction.

19.2.2.4 Derivation of Prediction Parameters Based on the *Scorpan* DSM Framework

Living organisms *(o* in *scorpan)*
ERDAS IMAGINE 2014 software was used to process the image and the derivatives. The attributes derived were landuse/landcover classes, the normalized difference vegetation index (NDVI), the optimized soil adjusted vegetation index (OSAVI) and the soil enhancement ratio (SER).

Unsupervised classification was performed on the RapidEye image using the iterative self-organizing data analysis technique (ISODATA) from which the study area was grouped into 36 land use/land cover classes.

The NDVI layer was calculated as:

$$NDVI = \frac{(NIR - R)}{(NIR + R)}, \tag{19.1}$$

where NIR is the spectral reflectance in the near-infrared band and R is the red band.

OSAVI was calculated as:

$$OSAVI = \frac{(NIR - R)}{(NIR + R + L)}, \tag{19.2}$$

where L = 0.16 is an optimal value to minimize the variation in soil background.

The SERs are calculated as b3/b2, b3/b7 and b5/b7, where b stands for band. Since a RapidEye image has bands 1–5, it was possible to calculate only b3/b2.

Topography *(r in scorpan)*
Derivatives were calculated from the ASTER DEM with a 30 m resolution and 1-Arc resolution SRTM DEM after the depression was filled using the Planchon and Darboux (2001) algorithm. Some DEM derivatives such as aspect were not used in this study because they were deemed of little contribution to soil formation due to the flatness of the study area. All the derivatives were calculated in Whitebox Geospatial Analysis Tool 3.2 software.

The slope gradient was estimated using Horn's (1981) third-order finite difference method. Plan curvature, profile curvature, tangential curvature and total curvature were calculated according to Gallant and Wilson's (2000) method.

The relative stream power index (RSP) was calculated as

$$RSP = A_s^p \; x \; \tan(B), \tag{19.3}$$

where A_s is the specific catchment area (the upslope contributing area per unit contour length), B is the local slope gradient in degrees and p is a user-defined exponent term that controls the location-specific relation between the contributing area and the discharge.The sediment transport index (STI) was calculated as:

$$STI = (m+1) \; x \; (A_s \; /22.13)^m \; x \; \sin(B/0.0896)^n, \tag{19.4}$$

where A_s is the specific catchment area (i.e., the upslope contributing area per unit contour length), B is the local slope gradient in degrees, with the contributing area exponent, m, set to 0.4 and the slope exponent, n, set to 1.4.

The wetness index (WI) was calculated as:

$$WI = \ln(A_s/\tan(S)), \tag{19.5}$$

where A_s is the specific catchment area, and S is the slope measured in degrees.

Deviation from the mean elevation was calculated as the difference between the elevation of each grid cell and the mean elevation of the centering local neighborhood normalized by the standard deviation. The difference from the mean elevation was determined as the difference between the elevation of each grid cell in an input DEM and the average elevation in the local neighborhood.

Other derived parameters included the topographic ruggedness index, which is a measure of local topographic relief, and the flow accumulation grid (i.e., the contributing area), which was generated using the D-infinity algorithm (Tarboton 1997).

19.2.2.5 Training Set Sampling

Training sets were extracted from the x, y locations of the 33 soil observations that were used to generate the 13 soil clusters. Two sets of training data were prepared:

one in which the topographic data were derived from ASTER DEM and the other from SRTM DEM. Both sets included the same satellite image derivatives and soil data sets.

19.2.2.6 Machine Learning

Two decision tree-based algorithms were used in machine learning to produce models that would predict soil clusters with an objective of picking the results from the algorithm that give a better predicted soil cluster map. The algorithms J48 and Random Forest (RF) were run in Machine Learning Workbench known as WEKA (Waikato Environment for Knowledge Analysis) as described in Witten et al. (2011).

J48 is an Open Source version of Quinlan's C 4.5 algorithm (Quinlan 1993) that uses the divide-and-conquer approach to learn decision trees. The Random Forest technique uses bootstrap samples to build multiple decision trees (Breiman 2001).

19.2.2.7 Mapping of the Predicted Soil Clusters

To map the predicted soil clusters, a 30 m SRTM DEM raster was converted to a points shape file. The XY coordinates for each point were extracted and stored in an Excel file. The XY points were used to extract the environmental correlates values from their sources (RapidEye satellite image derivatives, legacy soil map and SRTM and ASTER DEM derivatives) using Extract Values at the XY Coordinate tool in Whitebox GAT.

The J48 and Random Forest models were run using each training set, and the predictions of each model on the test set were obtained in a text file. The prediction output file for each model was then imported back into Excel and was queried against the original test area file containing the X and Y coordinates to get the corresponding locational information for each instance. The new file with the predictions and locations for each model was used to create point files in ArcGIS that were later converted to a prediction map for each model in the form of a raster grid.

19.2.2.8 Validation of the Predicted Soil Cluster Maps

A paired-sample t test was used to validate the results. A total of 24 validation points were randomly sampled and georeferenced. The soil samples from the validation points were analyzed for topsoil pH, organic carbon, available phosphorus and particle size distribution in the soil laboratory. A point shape file was created from these verification points and then was overlaid on the predicted soil cluster maps in ArcMap 10.1 (ESRI 2010). Using the identification tool in the software, the corresponding clusters for each verification point were recorded. The predicted topsoil values for each soil cluster were then extracted from the soil clustering output for each validation parameter. Thus, actual value and predicted value for each point were assembled for the predicted maps. The paired-sample t test was used to asses whether there was significant difference in the properties' values.

19.3 Results and Discussion

19.3.1 Ranking of Predictors

Before the data were trained for decision tree analysis, an information gain attribute evaluation was used to select and rank the input training data for the ASTER and SRTM DEM data sets in the WEKA application (Witten et al. 2011). The results showed that, of the 18 training attributes, the legacy map soil units (AU) attribute was highly ranked in the ASTER DEM training data set. The information gain for this attribute was 1.764, while for the other 17 attributes, each had a value of below 1. In the SRTM training set, AU was also highly ranked, again with an information gain value of 1.764. The difference from mean elevation (DFME) attribute was ranked second in the SRTM training set with an information gain value of 0.785. The rest of the attributes (16) in the data set had information gain values of less than 1.

The high ranking of the AU from both data sets suggests that they highly represent features that are result of the factors that influenced soil genesis in the study area. These units were delimited using aerial photograph interpretation. Low ranking of the other attributes suggests that these variables were poorly related to the factors that influenced soil genesis in these areas. Unsurprisingly, satellite image derivatives such as land cover classes, soil enhancement ratios and vegetation indices were not ranked high despite their fine resolution of 5 m. This could be because the major land use type in the study site is agriculture and the reflectance obtained could have just reflected differences in the growth stages of vegetation or land clearances at the time of exposure, not soil variability. For DEM-related derivatives, the explanation could be the scale from which the attributes were derived. For our relatively flat alluvial area, a 30 m resolution may not be the correct scale to capture soil–landscape relationships.

19.3.2 Comparing J48 and RF Predictions

Both training sets were run using the RF and J48 decision tree classifier algorithms to construct their respective models. The observations were not initially split into training and testing data sets for model accuracy testing because of fewer observations. Splitting observations into training and testing sets is a common practice (Tesfa et al. 2009; Subburayalu and Slater 2013), but when there are few observations, splitting may not be done (Brungard et al. 2015). Testing of the model accuracy was instead performed using the training sets. Evaluation of the model built using the ASTER training data set on RF showed that the algorithm classified all 33 instances correctly. In comparison, the J48 algorithm classified 20 instances (about 61 %) correctly. Using the SRTM training data, the model developed by the RF algorithm predicted 32 instances correctly (about 97 %), while the model by J48 predicted 19 instances correctly (about 58 %).

The comparison of the two learners suggests that Random Forest performed better than J48. In a study by Subburayalu and Slater (2013), Random Forest outperformed J48 in the prediction of soil series, and the authors suggested that the learner has a prospect of becoming a useful learning algorithm for digital soil mapping.

19.3.3 Soil Clusters' Prediction Outputs

Results showed that J48 learner predicted eight clusters out of possible 13 clusters. This was for both training data sets, ASTER and SRTM. Out of the eight predicted soil clusters, seven similar clusters were predicted on both data sets and one different for each data set. The predicted clusters are shown in Table 19.1.

The higher prediction rate of Random Forest compared to J48 could be attributed to how differently the two learners operate. Random Forest is an ensemble (forest)

Table 19.1 Soil clusters predicted for each training data set and learner

J48_ASTER	J48_SRTM	RF_ASTER	RF_SRTM
S56	S56	S56	S56
S58	S58	S57	S57
S59	S59	S58	S58
S60	S60	S59	S59
S61	S61	S60	S60
S62	S62	S61	S61
S65	S64	S62	S62
S68	S65	S63	S63
		S64	S64
		S65	S65
		S66	S66
		S67	S67
		S68	S68

Note S56, S57, ….S68 are arbitrary names of numerically classified soil clusters

J48_ASTER means soil clusters predicted by using J48 algorithm on ASTER DEM-based data set of bagged classification trees (typically 500–1000). The classification trees in Random Forest are independent, and the classification of samples does not depend upon previous trees in the ensemble (Kuhn and Johnson 2013). Because of this, the learner performs better in terms of prediction of the majority of soils and showing resistance to extreme changes in the data such as noise (Subburayalu and Slater 2013) compared to J48, which is a single tree classifier.

J48 learner tended to predict soil clusters that have higher constituent membership and leave those with few constituent members. The soil clusters that were not predicted by the learner are those that have either one or two members. The clusters with three or more constituent members were predicted by the J48 algorithm in the SRTM and ASTER data sets. As the number of members in a cluster also corresponded to the number of training sets in that soil cluster, it can therefore be suggested that J48 requires more training data than Random Forest to be able to predict a soil cluster.

19.3.4 Spatial Prediction of Soil Cluster Maps

The dominance of the legacy map soil units were demonstrated in all four predictions: ASTER (RF and J48) and SRTM (RF and J48). In all predictions, the predicted soil clusters generally followed boundaries prescribed by the legacy map soil units. The legacy map soil units are shown in Fig. 19.3. The dominance of the legacy map soil units may suggest the thoroughness of the work done to demarcate the soil units.

The dominance of well-demarcated soil units in decision tree–based predictions has been observed in other studies. For example, in a machine learning—based prediction, detailed soil maps were the strongest predictor of actual biological soil crust in the Canyonland National Park (Brungard and Boettinger 2012). As discussed earlier, the dominance can also suggest the inability of some of the derived attributes to explain the distribution of the soil clusters.

The results showed improvement in details when comparing the predicted map to the legacy soil map (Table 19.2). Soil units that were generalized as single units in the legacy map are shown to contain other soil types as complexes or inclusions in the map predicted by the DSM method. This information is very useful when deciding on site-specific type of crops and management option for climate change resiliency.

19.3.5 Choice of a Better Learner and DEM for Soil Cluster Prediction

As it was observed that J48 had a lower prediction rate compared to RF, and that J48 required more field observations for efficient prediction, we suggest using RF

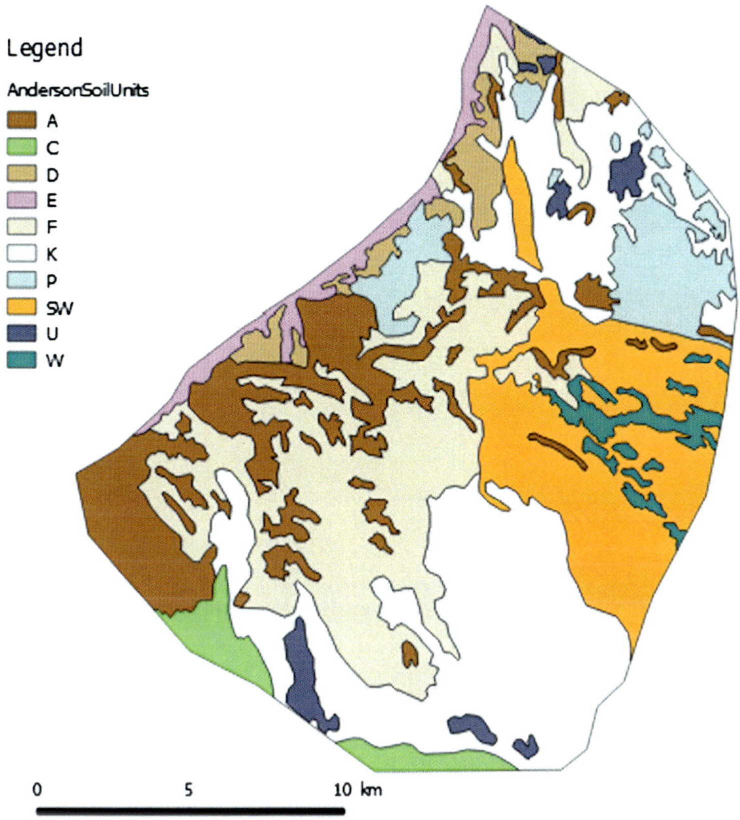

Fig. 19.3 Legacy map soil units boundaries. *Note* The Anderson soil units in the legacy map were described as: *A* sandy loams, including patches of sand, occupies about 15 % of total area. *C* clays affected by weathering since deposition, occupies about 3 % of total area. *D* sandy clays of enclosed basins, occupies about 3 % of total area. *E* soils following foot hill catena, occupies about 3 % of total area. *F* soils formed as a result of flood flow, occupies 22 % of total area. *K* recently deposited clays, subject to seasonal flooding, occupies about 28 % of the area. *P* pale sands without permanent water table and not developed from underlying sediment, occupies 6 % of the total area. *SW* Swamps, occupies about 16 % of the total area. *U* ferruginous sands, occupies about 2 % of the total area. *W* soils formed under ground-water forests, occupies about 2 % of the study area

for the study area and similar areas. Fewer field observations will save time and cost, thus making reliable and updated soil information cheaper and affordable.

The two data sets (ASTER and SRTM DEM derivatives) resulted in the prediction of soil cluster maps that contained almost the same level of detail. However, some differences were observed (Table 19.2). Decisions about which data set should be used are generally based on the clarity shown by the derivatives of the

Table 19.2 Predicted soil clusters (S56–S68) on Anderson's legacy soil map

Legacy map soil unit	Predicted soil clusters			
	SRTM training data set		ASTER training data set	
	J48 learner	Random forest learner	J48 learner	Random forest learner
A	Complex of S60, S58, few pixels of S61	S62 dominates; inclusions: S58, S56	Complex of S62, S58. More of S62	Complex of S62, S58. More of S62
C	Complex of S60, S59	S59	S59	S59
D	S60 dominant; S61, S65 minor components	S56 dominant, S61 minor inclusion	S65	S61 dominant, S65 inclusion
E	S60 dominant; S58, S61 minor component	S58	S58	S58
F	Complex of S59, S60, S63. More S60	Complex of S58, S59, S60, S63	Complex of S59, S60. More of S60	Complex of S57, S59, S63
K	Complex of S60, S63, S65	S65 dominant; S61, S63, S66 minor inclusions	S65 dominant; S59, S63 minor inclusions	S65 dominant; S58, S60, S61 minor inclusions
P	Dominantly complex of S60, S65. Minor inclusions: S58, S61	Dominantly S60. Few pixels: S58, S61, S66, S67	S60	S57 dominant; S 60 minor inclusions
SW	Complex of S60, S61, S63, S65. Comparatively more of S60	Complex of S59, S60. In lower proportions: S61, S63, S67	S65 dominant; few pixels of S61	Complex of S57, S59, S61, S63, S66. Much pixel coherence: S59
U	Complex of S56, S60, S61. Comparatively more of S60	S56	S56	S56
W	Dominantly S60; inclusions: S58, S65	S68 dominant; few pixels of S56	S68	Dominantly S68; few pixels of S58, S62

respective DEM. Visual analysis of the derivatives of the two DEMs showed that the ASTER DEM outputs were noisy for the flat surface of the study area compared to the SRTM outputs. Based on this, we suggest using SRTM as a data source for the environmental correlates needed to predict soil clusters in the study area and similar flat areas. The predicted soil cluster map using the SRTM data set and the RF algorithm is shown in Fig. 19.4.

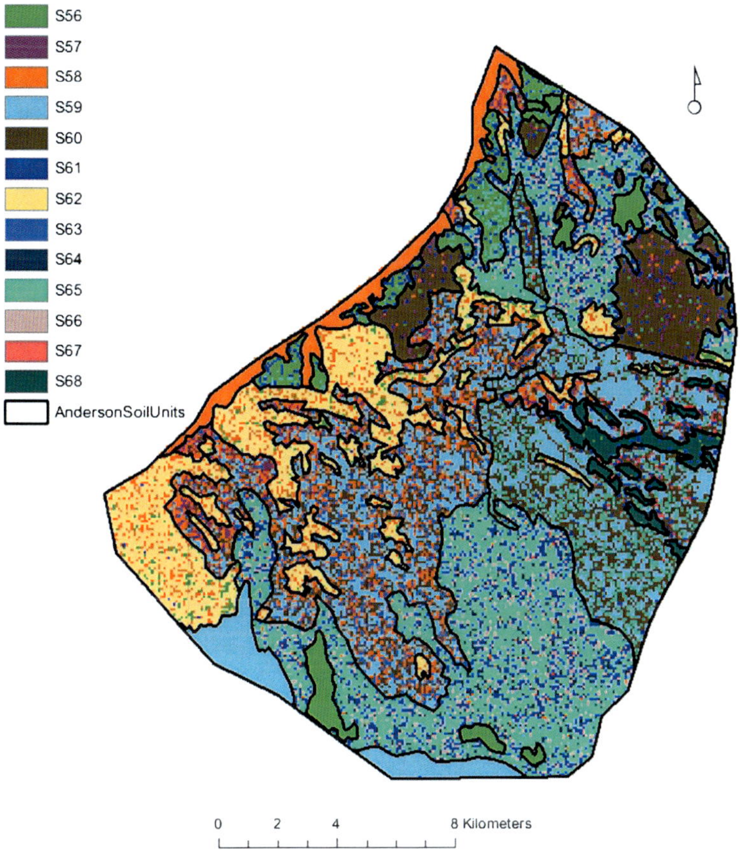

Fig. 19.4 Soil cluster map predicted using Random Forest algorithm on SRTM DEM-derived predictors. *Note* Cluster *S56* is represented by a pedon classified as Fluvaquentic Endoaquepts. It occupies 6 % of the total area. Cluster *S57* is represented by a pedon classified as Fluvaquentic Humaquepts. It occupies 3 % of the total area. Cluster S58 is represented by a pedon classified as Aquic Dystrudepts. It occupies 14 % of the total area. Cluster *S59* is represented by a pedon classified as Aeric Umbric Endoaqualfs. It occupies 16 % of the total area. Cluster *S60* is represented by a pedon classified as Umbric Endoaqualfs. It occupies 15 % of the total area. Cluster *S61* is represented by a pedon classified as Fluvaquentic Humaquepts. It occupies 8 % of the total area. Cluster *S62* is represented by a pedon classified as Fluvaquentic Hapludolls. It occupies 9 % of the total area. Cluster *S63* is represented by a pedon classified as Mollic Fluvaquents. It occupies 3 % of the total area. Cluster *S64* is represented by a pedon classified as Typic Endoaquepts. It occupies 3 % of the total area. Cluster *S65* is represented by a pedon classified as Aquic Udifluvents. It occupies 16 % of the total area. Cluster *S66* is represented by a pedon classified as Typic Endoaquerts. It occupies 4 % of the total area. Cluster *S67* is represented by a pedon classified as Oxyaquic Eutrudepts. It occupies 1 % of the total area. Cluster *S68* is represented by a pedon classified as Aeric Endoaquents. It occupies 2 % of the total area

19.3.6 Validation of Spatial Soil Cluster Prediction

To assess the correctness of the cluster predictions, a paired-sample t test was performed. A total of 24 georeferenced verification points were randomly sampled in the study area. Results show that there was no significant difference between the selected predicted and validation points for Random Forest and J48 learners when using the SRTM DEM data set (Table 19.3). The same results are shown when the ASTER DEM data set was used, except when available phosphorus was used for validation on the Random Forest learner.

However, these validation results should be interpreted with caution. This is because, visually, the four predicted soil cluster maps showed some obvious differences. For example, while RF predicted 13 soil clusters, J48 learner predicted only eight soil clusters for both test data sets.

Likely, sampling more validation points would have been more informative. In addition, although validation used only the topsoil attributes, the numerical clustering process did not take only the topsoil properties into consideration. The metrics involved all profile horizons to assign a soil profile to a given soil cluster. Thus, validation using information from the entire soil profile instead of the topsoil would probably give more informative results.

Table 19.3 T test of mean difference = 0 (vs. not = 0) for actual verses predicted cluster topsoil values for selected verification attributes

Predicting data source	Learner	Verification attribute	*P* value
SRTM	Random forest (RF)	Sand	0.303
		Clay	0.592
	J48	Sand	0.278
		Clay	0.147
		pH	0.834
ASTER	Random forest (RF)	Sand	0.113
		Clay	0.342
		pH	0.958
		Available phosphorus	0.000
	J48	Sand	0.307
		Clay	0.398
		pH	0.121

19.4 Conclusion

This paper assessed the performance of J48 and Random Forest (RF) learners in predicting soil clusters using two data sets, one based on derivatives from the ASTER DEM and the other using the same derivatives but from the SRTM DEM. Soil units from a legacy soil map and derivatives from a RapidEye satellite image were included in each data set.

RF was a better learner compared to J48 for predicting soil clusters in Kilombero Valley by showing a higher prediction rate. It was also found that J48 required more field observations than RF. Since the DSM aims at reducing the costs of soil surveys, the RF learner is preferred to J48. The study suggests the use of RF as a decision tree-based machine learner and predictors derived from the SRTM DEM for predicting soil clusters in Kilombero Valley. It was demonstrated that the DSM was able to add more detail to the legacy soil map units; thus, these methods are suggested in updating legacy soil maps in similar areas to assist farmers and planners in improving agricultural production as a method for dealing with the challenges caused by climate change.

References

Anwar MR, Liu DL, Macadam I, Kelly G (2013) Adapting agriculture to climate change: a review. Theor Appl Climatol 113:225–245

Anwar MR, Liu DL, Farquharson R, Macadam I, Abadi A, Finlayson J, Wang B, Ramilan T (2015) Climate change impacts on phenology and yields of five broadacre crops at four climatologically distinct locations in Australia. Agric Syst 132:133–144

Arslan A, McCarthy N, Lipper L, Asfaw S, Cattaneo A, Kokwe M (2015) Climate smart agriculture? Assessing the adaptation implications in Zambia. J Agric Econ 66(3):753–780

Baker RM (1970) The soils of Tanzania. FAO, Dar es Salaam

Baxter SJ, Crawford DM (2008) Incorporating legacy soil pH databases into digital soil maps. In: Hartemink A, McBratney A, Mendonça-Santos ML (eds) Digital soil mapping with limited data. Springer, Dordrecht, pp 311–318

Bishop TFA, McBratney AB, Laslett GM (1999) Modelling soil attribute depth functions with equal-area quadratic smoothing splines. Geoderma 91:27–45

Bray RH, Kurtz LT (1945) Determination of total, organic and available forms of phosphorus in soils. Soil Sci 59:39–45

Breiman L (2001) Random forests. Mach Learn 45:5–32

Bremner JM, Mulvaney CS (1982) Total nitrogen. In: Page AL, Miller RH, Keeney DR (eds) Methods of soil analysis, part 2; chemical and mineralogical properties, 2nd edn. American Society of Agronomy, Madison, pp 595–624

Brungard CB, Boettinger JL (2012) Spatial prediction of biological soil crust classes; value added DSM from soil survey. In: Minasny B, Malone BP, McBratney A (eds) Digital soil assessments and beyond: proceedings of the 5th global workshop on digital soil mapping. CRC Press, Sydney, pp 57–60

Brungard CW, Boettinger JL, Duniway MC, Wills SA, Edwards TC Jr (2015) Machine learning for predicting soil classes in three semi-arid landscapes. Geoderma 239–240:68–83

Bui EN, Morgan CJ (2001) Disaggregation of polygons of surficial geology and soil maps using spatial modelling and legacy data. Geoderma 103:79–94

Calton WE (1954) An experimental pedological map of Tanganyika. 2nd inter-African soils conference, pp 237–240
Cambule AH, Rossiter DG, Stoorvogel JJ, Smaling EMA (2015) Rescue and renewal of legacy soil resource inventories: a case study of the Limpopo National Park, Mozambique. Catena 125:169–182
Carré F, Jacobson M (2009) Numerical classification of soil profile data using distance metrics. Geoderma 148:336–345
Challinor AJ, Watson J, Lobell DB, Howden SM, Smith DR, Chhetri N (2014) A meta-analysis of crop yield under climate change and adaptation. Nat Clim Change 4:287–291
Cline W (2008) Global warming and agriculture. Financ Dev 45(1):23–27
Cook SE, Jarvis A, Gonzalez JP (2008) A new global demand for digital soil information. In: Hartemink AA, McBratney A, Mendonca-Santos ML (eds) Digital soil mapping with limited data. Springer, Dordrecht, pp 31–41
D'Hoore JL (1964) Soil map of Africa. Commission for technical cooperation in Africa, Lagos, Publication no. 23
de Pauw E (1984) Soil, physiography and agro-ecological zones of Tanzania. Consultants final report for crop monitoring and early warning systems project. KILIMO/FAO, Dar es Salaam
Diday E (1971) Une nouvelle méthode en classification automatique et reconnaissance des formes: la méthode des nuées dynamiques. Revue de Statistique Applique 19:283–300 In French
Eschweiler JA (1998) SOTER database—Tanzania. FAO, Rome
ESRI (2010) ArcGIS—a complete integrated system. Environmental Systems Research Institute Inc, Redlands
FAO (1961) The Rufiji basin Tanganyika. FAO Exp. Techn. Ass. Progr. No. 1269, vol 7, Rome
FAO (2006) Guidelines for soil description. FAO, Rome
Fridland VM (1974) Structure of the soil mantle. Geoderma 12:35–41
Gallant JC, Wilson JP (2000) Primary topographic attributes. In: Wilson JP, Gallant JC (eds) Terrain analysis: principles and applications. Wiley, New York, pp 51–86
Gee GW, Bauder JW (1986) Particle-size analysis. In: Klute A (ed) Methods of soil analysis: part 1—physical and mineralogical methods. SSSA book series 5.1 soil science society of America. American Society of Agronomy, Madison, pp 383–411
Hathout SA (1983) Soil atlas of Tanzania. Tanzania Publishing House, Dar es Salaam
Heuvelink GBM, Webster R (2001) Modelling soil variation: past, present, and future. Geoderma 100:269–301
Horn BKP (1981) Hill shading and the reflectance map. Proc IEEE 69(1):14–47
IPCC (2014) Summary for policy makers (SPM)—fifth assessment report (AR5). Working group (WG) II: impacts, adaptation and vulnerability. Intergovernmental Panel on Climate Change
Jacobson M, Carré F (2006) OSACA version 1.0 Land Management and Natural Hazards Unit, Institute for Environment and Sustainability. European Commission, Italy
Jenny H (1941) Factors of soil formation: a system of quantitative pedology. McGraw-Hill, New York
Kato F (2007) Development of a major rice cultivation area in the Kilombero Valley, Tanzania. Afr Study Monogr Supplement 36:3–18
Kempen B, Brus DJ, Heuvelink GBM, Stoorvoge JJ (2009) Updating the 1:50,000 Dutch soil map using legacy soil data: a multinomial logistic regression approach. Geoderma 151:311–326
Kuhn M, Johnson K (2013) Applied predictive modeling. Springer, New York
Lindsay WL, Norvell WA (1978) Development of a DTPA soil test for zinc, iron, manganese, and copper. Soil Sci Soc Am J 42:421–428
Malone BP, McBratney AB, Minasny B, Laslett GM (2009) Mapping continuous depth functions of soil carbon storage and available water capacity. Geoderma 154:138–152
Malone BP, Minasny B, Odgers NP, McBratney AB (2014) Using model averaging to combine soil property rasters from legacy soil maps and from point data. Geoderma 232–234:34–44
McBratney AB, Odeh IOA (1997) Application of fuzzy sets in soil science: fuzzy logic, fuzzy measurements and fuzzy decisions. Geoderma 77:85–113

McBratney AB, Mendonca-Santos ML, Minasny B (2003) On digital soil mapping. Geoderma 117:3–52

McBratney A, Field DJ, Koch A (2014) The dimensions of soil security. Geoderma 213:203–213

McLean EO (1986) Soil pH and lime requirement. In: Page AL, Miller RH, Keeney DR (eds) Methods of soil analysis, part 2; chemical and mineralogical properties, 2nd edn. American Society of Agronomy, Madison, pp 199–223

Milne G (1936) A reconnaissance journey through parts of Tanganyika Territory. J Ecol 35:192–265

Msanya BM, Magogo JP, Otsuka H (2002) Development of soil surveys in Tanzania. Pedologist 46(1):79–88

Nelson DW, Sommers LE (1982) Total carbon, organic carbon and organic matter. In: Page AL, Miller RH, Keeney DR (eds) Methods of soil analysis, part 2; chemical and mineralogical properties, 2nd edn. American Society of Agronomy, Madison, pp 539–579

Palm C, Sanchez P, Ahamed S, Awiti A (2007) Soils: a contemporary perspective. Annu Rev Environ Resour 32:99–129

Planchon O, Darboux F (2001) A fast, simple and versatile algorithm to fill the depressions of digital elevation models. Catena 46:159–176

Quinlan JR (1993) C4.5: programs for machine learning. Morgan Kaufmann, San Mateo

Rhoades JD (1982) Soluble salts. In: Page AL, Miller RH, Keeney DR (eds) Methods of soil analysis. Part 2, chemical and mineralogical properties, 2nd edn. American Society of Agronomy, Madison, pp 149–158, 167–180

Rogelj J, Hare W, Chen C, Meinshausen M (2011) Discrepancies in historical emissions point to a wider 2020 gap between 2 C benchmarks and aggregated national mitigation pledges. Environ Res Lett 6:1–9

Samki JK (1977) Provisional soils map of Tanzania. Survey and Mapping Division, Dar es Salaam

Samki JK (1982) Soils map of Tanzania. National Soil Service, Tanga

Schollenberger CJ, Simon RH (1945) Determination of exchange capacity and exchangeable bases in soils-ammonium acetate method. Soil Sci 59:13–24

Scott RM (1962) Soils of East Africa. In: Russel EW (ed) Natural resources of East Africa. East African Literature Bureau, Nairobi, pp 67–76

Scull P, Franklin J, Chadwick OA, McArthur D (2003) Predictive soil mapping: a review. Prog Phys Geogr 27:171–197

Subburayalu SK, Slater BK (2013) Soil series mapping by knowledge discovery from an Ohio county soil map. Soil Sci Soc Am J 77:1254–1268

Sulaeman Y, Minasny B, McBratney AB, Sarwani M, Sutandi A (2013) Harmonizing legacy soil data for digital soil mapping in Indonesia. Geoderma 192:77–85

Tarboton DG (1997) A new method for the determination of flow directions and upslope areas in grid digital elevation models. Water Resour Res 33(2):309–319

Tesfa TK, Tarboton DG, Chandler DG, McNamara JP (2009) Modeling soil depth from topographic and land cover attributes. Water Resour Res 45:W10438. doi:10.1029/2008WR007474

Thomas GW (1982) Exchangeable cations. In: Page AL, Miller RH, Keeney DR (eds) Methods of soil analysis. Part 2, chemical and mineralogical properties, 2nd edn. American Society of Agronomy, Madison, pp 595–624

U.S. Geological Society (2000) Shuttle radar topography mission, 1 Arc Second scene SRTM. http://earthexplorer.usgs.gov/. Accessed 24 Oct 2014

Vaysse K, Lagacherie P (2015) Evaluating digital soil mapping approaches for mapping GlobalSoilMap soil properties from legacy data in Languedoc-Roussillon (France). Geod Reg 4:20–30

Watanabe FS, Olsen SR (1965) Test of ascorbic acid method for determining phosphorus in water and $NaHCO_3$ extracts from soil. Soil Sci Soc Am J 29:677–678

Wickama JMW (1997) Pedological investigation and characterization in Kitanda Village, Mbinga District, Tanzania. M.Sc. Thesis, Sokoine University of Agriculture, Morogoro, Tanzania

Witten IH, Frank E, Hall MA (2011) Data mining: practical machine learning tools and techniques. Morgan Kaufmann, Burlington

You D, Hug L, Anthony D (2014) Generation 2030—Africa: child demographics in Africa. UNICEF, New York

Zinyengere N, Crespo O, Hachigonta S (2013) Crop response to climate change in southern Africa: a comprehensive review. Glob Planet Change 111:118–126

Chapter 20
Measuring Agricultural Sustainability in Agroforestry Systems

P.K. Ramachandran Nair and Gregory G. Toth

Abstract Sustainability is an intuitively understandable but difficult-to-measure concept. Despite numerous efforts over the years to measure and integrate the ecological, economic, and social aspects of sustainability, a set of universally acceptable standards for measuring sustainability does not exist. The prevailing ecology–economy conflict, in which ecologists consider economics as a subset of environment, while economists view the environment and its benefits as part of the economy, adds to the difficulty. Agroforestry systems (AFSs), considered paradigms of sustainability, are faced with these difficulties when it comes to measuring and comparing various AFSs with one another or with other land-use systems. In ecological terms, the best criteria and indicators of AFS sustainability are ecosystem services, such as soil-fertility improvement, climate-change mitigation through carbon sequestration, and biodiversity conservation. As an example of the variability of one of these measures across studies, estimates of carbon (C) stored in AFSs range from 30 to 300 Mg C ha^{-1} up to 1 m soil depth; additionally, 0.29–15.21 Mg C ha^{-1} $year^{-1}$ is estimated to be accumulated in aboveground biomass although most of it may not contribute to long-term C storage. In terms of economic sustainability, the principles and procedures of ecological economics and valuation of ecosystem services are useful approaches. Measurement of social sustainability, perhaps more challenging than measurement of the ecological and economic components, entails assessment of such social factors as policy, culture, and other socioeconomic indicators; a single measure of the combined manifestation of all these indicators is the adoption of improved practices by targeted land users. Standard procedures are available for measuring many of these indicators; however, most of them entail measurements taken over relatively long periods of time. Even if measurements and assessments are made rigorously, the ultimate benefit will depend on how sustainability is perceived and valued at all levels, from land users to national and international policy makers.

P.K.R. Nair (✉) · G.G. Toth
School of Forest Resources and Conservation, IFAS, University of Florida, Gainesville, FL 32611, USA
e-mail: pknair@ufl.edu

R. Lal et al. (eds.), *Climate Change and Multi-Dimensional Sustainability in African Agriculture*, DOI 10.1007/978-3-319-41238-2_20

Keywords Adoption · Biodiversity · Carbon sequestration · Ecological economics · Ecology–economy conflict · Ecosystem services · Policy · Nitrogen-fixing trees · Social values · Soil improvement

20.1 Introduction

At the outset, we acknowledge that the title of this chapter is somewhat presumptuous, because in a physical sense, the "measurement" of an entity implies that it can be measured and expressed in precise quantitative terms. But is sustainability measurable? Admittedly, this question has been asked many times before. The difficulty of measuring and expressing the ecological, economic, and social components of sustainability and thus capturing its scientific complexity has been taken as a challenge by various groups of academics, and numerous approaches have been suggested for measuring each of the three principal components of sustainability. These efforts have also contributed to the development of a new branch of science, appropriately termed "sustainability science." In this chapter, we focus on these developments in relation to measuring sustainability in agroforestry systems (AFSs) in developing countries.

20.2 Sustainability

Although the word "sustainable" has been used in European languages since the early Middle Ages, it was with the publication of the United Nations report *Our Common Future* (WCED 1987) that it was introduced into the agricultural and broader developmental vocabulary and became a commonly used term (de Vries 2012). In spite of the numerous definitions and explanations that have been proposed since that time, the World Commission on Environment and Development (WCED) definition of sustainability still encapsulates the concept and continues to be widely used: "meeting the needs of the present generation without compromising the ability of future generations to meet their own needs" (WCED 1987). To help policy makers decide what actions should be taken to make society sustainable, assessments of sustainability provide them with evaluations of integrated nature–society systems at both global and local scales and from both short- and long-term perspectives (Ness et al. 2007).

One of the oldest and most common meanings of the verb "to sustain" is to keep a person, a community, or the spirit from failing or giving way, to keep it at the proper level or standard. An English equivalent of this verb is "to last," meaning to go on existing or to continue. Thus, the concept of sustainable development has traditionally been framed as the balancing of this objective of preservation with economic advancement and well-being, acknowledging that economic advancement typically comes at a cost to the environment (MEA 2005). These contradictions manifest themselves in the apt term "ecology–economy divide."

20.2.1 *The Ecology–Economy Divide*

From the ecologist's perspective, the economy is a subset of the environment; all economic activity, indeed life, depends on the Earth's ecosystem. Inherent in this view is a realization of limits, often described in terms of the first two laws of thermodynamics ("conservation of matter and energy" and "entropy increases") and exemplified with phenomena such as energy flows through the food chain. Accordingly, this view recognizes resources as finite: water, gases, nutrients, and the cycles thereof that keep us alive are bound by constraints, which means that takings beyond regenerative capabilities equate to future deficits (Weiss 1992). Upon these grounds, ecologists call for "intergenerational equity," seeking to protect nature and natural resources for the benefit of future generations.

Traditional economists, on the other hand, view the environment and its benefits as part of the economy. This has major ramifications. For example, benefits derived from the environment are considered infinite and substitutable (Neumayer 2000); this translates into the belief that future generations are not affected by current environmental degradation and into the present-day undervaluation and/or degradation of natural resources. The traditional economic view reflects an adherence to an outdated model that defines the environment in terms of its potential for development and fails to internalize externalities. Externalities are market imperfections resulting from impacts of production, extraction, or consumption processes that typically affect third parties and are not compensated. Although these impacts can be positive, negative outcomes are equally possible and are unrecognized in the costs of the transaction. A prime example of a negative externality is biodiversity loss resulting from agricultural development.

20.2.1.1 Historical Examples of the Ecological Cost of Development

Food shortages caused by environmental destruction undermined several ancient civilizations to the point of collapse. Most of these declines can be traced to one or two damaging environmental trends. During the Sumerian civilization (which occupied a region in the lower valley of the Euphrates River in the Near East, fifth to third millennium BCE), rising salt levels in soils due to a flaw in the irrigation practices led to crop failures. In the Mayan Empire (Mexico, 2000 BCE to 600 CE), forest clearing led to soil erosion and loss of soil fertility. These examples illustrate how settlement and agricultural development that fail to account for environmental degradation can contribute to societal failure. Initially, as agriculture advanced, more people were fed and human survival rates increased. However, with increased survival rates, demand for food grew and was met with further agricultural expansion—at increasing expense to the surrounding environment. This expense came in various forms.

In Sumer, where wheat and barley were grown under heavy irrigation, high temperatures and an overfed water table led to a soil-salinity level beyond the tolerance threshold of these primary crops. Water-table elevation raises salt from throughout the soil profile up to the zone occupied by plant roots, where it can be further concentrated by evaporation. Given that desalination is a long process that was likely not understood at the time, little could be done to counteract the effect: a drop in crop yields of 42 % between 2400 and 2100 BCE and a continued loss of up to 65 % by 1700 BCE (Ponting 2007). Once this tipping point was reached, starvation quickly destabilized and ultimately led to the demise of Sumerian society.

Likewise, as the Mayan Empire sought arable land and fuelwood, it decimated the productivity of its soils through ever-expanding deforestation. Losses of tree cover led to increased erosion susceptibility, especially given the mountainous terrain of what is now Guatemala. Tree cover prevents erosion through several means, most prominently the anchoring of soil by roots and the reduction of rain and wind exposure (Khalilnejad et al. 2012). Moreover, decomposition of leaf litter and other senescent parts of vegetation helps to replenish soil. When these environmental benefits were eliminated through deforestation, malnutrition quickly ensued, leaving the society weak and increasing warfare over limited resources (Turner and Sabloff 2012). Whereas the Maya and many other ancient civilizations were encumbered by relatively few damaging environmental trends, today we have to deal with several.

20.2.1.2 Costly Side Effects of the Green Revolution

The Green Revolution of the 1970s to the early 2000s, brought about by technological advances in plant genetics, pesticides, and fertilizers, produced increases in crop production that helped eradicate large-scale hunger in many parts of the developing world. Moreover, like the settlement phase of early civilizations, it was accompanied by a corresponding population boom (Ehrlich and Daily 1993). Although the resultant increase in well-being was good for the contemporary population, the accompanying environmental degradation has led to concerns about intergenerational equity (Daily and Ehrlich 1996). To name just a few of the detrimental effects of agricultural expansion, there are shrinking forests, eroding soils, deteriorating rangelands, expanding deserts, rising atmospheric carbon dioxide, unpredictable water-table fluctuations, melting glaciers, and rising sea levels, each alarming in its own right (or weight). Although the Green Revolution contributed greatly to development, some of its methods are now clearly understood to be unsustainable.

20.2.1.3 Ecological and Environmental Economics

One of the ways in which traditional economics has attempted to adjust to the difficulties associated with the multidimensionality of development contexts is to classify the perspectives from which sustainability is measured. That is, sustainability is categorized into two forms: strong and weak sustainability. Those who perform assessments from the perspective of strong sustainability essentially give greater credence to the centrality of natural capital in the context of development. Natural capital is the totality of nature (resources, plants, and ecosystems usable by the Earth's inhabitants), one of the forms of capital typically distinguished by economists, the other forms of capital being social, man-made, and human capital. Strong sustainability maintains that there are no substitutes for this natural capital (Davies 2013). Weak sustainability, on the other hand, is the belief that such assets can be replaced with man-made capital with minimal ramifications.

As noted, when traditional economics, with its focus on cost efficiency, is applied to environmental concerns, it fails to account for the holistic essence of natural capital (Gasparatos et al. 2008). The subfield of environmental economics attempts to account for this shortcoming by valuing natural resources through contingent valuation and hedonic pricing methods, but it maintains a cost-efficiency focus. Contingent valuation is a mainstay of traditional economics that has been applied to environmental concerns in situations where it is difficult to observe behavior directly, such as in "non-use" or public goods, like the existence of a park or water quality. "Stated preference" is the core concept of contingent valuation. Such techniques work well in a developed context to price nature for consumption, but they do not take into account continued reliance on the resource or its role in producing other benefits within an ecosystem (weak sustainability). It is left to regulation and enforcement to address this "market failure," but because these are difficult to implement in a rural-development context, they have failed at all levels.

Ecological economics, alternatively, recognizes the finiteness and irreplaceability of nature (strong sustainability). This translates into greater recognition of natural capital and ultimately higher valuation of ecosystem services. Ecological economics, although still criticized for monetizing nature, does a better job than conventional economics of recognizing the importance of sustainability and attempts to account for externalities. Most criticism of ecological economics is based on the idea that the monetization of a particular ecosystem function will lead to the exploitation or abandonment of corresponding natural elements based on market shifts (Gómez-Baggethun et al. 2010). However, this criticism fails to consider the holistic nature of ecological economics in that any one element undoubtedly plays several roles within ecosystems and cannot be fully valued based on only one use, as elaborated in the next paragraph; many interactions are beyond our current comprehension, calling for greater application of the precautionary principle (UNESCO 2005).

A prime example of situations where caution is warranted is ecosystem services. An AFS, for example, provides a habitat for pollinators, water purification, and carbon sequestration, among many other benefits (Sect. 20.3.2). A technological advance may remove the need for the water-purification service of a particular system, but the other benefits would still be needed. Removal of the trees/AFS would not only damage these functions but also result in other ecological costs (some of which we may not yet understand). However, a noteworthy feature of the criticism of ecological economics is that although the latter recognizes the importance of social systems (as opposed to the individual rational actor at the center of traditional economics), it is not well suited to the sociological aspects of rural-agriculture-based development, because its focus continues to be on monetization, not interdependency, and it takes for granted the roles of property rights and other public policies. Interactions and conflicts, therefore, are realms in which ecological economics may fall short when human and animal habitats overlap, indigenous rights and protected species meet, or two purported sustainability efforts interfere with one another. Sometimes the interests of both parties coincide, but sometimes the interests of one must give way (perhaps allowing a further analogy with thermodynamics: there are elastic and inelastic collisions). For example, those supporting an ecotourism model for sustainable development in Costa Rica are in conflict with those in the country seeking clean energy from hydroelectric dams (Fletcher 2011).

Answers to these questions may be difficult to find. Indeed, there is no "correct" answer; there are only differing levels of willingness to both sacrifice and capitalize on present resources and different perspectives on what should be left for later generations. Identifying the best means of accomplishing such goals, if in fact they can be agreed upon, awaits further debate. The application of environmental and ecological economics to such problems often ends in stalemates. However, the inability (or inflexibility) of these varying forms of economics to find common ground on these issues led to the evolution of the concept of dynamism in what has become sustainability science (Weinstein et al. 2012).

20.2.1.4 Sustainability Science

Sustainability science can perhaps be viewed as an extension of ecological economics. It deals with the interactions between natural and social systems (institutions) and the measurement thereof, and it is especially significant for developing countries, whose inhabitants seek to improve their well-being. Numerous authors have suggested that the failure to agree on a collective vision of how to attain sustainability lies in the limitations and disconnections among disciplines (Kaufman and Cleveland 1995). The emerging field of sustainability science is not confined to the borders of traditional disciplines, but draws from sociology, ecology, and economics, among other disciplines, allowing for a dynamic approach to meeting the "needs of present and future generations while substantially reducing poverty and conserving the planet's life support systems" (PNAS 2015). Sustainability science arose from the realization that sustainable development is an aspiration to improve quality of life (development) in an enduring (sustainable) manner and that it can be accomplished only by acting across several scales of time and space; that is, it is a transdisciplinary approach that integrates and synthesizes the theory and practice of these quantitative (natural) and qualitative (social) aspects (de Vries 2012).

The salient characteristics of sustainability science are that it is use inspired and place based, hierarchical, and multidimensional and transdisciplinary (Wu 2012); it does not seek a broadly applicable "correct" decision. As the historical examples above illustrate, the ineffectual balancing of economy and environment can have disastrous results. This makes sustainability science—specifically, sustainable agriculture and the corresponding measurement of that sustainability—especially important.

20.3 Agroforestry and Ecosystem Services

20.3.1 Agroforestry

Over the past 35 years, agroforestry has been transformed from a vague concept into a robust, science-based, land-use discipline. Today, agroforestry is at the forefront of numerous development agendas, particularly in developing countries (Garrity 2012). The potential of agroforestry to sustain crop yields, diversify farm production, and provide ecosystem services has been well demonstrated in both the scientific literature and practical applications.

Various forms of agroforestry systems, such as silvopasture, intercropping, shaded perennials, riparian buffers, and forest farming, to name a few, are estimated to be practiced on over 1.6 billion ha globally (Nair 2012a, 2014). The underlying concept of the various forms of agroforestry is the beneficial role of on-farm and off-farm tree production in providing numerous products and services to support sustainable land-use and natural-resource management. Whereas the aboveground and belowground diversity provides more stability and resilience for the system at the site level, the system provides connectivity with forests and other landscape features at the landscape and watershed levels. These systems provide the ecosystem services and life-supporting functions of nutrient cycling, water-quality enhancement, and, in a self-perpetuating fashion, continued biological diversity (Hammond et al. 1995), which are recognized for their relevance in agriculture, biodiversity conservation, and natural-resource management (Heimlich 2003) as well as in food security, medical inputs, infectious-disease regulation, and climate-change mitigation (COHAB 2010). Although these functions interact with and promote one another, they can be categorized into the primary scales at which they operate: local (soil-productivity improvement), landscape (water-quality enhancement), regional (biodiversity conservation), and global (climate-change mitigation). The biophysical and ecological measurement of the sustainability of the systems will, therefore, depend on how each of these ecosystem services can be measured and quantified at various spatial levels (plot/farm → watershed → regional → global).

20.3.2 Major Ecosystem Services of Agroforestry

20.3.2.1 Soil Improvement

One of the tree-mediated benefits of considerable advantage in the tropics is that trees and other vegetation improve the productivity of the soil beneath them. Over the past three decades, research results have shown that three main tree-mediated processes determine the extent and rate of soil improvement in agroforestry systems: (1) increased nitrogen (N) input from nitrogen-fixing trees (NFTs) trees, (2) enhanced availability of nutrients resulting from production and decomposition of tree biomass, and (3) greater uptake and utilization of nutrients from deeper layers of soils by trees, the roots of which extend much deeper into the soil than roots of common crops.

Nitrogen-fixing trees and other "fertilizer trees" are a valuable resource in agroforestry systems. Farmland in many parts of the developing world generally suffers from the continuous depletion of nutrients, because farmers often harvest without fertilizing adequately or fallowing the land. One promising method for overcoming the acute problem of the low-nutrient status of soils, such as African soils in general, is to enable smallholders to use fertilizer-tree systems that increase on-farm food production. Nitrogen-fixing trees and shrubs, a large number of which are available (Table 20.1), are interplanted with food crops, the trees and shrubs are pruned periodically, and the biomass is added to the crops. The nitrogen-rich biomass decomposes rapidly, making the mineralized N and other nutrients available to the growing crop (Fig. 20.1). Additionally, the atmospheric N fixed by NFTs becomes available in the soil. Numerous estimates are available on the extent to which N is fixed by different NFTs under different conditions (Dubeux et al. 2015). Some widely held assumptions about their benefits could, however, be wrong or incomplete. Because of methodological difficulties in quantifying N_2 fixation, especially in older trees, our understanding of the extent of N_2 fixation, and therefore of the benefit that is actually realized by using NFTs in agroforestry systems, is unsatisfactory. Furthermore, it is not clearly understood how much of the N_2 fixed by an NFT is actually utilized or potentially made available to an associated crop during its growth cycle and how much goes into the soil's N store for eventual use by subsequent crops.

Biomass-decomposition patterns and therefore nutrient-release patterns from the decomposing biomass vary greatly among agroforestry tree species. Several biomass (litter)-quality parameters, based on the chemical composition of plant tissues,

Table 20.1 Biological nitrogen fixation: the family Leguminosae (Fabaceae) includes several, mostly tropical, N_2-fixing woody shrubs and trees

Subfamily	Genera ([a]) Number of species	N_2 fixation % N_2 fixers	Examples of common genera
Papilionoideae (T, S, H, C)	677 ([a]165) 12,000 spp.	High 90 (%)	*Erythrina, Flemingia Gliricidia, Sesbania*
Mimosoideae (T, S; tropical)	66 ([a]15) 2800 spp.	High to moderate 90 (%)	*Acacia, Calliandra, Leucaena, Prosopis*
Caesalpinioideae	256 ([a]84) 2800 spp.	Low 35 (%)	*Bauhinia, Parkinsonia, Tamerindus*

Source Compiled from various sources
Note The amount of N fixed by different species will vary widely depending on a number of factors, such as plant characteristics (species and age of plant), soil and climatic factors, and management issues (plant density and arrangement). Moreover, the amount reported will vary according to the method of estimation. Therefore, it is unrealistic and misleading to give estimates of nitrogen fixation under field conditions
T tree, *S* shrub, *H* herb, *C* climber
[a]Numbers in parentheses indicate genera not examined for N_2 fixation

Fig. 20.1 Fertilizer trees: fast-growing, nitrogen-fixing shrubs and trees, grown in association with agricultural crops, are pruned periodically; the succulent and easily decomposable tree biomass is returned to the cropped alleys as a source of nutrient for crops. Photo shows *Gliricidia sepium* grown with maize (*Zea mays*), a practice followed by many farmers in Eastern and Southern Africa. *Photo credit* ICRAF, Nairobi, Kenya

have been developed to interpret these patterns: ratios of C to N, polyphenols to N, lignin to N, and (polyphenols + lignin) to N. Using this information, management strategies can be developed to manipulate the decomposition of plant biomass in AFSs, thereby regulating the rates of nutrient release in the short term and, in the long term, improving soil fertility via improved soil-organic-matter status (Nair et al. 1999; Palm et al. 2001). Roots of the crops and NFTs also contribute biomass build-up in AFSs. Our knowledge of the dynamics of belowground biomass in AFSs, however, is much poorer than that of the dynamics of aboveground biomass.

Soil conservation is another major avenue of soil improvement in agroforestry. When properly designed and managed, agroforestry techniques can contribute to reducing water erosion and wind erosion and enhancing soil productivity (Fig. 20.2). Furthermore, under agroforestry, the presence of deep-rooted trees in the system can contribute to improved soil physical conditions and higher soil microbiological activities. About 2 billion ha of land—a third of total farmland—in developing nations are estimated to be degraded through erosion, salinity, and

Fig. 20.2 Soil conservation: contour hedgerows of trees and shrubs planted across slopes help arrest soil erosion in gently sloping lands. Depending on the trees and shrubs used, they could provide various products, such as nutrient-rich biomass, fodder for animals, fruits, and small timber. Photo shows hedgerows of *Leucaena leucocephala* in a maturing cowpea (*Vigna unguiculata*) field in Ibadan, Nigeria. *Photo credit* B.T. Kang, IITA, Nigeria (deceased)

fertility depletion (UNEP 2004). The potential of agroforestry to reduce the hazards of erosion and desertification as well as to rehabilitate such degraded land and to conserve soil and water has been widely recognized. The soil ameliorative potential of agroforestry systems has been demonstrated in the temperate zone as well (Schoeneberger et al. 2015).

20.3.2.2 Water-Quality Enhancement

The so-called safety-net effect of tree roots—the ability of deep-rooted trees to absorb nutrients that have leached below the rooting zone of agronomic crops, recycle them via leaf litter and fine-root turnover, and thus improve nutrient-use efficiency in the system as a whole—could have an important application in the heavily fertilized, sandy soils that have low nutrient-retention capacities. The capacity of tree roots to capture nutrients from the deeper soil horizons can enhance nutrient storage in the plant-soil system and thereby reduce the amount of nutrients that might otherwise be transported to ground and surface water through runoff and leaching. Research over the past decade has shown that riparian forest buffers can remove significant amounts of sediment, nutrients, and pesticides from both surface and subsurface waters and thus reduce the non-point-source pollution of water bodies in industrialized regions (Jose et al. 2012).

20.3.2.3 Biodiversity Conservation

The number and diversity of trees and shrubs present in AFSs help increase the ecosystem's "hospitality" to a greater number of organisms, such as pollinators, decomposers, herbivores, predators, and pathogens, both above- and belowground, thereby improving the efficiency and functionality of ecosystem services and food chains. For example, in a 7-year experiment, Zak et al. (2003) found that greater species diversity increased plant production by increasing biomass and modifying the composition of soil microbial communities. In combination with the trees themselves, these ecosystem services help promote the hydrological services of water filtration/purification, habitat preservation, seasonal flow regulation, and sediment and erosion prevention (Daily et al. 2001). In a unique experiment to determine the influence of agroforestry practices on biodiversity in an agricultural mosaic, Francesconi et al. (2013) studied the distribution of fruit-feeding butterflies in six different land-use systems in two agricultural landscapes in Central-West Brazil. They found that shaded coffee practices that represent long-term mixed tree-and-crop stands had better potential for conserving forest butterfly species compared to monoculture practices.

In addition to maintaining a healthier and biodiverse ecosystem, mixed-species AFSs could provide greater landscape connectivity in areas where landscapes are increasingly being fragmented and remaining patches of natural vegetation are reduced to isolated habitat islands. This can occur in at least three ways: (1) an intensification of AFSs that leads to reduced exploitation of protected areas, (2) increasing biodiversity in working landscapes through the expansion of AFSs into traditional farmlands, and (3) increasing the species diversity of trees in farming systems (Nair 2013). Where croplands occupy most of the landscape, riparian forest buffers and field shelterbelts can be essential for maintaining plant and animal biodiversity, especially under a changing-climate scenario. The trade-offs between ecosystem conservation and agricultural production can convincingly be addressed by shifting the focus from the plot scale to the landscape scale and integrating biodiversity-friendly land-use systems such as agroforestry into development strategies.

20.3.2.4 Carbon Sequestration and Climate-Change Mitigation

Agroforestry systems are perceived to have higher potential to sequester carbon (C) than comparable single-species crop or pasture systems. The underlying premise of this perception is the niche complementarity hypothesis, which states that a larger array of species in a system leads to a broader spectrum of resource utilization, which in turn makes the system more productive (Tilman et al. 1997); this hypothesis implies that plant species in a mixed system use resources in a complementary way. The estimates of carbon (C) stored in AFSs range from 30 to 300 Mg C ha^{-1} up to 1 m soil depth; additionally, 0.29–15.21 Mg C ha^{-1} $year^{-1}$ is estimated to be accumulated in aboveground biomass although most of it may not contribute to long-term C storage (Nair et al. 2010). Recent studies under various AFSs in diverse ecological conditions have shown that tree-based agricultural systems, compared to treeless systems, stored more C in deeper soil layers near the tree than away from the tree, and higher soil organic C content was associated with higher species richness and tree density. Furthermore, C3 plants (trees) have been found to contribute to more C in the silt + clay fractions (<53 μm diameter) that constitute more stable C than C4 plants (such as maize—*Zea mays*—and some other warm season grasses), in deeper soil profiles (Nair 2012a). The amount of C sequestered in an AFS depends to a great extent on environmental conditions and system management. Based on a comprehensive literature search, Nair and Nair (2014) estimated carbon sequestration rates for the different AFSs, as summarized in Table 20.2.

These are just a few examples of the ecosystem services provided by trees in general and AFSs in particular, on which some research data are available. Several other benefits have also been mentioned in the extant literature, and a great deal of undocumented information concerning such ecosystem services is said to exist in so-called indigenous/traditional knowledge.

Table 20.2 Estimates of carbon stocks and carbon-sequestration potential (CSP) in agroforestry systems in different ecological regions of the tropics

Agroforestry system subgroup	Major AF practices	Estimated C stock range (Mg ha^{-1})		CSP in new areas (Mg ha^{-1} $year^{-1}$)	
		Aboveground	Belowground (esp. soils)	Aboveground	Belowground (soils)
Multistrata systems	Home gardens	2–18	Up to 200	0.5–3	1.5–3.5
	Shaded perennials	5–15	Up to 300	1–4	1.0–5
Tree intercropping	Alleycropping	Up to 15	Very low to 150	0.5–4	1.5–3.5
	Multipurpose trees on farmlands	Up to 12	Very low to 150	0.2–2.5	1.5–3.5
Silvopasture	Tree fodder (Browsing, cut-and-carry)	1.8–3	Very low to 80	0.3–4	1.0–2.5
	Grazing under trees	1.5–8	Very low to 60	0.3–2	0.4–1.0
Protective systems	Windbreaks, shelterbelts, soil-conservation hedges, boundary planting	1.5–7	Very low to 60	0.7–2	0.4–1.0
Agroforestry tree woodlots	Woodlots for firewood, fodder, land reclamation	1.5–7	Very low to 60	1–5	1.0–6.0

Source Nair (2012a, b)

20.4 Policy and Institutional Aspects of Sustainability

20.4.1 Institutional Influence on Sustainability

Institutions are systems of established and prevalent social rules that structure social interactions (Hodgson 2006) and are formed through an iterative process involving a network of culture, policy, and socioeconomic factors (Holland 2007). As a network, all these factors influence one another as they interact to create an institutional environment. The resultant environment is in constant flux, and it serves as both a resource for (and constraint on) behavior in that it can mobilize information, social influence, resources, and social capital in highly differentiated ways (Ansell and Gash 2007). As would be expected, this flux situation influences the sustainability of agriculture; most importantly, it influences the perceived importance of ecosystem services and the incentives and ability to adopt the practices, such as agroforestry, that make possible the continued provision of those services.

Given that the biophysical underpinnings and impacts of agroforestry are well established, some may consider its adoption a measure of its perceived utility. Unfortunately, this alone does not translate into a measure of sustainability, because the financial and environmental incentives perceived by individual adopters cannot be distilled without closer scrutiny. Therefore, a large part of the sociological focus in the field currently revolves around determining perceived detriments and advantages to adoption—factors primarily controlled by the interaction of institutional influences and surrounding biophysical systems (Norgaard 1981). In ecological economic terms, technology availability and institutional structure determine the usefulness of any resource (Bromley 1991). Perceptions of practicality, in turn, are closely related to the knowledge potential adopters have about a technology (Meijer et al. 2014, 2015). The result of this causal chain is that culture, policy, and socioeconomic conditions are extensively explored in studies of technology (i.e., agroforestry) adoption and can be used to assess the environmental benefits of agroforestry in sustainable agriculture based on previous outcomes.

Identifying institutional factors is not difficult. Culture, and the social guidelines that define it, is easily ascertainable and for the most part well-defined for the majority of societies. Likewise, determining a particular household's socioeconomic status relies on indicators such as income, assets, and political position that require only cursory investigation. Moreover, even if policy is not clearly defined in writing, it can be identified through the rules it shapes and their effects. The difficulty lies in determining how these factors interact with one another to influence the adoption of agroforestry and thus the environmental sustainability of an agricultural setting. And this makes survey design and verification extremely important. Repetition has helped hone the quantification of these factors, and most agroforestry-adoption surveys today contain many of the same primary measures. Unfortunately, given the networked nature of these influences, it is inappropriate to use them individually for sustainability-assessment purposes. The existence of one factor may be ineffectual without the contributions of the other factors. Appraisals must be done holistically.

Given the appropriate cultural context and socioeconomic factors, accounting for all these influences can indicate the likelihood of the implementation of sustainable agriculture at the farm, or even community, level. In this sense, adoption rates and the policies that affect them can be used as part of the method for measuring potential agricultural sustainability. The clearest starting point for effecting change in an institutional environment is policy (Place et al. 2012). Policy measures include government programs instituted to support a particular technology (as this may contradict or complement agroforestry adoption), rules that govern markets for agroforestry products, extension programs, and land tenure. Policies can even affect culture, because incentives for certain genders and age groups can outweigh cultural motivations, and over time, the results can become solidified as norms (Stern 2000). Policies that reduce such risk and uncertainty, such as those that establish seed banks, nearby nurseries, and/or training, extension, and agroforestry subsidies, have positive effects on adoption (Pattanayak et al. 2002). Policies that raise awareness of the benefits of these technologies are also likely to instill optimistic perceptions regarding adoption (Ajayi et al. 2006).

Policy effects can also be extremely counterproductive to sustainable agriculture. The environmental impacts of poorly designed policy can be swift and long-lasting. For example, subsidies for inorganic fertilizers, common in southeast Africa, de-incentivize adoption of sustainable technologies and exacerbate the aforementioned downward cycle of environmental abuse. Although such policies benefit the politicians responsible for their propagation by temporarily increasing production, in combination with policies that neglect infrastructure, they lay the groundwork for perpetual food insecurity. This propensity stems from the fact that policy is often derived from economic concerns, making economic methodology determinative of environmental outcomes.

20.4.2 Influence of the Economic Perspective on Policy

To address the potential negative effects of policy, some economists have proposed the use of comprehensive institution-based analysis to assess policies concerning the sustainability of ecosystem function (Corbera et al. 2009). This relates closely to the many payment-for-ecosystem-service (PES) schemes that involve agroforestry, because sustainability assessment based on analysis of the institutional environment can be calibrated against the quality of the services produced by the corresponding ecosystem. In other words, indications regarding an institutional environment can be given by an assessment of whether the owners of a landscape who are purported to provide a hypothetical benefit have adopted practices intended to conserve this ability and whether this adoption has resulted in the continued provision of the benefit at an acceptable level of quality. If the benefit is not being produced and the ecological underpinnings remain constant, the sociological influences require closer scrutiny. Because the institutional context influences the coordination between policies that affect these influences, such as property rights, funding, and relationships between actors (Corbera et al. 2009), evaluating why the technology shown to create the ability was or was not adopted can then point to the relevant institutional issues. For example, failure to produce the hypothetical benefit could instigate a survey of current or potential PES participants that indicates that they chose not to adopt agroforestry because they did not feel secure in their property rights and, as such, could not justify the up-front investment costs in hopes of receiving a benefit in the long term.

Investigation of the institutional environment can potentially produce results that clarify the conservation issues that derive from sociological disconnections, for instance the causes of the differences between PES guidelines and implementation by land-use decision makers. Moreover, PES income effects, extension shortcomings, and influences that strengthen or weaken potential participants' interest in ecosystem conservation, as well as the underlying causes of deforestation that necessitated a PES program, can be identified. Such institutional measures can also be used to evaluate the collateral outcomes (both positive and negative) at the local level that result from PES (Corbera et al. 2009). Indeed, PES programs, such as Costa Rica's Programa de Pago por Servicios Ambientales (Payment for Environmental Services Program), often rely on land use for making program-enrollment and payment decisions. This relates back to the need for accurate measurement of the biophysical aspects of AFSs and other forms of sustainable agriculture as they are applied in the fields of economics. Of course, employing land use as a measurement is to rely on a proxy for an environmental service (i.e. it is not an actual output measurement). As such, a closer look at the concept of PES reveals a further opportunity for the advancement of sustainability science, because PES, along with the policies that must accompany it for successful functionality, encapsulates the difficulties involved in the measurement of ecological and sociological sustainability.

20.4.3 Difficulties in Sustainability Measurement of PES Schemes

A PES scheme is a "voluntary, conditional agreement between at least one 'seller' and one 'buyer' over a well-defined environmental service—or land use—presumed to produce that service" (Wunder 2007). Although this definition appears to be the most widely accepted, some researchers, such as Sommerville et al. (2009), have sought to "refine and refocus" the definition in order to highlight considerations of additionality, conditionality, and institutional contexts, while also contending that such agreements need not be voluntary. There is wide acknowledgement that because of the variety of local institutional contexts surrounding natural-resource management, pure PES approaches fulfilling all the criteria may not always be possible, or even preferable (Sommerville et al. 2009).

Additionality is a principal condition defined in the Kyoto Protocol that requires that benefits from proposed projects have real, measurable, and long-term effects in addition to any that would occur in the absence of the certified project activity (UNFCCC 1998). In short, in order for a project to be eligible to create a Certified Emission Reduction, the standardized and thus tradable unit under the Kyoto system, it must prove that the C being removed from the atmosphere is the result of an intentional effort by the project designers and not the by-product of another economically motivated activity (UNFCCC 1998). When applying this term to other PES situations, the focus shifts toward the "additional" environmental benefits (cobenefits) a PES scheme may provide and away from concepts of intentional design. Due to the difficulty in distinguishing between intended benefits and unintentional benefits resultant of some form of profit seeking Sommerville et al. (2009) feel additionality should be viewed as an aspiration and not a necessity of the Kyoto system.

Additionality and cobenefits, however, have led to considerable debate in the literature regarding how to treat "bundled" benefits (multiple services from the same system). The ramifications of "stacking" (being paid for more than one ecosystem service from an individual system) and "unbundling" (attempting to separate the intertwined services of an individual system in economic analysis) are being investigated from scientific and policy perspectives. Currently, the Wunder (2007) definition and much of the literature regard additionality simply as a PES-effectiveness indicator and not as a compensable construct, due to the difficulties associated with measurement and the possibility of the "leakage" or spatial shifting of an environmental pressure (Wunder 2008).

Although it appears straightforward, conditionality is open to interpretation, especially in light of additionality. The "conditions" for a PES scheme are as follows: the buyer pays the amount agreed upon at the agreed-upon interval, and the provider maintains practices that allow the environmental benefit to continue accruing to the buyer. Payment is "conditional" on provision of the service (Wunder 2005), not on intent. This basic understanding, however, can cause problems if all the terms are not clearly identified in the PES contract. For instance, what is to happen if the provider continues a practice that previously produced a particular result in the past but has ceased to produce that result despite continued effort, or the practice changes but the desired result remains the same, or the practice changes such that it discontinues a cobenefit but still provides the primary service. Designing institutions that address such issues while providing incentives for economic agents is an important part of the modern forms of economics discussed above (Laffont and Martimort 2002), and it is crucial to the appropriate recognition of environmental benefits through markets, reliable legal frameworks, and supporting governance.

PES schemes are constructed with the intention of providing incentives for conservation-oriented land-management practices. There is the added hope that these payments may eventually produce positive changes in attitudes toward conservation as participants experience the environmental benefits and the associated financial gains. Beyond the payment itself, participants using agroforestry could, for example, see financial gain from increased production and savings in inorganic-fertilizer and inorganic-pesticide costs. The intended effect on attitudes is necessary given that shifts in environmental conditions could require adjustments in the original provisions (e.g. a particular practice no longer producing the anticipated result or the possibility of buyers receiving facility from alternative sources). The ultimate goal is for these collective optimizations to heighten the recognition of ecosystem services at a macroscale (market). This relies on equal knowledge distribution for efficient functioning and continued existence, again attesting to the importance of appropriate social structures. Unfortunately, this recognition is not (yet) in place, nor does it seem to be forthcoming.

Low recognition of ecosystem services by markets is the result of three interrelated concepts: externalities, nonexcludability, and intangibility (Jindal and Kerr 2007). Proper valuation of the environment requires internalization of both external economies and diseconomies. The attainment of this condition is, however, complicated by the necessity of determining responsibility, which is inherently difficult due to the often abstract intangibility of such benefits and is complicated in places with weakly defined property rights. The term "intangibility" is used to signify either the current inability of science to identify the exact interactions that create specific environmental benefits or the degree to which such benefits can be realized through land-management practices such as agroforestry. Payments for ecosystem services, however, are primarily concerned with positive externalities. External economies are often not appropriately internalized because of the difficulty associated with excluding consumption by those who did not play a part in the production of the benefit (excludability—e.g., the oxygen produced by a tree plantation). A long history of such benefits being provided for free by nature has (in terms of incentive theory) dampened motivation: having received an external reward with minimal effort has conditioned an expectation that may be difficult to reverse (Singh 2015). With such conditions weighing against the possibility of attitude change regarding conservation, it is important that PES efforts be advanced carefully, with a focus on both the biophysical and institutional aspects of their implementation.

20.5 Measuring the Sustainability of Agroforestry Systems

20.5.1 Estimating Ecosystem Services of Agroforestry Systems

Having recognized what constitutes ecosystem services and the conservation potential of associated payment schemes (i.e., PES schemes), the next step in estimating the value of the service is to measure the parameters quantitatively using the most appropriate analytical procedures. Various analytical procedures can be used to measure the different parameters and entities of each of the ecosystem services discussed above. A discussion of the state of the knowledge regarding each of these is beyond the scope of this paper. A summary of the common procedures and methods available (Table 20.3) shows that, in general, the estimations/measurements are often unsatisfactory. This is due to one or a combination of several reasons, such as the lack of proper methods and rigorous procedures, the bias and errors in the assumptions based on which estimations are made, the extent of time and resources needed for long-term measurements of critical parameters, the lack of field validation of results generated by modeling, and so on. To illustrate this point, let us consider the situation regarding carbon-sequestration estimations under AFSs.

As mentioned above, numerous reports are available regarding the extent of carbon sequestration under AFSs. However, these studies exhibit enormous variability in terms of their nature, degree of rigor, and extent of detail, so that it becomes difficult to compare the datasets based on uniform criteria and hence to draw widely applicable conclusions. The reported values (Table 20.2) are mostly speculative, based on circumstantial and experiential rather than empirical and experimental evidence. The extreme site specificity of AFSs also contributes to the lack of uniformity in assessment methodologies. Even the systems in the same region vary considerably in structure (arrangements of components), function (expected outputs), species diversity (of crops and trees), management, and socioeconomics, such that no two agroforestry fields are identical. Consequently, the reports vary widely in the methods used and/or the extent of detail reported, making it difficult to subject such results to integrated analyses such as meta-analysis and other statistical tools. Furthermore, most published studies are of short duration and cannot be used to predict long-term consequences. The difficulty of modeling discontinuous multispecies stands adds to the problem. Most models used in forestry (for estimating stand volume, C content, growth patterns, etc.) have been developed for continuous, single-species stands, but agroforestry systems represent discrete stands of multiple species; therefore, applying available forestry models to AFSs results in a "round peg in a square hole" problem (Nair and Nair 2014). The extensive estimations of global forest biomass that are available are based on rough estimations, that is, measuring the volume of stem wood and multiplying it with species-specific wood density, and multiplying that number by 1.6 to get an estimation of whole-tree biomass. C content is assumed to be 50 % of the estimated whole-tree biomass, and root biomass is generally excluded (Nair 2012b; Malmer et al. 2010). Although the whole-tree harvesting method, which involves summing up the amount of harvested and standing biomass, has traditionally been used for more accurate estimations of tree biomass, the extremely tedious nature of the method limits its application to research purposes. Allometric equations developed based on biophysical properties of trees and validated by occasional measurements of destructive sampling are widely used in forestry for estimating volumes of standing forests. However, such allometric equations are seldom developed for trees common in AFSs. As far as soil carbon sequestration in

Table 20.3 Summary of procedures for estimating the ecological features of ecosystem services of agroforestry systems

Ecosystem service	Parameter	Commonly expressed as	Availability and clarity of methods
Soil improvement: soil fertility, soil conservation, soil properties	N_2 fixation by NFTs	N: kg ha^{-1} $year^{-1}$	Rigorous field-measurement-methods not available
	Nutrient cycling (nutrient release from biomass decomposition)	Biomass: Mg ha^{-1} $year^{-1}$ Nutrient turnover (individual nutrients): Mg ha^{-1} $year^{-1}$	Good
	Deep uptake of nutrients	Nutrient uptake (individual nutrients): Mg ha^{-1} $year^{-1}$	Fair
	Soil properties: chemical, physical, and microbiological (pH, CEC, salinity, organic matter, nutrient store, bulk density, porosity, aggregate stability, water-holding capacity, soil microbial composition, etc.)	Change over a period of time under specific land-use regime: chronosequence studies	Good methods are available, but long-term studies are tedious and rare; modeling is used, but is seldom validated by field testing
	Soil erosion	Soil loss: Mg ha^{-1} $year^{-1}$	
	Soil degradation/regeneration	Soil degradation/regeneration expressed in defined parameters	Vague/poor
Water-quality improvement	Common chemical and microbiological quality parameters in water bodies including irrigation water	Change over a period of time under specific land-use regimes: chronosequence studies	Fair
Biodiversity conservation	Various biodiversity and indicators; indices for species richness, abundance, and evenness	Life-cycle analyses; alpha diversity; species richness; conservation values	Vague, poor; no single indicator*
Carbon sequestration (climate-change mitigation)	C storage in long-lived pools—in soils and aboveground reservoirs	C sequestration (Mg ha^{-1} $year^{-1}$); C storage (Mg ha^{-1}); CO_2 equivalent; change over a period of time	Analytical methods are good, but estimations lack rigor due to wrong assumptions

concerned, the estimated values in AFSs vary greatly depending on biophysical and socioeconomic characteristics of the system parameters and because of the lack of uniformity in study procedures such as depth of sampling and soil analytical procedures. Many reports lack even the essential information for comparison and extrapolation of data, such as soil bulk density. The uncertainties arising from the lack of uniform methods for describing area under agroforestry is another difficulty in gauging the importance of agroforestry in carbon sequestration. Furthermore, the reported studies on carbon sequestration under AFSs are of a short-term nature (fewer than five years), even when a so-called "chronosequence approach" is used for soil sampling (Demessie et al. 2013).

Because changes in C stock are unlikely to be linear across time (Fig. 20.3), understanding the nature of the curve of C storage over time is important for identifying the periods when the most C is being sequestered. In addition, it is difficult to know whether the residence time of C that is sequestered initially in a system differs from that of C that is sequestered later. Are the cycles undergone by the initial C and later C additions the same? As Nair and Nair (2014) noted, many such questions need to be answered in order to realistically assess the impact of agroforestry on carbon sequestration.

Many aspects of the above analysis of the carbon-sequestration (and climate-change-mitigation) potential of AFSs apply to other ecosystem services as well. The perceptions regarding the potential of AFSs to render ecosystem services at a higher level compared with single-species stands of croplands and grazing lands are based on solid scientific foundations. The methods and procedures adopted in collecting or estimating the data, however, are inconsistent, such that the data lack scientific rigor, often cannot be compared, and are often inconclusive. Even if/when reliable quantitative estimates become available, the bigger question of the value that the society assigns to or is willing to accept for such services will be a major issue.

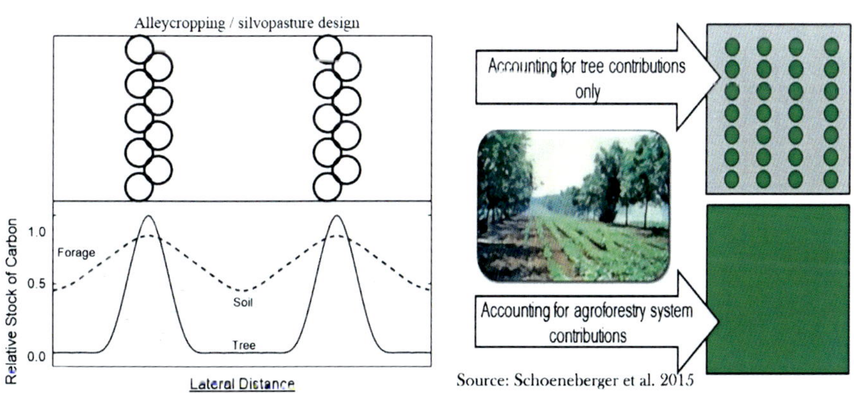

Fig. 20.3 Complexities of carbon-sequestration accounting in agroforestry systems. Agroforestry is not a "1 + 1 = 2" system, but rather a "1 + 1 maybe more, or less than 2" system. *Source* Schoeneberger et al. (2015)

20.5.2 *Institutional Measures of Sustainability*

As noted, an institutional environment is the interaction of culture, socioeconomic factors, and policy; the latter being the best entry point for stimulating change in an institutional environment. Given that policy's influence resonates through the casual chain that affects the use of sustainable agriculture, it can be utilized as an acceptable indicator of the potential for such use in a given community. This is because smallholder farmers view the influential factors of sustainable agriculture through the lens of an institutional environment, which policy helps to shape (Fig. 20.4). Applying this approach relies on an understanding of the connections between previous policy implementations and sustainability outcomes (Table 20.4). This understanding can then be compared with technology-adoption survey results and the functionality of schemes meant to incentivize sustainable-agriculture use, such as PES. A general sense of potential for sustainability can be gained if, in addition to policy, the cultural and socioeconomic elements described above are investigated properly, allowing for a summation of the manner in which drivers of sustainable agriculture are perceived by a community. Investigations of this nature are carried out through surveys, the results of which can then be calibrated against the results of the suggested biophysical-sustainability measurement technique in order to refine the process and produce a set of acceptable parameters.

Survey questions that focus on policy typically attempt to gauge participants' perceptions of policy effects rather than their knowledge of stated policies. For example, a policy that purports to solidify property rights for a given community may not result in its intended effect, and landowners could still feel uncomfortable about making long-term investments in their land. If a significant number of survey participants feel confident in their ownership, the stated policies are irrelevant from

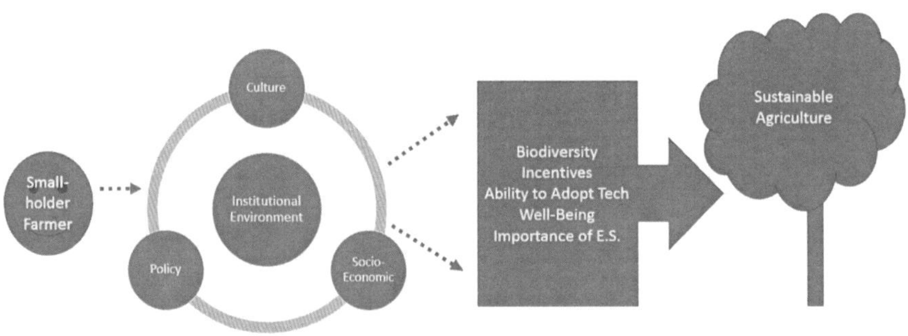

Fig. 20.4 Schematic presentation of how the institutional environment affects smallholder farmer perceptions. The institutional environment, which is the nexus of policy, culture, and socioeconomic conditions, affects farmers' perception (*dotted arrows*) of factors influencing adoption of sustainable agriculture (*solid arrow*), such as financial ability and incentives, benefits of biodiversity and ecosystem services, and their relationships to well-being

Table 20.4 Summary of measures for estimating adoption potential of agroforestry systems

Institutional environ.	Parameter	Influence on sustainability	Measurement/applicability	Ref.
Policy	Subsidies	Technology dependent, can be positive or negative	Typically not represented by stated policies but by perceptions (good b/c disconnect is common). Often quantified on a Likert scale using ordinal measures	11
	Property rights	Direct positive relationship		1
	Markets	Policies increasing access create blanket demand upturn		3
	Infrastructure	Provision of schools, medical, roads, etc., increase adoption		2
	Extension	Teaching and supporting tech use has large positive effect		9
	Tech available	Direct positive relationship		5
	Awareness	Direct positive relationship		2
Socioecon. factors	Resource access	Type of input acquirable can have positive/negative effect	Typically concrete, i.e., not perception. Often quantified through continuous measures denotable in intervals. This is good b/c it can highlight differences in population outcomes	13
	Property size	Often tied to soil quality; positive relationship		13
	Land tenure	Direct positive relationship		7
	Income/wealth	Direction of relationship often dependent on other factors		16
	Education	Mixed results; predominately positive esp. w/awareness		7
	Age	Inverse relationship		11
	Status	Mixes w/factors like subsidy creating positive effect		2

(continued)

Table 20.4 (continued)

Institutional environ.	Parameter	Influence on sustainability	Measurement/applicability	Ref.
Culture	Wealth meaning	If necessities met, value of added gain often still positive	No "typical" method. Difficult to quantify due to abstractness but has real effects. Responses can be represented through ordinal or interval measurement, making comparison across studies difficult	6
	Household roles	Stronger correlation with female household heads		12
	Communication	Direct positive relationship		8
	Marital residency	Variable depending on relation of resource manager to property owner, if one and the same influence is positive		12
	Family size	Often measure of available labor, with positive relationship		4
	Risk tolerance	Direct positive relationship		10
	Norm plasticity	Positive or negative relationship depends on other factors (e.g., policy)		15

(1) Ajayi and Place (2012), (2) Ajayi et al. (2006), (3) Bannister and Nair (2003), (4) Blatner et al. (2000), (5) Bromley (1991), (6) Fernandez and Fogli (2005), (7) German et al. (2009), (8) Kairuki and Place (2005), (9) Meijer et al. (2014), (10) Mercer (2004), (11) Pattanayak et al. (2002), (12) Place et al. (2009), (13) Sirrine et al. (2010), (15) Stern (2000), (16) Thangata and Alavapati (2003)

the perspective of sustainable-technology adoption. This is because it is the participants' perceptions that will ultimately determine their adoption decisions. Measuring these perceptions is typically accomplished by the use of ordinal measures quantified through a Likert scale (a five- or seven-point scale used in the social sciences to express the degree to which a survey respondent agrees or disagrees with a particular statement; Norman 2010). Continuing the property-rights example, participants might be asked to gauge their confidence that their land will remain under their control on a scale of 1–5, the larger numbers indicating greater confidence. This commonly used method accounts for disconnections between stated policies and their actual influence on adoptability, and it facilitates comparisons across adoption studies.

The quantification of socioeconomic factors is typically more straightforward than attempts at policy measurement, because most of these factors are represented by numbers, not perceptions. For example, property size can be physically measured, and in many societies a person's status is represented through clearly defined relationships with others in the community. The responses to questions gauging these factors are often collected as continuous measures and later assigned to intervals. For example, a question regarding age would yield ongoing varied responses that can then be placed into groupings. These groupings can then be used to uncover distinctions within the sample, such as the greater likelihood of participants in an age range of 18–25 years to adopt agroforestry. This measurement tendency has both good and bad attributes. Although it is good at revealing such features within a community, the intervals may appear in an inconsistent manner across studies. The effect of this potential inconsistency is not pronounced, however; the intervals and their related influences on sustainability align closely across the studies cited in this chapter.

Measurements of culture commonly used in technology-adoption studies have a low level of consistency, despite the strong influence of culture on adoption decisions. Techniques vary and can be based on the collection of nominal, ordinal, and interval data. Survey design, implementation, and interpretation rely heavily on anthropological considerations, necessitating the use of enumerators and consultants who belong to, or are very familiar with, the community being sampled. This reliance introduces another level of potential error in the measurement process, because the information is mediated by these persons' interpretations of the target community, most obviously when language translation is involved. In the same manner, a phrase can be translated from one language to another in multiple ways and with differing nuances, expression of the manner in which other cultural aspects appear can vary. For example, a culture-level propensity to tolerate risks (e.g. expenditures on a new sustainable technology) may not hold true for a consultant helping to design survey measures. Risk tolerance and the speed at which norms are prone to change within a community are highly influential (the prior having a positive relationship with propensity towards technology adoption and the latter functioning in conjunction with other considerations, such as policy). Therefore, inconsistencies in measurement design can have large effects on results.

The terms "typically" and "often" appear frequently in the descriptions of survey methodologies, because there is no standardized method for making such measurements, only common processes. Although this may appear to be a detriment, especially with respect to culture, such flexibility is required given the abstract nature of many of the concepts and the necessity of adapting to the complexities of different settings. When no standard exists for constructing indicators, issues of validity can arise. Validity is a fundamental property of good measurement that refers to the degree to which there is congruence between the operational definition and the concept the operational definition intends to measure. Because precise indicators of abstract concepts are especially important when measuring social phenomena, issues of validity require special attention. For example, when a study seeks to measure the influence of subsidies, there should be a precise understanding within the field about what constitutes a subsidy: is it a cash payment for a specific performance, the giving of implements conducive to the targeted technology, or something else? Although flexibility in measurement methods may help researchers capture difficult-to-define concepts, some of the distortion this creates can be compensated for with a concerted effort to improve validity.

Ultimately, analysis based on the influence of the varied aspects of policy, culture, and socioeconomic factors and how these affect sustainable agriculture can only provide a sense of the potential sustainability of an institutional environment at a macroscale. Despite some basic components identifiable as ubiquitous, "institutional contexts" that affect attitudes evolve over time and vary under the influence of a multitude of factors across societies (Corbera et al. 2009). Primarily, as motivation is shaped by the presence of incentives and disincentives, motivating behaviors (such as technology adoption) requires the creation of incentives and disincentives; this can be through law, monitoring, and financial frameworks that take form over iterative experiences in the development of a society (Weinstein and Turner 2012); and thus can vary widely. A review of the literature on ecosystem services found a consistent call for an improvement in the valuation of cultural ecosystem services, studies of culture in the context of bundling, and a better articulation of policy implications (Milcu et al. 2013); the review authors maintained that such a focus would help bridge gaps between academic disciplines, address real world problems, and foster new conceptual links between alternative logics relating to a variety of social and ecological issues.

20.6 Conclusions

For nearly the past four decades, the enigma of sustainability has appeared in almost all development agendas and paradigms as a leading item; yet, paradoxically, a clear definition of sustainability, let alone a well-defined set of criteria and indicators for measuring and expressing it quantitatively, has evaded researchers. The lack of such tools, however, has not dissuaded professionals from moving forward with a variety of sustainability-related programs, for which there is a growing demand. The American Society of Agronomy (www.agronomy.org; accessed 13 August 2015), for example, organizes and promotes webinars on measuring sustainability to help its "customers understand sustainability metrics and how to respond to downstream data requests for sustainable supply chain programs … and help farmers to measure environmental outcomes and provide opportunities to evaluate and adopt more sustainable practices." Thus, there is a contradiction between the lack of a clear set of criteria and indicators of sustainability measurements, on the one hand, and the demand for programs for educating practitioners on such measurements, on the other. The fact of the matter is that the demand for ensuring sustainability in agricultural operations is so overpowering that a combination of different measures and criteria, rather than any single one, is deemed acceptable as the measure of sustainability.

In agroforestry systems, as in other agricultural and natural-resource-management systems, these standards include measures of ecological, economic, and social sustainability. Among these three components, the one that stands out for agroforestry and sets it above other land-use disciplines is ecological sustainability, expressed in terms of ecosystem services. An important point in this context, however, is that the society at large must become more convinced and appreciative regarding the benefits of such ecosystem services for future generations, and norms and procedures (even legal mandates such as taxation) must be put in place. Until that happens, ecosystem services—and sustainability—will remain a mere talking point among academics.

References

Ajayi OC, Akinnifesi FK, Mullila-Mitti J, DeWolf JJ, Matakala PW, Kwesiga FR (2006) Adoption, profitability, impacts and scaling-up of agroforestry technologies in Southern African countries. ICRAF, Lilongwe

Ajayi OC, Place F (2012) Policy support for large-scale adoption of agroforestry practices: experience from Africa and Asia. In: Nair PKR, Garrity D (eds) Agroforestry: the future of global land use. Advances in agroforestry, vol 9. Springer, Dordrecht, pp 175–201. doi:10.1007/978-94-007-4676-3_12

Ansell C, Gash A (2007) Collaborative governance in theory and practice. J Public Adm Res Theory 18(4):543–571. doi:10.1093/jopart/mum032

Bannister ME, Nair PKR (2003) Agroforestry adoption in Haiti: the importance of household and farm characteristics. Agrofor Syst 57(2):149–157. doi:10.1023/A:1023973623247

Blatner KA, Bonongwe CSL, Carroll MS (2000) Adopting agroforestry: evidence from central and northern Malawi. J Sustain For 11(3):41–69. doi:10.1300/J091v11n03_03

Bromley DW (1991) Environment and economy: property rights and public policy. Blackwell, Oxford

Co-operation on Health and Biodiversity Initiative Secretariat (2010) The importance of biodiversity to human health. United Nations Convention on Biological Diversity, Galway

Corbera E, Sobera CG, Brown K (2009) Institutional dimensions of payments for ecosystem services: an analysis of Mexico's carbon forestry programme. Ecol Econ 68(3):743–761. doi:10.1016/j.ecolecon.2008.06.008

Daily GC, Ehrlich PR (1996) Socioeconomic equity, sustainability, and Earth's carrying capacity. Ecol Appl 6(4):991–1001

Daily GC, Ehrlich PR, Sanchez-Azofeifa GA (2001) Countryside biogeography: use of human-dominated habitats by the avifauna of southern Costa Rica. Ecol Applications 11: 1–13

Davies GR (2013) Appraising weak and strong sustainability: searching for a middle ground. Cons J Sustain Dev 10(1):111–124

Demessie A, Singh BR, Lal R (2013) Soil carbon and nitrogen stocks under chronosequence of farm and traditional agroforestry land uses in Gambo District, Southern Ethiopia. Nutr Cycl Agroecosyst 95(3):365–375. doi:10.1007/s10705-013-9570-0

de Vries BJM (2012) Sustainability science. Cambridge University Press, Cambridge

Dubeux JCB Jr, Muir JP, Nair PKR, Sollenberger LE, da Silva HMS, de Mello ACL (2015) The advantages and challenges of integrating tree legumes into pastoral systems. In: Evangelista AR, Avila CLS, Casagrande DR, Lara MAS, Bernardes TF (eds) International conference on forages in warm climates. Federal University of Lavras, Minas Gerais, pp 141–164

Ehrlich PR, Daily GC (1993) Food security, population and environment. Popul Dev Rev 19(1):1–32. doi:10.2307/2938383

Fernandez R, Fogli A (2005) Culture: an empirical investigation of beliefs, work, and fertility. Am Econ J Macroecon 1(1):146–177. doi:10.1257/mac.1.1.146

Fletcher R (2011) When environmental issues collide: climate change and the shifting political ecology of hydroelectric power. Peace Confl Rev 5(1):1–15

Francesconi W, Nair PKR, Levey DJ, Daniels J, Cullen L Jr (2013) Butterfly distribution in fragmented landscapes containing agroforestry practices in Southeastern Brazil. Agrofor Syst 87(6):1321–1338. doi:10.1007/s10457-013-9640-y

Garrity D (2012) Agroforestry and the future of global land use. In: Nair PKR, Garrity D (eds) Agroforestry: the future of global land use. Advances in agroforestry, vol 9. Springer, Dordrecht, pp 21–27. doi:10.1007/978-94-007-4676-3

Gasparatos A, El-Haram M, Horner M (2008) A critical review of reductionist approaches for assessing the progress towards sustainability. Environ Impact Assess Rev 28(4–5):286–311. doi:10.1016/j.eiar.2007.09.002

German G, Akinnifesi FK, Edriss AK, Sileshi GW, Masangano C, Ajayi OC (2009) Influences of property rights on farmers' willingness to plant indigenous fruit trees in Malawi and Zambia. Afr J Agric Res 4(5):427–437

Gómez-Baggethun E, de Groot R, Lomas PL, Montes C (2010) The history of ecosystem services in economic theory and practice: from early notions to markets and payment schemes. Ecol Econ 69(6):1209–1218. doi:10.1016/j.ecolecon.2009.11.007

Hammond A, Adriaanse A, Rodenburg E, Bryant D, Woodward R (1995) Environmental indicators: a systematic approach to measuring and reporting on environmental policy performance in the context of sustainable development. World Resource Institute, Washington, D.C.

Heimlich R (2003) Biological resources and agriculture. In: Heimlich R (ed) Agricultural resources and environmental indicators. U.S. Department of Agriculture, Washington, pp 1–17

Hodgson GM (2006) What are institutions? J Econ Issues 40(1):1–25

Holland J (2007) Tools for institutional, political, and social analysis of policy reform. World Bank, Washington, D.C.

Jindal R, Kerr J (2007) Basic principles of PES. United States Agency for International Development, Washington

Jose S, Gold MA, Garrett HE (2012) The future of temperate agroforestry in the United States. In: Nair PKR, Garrity D (eds) Agroforestry: the future of global land use. Advances in agroforestry, vol 9. Springer, Dordrecht, pp 217–245. doi:10.1007/978-94-007-4676-3_14

Kairuki G, Place F (2005) Initiatives for rural development through collective action: the case of household participation in group activities in the highlands of Central Kenya. International Food Policy Research Institute, Washington, D.C.

Kaufman RK, Cleveland CJ (1995) Measuring sustainability: needed—an interdisciplinary approach to an interdisciplinary concept. Ecol Econ 15(2):109–112

Khalilnejad A, Ali FHJ, Osman N (2012) Contribution of the root to slope stability. Geotech Geol Eng 30(2):277–288. doi:10.1007/s10706-011-9446-5

Laffont JJ, Martimort D (2002) The theory of incentives: the principal-model. Princeton University Press, Princeton, NJ

Malmer A, Murdiyarso D, Bruijnzeel LAS, Ilstedt U (2010) Carbon sequestration in tropical forests and water: a critical look at the basis for commonly used generalizations. Glob Change Biol 16(2):599–604. doi:10.1111/j.1365-2486.2009.01984.x

MEA (2005) Millennium ecosystem assessment: ecosystems and human well-being synthesis. Island Press, Washington

Meijer SS, Catacutan D, Ajayi OC, Sileshi GW, Nieuwenhuis M (2014) The role of knowledge, attitudes and perceptions in the uptake of agricultural and agroforestry innovations among

smallholder farmers in sub-Saharan Africa. Inter J Agric Sustain 13(1):30–54. doi:10.1080/14735903.2014.912493
Meijer SS, Catacutan D, Sileshi GW, Nieuwenhuis M (2015) Tree planting by smallholder farmers in Malawi: using the theory of planned behaviour to examine the relationship between attitudes and behaviour. J Environ Psychol 43:1–12. doi:10.1016/j.jenvp.2015.05.008
Mercer DE (2004) Adoption of agroforestry innovations in the tropics: a review. Agrofor Syst 61 (1):311–328. doi:10.1023/B:AGFO.0000029007.85754.70
Milcu AI, Hanspach J, Abson D, Fischer J (2013) Cultural ecosystem services: a literature review and prospects for future research. Ecol Soc 18(3):1–34. doi:10.5751/ES-05790-180344
Nair PKR (2012a) Carbon sequestration studies in agroforestry systems: a reality check. Agrofor Syst 86(2):243–253. doi:10.1007/s10457-011-9434-z
Nair PKR (2012b) Climate change mitigation and adaptation: a low hanging fruit of agroforestry. In: Nair PKR, Garrity D (eds) Agroforestry: the future of global land use. Advances in agroforestry, vol 9. Springer, Dordrecht, pp 31–67. doi:10.1007/978-94-007-4676-3_7
Nair PKR (2013) Agroforestry: trees in support of sustainable agriculture. In: Elias SA (ed) Earth systems and environmental sciences. Elsevier, Philadelphia, pp 1–15. doi:10.1016/B978-0-12-409548-9.05088-0
Nair PKR (2014) Agroforestry: practices and systems. In: van Alfen N (ed) Encyclopaedia of agriculture and food systems, vol 1. Elsevier, San Diego, pp 270–282
Nair PKR, Nair VD (2014) Solid-fluid-gas: the state of knowledge on carbon-sequestration potential of agroforestry systems in Africa. Curr Opin Environ Sustain 6:22–27. doi:10.1016/j.cosust.2013.07.014
Nair PKR, Buresh RJ, Mugendi DN, Latt CR (1999) Nutrient cycling in tropical agroforestry systems: myths and science. In: Buck LE, Lassoie JP, Fernandes ECM (eds) Agroforestry in sustainable agricultural systems. CRC Press, Boca Raton, pp 1–31
Nair PKR, Nair VD, Kumar BM, Showalter JM (2010) Carbon sequestration in agroforestry systems. In: Sparks D (ed) Advances in agronomy, vol 108. Elsevier, San Diego, pp 237–307
Ness B, Urbel-Piirsalu E, Anderberg S, Olsson L (2007) Categorising tools for sustainability assessment. Ecol Econ 60(3):498–508. doi:10.1016/j.ecolecon.2006.07.023
Neumayer E (2000) Scarce or abundant: the economics of natural resource availability. J Econ Surv 14(3):307–329. doi:10.1111/1467-6419.00112
Norgaard RB (1981) Socio-system and ecosystem co-evolution in the Amazon. J Environ Econ Manag 8(3):238–254
Norman G (2010) Likert scales, levels of measurement and the "laws" of statistics. Adv Health Sci Educ Theory Pract 15:625–632. doi:10.1007/s10459-010-9222-y
Palm CA, Gachengo CN, Delve RJ, Cadisch G, Giller KE (2001) Organic inputs for soil fertility management in tropical agroecosystems: application of an organic resource database. Agric Ecosyst Environ 83(1–2):27–42. doi:10.1016/S0167-8809(00)00267-X
Pattanayak SK, Mercer DE, Sills E, Yang JC (2002) Taking stock of agroforestry adoption studies. Agrofor Syst 57(3):173–186
Place F, Roothaert R, Maina L, Franzel S, Sinja J, Wanjiku J (2009) The impact of fodder trees on milk production and income among smallholder dairy farmers in East Africa and the role of research. ICRAF, Nairobi
Place F, Ajayi OC, Torquebiau E, Detlefsen G, Gauthier M, Buttoud G (2012) Improved policies for facilitating the adoption of agroforestry. In: Koanga ML (ed) Agroforestry for biodiversity and ecosystem services: science and practice. InTech, Rijeka, pp 113–128
PNAS (2015) Sustainability science. In: Proceedings of the National Academy of Sciences. http://sustainability.pnas.org/page/about. Accessed 21 Sept 2015
Ponting C (2007) A new green history of the world: the environment and the collapse of great civilizations. Penguin, Westminster
Schoeneberger M, Domke G, Nair PKR, Marlen E, Franzluebbers A, Woodall C, Patel-Weynand T, Brandle J, Ballesteros W (2015) Greenhouse gas mitigation and accounting. In: Patel-Weynand T, Bentrup G, Schoeneberger M (eds) Agroforestry and climate change:

reducing threats and enhancing resiliency in agricultural landscapes. U.S. Department of Agriculture, Forest Service, Washington
Singh NM (2015) Payments for ecosystem services and the gift paradigm: sharing the burden and joy of environmental care. Ecol Econ 117:53–61. doi:10.1016/j.ecolecon.2015.06.011
Sirrine D, Shennan C, Sirrine JR (2010) Comparing agroforestry systems' ex ante adoption potential and ex post adoption: on-farm participatory research from Southern Malawi. Agrofor Syst 79(2):253–266. doi:10.1007/s10457-010-9304-0
Sommerville MM, Jones JPG, Milner-Gulland EJ (2009) A revised conceptual framework for payments for environmental services. Ecol Soc 14(2):34–48
Stern PC (2000) Toward a coherent theory of environmentally significant behavior. J Soc Issues 56 (3):407–424. doi:10.1111/0022-4537.00175
Thangata P, Alavapati JPR (2003) Agroforestry adoption in Southern Malawi: the case study of *Gliricidia sepium* and maize. Agric Syst 78(1):57–71. doi:10.1016/S0308-521X(03)00032-5
Tilman D, Lehman CL, Thomas KT (1997) Plant diversity and ecosystem productivity: theoretical considerations. Proc Natl Acad Sci 94(5):1857–1861
Turner BL, Sabloff JA (2012) Classic Period collapse of the central Maya lowlands: insights about human–environment relationships for sustainability. Proc Natl Acad Sci 109(35):13908–13914. doi:10.1073/pnas.1210106109
UNEP (2004) GEO—global environment outlook 3: past, present and future perspectives. United Nations Environment Programme, Nairobi
UNESCO (2005) The precautionary principle. World Commission on the Ethics of Scientific Knowledge and Technology, Paris
UNFCCC (1998) Kyoto protocol, Article 12-5. http://unfccc.int/resource/docs/convkp/kpeng.pdf. Accessed 21 Sept 2015
WCED (1987) Our common future. World Commission on Environment and Development. Oxford University Press, Oxford
Weinstein MP, Turner RE (2012) Sustainability science. In: Weinstein MP, Turner RE (eds) Sustainability science: the emerging paradigm and the urban environment. Springer, New York, pp vii–xii. doi:10.1007/978-1-4614-3188-6
Weinstein MP, Turner RE, Ibanez C (2012) The global sustainability transition: it is more than changing light bulbs. Sustain Sci Pract Policy 9(1):4–15
Weiss EB (1992) Intergenerational equity: a legal framework for global environmental change. In: Weiss EB (ed) Environmental changes and international law: new challenges and dimensions. United Nations University, Tokyo, pp 1–66
Wu JJ (2012) A landscape approach for sustainability science. In: Weinstein M, Turner R (eds) Sustainability science: the emerging paradigm and the urban environment. Springer, New York, pp 59–78
Wunder S (2005) Payments for environmental services: some nuts and bolts. CIFOR, Bogor
Wunder S (2007) The efficiency of payments for environmental services in tropical conservation. Conserv Biol 21(1):48–58. doi:10.1111/j.1523-1739.2006.00559.x
Wunder S (2008) How do we deal with leakage? In: Angelsen A (ed) Moving ahead with REDD: issues, options, and implications. CIFOR, Bogor, pp 65–76
Zak DR, Holmes WE, White DC, Peacock AD, Tilman D (2003) Plant diversity, soil, microbial communities, and ecosystem function: are there any links? Ecology 84(8):2042–2050

Chapter 21
Experiences with Adopting the Catchment Approach in Sustainable Land Management: The Case of Kagera TAMP Tanzania

Fidelis Kaihura and Stefan Schlingloff

Abstract A regional project as part of the Transboundary Agroecosystems Management Project (Kagera TAMP) was implemented in the Kagera River basin covering Burundi, Rwanda Tanzania and Uganda. The project was implemented by the Food and Agriculture Organization (FAO) of the United Nations with financial support from the Global Environment Facility (GEF) and the governments of the riparian countries. In Tanzania, the project was conducted in four districts (Bukoba, Karagwe, Missenyi and Ngara) in the Kagera region. The project was implemented to address the causes of land degradation, restore ecosystem health and function and generate a range of global environmental benefits across the Kagera basin. In Tanzania, 10 micro-catchments were selected for project implementation involving 14,282 farmers covering about 3500 ha of land. The type, degree and direct and indirect causes of land degradation in the basin were assessed for different land use systems. Similarly, good practices of sustainable land management and their extent, effectiveness and trends were evaluated. A range of maps were developed and used to select land degradation priority areas for project intervention (hotspots). Successful sustainable land management (SLM) practices were introduced into the curriculum of the Farmer Field School (FFS), and a holistic catchment approach was adopted to address SLM in an integrated and multi-sectoral way. Through adopting this catchment approach (see also a long version of the definition on p. 15), drying water sources were regenerated that provide a reliable water supply for domestic and livestock use. Furthermore, training of extension staff and facilitators for the farmer field schools improved smallholder farmers' knowledge and skills in applying sustainable land management practices. Cropland management increased the crop yield of bananas, maize, beans, and cassava from 20 % to

F. Kaihura (✉)
Agriculture Research Institute Maruku, P.O. Box 127, Bukoba, Tanzania
e-mail: fkaihura@yahoo.com

S. Schlingloff
Land and Water Division, Food and Agriculture Organization of the United Nations, Viale delle Terme di Caracalla, 00153 Rome, Italy
e-mail: Stefan.Schlingloff@fao.org

R. Lal et al. (eds.), *Climate Change and Multi-Dimensional Sustainability in African Agriculture*, DOI 10.1007/978-3-319-41238-2_21

sometimes more than 60 %. Planting of *Cajanus cajan* and other types of agroforestry technologies, application of farmyard manure and contouring were superior for improving soil quality and reducing soil erosion by more than 20 % of the covered area. Construction of physical contours and their strengthening with vegetative materials, such as fruit trees, grasses and shrubs, including *Cajanus cajan*, pineapples, vetiver and lemon grass, improved the soil quality. Improved soil quality also increased household income and improved household and community nutrition, which are short-term SLM benefits. These benefits were powerful incentives that led to wider adoption of SLM practices in all catchments. Soil organic carbon sequestration exceeded 20 %. Grassland carrying capacity was improved from 2.5 to 0.7, 5.0 to 0.4 and 3.3 to 0.7 ha/tropical livestock units (TLU). Grassland biodiversity regenerated with the reappearance of endangered trees, shrubs and grasses that have medicinal, socio-economic and cultural value. A total of 31 technical and three policy recommendations were developed and submitted to policy and decision makers for inclusion in development plans. The main challenges included the low priority given by national and local governments to land degradation issues and the lack of financing for and investment in upscaling of sustainable land management practices, limited attention given to land use conflicts and their management, and land tenure and addressing SLM on a sector basis instead of a multi-sector basis. With the benefits demonstrated in implementing the Kagera TAMP project, communities have been empowered to demand further improvement of their natural resources base.

Keywords Transboundary agro-ecosystem management · Tanzania · Catchment approach · Sustainable land management · Farmer field schools · Incentives

21.1 Introduction

A Project on Transboundary Agroecosystems Management with a focus on land degradation (LD) and sustainable land management (SLM) was implemented jointly by the Food and Agriculture Organization (FAO) in selected districts in the Kagera River basin in four riparian countries, Burundi, Rwanda, Tanzania and Uganda. The basin is faced with increasing human and animal pressures that have led to intensification of land use and the adoption of unsustainable practices, including the following: (i) overstocking and overgrazing of pastures and rangelands, excess bush burning; (ii) continuous cropping, with reductions in fallow and rotations, reduced crop diversity in response to markets (food and forage species/varieties), repetitive tillage and soil nutrient mining (lack of nutrient restoration practices); (iii) encroachment of subsistence cropping into more fragile, drier areas, previously used/reserved for pasture and grazing, and in wetlands; (iv) over-exploitation of forests and woodland, especially the loss of riverine forest, and unsustainable harvesting (timber, fuelwood, charcoal, brick making etc.); and (v) over-exploitation and degradation of communal areas, such as forested highland and riverine areas, grazing lands, riverbanks and cultivated steep slopes. The project was implemented to address the causes of land degradation, to restore ecosystem health and functions and to generate a range of global environmental benefits across the Kagera basin through the introduction of adapted agro-ecosystem management approaches, and contribute to reducing the poverty of rural communities in the Kagera basin through more productive and sustainable resource management practices. Implementation followed a "catchment approach" by first identifying the existing land use systems as well as the types, extent and impact of land degradation and ongoing management practices to mitigate the problems followed by introducing and/or improving existing ecosystems management technologies. A total of 10 land degradation hotspots or micro-catchments were selected that involved 14,282 farmers covering 300 ha of testing and demonstrating and 3128 ha through the adoption of successful technologies. A number of stakeholders, including stakeholder ministries, district councils, NGOs, research and training institutions and the private sector, as well as community beneficiaries, were involved. To address transboundary resource degradation issues, regional workshops and meetings were organized especially for policy harmonization and information-sharing. The project was implemented for the period 2010–2015 with effective technologies testing in the field for two and half years.

21.2 Catchment Delineation and Selection of Project Implementation Sites

The catchment approach as utilized in this paper is a comprehensive, integrated and multisectoral approach to sustainable land management. The main principles underlying the approach are the following: (1) involves and gives land users/communities responsibility at all stages, that is, is participatory and collaborative, (2) is multilevel and multi-stakeholder and places people and supportive institutions at the center for maximum sustainability (3) and is partnership-based in a collaborative approach the role of partnerships is to mobilize scientific knowledge for agricultural investments that are pro-poor, pro-growth and pro-environment, to have more equitable partnerships by blending scientific and traditional knowledge, achieve a common vision for SLM, provide the right framework to work together to develop policy, govern programs and share information and to target a broad spectrum of stakeholders (policymakers, civil society (NGOs), land users/owners, community-based organizations, research institutions, mass media and the private sector).

21.2.1 Identification of Catchments

A catchment is an area of land where water collects when it rains, often bound by hills. As the water flows over the landscape, it finds its way into streams and down into the soil, eventually feeding the river. Catchments can range greatly in size from sub-catchments to massive catchments. Due to resource limitations, SLM interventions were implemented starting at the smaller parts of the catchments and planned to gradually extend and spread to the entire catchment.

Catchment selection started by establishing a database and maps for land use systems and land degradation types in the entire basin. For Tanzania, the identified land use systems included protected natural forest; natural forest; protected forest plantation; forest plantation moderate livestock; protected savannah; savannah with livestock; wetland with livestock; protected crops (perennial, seasonal irrigated); perennial crops with livestock; seasonal crops high livestock; irrigated crops with livestock; urban areas and protected surface water. In terms of land degradation, four main types were described, including physical, chemical, biological and water degradation. Resource degradation was evaluated in terms of the rate, extent and impact using the World Overview of Conservation Approaches and Technologies (WOCAT) procedures. Existing SLM technologies were also documented. Figure 21.1 illustrates the types of land degradation, their severity and existing SLM technologies in Tanzania.

Land use is the driving force of land degradation. Therefore, with the help of Geographic Information System (GIS data, land use systems maps have been

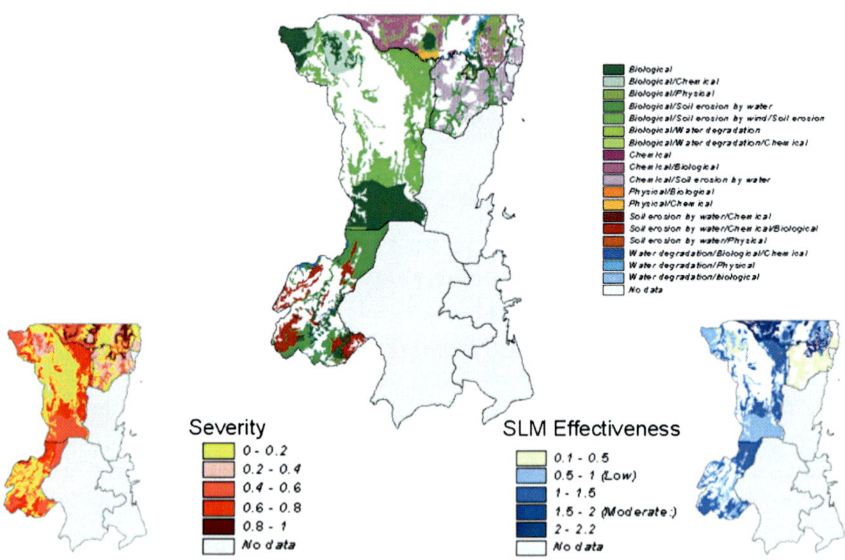

Fig. 21.1 Land degradation types, their severity and SLM effectiveness in the Kagera basin of Tanzania (Kagera TAMP WOCAT QM maps, 2012)

developed, based on which a multi-disciplinary team of national experts and experienced staff from local/district agricultural authorities have defined (in a workshop with a structured questionnaire) the type, degree and extent of land degradation, as well as the extent, effectiveness and trends of sustainable land management practices, to address these problems.

21.2.2 Identification of Implementation Micro-Watersheds or Villages Within Districts

Project pilot sites and micro-watersheds within districts were identified using the maps and applying a set of 10 selection criteria. Table 21.1 summarizes the selection process and established criteria.

The district experts then selected one or two villages based on the following criteria: (a) the extent and impact of land degradation, (b) possibilities of success, (c) readiness and capability of farmers, (d) relationships with village facilitators (e) and expected cooperation from the village communities. Minziro and Kigazi villages were selected.

21.2.3 Implementation Following the Catchment Approach

The catchment approach was set as comprehensive, integrated and multi-sectoral. The main principles underlying the approach included the following: (i) involving and giving land users/communities responsibility at all stages (i.e., is participatory and collaborative), (ii) placing people and supportive institutions at the center for maximum sustainability and (iii) involving many stakeholders and ensuring the mobilization of scientific knowledge for agricultural investments that are pro-poor, pro-growth and pro-environment. Implementation ranged between milli-watersheds (100–1000 ha) and micro-watersheds (<100 ha to farm size) in the Bukoba, Missenyi, Karagwe and Ngara districts in the Kagera basin of Tanzania. Figure 21.2 shows the Kyazi micro-catchment, one of the SLM intervention sites. Characteristics of micro-watersheds include natural springs, permanent rivers, cropland, pasture land, natural forest, pine tree plantations and settlements.

Activities to be carried out were developed after a participatory rural appraisal (PRA) was conducted for each selected micro-watershed. Biophysical and socio-economic aspects of the land degradation of the soil, vegetation and water, as well as poverty, education, land tenure and access to capital assets, were evaluated. Identified land degradation issues were prioritized and the highest ranking considered for the development of an action plan and activities. During this process, community members, other land management stakeholders in the district and land

Table 21.1 Scoring and selection for natural forest high livestock in Missenyi District

Selection criteria	Ward					Selected ward and village	
	Bugorora	Minziro	Kassambya	Nsunga	Mtukula	Ward	Village
Contribution to project outputs, esp. transboundary issues	1	3	2	2	3	MINZIRO	KIGAZI + MINZIRO
Presence of development partners	3	3	2	3	3		
Degree and type of LD	3	3	2	3	2		
Impact of LD	2	2	2	2	2		
District priorities in SLM	3	3	1	3	3		
SLM types on the ground	2	2	1	2	2		
Probability of success	2	3	2	2	2		
Access and visibility	2	2	2	2	2		
Activities in relation to national plans	2	3	2	2	2		
Acceptance by majority of stakeholder	2	3	2	2	2		
Total score	22	27	18	23	23		

NB: Score ranking: *3* high; *2* medium and *1* low (*Source* Kaihura 2012)

Fig. 21.2 Kyazi micro-catchment image: the Kyazi micro-catchment covers mainly the Rubaya and Rubumba sub-villages. It water drains westward in Ngono River. The predominant land use types include Kibanja (settlements associated with plots of permanent crops, mainly bananas and coffee), Kikamba (abandoned Kibanja used for growing annual crops such as cassava, sweet potatoes and ground nuts), Rweya (mainly used for communal grazing, grass cut for mulch while crop production is mostly limited to Bambaranuts) and natural and planted forests (Rwazo 2012)

management consultants were involved in the development of action plans and activities, with variable benefits.

Table 21.2 summarizes community participation and area coverage over a period of 4.5 years. For each catchment, activities were in four basic categories, including (a) SLM-FFS learning-by-doing methodology, (b) community-based SLM activities mainly on degraded open access resources, (c) demonstrations of successful SLM technologies at easily accessible sites, including Ward Agricultural Resource Centers (WARCs), health centers and/or primary schools and (d) carrying out quick-win Income Generating Activities (IGAs), including the use of

Table 21.2 Catchments area coverage under project implementation

S/N	Summary	Male	Female	Total
1	Number of active farmers	3831	3179	7010
2	Number of adopters	3667	3605	7272
3	Total number of farmers	7498	6784	14,282
4	Area contributed by active farmers			300
5	Area contribute by adopters			3128
6	Total area conserved			3428
7	Total catchments area			64,947

micro-finance institutions. Overall, 14,282 farmers (7498 men and 6784 women) were involved in implementing the project covering a total of about 65,000 ha of land.

To create an enabling environment and sustainable adoption of the tested and demonstrated technologies, institutional and policy frameworks, rules, regulations and by-laws were reviewed and improved and/or mechanisms of their reinforcement proposed. Institutional capacity was also improved through (i) development of land user group management and organizational capacities and formation of catchment committees, (ii) identification of key issues hindering or promoting SLM and ways to improve them, developing useful and workable technical policy recommendations suitable for promoting and sustaining SLM and (iii) fostering collaboration and networking for collective action.

21.3 Catchment Approach Adoption Benefits of Sustainable Land Management

21.3.1 Pasture and Biodiversity Improvement

Grasslands were one of the most degraded land use types mostly due to overgrazing, cattle tracks due to livestock movements in search of pastures and water, the outbreak of fires mostly for killing teaks and pasture regeneration, which, in turn, caused soil erosion and compaction. Most traditional medicine grasses, shrubs and trees became extinct. New technologies, including area closure, introduction of new pasture species, controlled fire outbreaks, review and enforcement of bylaws on grasslands/rangelands resulted in a number of improvements in the grasslands. Table 21.3 indicates changes in forage yields, stock rates and carrying capacity of grasslands in the districts in which the project was implemented. Grassland carrying capacity was improved from 2.5 to 0.7, 5.0 to 0.4, and 3.3 to 0.7 ha/TLU in Ngara, Karagwe and Missenyi districts, respectively. Pasture quality was improved through planting improved pastures that included *Chloris gayana, Clitoria ternatea, Setaria sphacelata, Cenchrus ciliaris, Desmodium uncinatum, Centrocema pubescens, Stylosanthes scraba and Macroptilium atropulpureum.*

In addition to improving pasture quality, recorded extinct and endangered biodiversity of grasses, shrubs and trees were observed in order to allow for possible regeneration of degraded species.

Trees and shrubs that have cultural, socio-economic value, including medicinal plants, poles for construction and thatching grass species, reappeared. Figure 21.3 shows a youth carrying home previously extinct medicinal plants. A variety of insects and ants started working the soil and bringing it to life after years of compaction.

Table 21.3 Forage yield, estimated stocking rate and carrying capacity under managed and degraded grasslands

District	Managed/degraded grassland	Usable forage yield (kg DM/ha)	Monthly stocking rate (AUM)	Carrying capacity (TLU/ha/year)	(ha/TLU)
Ngara	Managed	4000	17.8	1.5	0.7
	Degraded	1000	4.4	0.4	2.5
Karagwe	Managed	7030	31.2	2.6	0.4
	Degraded	570	2.5	0.2	5
Missenyi	Managed	3750	16.7	1.4	0.7
	Degraded	800	3.6	0.3	3.3

Source Kizima (2014)

Fig. 21.3 Recovered biodiversity of medicinal plants collected for local treatments

21.3.2 Recovery of Drying Water Sources

In almost every catchment, water shortage for human and livestock use was recovered through closing the areas against human and livestock activities, for example, cultivation, trees cutting, fire burning and grazing.

The closed areas were planted with water-loving trees such as the Ficus and Markhamia species. In seven micro-catchments, drying water sources were recorded as recovering with clean and reliable water supply for 2–4 months longer during the dry season and for the entire dry season in a few sites. Improved water availability saved many women from traveling long distances in search of water and young children could collect water from wells only a short distance away. Figure 21.4 illustrates a young girl collecting water in the dry season from a well previously dry at such times.

Fig. 21.4 Water collection from a previously dry water source for home consumption in Karagwe district

21.3.3 Increased Soil Quality and Crop Production

The crop yield of bananas, maize, beans and cassava increased from 20 to 60 % compared to normal yields. Beans, for example, increased from an average of 400–650 kg/ha, maize from 300 to 500 kg/ha and bananas from 12,000 to 21,000 kg/ha for farms that were owned by farmers who adopted recommended practices in areas of project implementation. Increases in crop yield were associated with applications of organic fertilizers, including manure and mulch, and inorganic fertilizers (NPK). Other technologies included improved agroforestry activities, such as planting nutrient improvement trees, boundary trees, biological nitrogen fixation and nutrient recycling, as well as fruit trees. Agroforestry systems were also combined with livestock production for manure and income generation. Other technologies included application of soil and water conservation by combining contouring, construction of cutoff drains and construction of sunken beds especially with banana production in low-rainfall areas. In most cases, the contour structures were strengthened by vegetation materials such as *Cajanus cajan*, Lemon grass, Vetiver grass and fruit trees like pineapples. Contours in grasslands were strengthened with Guatemala and Euphobia plants. Evaluation of the soil properties of the tested management technologies indicated a general low content of nutrients in most soils especially for Nitrogen (N), Potassium (P), Phosphorus (K), Organic Carbon (OC), Calcium (Ca) and Magnesium (Mg).

The applied technologies did not have a significant effect on bulk density (BD) and available water capacity (AWC). Figure 21.5 summarizes the soil properties and the changes. Estimates of the level of organic carbon sequestration (Spiotta and Sharma 2013) were considerably higher than 20 %, a project target, probably due to the soil-sampling procedures used.

Planting *Cajanus cajan* as a vegetative contour or within agroforestry systems and the application of different types of agroforestry technologies were superior in improving soil quality, including nitrogen and pH levels. Twenty-one technologies

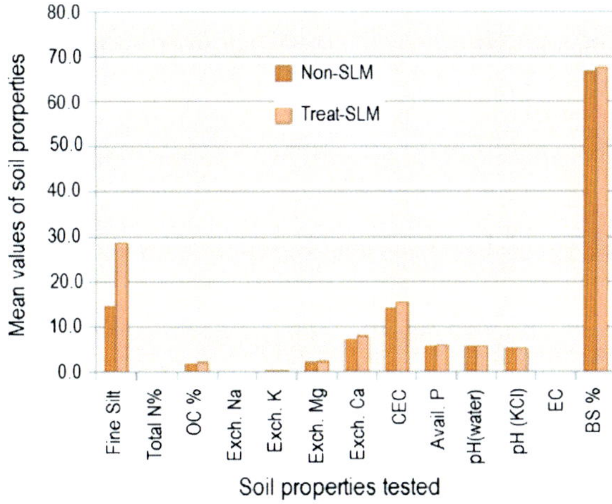

Fig. 21.5 Influence of sustainable land management technologies on soil properties

were analyzed using linear discriminant analysis. The highest score coefficients for key soil indicators were selected for best-performing technologies. They are presented in the following table (Table 21.4). The *Cajanus cajan* and Elephant grass combination was the best in improving soil nitrogen (coeff. 364.59) probably through N_2 fixation. Other *Cajanus cajan* combinations with the highest coefficients were *Cajanus cajan* and Elephant grass (32.82) and Mulching and *Cajanus cajan* and castor (26.28). *Cajanus cajan* also best increased soil phosphorus in grasslands (0.63). Farmers also like *Cajanus* for other benefits apart from soil quality, including food, medicine, firewood and erosion control. *Cajanus* technologies were therefore found to be preferred by farmers and technicians for scaling up and out in the catchments. The tested technologies were applied, however, over a period of 2–2.5 effective years, a relatively short period to have a desirable impact on soil quality.

21.3.4 Recovery of Degraded Hilltops and Sloping Lands

Bare hilltops and eroded sloping lands are a common phenomenon in Karagwe district mainly due to overgrazing, tree cutting, fire outbreaks and the absence of land use plans. Most hilltops have surface cover of less than 20 % and exposed sub-soils due to erosion. Establishment of woodlots on hilltops, mulching and contouring of steep sloping lands have been used to recover degraded and abandoned fields with high yields of planted crops. More than 100 ha abandoned land was recovered. The areas also support planned grazing of cattle and goats. Figure 21.6 illustrates SLM-FFS members in Kibingo village eating pineapples planted on fanya chini contours. In addition to seating, pineapples are sold for household income generation. The fanya chini contours are also strengthened by

Table 21.4 Summary of coefficient of discriminant analysis for different SLM practices

SLM practices	Discriminant coefficient								
	Fsilt	TN	OC	K	P	Ca	Mg	pHw	CEC
Bare land	2.65	230.91	23.44	32.02	0.01	6.75	15.16	38.98	4.02
Bee keeping (tree farm)	2.94	279.52	18.75	27.91	0.01	2.72	12.64	40.3	1.73
Cajanus (grassland)	3.38	218.76	15.91	20.11	0.63	1.27	4.164	35.02	0.96
Cajanus + elephant (pasture)	1.47	364.59	32.82	27.91	0.10	4.81	15.49	38.98	3.30
Cajanus + *Euphobia* spp. + E. grass (pasture)	3.23	279.52	17.10	17.24	0.006	3.31	14.51	39.64	2.22
Grassland	2.5	279.52	22.68	29.55	0.013	1.60	8.79	34.02	1.43
Mulching (banana)	2.65	255.21	20.66	19.7	0.021	5.72	13.78	40.3	3.39
Mulching + agroforestry (cropland)	2.06	267.37	16.44	105.0	1.105	3.18	11.55	44.26	2.07
Mulching + Cajanus + castor (banana)	2.06	255.21	26.28	30.38	0.027	4.2	13.42	40.96	2.29
Mulching + compost + Manure (banana + maize)	2.35	243.06	22.0	27.09	0.001	1.37	4.37	32.37	1.28
Mulching + L. + vertiver grass (cropland)	2.35	279.52	24.3	20.52	0.023	4.15	13.3	42.28	2.45
Stoneline + agroforestry (cropland)	2.35	315.98	21.9	29.55	0.02	1.29	4.02	31.71	1.32
Vertiver grass (cropland)	3.53	267.37	17.6	31.20	0.00	2.65	14.05	38.98	1.89

Fig. 21.6 Sustainable land managementSustainable land management benefits through planting pineapples as micro-contours

planted *Cajanus cajan* on top. This species fertilizes the soil by fixing nitrogen and adding other nutrients. The peas are eaten as food while the leaves are used as medicine. Although recovery from erosion is gradual or long term, eating and selling pineapples and peas are immediate or short-term benefits of SLM, which are incentives for adopting and sustaining SLM in catchments.

21.3.5 Improved Knowledge and Skills in Sustainable Land Management

Low crop production has always been associated with use of indigenous crop varieties by farmers and extension staff. Soil erosion has not been considered a problem as there is some soil left after erosion and one can go elsewhere if one's farmland turns into rocky outcrops. Through learning by doing land management activities of FFS-SLM groups, farmers have now gained knowledge about the effects of soil erosion and have gained skills to control soil erosion. Farmers active in FFS-SLM groups can make correct decisions and actions to control soil erosion in their plots and other people's fields. After training in contouring, several farmers can now make their own contours in their fields, as well as neighbors' fields.

21.3.6 Policy and Legislation Issues

A number of existing land-related legislation and bylaws were reviewed, and at least two bylaws per micro-catchment were developed in support of the developed SLM practices. Recommendations for improving existing natural resources policies and technical recommendations for addressing identified natural resources mismanagement issues were developed and presented to policy and decision makers for action.

Specifically, the technical recommendations addressed the following: common property resource management in catchments (7); land- and natural resource–related conflicts (3); land use planning and management (4); and trans-boundary issues of land degradation and conflicts related to livestock management and movement (10) and beach management (7). Three policy recommendations were developed that addressed land use planning and management (2) and priority trans-boundary conflicts management and policy (1). The numerals inside the parentheses are the number of recommendations developed for each key issue.

21.4 Lessons Learned

- The catchment and SLM-FFS approaches were well received by research and extension as holistic and appropriate for targeting technologies, knowledge and skills related to natural resource management.
- Despite the sensitization of decision makers and policy makers to the extent and impact of LD during project implementation, prioritization and financing remain low
- Although soil erosion and fertility decline were priority issues of LD in the past, water shortage has become the primary critical limitation in rural and urban communities
- Although there were few improved SLM technologies on the ground, there are several indigenous and effective undocumented SLM technologies for dissemination
- The land use system for managing rangeland and pastures needs closer attention as the system has been neglected
- Land use conflicts and management and land tenure systems have been identified as an important issue but with limited attention and investment and should be addressed for successful SLM
- Limited use of GIS at the district level is a major limitation to monitor LD and develop targeted and timely SLM programs to address land degradation in catchments
- Income-generating activities are an incentive measure for rural communities to invest and adopt SLM. Most such activities should also be aimed at being inputs in terms of materials and cash.

21.5 Challenges

(i) Low prioritization of LD/SLM issues from the national to district levels. Due to the low prioritization of SLM, there is hardly any financial investment to control land and other natural resource degradation. Sensitization of policy and decision makers to invest in LD/SLM was one of the key project activities

(ii) Site facilitators are unfamiliar with LD/SLM issues and their effective management. Extension staff are also inadequate to cover areas that require extension services. Custom training that addresses land management is uncommon. The project conducted many SLM-related trainings for the majority of the extension staff as well as farmer animators to address the problem

(iii) The degradation of natural resources are addressed on an individual sector basis, but the problems are multi-sectoral. The Kagera TAMP project established multi-sector district teams to address LD/SLM following the catchment approach. A pilot multi-sector SLM team has been established, but its facilitation in the absence of a follow-up project is uncertain.

21.6 Way Forward

To establish a National SLM Team and, finally, a National Strategic Investment Framework that emphasizes a multi-sector, multidisciplinary and catchment/watershed approach to achieve sustainable land management. In addition, to follow up on processes to mainstream developed technical and policy recommendations into national and district development plans.

References

Kaihura FBS (2012) Sites selection criteria and process in Tanzania. Kagera TAMP Project. Agriculture Research Institute, Bukoba

Kizima JB (2014) Land degradation and natural resources conflicts under livestock management and movements and strategies and options for sustainable and equitable land and livestock management

Rwazo J (2012) Kyazi micro-catchment characterization. In: Sites characterization and management plans for the Kagera Transboundary Agro-ecosystems Management Project. Land Use Planning and Management Division, Ministry of Agriculture Food Security and Cooperatives, United Republic of Tanzania

Spiotta EM, Sharma S (2013) Carbon storage in succession and plantation forest soils: a tropical analysis. Global Ecol Biogeogr 22:105–117

Chapter 22
The Role of Forages in Sustainable Intensification of Crop-Livestock Agro-ecosystems in the Face of Climate Change: The Case for Landscapes in Babati, Northern Tanzania

F. Kizito, B. Lukuyu, G. Sikumba, J. Kihara, M. Bekunda, D. Bossio, K.W. Nganga, A. Kimaro, H. Sseguya, B. Jumbo and P. Okori

Abstract Agro-ecosystem productivity is highly dependent on soil moisture fluxes yet climate change induces unpredictable dynamic interactions on water and nutrient resources. This study assessed on-farm seasonal productivity, runoff and soil moisture storage estimates within forage grass and forage legume intercrops at the Long site in Babati District of Northern Tanzania and how these would be impacted by climate change. The WaterWorld model was used to ascertain the impact of climate change on temperature and moisture fluxes at landscape level within these agro-ecosystems. Study results revealed a steady increase in temperature and a projected increase in rainfall over the next 40 years to the 2050s with an average future precipitation of 1300 mm yr^{-1} compared to the current baseline of 960 mm yr^{-1}. On-farm seasonal water balance estimates within forage grass–forage legume intercrops revealed that with the 645 mm of rainfall received in the 2014 rainy season, evapotranspiration (ET) was the predominant factor accounting for about 75 % of the fluxes. We demonstrate that compared to the control trials, runoff

F. Kizito (✉) · J. Kihara · D. Bossio · K.W. Nganga
International Center for Tropical Agriculture (CIAT), P.O. Box 823-00621, Nairobi, Kenya
e-mail: f.kizito@cgiar.org

B. Lukuyu · G. Sikumba
International Livestock Research Institute (ILRI), Nairobi, Kenya

A. Kimaro
International Center for Research in Agroforestry-World Agro-Forestry Center (ICRAF), Nairobi, Kenya

M. Bekunda · H. Sseguya
International Institute of Tropical Agriculture (IITA), Ibadan, Nigeria

B. Jumbo
International Maize and Wheat Improvement Center (CIMMYT), El Batan, Mexico

P. Okori
International Centre for Research in Semi-Arid Tropics (ICRISAT), Patancheru, India

R. Lal et al. (eds.), *Climate Change and Multi-Dimensional Sustainability in African Agriculture*, DOI 10.1007/978-3-319-41238-2_22

levels were significantly lower in areas with forage grass–legume intercrops which translated to 20 % lower runoff levels; there was higher soil moisture storage with an average of about 25 mm (30 % higher) in areas with forage grass–forage legume intercrops than the bare plot control areas. The Napier-Desmodium and Napier-Lablab combinations had about 15 % higher soil moisture storage and 30 % higher water productivity compared to the sole Napier accessions. The sole forage grasses depicted about 15–50 % higher runoff levels compared to the Napier-Desmodium and Napier-Lablab combinations. In doing so, a combination of perennial forages (grasses and legumes) improves the sustainability of farming systems through erosion control and soil moisture retention beyond serving as feed resources. Using both qualitative and quantitative metrics from this study, we draw on the sustainable intensification indicators framework to illustrate explicit linkages on synergies and tradeoffs associated with forage interventions within smallholder farming systems. Sustainable intensification within these landscapes will thus require more innovative solutions that incorporate establishing different types of alternative forage grass–forage legume combinations coupled with other improved agronomic practices into a compendium package of interventions that allows for sustainable land use to cope with climate change and variability.

Keywords Sustainable intensification · Climate change · Adaptation · Farmer options · Innovative solutions

22.1 Introduction

Historically, agroecosystems the world over have responded rather resiliently to the increasing pressure for producing food for an expanding human population (Robertson et al. 2014). As a result, it is not surprising that recent years have witnessed a gradual but steady increase in urbanization and prominent rise in incomes of emerging economies (Cohen 2006), with shifting of human diets toward higher consumption of calories, fats, and animal products (Nair 2014). This therefore calls for exploring novel and sustainable ways of intensifying agro-ecosystems to ensure higher crop and forage productivity that reduces competition between man and livestock for food and feed respectively. This is more pertinent than ever because climate change is among the plethora of factors affecting crop and livestock productivity resulting in negative impacts on livelihoods in semiarid landscapes as evidenced in portions of central and northern Tanzania.

Climate change is further expected to exert more pressure on water and agriculture with potential negative impacts on livelihoods. The vulnerability of Northern Tanzania is high due to the large number of households that depend on the natural resource base for their livelihood. Consequently, there is a growing need for 'anticipatory adaptation', in a more proactive rather than reactive management of climate change risk. The productivity of agro-ecosystems in the region is controlled primarily by water dynamics an aspect that is intrinsically linked with the amount and distribution of rainfall. This also affects agricultural productivity among smallholder farmers in SSA, namely crop enterprises, cropping calendars, incidence and growth of weeds, crop pests and diseases. This erratic variability of climate exacerbates environmental vulnerabilities which in turn affect the poorest segments of society. Recent studies indicate that 40 and 26 % of agro-pastoralists in Kiteto and Longido districts respectively identified climate variability and extreme climate events, especially, as the major challenge to sustained livestock and agricultural productivity. In particular, frequent and prolonged drought and insufficient pasture of good quality and quantity were noted as results of climate variability (Coulibaly et al. 2015) that impact sustainable intensification of crop-livestock mix agro-ecosystems. As a result of these climate induced seasonal changes, livestock death and crop failure are frequent in the two districts.

Sustainable intensification innovations, such as integrated land and water management practices and agroforestry practices, can provide win–win solutions through improving yields and land and animal productivity; hence food security. Other associated benefits include improved ecosystem services and socioeconomic benefits, and increased resilience to climate change and associated extreme weather events, such as water scarcity, intense rainfall, or droughts. These benefits occur as a result of increase in soil organic matter, improved soil structure, reduced soil erosion, increased water filtration and efficiency of water use, replenishing of soil nutrients, and increased efficiency of nutrient uptake (Winterbottom et al. 2013). For instance, in situ rainwater harvesting complemented with agroforestry and/or nutrient management practices such as micro dosing has been known to double or triple crops yields in the Sahel (Winterbottom et al. 2013), a region with similar climatic conditions to semiarid central Tanzania. These practices are currently being promoted and scaled up within Africa RISING sites in Babati, Kongwa and Kiteto to sustainably intensify farming systems to increase yields, reduce land degradation and increase community resilience through diversified production and income options (Okori 2014).

This study (1) Assessed forage water productivity within forage grasses–forage legume intercrops compared to sole forage grass monocrops and bare control plots; (2) Determined on-farm erosion, runoff and soil moisture storage dynamics within forage grasses–forage legume intercrops compared to sole forage grass monocrop; and bare control plots; (3) Projected regional climatic trends that impact on both farm-scale and catchment-scale water management in Northern Tanzania over the next 40 years to the 2050s; (4) Assessed study results against the sustainable intensification indicators framework to discern synergies, tradeoffs and minimize unintended negative consequences in future work. We posit that where applicable, incorporation of forage grass and forage legume combinations into smallholder farming systems (from farm-scale to landscape level) will play a critical role towards higher crop and forage water productivity, increased soil retention and nutrient composition and improved agricultural soil moisture management.

22.2 Materials and Methods

22.2.1 *Site Characteristics*

The study was conducted in the Babati district of Northern Tanzania (Fig. 22.1), located between the latitudes 3° and 4° south and the longitudes 35° and 36° with an altitude between 1650 and 2250 m above sea level. The Region is a part of the Great Rift Valley and the landscape is characterized by mountains, undulating hills and plains. The precipitation varies with the altitude and ranges from 1200 mm/year in the highlands down to 500 mm/year in the lowlands. The rains are predominantly unimodal with the major rains of the growing season between February and May (Bishop-Sambrook 2004). Based on description given by Kihara et al. (2014), the area is characterized by low fertilizer use and has one lengthy growing season between November and June. Maize is mainly grown as an intercrop with a late maturing pigeon pea (*Cajanus cajan* L. Millsp.) cultivar. The soils are mainly of volcanic origin and range from sandy loams to clay alluvial soils. The content of organic material and availability of phosphorus is generally low across the district (Jonsson 1996). Many farmers in Babati District are agro-pastoralists and the number of livestock in the area is high, livestock rearing constitutes about 35 % of the overall land use in the district (Shetto and Owenya 2007). In some areas, farmers practice traditional post-harvest grazing which is not compatible with systems where soil cover is desired or where contour bunds are practiced.

22.2.2 *Experimental Setup*

A total of three Napier grass accessions (KK1, KK2, and ILRI 16837) were grown and harvested every 6 weeks at an on-farm trial replicated three times (Fig. 22.2). The replications were a combination of Napier grass (*Pennisetum purpureum*) accessions with Desmodium (*Desmodium uncinatum*), Lablab (*Lablab purpureus*) and the sole Napier grass of each accession (KK1, KK2, and ILRI 16837). The choice of these forage combinations were a result of prior participatory variety assessments involving 77 farmers on the field trials using a rating and voting exercise where farmers identified and ranked their preferred characteristics. The main characteristics identified by farmers included the number of leaves and shoots, tolerance to drought, rapid regeneration and length of stem after harvest. In addition, control plots that had neither sole forage grass nor forage grass–forage legume combinations were used to discern soil moisture flux differences.

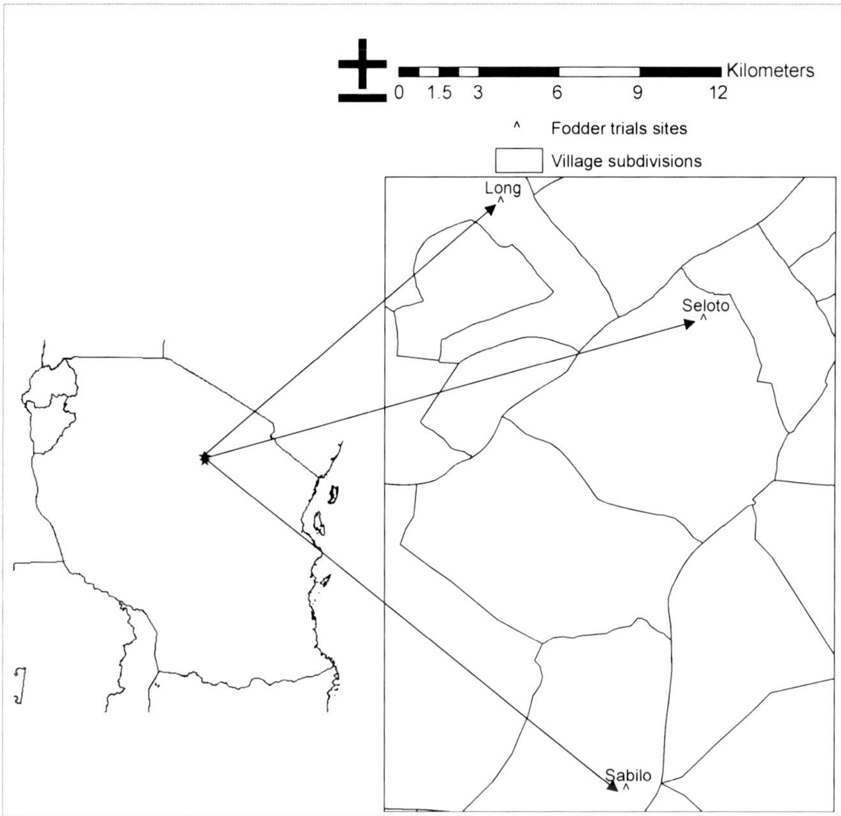

Fig. 22.1 Location of forage grass–forage legume trials in the villages of Long, Seloto and Sabillo in Babati district, Manyara region of Tanzania (*Note* that this paper only reports results from Long site)

22.2.3 *Micro-Climatic Data Collection for Forage Water Productivity Estimates*

All micro-climatic parameters were measured using an automated weather station (Spectrum 9 Technologies) at hourly intervals. Rainfall was monitored with a tipping bucket rain gauge (0.5 mm per tip) and evapotranspiration was estimated using the modified FAO Penman–Monteith approach at hourly intervals. Daily reference crop evapotranspiration (ETo) was computed from measured meteorological data; namely solar radiation, air temperature, relative humidity and wind speed. The FAO Penman–Monteith equation (Allen et al. 1998) used for hourly time steps (for a well-watered crop) in this study Eq. (1) is:

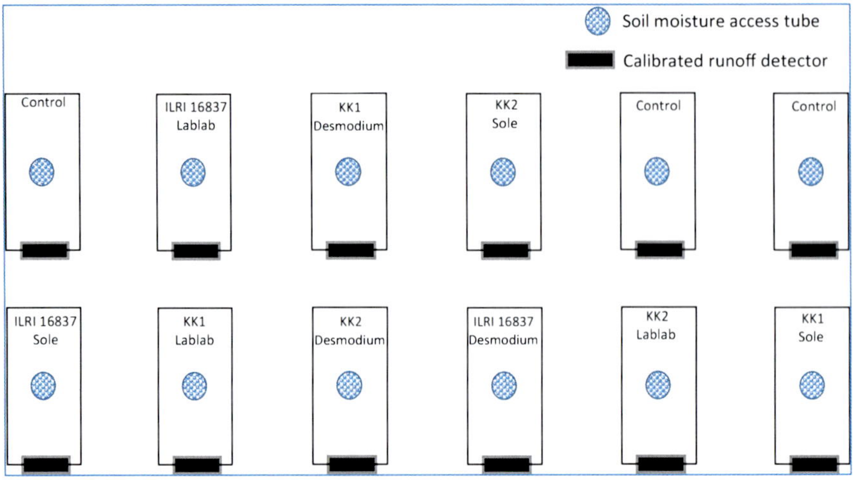

Fig. 22.2 Experimental set up of forage-grass and forage-legume interactions showing soil moisture access tubes and runoff soil trap detectors with each plot measuring 10 m × 5 m

$$ETo = \frac{0.408\Delta(R_n - G) + \gamma\left(\frac{37}{T_{hr}+273}\right)u_2(e_s - e_a)}{\Delta + \gamma(1 + 0.34u_2)} \tag{22.1}$$

where ETo is the reference crop evapotranspiration (mm h^{-1}), Rn the net radiation (MJ m^{-2} h^{-1}), G the soil heat flux density (MJ m^{-2} h^{-1}), T_{hr} is the mean hourly air temperature (°C), (e_s–e_a) the hourly vapor pressure deficit of the air (kPa), Δ the slope of the saturation vapour pressure function (kPa $°C^{-1}$), γ the apparent psychrometric constant (kPa $°C^{-1}$), u_2 is the average hourly wind speed (m s^{-1}) measured at 2 m above the soil surface.

Forage water productivity is the amount of water required (crop evapotranspiration, ETc) per unit of biomass yield (Amede et al. 2009) and is a vital parameter to assess the performance of agricultural systems for targeted integrated water resources management.

$$Forage\ Water\ productivity = \left(\frac{Forage\ yield(Y)}{Forage\ Evapotranspiration\ (ET_c)}\right) \tag{22.2}$$

Forage water productivity will vary greatly according to the specific conditions under which the crop is grown. There are standard procedures used to assess forage water productivity in the context of the framework of water management applications and practices (FAO 2006). A suite of these procedures was used in combination with the seasonal forage yield averages from primary field data to estimate forage water productivity at the study site.

22.2.4 *On-Farm Runoff and Soil Moisture Storage Dynamics Within Forage Trials*

Within the Napier grass accessions, soil moisture measurements were conducted using a Diviner 2000 Probe Series. Measurements were conducted every week over a 2-year period (2014–2015) within the forage trials. The Diviner 2000 probe soil moisture data was calibrated gravimetrically under field conditions. For the vertical profile study, measurements were conducted at 0.10 m depth increments to a depth of 1.0 m. Profile stored water was calculated on a depth basis as the product of volumetric water-content and the depth interval (0.10 m) and expressed as millimeters of water. In this study, we present a mean value of soil moisture storage for the 0–40 cm depth range. Erosion assessments were conducted with flexible corrugated iron cubic boxes of 15 cm dimension providing a total cubic volume of 3375 cm^3 as soil traps.

22.2.5 Climate Change Assessment and Projections

WaterWorld is a support modeling platform for simulation of hydrological systems and human impacts upon natural resources. The model is designed for application by stakeholders at the local to international scale in order to understand the baseline distribution of water and the impact of land use, land management and climate change upon the natural resource base (Mulligan et al. 2010). Within the modeling platform, Babati District was defined for the study analysis by using a one degree tile (high resolution) covering a 1 hectare resolution. The climate change simulation was for the tile with boundaries 10.0 (to the N), 9.0 (to the S), −1.0 (to the E and) −0.0 (to the W). The extreme west of the District (about 5 %) fell outside the designated tile while 95 % of the Region was captured. A baseline scenario was run which showed the current state of the system then an alternative run for water balance dynamics and climate change scenarios in the Region was conducted. The baseline run yielded mean monthly air temperature and total precipitation for each month of the year. However, only results for selected months are presented herein to highlight major trends within the annual cycle.

For each baseline, an alternative scenario was generated for water balance and climate change scenarios. The scenario characteristics for water balance estimates were based on global hydrological data sets while the climate change scenarios were based on IPCC assessments downscaled for various regions. The scenario chosen was the 'AR4' upgrade which includes the 2000 IPCC Special Report Emission Scenarios (SRES): uncertainty of future GHG emissions given a wide range of driving forces; no climate policies; complemented by storylines/narratives of the future; open process involving many different modeling teams (IPPC 2000). Thins study used emission scenario 'A2 emission scenario' which is based on the hypothesis that 'the world evolves in a very heterogeneous way, the world

Table 22.1 Climate change scenario properties used for Babati, Tanzania

Variable	Value
IPCC assessment report	AR4
Emissions scenario	A2a
Downscaled by	CIAT
GCM name	Mean of all models plus one standard deviation
Projection year	2050s

population reaches 15 billion people in 2100, and rising, economic growth and the spreading of new efficient technologies are very different depending on the region of the world'. The GCM platform used was the mean of all models plus one standard deviation and the scenarios were projected to 2050. A summary of the scenario attributes used in this study are presented in Table 22.1.

22.2.6 Data Analysis

Forage yields, forage water productivity, runoff and soil moisture storage data were statistically analyzed with SAS V8 (2001) for two treatments factorial random block design. Since sampling was conducted on the same individuals over time (forage grasses, forage legumes and soil moisture) data were analyzed using a repeated measures model. Two factorial ANOVAs with replication were conducted to ascertain the interactions between the forage grasses and forage legumes and test if the mean values for forage water productivity, runoff and soil moisture storage were significantly different at $P = 0.05$.

22.3 Results

22.3.1 Forage Biomass and Water Productivity Trends

Farmers ranked the accessions in the following order: KK2, ILRI 16837 and KK1 as first, second and third best accessions on overall preference respectively. Among the three accessions, ILRI 16837 produced the highest yield (mean = 1.77 t ha^{-1} (DM); sd = 0.93). The number of tillers showed a significant ($P < 0.05$) positive relationship with dry matter yield for all the 3 accession forage grass–forage legume combinations. A two way factorial ANOVA analysis revealed that there were significant differences in overall dry matter results of the three forage grass and forage legume combinations.

As depicted in Table 22.2, considering the forage legumes analysis, the F distribution results revealed that $F(2,18) = 10.58$, $P < 0.05$ and P value (0.001) is <0.05 hence we showing that forage legumes had a significant difference on the

Table 22.2 ANOVA results for forage grass–forage legume dry matter harvest combinations for 2014–2015

Source of variation	SS	df	MS	F	*P* value	F critical
Forage legumes	34.991	2	17.496	10.582	0.001	3.555
Forage grasses	46.796	2	23.398	14.153	0.0002	3.555
Interaction	14.524	4	3.631	2.196	0.110	2.928
Within	29.758	18	1.653			
Total	126.070	26				

outcome of the dry matter biomass. The forage grasses analysis showed F distribution results of $F(2,18) = 14.15$, $P < 0.05$ and P value (0.0002) is <0.05 hence showing that forage grasses also had a significant difference on the outcome of the dry matter biomass with less than 0.09 % chance of getting these values by a random chance. The forage grass–forage legume interactions the F distribution results revealed: $F(4, 18) = 2.196$, $P < 0.05$ and P critical value (0.110) is >0.05, additionally, the F critical value (2.928) is >than the F value (2.196) (Table 22.2); revealing that forage grasses–forage legume interactions did not have a significant effect on the outcome of the dry matter biomass yields.

Water productivity statistical analysis (Table 22.3) for forage legumes analysis, the F distribution results revealed: $F(2,24) = 109.64$, $P < 0.05$ and P value is <0.05; F critical value (3.403) is less than the F value showing that forage legumes had a significant effect on the outcome of the water productivity results. The trends in Fig. 22.3 revealed that both KK2 and ILRI 16827 were superior to KK1 with the Desmodium legume combinations. Water productivity statistical analysis (Table 22.3) indicate that forage grasses and the forage grass–forage-legume interactions had significant influence on the water productivity results. Clearly graphical trends (Fig. 22.3) depict that KK2 and KK1 were superior to ILRI 16827 with both the Lablab and sole components over the two year period. On the overall, the Napier-Desmodium combination performed better than the Napier-Lablab combination which in turn outperformed the sole forage grass.

Table 22.3 ANOVA results for forage grass–forage legume water productivity combinations for 2014–2015

Source of variation	SS	df	MS	F	*P* value	F critical
Forage legumes	0.754	2	0.377	109.636	8.5E−13	3.403
Forage grasses	3.613	3	1.204	350.240	6.13E−20	3.009
Interaction	0.276	6	0.046	13.362	1.26E−06	2.508
Within	0.083	24	0.003			
Total	4.725	35				

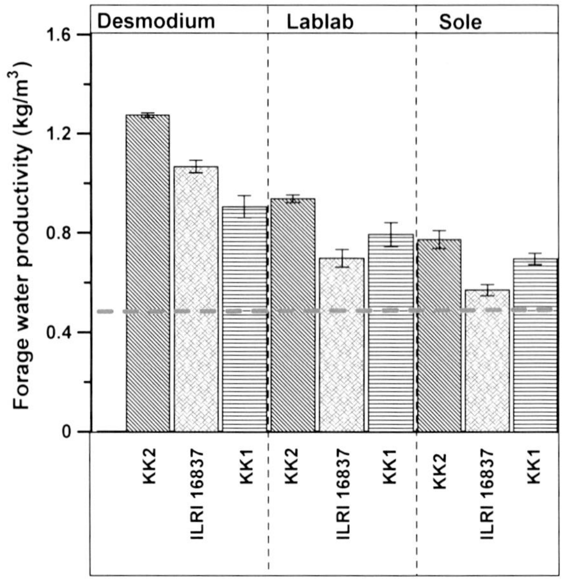

Fig. 22.3 Mean water productivity trends among forage grass–forage legume combinations for 2014 and 2015. Note the *green dotted line* as the global average threshold for forage water productivity (Sala et al. 1988)

22.3.2 *Runoff and Soil Moisture Storage Dynamics in Forage Trials*

Runoff results (Fig. 22.4) indicated that there were significant differences between the forage grass–forage legume combinations and the control. The control had significantly higher runoff regimes (>60 %) than the grass–legume combinations over the 2 year period. Likewise, sole Napier accessions showed significantly higher runoff levels than the Napier-Desmodium and Napier-Lablab combinations. The differences between Napier-Desmodium and Napier-Lablab (Fig. 22.4) in runoff control were not easily discernible though Desmodium registered slightly lower runoff values.

The two-way ANOVA analysis (Table 22.4) for forage legumes had F distribution results with $F(2,24) = 118.56$ at $P < 0.05$ yet the P value is much smaller than 0.05 and the F critical is less than the F value hence revealing that forage legumes had a significant effect on the outcome of the mean annual runoff. Considering the forage grasses analysis, the F distribution results revealed: $F(2,24) = 3799$ at $P < 0.05$ yet the P value is much smaller than 0.05 and the F critical is less than the F value hence revealing that forage grasses too had a significant effect on the outcome of the mean annual runoff. The forage grasses–forage legume interactions as well depicted that they had a significant effect on the mean annual runoff.

There were differences observed in soil moisture storage among the forage grass–forage legume combinations. Results indicate that the combination of KK2-Desmodium had significantly higher soil moisture storage than the other

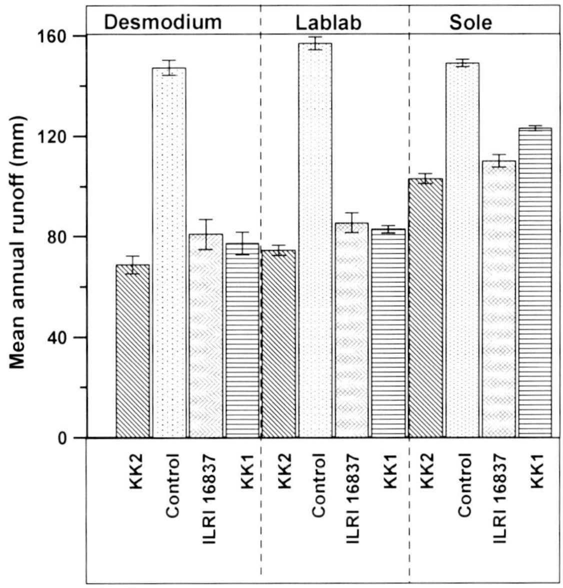

Fig. 22.4 Runoff trends among forage grass–forage legume combinations over 2 years 2014 and 2015

Table 22.4 ANOVA results for forage grass–forage legume mean annual runoff for 2014–2015

Source of variation	SS	df	MS	F	*P* value	F critical
Forage legumes	683.298	2	341.649	118.562	3.63E−13	3.403
Forage grasses	32,845.070	3	10,948.360	3799.406	2.98E−32	3.009
Interaction	320.849	6	53.475	18.557	6.15E−08	2.508
Within	69.158	24	2.882			
Total	33,918.380	35				

combinations. The two-way ANOVA results (Table 22.5) for forage legumes analysis had F distribution results with $F(2,24) = 75.48$ at $P < 0.05$ yet the P value is much smaller than 0.05 and the F critical is less than the F value hence revealing that forage legumes had a significant difference on the mean soil moisture storage. Considering the forage grasses analysis, the F distribution results revealed: $F(3,24) = 1342$ at $P < 0.05$ yet the P value is much smaller than 0.05 and the F critical is less than the F value indicating that forage grasses too had a significant effect on the soil moisture storage. The forage grasses–forage legume interactions as well depicted that they had a significant impact on the mean soil moisture storage.

Table 22.5 Two way ANOVA with replication for forage grass–forage legume soil moisture storage for 2014–2015

Source of variation	SS	df	MS	F	*P* value	F critical
Forage legumes	150.967	2	75.483	39.559	2.53E−08	3.403
Forage grasses	4026.958	3	1342.319	703.489	1.64E−23	3.008
Interaction	319.973	6	53.328	27.948	1.06E−09	2.508
Within	45.794	24	1.908			
Total	4543.692	35				

22.3.3 Climate Change Assessments

Both temperature (Fig. 22.6) and rainfall trends (Fig. 22.7) for the region revealed that there was a significant increase in the regional temperature and rainfall amounts respectively. For example, model results revealed that the total monthly rainfall is projected to have a 10 % increase in February and comparisons between the annual results for the alternative 2050s scenario and baseline conditions revealed that there would be a mean precipitation increase of about 360 mm/year. Beside the increment in amount for each month, there was about 15 % higher increment reported for non-conventional rainfall months (Fig. 22.7).

Results of the General Circulation Models (GCM) used by OECD indicated that the temperature will rise by 2 °C by 2050. The highest increase in temperature will be during the cooler period, June–August and lower in the warmer period Dec–Feb as depicted in the Table 22.6. Initial assessments by the Tanzania Adaptation Team indicate that there will be an increase in daily mean temperature by 3–5 °C throughtout the country and an average annual mean increase by 2–4 °C (Tanzania Adaption Team 2006).

Predictions of changes in rainfall are less certain with very pronounced differences among the different GCM models. However an increase of about 10 % is the most commonly accepted value. According to OECD, the distribution will also be uneven, with a 6 % predicted decrease in Jun–Aug and a 17 % increase in Dec–Feb. Changes will not be distributed accordingly over the whole country however, some parts will receive an increase while other parts a decrease. Changes will not occur in the same time and timing and intensity of rains will be less predictable (Häckner 2009). Changes in rain season patterns could also be significant, in the northern parts, the amount of rain during the short rain period could increase by 25–

Table 22.6 Estimated temperature changes in Babati based on GCM (Agrawala et al. 2003; Maddison 2007)

Year	Temperature changes				
	Annual	Jun–Aug	Sept–Nov	Dec–Feb	Mar–May
2030	0.9	1.0	0.8	0.8	0.9
2050	1.3	1.5	1.2	1.1	1.3

60 % and the amount in the long rain period by 20–45 %. The distribution of increased rain may also be uneven with an increase during the long rain period and a decrease of the short rains (Häckner 2009).

22.4 Discussions and Conclusion

The overarching message of this study is that the forage-environment-human nexus is important but under-researched and that huge opportunities exist to improve the productivity of water associated with forage production. Peden et al. (2007) illustrated that water that is used to produce 1 kg of dry animal feed through evapotranspiration is highly variable, ranging from about 0.5 to 8 kg m^{-3}. Many factors affect the amount of water depleted through evapotranspiration, including the vegetative leaf area index, root depth, rainfall, plant genetics, soil structure, moisture, and soil nutrient composition. The forage yield and forage productivity results (Tables 22.2 and 22.3) indicated that both grasses and legume combinations with Napier had a significant contribution to overall biomass yield and productivity. Sala et al. 1988 analyzed 9500 sites throughout the central United States and found that the water productivity of diverse temperate grasslands receiving 200–1200 mm of annual rainfall was similar, at about 0.5 kg of aerial biomass per cubic meter of evapotranspiration, with productivity slightly higher in wetter sites than in drier ones. The forage water productivity in this study was above the 0.5 kg m^{-3} threshold, the higher levels of water productivity are potentially because the cumulative evapotranspiration was measured only during plant growth without representing year-round calculations of evapotranspiration (Table 22.7).

In Babati, an area with inherently low biomass landscapes, the ability to produce sufficient forage products while simultaneously sustaining the natural resource base (soil, water, air and biodiversity) is a key issue confronting the future farming practices. Thus improving productivity and reducing the existing wide gap between actual and maximal forage yields will contribute towards alleviating food insecurity through enhancing forage production with suitable forage grass–forage legume combinations.

Drawing on results from field measurements, we demonstrate that forage–water interactions serves as an entry point to better understand the wider dimensions and complexity of agricultural water use in resource scarce landscapes. We thus invoke

Table 22.7 Estimated rainfall changes in Babati based on GCM (Agrawala et al. 2003; Maddison 2007)

Year	Precipitation changes				
	Annual	Jun–Aug	Sept–Nov	Dec–Feb	Mar–May
2030	4.1	−2.4	3.9	6.6	2.2
2050	5.9	−3.5	5.6	9.6	3.1

the Sustainable Intensification Indicators framework using both quantitative study results as well as qualitative assessments to deduce synergies and associated tradeoffs in a bid to minimize any unintended negative consequences in future work. The in-depth understanding of these interactions if explored with the sustainable intensification indicators framework will help to explore alternative options for improving the use of scarce water, soil and feed resources. Because forage water productivity is a function of both forage biomass yield and water input, there is a need to consider practical avenues for enhancing forage biomass alternatives along with water use efficiency in a manner that is more compatible to the specific local contexts.

The sustainable intensification indicators framework aims at providing a synthesized list of sustainable agricultural intensification (SI) indicators and metrics, categorized into five domains (economic, human condition, environmental, social and productivity) (Fig. 22.8) and three scales (field farm/households, and landscape). Regardless of the size of the land area covered, water enters an agricultural system in the form of rain or surface inflow. Water is depleted or lost through transpiration, evaporation, and runoff and cannot be readily used again. Runoff results (Fig. 22.4) and soil moisture storage trends (Fig. 22.5) clearly demonstrate that introducing management practices such as cover crops (Desmodium and lablab) that promote beneficial evapotranspiration or infiltration of available water will likely increase forage water productivity (Fig. 22.3). The forage legumes not only target rapid early growth to shade the soil and reduce evaporation but also improve

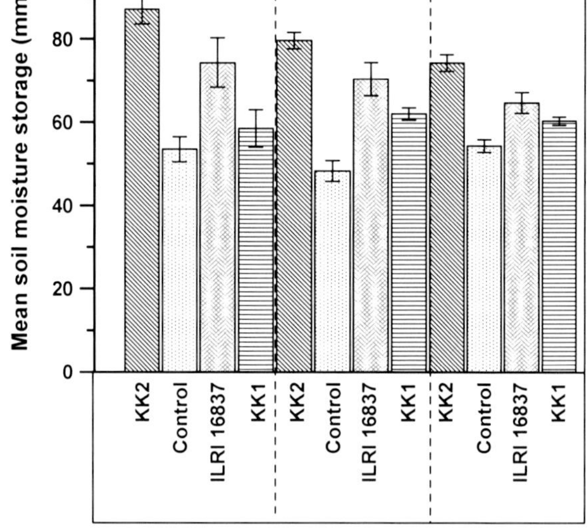

Fig. 22.5 Mean soil moisture storage trends among forage grass–forage legume combinations over 2 years 2014 and 2015

the nutritional quality of forage (Descheemaeker et al. 2009). Nyambati et al. (2003) and Kabirizi et al. (2007) reported that these forages contribute to soil fertility through the fixation of atmospheric N while serving as an excellent food source.

The case example drawn from Babati highlights a specific production system and the need for integrated site-specific interventions to ensure that agricultural production contributes to sustainable and productive use of water resources and to improved livelihoods of the poor. Study results demonstrated that lessening non-productive evaporation is possible through the use of forage legumes acting as cover crops, enhancing soil infiltration and increasing soil storage thus reducing irrecoverable deep percolation and surface runoff (Figs. 22.4 and 22.5; Tables 22.4 and 22.5). For this study, both the productivity and environmental domains of the SI indicators pentagon showed synergistic linkages. The forages are easy to establish and fast growing hence not only provide sufficient biomass for fodder but also have the capacity to stabilize land and gullies (Magcale-Macandog et al. 1998), thereby leading to water conservation as well. The aforementioned quantitative data on forage grass–forage legume combinations in relation to biomass and water productivity depicts that the productivity component of the SI indicator framework is strong (Fig. 22.8). Similarly, the runoff and soil moisture storage trends clearly depict the strength of having forage interventions within farming landscapes to reduce on soil erosion losses and downslope sedimentation while enhancing soil water infiltration, aspects that are strong in the environmental domain of the SI indicators pentagon (Fig. 22.8).

We surmise that if farmers conduct forage production with a business lens, then the economic domain could be more pronounced and would follow the green trajectory depending on the external prevailing factors such as policy, market structures and cultural preferences (Fig. 22.8). Additionally, innovative use of dual purpose cover crops such as cowpea (*Vigna ungulata*) (Tarawali et al. 1997; Singh et al. 2003) could provide higher nutritional benefits for household consumption in addition to serving as fodder for livestock. This would potentially follow the red trajectory (Fig. 22.8) to increase the human domain of the SI indicators pentagon. The role that forage–grass and forage legumes combinations (Napier-Desmodium and Napier-Lablab) play towards improving the nutritive value of fodder cannot be underestimated (Zhang et al. 2009). Adding these sources of nutritive fodder to Napier grass as animal diets improve feed conversion and increase digestibility (Descheemaeker et al. 2009) hence reducing methane emissions from enteric fermentation (Herrero et al. 2008) thus providing positive outcomes on the environmental domain of the SI framework (Fig. 22.8) through climate change mitigation. This may in turn increase resilience and adaptive capacity of smallholder farmers (an aspect that would enhance the under-represented social domain in Fig. 22.8).

In semi-arid environments up to 90 % of rainfall evaporates back into the atmosphere, leaving just 10 % for productive transpiration. Micro- and macro-catchment management techniques that can capture more of this water such as use of forage grass and forage legume cover crop combinations for subsequent crop use before it evaporates, increase beneficial rainwater available for transpiration to 20–50 % (Oweis et al. 1999). Agricultural water management practices can provide multiple ecosystem services beyond food production. For example, the value of forage legume and forage grass combinations is underestimated unless its multifunctional roles are taken into consideration. These practices reduce environmental costs and enhance ecosystem services increase the value derived from agricultural water management (Matsuno et al. 2006). In rangelands, especially dry ones, forage water productivity is low, but there are few alternate uses of agricultural water, only a small part of the evapotranspiration typically attributed to pasture production is actually used by grazing animals. Typically, about half of plant biomass production takes place below ground. In well managed pastures only about half of the above biomass is consumed by grazing animals. Of the amount consumed only about half is digested, with the remainder being returned to the soil. Thus, only about one–eighth of depleted evapotranspiration contributes to animal production. The rest contributes to maintaining the pasture ecosystem and providing ecosystem services like soil health attributes (improved nutrient composition, improved soil structure, better soil moisture storage and reduced erosion impacts). These services either directly or indirectly influence the 5 domains of the SI indicators pentagon with numerous permutations of synergies and tradeoffs depending on the context at hand.

Study results indicate a steady increase in temperature and a projected increase in rainfall over the next 40 years to the 2050s (Figs. 22.6 and 22.7). A warming climate will inevitably place additional stresses on water resources, whether or not future rainfall is significantly altered. Increments in regional rainfall amounts will call for more concerted management of water resources in order to optimize agricultural productivity within the cropping cycles. It also creates opportunities for potential storage options as a coping mechanism and adaptation to climate change. Though predictions pertaining to future warming are robust, there remains

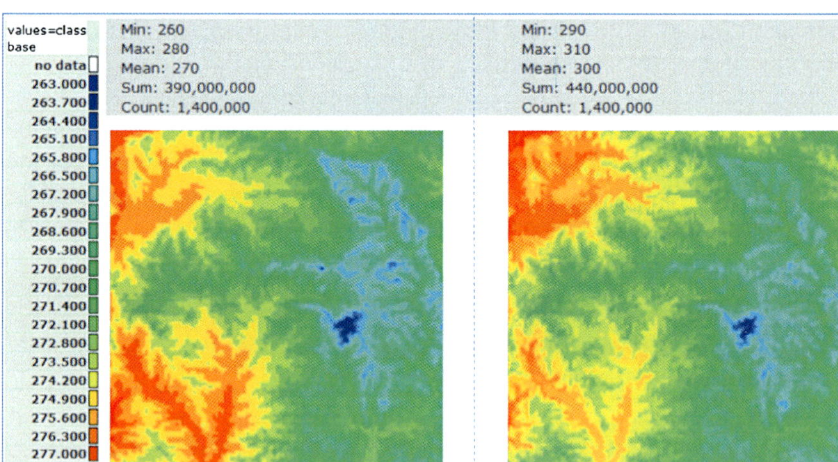

Fig. 22.6 Baseline (*left*) and 2050s scenario (*right*) mean monthly temperture (°C). *Upper greyed* tabs provide monthly statistical summaries for each simulation

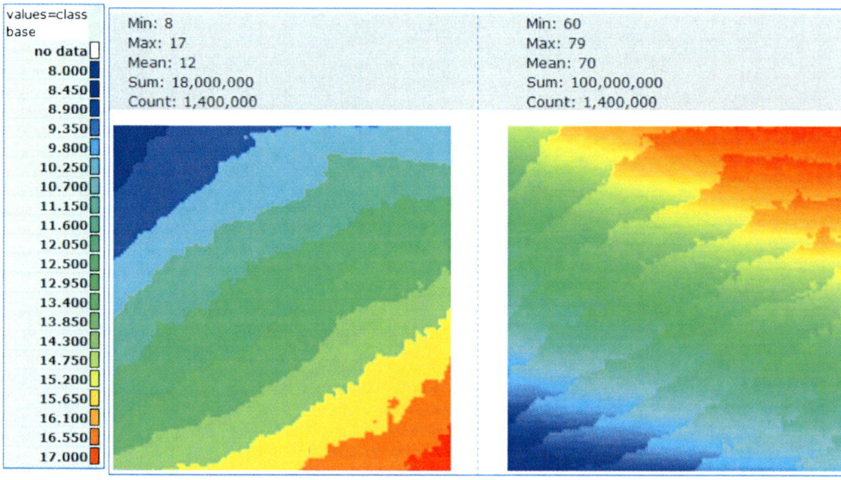

Fig. 22.7 Baseline (*left*) and 2050s scenario (*right*) total monthly rainfall for November (mm/month). *Top grey* tabs depict monthly statistical summaries for each simulation

significant uncertainty about the magnitude and direction of regional rainfall changes for the most of Africa. In their work on African climate change, (Hulme et al. 2005) surmise that there is a rather ambitious representation in most GCMs of the El Nino Southern Oscillation (ENSO)-type climatic variability in the tropics (a key determinant of African rainfall variability). This is further coupled by the omission of any representation of dynamic land cover-atmosphere interactions and dust and biomass aerosols. These relationships and interactions have been suggested to be critical in determining African climate change.

If rainfall is received in higher amounts at greater intensities over short durations, it may translate into an extreme event in an area that is prone to flooding. The impacts of extreme events on many developing countries have been reported to likely be negative (Low 2005). Therefore, efforts should be directed towards reducing the rate of change (mitigation) or manage its consequences (adaptation). Depending on how climatic changes unfold, and how local communities in Tanzania mitigate or adapt to these changes, a significant number of people could be at risk from extreme events such as floods which may further lead to negative social externalities and hunger. The identification of pathways for adaptation should form a key feature of the development landscape. Identification of local, institutional, knowledge and policy gaps that may constrain effective response to climate change and how the use of science, technology and innovation may be targeted to bridge these gaps in future and enhance community adaptation strategies. These would strengthen the social domain in the SI indicator pentagon presented in Fig. 22.8. Finally, a deeper understanding of the ecological consequences of more extreme intra-annual precipitation patterns will also strengthen our knowledge of vegetation–climate relationships and how forage legumes and forage grass

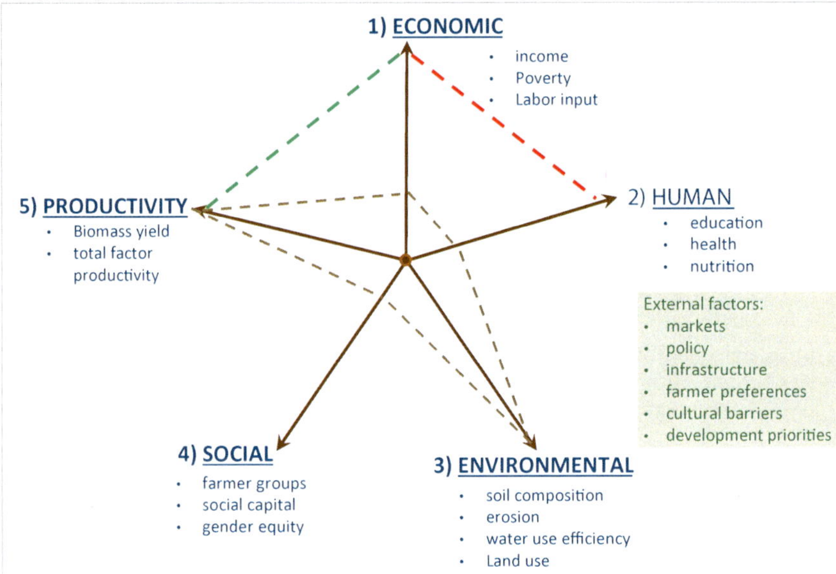

Fig. 22.8 Representation of the forage system synergies and tradeoffs (*dotted brown lines*) along a sustainable intensification indicators framework with five core domains and some indicators for each domain. Potential trajectories of change are shown in *green* and *red dotted lines* (adapted from Africa RISING Sustainable Intensification Workshop, Accra, July 2013)

combinations can help reduce some negative impacts associated with climate change at farm level and catchment scales.

Additional research is needed to further scale and test the findings highlighted in this study to fill critical knowledge gaps in our understanding of ecological responses from farm level to landscape scales in the context of water balance dynamics within forage systems. We suggest that future research focuses on the need for (a) enhanced documentation and projection of intra-annual precipitation patterns at local and regional scales; (b) greater insight into the direct effects of these modified rainfall delivery patterns on agricultural productivity, ecosystem structure and function, as well as interactions with other regional and global change drivers; and (c) greater understanding of how modifying the dynamics of the ecosystem water balance may impact forage production and vice versa. There is a clear need for field experimentation combined with systems modeling to address these under-studied components. Key to these experiments is greater knowledge of exactly how precipitation regimes are changing and how much they can be expected to change in the future and their impact on agroecosystem productivity including forages. Finally, a deeper understanding of the ecological consequences of more extreme intra-annual precipitation patterns will also strengthen our knowledge of vegetation-climate relationships and anthropogenic feedbacks at both farm level and catchment scales.

Acknowledgments This study was sponsored by the USAID Feed the Future Africa RISING Program. We recognize extension officers at Long village Mr. Zahoro Madongo and Mr. Sambali Bernard for their role in data collection; Mr. Gilbert Mbesere from the Babati District Council for his dedicated efforts in soil moisture measurements throughout the study period.

References

Agrawala S, Moehner A, Hemp A, van Aalst M, Hitz S, Smith J, Meena J, Mwakifwamba SM, Hyera T, Mwaipopo OU (2003) Development and climate change in Tanzania: focus on Mount Kilimanjaro. OECD, Paris

Allen RG, Smith M, Perrier A, Pereira LS (1998) An update for the definition of reference evapotranspiration. ICID Bull 43(2):1–34

Amede T, Geheb K, Douthwaite B (2009) Enabling the uptake of livestock water productivity interventions in the crop-livestock systems of sub-Saharan Africa. Rangeland J 31:223–230. doi:10.1071/RJ09008

Bishop-Sambrook C, Kienzle J, Mariki W, Owenya M, Ribeiro F (2004) Conservation agriculture as a labour-saving practice for vulnerable households: a study of the suitability of reduced tillage and cover crops for households under labour stress in Babati and Karatu districts, Northern Tanzania. IFAD and FAO, Rome

Cohen B (2006) Urbanization in developing countries: current trends, future projections and key challenges for sustainability. Technol Soc 28:63–80

Coulibaly YJ, Kundhlande G, Amosi N, Tall A, Kaur H, Hansen J (2015) What climate services do farmers and pastoralists need in Tanzania? Baseline study for the GFCS adaptation program in Africa. CCAFS working paper no. 110. CGIAR Research Program on Climate Change, Agriculture and Food Security (CCAFS). Copenhagen. www.ccafs.cgiar.org

Descheemaeker K, Amede T, Haileslassie A (2009) Livestock and water interactions in mixed crop-livestock farming systems of sub-Saharan Africa: interventions for improved productivity. (IWMI working paper 133). Colombo, Sri Lanka: International Water Management Institute

FAO (2006) FAO Irrigation and drainage paper No. 56. Crop evapotranspiration (guidelines for computing crop water requirements)

Häckner L (2009) Climate change and agriculture in Babati: awareness, strategies and constraints. Södertörns University College, School of Life Sciences, thesis 15 ECTS 2009

Herrero M, Thornton PK, Kruska R, Reid S (2008) Systems dynamics and the spatial distribution of methane emissions from African domestic ruminants to 2030. Agric Ecosyst Environ 126:122–137

Hulme M, Doherty R, Ngara T, New M (2005) Global warming and African climate change: a reassessment. In: Low PS (ed) Climate change and Africa. Cambridge University Press, Cambridge

IPPC (2000) IPPC: special report on emission scenarios. IPCC, Rome

Jonsson LO (1996) Rainwater management to avoid drought, IRD Currents no. 12, Aug 1996, pp 56–64

Kabirizi J, Mpairwe D, Mutetikka D (2007) The effect of integrating forage legumes in smallholder crop/livestock farming systems on food, fodder and animal performance. Paper presented at Tropentag. Witzenhausen, 9–11 Oct 2007

Kihara J, Tamene LD, Massawe P, Bekunda M (2014) Agronomic survey to assess crop yield, controlling factors and managementimplications: a case-study of Babati in northern Tanzania. Nutrient Cycling in Agroecosystems. Nutr Cycl Agroecosyst. doi: 10.1007/s10705-014-9648-3

Low PS (2005) Climate change and Africa. Cambridge University Press, Cambridge

Maddison D (2007) The perception of and adaption to climate change in Africa. The World Bank, Washington

Magcale-Macandog DB, Predo CD, Menz KM, Predo AD (1998) Napier grass strips and livestock: a bio-economic analysis. Agrofor Syst 40:41–58
Matsuno Y, Nakamura K, Masumoto T, Matsui H, Kato T, Sato Y (2006) Prospects for multi-functionality of paddy rice cultivation in Japan and other countries in Monsoon Asia. Paddy Water Environ 4(4):189–197
Mulligan M, Rubiano J, Hyman G, White D, Garcia J, Saravia J, Gabriel Leon J, Selvaraj JJ, Guttierez T, Saenz-Cruz LL (2010) The Andes basins: biophysical and developmental diversity in a climate of change. Water Int 35(5):472–492
Nair RPK (2014) Grand challenges in agroecology and land use systems. Front Environ Sci. doi:10.3389/fenvs.2014.00001
Nyambati EM, Sollenberger LE, Kunkle WE (2003) Feed intake and lactation performance of dairy cows offered Napier grass supplemented with legume hay. Livestock Prod Sci 83:179–189
Okori, P. 2014. Africa rising report archives for Kongwa and Kiteto, Tanzania
Oweis T, Hachum A, Kijne J (1999) Water harvesting and supplementary irrigation for improved water use efficiency in dry areas. SWIM paper 7. International Water Management Institute: Colombo, Sri Lanka
Peden D, Tadesse G, Misra A (2007) Water and livestock for human development. In: Water for food, water for life: a comprehensive assessment of water management in agriculture. London, UK: Earthscan, and Colombo, Sri Lanka: International Water Management Institute. pp 485–514
Robertson GP, Gross KL, Hamilton SK, Landis DA, Schmidt TM, Snapp SS, Swinton SM (2014) Farming for ecosystem services: an ecological approach to production agriculture. Bioscience. 64(5):404–415. http://www.ncbi.nlm.nih.gov/pmc/articles/PMC4776676/ Accessed Nov 2015
Sala O, Parton W, Joyce A, Lauenroth W (1988) Primary production of the central grasslands of the United States. Ecology 69(1):40–45
Shetto R, Owenya M (eds) (2007) Conservation agriculture as practised in Tanzania: three case studies. African Conservation Tillage Network, Centre de Coopération Internationale de Recherche Agronomique pour le Développement, Food and Agriculture Organization of the United Nations. Nairobi
Singh B, Ajeigbe H, Tarawali S, Fernandez-Rivera S, Abubakar M (2003) Improving the production and utilization of cowpea as food and fodder. Field Crops Res 84:169–177
Tanzania Adaption Team (2006) Poverty and climate change a South North Collaboration, summary of Tanzania vulnerability and adaption to climate change, variability and extreme events. The Centre for Energy, Environment, Science and Technology, Dar es Salaam
Tarawali SA, Singh BB, Peters M, Blade SF (1997) Cowpea haulms as fodder. In: Singh BB, Mohan Raj DR, Dashiell K, Jackai LEN (eds) Advances in Cowpea research. Ibadan, International Institute of Tropical Agricultural Sciences (IITA) and Japan International Research Center for Agricultural Sciences (JIRCAS)
Winterbottom R, Reij C, Garrity D, Glover J, Hellums D, McGahuey M, Scherr S (2013) Improving land and water management. Working paper, installment 4 of creating a sustainable food future. World Resources Institute: Washington. http://www.worldresourcesreport.org
Zhang X, Gu HR, Ding CL, Xu NX, Ran JS (2009) Variation and cluster analysis of yield and biological characters of *Pennisetum purpureum*. Chin J Grassland 31(1):58–63

Chapter 23
Smart Strategies for Enhanced Agricultural Resilience and Food Security Under a Changing Climate in Sub-Saharan Africa

Kennedy Were, Aweke Mulualem Gelaw and Bal Ram Singh

Abstract Africa's population, growing at a rate of 2.7 %, reached 1.1 billion as of mid-2015 and is projected to be 1.7 billion in 2030, 2.5 billion in 2050 and 4.4 billion in 2100. This population depends on agricultural outputs from smallholder farmers who cultivate small parcels of, mostly degraded, land and have no access to reliable irrigation, affordable inputs, financial credit services, output markets and agricultural information. Thus, food security remains a great concern with ~220 million people (23.2 %) having been unable to consume enough food to lead active and healthy lives in 2014–2016. This is an increase of ~44 million people from 1990–1992. Ensuring food security without compromising sustainability of land resources under a rapidly growing population and changing climate is among the major challenges of this era. In this chapter, we present climate-smart agriculture (CSA) as an approach that can be adopted to increase agricultural productivity and incomes in environmentally and socially sustainable ways, enhance farmers' resilience and mitigate climate change in sub-Saharan Africa (SSA). Specifically, we describe the array of proven practical techniques that underpin CSA, highlight soil as a limited resource and emphasise the importance of its sound management for present and future use. As none of the CSA technologies individually offers a magic bullet solution to the foregoing challenges and most of the promising technologies are founded on local knowledge, local and scientific knowledge must be integrated when choosing the most suitable climate-smart technologies and practices for any given agro-ecology. We thus recommend creating policies and multi-sectoral and

K. Were (✉)
Kenya Agricultural and Livestock Research Organisation, Kenya Soil Survey, P.O. Box 14733-00800, Nairobi, Kenya
e-mail: kenwerez@yahoo.com

A.M. Gelaw
World Agroforestry Centre (ICRAF), Ethiopia Office, C/O ILRI, P.O. Box 5689, Addis Ababa, Ethiopia

B.R. Singh
Department of Environmental Sciences, Norwegian University of Life Sciences, P.O. Box 5003, 1432 Ås, Norway

R. Lal et al. (eds.), *Climate Change and Multi-Dimensional Sustainability in African Agriculture*, DOI 10.1007/978-3-319-41238-2_23

multi-agency approaches which foster partnerships between governments, the private sector and non-governmental organisations to achieve CSA in SSA.

Keywords Climate-smart agriculture · Climate-resilience · Climate change · Food security · Environmental sustainability · Sub-Saharan Africa

23.1 Introduction

23.1.1 The Sub-Saharan Africa Context

Sub-Saharan Africa (SSA), which lies south of the Sahara desert, is a huge land area ($\sim$24.2 million km^2) rich in cultures, soils, minerals, agro-ecological climates, biomes, flora and fauna. The climate varies from warm tropics with high rainfall in the Congo region and very low rainfall in the Sahel to the cool tropics in the Afro-montane region (Fig. 23.1). Similarly, the vegetation ranges from humid forests in the Congo basin to the arid grasses and shrubs in the Sahel. Most of the economies in SSA are agro-based driven by small-scale crop and livestock production under rain-fed conditions. Agriculture feeds $\sim$1 billion people, employs $\sim$65 % of the workforce and contributes $\sim$32 % of the gross domestic product (GDP) and $\sim$40 % of exports [Alliance for a Green Revolution in Africa (AGRA) 2013]. As the primary producers of agricultural outputs, smallholder farmers account for $\sim$80 % of all farms in the region (AGRA 2014). They cultivate small parcels of land, which are often degraded, and have limited access to reliable irrigation, affordable inputs (e.g. fertiliser), financial credit services, output markets and agricultural information. That is, the farmers practice low-input and low-yield subsistence agriculture. The average yields fall below global averages, almost irrespective of the crop being grown. Labour productivity and incomes from

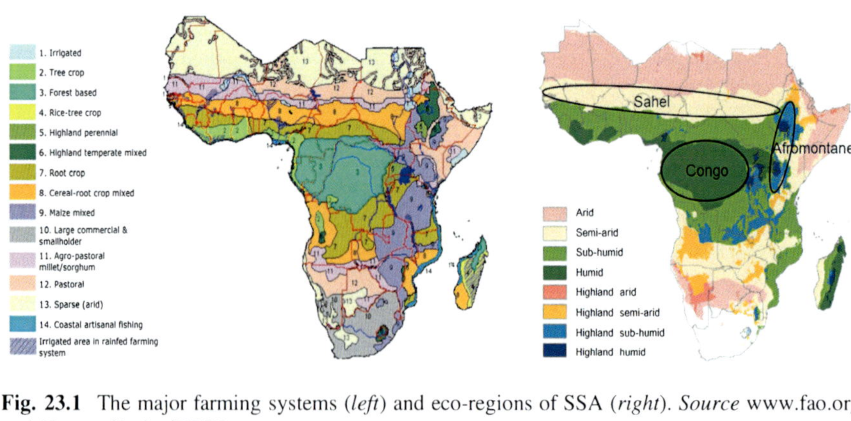

Fig. 23.1 The major farming systems (*left*) and eco-regions of SSA (*right*). *Source* www.fao.org and HarvestChoice/IFPRI

agriculture are also very low relative to global averages (~USD 2.00 per day or less), and the farmers typically spend ~60 % of their incomes on food (AGRA 2014). Of all crops grown in SSA, the major farming systems consist of maize-mixed, irrigated, cereal-root crop mixed, tree crop and agro-pastoral/ millet/sorghum systems (Fig. 23.1) (FAO 2001).

Besides agricultural challenges, SSA is also grappling with a high population growth rate of 2.7 % per year and the attendant implications. According to the official United Nations population estimates and projections, Africa's population reached 1.1 billion as of mid-2015 and is projected to be 1.7 billion in 2030, 2.5 billion in 2050 and 4.4 billion in 2100 (UN 2015). The escalating demand for food, fibre, fuel and shelter associated with the demographic changes has culminated in food and nutritional insecurity, in addition to rapid and large-scale conversion of fragile natural ecosystems (i.e. forests, woodlands, savannahs, grasslands and steppe) to agro-ecosystems (Fig. 23.2). Prominent conversions have been observed in the sub-tropical dry Miombo forests in southern Africa (Lepers et al. 2005). The FAO (2015) also reported that ~220 million people, or 23.2 % of the population, in SSA were hungry and unable to consume enough food to lead active and healthy lives in 2014–2016.

Subsequently, agricultural expansion has contributed to widespread land degradation and climatic changes through the emission of greenhouse gases (GHGs) from sources, such as soils, biomass burning, enteric fermentation, rice cultivation, fertilisers and manure management. Climate change and land degradation are intricately linked because of the feedback between land degradation and climatic elements. Climate change aggravates land degradation, especially soil degradation, by altering the spatial and temporal patterns of temperature, rainfall,

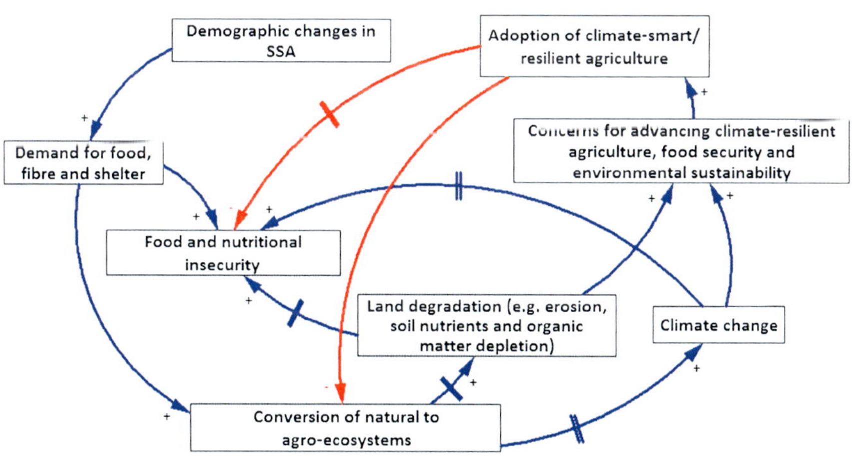

Fig. 23.2 Complex interactions and feedbacks between demographic changes, food security, and environmental sustainability. Each *arrow* indicates a causal relationship, which can be large or small, immediate or delayed [||], and an increasing [+] or decreasing [−] effect

solar radiation and winds. It is expected that climate change will increase soil erosion in parts of SSA through heavy rainfall and increased wind speed. Soil properties and processes, including organic matter decomposition, leaching and soil water regimes, will also be affected by rising temperatures. The resultant soil loss and degradation will, in turn, reinforce the detrimental effects of temperature rise on agricultural yields. Therefore, climate change and soil degradation remain priority issues for SSA because of their cross-cutting impacts on ecosystem services (e.g. food security) and other resources. Soil is also a limited resource whose degradation is unrecoverable within the human lifespan. In the following sections, we discuss climate change in the context of agriculture in SSA (i.e. climatic projections and impacts) and present climate-smart agriculture (CSA) as a potential approach for enhancing agricultural resilience in the region.

23.1.2 Climate Change and Agriculture in SSA

The climatic changes resulting from unsustainable human activities (i.e. agricultural expansions) render SSA the most vulnerable region in the world because of its heavy reliance on rain-fed agriculture, widespread poverty and low adaptive capacity. The changes exacerbate the challenges facing the smallholder farmers in their efforts to produce enough food for the region's growing population. Generally, the region is confronted with several climatic risks that could have far-reaching consequences to its agricultural systems in the future. Climatic changes and variability will threaten food production, lead to food price shocks, increase the vulnerability of smallholders and accentuate rural poverty. Greater warming is expected across all seasons in the twenty-first century with the temperature increases exceeding the global mean increase of 2.5 °C (i.e. ~3–4 °C rise by 2099), and with more intensity in central and southern Africa and the semi-arid tropical margins of the Sahara (Cairns et al. 2013). Under 2 °C warming, the existing variations in water availability across the region could become more pronounced (World Bank 2013). With regard to precipitation, rainfall patterns are projected to change at all locations, although with varying magnitude and directions. For example, in the western margins of southern Africa, annual precipitation is likely to decrease by up to 30 % under a 4 °C warming scenario (AGRA 2014), while in East Africa, rainfall is predicted to increase by 5–20 % in the December–January–February season and decrease by 5–10 % in the June–July–August season by 2050 (Christensen et al. 2007). This could substantially reduce the groundwater recharge rates and increase the risk of drought. The 2011 Horn of Africa drought, which was particularly severe in Kenya, Somalia and Ethiopia, is consistent with an increased probability of long-rains failure due to climate change (World Bank 2013). The length of growing period (LGP), which indicates the adequacy of moisture availability, temperature and soil conditions for crop growth, is projected to decrease by up to 20 % for most parts of SSA by 2050 (Thornton et al. 2011; Sarr 2012). This implies rapid maturation of plants and shorter periods of grain-filling leading to low yields. From their modelling studies, Ringler et al. (2010) found a largely negative impact of climate change on crop yield in the region. Among the cereal crops, the largest negative yield impact by 2050 will be for wheat (−22 %) followed by maize (−5 %) and rice (−2 %). By contrast, millet and sorghum yields will slightly increase by 1 and 2 %, respectively, the increment of which can be ascribed to their higher tolerance to elevated temperatures and drought stress. Adhikari et al. (2015) also reported similar findings.

Moreover, various future climate scenarios depict limited diversification options and livelihood transitions for agro-pastoral systems. This is because climate change will reduce the carrying capacity of grazing (or range) lands and livestock productivity following a decline in the quantity, quality, productivity and distribution of pastures and watering points. For example, the projected shifts in the extent of savanna grasslands could lead to a reduction in the availability of forage for grazing animals. Additionally, higher temperatures could affect the food intake of animals and impair their reproductive success. Most livestock species thrive in temperatures between 10 and 30 °C; beyond this, animals reduce their feed intake by 3–5 % for each degree Celsius rise in temperature (Thornton and Cramer 2012). In a similar vein, climate change will reduce the area suitable for agriculture by ~3 % (because of increased aridity, inundation and land degradation), with most of the reduction occurring in the Sahelian belt and southern Africa (Lane and Jarvis 2007). Lastly, climate change might also influence the patterns, geographic range, intensity, severity and incidence of pests, diseases and weeds. For example, it has been established that the migration pattern of locusts in SSA is influenced by rainfall trends (Gornall et al. 2010) and outbreaks of the African army worm moth follow prolonged drought periods (Gachene et al. 2015).

Owing to the linkages between climate change, agriculture and food security, SSA is faced with the challenge of advancing agricultural productivity and food security under a rapidly growing population, while mitigating the contributions of agriculture to climate change and maintaining soil resources for future use. This calls for revolutionary approaches for the development, identification and adoption of appropriate agronomic technologies, innovations and practices that offer triple wins for sustainability of land resources, food security and climate change mitigation. The global scientific community considers that CSA offers an integrated and systemic response to this challenge. CSA, which was first presented by FAO at the Hague conference on agriculture, food security and climate change in 2010, aims to increase agricultural productivity and incomes in environmentally and socially sustainable ways, adapt and build farmers' resilience to climate change and contribute to climate change mitigation by reducing GHG emissions and increasing carbon (C) storage on farmland (FAO 2013). CSA comprises proven practical techniques, such as mulching, intercropping, conservation agriculture (CA), agroforestry, crop rotation, integrated crop and livestock management, improved grazing, soil and water management, weather forecasting, early warning systems, risk insurance and livelihood diversification.

A wide range of CSA technologies and practices are currently in use in many SSA countries, including Malawi, Mozambique, Zambia, Zimbabwe, Rwanda,

Niger, Kenya and Ethiopia (Lima 2014). The technologies and practices that have been tried are mainly based on options that promote adaptation and resilience, with mitigation as a co-benefit. In Southern Africa, a region identified as the most suitable for rapid scaling up of these approaches, the CSA technologies and practices most commonly encountered are CA, agroforestry, mixed livestock and cropping systems, and improved crop varieties (UNDP 2013). Generally, smallholder farmers across the continent have begun to embrace climate-smart farming, but as the impacts of climate change become increasingly evident, they may need to adapt more quickly and comprehensively.

23.2 Climate-Smart Options for Enhancing Agricultural Resilience in SSA

Sub-Saharan Africa has a great agro-ecological diversity, and the uniqueness of the agro-ecologies imply that they are impacted differently by climate change. Thus, no single specific agricultural technology or practice can be universally applied to achieve sustainable agriculture, climate change mitigation and climate resilience in the region. CSA requires an objective assessment to identify the most suitable agricultural production technologies and practices for any given agro-ecology, taking into consideration site-specific biophysical and socio-economic factors. The objective assessment is also important because CSA technologies and practices have their own merits and demerits; hence, trade-offs must be made (Lal 2015). This means there are multiple pathways to CSA (Fig. 23.3). However, the purpose of each pathway is to make agricultural systems more productive, efficient in their use of resources and resilient to climatic risks and shocks. Thus, CSA technologies and practices are simply better ways of managing water, soil, energy, genetic resources, livestock and crops to ensure efficiency in their use, and enhancing agricultural productivity and resilience in the face of climate change. Increased efficiency in the use of inputs is critical because climate change will affect input availability, especially water. Most of the emerging technologies and practices are designed to alter the use of farm inputs in ways that reduce GHG emissions and the impacts of weather on agricultural production. These technologies and practices are manifold and can be instrumental in mitigating climate change and building resilient agriculture in SSA.

23.2.1 Soil Management

As previously mentioned in Sect. 23.1.1, soil is a limited but precious resource that delivers ecosystem services (e.g. food, fodder, fibre, renewable energy and raw materials) to all life forms on Earth. Soil health is the cornerstone of ecological stability and sustainability, food security, viable rural livelihoods and agricultural

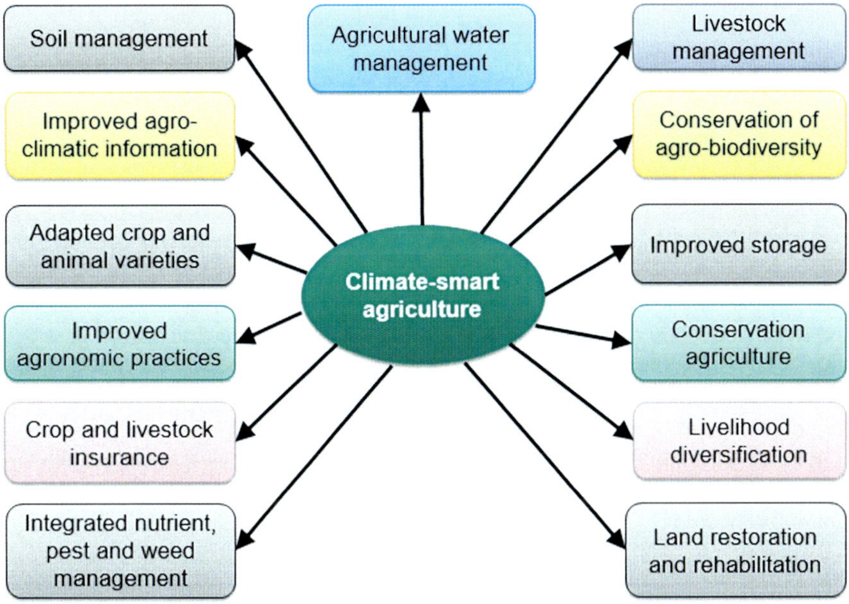

Fig. 23.3 An illustration of the multiple pathways to CSA

resilience under a changing climate. Productive soils constitute a small fraction of the total land area in SSA and face great pressure from competing land uses, such as agriculture, grazing, forestry, infrastructural development and energy production. This often enhances soil degradation (e.g. erosion, nutrient and organic matter depletion), which is unrecoverable within the human lifespan. In fact, the success of any soil restoration efforts being made today will only be known by those who are on Earth in 2100. About 180 million people are affected by soil degradation in SSA and the attendant economic losses are estimated at $68 billion per year (Glatzel et al. 2014). Thus, the proper care, management, and protection of soils for present and future use is needed. Numerous soil-smart technologies, practices and interventions can be organised and applied to help farmers adapt to the likely adverse impacts of climate change and variability, and to reduce GHG emissions from agriculture, build resilience in farming systems and advance food security.

23.2.1.1 Soil Carbon Sequestration

Re-carbonisation of soil through the uptake and transfer of atmospheric CO_2 into the soil reservoir by plants, plant residues and other organic solids is an integral strategy towards adapting to and mitigating climate change, advancing food security and improving the environment (Lal 2014). Soil organic carbon (SOC) pool plays a significant role in the global carbon cycle and is a key determinant of the physical, chemical and biological properties requisite for proper functioning of the soil system. The world soils contain ~1500 Pg C to 1 m depth (1 Pg = 10^{15} g), which is twice the amount of C in the atmospheric pool (~750 Pg C) and almost three times the amount in the biotic pool (~610 Pg C) (Lal 2004; Smith 2004, 2008). Thus, any changes in the SOC pool are bound to instigate environmental problems such as climate change. Many studies in SSA and beyond have found that converting native systems to agricultural lands diminishes the SOC pool (Brown and Lugo 1990; Braimoh and Vlek 2004; Lemenih et al. 2005; Yimer et al. 2007; Girmay and Singh 2012; Demessie et al. 2013; Were et al. 2015), which is attributed to the disrupted balance between the inputs of carbon through litter fall, dead roots, below ground biomass and root exudates and the outputs through leaching, decomposition and erosion (Detwiler 1986; Eaton et al. 2008). Some of the agricultural practices that contribute to the disruption include the removal of ground cover, continuous cultivation and grazing, bare fallows, and biomass burning.

Re-carbonisation to address SOC loss can be achieved by identifying, selecting and adopting best management practices (BMPs), which can create positive carbon budget in the soils. The BMPs are wide-ranging, and include conservation tillage, agroforestry, mulching, integrated nutrient management, optimal stocking, and the use of crop residues, cover crops, biochar, improved plant varieties with greater root mass, crop rotations, and soil and water conservation structures (e.g. terraces, tied ridges and windbreaks), among others. Soil carbon sequestration offers a myriad of benefits, such as improved soil and water quality, increased water and nutrient retention capacity, efficiency in the use of inputs, decreased vulnerability to extreme climatic events (e.g. drought) and susceptibility to soil degradation, as well as sustained agronomic productivity and food security.

23.2.1.2 Restoration of Degraded Soils

Restoration of degraded soils is tightly linked to soil carbon sequestration, water retention, risk reduction in rain-fed agriculture, biodiversity and food security (Lal 2008). Degraded soils (Fig. 23.4) have the highest potential for carbon

Fig. 23.4 Deep gullies in West Pokot, Kenya *Photo* Wilson Ng'etich

sequestration (FAO 2013; Lal 2004). Such soils have lost a large fraction of the antecedent SOC pool, which can be restored by adopting a range of BMPs. The BMPs for carbon sequestration in degraded soils are synonymous with those presented in Sect. 23.2.1.1. The Global Partnership on Forest Landscape Restoration estimates that over 400 million hectares of degraded forest landscapes in Africa offer opportunities for restoring or enhancing the functionality of 'mosaic' landscapes that mix forest, agriculture and other land uses. Restoration of degraded ecosystems can improve human livelihoods, repair ecosystems and increase the resilience of people and landscapes to climate change. Such restoration can generate private and public benefits, and thus constitutes a potentially important means of generating 'win–win' solutions to address poverty, food insecurity and environmental issues (Scherr et al. 2012). For example, restoration of degraded ecosystems increases resilience of local communities by providing food, livestock feeds and fuel wood (Milder et al. 2011). Restoration also improves ecological resilience in terms of watershed functions, habitat for wildlife, biodiversity conservation, reducing soil erosion, and carbon sequestration (Bernazzani et al. 2012; Scherr and Shames 2009).

23.2.1.3 Soil Water Storage

Soil management can influence rainwater infiltration and the soil's capacity to reduce evaporation and store water. By increasing water infiltration and storage in soils and reducing evaporation, many existing soil and crop management technologies and practices (e.g. mulching, CA, deep tillage, manure application, terracing, soil bunds, use of crop residues and cover crops) will help land managers in areas with low levels of precipitation to adapt to climate change. Other beneficial effects of such practices include improved soil surface conditions, SOC content, soil structure, porosity, aeration, bulk density, reduced rates of erosion and increased agricultural productivity.

23.2.1.4 Prevention of Land Conversions and Degradation

This strategy entails preventing the expansion of intensive land uses (e.g. agriculture) into areas where SOC stocks are less resilient; for example, the conversion of semi-arid savannahs, grasslands, tropical rainforests or peatlands into arable land. Brown and Lugo (1990) found that cultivation of tropical forest soils causes losses of more than 60 % of original SOC stocks in just a few years, which has implications for climate change.

Further, the strategy involves deterring soil degradation (e.g. soil compaction caused by the use of heavy machinery, trampling of livestock and frequent ploughing). It has been established that degraded soils are more vulnerable to climate change impacts owing to the serious losses of SOC and soil biodiversity, greater compaction and increased rates of erosion (Lal 2004). These soils also influence climate change.

23.2.2 Crop Management

23.2.2.1 Plant Breeding for Adapted Varieties and Traits

As for crops, developing and adopting new varieties and traits (Fig. 23.5) to overcome abiotic stresses can be an effective means of increasing crop yields, while simultaneously mitigating and adapting to climate change. This means that genetic resources for food and agriculture are fundamental for building resilient agriculture and providing suitable crop varieties for adapting agricultural production to future climates (FAO 2013). Thus, agro-biodiversity (especially genetic resources) must be conserved and used sustainably. The roles of conventional breeding and biotechnology are also of paramount importance. For instance, biotechnology can contribute to improvements in crop yields and reductions in production costs and intensity of the use of farm inputs (e.g. fertilisers). Improved yields can thus boost climate change mitigation efforts through foregone land cover conversions associated with GHG emissions (Lybbert and Sumner 2010). Moreover, new crop varieties (that are early-maturing; efficient in using nitrogen; high yielding; nutritionally improved; tolerant to drought, heat, fire and salinity; and resistant to pests, parasites and diseases) can provide farmers with a wide range of options for adapting to climate change and reduce their vulnerability. Their contribution to reduction in pesticide demand and the number of in-field applications also shows that new varieties can play a role in reducing the GHG cost of production in the agricultural sector.

Fig. 23.5 A new sorghum variety growing at a trial site in Western Kenya *Photo* Authors

23.2.2.2 Agronomic Technologies and Practices

Besides new crop varieties, improved agronomic techniques and practices can also be significant in increasing agricultural productivity and advancing food security in the face of climate change. In this category, CA is a key option for the smallholder farmers, although its uptake can be constrained by biophysical and socio-economic factors, such as declining soil quality (Lal 2015). CA is based on the principles of minimum soil disturbance, maintenance of at least 30 % permanent organic mulch on soil surface and a diversified cropping system through intercropping or crop rotation (Aune and Coulibaly 2015; FAO 2013). It has a wide range of benefits, including (1) increased infiltration capacity, soil water content, soil organic matter (carbon storage), soil biological activity and flexibility in planting and harvesting; (2) reduced weeds, soil erosion, soil surface temperatures, soil compaction, emission of soil carbon to the atmosphere and fossil fuel consumption by avoiding ploughing; and (3) improved soil tilth and fertility.

Furthermore, applying agronomic techniques and practices that aim to reduce N_2O and CH_4 emissions can boost the efforts to enhance climate change mitigation. In field operations, greater N_2O emissions occur when nitrogen (N) fertilisers are applied. Thus, practices that improve nitrogen-use efficiency, such as N field-testing, precision farming (i.e. applying fertiliser based on a precise estimation of crop needs), proper timing of fertiliser applications (i.e. often just prior to plant uptake), precise placement of N in the soils to make it more accessible to the roots, improvement of field drainage, the use of cover crops and manures, composting, N-fixing crop rotations, the use of nitrification inhibitors with fertiliser, integrated pest management, and feeding the animals on more proteins, can mitigate N_2O emissions (Schahczenski and Hill 2009; Branosky and Greenhalgh 2007; Smith et al. 2007; UNFCCC 2008). By contrast, CH_4 emissions in the agricultural sector are attributed to rice cultivation mainly through stimulation of anaerobic decomposition of plant residues. Branosky and Greenhalgh (2007) and Smith et al. (2007) mentioned cultivation of rice cultivars with low exudation rates and improved water management practices (e.g. draining wetland rice once, or several times, during the growing season and keeping the soil dry during the off-rice season) as some of the good practices for reducing CH_4 emissions from rice cultivation. The System of Rice Intensification (SRI) strategy could also reduce CH_4 and N_2O emissions because it aims to increase the productivity of irrigated rice by changing the management of the plants, soil, water and nutrients. For example, SRI reduces the amount of flooding of irrigated rice and saves water.

Finally, the literature has also cited agro-forestry, terracing, shelterbelts and windbreaks, direct seeding, rotational cropping and grazing, mulch cropping, cover cropping, integrated pest management, integrated weed management, integrated soil fertility management (ISFM), the use of perennial forage crops, proper straw management, the reduction of bare fallows, and organic systems of production as climate-smart practices on rain-fed systems which can promote soil carbon sequestration (FAO 2013; Schahczenski and Hill 2009; Smith et al. 2007). Nevertheless, the use of pesticides, herbicides and fertilisers is only defensible if it increases land use intensity and agricultural yields; hence, averting land cover conversions.

23.2.3 *Livestock Management*

Ruminant animals are important sources of CH_4 and N_2O through enteric fermentation and manure disposals. Enteric fermentation produces CH_4 when micro-organisms break down complex carbohydrates into simple sugars that can be assimilated into the animal's body. Up to 7 % of an animal's feed can be lost as CH_4; hence, attempts to increase animals' digestive efficiency can lessen CH_4 emissions (Mushi et al. 2015). This can also reduce feed costs. Various climate-smart options exist to minimise the levels of CH_4 and N_2O emissions in the livestock sector.

23.2.3.1 Grazing Land Management

One option for climate-smart livestock is better management of grazing systems. This can be accomplished through rotational grazing, increasing livestock mobility, reducing grazing pressure and improving the soil's physical condition through drainage to prevent soil degradation or restore grazing lands. Wet soils are compacted easily by grazing animals, which increases their anaerobicity, denitrification conditions and potential for N_2O emissions (Mushi et al. 2015). Restoring degraded grasslands enhances soil health, carbon sequestration and water retention, and establishes climate-resilient grazing systems.

23.2.3.2 Pasture Management

Practices for managing pasture to lower GHG emissions by livestock include fertilisation and cutting regimes, irrigation, controlled grazing regime, the introduction of earthworms, fire management and the use of improved grass species with regard to yields, environmental adaptation and digestibility (e.g. perennial fodders, pastures and legumes). Increasing the digestibility of feeds by improving the quality of grass and crop residues, or supplementing diets with concentrates reduces CH_4 emissions (Lal 2004).

23.2.3.3 Animal Breeding

Breeding can also be done to select more productive, drought-tolerant and disease-resistant animals that are low CH_4-emitting and use dietary protein more efficiently (Fig. 23.6) (Mushi et al. 2015). Compared to the indigenous breeds,

Fig. 23.6 Improved cattle breed *Photo* www.kalro.org

animals bred for efficient feed conversion direct more of their intake into production and less into nitrogen excretion through urine, thereby reducing N_2O emissions; however, this does not imply that indigenous breeds have no merits. In SSA, for instance, most pastoralists keep local breeds. The breeds may be less productive than the improved high-yielding ones, but they are well adapted to their harsh environments and can produce in such conditions. The indigenous breeds are also more disease-resistant, drought-tolerant and crucial to the effective management of their environments. International Union for Conservation of Nature (IUCN) (2010) argued that without resilient livestock that can cope with the rigors of transhumance, rangelands systems collapse and environmental degradation often follows. Therefore, as local breeds are crossed with genetically improved breeds to produce crossbreeds that are hardy and high yielding, deliberate efforts must be made to avoid wiping out the local breeds (Mushi et al. 2015). Some local breeds must be maintained as a genetic resource base.

23.2.3.4 Nutrition

Mushi et al. (2015) stated that emission per animal and per unit of product is normally higher when the animal is fed a poor diet. Therefore, better nutrition can improve efficiency by reducing CH_4 emissions per unit of animal product, in the same way that better animal husbandry and maintenance of health can improve productivity and reproductive rates and reduce mortality rates of animals. Some of the measures that can increase the amount of output for a given level of emissions include feeding animals on young pastures (i.e. harvested early before lignification) or feed resources with a low carbon footprint (e.g. feed crops that have been produced through CA practices or grown in cropping areas that have not been recently extended from forests or natural pastures), mixing legumes in pure pasture stands and applying chemical treatment to crop residues (e.g. wheat straw).

23.2.3.5 Manure Management

Manure emits CH_4 and N_2O as it breaks down in the soil. Similarly, manure stored in central tanks or lagoons releases CH_4 during anaerobic decomposition. Innovative technologies such as anaerobic digestion facilitate the capture of CH_4 produced in such a manner and its conversion to energy (biogas) or direct usage. This offsets GHG emissions from burning fossil fuels; provides energy for electric generators, heating and lighting; reduces odour from animal manure; and lowers labour costs associated with removal of manure from the barns (Schahczenski and Hill 2009; Branosky and Greenhalgh 2007; UNFCCC 2008). Other good practices exist to manage manure and reduce N_2O and CH_4 emissions, such as solid disposal of manure, minimal compaction of manure heaps, regular and complete removal of manure from the barns, and the application of slurry from manure–rather than fresh manure–to wet grassland soils.

23.2.4 Agricultural Water Management

Climate change is likely to result in hotter temperatures, high evaporation rates, increased droughts and reduced or erratic rainfall over large areas, especially in SSA (Christensen et al. 2007; AGRA 2014). These effects have negative implications for water supply, quantity and distribution in agro-ecosystems, which in turn reduce nutrient-use efficiency and agricultural productivity. The resultant water shortages may also intensify the current water competition and heighten internal social conflicts. Therefore, improvements in water-use efficiency through measures, such as drip irrigation and other systems that maximise water use and cultivation of water-efficient and drought-tolerant crops are increasingly needed. Such water-smart measures can essentially reduce the amount of water and N applied to the cropping system, thereby mitigating the emissions of N_2O and water withdrawals (Schahczenski and Hill 2009). In irrigated areas, climate change and variability may strain irrigation capacity due to a water shortage; therefore, improving the efficiency of existing irrigation systems and extending the irrigation infrastructure may be indispensable in such areas. Lal (2004) pointed out that irrigation in such areas can enhance biomass productivity, increase the amount of aboveground biomass and root biomass returned to the soil, and ultimately improve soil quality. In non-irrigated (rain-fed) areas, seasonal rainwater harvesting, conservation technologies (e.g. dams, pans, tanks, ponds, roof-catchment systems (guttering), minimum tillage, contouring and terracing) and supplementary irrigation may be appropriate (Fig. 23.7).

Fig. 23.7 Examples of climate-smart water management technologies: a small-scale irrigation system and a small dam *Photo* Wilson Aore

23.2.5 Energy Management

The movement of agricultural produce (e.g. milk, vegetables and flowers) from the farm gate to the market is a major contributor to GHG emissions. Thus, energy-smart measures, such as improving transport efficiency, can be useful in mitigating post-harvest GHG emissions. The emissions are attributed more to transport inefficiencies than mileage (Lybbert and Sumner 2010). Within the farm gate, the use of renewable and carbon-neutral forms of energy, such as wind, solar, biogas and biofuels for heating, lighting and supplying water offer excellent opportunities for climate change mitigation. For example, biofuels emit CO_2 of recent atmospheric origin when burned and replaces the CO_2 which would have otherwise come from combustion of fossil fuels. However, the production of biofuels crops (e.g. *Jatropha curcas*) has negative implications for food production. Carbon-neutrality of some biofuels from oilseeds (biodiesel), feed corn (ethanol) and cellulosic sources has also been questioned in the scientific sphere (Schahczenski and Hill 2009). To address this, careful life cycle analysis (LCA) of specific biofuels is important.

23.2.6 Land Restoration and Rehabilitation

The aims of CSA can also be achieved through land restoration and land use changes that conserve and improve soil, water and air quality. As an example, conversion of marginal (or degraded) croplands to natural vegetation systems, forest plantations, improved pastures or other perennial land uses can result in the substantial accrual of SOC because of minimal disturbance to soil and removal of carbon by harvested products. Natural vegetation may also emit less N_2O owing to lower N inputs and higher rates of CH_4 oxidation compared to croplands. Similarly, restoration of drained wetlands (i.e. organic or peaty soils) can enhance soil carbon

Fig. 23.8 Degraded land (*left*) and its condition after rehabilitation (*right*) *Photo* www.ser.org

sequestration because of the slow rate of decomposition under anaerobic conditions; however, such conditions may also be conducive for CH_4 emissions. Planting trees (afforestation), or improved grass, and agro-forestry also have higher potential to sequester carbon because they yield higher aboveground biomass carbon than equivalent treeless land uses (Smith et al. 2007). Part of the biomass also enters the soil carbon pool through humification. For instance, agro-forestry systems in Africa accumulate $\sim$1–18 tons C ha^{-1} in aboveground biomass and $\sim$200 tons C ha^{-1} in soils, especially where leguminous trees such as *Faidherbia* are planted (Galtzel et al. 2014). Observations about the effects of afforestation, grassland establishment and agro-forestry on SOC have been reviewed by Post and Kwon (2000) and Lorenz and Lal (2014). In degraded soils, carbon sequestration can be achieved through re-vegetation (Fig. 23.8) and the adoption of practices that improve the structure and functioning of the soils (e.g. water storage capacity) and reduce GHG emissions, such as the application of nutrient amendments and organic substrates (e.g. manures, bio-solids and composts), minimal tillage, retention of crop residues and water conservation.

23.2.7 Climate Forecasting and Early Warning Systems

Improvements in the forecasts of weather events and inter-seasonal, or inter-annual, weather probabilities can enhance farmers' preparedness for weather variations and their capacity to make short-term decisions, which can either minimise the negative impacts of climatic shocks (e.g. droughts) or make the most of good years. This is quite significant because the frequency and intensity of climatic extremes are expected to increase. Lybbert and Sumner (2010) and Vermeulen et al. (2010) stated that farmers with foreknowledge of extreme events can respond accordingly by planting more appropriate crops, adopting improved technology, intensifying production, replenishing soil nutrients, investing in more profitable enterprises when climatic conditions are favourable and protecting their families and farms against the long-term consequences of adverse climatic extremes. Therefore, technology is being used to improve the quality and accessibility to climatic information and weather-based agro-advisories (see Box 1). Some of the advances made so far include better remote sensing of weather phenomena (e.g. rainfall and land surface temperatures), validation of different land use products and dissemination of weather-based, locale-specific agro-advisories and early warnings to farmers through radios, mobile phones and the Internet (Vermeulen et al. 2010). Major innovation is also taking the form of improved micro-climate modelling to enable more accurate interpolations between actual weather stations and, ultimately, create virtual weather stations for nearly all locations. These improved interpolations can lead to improved short-term forecasts, which can then be disseminated via short message service (SMS) using the rapidly spreading cell phone networks in SSA.

Box 1. Agro-Weather tools for CSA in Kenya and Ethiopia
Through RMSI Company, the World Bank Group launched a pilot project in September 2012 in Embu County (Kenya) and Adaá District (Ethiopia) to improve access to information on weather forecasts and climatic patterns with a view to enhancing the adaptive response of smallholder farmers. The project is developing web- and mobile phone-based agro-weather tools that incorporate climate information and suitable agronomic management recommendations to help farmers manage weather risks, maximise productivity and minimise the environmental impacts of farming practices. Essentially, climate information and agro-advisories are delivered to farmers through information communication technologies (ICTs), such as agro-websites, Interactive Voice Response System (IVRS), SMS, and smart phone applications, as well as through more conventional media, such as radio messages. Farmers register online or through mobile phones to obtain information on crops (e.g. tea, coffee, sorghum, maize, beans, wheat, *teff* and chick peas) and weather risks, including seasonal, monthly and five-day forecasts. In addition, the farmers receive agro-advisories based on the weather forecasts for proper crop management. The agro-advisories focus on future climate outlooks, land preparation and crop sowing dates, the selection of crop type and variety, irrigation and fertiliser application, harvesting, and pests and diseases. (Source: http://www.rmsi.com; http://www.eiar.gov.et).

23.2.8 Crop and Livestock Insurance

In many parts of SSA, the rain-fed agriculture being practiced is an extremely risky business. Development of innovative insurance products is among the climate-smart approaches that can improve farmers' capacity to manage climate risks and cushion them against losses, especially in the dry lands which are prone to greater variability and frequent extreme events. Presently, index-based insurance products are being piloted in many developing countries. These innovations trigger indemnity pay-outs (e.g. via mobile phones) based on meteorological, remotely-sensed or area yield indices (e.g. the normalised difference vegetation index (NDVI)) that have a correlation with agricultural losses or livestock mortality. In Kenya, for instance, the International Livestock Research Institute (ILRI), under the auspices of USAID, and in collaboration with Cornell University, University of Wisconsin-Madison and Syracuse University, developed an NDVI-based livestock insurance contract in 2010 to help the vulnerable pastoralists in the northern parts cope with severe droughts by reducing the risk of asset loss and food shortage owing to livestock mortality (Vermeulen et al. 2010). Similarly, the World Food Programme (WFP) with technical assistance from the World Bank launched an index-based crop insurance contract in Ethiopia in 2006 (AGRA 2014).

23.2.9 Livelihood Diversification

Livelihood diversification encompasses diversity of on-farm activities (e.g. mixed cropping, agroforestry, integrated crop-livestock farming, value addition to agricultural produce, and apiculture) and diversity of off-farm activities (e.g. employment and small businesses). It deals with climate uncertainty and vulnerability by reducing the burden of productivity from agriculture (i.e. spreading the risk) and mitigating climate-related risks (e.g. soil degradation). For instance, livelihood diversification through mixed cropping can contribute to balancing soil nutrients; preventing weeds and pests; reducing plant diseases; increasing the overall efficiency and productivity of the land; improving health and nutrition through more diverse, nutritious and fresh diets; increasing incomes; and reducing crop failures (Chung and Billingsley 2012). In addition, diversification through agroforestry can reduce soil erosion, improve soil fertility, and increase and diversify household income. Lastly, integrating crops and livestock farming and enhancing animal fodder production can ensure accessibility to animal manure and availability of organic materials to cover the soils.

23.2.10 Food Storage

Building the farmers' resilience and advancing food security under a changing climate also calls for the development of proper food storage facilities to cater for surplus harvest and reduce post-harvest losses caused by pests and contamination (e.g. aflatoxins in maize). Climate-smart options, such as the metal silos being used to store grains in Kenya (Fig. 23.9), promote both traditional and modern food preservation methods. The hermetic sealed metal silos protect grains against rodents

Fig. 23.9 Metal silos for improved grain storage *Photo* CIMMYT

and weevils (large grain borers), and their capability to kill pests through oxygen deprivation obviates the usage of pesticides which usually affect human health.

23.3 Summary and Conclusions

In a nutshell, population dynamics in SSA pose the challenge of increasing agricultural productivity for food security without compromising the finite land resources and ability of posterity to use them. This challenge is compounded by the spectre of climate change and variability. We have described how these challenges can be partly solved by CSA. CSA comprises an array of promising technologies and practices that aim to increase agricultural productivity and incomes in environmentally and socially sustainable ways, build farmers' resilience to climate change and contribute to climate change mitigation. Climate-smart options for soil management have been highlighted in particular because of the value of soils in the ecosystem, and because soil degradation is unrecoverable within the human life span. However, no single specific technology or practice individually offers a panacea or magic bullet solution. Additionally, most of the promising practices are grounded in local knowledge and innovations. Thus, traditional and scientific knowledge must be integrated to choose the most suitable climate-smart technologies and practices for any given agro-ecology. Some of the technologies and practices are readily available, while others need further applied research (e.g. for protecting against devastating pests and diseases). Finally, achieving CSA in SSA requires the creation of policies and multi-sectoral and multi-agency approaches, which foster partnerships between governments, the private sector and non-governmental organisations. An example of such a policy is the African Ministerial Conference on the Environment (AMCEN) climate change policy, which stresses adaptation as a priority intervention, but with mitigation as a co-benefit for SSA agriculture (AGRA 2014). Policies should remove the existing barriers (e.g. financial, socio-cultural and technological) to implementing CSA and create synergies with alternative options.

Acknowledgments The authors wish to thank the reviewers for their constructive comments which greatly improved the original manuscript of this work.

References

Adhikari U, Nejadhashemi AP, Woznicki SA (2015) Climate change and eastern Africa: a review of impact on major crops. Food Energy Secur 4(2):110–132

AGRA (2013) Africa agriculture status report: focus on staple crops. Nairobi, Kenya. http://www.agra.org/our-results/agra-status-reports/. Accessed 13 Mar 2016

AGRA (2014) Africa agriculture report: climate change and smallholder agriculture in sub-Saharan Africa, No. 2. Nairobi, Kenya. http://www.agra.org/our-results/agra-status-reports/. Accessed 13 Mar 2016

Aune JB, Coulibaly A (2015) Microdosing of mineral fertilizer and conservation of agriculture in sustainable agricultural intensification in sub-Saharan Africa. In: Lal R, Singh BR, Mwaseba DL et al (eds) Sustainable intensification to advance food security and enhance climate resilience in Africa. Springer, Berlin, pp 223–234

Bernazzani P, Bradley BA, Opperman JJ (2012) Integrating climate change into habitat conservation plans under the U.S. Endangered Species Act. Environ Manag 49:1103–1114

Braimoh AK, Vlek PLG (2004) The impact of land cover change on soil properties in northern Ghana. Land Degrad Dev 15:65–74

Branosky E, Greenhalgh S (2007) Agriculture and climate change: greenhouse gas mitigation opportunities and the 2007 Farm Bill. World Resources Institute, Washington DC, Policy note No. 2. http://www.wri.org/publication/agriculture-and-climate-change-0. Accessed 13 Mar 2016

Brown S, Lugo AE (1990) Effects of forest clearing and succession on the carbon and nitrogen content of soils in Puerto Rico and US Virgin Islands. Plant Soil 124:53–64

Cairns JE, Hellin J, Sonder K (2013) Adapting maize production to climate change in sub-Saharan Africa. Food Secur 5:345–360

Christensen JH, Hewitson B, Busuioc A et al (2007) Regional climate projections. In: Solomon S, Qin D, Manning M et al (eds) Climate change 2007: the physical science basis—contribution of working group I to the fourth assessment report of the intergovernmental panel on climate change. Cambridge University Press, Cambridge, United Kingdom and New York, NY, USA

Chung YB, Billingsley C (2012) Climate-resilient sustainable agriculture: a real alternative to false solutions (a backgrounder). Actionaid. http://www.actionaid.org/sites/files/actionaid/crsa_backgrounder_june_2012_design.pdf. Accessed 13 Mar 2016

Demessie A, Singh BR, Lal R (2013) Soil carbon and nitrogen stocks under chronosequence of farm and traditional agro-forestry uses in Gambo district, southern Ethiopia. Nutr Cycl Agroecosyst 95:365–375

Detwiler RP (1986) Land use change and the global carbon cycle: the role of tropical soils. Biogeochemistry 2:67–93

Eaton JM, McGoff NM, Byrne KA, Leahy P, Kiely G (2008) Land cover change and soil organic carbon stocks in the Republic of Ireland 1851–2000. Clim Change 91:317–334

FAO (2001) Farming systems and poverty: improving farmers' livelihood in a changing world. FAO, Rome. http://www.fao.org/3/a-ac349e.pdf. Accessed 13 Mar 2016

FAO (2012) The state of food insecurity in the world 2012. FAO, Rome. http://www.fao.org/docrep/016/i3027e/i3027e.pdf. Accessed 13 Mar 2016
FAO (2013) Climate-smart agriculture sourcebook. http://www.fao.org/docrep/018/i3325e/i3325e.pdf. Accessed 13 Mar 2016
FAO (2015) The state of food insecurity in the World 2015: meeting the 2015 international hunger targets: Taking stock of uneven progress. FAO, Rome. http://www.fao.org/3/a-i4646e.pdf. Accessed 13 Mar 2016
Gachene CKK, Karuma AN, Baaru MW (2015) Climate change and crop yield in sub-Saharan Africa. In: Lal R, Singh BR, Mwaseba DL et al (eds) Sustainable intensification to advance food security and enhance climate resilience in Africa. Springer, Berlin, pp 165–183
Girmay G, Singh BR (2012) Changes in soil organic carbon stocks and soil quality: land use system effects in northern Ethiopia. Acta Agric Scand Sect B, Soil Plant Sci 62(6):519–530
Glatzel K, Conway G, Alpert E, Brittain S (2014) No ordinary matter: Conserving, restoring and enhancing Africa's soils. A Montpellier panel report. http://ag4impact.org/wp-content/uploads/2014/12/MP_0106_Soil_Report_LR1.pdf. Accessed 13 Mar 2016
Gornall J, Betts R, Burke E, Clark R, Camp J, Willett K, Wiltshire A (2010) Implications of climate change for agricultural productivity in the early twenty-first century. Philos Trans R Soc B 365:2973–2989
IUCN (2010) Building climate change resilience for African livestock in sub-Saharan Africa. Nairobi, Kenya. https://cmsdata.iucn.org/downloads/resilence2.pdf. Accessed 13 Mar 2016
Lal R (2004) Soil carbon sequestration to mitigate climate change. Geoderma 123:1–22
Lal R (2008) Soils and sustainable agriculture: a review. Agron Sustain Dev 28(1):57–64
Lal R (2014) Climate-resilient agriculture and soil organic carbon. Indian J Agron 58(4):440–450
Lal R (2015) Sustainable intensification for adaptation and mitigation of climate change and advancement of food security in Africa. In: Lal R, Singh BR, Mwaseba DL et al (eds) Sustainable intensification to advance food security and enhance climate resilience in Africa. Springer, Berlin, pp 3–17
Lane A, Jarvis A (2007) Changes in climate will modify the geography of crop suitability: agricultural biodiversity can help with adaptation. J Semi-Arid Trop Agric Res 4:1–12
Lemenih M, Karltun E, Olsson M (2005) Assessing soil chemical and physical property responses to deforestation and subsequent cultivation in smallholders farming system in Ethiopia. Agric Ecosyst Environ 105:373–386
Lepers E, Lambin EF, Janetos AC, DeFries R, Achard F, Ramankutty N, Scholes RJ (2005) A synthesis of information on rapid land-cover change for the period 1981–2000. Bioscience 55 (2):115–124
Lima MG (2014) Policies and practices for climate-smart agriculture in sub-Saharan Africa: a comparative assessment of challenges and opportunities across 15 countries. Pretoria, Food, Agriculture and Natural Resource Policy Analysis Network (FANRPAN). http://www.fanrpan.org/documents/d01827/Regional_Synthesis_Report_Comprehensive_Scoping_Assessment_of_CSA_Policies.pdf. Accessed 13 Mar 2016
Lorenz K, Lal R (2014) Soil organic carbon sequestration in agroforestry systems: a review. Agron Sustain Dev 34:443–454
Lybbert T, Sumner D (2010) Agricultural technologies for climate change mitigation and adaptation in developing countries: policy options for innovation and technology diffusion, ICTSD–IPC platform on climate change, agriculture and trade, Issue brief, No. 6. International Centre for Trade and Sustainable Development, Geneva, Switzerland and International Food and Agricultural Trade Policy Council, Washington. http://www.kacst.edu.sa/en/research/agrt/Documents/pdf1.pdf. Accessed 13 Mar 2016
Milder JC, Majanen T, Scherr SJ (2011) Performance and potential of conservation agriculture for climate change adaptation and mitigation in sub-Saharan Africa. Ecoagriculture discussion paper No. 6. EcoAgriculture Partners, Washington. http://ecoagriculture.org/wp-content/uploads/2015/08/PerformancePotentialofConservationAg.pdf. Accessed 13 Mar 2016
Mushi D, Eik LO, Bernués A, Ripoll-Bosch R, Sundstøl F, Mo M (2015) Reducing GHG emissions from traditional livestock systems to mitigate changing climate and biodiversity. In:

Lal R, Singh BR, Mwaseba DL et al (eds) Sustainable intensification to advance food security and enhance climate resilience in Africa. Springer, Berlin, pp 343–365

Post WM, Kwon KC (2000) Soil carbon sequestration and land use change: processes and potentials. Glob Change Biol 6:317–327

Ringler C, Zhu T, Cai X et al (2010) Climate change impacts on food security in Sub-Saharan Africa: Insights from comprehensive climate change scenarios. IFPRI discussion paper No. 1042. International Food Policy Research Institute, Washington. http://www.cosv.org/download/centrodocumentazione/climate%20change%20impacts%20on%20food%20security.pdf. Accessed 13 Mar 2016

Sarr B (2012) Present and future climate change in the semi-arid region of west Africa: a crucial input for practical adaptation in agriculture. Atmos Sci Lett 13:108–112. doi:10.1002/asl.368

Schahczenski J, Hill H (2009) Agriculture, climate change and carbon sequestration. ATTRA: National Sustainable Agriculture Information Service. https://attra.ncat.org/publication.html. Accessed 13 Mar 2016

Scherr SJ, Shames S (2009) Mitigating climate change through food and land use: worldwatch report No. 179. Worldwatch Institute, Washington. http://www.worldwatch.org/system/files/179%20Land%20Use.pdf. Accessed 13 Mar 2016

Scherr SJ, Shames S, Friedman R (2012) From climate-smart agriculture to climate-smart landscapes. Agric Food Secur. doi:10.1186/2048-7010-1-12

Smith P (2004) Soils as carbon sinks: the global context. Soil Use Manag 20:212–218

Smith P (2008) Land use change and soil organic carbon dynamics. Nutr Cycl Agroecosyst 81:169–178

Smith P, Martino D, Cai Z, Gwary D, Janzen H, Kumar P, McCarl B, Ogle S, O'Mara F, Rice C, Scholes B, Sirotenko O (2007) Agriculture. In: Metz B, Davidson OR, Bosch PR et al (eds) Climate Change 2007: mitigation—contribution of working group III to the fourth assessment report of the intergovernmental panel on climate change. Cambridge University Press, New York

Thornton P, Cramer L (eds) (2012) Impacts of climate change on the agricultural and aquatic systems and natural resources within the CGIAR's mandate: CCAFS Working paper 23. CGIAR Research Program on Climate Change, Agriculture and Food Security: Copenhagen, Denmark. https://ccafs.cgiar.org/publications/impacts-climate-change-agricultural-and-aquatic-systems-and-natural-resources-within#.VuUtv7pJnIU. Accessed 13 Mar 2016

Thornton PK, Jones PG, Ericksen PJ, Challinor AJ (2011) Agriculture and food systems in sub-Saharan Africa in a 4 °C+ world. Philos Trans R Soc A 369:117–136

UNDP (2013) Climate-smart agriculture in Africa, e-discussion summary report. http://www.ipc-undp.org/pressroom/files/ipc819.pdf. Accessed 13 Mar 2016

UNFCCC (2008) Challenges and opportunities for mitigation in the agricultural sector. Technical paper GE-08-64351. http://unfccc.int/resource/docs/2008/tp/08.pdf. Accessed 13 Mar 2016

United Nations, Department of Economic and Social Affairs, Population Division (2015) World population prospects: the 2015 revision, key findings and advance tables. Working paper No. ESA/P/WP.241. http://esa.un.org/unpd/wpp/publications/files/key_findings_wpp_2015.pdf. Accessed 13 Mar 2016

Vermeulen SJ, Aggarwal PK, Ainslie A, Angelone C, Campbell BM, Challinor AJ, Hansen J, Ingram JSI, Jarvis A, Kristjanson P, Lau C, Thornton PK, and Wollenberg E (2010) Agriculture, food security and climate change: outlook for knowledge, tools and action. CCAFS Report 3. Copenhagen. https://ccafs.cgiar.org/publications/agriculture-food-security-and-climate-change-outlook-knowledge-tools-and-action#.VuUvcLpJnIU. Accessed 13 Mar 2016

Were KO, Singh BR, Dick ØB (2015) Effects of land cover changes on soil organic carbon and nitrogen stocks in the Eastern Mau Forest Reserve, Kenya. In: Lal R, Singh BR, Mwaseba DL et al (eds) Sustainable intensification to advance food security and enhance climate resilience in Africa. Springer, Berlin, pp 113–133

World Bank (2013) Turn down the heart: climate extremes, regional impacts and the case for resilience. Washington. http://www.worldbank.org/content/dam/Worldbank/document/Full_

Report_Vol_2_Turn_Down_The_Heat_%20Climate_Extremes_Regional_Impacts_Case_for_Resilience_Print%20version_FINAL.pdf. Accessed 13 Mar 2016

Yimer F, Ledin S, Abdelkadir A (2007) Changes in soil organic carbon and total nitrogen contents in three adjacent land use types in the Bale Mountains, south-eastern highlands of Ethiopia. For Ecol Manag 242:337–342

Part IV
Outreach and Extension

Chapter 24
Globalizing Environmental Sustainability: "2015 International Year of Soil" Transitioning to "2015–2024 International Decade of Soil"

Rattan Lal

24.1 Introduction

International Union of Soil Science (IUSS), formerly called International Society of Soil Science (ISSS), was created in 1924 (Hartemink 2015, Hartemink and McBrantey 2008). Since its inauguration, 20 World Congresses of Soil Science have been held. The 20th Congress was held at Jeju, Korea in 2014, the 21st will be held at Rio, Brazil in 2018 and the 22nd at Glasgow in U.K. in 2022. The IUSS, a link to the world's 60,000 soil scientists, is a member of the International Council of Science (ICSU). The secretariat of IUSS is located in Vienna, Austria, and the office is managed by a Secretary in consultation with Presidents (current, past, and incoming), and other elected members of the Executive Committee. The Mission of the IUSS is to promote soil science and all its activities, engage with global stakeholders, stimulate soil science initiatives inside and outside of the IUSS, and improve communication with other scientific disciplines and the general public.

In accord with its mission, IUSS has been proactive in celebrations of the U.N. 2015 International Year of Soil (2015 IYS), and has launched 2015–2024 International Decade of Soil (2015–2024 IDS).

The objectives of this chapter are (1) to explain the significance of the "2015 IYS;" (2) to outline major soil and environmental degradation issues; (3) to describe the adverse impacts of an excessive depletion of soil organic carbon (SOC) pool on soil degradation: (4) to explain the importance of soil conservation and restoration for development of climate-resilient agro-ecosystems; and (5) to describe the goals of "2015–2024 IDS".

R. Lal (✉)
International Union of Soil Sciences, Carbon Management and Sequestration Center,
The Ohio State University, Columbus, OH 43210, USA
e-mail: lal.1@osu.edu

R. Lal et al. (eds.), *Climate Change and Multi-Dimensional Sustainability in African Agriculture*, DOI 10.1007/978-3-319-41238-2_24

24.2 2015 International Year of Soil

After 2 years of intensive diplomacy, 2015 was declared the International Year of Soils by the 68th U.N. General Assembly (A/RES/68/232). The 2015 IYS was aimed to be a platform for raising awareness of the importance of soils for food security and essential eco-system functions and services.

The objectives of the 2015 IYS were to:

1. Create full awareness of civil society and decision makers about the fundamental roles of soils for human's life;
2. Achieve full recognition of the prominent contributions of soils to food security, climate change adaptation and mitigation, essential ecosystem services, poverty alleviation and sustainable development;
3. Promote effective policies and actions for the sustainable management and protection of soil resources;
4. Sensitize decision-makers about the need for robust investment in sustainable soil management activities aiming at healthy soils for different land users and population groups;
5. Catalyze initiatives in connection with the Sustainable Development Goals (SDG) process and Post-2015 agenda; and
6. Advocate rapid enhancement of capacities and systems for soil information collection and monitoring at all levels (global, regional and national).

In addition to the objectives outlined above for 2015 IYS, IUSS has and will also promote the SDGs of the U.N. launched in September 2015. Specifically, three SDGs closely linked with soil functions are the following:

> SDG.2: "End hunger, achieve food security, improve nutrition and promote sustainable agriculture." This SDG is mainly linked to soils and their functioning in natural environments.
>
> SDG.6: "Ensure availability and sustainable management of water and sanitation for all." This SDG is closely linked to soil functions of denaturing and filtering of pollutants.
>
> SDG.7: "Ensure access to an affordable, reliable, sustainable and modern energy for all." This SDG is associated with an appropriate use of agriculturally marginal soils for growing bioenergy crops, such as miscanthus, switchgrass, and short rotation woody plantations like poplar or willow.

Despite the momentum and success of the 2015 IYS, a great need remains to continue to enhance awareness about the importance of soil to human wellbeing and nature conservancy. Soils are critical to addressing major global issues of the 21st century. Among them are:

- Food insecurity (affecting 795 million in 2015, FAO 2015);
- Hidden hunger (deficiency of 17 essential micronutrients such as Fe, Zn, I, Mo, B, Se (and protein) affecting about 2 billion people (Muthayya et al. 2013);

- Water scarcity that affected 1.2 billion people in 2000, another 500 million tomorrow, another 1.6 billion people in the near future (IWMI 2001, UNDP 2014), and another 3.5 billion (~48 % of the world population) by 2025;
- CO_2–C emissions from fossil fuel combustion, cement production and land use conversion (deforestation) estimated at ~11 Pg (Pg = petagram = 10^{15} g = 1 billion metric ton = 1 gigaton = 1 Gt) in 2015 and increasing (Le Quéré et al. 2015);
- Energy use estimated at 570 EJ in 2013 and increasing at the rate of ~3 % per annum and 865 EJ by 2010 (EIA 2013a, b);
- A world population of 7.3 billion (B) that is increasing by ~75 million per annum, and is estimated to be 9.7 B by 2050 and 11.2 B by 2100 (U.N. 2015);
- Deforestation of tropical rainforests at ~13 million ha (Mha) per annum between 2000 and 2010 (FAO 2012) with a strong impact on biodiversity and climate change;
- Global land degradation affecting ~24 % of the Earth's land area, 2.4 billion people, and large areas of soils of agroecosystems (Bai et al. 2008);
- Urban encroachment and surface sealing diverting 3 Mha of agricultural land per annum to non-agricultural uses, and covering 152 Mha or 10 % of the current cropland area of the world; and
- Water pollution eutrophication, and loading of P from soils of agroecosystems exacerbating the problems of algal blooms in Great Lakes in North America, Gulf of Mexico, Chesapeake Bay in the U.S., Lake Taihu in China, the Baltic Sea in Europe, and numerous other water bodies throughout the world.

The answer to these and numerous other global issues lies in sustainable management and protection of the limited soil resources. Indeed, soils matter. The finite, but essential, soil resource must never be taken for granted.

Thus, the urgent need to enhance awareness about soils (among land managers, general public, and policy makers) cannot be over-emphasized. And the U.N. declaration of the 2015 IYS was an important political milestone.

24.3 Soil and Environmental Degradation

Environmental degradation (Fig. 24.1), involving degradation of land and air by natural factors and anthropogenic activities, is a global issue impeding sustainable management of natural resources and a challenge to achieving SDGs of the U.N. There are several interactive and reinforcing (positive feedback) components of environmental degradation (Fig. 24.1). The term land degradation, erroneously used synonymously with soil degradation, comprises of three different but inter-related components: soil degradation, water pollution, and biodiversity decline. Soil degradation, one of the components of land degradation, implies a decline in soil quality and its reduced capacity to provide ecosystem services and functions of importance to human wellbeing and nature conservancy. In general, there are four types of soil degradation: physical, chemical, biological and ecological. Soil physical degradation is set in motion by a decline in structure and tilth due to reductions in amount and strength of aggregates, and declines in porosity and continuity of pores. Declines in soil structure are set-in-motion by the conversion of natural to agricultural ecosystems through deforestation, biomass burning, drainage and exposure of the fragile soil to vagaries of climate. Soil physical degradation includes crusting, compaction, hard setting, decreases in water infiltration rates, increases in water runoff, accelerated erosion by water and wind, water imbalance (drought, inundation), and adverse changes (e.g., supra-optimal, sub-optimal) in soil temperature regimes.

Soil chemical degradation implies a decline in soil chemical quality. It includes numerous processes including acidification, salinization and sodication, elemental imbalances as characterized by toxicity of some (e.g., Al, Mn, Fe) and deficiency of others (e.g., P, Ca, Mg, N), declines in cation exchange capacity (CEC), and depletions of essential plant nutrients by perpetual use of extractive farming practices.

In comparison, soil biological degradation implies a reduction in activity and species diversity of soil fauna and flora, especially that of microbes, and some invertebrates. A decline in the activity of macrofauna, such as earthworms, termites, centipedes and millipedes, can adversely impact bioturbation and reduce the amount and continuity of pores. Biological degradation is characterized by a decline in the soil organic carbon (SOC) pool, microbial respiration, and microbial biomass C (MBC) with attendant adverse impact on biochemical transformations.

Processes moderating degradation of soil physical, chemical and biological quality strongly interact and reinforce one another. It is this mutual reinforcement, exacerbated by anthropogenic and natural perturbations, which leads to soil ecological degradation (Fig. 24.1). The latter is characterized by a disruption in coupled cycles of water and nutrients/elements (C, N, P, S), anaerobiosis, a decline in net primary production (NPP), and a reduction in the use efficiency of inputs.

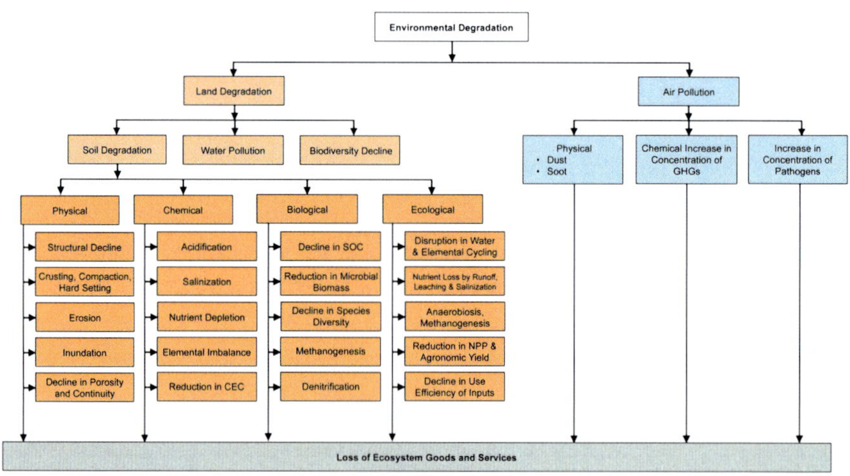

Fig. 24.1 Types of environmental degradation

The overall impact of environmental degradation is a downward spiral leading to loss of ecosystem goods and services. Thus, it is essential that these degradation trends are reversed by understanding the scientific principles governing the cause-effect relationship. Adoption of preventative measures, through judicious land use and adoption of recommended management practices (RMPs), is a prudent strategy. In general, curative measures are too expensive, too late, and often ineffective.

24.4 Importance of Soil Organic Carbon Pool to Soil Quality

While numerous soil properties play a critical role in moderating soil quality and functionality, SOC pool and its composition are major factors which affect numerous other properties and processes. Indeed, there is a critical level of SOC pool below which soil functionality is strongly jeopardized. The critical level may differ among soils of temperate and tropical climates, land uses and management. In general, the critical level of SOC in the rootzone (top 20–30 cm) is estimated to be 1.1 % for soils of the tropics (Aune and Lal 1997) and ~2 % for soils of the temperate climates (Loveland and Webb 2003). The critical level may be lower in soils receiving external inputs, such as compost, manure and chemical fertilizers, than in those managed by low external inputs.

The SOC pool is strongly depleted in soil of agroecosystems, especially in croplands managed by extractive farming practices, such as residue removal, biomass burning, little or no input of fertilizers and organic amendments. The relative magnitude of depletion of SOC pool is greater in soils of the tropics than in those in temperate climates, in sandy rather than clayey soils, and in soils vulnerable to erosion by water and wind and the rates of erosion exceed the tolerable limit (Lal 2004a, b, 2010, Lal et al. 2004). Accelerated soil erosion also exacerbates the mineralization of SOC because of changes in soil temperature and moisture regimes and the redistribution of SOC-enriched sediments over the eroded landscape (Lal 2003a, b). While some of the SOC is buried in depressional sites, and eventually carried into aquatic ecosystems (Van Oost et al. 2007; Amundson et al. 2015), the SOC transported with run-on or aelian sediments does not represent at sequestration because it is not photosynthesized by plants grown on the same landscape unit (Olson et al. 2014).

Strong depletion of the SOC pool adversely impacts soil physical, chemical, biological and ecological quality (Fig. 24.2). In addition to a decline in ecosystem goods and services, soil degradation also aggravates numerous disservices, such as the pollution of water and air, reduce biodiversity, a decline in disease-suppressive attributes of soil, emission of greenhouse gases (CO_2, CH_4, N_2O), and a decline in NPP (Fig. 24.2). Therefore, restoration of SOC pool is important in order to

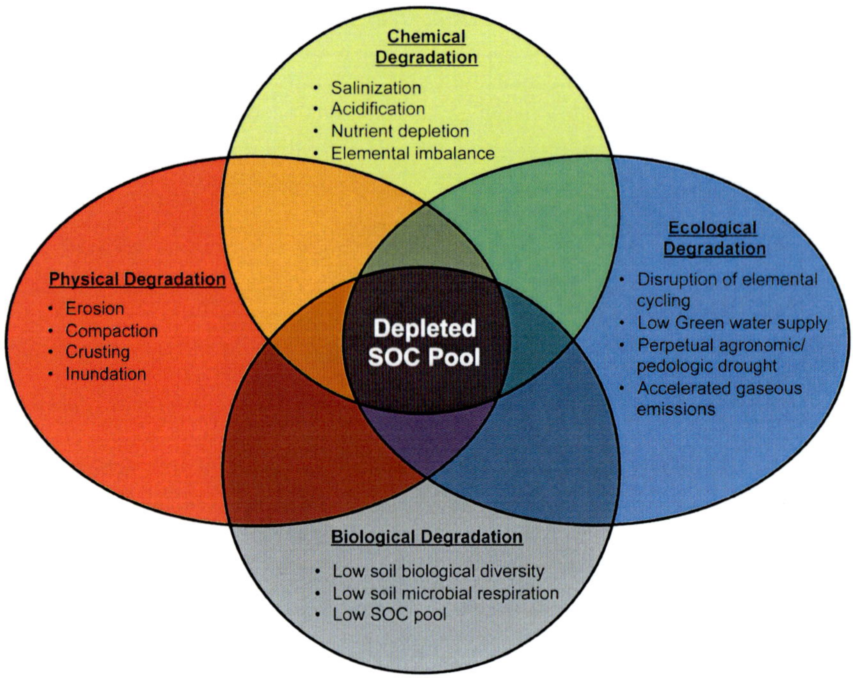

Fig. 24.2 Depletion of the soil organic carbon (SOC) pool at the nexus of sol degradation processes

enhance ecosystem goods and services, improve use efficiency, and increase environmental quality and sustainability (refer to Chap. 1).

It is precisely for this reason that the COP-21 Climate Summit in Paris recommended the "4 Per Thousand" Program. This program advocates for increasing the global SOC pool by 0.4 % per annum to 40-cm depth. The global SOC pool to 40-cm depth is estimated at ~850 PgC (Batjes 1996). Therefore, increasing the SOC pool by 0.4 %/year amounts to sequestration of atmospheric CO_2 in world's soil at the rate of 3.4 PgC/year. Implementation of this concept enhances the importance of soils and SOC pool as a means to address global environmental threats listed in Introduction of this chapter. SOC sequestration is a win-win-win option because it mitigates climate change, advances food and nutritional security, and improves the environment. However, implementation of the "4 per Thousand" program has numerous challenges (Lal 2016), especially in relation to its adoption by the resource-poor and small landholders of the developing countries. Nonetheless, the "4 per Thousand" program is an historic, bold initiative, and it is a landmark concept which emphasizes that agricultural ecosystems and world soils are an integral component of all agendas to adapt and mitigate climate change and to achieve food and nutritional security. Rather than deliberating a specific number, such as the 0.4 % per annum increase in SOC pool, the focus should be on the concept of soil restoration through an increase in its SOC pool because it has numerous co-benefits. Increasing the SOC pool, even at a lower rate of 0.1 % per annum, would have numerous co-benefits. While there is no panacea or a silver bullet to address the huge problem of anthropogenic climate change, even the incremental drawdown in atmospheric concentration of CO_2 by SOC sequestration can have strong environmental and agronomic impacts. Thus, the importance of restoring the SOC pool to reverse the trends in soil and environmental degradation, at whatever level, cannot be over-emphasized. The societal value of soil C must be recognized (Lal 2014).

24.5 Soil Conservation and Climate Resilient Agro-ecosystems

The use of conservation-effective measures to reduce risks of soil erosion is an important option among RMPs to adapt and mitigate climate change. Similarly, restoration of eroded and degraded soils, such as salinized, nutrient-depleted, low SOC, or low pH, can result in C sequestration and the mitigation of climate change. There is no silver bullet or a panacea. However, basic principles of site-specific RMPs include: (1) creating a positive soil/ecosystem C budget, (2) reducing losses of soil, water, nutrients and SOC by accelerated erosion, (3) improving soil biodiversity, (4) maintaining a continuous ground cover, (5) adopting complex rotations and farming systems, and (6) using integrated nutrient management (INM) strategies. These principles are imbedded in the concept of conservation agriculture (CA) (Lal 2015a, b, c, d), agroforestry systems, etc.

The adoption of soil conservation measures, that are based on a judicious combination of biological and engineering techniques, strengthens the climate-resilience of agroecosystems. Principal aspects of climate-resilience through soil conservation include (Fig. 24.3): (1) enhancing C sequestration in biomass and soil; (2) promoting emission avoidance by reducing the use of diesel and other C-intensive inputs (agro-chemicals); (3) increasing tolerance to agronomic/pedologic drought by increasing plant available water capacity and generally improving soil edaphological attributes; (4) moderating soil temperatures and decreasing diurnal amplitudes, and (5) increasing vegetative cover while providing continuous inputs of biomass-C to the soil surface. Overtime, perpetual use of conservation-effective measures, such as CA, improving and sustaining productivity, and buffering against extreme events, such as the drought-inundation cycle), can facilitate adaptation to and mitigation of anthropogenic climate change (Fig. 24.3).

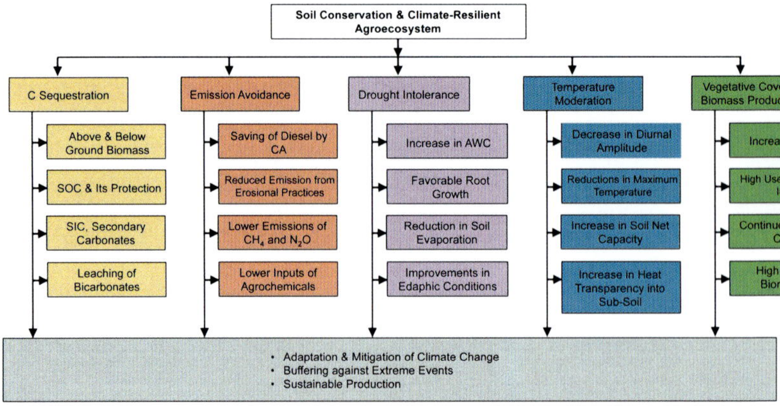

Fig. 24.3 Adoption of conservation-effective soil and crop management systems leads to climate-resilient agro-ecosystem

24.6 2015–2024 International Decade of Soil

In view of the basic principles of soil management outlined above and encouraged by the achievements of 2015 IYS, IUSS proclaimed the "2015–2024 IDS". The proclamation was made by the Executive Committee of IUSS at its meeting on 8th December 2015 in Vienna, Austria. The decade long celebrations will culminate in 2024 with the centennial celebrations in Rome, Italy where the IUSS was inaugurated in 1924. Maintaining the momentum generated by activities of 2015 IYS, IUSS has identified several priority activities that will need to be addressed, including the following:

1. Evaluating the effects of anthropogenic activities on soil and the environment at the landscape scale using a multidisciplinary approach;
2. Assessing all dimensions of soil security as it relates to food security, water security, energy security and other emerging demands of the growing and increasingly affluent world population,
3. Promoting the role of soils in adapting to and mitigating climate change, and integrating soil into all agendas addressing climate change,
4. Protecting soils against urban encroachment, surface sealing, accelerated erosion, compaction, salinization, SOC and nutrient depletion and other degradation processes, and
5. Enhancing awareness of the importance of soil as a provider of numerous ecosystem services pertinent to human and nature conservancy.

During 2015–2024 IDS, IUSS will organize three Congresses: (1) the 21st WCSS at Rio in 2018; (2) the 22nd WCSS at Glasgow in 2022, and (3) the Centennial Celebrations at Rome in 2024. In addition, IUSS may also publish thematic books relating soil science to topics such as Food and Nutritional Security, Adaptation and

Mitigation of Climate Change, Pharmaceuticals and Antibiotics, Biodiversity, Denaturing and Filtering of Pollutants, and Biofuel Production. The Executive Committee of IUSS will encourage its Divisions, Commission and Working Groups to promote activities relevant to thematic topics of interest to their mission and goal.

Implementation of these and other activities of IUSS during 2015–2024 IDS will be undertaken in close cooperation with prominent international organizations such as CGIAR, FAO, IAEA, UNEP, EU, etc.

24.7 Conclusion

The 2015 IYS enhanced awareness about the importance of soils as providers of essential services including: (1) the production of food, feed, fuel, and fiber, (2) the filtration and purification of water, (3) serving as the foundation for homes, roads, dams and other civil structures, (4) the storage of carbon and moderation of climate, (5) the archiving of human and planetary history, (6) the preservation of germplasm and seeds, (7) a habitat for biodiversity, (8) a source of raw material, (9) an inspiration for arts and culture, and (10) the perpetuation of faith and spiritualism. Indeed, the next green revolution will need to be soil-centric for its impact must last for centuries if 11 billion or more inhabitants on the earth are to be fed.

The momentum gained during 2015 IYS must be continued and strengthened. In this context, IUSS will need to strengthen its linkages with diverse disciplinary organizations and professions. Important among these are climatology, hydrology, geology (glaciology), pharmacology, astronomy, biology, and anthropology. Because it is the essence and substance of all terrestrial life, the study of soil science has a bright future.

References

Amundson R, Berhe A, Hopmans J, Olson C, Sztein A, Sparks D (2015) Soil and human security in the 21st century. Science 348:647

Aune J, Lal R (1997) Agricultural productivity in the tropics and critical limits of properties of Oxisols, Ultisols and Alfisols. Trop Agric (Trinidad) 74:96–103

Bai Z, Dent D, Olsson L, Schaepman M (2008) Proxy global assessment of land degradation. Soil Use Manag 24:223–234

Batjes N (1996) Total carbon and nitrogen in the soils of the world. Eur J Soil Sci 47:151–163

EIA (2013a) Today in energy. U.S. Energy Information Administration. http://eia.gov/todayinenergy/detail.cfm?id=12551

EIA (2013b) International energy statistics. U.S. Energy Information Administration http://www.eia.gov/cfapps/ipdbproject/IEDIndex3.cfm?tid=44&pid=44&aid=2

FAO (2012) World agriculture towards 2030/2050. ESA Working Paper No.10–30. FAO, Rome, Italy

FAO (2015) The state of food insecurity in the world. FAO, Rome

Hartemink A (2015) 90 years IUSS and global soil science. Soil Sci Plant Nutr 61:579–586

Hartemink A, McBratney A (2008) A soil science renaissance. Geoderma 148:123–129
IWMI (2001) Projected water scarcity in 2025. IWMI, Colombo
Lal R (2003a) Global CO_2 emissions by restoration of degraded soils and intensification of world agriculture and forestry. Land Degrad Dev 14:309–322
Lal R (2003b) Soil erosion and the global carbon budget. Environ Int 29:437–450
Lal R (2004a) Carbon emission from farm operations. Environ Int 30:981–990
Lal R (2004b) Soil carbon sequestration impacts on global climate change and food security. Science 304:1623–1627
Lal R (2010) Managing soils and ecosystems for mitigating anthropogenic carbon emissions and advancing global food security. Bioscience 60:708–721
Lal R (2014) Societal value of soil carbon. J Soil Water Conserv 69:186A–192A
Lal R (2015a) A system approach to conservation agriculture. J Soil Water Conserv 70(4):82A–88A
Lal R (2015b) Cover cropping and the "4 per thousand proposal". J Soil Water Conserv 70:141A
Lal R (2015c) Sequestering carbon and increasing productivity by conservation agriculture. J Soil Water Conserv 70(3):55A–62A
Lal R (2015d) Soil carbon sequestration and aggregation by cover cropping. J Soil Water Conserv 70:329–339
Lal R (2016) Beyond COP21: potential and challenges of the "4 per Thousand" initiative. J Soil Water Conserv 71:20A–25A
Lal R, Griffin M, Apt J, Lave L, Morgan MG (2004) Managing soil carbon. Science 304:393
Le Quere C, Moriarty R, Andrew R, Peters G, Ciais P, Friedlingstein P, Jones S, Sitch S, Tans P, Arneth A, Boden T, Bopp L, Bozec Y, Canadell J, Chini L, Chevallier F, Cosca C, Harris I, Hoppema M, Houghton R, House J, Jain A, Johannessen T, Kato E, Keeling R, Kitidis V, Goldewijk K, Koven C, Landa C, Landschutzer P, Lenton A, Lima I, Marland G, Mathis J, Metzl N, Nojiri Y, Olsen A, Ono T, Peng S, Peters W, Pfeil B, Poulter B, Raupach M, Regnier P, Rodenbeck C, Saito S, Salisbury J, Schuster U, Schwinger J, Seferian R, Segschneider J, Steinhoff T, Stocker B, Sutton A, Takahashi T, Tilbrook B, Van Der Werf G, Viovy N, Wang Y, Wanninkhof R, Wiltshire A, Zeng N (2015) Global carbon budget 2014. Earth Syst Sci Data 7:47–85
Loveland P, Webb J (2003) Is there a critical level of organic matter in the agricultural soils of temperate regions: a review. Soil Tillage Res 70:1–18
Muthayya S, Rah JH, Sugimoto JD, Roos FF, Kraemer K, Black RE (2013). The global hidden hunger indices and maps: an advocacy tool for action. PlosOne 8(6). doi:10.1371/journal.pone.0067860
Olson K, Al-Kaisi M, Lal R, Lowery B (2014) Experimental consideration, treatments, and methods in determining soil organic carbon sequestration rates. Soil Sci Soc Am J 78:348–360
U.N. (2015) World population prospects: 2015 Revision—The 2015 revision, key findings and advance tables. Working paper no. ESA/P/WP.241
UNDP (2014) Human development report. United Nations Development Programme, New York
Van Oost K, Quine T, Govers G, De Gryze S, Six J, Harden J, Ritchie J, Mccarty G, Heckrath G, Kosmas C, Giraldez J, Da Silva J, Merckx R (2007) The impact of agricultural soil erosion on the global carbon cycle. Science 318:626–629

Chapter 25
From Ujamaa to Big Results Now: Sustainable Transformation of Tanzanian Agriculture in the Frame of Climate Change

Ruth Haug

Abstract The relationships of frame conditions to the sustainable transformation of agriculture and the adoption of new technologies are discussed in this paper. Emphasis is given to identifying the place for a successful agricultural extension in this regard. Persistent rural poverty, chronic undernourishment, low agricultural productivity, and uncertainties surrounding the future impacts of climate change on food production are the challenges faced by the Tanzanian government. The objective of this research is to assess why different political actions and technological innovations have only had limited impacts on the planned transformation of Tanzanian agriculture. The main finding in relation to achieving the overall aim of the sustainable transformation of Tanzanian agriculture is the need to involve farmers and their organizations in policy formulation and implementation. As well, trust needs to be restored in public institutions and the many, changing initiatives put in place by the government and donors since independence. Predictability in relation to stable, conducive frame conditions and risk-cutting measures are important factors in both female and male farmers' willingness to adopt new technologies. These factors also affect the ability of extension services to contribute to sustainable transformation by providing advice relevant to various local conditions and the impacts of climate change.

Keywords Sustainable transformation · Agriculture · Climate change · Extension · Tanzania

R. Haug (✉)
Department of International Environment and Development Studies,
Norwegian University of Life Sciences, Ås, Norway
e-mail: ruth.haug@nmbu.no

R. Lal et al. (eds.), *Climate Change and Multi-Dimensional Sustainability in African Agriculture*, DOI 10.1007/978-3-319-41238-2_25

25.1 Introduction

Tanzania is a country with huge potential for agricultural development. However, despite the country's promising potential, the agricultural sector has not performed according to expectations for productivity increase and poverty reduction (Lintelo et al. 2014; World Bank 2015a). The country is more than self-sufficient in food production in most years. However, production has kept up with population growth largely through the expansion of cultivated land rather than increased productivity (Cooksey 2012; Mtengeti et al. 2014).

Average cereal grain yields have remained at the same low level of approximately 1300 kg/ha for the past 30 years (World Bank 2015b). Limited growth in productivity is regarded as an important reason why rural poverty remains high in the country (Lintelo et al. 2014). The agricultural sector in Tanzania employs around 75 % of the population and accounts for 30 % of all gross domestic product value added (World Bank 2015c). Approximately 70 % of the population lives on less than US$2 per day (World Bank 2015a). Persistent rural poverty, chronic undernourishment, low agricultural productivity, and uncertainties surrounding the future impacts of climate change on food production are major challenges faced by the government.

The link between supportive political frame conditions, agrarian change, and the adoption of new technologies is a focus of this paper. It also includes discussion of the place for a successful agricultural extension. During the past decade, Tanzania has performed well in economic growth, but the growth has not yet resulted in much improvement of rural poverty or malnutrition (WB 2015a). Small-scale farmers are responsible for 90 % of all rice and maize production, with farm size ranging from 0.5 to 2.0 ha (Mtengeti et al. 2014). The 2007–2008 agricultural census record 1006 large-scale farmers and approximately 6 million small-scale agricultural households (NSB 2007/08). In 2011, the number of small-scale farmers was around 8 million (URT/MFEA 2011). The proportion of household income that farmers obtain from agriculture declined from 60 % in 2000–2001 to 50 % in 2007, according to the 2007 household baseline survey (URT/MFEA 2009). Future possibilities for continued land expansion are limited. Consequently, national food security might become a problem if the current population more than doubles by 2050, as is forecasted (UN 2015). Another major concern is the lack of employment opportunities for youth and the creation of create jobs for the rapidly growing population of young people.

25.1.1 Approach and Analytic Framework

The major questions guiding the discussion in this paper are as follows:

- Why have technologies that have been pushed for decades only been adopted to a limited degree?

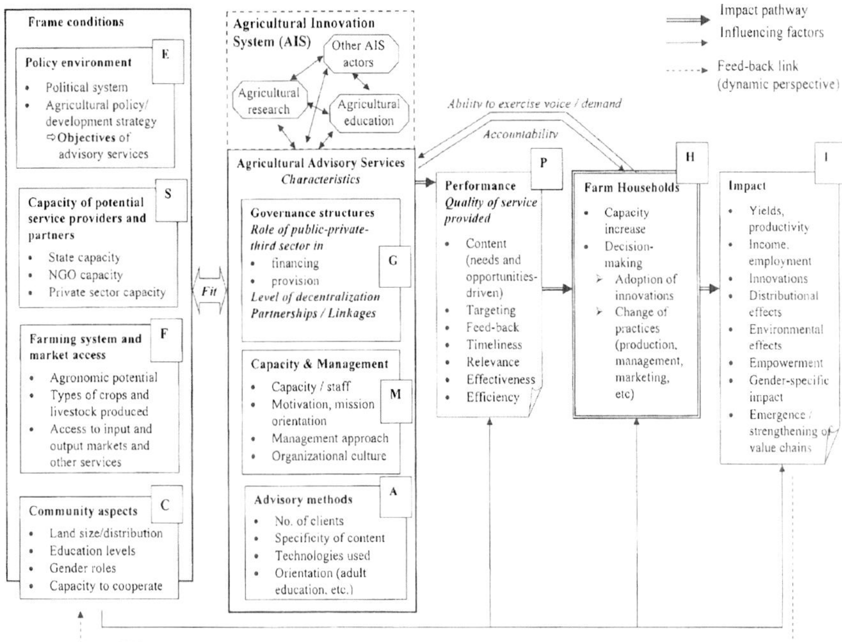

Fig. 25.1 Framework for designing and analyzing agriculture advisory services (Birner et al. 2006)

- To what extent are existing political frame conditions perceived as conducive to agricultural development?
- What role can an extension service play in the sustainable transformation of agriculture in Tanzania, including adaptation to climate change?

The content of this paper is based on an extensive literature review and qualitative interviews conducted over the past five years with Tanzanian key informants, such as women and men farmers, extension agents, agro-dealers, NGOs, private-sector representatives, researchers, and other civil district and national service officers. The framework developed by Birner et al. (2006) to design and analyze agriculture advisory services was selected as a guide for the research. This framework provides a holistic overview of the agricultural innovation systems, including frame conditions and impact factors, aligned with the sustainable transformation of agriculture (Fig. 25.1).

25.1.2 Sustainable Transformation of Agriculture

What is sustainable transformation of agriculture? With transformation, we might think of a radical change that leaves no way of return to what was before.

Transformation can be both good and bad, but we usually think of transformation for the better. However, different actors might have different views on what "for the better" means in practice as well as for whom "for the better" should be a reality. Transformation of agriculture is often seen as more or less the same as modernization. Staatz (1998:1) defines agricultural transformation as *the process by which individual farms shift from highly diversified, subsistence-oriented production towards more.*

Transformation implies radical change that limits the possibilities of returning to past systems and practices. In theory, transformation can be both good and bad, but transformation normally implies changing for the better. Different actors might have different views of what *for the better* means in practice and for whom *the better* should be a reality. Staatz (1998, p. 1) defines agricultural transformation as "the process by which individual farms shift from highly diversified, subsistence-oriented production towards more specialized production oriented towards the market or other systems of exchange." This definition suggests that agricultural transformation is somewhat analogous to modernization. Using a more popular approach, Kahunga (2015) describes agricultural transformation in East Africa as a continuum between the two extremes: (a) the *get-big-or-get-out school* involving large-scale foreign investors that demand tax exemptions, cheap labor, uncontrolled use of chemicals, free access to huge areas of land, and other inputs; and (b) the *food security activist school* that claims that farmers in Africa should be left alone with their age-old farming practices and culture. Neither of these two extreme schools as described by Kahunga (2015) necessarily represents transformation *for the better.*

Describing *transformation* as *sustainable* implies the need to account for social, economic, and environmental factors during the transition. In other words, the addition of *sustainable* implies the need for the transition to be *good* rather than *bad*. However, use of the term sustainable also opens the type of transformation to be promoted to different interpretations. Many different definitions of sustainability exist. According to the World Commission on Environment and Development (UN 1987, p. 16), "sustainable development is a process of change in which the exploitation of resources, the direction of investments, the orientation of technological development, and institutional change are made consistent with future as well as present needs." Sustainability is about how balancing social, economic, and environmental objectives in ways that meet the needs of the present generation without compromising the ability of future generations to meet their needs. The list of expected impacts from extension efforts in the framework found in Fig. 25.1 can be defined as the outcomes of sustainable transformation (e.g., increased yield and productivity) and as environmentally friendly- and climate-smart agriculture, as well as improved incomes, increased food security, reduced poverty, and greater social equality.

National states are responsible for defining policies and development strategies that lead to the sustainable transformation of their agricultural sectors. China is an example of a country that has transformed its agricultural sector over the past three decades. Smallholder farmers have driven the agricultural transformation and contributed toward huge improvements in productivity and food security. Public policies, such as land reforms and subsidized inputs, were powerful incentives for family farming and resulted in great productivity gains. According to the China-DAC Study group (2010), the three actors transforming Chinese agriculture since 1978 have been the state, the market, and small-scale farmers. However, the transition has not been equally successful from environmental or climate-change perspectives. Whether the transition can claim to be sustainable is questionable.

25.2 From Ujamaa to Big Results Now

According to Havnevik (2010), Tanzania is a special case because of its frequent, radical attempts to change policy and development strategy. The changes, instituted during the first decades after independence, are closely related to President Julius Nyerere's tenure from 1961 to 1985. The first years after independence, 1961–1967, are regarded as the post-independence open-market period, during which the focus was on economic modernization (Haug and Hella 2013). This period of open-market and modernization policy resulted in greater economic and social differences that were not in accordance with Nyerere's vision for Tanzania (Havnevik 2010). In the famous Arusha Declaration of February 1967, President Nyerere introduced a shift in policy toward what he defined as *African socialism and self-reliance*. In the agricultural sector, Ujamaa villages were established, cooperatives were formed, and farmers were told to engage in communal farming (collectivization). The Ujamaa period lasted from 1967 to 1985. The Ujamaa villages were gradually implemented, and by 1976, about 70 % of the population had been moved from their homes, either by free will or by force (Simonsen 2010). Villagization had several aims, including increasing the government's ability to provide public services in rural areas (Bryceson 2010). Nyerere's alternative path to development focused on social equity and distributive justice and received considerable external support, including from the Nordic countries (Ibhawoh and Dibua 2003).

The economic crisis of the early 1980s arose from food insecurity, poor performance in the agricultural sector, and decreased export earnings from agriculture. It resulted in the International Monetary Fund and the World Bank demanding economic reforms as a condition for continued support (Simonsen 2010). Accordingly, in 1986, the Ujamaa period was replaced by a period of liberalization through market-oriented economic policy reforms. The period saw the introduction of major structural adjustments to the Tanzania economy. While the Ujamaa period had focused on agriculture and farmers, the early phase of economic liberalization set in motion processes of deagrarianization and depeasantization in motion (Bryceson 2010). The structural adjustment program was intended to provide stimuli to Tanzanian agriculture. However, it forced smallholders to seek income diversification through non-farm employment and to view agriculture as a subsistence fall-back in order to reduce risks and secure a livelihood (Skarstein 2010).

Neither the collectivization of agriculture through Ujamaa nor market liberalization has transformed Tanzanian agriculture in the direction of modernization in the way that various donors have intended (Cooksey 2011). During the past decade, numerous strategies and plans have assigned priority to transforming agriculture in a direction that should result in increased production and productivity, higher income for farmers, reduced rural poverty, and improved food and nutrition security. In 2004, the Poverty Reduction Strategy Paper (MKUKUTA) was introduced, and in 2006 came the Agricultural Sector Development Program (ASDP), including the National Agricultural Input Voucher Scheme (NAIVS). Then, in 2009, President Jakaya M. Kikwete announced a new initiative for agriculture, Kilimo Kwanza (Agriculture First). This initiative focused more on the role of the private sector in agricultural development than the more public-sector oriented ASDP. The goal of Kilimo Kwanza was to commercialize and modernize agriculture in Tanzania (TNBC 2009; Kilimo Kwanza 2009). An important component of Kilimo Kwanza was the Southern Agriculture Growth Corridor (SAGCOT), which was launched in 2010. SAGCOT promoted the private sector and emphasized the importance of large-scale commercial farms as engines of growth in this high-potential agricultural area of the country (Coulson 2010; Kaarhus et al. 2010; Cooksey 2012).

The last initiative promoted by President Kikwete before he stepped down in November 2015 was Big Results Now (BRN), which drew upon the Malaysian model of development and identified agriculture as one of six priority areas (President's Office 2014; Kikwete 2014). In the agricultural sector, the goal of BRN was to increase maize productivity from 1.3 to 2.5 tons per ha. It also proposed establishing 25 commercial rice and sugarcane farms, 78 collective rice-irrigation and marketing schemes, and 275 collective warehouse-based marketing schemes (BRN 2013). The BRN approach sets quantitative targets and focuses on implementation and effective delivery of them, as well as on governance and stakeholder commitment. It is too early to evaluate the results of BRN and to what degree the initiative will continue to be a priority under newly elected President John Magufuli.

25.2.1 Extension and Sustainable Transformation

The role that extension services can play in the sustainable transformation of agriculture in Tanzania remains to be clarified. Extension services, which some prefer to call *agricultural advisory services,* are important for technology to be put into use and scaled-up. Extension services often receive the blame when farmers fail to adopt technologies whose value has been verified through research trials. To define the expected results from extension officer efforts, it is important to assess these services as part of the wider system in which knowledge and innovations are generated, disseminated, and used by farmers (Birner et al. 2006). In Fig. 25.1, the framework for analyzing agricultural advisory services developed by Birner et al.

(2006) emphasizes the frame conditions required for an extension to succeed and the ideal characteristics of extension systems, including the quality of the services provided and household-level factors affecting the situation. How to manage uncertainty, unpredictability, and uncontrollability are important issues considered both by farmers when deciding whether to try new technology and by extension officers when providing advice to farmers (Christoplos 2010).

The ultimate aim of extension services is to contribute to reduced poverty and increased yield, productivity, income, food security, social equality, and gender equality through environmentally friendly and climate-smart agriculture (Birner et al. 2006). New technologies might affect men and women differently due to differences in gender roles, responsibilities, and involvement in agriculture (Meinzen-Dick et al. 2014). The advice provided by extension services should lead to the adoption of innovations that contribute to a change for the better in accordance with the listed impact factors. In this way, an extension serves a tool in the efforts to achieve the sustainable transformation of agriculture.

In Tanzania, the public agricultural extension service predominates, with a few alternatives provided by the private sector, NGOs, CBOs, and participatory research operations. Since the establishment of subsidized inputs, private agro-dealer shops have spread around the country. These agro-dealers usually also provide agricultural advice to customers. The public extension system is decentralized, and district councils are expected to play an active role in the provision and management of the public extension service (Nchimbi-Msolla et al. 2015). The public extension service is perceived as performing poorly in its impact on farmers' uptake of new technology. Its funding has been reduced in real terms since the introduction of structural adjustment measures in the 1980s (Hella 2013). A whole range of different extension methods have been tried: demonstration plots, farm visits, farmer-training center, lead farmers, farmer field schools, farmer groups, farmer exchange visits, farmer-to-farmer extensions, women's groups, female extension officers, and participatory research. Mobile phones provide a new channel for the extension service to reach out to more farmers and for farmers to more frequently contact extension officers. With the high number of mobile phone subscribers in Tanzania, information and communication technology (ICT) provides an exciting opportunity for the extension service to achieve increased coverage, efficiency, and impacts (Sanga et al. 2014).

25.2.2 Transformation and Adoption of New Technology

The rate of technology adoption by farmers, along with productivity increases, has been low, despite decades of efforts to promote green revolution technologies, such as improved seeds and fertilizers, and agro-ecological practices, such as conservation agriculture. The average yield level for maize has stayed at the same level of around 1.3 tons per ha for the past 30 years (World Bank 2015b). Production has kept up with population growth through the expansion of areas under cultivation, and the country has remained food secure at the national level in most years (Mtengeti et al. 2014). Most Tanzanian farmers do not use fertilizer, improved seeds, or chemicals (Cooksey 2012). According to BRN (2013), 17 % of farmers use improved seeds, the average fertilizer use per ha is among the lowest in Africa, and 62 % of farmers use only hand hoes to cultivate the land. The government of Tanzania tends to explain low agricultural productivity by lack of access to technology, such as improved seeds and fertilizers, mechanization, and irrigation (URT/MAFC 2011; URT/TAFSIP 2011; URT/MAFC/NAP 2013; BRN 2013). Extension services often receive the blame for farmers not adopting new technology. The performance of the extension service has been questioned, and its capacity to deliver results with severely limited funding and an inadequate workforce is regarded as problematic. The public extension service in Tanzania has shrunk, when measured by the number of extension officers and amount of funding (BRN 2013; Hella 2013).

The government and donors collaborated in 2003 and 2004 to establish the ASDP in response to the low uptake of fertilizers and improved seeds. ASDP included an important input subsidy scheme, which was further developed as NAIVS in response to the 2007–2008 food-price crisis. NAIVS was designed to promote adoption of fertilizers and boost productivity (WB 2014). It provides targeted farmers with one bag of basal fertilizers, one bag of top dressing, and improved maize or rice seeds at half price for one half ha of their land. Farmers are expected to pay the other half of the cost for the fertilizer and the seeds themselves (Aloyce et al. 2014; WB 2014).

The overall results and lessons learned from the input subsidy program have been contested. According to the World Bank (2014), NAIVS has reached 2.5 million smallholders and increased agricultural production by a promising 2.5 million tons due to the subsidies. The logistical challenges of organizing and distributing the vouchers have been substantial, and the vouchers have often arrived late at the district level or been too few (WB 2014; Aloyce et al. 2014). The World Bank (2014) reports a low level of corruption problems, but Pan and Christiansen (2012) found that village elites captured 60 % of the vouchers. Rich farmers benefit more than poor farmers because they can afford to pay the 50 % top-up, while poor farmers who need cash give vouchers to their richer relatives or sell them cheaply to agro-dealers (Cooksey 2012; Aloyce et al. 2014). The World Bank (2014) has acknowledged that the goal of NAIVS was never to target the poorest. Rather, it was intended to reach farmers who can afford to pay a share of the costs for seeds and fertilizers.

The targeted farmers are supposed to graduate from the subsidy program after 3 years of participation. They are then expected to continue to buy inputs on their own (WB 2014; Aloyce et al. 2014). The World Bank (2014) documented that 47 % of the graduates who had not used these inputs before entering the program continued to buy improved seeds, while only 19 % continued to purchase fertilizers. The World Bank (2014) highlighted the low profitability of fertilizer use for most farmers due to high input costs at the farm-gate relative to the low farm-gate prices of produce sold to traders. Farmers cannot be expected to continue to use fertilizers when their use is not profitable. Access to credit will not be a solution as long as the low profitability of farming makes it impossible to repay loans. In fact, farmers might end up being worse off after taking out credit to purchase fertilizers. In theory, it is possible to buy improved seeds without also buying fertilizers, but the improved maize seeds do not perform well without fertilizers (Nchimbi-Msolla et al. 2015). Nchimbi-Msolla et al. (2015) found that the main reasons why few farmers use improved seeds are their high cost relative to the high risks and low profitability associated with farming. As well, farmers are concerned about the uncertainties regarding seed quality, especially the danger of purchasing low-quality seeds. The presence of low-quality or fake seeds in the market is a problem that discourages farmers from using improved seeds (Amugune 2014; Nchimbi-Msolla et al. 2015). Few farmers reported a lack of awareness or availability of improved seeds as a reason for not buying improved seeds, although availability was limited in more remote areas (Nchimbi-Msolla et al. 2015). Awareness of the benefits of using quality seeds is prevalent, which suggests that the extension service has done a good job of making farmers aware of the productivity implications of using improved seeds. Regarding climate change, improved seeds are an important adaptation measure. Most farmers are aware of the potential benefits of improved seed, so the uptake of climate-smart seed could be successful if the seeds are within the affordable reach of small-scale farmers.

If the use of non-subsidized fertilizers is not profitable for most farmers (WB 2014), the appropriateness of the message provided by extension agents to farmers needs to be questioned. Alternative practices that are less costly than buying fertilizers might be more profitable for many farmers and could be promoted. Although donors, NGOs, and CBOs have supported agro-ecological approaches for almost as long as green revolution technologies have existed. Support for conservation agriculture (CA) was started in Tanzania during the 1980s by the Soils Conservation and Agroforestry Project Arusha. It was succeeded by other projects, such as those introduced by the Research, Community and Organizational Development Associates; Women in Agricultural Development and Environmental Conservation; World Vision; CARE International, and the Food and Agriculture Organization (Kahimba et al. 2014). In Tanzania, CA takes different forms, such as terracing, conservation tillage, pit and trench farming, and micro catchment water harvest systems (Kahimba et al. 2014). The low availability of inputs and equipment at affordable costs has constrained the adoption of CA (e.g., rippers, no-toll seeders and seeds), as has the high labor demand of some CA practices (Umar et al. 2012; Kahimba et al. 2014). In addition, CA requires a well-functioning extension service with officers who possess adequate capacity, motivation, and understanding of CA principles. Regarding promoting different kinds of climate-smart agriculture, there has been much attention to agricultural technologies and practices (e.g., livestock and crop management) but less attention to the advisory service that can facilitate the use of such climate-smart practices (Richards et al. 2015).

The degree to which technologies, such as improved seeds and fertilizers, and agro-ecological approaches, such as CA, are the right technological means for transforming Tanzanian agriculture along a path of sustainable intensification is difficult to say. Tanzania is a large country with great variations in infrastructure and agro-ecological zones. Thus, a need exists to tailor technology to local conditions, such as high-potential areas or drylands, and to optimize access to roads and markets. The semi-arid areas of Tanzania are mostly food-deficit areas and contain approximately 40 % of the population. They need different approaches to improve agricultural production and related improvements in livelihoods than high-potential areas (Hella et al. 2011). It is important for the extension service to achieve impacts and contribute to the transformation of agriculture, not the least in light of the predicted reductions in crop yields due to climate change.

The advice given by extension officers needs to be for the better so that farmers profit from following it. Improvements could be increased production or productivity, higher income, better food security, a healthier diet, reduced labor demands, lower risks, or other factors of importance to both female and male farmers. If the extension message, whatever it is, does not prove helpful to farmers, transformation of agriculture will be difficult. Currently, the trend appears to be that farmers, increasingly and with limited investments, produce for home consumption and seek their cash income elsewhere (URT/MFEA 2009; Skarstein 2010; Cooksey 2012). New technologies presented under any umbrella, whether sustainable intensification or agro-ecological approaches, will require the right incentives for farmers to adopt them.

25.2.3 Transformation and the Role of Policy and Market Access

The analytic framework in Fig. 25.1 developed by Birner et al. (2006) illustrates the importance of supportive frame conditions to achieve the expected impacts of extension services. Agricultural policy and market access are important frame conditions. Agricultural price policy has proved to be of great importance in the conduciveness of present policies. In 1986, structural adjustment, liberalization, and the market economy came to Tanzania. However, a full market economy has not been implemented because restrictions on both in-country and cross-border trade have been in place (EAC 2012; Haug and Hella 2013). Periodic export bans on maize have been used to keep national food prices relatively low, while prices in neighboring countries have been much higher (EAC 2012). There is no explicit policy on when to use export bans, but the bans have been put in place by the government for national food-security reasons when food prices have increased significantly. Export bans can be good for food security in the short term but have dramatically reduced farmers' incomes, providing disincentives for future production.[1] Export bans have led to maize being smuggled out of the country to Malawi, Zambia, Kenya, Sudan, Somalia, and Ethiopia. Smuggling, however, is a risky

[1]For example, when an export ban on maize was imposed in July 2011, farm-gate prices in Rukwa were cut by more than half overnight (Haug and Hella 2013).

undertaking, and most of the profits go to the traders, rather than the farmers (KI 2011, Hella et al. 2011; EAC 2012). Regarding rice, imports of cheap rice from Asia have contributed to keeping rice prices low in the country. Tanzania has great potential for increased rice production, but in recent years, low farm-gate prices for rice due to both official and unofficial rice imports have created disincentives for both small and big national rice producers.

In the short run, keeping food prices down through export bans on maize and the import of low-priced rice might appear an attractive and relatively cheap measure to achieve national food security. However, as stated by Coulson (2010, p. 6), "if the Tanzanian government intervenes to hold down food prices, it can easily discourage production and make the problem it is trying to solve even worse." If farmers perceive that it is their responsibility to pay the price for food security in the country, overall production might suffer. There might be better ways of assisting food-insecure people, for example, through social protection program (Haug and Hella 2013). In rural areas, low food prices might have a negative impact on both net sellers and net buyers of food because low incomes for farmers mean low incomes for everybody in rural areas where most economic activities are closely tied to farming (Aksoy and Isik-Dikmelick 2008). Diao et al. (2013) also found that export bans in Tanzania hurt poor rural households and are thereby increasing poverty in the country.

Despite variations, a lack of market access for farmers is a big problem in Tanzania (IFPRI 2012). Before liberalization, the national Milling Corporation was a monopoly grain purchaser, though a parallel, private grain trade did also exist (Cooksey 2011). Neither state authorities nor current middlemen or private traders have solved the problem of connecting farmers to the market in an effective way. Inefficient, long value chains, including poor roads, long transport distances, numerous roadblocks, tax collections, and traffic checks, make marketing cumbersome and expensive (Haug and Hella 2013). Small-scale farmers produce limited volumes and are not organized collectively to transport produce to possible pick-up centers or storage facilities. Tanzania also lacks storage capacity. According to BRN (2013), less than 1 % of Tanzanian farmers have access to storage facilities. Tanzania experienced a bumper harvest in 2013–14, which represented great potential for boosting farmers' income (Muchoki 2014). However, as stated by Agnes Kalibata, president of the Alliance for the Green Revolution in Africa, in 2014,

> Tanzania's farmers produced one of the biggest maize crops the country has ever seen, but the bumper crop was a benefit mainly to farmers who had access to storage facilities and links to market opportunities. Too many farmers were left singing about "the problem of a good year"—lots of surplus produce rotting before they see a market. (Kalibata 2015, p. 1)

Lack of market access and unpredictable market conditions contribute to Tanzanian farmers' preference to produce only for home consumption (BRN 2013). Under such frame conditions, it will be difficult for extension officers to contribute toward farmers adopting technology that requires monetary investments.

25.2.4 *Sustainable Transformation of Tanzanian Agriculture?*

The need for conducive frame conditions, such as policy and market access, for technology to be adopted and extension services to be successful has been discussed. The high risks and low profitability of farming discourage many farmers from investing in agriculture. As noted by the World (2014), the majority of farmers targeted by input subsidy programs find it unprofitable to buy fertilizers and improved seeds if they are not subsidized. Without predictable and conducive frame conditions for agricultural development, the expected impact of efforts by extension officers to promote sustainable transformation and climate change adaptation will be hard to achieve. The extension service needs to adjust its technology advice to help farmers manage the risks resulting from unpredictable policies and markets and the uncertainties regarding weather and climate change (Christoplos 2010). Climate change is expected to increasingly affect yields, especially from 2030 and onward, and might already be doing so (Challinor et al. 2014). Adaptation will be a demanding task at all levels in both the short- and the long-term perspective. According to Hallegatte et al. (2015), by 2030, the only way of reducing the negative impact of climate change will be by lowering socio-economic vulnerability.

Since independence, from Ujamaa to BRN, Tanzanian governments have introduced different policies and strategies to transform agriculture. The following list of contradictions has been present in the government's efforts to formulate effective policies and strategies.

- Socialism and self-reliance versus market liberalization
- Urban versus rural focus
- The state versus the private sector
- High-potential areas versus the drylands
- Small-scale versus large-scale farmers
- Low food prices versus decent farm-gate prices
- National food security versus exports of maize
- Green revolution technology versus agro-ecological practices
- Input subsidies versus no input subsidies
- Gender-neutral versus women-directed interventions
- Public extension service versus extension services from the private sector and NGOs
- National ownership versus donor-driven approaches
- Top-down versus empowerment of small-scale farmers.

According to (Cooksey 2012), there is limited knowledge on the manner in which various interest groups inside and outside the country, public and private, influence policy formulation and implementation. Small-scale farmers lack political power and have little direct influence on national policy formulation (Maghimbi et al. 2011). However, influence on policy formulation and implementation is important for farmers to ensure that their concerns and interests are addressed. Cooksey (2012) points out that problems, such as patronage and rent-seeking, undermine official policies meant to deliver results. Cooksey (2012) also questions the government's ability to formulate policies which the state has the resources, both human and financial, to implement. If Tanzanian farmers are not empowered to hold the state accountable for failing to deliver on its policies and strategies, any policy approach will probably be unsuccessful (Maghimbi et al. 2011). Farmers and their organizations need to be involved in policy formulation and implementation. Engagement in this process will also help restore trust in the government. This trust has been eroded by the failure of too many government initiatives to deliver the promised results.

25.3 Conclusion

The role of frame conditions in the sustainable transformation of agriculture and the adoption of new technologies have been addressed in this paper. This discussion has included the role of public agricultural extension efforts. The main finding is that achieving the overall aim of sustainable transformation of Tanzanian agriculture requires involving farmers and their organizations in policy formulation and implementation and restoring trust in public institutions and the many changing initiatives put in place by the government and donors over the past five decades. Predictability resulting from stable frame conditions and risk-reduction measures is important for both female and male farmers. Predictability affects the willingness of farmers to adopt new technologies and the ability of extension services to contribute to sustainable transformation by providing advice relevant to various local conditions and farmers' needs. The predicted reductions in crop yields due to climate change Call for a fresh look at the role of agricultural advisory services in facilitating uptake of climate-smart practices.

References

Aksoy MA, Isik-Dikmelick A (2008) Are low food prices pro-poor? Net food buyers and sellers in low-income countries. Policy research working paper 4642. World Bank Development Research Group, World Bank, Washington

Aloyce GM, Gabagambi DM, Hella JP (2014) Assessment of operational aspects of the input supply chain under national agriculture input voucher schemes (NAIVS) in Tanzania. J Dev Agric Econ 6(3):94–104

Amugune A (2014) Alarm raised on fake seed. http://www.mediamaxnetwork.co.ke/thepeople/74824/alarm-raisedfake-seeds. Accessed June 2014

Big Results Now (2013) National key results area: agriculture lab. Ministry of Agriculture, Food Security and Cooperatives, Dar es Salaam

Birner R et al (2006) From best practice to best fit. A framework for analyzing pluralistic agricultural advisory services worldwide. International Food Policy Research Institute (IFPRI) discussion paper. IFPRI, Washington

Bryceson DF (2010) Agrarian fundamentalism or foresight? revisiting Nyerere's vision for rural Tanzania. In: Havnevik K, Isinika AC (eds) From Nyerere to Mkapa. Mkuki na Nyota Publishers, Dar es Salaam, pp 71–99

Challinor AJ et al (2014) A meta-analysis of crop yield under climate change and adaptation. Nat Clim Chang 4:287–291

China-DAC Study Group (2010) Agricultural transformation, growth and poverty reduction. IPRCC and DAC, Bamako

Christoplos I (2010) Mobilizing the potential of rural and agricultural extension: FAO—office of Knowledge Exchange, Research and Extension. The Global Forum for Rural Advisory Service, Rome

Cooksey B (2011) Marketing reform? the rise and fall of agricultural liberalism in Tanzania. Dev Policy Rev 29:57–81

Cooksey B (2012) Politics, patronage and projects: the political economy of agricultural policy in Tanzania. Futures agricultures working paper 40. http://www.future-agricultures.org

Coulson A (2010) Kilimo Kwanza: a new start for agriculture in Tanzania? http://www.btsociety.org/app/images/events/kilimo-kwanza_paper.pdf. Accessed 8 July 2011

Diao X et al (2013) Economy wide impact of maize export bans on agricultural growth and household welfare in Tanzania: a dynamic computable general equilibrium model analysis. IFPRI discussion paper 1287. International Food Policy Research Institute, Washington

EAC (2012) East African Community. http://www.pma.go.ug/docs/Analysis€20for%%20Tanzania%Maize%20Export%Ban_%20January%202012. Accessed 3 July 2012

Hallegatte S et al (2015) Shock waves: managing the impacts of climate change on poverty. Climate change development series. World Bank, Washington

Haug R, Hella JP (2013) The art of balancing food security: securing availability and affordability of food in Tanzania. Food Secur J 5:415–426

Havnevik K (2010) A historical framework for analyzing current Tanzanian transition: the post-independence model, Nyerere's ideas and some interpretations. In: Havnevik K, Isinika AC (eds) From Nyerere to Mkapa. Mkuki na Nyota Publishers, Dar es Salaam, pp 19–57

Hella JP (2013) Study to establish return to investment in agricultural extension services in Tanzania. Ministry of Agriculture, Food Security and Cooperatives, Dar es Salaam

Hella JP et al (2011) High global food prices: crisis or opportunity for smallholder farmers in Tanzania. Dev Pract 21(4–5):652–665

Ibhawoh B, Dibua JI (2003) Deconstructing Ujamaa: the legacy of Julius Nyerere in the quest for social and economic development in Africa. Afr J Polit Sci 8(1):59–83

International Food Policy Research Institute (2012) www.foodsecurityportal.org/tanzania/resources. Accessed 5 Jan 2013

Kaarhus R et al (2010) Agro-investment in Africa: impact on land and livelihoods in Mozambique and Tanzania. Noragric report no 53. Norwegian University of Life Sciences, Ås, Norway

Kahimba FC et al (2014) Adoption and scaling-up of conservation agriculture in Tanzania: case of Arusha and Dodoma regions. Nat Res 5:161–176

Kahunga M (2015) Agricultural transformation: which way for East Africa? http://www.newtimes.co.rw/section/article/2015-03-03/186507/. Accessed 24 June 2015

Kalibata A (2015) Africa rise to economic power? http://www.theguardian.com/global-development-professionals-network/2015/jun/03/agriculture-africa-rise-to-economic-power?CMP=share_btn_link. Accessed 24 June 2015

Kikwete JM (2014) Tanzania's transformation and vision 2025: governing economic growth for social gain. http://www.chathamhouse.org/sites/files/chathamhouse/home/chatham/public_html/sites/default/files/20140331TanzaniaKikwete.pdf. Accessed 12 Dec 2014

Kilimanjaro International (2011) Consultancy for undertaking the study on the impact of domestic and cross border trade restrictions in Tanzania. Final report submitted to Chief Executive Officer, Rural Livelihood Development Company, Kilimanjaro International, Dodoma

Kilimo Kwanza (Agriculture First) (2009) Kilimo Kwanza http://www.tnbctz.com/index2.php?option=com_docman. Accessed 10 Oct 2009

Lintelo te DJH et al (2014) The hunger and nutrition commitment index (HANCI 2013): measuring the political commitment to reduce hunger and undernutrition in developing countries. IDS evidence report 78. IDS, Brighton

Maghimbi S et al (2011) The agrarian question in Tanzania? A state of the art paper. Current African issues 45. Nordiska Afrikainstitutt, Uppsala

Meinzen-Dick R et al (2014) A system that delivers: integrating gender into agricultural research, development and extension. In: Quisumbing AR, Meinzen-Dick R, Raney TL, Croppenstedt A, Behrman JA, Peterman A (eds) Gender in agriculture: closing the knowledge gap. Springer, Berlin, pp 373–393

Mtengeti EJ et al (2014) Sustainable intensification of maize and rice in smallholder farming systems under climate change in Tanzania. In: Lal R, Singh BR, Mwaseba DL, Kraybill D, Hansen DO, Eik LO (eds) Sustainable intensification to advance food security and enhance climate resilience in Africa. Springer, Berlin, pp 441–467

Muchoki A (2014) Farmers fear maize losses could reach catastrophic levels. http://www.agra-alliance.org/media-centre/news/farmers-fear-maize-losses-could-reach-catastrophic-levels. Accessed 30 Oct 2014

National Bureau of Statistics (2007/08) National sample census of Agriculture 2007/2008. NBS, Ministry of Finance and Economic Affairs, Dar es Salaam

Nchimbi-Msolla S et al (2015) Assessment of the impact of seed policies on availability and accessibility of improved seeds of staple crops by smallholder farmers in Tanzania. EPINAV research report. Sokoine University of Agriculture, Morogoro

Pan L, Christiaensen L (2012) Who is vouching for the input voucher? decentralized targeting and elite capture in Tanzania. World Dev 40:1619–1633

President's Office (2014) Big results now (BRN). http://www.pmoralg.go.tz/quick-menu/brn/. Accessed 10 Dec 2014

Richards M et al (2015) How countries plan to address agricultural adaptation and mitigation: the Montpellier statement. Climate-smart agriculture: Towards sustainable landscapes and food systems. CGIAR–CCAFS, Montpellier, France

Sanga C et al (2014) On the development of the mobile based agricultural extension system in Tanzania: a technological perspective. Int J Comput ICT Res 8(1):49–67

Simonsen J (2010) The Norwegian-Tanzanian aid relationship: a historical perspective. In: Havnevik K, Isinika AC (eds) From Nyerere to Mkapa. Mkuki na Nyota Publishers, Dar es Salaam, pp 57–71

Skarstein R (2010) Smallholder agriculture in Tanzania: can economic liberalization keep its promises? In: Havnevik K, Isinika AC (eds) From Nyerere to Mkapa. Mkuki na Nyota Publishers, Dar es Salaam, pp 99–131

Staatz JM (1998) What is agricultural transformation? USAID. Workshop on structural transformation in Africa. http://fsg.afre.msu.edu/ag_transformation/Def_Trans.htm. Accessed 25 June 2015

Tanzania National Business Council (TNBC) (2009) Kilimo Kwanza: towards Tanzanian green revolution. Tanzania National Business Council, Dar es Salaam

Umar BB et al (2012) Are smallholder Zambian farmers economists? A dual-analysis of farmers' expenditure in conservation and conventional agriculture systems. J Sustain Agric 36:908–929

United Nations (1987) World Commission on Environment and Development. Brundtland Commission, New York

United Nations (2015) Population estimates. http://esa.un.org/wpp/unpp/p2k0data.asp. Accessed 24 June 2015

URT/MFEA (2009) Poverty and human development report. Ministry of Finance and Economic Affairs, Dar es Salaam

URT/MAFC (2011) Evaluation of the performance and achievements of the agricultural sector development program (ADSDP). United Republic of Tanzania Ministry of Agriculture, Food Security and Cooperatives, Dar es Salaam

URT/MFEA (2011) MKUKUTA annual implementation report 2010/11: delivering on commitments and implementation. United Republic of Tanzania Ministry of Finance and Economic Affairs, Dar es Salaam

URT/MAFC/NAP (2013) National agricultural policy: United Republic of Tanzania (Ministry of Agriculture, Food Security and Cooperatives, Dar es Salaam

URT/TAFSIP (2011) Tanzania agriculture and food security investment plan 2011–12 to 2020–21. United Republic of Tanzania, Dar es Salaam

World Bank (2014) Tanzania—strengthening national comprehensive agricultural public expenditure in Sub-Saharan Africa: national agricultural input voucher scheme (NAIVS). Public expenditure review. World Bank, Washington. http://documents.worldbank.org/curated/en/2014/02/19456105/tanzania-strengthening-national-comprehensive-agricultural-public-expenditure-sub-saharan-africa-national-agricultural-input-voucher-scheme-naivs. Accessed 24 June 2015

World Bank (2015a) United Republic of Tanzania mainland poverty assessment: Report No: AUS6819. World Bank, Washington, DC

World Bank (2015b) Cereal yield kg/ha Tanzania. http://data.worldbank.org/indicator/AG.YLD.CREL.KG. Accessed 10 Jan 2015

World Bank (2015c) Agriculture: value added GDP. http://data.worldbank.org/indicator/NV.AGR.TOTL.ZS. Accessed 24 June 2015

Part V
Private Sector Solutions

Chapter 26
Effect of Improved Plant Nutrition on Maize (*Zea mays*) and Rice (*Oriza sativa*) Grain Chemical Nutrient Content Under Smallholder Farming Systems in Tanzania

Ephraim J. Mtengeti, Eva Mtengeti, Frank Brentrup, Lars Olav Eik and Ramadhani Chambuya

Abstract Maize and rice are the primary cereal crops constituting more than 50 % of the dietary energy for the population of Tanzania. The current increased demand for food to feed an increasing population in the country has called either for an expansion of cultivated land or intensification of these two crops. Expansion of cultivated land is limited by high land use pressure and the concern over natural resources conservation. The only way to grow more food is then through agricultural intensification by improving plant nutrition and protection. Smallholder farmers, however, lack information on appropriate use of agro-inputs and the effect of inorganic fertilizers on these cereals' grain quality. This has led to either improper or disproportionate use of inorganic fertilizers resulting in disappointing low yield and frequent household food insecurity. To address this matter, a public–private partnership comprising two public universities and multinational companies dealing with fertilizer and crop protection was initiated in December 2010, aiming at demonstrating the effect of appropriate inorganic fertilizer use on the yield and chemical composition of maize and rice grains. In total, four farms of maize and three of rice crops in different villages and districts were selected for the demonstration that was carried out from 2011 to 2014. The demonstrated treatments were farmers' practice (FP) and appropriate use of inorganic fertilizers (YSS). Maize and rice grains were harvested, oven dried and analyzed for

E.J. Mtengeti (✉) · E. Mtengeti
Sokoine University of Agriculture, P.O. Box 3004, Morogoro, Tanzania
e-mail: emtengeti@yahoo.co.uk

F. Brentrup
Research Centre Hanninghof, Yara International, Duelmen, Germany

L.O. Eik
Department of International Environment and Development Studies/Noragric, Norwegian University of Life Sciences (NMBU), P.O. Box 5003, 1432 Ås, Norway

R. Chambuya
Syngenta International AG, Schwarzwaldalle e 215, CH-4058 Basel, Switzerland

R. Lal et al. (eds.), *Climate Change and Multi-Dimensional Sustainability in African Agriculture*, DOI 10.1007/978-3-319-41238-2_26

Nitrogen (N), Phosphorus (P), Potassium (K), Magnesium (Mg), Sulphur (S), Calcium (Ca), Boron (Bo), Copper (Cu), Iron (Fe), Manganese (Mn), Molybdenum (Mo) and Zinc (Zn). The concentrations of N, P, K, Mg and S were not significantly ($P > 0.05$) different between the farmers' and appropriate inorganic fertilizers use practices and ranged from 1.21 to 1.69, 0.18 to 0.34, 0.24 to 41, 0.08 to 0.13 and 0.09 to 0.12 % for maize grain and from 0.97 to 1.19, 0.26 to 0.31, 0.28 to 0.41, 0.09 to 0.12 and 0.07 to 0.10 % for rice grain, respectively. Comparable nutrients in the two agronomic practices are related to nutrient dilution under YSS due to high biomass production since grain yields under YSS were significantly ($P < 0.05$) higher than under FP, and thus, the total nutrient content produced per unit area was higher under YSS than FP. Since cereal grains from improved plant nutrition did not differ significantly ($P > 0.0.5$) from those from farmers' practice in terms of nutrient concentration, it can be concluded that farmers' practice grains obtained nutrients mainly from soil reserve and thus soil fertility mining with crop harvest. There is an urgent need, therefore, to train farmers on the appropriate improvement of plant nutrition to reduce depletion of soil fertility. The calcium concentration in maize grain regardless of farming practice was very low; thus, maize diets should include Ca-rich foods.

Keywords Agricultural intensification · Inorganic fertilizers · Macronutrients · Micronutrients

26.1 Introduction

26.1.1 Maize and Rice Grains Nutrient Concentrations

Maize and rice are the most widely cultivated and consumed cereal crops globally. Maize is the third most important cereal grain after wheat and rice in the world, providing nutrients for humans and animals and serving as the basic raw materials for the production of starch, oil and protein, alcoholic beverages, food sweeteners and, more recently, fuel. The green plant when made into silage has been used successfully in the dairy and beef industries. Protein of the whole maize grain ranges from 7.7 to 14.6 % (Ijabadeniyi and Adebolu 2005; Ullah et al. 2010). The maize grain germ is relatively rich in minerals, with an average value of 11 % compared with less than 1 % in the endosperm (Food and Agriculture Organization (FAO) 1992). The germ provides about 78 % of the whole kernel minerals. The most abundant mineral in maize grain is phosphorus followed by potassium, but as with most cereal grains, maize grain is low in calcium and micronutrients such as zinc, iron, manganese and copper (Enyisi et al. 2014).

Rice is the second major staple food for 75 % of the global population and provides 60 % of the food intake in Southeast Asia (Anjum et al. 2007). Potassium is the most abundant mineral in rice (brown rice) followed by magnesium and calcium. Unfortunately, rice is a poor source of many essential micronutrients such as iron and zinc. Thus, iron, zinc, and vitamin A deficiencies are common in rice-consuming regions (Rivero-Huguet 2007). Batista et al.'s (2012) findings demonstrated that rice can contribute significantly to the intake of molybdenum and potassium, but rice cannot be considered an important source of Fe and Ca. Rice has relatively low protein content ranging from 7 % for milled to 8 % for brown rice (Anjum et al. 2007).

Plant mineral nutrition affects the protein content of the rice grain such that soil organic matter, total nitrogen, exchangeable calcium, available copper and molybdenum tend to increase the grain protein content. Hussaini et al. (2008) showed that nitrogen fertilizer application up to 60 kg N ha^{-1} significantly increased the concentration of nitrogen, phosphorous, magnesium and potassium in maize grain. Therefore, in addition to the genetic factor, the mineral content of the cereal grain is affected by the mineral content of the soil where the cereal grain has been planted.

26.1.2 Strategies for Improving Maize and Rice Grain Quality

Maize and rice cereals are basic staple foods for a large population of Tanzania (Water Efficient Maize for Africa (WEMA) (2010); Private Agricultural Support (PASS) (2012); Food and Agricultural Organization Statistical base (FAOSTAT) (2012). Therefore, any move to improve the productivity of these two cereal crops must also make sure the quality (especially protein content and mineral concentration) is not overlooked. The concern over their low nutritional value, mainly for protein, and trace minerals have been discussed elsewhere through three approaches: genetic manipulation, improved agronomic practices and processing and fortification of harvested grain. Although the literature on processing and fortification of harvested grain is scant, breeding for higher concentrations of minerals in food crops has been used as an option for improving the health of humans who suffer from protein and mineral deficiencies. The plant breeding approach, however, requires the varietal differences be stable across different environmental conditions. However, the environmental conditions are not always stable especially with the current effects of climate change scenarios. Recurrent drought and high temperatures caused by climate change often hinder efforts made by crop breeders to improve crop productivity and quality properties.

To meet the food requirements of the country's growing population, intensification of crop production remains a viable option because of land use pressure and the continued decrease in arable land. Most soils are highly weathered and deficient in major and trace elements for crop production. Therefore, to achieve the required yield or physical quality standards (appearance of the crop product in the market), farmers sometimes or often apply high doses of nitrogen, for example, irrespective of the adverse impacts on the environment and the nutritional quality of the crop product. Excessive and inappropriate application of inorganic fertilizers can result in severe environmental and ecological problems. For example, high use of nitrogen fertilizers can lead to nitrate pollution in groundwater (Ju et al. 2006; Thorburn et al. 2003), eutrophication of water sources and greenhouse gases emissions that contribute to global warming (Scheer et al. 2008). An inappropriate balance of fertilizer elements may lead to other elements not available to the crops and thus reduce the crop product quality for the consumer. Excessive use of nitrogen and phosphorous fertilizers is common in growing high-yield varieties of crops but has been found to decrease the uptake capability of plants for micronutrients. Alamadari and Mobasser (2014) noted that smallholder rice farmers in Asia do not apply all the nutrients required by the crop, even though intensive modern agriculture has a soil fertility–depleting effect. These farmers in Asia, and even smallholder cereal crop farmers in Tanzania (Mtengeti et al. 2015), tend to apply nitrogen only, a small amount of phosphorus and no micronutrients at all. With each cropping season, the micronutrients are depleted out of the arable land with the harvested crop and are not being replaced. The result of this mining of nutrients is a widespread micronutrient deficiency problem in almost all arable lands worked by smallholder farmers in Asia and Africa (Alamadari and Mobasser 2014). Application of high-nitrogen fertilizer reduced the concentrations of Ca and Zn and increased the concentration of Mn in maize grains (Feil et al. 2005). However, through seed priming or coating with zinc fertilizer the rice grain zinc concentration may be increased substantially (Johnson et al. 2005; Shivay et al. 2008). Other strategies for nitrogen efficiency use have been advocated, such as postponing part of the nitrogen fertilization to more advanced growth stages (three to eight expanded leaves in maize and even to silking and grain filing for maize hybrid varieties), when plants have a greater capacity to take nutrients (da Siliva et al. 2005).

Effective nutrient management in agricultural intensification, therefore, requires an accurate accounting of nutrients the plant can take at that age of growth and nutrients removed from soils in the harvested portion of a crop (Herkman et al. 2003). Effective plant nutrient management among smallholder farmers who account for nearly 90 % of the production of maize and rice cereal crops in Tanzania (Regional Agricultural Trade Expansion Support (RATES) (2003); US Department of Agriculture (USDA) (2012) requires skill and capital. A strong partnership between public technical advisors (researchers and extension) and agro-industries is the appropriate vehicle for propagating the appropriate skills for effective plant nutrient management under agricultural intensification. This will enhance eco-friendly crop productivity and quality and thus reduce the vulnerability of rural communities, especially to the effects of climate change. To address the need for such a vehicle for propagating appropriate skills for an effective intensification of maize and rice growing in the country, a public–private partnership (PPP) between two public universities (Sokoine University of Agriculture and Norwegian University of Life Sciences) and two international agro-input companies (Yara, an international fertilizer company, and Syngenta, an international plant protection inputs company) was initiated in December 2010. The PPP aimed at conducting research and demonstrating how to achieve sustainable agricultural intensification under smallholder maize and rice farmers through effective plant nutrient management and protection for increasing crop productivity and crop product quality while preserving the environment. This paper presents part of the results of the PPP implementation dealing with investigating the effects of improved plant nutrition on the chemical nutrient concentrations of maize and rice grains under smallholder farmers in the Njombe, Mvomero, Morogoro and Kilombero districts.

26.2 Materials and Methods

26.2.1 Description of the Study Area

Maize crop trials were established under smallholder farmers in Njombe district and at the Sokoine University of Agriculture (SUA) farm in Morogoro district. Njombe district is located in Southern Highlands of Tanzania, 1700–1800 masl, and experiences a unimodal type of rain from late November to April. The total annual rainfall amount is 1000 to 2000 mm. The temperature ranges from 22 to 30 °C maximum and 15 to 20 °C minimum. The study was conducted in three villages in Njombe district: Matiganjola (1791 masl and S 09 13 883; E 034 53 452), Welela (1793 masl and S 09 01 233; E 034 48 233) and Ibumila (1819 masl and S 09″ 06.620, E 034″50). The Sokoine University of Agriculture farm is located along the western foot of the Uluguru Mountains and receives average annual rainfall of 800 mm. The farm is at 540 masl and S 06 50 870; E 037 39 270 with temperature ranging from 20 to 33 °C. The soil parent material is alluvium derived from intermediate metamorphic rock from the Uluguru Mountains and is classified as isohyperthermic, very fine, kaolinotic, kanhaplic, Haplutults (Soil taxonomy) or Chromic Acrisol according to the World Reference Base resource (Msanya et al. 2003; Szilas 2002).

Rice trials were established at smallholder farmers' fields in Dihombo village (360 masl and S 06 15 749; 037 32. 357) and the Dakawa Rice Research Institute farm (370 masl and S 06 25 236: E 037 32 476) in Mvomero district and in Mkula village (300 masl and S 074 78 826 E 36 54 700) in Kilombero district. These rice-growing areas are the floodplains of the Wami and Ruaha rivers, respectively, and have vertisols, fluvisols and complexes of vertisols and fluvisol (Msanya et al. 2003). The average annual rainfall in these bimodal rainfall rice-growing areas is 1000 mm, and the average annual temperature range is 24–32 °C.

According to Mtengeti et al. (2015), the soils at the Njombe study sites (i.e., Ibumila, Matiganjola and Welela villages) were acidic with pH < 4.4, very low plant available P Bray 1 *P* < 1.4 mg/100 g soil in all sites, low to very low potassium >11.2 mg/100 g and very low mineral sulphur >6 kg/ha (Landon 1991). Mineral N varied in Njombe sites ranging from 13 to 40 kg/ha where the highest level was found at Matiganjola and Ibumila. Macronutrients and micronutrients were low in all sites. Soils at SUA were moderate acid with P deficient <0.9 mg/100 g. Mineral N and sulphur at SUA were low. The micronutrients, especially Zn and Bo were at sufficient levels, and also potassium and magnesium were high.

Rice sites at Dakawa, Dihombo and Mkula were located in floodplains. The pH of the rice sites ranged from 5.5 to 6. These rice-growing site soils had sufficient levels of K, N, S, boron and zinc, but the levels of Bray 1 P were low, 1.3 mg/100 g. However, the soils at Dihombo had high pH (i.e., 7.6), low Bray 1 P and mineral N, very low K and S while Bo and Zn were at sufficient levels. The Mkula soils were acidic, low in mineral N, very low K and Bray 1 P, while S was at sufficient levels.

26.2.2 Treatments

Multi-location trials were conducted for three years (2012–2014) consecutively. Two treatments were included in each trial: FP and the Yara/SUA/Syngenta (YSS) practice. The YSS entails application of fertilizer depending on crop requirements to achieve optimum yields. The initial application was based on the initial nutrients present in the soils after soil analysis. Subsequent fertilizer application was based on replenishment of the amount removed through crop harvest. Additionally, the amount and type of fertilizer were applied according to the crop growth stage (Tables 26.1 and 26.2). Maize trials were planted in the beginning of the long rainy season in December in Njombe and March at SUA in Morogoro for the three consecutive years of this study. Planting spaces were 90 cm by 30 cm for the long maturing variety (120–150 days) planted in Njombe and 75 cm by 30 cm for the medium maturing variety (90–110 days) in Morogoro. Rice trials were planted twice per year in Dihombo and Mkula in August and March and harvested in December and June, respectively; and in Dakawa, rice was planted only once in March and harvested in July. Rice in all trials was planted in 20 cm by 20 cm spaces.

26.2.2.1 Input Application for Maize Crop

Fertilizer in FP was applied only twice to the maize crop (Table 26.1). Phosphorus in combination with nitrogen as 62 kg/ha of DAP (46P-18N) was applied during planting and only nitrogen as 123 kg/ha of urea (46N) in the fifth week after planting when the maize plant was at knee height. Therefore, the farmers applied only two macronutrients (N and P) and normally at a low rate that lead to continuous soil mining of other macronutrients and all micronutrients with crop harvests every year (Alamadari and Mobasser 2014) and, thus, a rapid decline in soil fertility. Therefore, the only way to meet household food demand under such a farming practice is to expand the farm.

Table 26.1 Inputs application for maize in YSS and farmers' practices (treatments)

Activities	Inputs		Period of application
	Yara/SUA/Syngenta practice (YSS)	Farmers' practice (FP)	
Seed treatment	Apron Star at 10 g/4 kg seed	No seed treatment	Seed preparation during planting
1st fertilizer application	YaraMila cereal @ 350 kg/ha	(DAP) @ 62 kg/ha	During planting
Pre-emergence herbicide application	Primagram Gold 3 lts/ha	No application. But 1st weeding 3rd week after planting	Just after planting
1st insecticide application	Karate 5 EC for control of stalk borers @ 395 mls/acre	Karate 5 EC for control of stalk borers @ 395 mls/ha	3rd week after planting if symptoms of attack occur
2nd fertilizer application. Sprayed on the leaves	YaraVitaTracel^{TM} BZ @ 2 kg/ha	No application	3rd week after planting (4–6 leaves)
3rd fertilizer application	YaraMila cereal 200 kg/ha	Application of Urea fertilizer @ 123 kg/ha	5th week after planting (knee height)
4th fertilizer application	YaraMila Java 52 kg/ha	No application	7 weeks after planting (tasseling)
2nd insecticide application	Karate 5 EC @ 395 mls/ha	No application	8th week after planting If symptom of attack occur
Herbicide application	Gramoxone @ 1240 mls/ha	Weeding by use of draft animals	10th week after planting

1. Yara mila cereal = 23N − 10P(P_2O_5) − 5K_2O − 2MgO − 3S − 0.3Zn
2. Yara Vita ^{TM}Tracel BZ = 5N − 7.5P_2O_5–5 K_2O − 5MgO − 5S − 5Zn − 5Bo − 0.1Cu − 0.1Fe − 0.1Mn − 0.1Mo
3. Yara Mila Java = 22N − 6P_2O_2 − 12K_2O − 2CaO − 1MgO − 3S − 0.2Bo − 0.2Zn
4. DAP = 18 % N − 46 % P_2O_5 (20 % P)
5. Urea – 46 % N

In improved agronomic practice (YSS), fertilizer was applied four times to the maize crop (Table 26.1). During planting, all macronutrients were applied except calcium, and again N, P, K, Mg and S were sprayed on the plants at four to six leaves stage of growth. Thereafter, the same macronutrients were top dressed when the plants were at knee height, and all six macronutrients were top dressed when the plants were tasseling. Only one micronutrient (Zn) was applied during planting, six (Zn, B, Cu, Fe, Mn and Mo) were applied at four to six leaves and two (Zn and B) were added at the tasseling growth stage. This fertilizer application regime seemed to match that of plant needs for the macro- and micronutrients, and with the split application, the fertilizer was offered when the uptake capacity of the plant was

Table 26.2 Inputs application for rice in YSS and farmers' practices (treatments)

Activities	Inputs		Period of application
	Yara/SUA/Syngenta practice (YSS)	Farmers' Practice (FP)	
Seed treatment	Apron Star at 10 g/4 kg seed	No seed treatment	Seed preparation during planting
Fertilizer application in rice nursery	YaraMila cereal 7.4 kg/100 m^2	Fertilizer application (urea @ 2.47 kg/100 m^2)	For the nursery seedbed
Herbicide application	Touchdown Forte @ 2.47 lt/ha	No treatment	Clear weeds before paddling
1st fertilizer application	YaraMila cereal @ 200 kg/ha	No application	During transplanting planting
Herbicide application	Solito 320 EC @ 1482 mls/ha	Hand weeding	2–3 weeks after transplanting
2nd fertilizer application on the leaves	YaraVita^{TM} Tracel BZ @ 2 kg/ha	Urea @ 123 kg/ha	4 weeks after transplanting (4–6 leaves)
3rd fertilizer application	YaraMila Java @ 247 kg/ha	No fertilizer	5th week after transplanting (tillering/booting)
4th fertlizer application	YaraLiva Nitrabor @ 61.7 kg/ha	No fertilizer	Booting
Insecticide application	Karate 5 EC @ 395 mls/ha	Application of Karate 5 EC @ 395 mls/ha	4–5 weeks after transplanting. If symptom of attack noted
Fungicide application	Artea 330 EC @ 494 mls/ha	No application of fungicide	4–5th week after transplanting. If symptom of attack noted

1. YaraMila cereal = 23N − $10P_2O_5$–$5K_2O$ − 2MgO − 3S − 0.3Zn
2. YaraVita ^{TM}Tracel BZ = 5N − $7.5P_2O_5$–5 K_2O − 5MgO − 5S − 5Zn − 5Bo − 0.1Cu − 0.1Fe − 0.1Mn − 0.1Mo
3. YaraMila Java = 22N − $6P_2O_2$ − $12K_2O$ − 2CaO − 1MgO − 3S − 0.2Bo − 0.2Zn
4. YaraLiva Nitrabor = 15.4N − 25.5CaO − 0.3Bo
5. Urea = 46 % N

optimum (da Siliva et al. 2005) thus leaving no more nutrient in the soil to pollute the environment and optimize profit over fertilizer use.

26.2.3 Input Application for Rice Crop

In farmers' practice, fertilizer was applied once in the rice field at four to six leaves stage of growth. The fertilizer applied was 62 kg urea (46N)/acre. Few farmers applied fertilizer during planting depending on the price of rice grain and availability of the fertilizer. Rice farmers, therefore, applied only one macronutrient and no micronutrients. The farmers relied on the alluvial floodplain soils to grow their rice with negligible external plant nutrient inputs. This practice, however, continues to deplete the soil fertility year after year, and farmers may need to expand their farms to meet the required rice grain demand.

In improved practice, fertilizer was applied four times in the rice field. The first fertilizer applied during planting had five macronutrients (N, P, K, Mg, S) and one micronutrient (Zn), and the second fertilizer application contained the same five macronutrients and six micronutrients (Zn, B, Cu, Fe, Mn and Mo) was sprayed on the plants at four to six leaves stage of growth. The third fertilizer containing six macronutrients (the same as the first and second but with an addition of Ca) was applied at the tillering growth stage. The last treatment of N, Ca and B was applied at the booting growth stage. Therefore, as in the maize crop fertilizers in the YSS practice, plant nutrients were applied strategically when the plant had optimum uptake capacity, thus preserving the environment from excess nutrients leaching to the streams.

26.3 Crop Harvesting and Grain Chemical Composition Analysis

Each treatment plot in each trial was demarcated into three sub-plots during the crop harvesting period. Two sampling units were then located in the middle of each sub-plot making a total of six sampling units per treatment. The maize sampling units were lines 4 m long, and the rice units were 1 m^2. Farmers continued to harvest their crops after sampling. After each crop sampling, two soil samples at 0–20 and 20–40 cm deep were collected for physical and chemical properties analysis. The soils, crop residues (maize stover and rice straw) and grains samples were sent to Research Centre Hanninghof, YARA International, Duelmen, Germany, for analysis of the macro- and micronutrients. This helped to understand the influence of improved plant nutrition on the chemical nutrient contents of the maize and rice grains and to calculate the nutrient removal with the crop harvest. The crop agronomic data recorded for maize were spacing of plants, number of plants per 4 m row (sampling unit), plant height, cob weight, cob length, grain yield (t/ha at 14 % MC), grain-specific weight (1000 seed wt), stover biomass (tDM/ha) and weed biomass/sampling areas, while the rice crop parameters recorded were number of plants/m^2, number of tillers/m^2, tiller height, number of panicles/m^2, grain yield (t/ha at 14 % MC), grain-specific weight (1000 seed wt) and weed biomass/m^2. In this chapter, however, only grain chemical nutrient contents in terms of the macro- and micronutrients and the dry matter yield of the crops are presented and discussed.

26.4 Data Analysis

The data was handled and analyzed using Excel, and a t-test was used to check if the difference between the FP and the improved (YSS) practices was significant at the probability level of $P < 0.05$. The data for each site and each season was analyzed individually because of the seasonal variation and environmental heterogeneity of the site.

26.5 Results and Discussion

26.5.1 Effects of Improved Plant Nutrition on Whole Maize Grain Nutrients Concentration and Dry Matter Yield

The mean macronutrient concentrations in maize grain from different demonstration sites are shown in Table 26.3. At Ibumila village, the whole maize grains from FP in 2013 and 2014 had significantly ($P < 0.05$) higher P and Mg concentrations than YSS, but at the same site, this was true only for K in 2014. However, the N concentrations of the whole maize grains at Matiganjola village were significantly ($P < 0.05$) higher in YSS than FP in 2012 and 2013, and the same difference was shown in S at the same site in 2013 while the Mg content was higher in YSS than FP only in 2014. At SUA, the YSS maize grains had a higher ($P < 0.05$) N concentration than FP in all years of the study, P in 2013 and 2014 and K in only 2013. At Welela village, the FP maize grains had higher N in 2013, higher Ca in 2012 and 2013 and lower K and S in 2012 than YSS.

In this study, the N concentration in the whole maize grains ranged from 1.21 to 1.63 % and from 1.28 to 1.81 % in the FP and YSS practices, respectively, and fell within the range reported by other researchers (Ijabadeniyi and Adebolu 2005; Ullah et al. 2010). Elsewhere, the application of 60 kg N/ha increased N, P, Mg and K significantly ($P < 0.05$) (Hussaini et al. 2008). The abundance of P and K and very low concentrations of Ca in the maize grain observed in this study are in agreement with the findings by Feil et al. (2005) and Enyisi et al. (2014).

In most cases, however, the macronutrient concentrations of the whole maize grain did not differ significantly ($P > 0.05$) between the YSS and the FP. This could be due to the diluting effect of the higher dry matter yield in YSS compared to FP. YSS had statistically significantly ($P < 0.05$) higher grain dry matter yield than FP at all sites and years except for 2012 and 2014 at Ibumila village. Therefore, there was a higher yield of macronutrients per unit of land in YSS than in FP. Thus, the disadvantage of the farmers' practice is not only the low productivity but also the soil mining of most of these macronutrients with harvest since only two elements (i.e., nitrogen and phosphorus) are applied to the plants and at a very low rate.

The mean micronutrient concentrations in the maize grain from different trial sites are shown in Table 26.4. At Ibumila village, the whole maize grain in YSS had

Table 26.3 Mean macronutrient concentration and dry matter yield of maize grain from different demonstration sites in Tanzania

Nutrients (%)	Practices	Ibumila village			Matiganjola village			SUA			Welela village		
		2012	2013	2014	2012	2013	2014	2012	2013	2014	2012	2013	2014
N	FP	1.25	1.28	1.34	1.22^{b}	1.21^{b}	1.32	1.55^{b}	1.63^{b}	1.59^{b}	1.46	1.61^{a}	1.50
	YSS	1.31	1.35	1.34	1.34^{a}	1.34^{a}	1.28	1.81^{a}	1.69^{a}	1.68^{a}	1.47	1.51^{b}	1.47
P value		0.340	0.370	0.980	0.030	0.000	0.610	0.000	0.030	0.010	0.900	0.050	0.480
P	FP	0.23	0.27^{a}	0.28^{a}	0.19	0.20	0.25^{a}	0.25	0.18^{b}	0.19^{b}	0.16^{b}	0.18	0.19
	YSS	0.24	0.22^{b}	0.19^{b}	0.18	0.20	0.21^{b}	0.25	0.22^{a}	0.21^{a}	0.21^{a}	0.21	0.20
P value		0.870	0.010	0.000	0.401	0.810	0.050	0.320	0.010	0.051	0.000	0.210	0.540
K	FP	0.32	0.34	0.32^{a}	0.30	0.32	0.32	0.30	0.26^{b}	0.26	0.24^{b}	0.24	0.26
	YSS	0.32	0.28	0.27^{b}	0.28	0.30	0.29	0.28	0.28^{a}	0.26	0.29^{a}	0.25	0.25
P value		0.860	0.081	0.030	0.340	0.310	0.072	0.072	0.030	0.650	0.000	0.360	0.710
Mg	FP	0.09	0.10^{a}	0.10^{a}	0.08	0.08	0.09^{a}	0.09	0.09	0.09	0.08	0.08	0.08
	YSS	0.09	0.09^{b}	0.08^{b}	0.08	0.08	0.08^{b}	0.09	0.10	0.09	0.08	0.09	0.08
P value		0.270	0.040	0.000	0.301	0.550	0.030	0.100	0.070	0.260	0.401	0.060	0.090
S	FP	0.11	0.09	0.10	0.10	0.09^{b}	0.09	0.12	0.12	0.12	0.10^{b}	0.09	0.09
	YSS	0.11	0.09	0.09	0.11	0.10^{a}	0.09	0.13	0.12	0.12	0.20^{a}	0.10	0.09
P value		0.350	0.830	0.280	0.070	0.010	0.030	0.441	0.581	0.430	0.010	0.330	0.101
Ca	FP	0.004	0.008	0.01	0.005	0.007	0.010	0.005	0.006	0.003	0.007^{a}	0.007^{a}	0.004
	YSS	0.004	0.008	0.01	0.005	0.006	0.010	0.003	0.005	0.002	0.005^{b}	0.006^{b}	0.005
P value		0.360	0.621	0.040	0.680	0.370	0.000	0.000	0.480	0.280	0.000	0.000	0.441
YieldtDM/ha	FP	2.19	0.95^{b}	3.35	2.17^{b}	1.98^{b}	1.57^{b}	2.29^{b}	4.14^{b}	4.25^{b}	1.56^{b}	2.73^{b}	3.78^{b}
	YSS	2.50	3.67^{a}	3.67	5.78^{a}	4.78^{a}	3.67^{a}	5.30^{a}	5.11^{a}	6.00^{a}	4.12^{a}	7.67^{a}	7.76^{a}
P value		0.658	0.027	0.680	0.006	0.003	0.20	0.000	0.017	0.002	0.058	0.031	0.018

FP Farmers' practice, *YSS* Yara/SUA/Syngenta improved practice. Values in the same column and nutrient followed by different superscript are significantly different at $P < 0.05$

Table 26.4 Mean micronutrients concentration of maize grain in different demonstration sites

Nutrients (mg/kg)	Practice	Ibumila village			Matiganjola village			SUA			Welela village		
		2012	2013	2014	2012	2013	2014	2012	2013	2014	2012	2013	2014
B	FP	2.45	2.43	2.99^{a}	2.90	2.47	2.92^{a}	2.28	2.03	2.73	2.68	1.61	1.83
	YSS	2.36	2.73	2.41^{b}	2.64	3.13	2.50^{b}	2.30	1.73	2.48	2.35	1.65	1.46
P value		0.690	0.120	0.350	0.390	0.830	0.020	0.97	0.645	0.150	0.500	0.960	0.060
Cu	FP	1.69	12.95	1.87	1.36	16.42	2.12^{a}	2.06	2.36	2.32a	2.16	11.64a	1.63
	YSS	1.64	12.61	1.64	1.42	14.96	1.50^{b}	2.07	2.24	2.00b	2.22	11.09b	1.29
P value		0.660	0.030	0.220	0.680	0.560	0.001	0.950	0.650	0.000	0.820	0.050	0.161
Fe	FP	16.39^{b}	16.76	15.24	15.38	17.72	13.05^{b}	17.44	20.52	17.95	16.08	15.59	14.64
	YSS	26.01^{a}	18.47	16.63	15.74	17.34	13.96^{a}	18.22	20.53	17.29	16.27	16.64	14.72
P value		0.322	0.151	0.310	0.780	0.820	0.003	0.351	0.982	0.351	0.862	0.330	0.343
Mn	FP	5.98	8.81	7.52^{a}	2.66	5.12	6.46^{a}	5.17	6.05	5.67	8.44^{a}	15.09a	11.94
	YSS	7.01	8.09	6.14^{b}	2.09	5.28	5.14^{b}	5.33	6.09	6.20	3.93^{b}	7.47b	6.51
P value		0.161	0.462	0.010	0.330	0.771	0.030	0.951	0.050	0.100	0.050	0.041	0.050
Mo	FP	0.10	0.69	0.14	0.04	0.69	0.10	0.51^{a}	0.13	0.04	0.08	0.65	0.19
	YSS	0.10	0.41	0.11	0.17	0.61	0.22	0.09^{b}	0.16	0.12	0.33	0.24	0.33
P value		0.970	0.482	0.691	0.320	0.780	0.060	0.030	0.791	0.191	0.511	0.101	0.091
Zn	FP	15.46	18.26	20.00^{a}	15.20	15.82	15.63^{b}	20.57^{b}	20.38^{a}	18.89	15.51^{a}	18.26	15.40^{a}
	YSS	15.42	14.70	15.32^{b}	15.02	17.27	15.76^{a}	23.84^{a}	17.93^{b}	16.97	12.69^{b}	15.31	13.09^{b}
P value		0.971	0.062	0.010	0.811	0.172	0.922	0.010	0.170	0.010	0.010	0.081	0.041

FP farmers' practice, *YSS* Yara/SUA/Syngenta improved practice. Values in the same column and nutrient followed by different superscript are significantly different at $P < 0.05$

significantly ($P < 0.05$) higher Fe than FP in 2012 and higher ($P < 0.05$) B, Mn and Zn in FP than YSS in 2014. The mean concentrations of B, Cu and Mn of maize grain were statistically significantly ($P < 0.05$) higher while Fe and Zn were significantly ($P < 0.05$) lower in FP than in YSS at Matiganjola village in 2014. In 2012, the maize grain at SUA had higher ($P < 0.05$) Mo and lower Zn in FP compared to YSS. The most variable micronutrients in maize grain were Cu and Zn. Improved plant nutrition did not improve ($P > 0.05$) Cu significantly at Matiganjola, SUA and Welela sites in 2013 and 2014, but in the same years, it improved significantly ($P < 0.05$) Zn at SUA and Welela. Improved plant nutrition had also no significant ($P < 0.05$) effect on the micronutrients of the whole maize grain at Matiganjola village in 2012 and 2013. At all Njombe sites, regardless of practice, copper was higher in the second year of the study than in the first and third years, a trend that was not observed at the SUA site. However, a comparison across season/years may not be very practical due to seasonal variations. The zinc, iron and copper concentrations in the maize grains observed in this study were in agreement with those reported by Ullah et al. (2010).

26.5.2 Effects of Improved Plant Nutrition on Whole Rice Grain Nutrients Concentration

The mean macronutrient concentrations of the rice grain are shown in Table 26.5. Nitrogen was significantly ($P < 0.05$) higher in YSS than FP in 2014 during the long rainy seasons in Dihombo and Mkula in 2012 and 2013 and was significantly lower in YSS during the short rainy season in 2012 at Mkula village. These results are nearly similar to those reported by Anjum et al. (2007). The N concentrations measured in this study were in agreement with those reported in a long field experiment in Asia (Dobermann and Fairhurst 2000). In all sites and seasons, none of the farmers' practice had N% lower than 1 % while in YSS out of 13 assessments six had levels below 1 %. The relatively lower N concentrations could be attributed to the higher grain yield in YSS than FP that led to a dilution effect because grain N removal was higher in YSS than FP (Mtengeti et al. 2015). In long rainy seasons, the mean dry matter yield of rice grain was significantly ($P < 0.05$) higher in YSS compared to FP, but there were no significant ($P > 0.05$) changes in yield in the short rainy season. This could be due to the use of controlled water through irrigation during the short rainy season that did not wash away the applied plant nutrients as in the case of the long rainy season.

Improved plant nutrition did not have any significant ($P > 0.05$) effect on the P concentration in whole rice grain except for 2013 and 2014 for the short and long rainy seasons in Mkula village and the Dakawa Research Institute. In the long rainy seasons of 2013 and 2014, the rice grain K concentrations were significantly ($P < 0.05$) different between YSS and FP. However, the P and K concentrations in the rice grain in this study were in agreement with those reported by Dobermann and Fairhurst (2000) in Asia. The rice grain Mg, S, and Ca concentrations varied

Table 26.5 Mean macronutrients concentration of rice grain from different demonstration sites

Nutrients (%)	Practice	Long rainy season									Short rainy season			
		Dakawa Research Institute			Dihombo village			Mkula village			Dihombo village		Mkula village	
		2012	2013	2014	2012	2013	2014	2012	2013	2014	2013	2014	2012	2013
N	FP	1.11	1.01	1.00	1.13	1.02	1.10^{b}	1.15^{b}	1.06^{b}	1.18	1.01	1.00	1.21^{a}	1.00
	YSS	1.19	0.99	0.98	1.20	0.97	1.18^{a}	1.17^{a}	1.17^{a}	1.16	0.97	0.99	1.16^{b}	0.96
P value		0.021	0.055	0.047	0.017	0.019	0.003	0.076	0.000	0.048	0.024	0.082	0.005	0.017
P	FP	0.26	0.28	0.23^{b}	0.27	0.26	0.32	0.30	0.31	0.28	0.30	0.30	0.31	0.28^{a}
	YSS	0.27	0.27	0.26^{a}	0.28	0.27	0.31	0.30	0.31	0.28	0.29	0.29	0.29	0.27^{b}
P value		0.058	0.055	0.001	0.032	0.025	0.051	0.054	0.083	0.044	0.011	0.010	0.068	0.004
K	FP	0.33	0.38^{a}	0.32^{b}	0.33	0.34	0.41^{a}	0.34	0.30	0.28	0.36	0.39	0.39	0.33
	YSS	0.33	0.34^{b}	0.35^{a}	0.33	0.35	0.37^{b}	0.30	0.30	0.28	0.36	0.39	0.35	0.33
P value		0.084	0.000	0.000	0.084	0.042	0.000	0.038	0.0196	0.042	0.052	0.095	0.054	0.098
Mg	FP	0.10	0.11	0.09	0.11	0.10	0.11	0.12	0.12	0.10	0.12	0.11	0.14	0.11
	YSS	0.11	0.11	0.09	0.11	0.10	0.11	0.12	0.12	0.11	0.11	0.11	0.13	0.11
P value		0.051	0.015	0.004	0.023	0.013	0.020	0.064	0.094	0.034	0.005	0.010	0.070	0.001
S	FP	0.09	0.09	0.10	0.09	0.07	0.09^{b}	0.09	0.09	0.09	0.08	0.08	0.10	0.09
	YSS	0.09	0.09	0.10	0.10	0.07	0.10^{a}	0.09	0.09	0.09	0.08	0.08	0.09	0.08
P value		0.049	0.093	0.002	0.047	0.033	0.002	0.070	0.052	0.062	0.012	0.093	0.061	0.015
Ca	FP	0.021	0.019^{b}	0.023	0.018	0.02	0.024^{a}	0.017	0.022^{a}	0.022	0.020	0.022	0.020	0.023
	YSS	0.021	0.023^{a}	0.023	0.019	0.02	0.022^{b}	0.018	0.017^{b}	0.021	0.020	0.022	0.018	0.020
P value		0.050	0.000	0.041	0.074	0.024	0.000	0.067	0.000	0.070	0.029	0.020	0.047	0.001
Yield t/ha	FP	4.56^{b}	6.63^{b}	4.22^{b}	5.14^{b}	5.74^{b}	4.67^{b}	3.72^{b}	4.75^{b}	8.16	7.49	6.03	4.10	7.11
	YSS	6.25^{a}	7.13^{b}	7.32^{a}	6.37^{a}	6.03^{a}	5.19^{a}	4.52^{a}	7.05^{a}	8.51	7.51	6.06	4.52	7.55
P value		0.061	0.158	0.051	0.010	0.620	0.021	0.612	0.000	0.897	0.989	0.978	0.347	0.879

FP Farmers' practice, *YSS* Yara/SUA/Syngenta improved practice. Values in the same column and nutrient followed by different superscript are significantly different at $P < 0.05$

slightly between FP and YSS and were lower than those reported by Dobermann and Fairhurst (2000) in Asia (0.15, 0.1 and 0.05 %), respectively. The lower concentrations of these nutrients in the studied sites are attributed to lower concentrations of these minerals in the soils under study (Mtengeti et al. 2015).

The mean micronutrient concentrations of the rice grain are shown in Table 26.6. In the long rainy seasons of 2013 and 2014, the rice grains at Dakawa and Dihombo had significantly ($P < 0.05$) higher B in YSS than in FP. The only significant ($P < 0.05$) change in the Cu concentration in rice grain due to improvement of plant nutrition was at Dakawa during the long rainy season in all years of the study. On average, the B and Cu concentrations were lower in rice grains produced during the short rainy season. This could be caused by reduced mobility of Bo and Cu due to low soil moisture because of intermittent and scarce water for irrigation. In the 2013, the Cu concentration in rice grain grown during the long rainy seasons at Dihombo and Mkula villages was exceptionally higher (>10 mg/kg) than in all other years. This could just be due to a favorable season for plant growth and uptake of nutrients. In this study, Cu was sprayed on the rice crop once at the four to six leaves stage of growth and probably the absorption through leaves was poor, thus leading to the low concentration of the element in rice grain.

Regardless of management practice and site, iron levels in rice grains ranged from 25 to 612 mg/kg, which was comparable to those obtained elsewhere (Dobermann and Fairhurst 2000) suggesting iron availability to plants were sufficient and did not limit crop growth. In Dihombo village, in 2013 the rice grain Mn concentration in YSS was significantly lower ($P < 0.05$) than in FP during the long rainy season but higher during the short rainy season. Zn was the most variable micronutrient in rice grain. It was significantly ($P < 0.05$) higher in FP than YSS in 2013 during the short rainy season but lower during the long rainy seasons at Mkula in 2013. In the short rainy season at Mkula village in 2013, the rice grain had significantly higher Fe, Mn, Mo and Zn concentrations in YSS than in FP.

Under smallholder, as in maize production and rice farmers' practice, fertilizers are not only inadequately applied, but there was a lack of effective nutrient management and thus an inability to cope with environmentally friendly agricultural intensification in the country. The farmers applied only two macronutrients (N and P) and normally at a low rate that led to continuous soil removal of other macronutrients and all micronutrients with the crop harvest every year. For example, rice farmers relied heavily on the alluvial floodplain soils to grow their crops year after year with almost negligible external plant inputs. Whether for maize or rice, the plants under farmers' practice were not fed according to their phonological needs, especially during flowering and grain filling. When plants are supplied with high amount of nutrients when their uptake capacity is still low especially in the early stages of growth, portions of the supplied nutrients are leached and pollute the underground water (Thorburn et al. 2003). Therefore, as smallholder farmers are urged to adopt agricultural intensification as a means of improving household food demand with small farmed areas, training on how to feed the crop plants appropriately for sustainable productivity and healthy environment is very important.

Table 26.6 Mean micronutrients concentration of rice grain from different demonstration sites

Nutrients (mg/kg)	Practice	Long rainy season									Short rainy season			
		Dakawa research institute			Dihombo village			Mkula village			Dihombo village		Mkula village	
		2012	2013	2014	2012	2013	2014	2012	2013	2014	2013	2014	2012	2013
B	FP	3.03	3.87[b]	2.57[b]	2.67	1.34	2.82[b]	2.43	3.25	1.51	1.85	2.33	2.85	1.79
	YSS	3.00	4.25[a]	2.96[a]	3.13	1.64	3.15[a]	2.33	2.47	1.65	2.39	2.29	2.34	1.36
P value		0.995	0.344	0.030	0.250	0.512	0.270	0.729	0.021	0.400	0.241	0.863	0.420	0.090
Cu	FP	1.73[b]	2.28[b]	4.38[a]	1.92	11.87	2.36	2.29	11.38	4.24	2.35	3.08	3.01	4.08
	YSS	2.20[a]	7.62[a]	3.52[b]	2.53	12.51	2.80	2.31	11.93	4.57	2.76	3.70	2.50	3.46
P value		0.015	0.020	0.020	0.141	0.130	0.020	0.413	0.000	0.320	0.340	0.080	0.371	0.170
Fe	FP	38.7	53.6	288.8	210.7	52.6	524.1	399.6	72.4[b]	554.0	294.5	530.9	173.7	612.2[a]
	YSS	25.6	31.3	218.0	159.7	70.99	556.5	354.9	251.4[a]	650.7	304.3	696.3	185.0	323.2[b]
P value		0.331	0.413	0.260	0.33	0.390	0.480	0.641	0.220	0.220	0.700	0.311	0.830	0.031
Mn	FP	56.0[b]	70.1	79.9	79.8	79.5[a]	81.0	46.5	45.3	51.6	80.9[b]	88.4	49.4	82.5[a]
	YSS	79.8[a]	82.0	135.1	76.9	75.4[b]	86.3	51.2	47.0	57.9	114.1[a]	97.1	44.3	70.8[b]
P value		0.003	0.078	0.000	0.540	0.361	0.240	0.403	0.260	0.121	0.000	0.352	0.540	0.031
Mo	FP	0.42	0.19	0.38	0.11	1.38	1.96	0.90	0.88	1.95	0.51	1.77	1.40	2.05[a]
	YSS	0.61	0.29	0.55	0.26	0.79	1.60	0.69	1.14	2.11	0.51	1.69	1.20	1.29[b]
P value		0.608	0.667	0.260	0.351	0.040	0.140	0.634	0.320	0.371	0.990	0.733	0.660	0.04
Zn	FP	18.0	18.2	18.2	20.1	18.5	22.2[a]	20.5	23.1[b]	21.7	20.6[a]	19.5	28.4	19.4[a]
	YSS	16.7	17.7	17.8	19.9	19.0	20.9[b]	20.6	25.0[a]	22.2	19.3[b]	20.0	25.4	17.8[b]
P value		0.306	0.304	0.290	0.121	0.530	0.000	0.431	0.130	0.410	0.020	0.700	0.520	0.031

FP farmers' practice, *YSS* Yara/SUA/Syngenta improved practice. Values in the same column and nutrient followed by different superscript are significantly different at $P < 0.05$

26.6 Conclusions and Recommendations

The results from this study have shown (1) crop plants under smallholder maize and rice farmers' practice were not fed adequately, but the plants' grain nutrient content did not differ appreciatively with improved agronomic management practice. (2) Therefore, there is an urgent need to train smallholder farmers on effective plant nutrient management, that is, appropriate feeding of the crop plants according to their demand and capacity for nutrient uptake. This will enhance agricultural intensification and thus no expansion of arable land, increased conservation of nature and reduced effects of climate change to the household. (3) Even with optimal and appropriate application of the required plant nutrients, the calcium content of the maize grains was poor during all years of the study. Farmers whose daily diet is mainly composed of maize should be urged to include calcium-rich food items in their diet; otherwise, they could suffer calcium deficiency symptoms such as rickets in children and osteomalacia (weak bones) in adults. (4) Improved plant nutrition did not affect the grain nutrient concentration appreciably but increased the amount of harvested nutrients with increased crop yield compared to the low crop yield in smallholder farmers' practice. This means that with appropriate maize and rice intensification the aim may not entirely be to increase the concentration of nutrients in their grain but to increase the amount of nutrients harvested per unit area and thus reduce household essential nutrient insecurity.

Acknowledgments The authors of this article would like to thank NORAD, Yara International and Sygenta International for their financial support and Sokoine University of Agriculture for logistical support during the period of this study. The authors are also grateful to the district council's extension workers for their support in data collection and smallholder maize and rice farmers in Njombe, Kilombero and Mvomero districts for allowing portion of their farms be used in this study.

References

Alamadari MQ, Mobasser HR (2014) The macro- and micro-nutrient fertilizer on yield and yield attributes of rice in a calcareous soil. Am J Exp Agric 4:1604–1615

Anjum FM, Pasha I, Bugti A, Butt MS (2007) Mineral composition of different rice varieties and their milling fractions. Pak J Agric Sci 44:332–336

Batista BL, Nacano LR, de Freitas R, de Oliveira-Souza VC, Barbosa F (2012) Determination of essential (Ca, Fe, I, K, Mo) and toxic elements (Hg, Pb) in Brazilian rice grains and estimation of reference daily intake. Food Nutr Sci 3:129–134

da Siliva PRF, Strieder ML, da Silva Coser RP, Rambo L, Sangoi L, Argenta G, Forsthofer EL, da Silva AA (2005) Grain yield and kernel crude protein content increases of maize hybrids with late nitrogen side-dressing. Sci Agric 62:487–492

Dobermann A, Fairhurst TH (2000) Rice: nutrient disorders and nutrient management handbook series. Oxford: Potash & Phosphate Institute (PPI), Potash and Phosphate Institute of Canada (PPIC) and International Rice Research Institute (IRRI)

Enyisi IS, Umoh VJ, Whong CMZ, Abdullahi IO, Alabi O (2014) Chemical and nutritional value of maize and maize products obtained from selected markets in Kaduna State, Nigeria. Afr J Food Sci Technol 5(4):100–104

FAO (1992) Maize in human nutrition. FAO Food and nutrition series, No. 25. FAO, Rome

FAOSTAT (2012) The Food and Agricultural Organisation's Statistical Database. Land Use Area Database. http://faostat.fao.org/site/377/DesktopDefault.aspx?PageID=377#ancor. Accessed 21 Oct 2013

Feil B, Moser SB, Jampatong S, Stamp P (2005) Mineral composition of the grains of tropical maize varieties as affected by pre-anthesis drought and rate of nitrogen fertilization. Crop Sci 45:516–523

Herkman JR, Sims JT, Beegle DB, Coale FJ, Herbert SJ, Bruulsema TW, Bamka WJ (2003) Nutrient removal by corn grain harvest. Agron J 95:587–591

Hussaini MA, Ogunlela VB, Ramalan AA, Falaki AM (2008) Mineral composition of dry season maize (*Zea mays* L.) in response to varying levels of nitrogen, phosphorus and irrigation at Kadawa, Nigeria. World J Agric Sci 4:775–780

Ijabadeniyi AO, Adebolu TT (2005) The effect of processing methods on the nutritional properties of produced from three maize varieties. J Food Agric Environ 3:108–109

Johnson SE, Lauren JG, Welch RM, Duxbury JM (2005) A comparison of the effects of micronutrient seed priming and soil fertilization on the mineral nutrition of chickpea (*Cicerarietinum*), lentil (*Lens culinaris*), rice (*Oryza sativa*) and wheat (*Triticum aestivum*) in Nepal. Exp Agric 41:427–448

Ju XT, Kou CL, Zhang FS, Christie P (2006) Nitrogen balance and groundwater nitrate contamination: comparison among three intensive cropping systems on the North China Plain. Environ Pollut 143:117–125

Landon JR (1991) Booker tropical soil manual: a handbook for soil survey and agricultural land evaluation in the tropics and subtropics. Wiley, New York

Msanya BM, Kaaya AK, Araki S, Otsuka H, Nyadzi GI (2003) Pedological characteristics, general fertility and classification of some benchmark soils of Morogoro District, Tanzania. Afr J Sci Techn 4(2):101–112

Mtengeti EJ, Brentrup F, Mtengeti E, Eik LO, Chambuya R (2015) Sustainable intensification of maize and rice in smallholder farming systems under climate change in Tanzania. In: Lal R, Singh BR, Mwaseba DL, Kraybill D, Hansen DO, Eik LO (eds) Sustainable intensification to advance food security and enhance climate resilience in Africa. Springer, Heidelberg, pp 441–465

PASS (2012) Private agricultural sector support: investment potential in Grain. http://www.pass.ac.tz/investmentpotential.html

RATES (2003) Maize market assessment and baseline study for Tanzania. Regional Agricultural Trade Expansion Support Program (RATES). United States Agency for International Development, Nairobi

Rivero-Huguet M (2007) Rice (*Oryza sativa* L.) as a source of microelements and toxic contaminants. http://www.eoearth.org/view/article/155736. Accessed 20 May 2015

Scheer C, Wassman R, Kienzler K, Ibragimov N, Eschanov R (2008) Nitrous oxide emissions from fertilized, irrigated cotton *(Gossypium hirsutum* L.) in the Aral Sea Basin, Uzbekistan: Influence of nitrogen applications and irrigation practices. Soil Biol Biochem 40:290–301

Shivay YS, Kumar D, Prasad R, Ahlawat LPS (2008) Relative yield and zinc uptake by rice from zinc sulphate and zinc oxide coatings onto urea. Nutr Cycl Agroecosyst 80:181–188

Szilas C (2002) The Tanzania Minjingu phosphate rock: possibilities and limitation for direct application. Thesis for award of Ph.D. at Chemistry Department Royal Vetinary and Agricultural University Copenhagen, Denmark

Thorburn P, Biggs JS, Weier KL, Keating BA (2003) Nitrate in ground waters of intensive agricultural areas in coastal Northeastern Australia. Agric Ecosyst Environ 94:49–58

Ullah IU, Mohammed A, Arifa F (2010) Chemical and nutritional properties of some maize (Zeamays L.) varieties grown in NWFP, Pakistan. Pak J Nutr 9(11):1113–1117

USDA (2012) United States Agency for International Development review of policy constraints on competitive East Africa Community rice production. http://agritrade.cta.int/en/layout/set/print/Agriculture/Commodities. Accessed 21 Oct 2013

WEMA (2010) Mitigating the impact of drought in Tanzania: the WEMA intervention. Policy brief. Water Efficient Maize for Africa (WEMA). http://www.aatf-africa.org/userfiles/WEMA-Tz-policy. Accessed 20 Oct 2013

Chapter 27
Public-Private Partnership for Sustainable Production and Marketing of Goat's Milk in Light of Climate Change

G. Msalya, H. Lie, V. Mfinanga, A.S. Ringheim, R. Sandvik, M. Åsli, O.A. Christophersen, A. Haug, D.E. Mushi, D. Mwaseba, G.C. Kifaro and Lars Olav Eik

Abstract In Tanzania, goat's milk has a high market value due to its desirable nutritional profile and cultural recognition that it is beneficial to human health. A joint initiative between Sokoine University of Agriculture (SUA) and the Norwegian University of Life Sciences (NMBU) that introduced dairy goats to Tanzania has resulted in approximately 400,000 goats in the region to this day,

G. Msalya (✉) · D.E. Mushi · D. Mwaseba · G.C. Kifaro
Sokoine University of Agriculture (SUA), Morogoro, Tanzania
e-mail: msalya@suanet.ac.tz

D.E. Mushi
e-mail: danielmushi@gmail.com

D. Mwaseba
e-mail: dilmwase@hotmail.com

G.C. Kifaro
e-mail: gkifaro@yahoo.com

H. Lie · A.S. Ringheim · R. Sandvik · M. Åsli · O.A. Christophersen
A. Haug · L.O. Eik
Norwegian University of Life Sciences (NMBU), Ås, Norway
e-mail: helene.lie@nmbu.no

A.S. Ringheim
e-mail: alexandersolstad@gmail.com

R. Sandvik
e-mail: ritah.sandvik@nmbu.no

M. Åsli
e-mail: magnus.aasli@gmail.com

O.A. Christophersen
e-mail: olav.albert.christophersen@gmail.com

A. Haug
e-mail: anna.haug@nmbu.no

R. Lal et al. (eds.), *Climate Change and Multi-Dimensional Sustainability in African Agriculture*, DOI 10.1007/978-3-319-41238-2_27

providing households and communities with milk and various other animal byproducts. In areas such as the highlands of Mgeta in the Morogoro region, where there was previously no production of milk, dairy goats have achieved an average milk production of 1.4 L of milk per animal per day. This led to a rise in milk consumption from 0 L per household per day in 1988 to 1.6 L per household per day in 2012. Although the joint initiative between SUA and the NMBU was successful in diversifying the diets and improving the livelihoods of the poor, farm group efforts to distribute surplus milk to larger and more distant markets has remained a challenge. Suboptimum feeding practices and low education levels in milk handling and entrepreneurship may explain this lack of progress. In order to expand milk distribution beyond the local market, the involvement of an established dairy company is required. This paper will look at collaborations between farmers, private-milk-processing enterprises such as Shambani Graduates Ltd. (SGL) and a research institution (SUA) to discuss the value of milk in human nutrition and the feasibility of a pro-poor value chain for climate-smart goat's milk processing and dairy goat maintenance.

Keywords Goat's milk · Human nutrition · Public-private partnership · Climate-smart · Shambani Graduates Ltd. · Tanzania

27.1 Introduction

For the majority of the global population, livestock plays a significant role in the diversity of diets. Protein is vital to the human diet, and over 80 % of protein-rich food is primarily derived from the meat and milk of livestock. Milk provides fat, protein, and essential micronutrients, such as iron, zinc, and vitamin A, making it a desired food product that is especially sought after in developing countries (FAO 2013b). Globally, at least 150 million households depend on milk production for their livelihood, with the majority being smallholder farmers (FAO 2013a, 2015). Looking ahead, it is projected that the consumption of milk and milk products in African countries will continue to increase from their current levels (Tschirley et al. 2014).

In Tanzania, the livestock sector contributes significantly to the national food supply and to the livelihoods and food security of smallholder farmers. Livestock keeping is common throughout the country and includes 15.2 million goats and 22.8 million cattle that are kept by 2.3 million households. Furthermore, at least 70 % of these households are smallholder farmers (Njombe and Msanga 2011;

L.O. Eik
e-mail: lars.eik@nmbu.no

V. Mfinanga
Shambani Graduates Limited, Morogoro, Tanzania
e-mail: victor.mfinanga@shambani.co.tz

URT 2012). The increasing number of dairy animals (goats and cattle) and growing investments in milk production and processing indicate that the expansion of Tanzania's dairy sector is ongoing. Between 1995 and 2007, milk production increased from 555 million liters to 1640 million liters, contributing to 30 % of the gross domestic product (GDP) for livestock (MLFD 2011; Njombe and Msanga 2011).

More than any other farm animal, goats supply milk, meat, and fiber to rural people. In many cases, the ability for goats to adapt to a wide range of climates and management practices make them irreplaceable. Their contribution to the economy is also noteworthy because they can thrive in arid and semi-arid lands that are considered unfavorable for the production of agriculture and other purposes (Zervas and Tsiplakou 2013). In Tanzania, where goats are traditionally kept for meat, dairy goats have become increasingly popular among smallholder farmers. Currently, approximately 400,000 of these animals are kept nationwide (NBS 2012). For smallholder farmers, it is often easier to buy a goat than a cow due to the substantial difference in cost. Compared to cows, goats are relatively inexpensive to keep and require less fodder and a smaller plot of land for grazing (Peacock 2007). This is particularly beneficial for smallholder Tanzanian farmers, who typically own smaller plots of land. Due to their small body size and because they produce fewer emissions than cattle goats are also likely to fit well into integrated and climate-friendly farming systems (Nziku et al. 2015).

Consumers in Tanzania and elsewhere consider goat's milk to be healthy. The milk is easier to digest than cow's milk and is also better suited to the sick, elderly, children, nursing mothers, and the lactose intolerant. Furthermore, goat's milk has proven to be a viable milk substitute for infants with mothers who are HIV-positive and in cases where few or no alternative infant formulas exist. These factors that highlight the value of dairy goats in the poor and rural households of Sub-Saharan Africa (Zervas and Tsiplakou 2013).

In Mgeta, a set of rural communities located in the Uluguru Mountains of Morogoro, Tanzania, 430 farmers currently keep 2000 dairy goats. Goat's milk is an important source of income, and both goat's milk and meat is a valuable nutritional addition to household diets (Fig. 27.1). While goat farmers primarily sell goat's milk to their neighbors, a small amount of the milk is collected by Twawose (the local dairy goat farmers association), who process the milk into drinking yogurt and sell it at the local markets. While Twawose has made several attempts to distribute excess milk to larger markets, these efforts have proven unsuccessful due to various challenges that include poor infrastructure such as poor condition of roads to and from Mgeta as well as a processing building which is not well furnished or equipped, limited access to credit and finance, and insufficient knowledge of marketing and business management (Lie et al. 2012). Recently, contract farming with established market actors has increased to the point that smallholder farmers can gain access to larger markets (FAO 2013a). Contract farming between Twawose and established dairy companies could therefore allow the smallholder farmers through their association (Twawose) to have access to regional markets and increases the base of consumers who can purchase healthy milk.

Fig. 27.1 Photographs courtesy of the Mahenge Family (*L*) and the Kulinywangwa Family (*R*). Upon their introduction to the highlands of Mgeta in the Morgoro region of Tanzania, these were the first farmers to receive Norwegian dairy goats. These individuals have sustained their business in goat farming for over 25 years and have witnessed the improvements that goat farming has brought to household income and nutrition, in addition to the reductions it has brought to vulnerability. Photos by Asle Olav Rønning

Shambani Graduates Limited (SGL), also known as Shambani Limited (Shambani Ltd.), is a Morogoro-based dairy processor that was established by former students of Sokoine University of Agriculture (SUA) in 2003. The company collects and processes cow's milk that is predominantly sourced from Maasai pastoralists. Strong collaboration between pastoralists and the dairy company has ensured positive revenue and a stable supply of milk. Within the pastoralist sector, a market-oriented system has emerged due to the profitability of increased milk and meat outputs. SGL has recognized the benefits of goat's milk due to its nutritional value, capacity as a niche product, and potential to prove profitable in Tanzania's increasingly competitive dairy sector.

This paper discusses the collaboration between SUA and Twawose and the involvement of private sectors, such as SGL, in the commercialization of goat milk in Tanzania. In terms of improved income, nutrition, and job creation, this public-private partnership (PPP) has the potential to significantly improve the benefits of regional and household dairy goat keeping. A PPP is a collaboration between the public and private sector that has clearly defined goals, builds on each partners' expertise, and leads to an appropriate allocation of risks, rewards, and resources (Narrod et al. 2009).

This paper is based on inter-disciplinary research that was conducted in Mgeta over the past eight years. The current research had three specific objectives and included value chain analyses developmental work, such as goat's milk collection and processing, an exploration of marketing opportunities, an analysis of the nutritional value of goat's milk, and goat feeding and breeding for increased milk production. First, Tanzania's dairy sector will be described and followed by an in-depth description of the three main benefits of keeping dairy goats: income potential, nutritional value, and climate-smartproduction practices (objective 1). Second, the paper will explore alternative methods for increasing the production of goat's milk in Mgeta and the PPP's potential to improve goat's milk collection and processing so that it can reach new markets (objective 2). Finally, a conclusion will be drawn and the possible division of the three actors' roles in the PPP will be recommended (objective 3).

27.2 The Tanzanian Dairy Sector

27.2.1 Milk Production

While about 30 % of Tanzanian cow's milk is obtained from commercial dairy breeds, the remaining 70 % is derived from the indigenous Tanzanian shorthorn Zebu (TSZ) species. Out of the 22.8 million cattle in Tanzania, only 700,000 are dairy-cattle breeds and crossbreeds, which are known for their higher potential for productivity (MLFD 2011; NBS 2012). While comprehensive, on-farm studies about the milk yield and milk content of the TSZ species are lacking, it is estimated that 20 % of these indigenous cattle herds produce an average of 160 L per cow per lactation period, or 500–800 L of milk per year (excluding milk that is consumed by calves) (Msanga et al. 2009). In the southern highlands of Tanzania, Mwambene et al. (2012) found that the peak yield of milk per cow was 3 L per day (excluding suckled milk).

In many countries, small ruminants (sheep and goats) contribute to the production of milk. Globally, sheep and goat's milk contribute to approximately 3 % of the annual milk that is consumed by humans (FAO 2015). Smallholder farmers in Tanzania have recognized the benefits of dairy goats since the 1960s, due to the lower capital input of goats than cattle and the reduction of malnutrition in poor households. Until the early 1980s, the Ministry of Agriculture and Livestock Development (MALD) imported exotic dairy goat breeds, such as Saanen and Toggenburg, from Europe to crossbreed with indigenous goats. However, because the impact of the crossbreeding program was insubstantial, few goats provided higher yields of milk.

In 1983, SUA collaborated with the Norwegian University of Life Sciences (NMBU) to initiate a dairy goat crossbreeding program that utilized the Norwegian Land Race (NLR). The project was based at SUA (within Morogoro municipality) and was implemented in Mgeta, a set of rural communities that are located in the Uluguru Mountains. While most farmers were new to raising dairy goats, they soon embraced the innovation. The initiative, which started with 84 goat children that were 2-weeks old, resulted in 2000 animals for more than 430 farmers in 7 Mgeta villages. Since the initiative, about 400,000 animals have been estimated to populate the country, and a pure-base herd has been established at Mulbadaw Farm Ltd. in Northcentral Tanzania (Kifaro et al. 2008; NBS 2012). In Mgeta, an average production of 1.4 L of milk per goat per day, average sales of up to 2 L of milk per household per day, and an average consumption of 1.2 L of milk per household per day have been achieved (Table 27.1; Kifaro et al. 2012). By comparison, the

Table 27.1 Sales and consumption information from Mgeta, Morogoro and Norwegian dairy goats' milk production during the wet and dry seasons

Ward	No. of goats	Wet season (L/day)	Dry season (L/day)	Consumption (L/day)	Sales (L/day)	Neighbor (TZS/L)	Collection center (TZS/L)
Tchenzema	46	60.0 (1.3)	72.0 (1.6)	1.4	2.0	960	875
Nyandira	71	76.8 (1.1)	100.8 (1.4)	1.2	1.2	960	875
Kibungo Juu	1	0.5 (0.5)	0.5 (0.5)	0.0	0.0	0.0	0.0

(): average production per goat; L: liters of milk; TZS: Tanzanian Shilling (As of February, 2016, $1 = TZS 2200)

average milk yield for the NLR, when limited to Norway, is 1.6 L of milk per day (600 kg). In Tanzania, the dominant goat species is the local Small East African (SEA) goat, which is primarily kept for meat. Currently, dairy goats are less than 2 % of all goats in Tanzania (MLFD 2011; NBS 2012).

27.2.2 *Milk Marketing and Processing*

Most Tanzanian goat's milk is consumed on-site within farming households and a fair amount is left for the calves. In practice, only about 10 % of the country's milk reaches the market, most of which includes milk that is sold to neighbors. The milk that reaches the commercial market is moved to 1 of the country's 62 dairy processing facilities (MLFD 2010). The lack of sales to commercial markets is chiefly due to the remoteness and poor infrastructure of village production systems, which makes the marketing and collection of milk a challenge (Njombe and Msanga 2011; RLDC 2010). Most processors in Tanzania are small- or micro-dairies[1] that produce less than 1000 L of milk per day, and there are only a few medium-sized commercial operations (Njombe and Msanga 2011). Overall, every processing plant processes milk at about 30 % of their capacity. This demonstrates the significant potential for increased milk collection and production.

The reasons behind the poor performance of national dairy processing include poor infrastructure and inadequate access to credit, in addition to the high cost of milk collection, processing, and transportation due to the high costs of equipment, machinery, and packaging materials. The high price of conducting business and limited marketing of processed products also increase costs (MMA 2008). The low

[1]In Tanzania, large-scale dairy processors produce more than 5000 L of milk per day, medium-scale dairy processors produce between 1000 and 5000 L of milk per day, small-scale dairy processors produce between 500 and 1000 L per day, and micro-dairy processors produce less than 500 L of milk per day.

processing performance of milk is additionally impacted by competition with imported dairy products. Imports of processed milk and dairy products account for about 48 % of processed dairy products in Tanzania, which indicates opportunity for the national dairy industry to expand (RLDC 2010).

Although Tanzania has a higher number of cattle and dairy goats than their East African neighbors, Tanzanians consume smaller amounts of milk. In 2009–2010, Tanzanians consumed an average of about 43 L of milk per year per capita. While this was an increase from about 26 L of milk per year per capita in 2003, it was still significantly less than the consumption of 200 L of milk per year per capita that was advised by the World Health Organization (WHO) and lower than the averages of neighboring countries, such as Kenya (100 L per year) and Uganda (50 L per year) (MMA 2008; Njombe and Msanga 2011). Approximately 50 % of Tanzanians consume milk regularly, with higher levels of consumption occurring primarily within urban areas (RLDC 2010). Typically, goat's milk is only made available in local communities, such as Mgeta, where dairy goat keeping is common.

Tanzania's low rate of milk consumption is primarily caused by cultural beliefs, low purchasing power of many Tanzanians, and perceptions that it is a children's drink. The most popular milk products are fresh milk and fermented milk products, like mtindi and yogurt (RLDC 2010). The Tanzanian government currently has aims to raise the annual per-capita consumption of milk to at least 80 L. To meet this demand for milk, their goal is to increase the processing capacity of milk processing plants from 30 % to at least 75 % by applying beneficial tax policies for processors, instructing consumers about the nutritional benefits of milk, offering diverse policies that focus on improved feeding and breeding practices, and establishing a greater number of milk collection centers and dairy production organizations (Njombe et al. 2012).

27.3 Dairy Goats: The Untapped Potential of the Poor Man's Cow

27.3.1 Milk in Human Nutrition

Milk contains low levels of ascorbate and all essential vitamins and minerals except Iron (Fe), copper (Cu), and manganese (Mn) (Koivistoinen 1980). This nutritional profile makes milk particularly suitable for vulnerable individuals, such as the sick, children, the elderly, and nursing mothers. Likewise, when combined with unrefined fruits, cereals, and vegetables, milk meets most people's dietary requirements (Koivistoinen 1980). Low intake of animal-based products has been connected to a range of health problems that include anemia and stunted growth in children (Haug et al. 2010; Hotz and Gibson 2001). A study in rural Tanzania revealed that 25 % of children under the age of 5 were underweight and 52 % of children under the age of 5 showed stunted growth (Kinabo et al. 2003). These findings were linked to diet, which in rural areas, typically consisted of plant- and limited animal-based products (MLFD 2011).

Goat's milk is considered by some to be healthier than cow's milk. According to Zervas and Tsiplakou (2013), the difference between the two milks may be due to the lower n-6: n-3 ratios, lower allergenicity, higher digestibility, low levels of a_{s1}-casein, smaller-sized fat cells, higher levels of α_{s2}-casein and β-casein, and higher proportion of medium chain fatty acids in goat's milk. Furthermore, goat's milk contains high levels of taurine (Manzi and Pizzoferrato 2013; Prosser et al. 2008; Sarwar et al. 1998), a nutrient which is essential to the optimal health and development of infants (Sturman 1993). When compared to the milk proteins of other dairy goat breeds, the proteins in NLR goat's milk has a higher prevalence of the *alpha-S_1* gene (Hayes et al. 2006), which results in fewer stable fat cells. Milk of this quality is easy to digest and is particularly beneficial to the diets of infants and sick people (Sturman 1993). Goat's milk has also been considered to be superior to the milk of other animals due to its higher percentages of ash, solids, proteins, and milk fat (Table 27.2). In a comparative study that was held in Tanzania, SEA goats showed significantly higher percentages of mean milk fat (6.95 %) than three breeds of cattle (Jersey, Friesian, and Ayrshire), whose values ranged between 4.09 and 4.83 % (Ryoba and Hansen 1988).

Milk from species other than goats or cattle is rarely consumed and largely unavailable. The consumption of fresh milk can cause some consumers to experience digestive problems, such as hypolactasia (lactose intolerance) and diarrheal diseases that include symptomatic *Giardia lamblia* infections and general diarrhea in malnourished measles patients (Christophersen 1977; Mantovani et al. 1989; Montgomery et al. 1991). Lactose intolerance is the inability to digest lactose due to a lack of the lactose-degrading enzyme lactase in the intestine. Intolerance of lactose may be caused by genetic disorders (primary lactose intolerance) or malnutrition and other diseases in persons who are not genetically intolerant (secondary lactose intolerance). Secondary lactose intolerance is often combined with sucrose intolerance, which is a complication of severe protein-energy malnutrition disorders, such as kwashiorkor (Hansen 1968; Prinsloo et al. 1969; Wharton 1968). For individuals with giardiasis, milk consumption can also result in the depression of

Table 27.2 Constituents of milk from different species

	Cows	Goats	Sheep	Buffalo
% Milk fat	3.85	3.93	6.86	7.96
% Milk protein	3.50	3.56	6.00	4.16
% Milk sugar	4.72	4.65	4.91	4.86
% Milk ash	0.72	0.81	0.89	0.78
% Milk solids	12.79	12.95	18.66	17.76

Source Rasic and Kurmann (1978)

intestinal disaccharides, such as lactase and sucrase (Holzel 1968). Primary lactose intolerance should not be considered a disease, as it is a phylogenetically more primitive condition in humans and is still present in most of the global adult population (Montgomery et al. 1991).

The content levels of lactose in milk can be significantly reduced by sour milk bacteria, and much of the remaining lactose can be degraded by bacterial lactase after ingestion (Hertzler and Clancy 2003; Kolars et al. 1984; Marteau et al. 1990). In Mgeta, allergic reactions to milk proteins are rare among people who consume goat milk (Mushi 2014). Nevertheless, infants with HIV-positive mothers should be monitored for the development of food allergies. Due to the considerable medical and nutritional advantages that have been presented, goat's milk and goat's milk products, such as fermented milk, should be introduced to the Tanzanian market.

27.3.2 Climate-Smart Goat Keeping Practices

Climate change causes extreme weather in the form of floods, droughts, landslides, and erratic rainfall. Agricultural practices that can manage these challenges is therefore crucial, especially among smallholder farmers who are exposed and vulnerable to climate change. Because only minor portions of the energy and protein in animal feed can be recycled into edible products, the production of food from livestock demands numerous resources. This is especially true for ruminant-based meat production, where less than 10 % of the energy in feed is retained in food for human consumption (Syrstad 1993). Ruminants have a direct impact on the global emission of greenhouse gases (GHG), specifically due to the GHG methane (CH_4) and nitrogen oxide (N_2O) that is produced from their rumen and manure (Broucek 2014; Smith et al. 2012). This disadvantage to ruminant meat production should be weighed against the advantages that it can bring to human health. For example, when they reside within agroforestry systems, ruminants can play an important role in the restoration of local ecosystems. In hilly terrains, the groundwater storage capacity for CO_2 can be enhanced through the construction of terraces (Chepstow-Lusty and Jonsson 2000; Mitiku et al. 2006), especially when they have been planted with trees. Trees and forests contribute to the reduction of water-related risks, such as droughts, landslides, and local floods, and can help to prevent salinization and desertification (FAO 2015).

As natural browsers, goats have a higher preference for trees and shrubs than cows. As a result, goat farmers are able to use a greater quantity of leaves and branches in their stall-feeding systems. Compared to fresh fodder from grasses, fresh fodder from trees can maintain its nutritive value into the dry season and for longer periods of time. During the production of animal-based products, there is argument for both modern agriculture techniques in the form of feed and edible plant production and an integrated and environmentally-friendly agricultural system of production.

By reducing disease losses and improving the feed-conversion ratio, it is also possible to decrease GHG emissions for each kg of animal protein that is produced from ruminants (Hristov et al. 2013a, b). Due to the low reproductive rates of ruminant females, and because cows in beef herds receive nearly half of Tanzania's feed, the output-input ratio for ruminants that are kept solely for meat is poor (Syrstad 1993). From a resource-efficiency point of view, keeping ruminants for meat production alone can be justified only when it is based almost exclusively on feeds, which have no alternative benefit but to nourish livestock. The proportion of feed energy that is retained in edible products is usually considerably higher during milk production than it is during meat production. While the efficiency of meat production depends on the quantity of production, even at moderate quantity of production, milk is typically more efficient. For an animal that produces 5 times more milk than its bodyweight, this proportion (milk production/animal bodyweight) increases to about 30 %. It should be noted that quantity that is increased from low to medium is much more effective than levels that are increased from medium to high (Syrstad 1993).

Increased animal productivity can be achieved by improving the genetic potential of animals through planned crossbreeding or breed selection. Genetic potential can be further advanced by combining breeding with proper nutrition, good animal health, and improved reproductive lifespan and reproductive efficiency (Hristov et al. 2013b). Because a single dairy goat can provide the same amount of edible animal protein as 33 SEA goats, there is also a high potential to reduce the emission of GHG per kg of animal protein that is produced (Eik et al. 2008). By implementing optimal feeding in the diets of Tanzanian dairy goats, the emission of GHG per kg of produced animal protein can therefore be improved by a value of 3.

27.3.3 Income-Generation from Goat's Milk

When resources for cattle keeping are insufficient, the small sizes of goats make them suitable for small-scale (subsistence) production. Even in cases where cattle can be kept, the level of risk is reduced when investments are dispersed over a greater number of animals. Other advantages to keeping goats include shorter intervals between generations and higher rates of twinning in crossbreeds with local goats. These factors explain why the expansion and reproduction of dairy goats occurs more quickly than it does for cows. Goats are also relatively inexpensive to keep, as they require relatively little maintenance. Additionally, they need less fodder and smaller plots of land for grazing. Practically speaking, goats are also easier for women and children to handle, who traditionally do most of the work with livestock in several developing countries (Devendra 1999). Milk also provides smallholder farmers with income throughout the year, which is especially important for smallholder farmers who face large seasonal variations in the income they earn from crops. When compared to the price of cow's milk, the price for goat's milk can be relatively high due to its high demand and nutritious characteristics.

Although dairy products can generate substantial income for farmers, access to markets can be difficult for smallholder farmers to achieve. Some of the challenges that smallholder farmers face include limited access to credit, poor financial management, low education levels that leave many illiterate, and inadequate knowledge in marketing and entrepreneurship. Community milk processing facilities are also often located in areas with poor infrastructures, limited power supplies, and inferior road conditions that are impacted by the rainy season (Arias et al. 2013).

To conclude, the potential to commercialize goat's milk in a climate-friendly way, which will ultimately lead to increased income for smallholder dairy-goat farmers due to the positive nutritional characteristics of goat's milk, is high. In order to reach additional global and national markets, smallholder farmers need to collaborate with private sector actors so that they can increase their knowledge and experience in milk marketing, collection, and processing.

27.4 Increasing the Marketing and Production of Tanzanian Goat's Milk

27.4.1 Shambani Graduates Limited

In 2003, SGL was established with limited capital from its shareholders. While the company was initially only capable of processing a single milk-based product from 30 L of milk per day, it has since grown into a large and trusted entity with a processing plant that can manufacture as many as 2500 L of milk per day. As of 2015, the company's production capacity has increased to 4000 L of milk per day and has been divided into a total of 5 different types of milk products. The company receives milk from approximately 250 trained and equipped suppliers, who work within 90 km of the company's processing plant in the Morogoro region. In 2003, the company employed only 2 people, but today, the company employs as many as 14 people. In regards to sales and markets, the company currently sells its product in three Tanzanian regions. These regions include Dodoma, Morogoro, and Dar es Salaam. The company also has a positive turnover. In 2014 for instance, collaborating pastoralists secured delivery for their milk and earned suppliers approximately \$90000.00. At least 87 % of the company's suppliers are Maasai pastoralists who primarily own TSZ cattle, while the remaining 13 % are smallholder farmers that keep dairy cows. The suppliers bring milk to two collection centers in Kilosa and Morogoro, which are in close proximity to the milk's producers. Once it has been transported to a collection center, the quality and quantity of the product is recorded by a SGL employee.

The supply of milk to processing plants can be impacted by various factors, such as season, weather, and inconsistent milk production. These factors negatively influence SGL's ability to efficiently use its machinery and maintain a consistent supply of milk. The production of cow's milk is minimal during the dry season due to the general lack of supplementary feeding practices among pastoralists. To overcome this limitation, SGL has begun to source its milk from large scale farms. By comparison, more goat's milk is produced during the dry season due to the higher number of goat kids during the late-rainy and early-dry seasons. Therefore, working with both types of milk may improve SGL's potential to meet its processing capacity.

By collaborating with pastoralists and establishing contract dairy farming that increases income across generations and improves the livelihoods of smallholder farmers, SGL can play a significant role in helping smallholder dairy farmers to realize increased financial gains. goat's milk can provide SGL with a niche product that can be marketed both to the wider population and to individuals with weaker immune systems, such as the elderly, infants and children, and the lactose intolerant. Because no other Tanzanian processing company currently offers goat's milk products, the introduction of goat's milk could provide SGL with a competitive advantage. In Kenya, goat' milk products (Kibidav Ltd. 2015) and information about dairy goat keeping and the nutritional value of goat' milk (Dairy Goat Kenya 2015; NAFIS 2015) are provided in supermarkets.

27.4.2 *Potential for Production of Goat Milk*

Due to the growing number of dairy goats and increasing sizes of goat herds in farms, the production of goat's milk in Tanzania is expected to expand considerably (Kifaro et al. 2012). The dry season carries the highest potential for the delivery of milk, and as natural browsers, goats are capable of finding quality foraging through the duration of the season. By comparison, TSZ cattle's milk is more likely to be produced during the rainy season, when grass is abundant. While the number of infant goats is highest in Mgeta between April and October, and peaks between August and September, it is at its lowest between February and March. Consequently, goat keepers may plan for the production of goat kids to meet its peak in the dry season, when the supply of cow's milk is low. However, because goats are prone to diseases (such as children's pneumonia) and parasites, health management is recommended (Eik et al. 1985).

In 2008, 58 % of Mgeta's total herd were milking goats. Over a period of 10 months of lactation, the average milk yield was estimated to be 1.0 L of milk per day for purebred NLR goats, 0.9 L of milk per day for 75 % of NLR-SEA crossbreeds, and 0.7 L of milk per day for 50 % of NLR-SEA crossbreeds (Eik et al. 2008). In Tanzania, the birthweights of all genotypes are estimated to be 2.6 kg, and the intervals between the birth of goat kids is projected to be around 11 months. These numbers are significantly lower than Norwegian estimates, which are 3.0 kg and 12 months, respectively, and also indicate sub-optimal feeding practices and intervals that are too brief between the birth of goat kids among Mgeta farmers (Eik et al. 2008).

Improved feeding and management practices have increased milk yields and improved the production of meat from NLR goats and NLR-SEA crossbreeds.

NLR-SEA crossbred goats are sold at rates of between $85.00 and $170.00. Studies have revealed that farmers find it more profitable to keep multiple goat species, as this allows them to benefit from the sale and breeding of purebred bucks and pure or crossbred does and kids (Hango et al. 2007; Safari et al. 2009). By comparison, SEA goats provide an average income of around $18.00 a head and are sold at a later age than their NLR and crossbred counterparts. These positive results have produced a high cropping rate in areas like Mgeta and have encouraged farmers to opt for live sales to neighbors and markets in other Tanzanian regions, rather than sales through milk marketing, production, and processing (Kifaro et al. 2012). The on-farm modelling of these integrated dairy goat systems are also indicative of high economic returns. Therefore, a shift in focus from seasonal vegetables to dairy goats, permanent grass, and multipurpose fodder trees could increase the gross margin of goat's milk sales by about 14 %. The gross margin of dairy goat sales are also more resistant to climate change. While declines in sales due to seasonal changes are estimated to be only 3.5 % for farmers with dairy goats, the estimated decline in sales for farmers without dairy goats is estimated to be 9.6 % (Nziku et al. 2015).

Although the potential for increased milk production is high, feeding is considered to be a major limitation. In Norway, 1 kg of concentrate per animal per day has been proven to increase the yield of goat's milk by as much as 2.5 L per day (Eik et al. 1985), and at the Yole Agricultural Research Station in Tanzania, 2 kg of concentrate per animal per day increased milk production by as much as 1 L per day (N. A. Urio 2015, Per. Comm.). In Tanzania, and Mgeta in particular, it is projected that proper quantities of concentrate can triple the daily milk production of both purebred-NLR and crossbred goats. However, the cost of investment for introducing supplemental feed is high. At a price of $0.2 per kilo, supplementation is considered to be expensive by most farmers. These individuals probably misunderstand the economics of proper feeding and are likely unaware that 1 kilo of concentrate, at an expense of $0.2, will increase their milk yield by up to 2 L at a market price of $1.2 per kilo. Furthermore, improved feeding can increase the birth weights of goat kids, which in turn can reduce mortality and increase sales from breeding. For these calculations to be accurate, a well-functioning local milk market must first be established.

Other existing options for improved concentrate supplementation include (1) an optimal concentrate formula that consists of lime (0.5 %), bone meal (0.5 %), maize bran (72.5 %), cotton seed cake (12 %), and sunflower seed cake (14.5 %) (farmers can more easily adopt concentrate supplementation if they communicate the results of cattle production to rural goat farmers); (2) conservation farming practices where Mgeta farmers use manure to produce more fruits and vegetables and buy cheap concentrates at marketplaces in Morogoro and Dar es Salaam (trucks used for this transport can also bring concentrate up from the lowlands); and (3), balanced and annual distributions of goat's milk that are based on established pro-poor contract farming (at $0.2 per liter during payment of milk) or other loans that allow SGL to supply concentrate to farmers.

The number of dairy goat farmers has increased with the growth in the milk yields of goats per lactation period, and the high local sales rate of dairy goats has increased the number of dairy goat keepers in Mgeta. There is even a program in Mgeta that trains orphans in goat keeping and allows its 60 enrolled youths to share their experiences. In order to improve the nutrition of school children and teach students good agricultural practices, dairy goats have been given to three schools that are located inside of Mgeta but outside of the primary goat keeping villages. This training is offered to the local community and conducted by experienced dairy goat farmers and the orphans that were trained in the aforementioned program. As a result of these numerous innovations, the number of dairy goat keepers is expected to increase and contribute significantly to Mgeta's current milk supply.

27.4.3 The Potential for the Marketing, Collection, and Processing of Goat's Milk

The large increase of dairy goats in Mgeta has resulted in an excess supply of goat's milk. Over the years, the focus of goat's milk production has moved from the consumption of milk at home to the sale of milk to neighbors and the production and distribution of drinking yogurt to local markets by Twawose. To add value to milk and increase its shelf life, Twawose, which currently has 64 members, began processing drinking yogurt in early 2010. Currently, sales at the local market are limited due to low purchasing power and the low frequency of markets. In Mgeta, the communal market (where traders and farmers meet in a single village to conduct business) opens twice weekly (two market days). This limits the meeting between the two parts and therefore limits the sales.

In 2012, Twawose established a school-based yogurt program to sale drinking yogurt to two primary schools in Mgeta. This program provided Twawose with a stable, accessible, and more predictable local market. While the supply of yogurt from goat's milk regularly encountered challenges, it also provided Twawose with valuable experience in the supply of products and the management of associations (Lie 2015, Unpublished). In 2013 and 2014, Twawose collected an average of 30 L of goat's milk per day (between 100 and 600 L of goat's milk per month) (Fig. 27.2), a figure which was 10 times higher than the amount of milk that was collected when the center was established (Lie 2015, Unpublished). Because the supply for local milk is greater than the demand, Twawose has been forced to restrict its collection of milk by reducing its overall number of milk collection days each year.

Because Twawose consistently seeks new marketing options, potential outlet points for goat's milk and yogurt distribution have been identified on several occasions. These outlets include street shops, restaurants, supermarkets, and mobile traders (Lie et al. 2012). However, the risk for Twawose to penetrate nearby town markets is high due to unknown demand, increased distances between production areas and the prospective market places, potential loss of quality, competing dairy products, and high transportation costs. Likewise, Twawose lacks the necessary

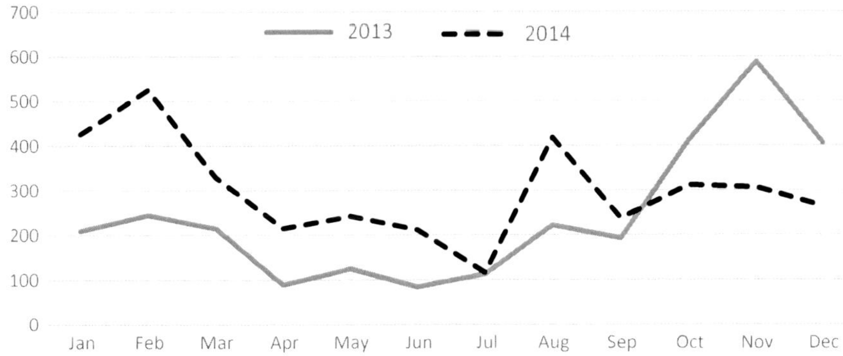

Fig. 27.2 The 2013 and 2014 average monthly collection of goat's milk (liters) at the milk collection and processing center in Mgeta

business skills to minimize risks, attract investments, and devise a marketing plan. Low education levels, limited access to information, and poor roads and power infrastructures also contribute to Twawose's low rates of penetration. Furthermore, poor coordination between milk suppliers has caused the demand for the product to fluctuate and has created issues with communication and the management of the association (Lie et al. 2012).

With the many farmers who are interested in becoming goat's milk suppliers and an estimated 1000 L of goat's milk available for collection every day, the potential to initiate a contracted supply of goat's milk with a commercial dairy actor, such as SGL, is high. At a distance of approximately 50 km from SGL's headquarters in Morogoro, Mgeta falls under the company's existing collection region. Twawose operates one collection center and two collection points in the nearby villages of Tchenzema and Mwarazi. Collection now covers four of the seven villages where the project of upscaling of dairy goat technologies operates and where farmers have previously been educated about dairy goats and goat's milk. Electricity was first installed in Mgeta at the end of 2014. With the installation of cooling systems and other equipment that ensures high quality products during the storage and collection of goat's milk, this advancement will allow for the improved collection of goat's milk.

To increase the collection of goat's milk to an amount that is necessary for contract farming, Mgeta dairy goat farmers must first expand their production capacity. Focus should be placed on transportation, proper collection, storage in households, hygienic milk handling practices, storage and bulking routines by Twawose, record keeping for both milk and animals, and increasing the volume of milk production. Special attention should also be paid to female dairy goat keepers. Through training and follow-up, SGL, SUA, and Twawose can guarantee that these factors are met. SGL's experienced staff should be used as a model for the value of hiring trained and experienced outsiders to run milk collection centers, as outsiders are less likely to be impacted by social relationships with milk suppliers that can coerce them into accepting poor quality milk. Because growth in the number of goat's milk suppliers in the seven villages is necessary to ensure economy of scale and to manage the seasonal fluctuations of milk production, SGL's history of coordinating a large number of milk suppliers will prove especially valuable. Bad roads and power outages are the two primary challenges to milk collection that SGL will need to overcome. The transportation of milk from Mgeta can be difficult due to its winding and steep roads, and during the rainy season, products are frequently reported to be damaged.

The price for fresh goat's milk in Mgeta is \$0.6 per liter throughout the year. In spite of Morogoro's \$0.6 per liter price for cow's milk, SGL collects cow's milk for \$0.4 per liter. If goat's milk were to be marketed as a niche product, its price premium could cover this difference in cost. Regardless of any small reductions in the price of milk, Mgeta's dairy goat farmers can increase their daily income by supplying larger amounts of milk.

A contract collaboration between SGL and interested dairy farmers would increase the incomes of farmers and the supply of goat's milk. In 2014, at least 50 Mgeta goat's milk suppliers earned as much as \$46.00 per month. While this figure does not include any of the expenditures of goat's milk production, it reveals the potential income for farmers who follow recommended goat management practices. In 2014, Mgeta farmers supplied an average of 2.5 L of milk per day. However, a number of productive farmers with large herds supplied as much 9 L of milk per day. Because 50 % of the 430 dairy goat farmers in Mgeta supplied 2.5 L of milk per day, at least 540 L of milk per day were collected in total. This means that farmers earned an average monthly income of \$43.00, which is a 90 % increase from the 2014 average. With the high demand for milk, the likelihood for farmers to invest in improved goat management practices that boost milk production and increase the size of their dairy goat herds has grown. Other farmers may also consider becoming dairy goat keepers so that they can seize the increased opportunity for income.

During collaboration with SGL, safe payment practices can be introduced by using the mobile banking services such as M-Pesa. Although SGL does not currently use this technology, Mgeta farmers and SUA researchers have discussed it as a viable banking solution for Tanzania. SGL offers small, interest free loans to committed milk suppliers that enable them to purchase materials, such as bikes or milk buckets, and that can be paid through the supply of milk. As part of the collaboration to ensure high quality milk and increased milk production, small loans can be introduced to dairy goat farmers so that may purchase medicine, milk buckets, supplementary feeds, and other goods that will help them to farm.

SGL can also use its existing sales and distribution channels to boost the sales potential of goat's milk and goat's milk products. As a result, consumers in Dodoma, Morogoro, and Dar es Salaam, who have previously not had access to goat's milk, will be able to benefit from its numerous healthy characteristics. However, because little is known about the demand for goat's milk in these markets, competition with cow's milk will likely remain high. To curve these factors, Tanzania's general public must be educated about the beneficial characteristics of goat's milk. The SUA, with its in-depth research about the nutritional advantages of goat's milk, should therefore contribute to the marketing campaigns (which can be run by SGL) of goat's milk related products.

In sum, there are three primary reasons why a PPP should be established within the Morogoro region: (1) nutritious goat's milk is currently available in Mgeta for commercialization, (2) a professional dairy company is nearby, and (3) a highly regarded university is located between the two. In order to achieve a milk surplus, market surveys and economic analyses that are based on investment costs will be required to calculate the profitability and amount of goat's milk that will need to be produced. Market surveys should pay special attention to fermented goat's milk products, such as mtindi (sour milk), which is popular in Tanzania due to its health benefits. The attraction of investment capital will also be critical to the success of the collaboration's ability to combine research about storage, dairy goat breeding, increased production for dairy goat farmers, the marketing of niche products in new markets, optimizing feeds for increased animal productivity, and the introduction of equipment that ensures proper milk collection.

27.5 Conclusions and Recommendations

The proposed PPP among SUA, the privately owned dairy firm, SGL, and the association of dairy goat farmers in Mgeta, Twawose, will help to provide nutritious goat's milk to Tanzanian consumers and offer valuable income to smallholder dairy farmers who follow climate-smart production practices. It will also allow SGL to introduce a new niche product in goat's milk and give SUA the opportunity to collect valuable data from goat farms and research about the development of goats.

By expanding the production of goat's milk in Tanzania, the diets of people in rural and urban areas will improve. The introduction of goat's milk will prove especially beneficial to the diets of people with impaired or underdeveloped immune systems. With a close collaboration between small-scale dairy farmers and commercial dairy manufacturers, farmers will sell more milk, gain access to larger markets, and receive higher annual incomes. Similarly, dairy manufacturers will be able to use goat's milk to expand their production capacity and contribute to the Tanzanian government's goal of efficiently filling the capacities of milk processing plants. Compared to other types of livestock, dairy goats produce lower amounts of GHG per emission per kg of protein. Their environmental impact can be further reduced by establishing terraces with trees, which can be used for fodder and thereby encourage the integration of agricultural systems.

The collaborative actors in the PPP will have defined roles. The Mgeta dairy goat farmers (Twawose) will supply goat's milk, receive training, and access new services; SGL will gain a new, niche product in goat's milk and market, collect, and process it; and to ensure the high quality and quantity and of milk, in addition to a long-lasting collaboration with Mgeta dairy goat farmers, the SUA will contribute their knowledge of integrated dairy farming practices, the nutritional characteristics of goat milk, which can be important in marketing, and optimal goat management practices that include feeding, breeding, and milk handling.

Acknowledgments The dairy goats upscaling project was financed by the Enhancing Pro-poor Innovation in Natural Resources and Agricultural Value Chains (EPINAV) program, a research collaboration between Sokoine University of Agriculture (SUA) and the Norwegian University of Life Sciences (NMBU).

References

Arias P et al (2013) Smallholder integration in changing food markets. Food and Agriculture Organization of the United Nations, Rome

Broucek J (2014) Production of methane emissions from ruminant husbandry: a review. J Environ Prot 5:1482–1493. doi:10.4236/jep.2014.515141

Chepstow-Lusty A, Jonsson P (2000) Inca agroforestry: lessons from the past. AMBIO: J Hum Environ 29(6):322–328

Christophersen O (1977) Fiskeproteinkonsentrat—et næringsmiddel for fremtiden? [Fish protein concentrate—a food for the future?] [Article in Norwegian]. Inter Medicos 23:53–67

Dairy Goat Kenya (2015) Ongata Rongai dairy goat resource center. https://goatkenya.blogspot.no/p/about.html. Accessed 09 Apr 2015

Devendra C (1999) Goats: challenges for increased productivity and improved livelihoods. Outlook Agric 28(4):215–226

Eik L, Mtenga L, Olsen O (1985) Preliminary results on some economic traits in Norwegian and Tanzania Saanen goats. Tanzan Soc Anim Prod 12:129

Eik L et al (2008) Productivity of goats and their contribution to household food security in high potential areas of East Africa: a case of Mgeta, Tanzania. Afr J Food Agric Nutr Dev 8(3):278

FAO (2013a) Contract farming for inclusive market access. Food and Agriculture Organization of the United Nations (FAO), Rome

FAO (2013b) Milk and dairy products in human nutrition. Food and Agriculture Organization of the United Nations, Rome

FAO (2015) FAO news article: forests are key for high quality water supply. Better forest management needed to maximize water-related benefits from forests. https://www.fao.org/news/story/en/item/53391/icode. Accessed 01 Feb 2016

Hango A, Mtenga LA, Kifaro GC, Safari J, Mushi DE, Muhikambele VRM (2007) A study on growth performance and carcass characteristics of Small East African goats under different feeding regimes. Livest Res Rural Dev 19:130. http://www.lrrd.org/lrrd19/9/hang19130.html. Accessed 15 July 2015

Hansen J (1968) Features and treatment of kwashiorkor at the Cape. In: Bender AE (ed) Calorie deficiencies and protein deficiencies. Churchill, London, pp 33–48

Haug A et al (2010) Use of dried Kapenta (Limnothrissa miodon and Stolothrissa tanganicae) and other products based on whole fish for complementing maize-based diets. Afr J Food Agric Nutr Dev 10(5):2478–2500

Hayes B et al (2006) Effects on production traits of haplotypes among casein genes in Norwegian goats and evidence for a site of preferential recombination. Genetics 174(1):455–464

Hertzler SR, Clancy SM (2003) Kefir improves lactose digestion and tolerance in adults with lactose maldigestion. J Am Diet Assoc 103(5):582–587

Holzel A (1968) Defects of sugar absorption. Proc R Soc Med 61(11 Pt 1):1095

Hotz C, Gibson R (2001) Complementary feeding practices and dietary intakes from complementary foods amongst weanlings in rural Malawi. Eur J Clin Nutr 55(10):841–849

Hristov A et al (2013a) Special topics—mitigation of methane and nitrous oxide emissions from animal operations: I. A review of enteric methane mitigation options. J Anim Sci 91(11):5045–5069

Hristov A et al (2013b) Special topics—mitigation of methane and nitrous oxide emissions from animal operations: III. A review of animal management mitigation options. J Anim Sci 91(11):5095–5113

Kibiday Ltd (2015) Toggs goat milk products for a healthier life. https://www.toggskenya.com/. Accessed 09 Apr 2015

Kifaro GC et al (2008) The potential use of artificial insemination in sustainable breeding of dairy goats in developing countries: a case study of Norwegian dairy goats in Tanzania. Tanzan J Agric Sci 8(1):19–24

Kifaro GC et al (2012) Status of Norwegian dairy goats in two districts of Morogoro region at the start of an up-scaling project. Paper presented at the 1st annual conference of the EPINAV programme, Morogoro, Tanzania, 18–19 Dec 2012

Kinabo J, Msuya J, Nyaruhucha CN, Mamiro P, Mnkeni A, Ishengoma J (2003) Nutritional status of children and adults in Morogoro and Iringa regions. In: Collaborative research for food security (TARPII & MAFS) held in Morogoro, Tanzania. Sokoine University of Agriculture, Morogoro, pp 28–30

Koivistoinen P (1980) Mineral element composition of Finnish foods: N, K, Ca, Mg, P, S, Fe, Cu, Mn, Zn, Mo, Co, Ni, Cr, F, Se, Si, Rb, Al, B, Br, Hg, As, Cd, Pb and ash. Acta Agric Scand Suppl 22:171. Scandinavian Association of Agricultural Scientists and the Royal Swedish Academy of Agriculture and Forestry, Stockholm

Kolars JC et al (1984) Yogurt an autodigesting source of lactose. N Engl J Med 310(1):1–3

Lie H, Rich RM KM, Kurwijila LR, Jervell AM (2012) Improving smallholder livelihoods through local value chain development: a case study of goat milk yogurt in Tanzania. Int Food Agribusiness Manage Rev 15:55–85

Mantovani MP et al (1989) Lactose malabsorption in children with symptomatic Giardia lamblia infection: feasibility of yogurt supplementation. J Pediatr Gastroenterol Nutr 9(3):295–300

Manzi P, Pizzoferrato L (2013) Taurine in milk and yoghurt marketed in Italy. Int J Food Sci Nutr 64(1):112–116

Marteau P et al (1990) Effect of the microbial lactase (EC 3.2. 1.23) activity in yoghurt on the intestinal absorption of lactose: an in vivo study in lactase-deficient humans. Br J Nutr 64(01):71–79

Mitiku H, Herweg K, Stillhardt B (2006) Sustainable land management: a new approach to soil and water conservation in Ethiopia. Mekelle University and University of Berne, Mekelle

MLFD (2010) Livestock sector development strategy. Ministry of Livestock and Fisheries Development, Dar es Salaam

MLFD (2011) Livestock sector development programme. Ministry of Livestock and Fisheries Development, Dar es Salaam

MMA (2008) Dairy sector quick scan and selective value chain analysis Tanzania. Netherlands Development Organization. Dutch Ministry of Economic Affairs

Montgomery RK et al (1991) Lactose intolerance and the genetic regulation of intestinal lactase-phlorizin hydrolase. FASEB J 5(13):2824–2832

Msanga YN et al (2009) Cattle improvement in Tanzania with emphasis on management of field artificial insemination services. In: The proceeding of the Tanzania society of animal production (TSAP), vol 36. Arusha, Tanzania, pp 238–255

Mushi PM (2014) A study on lactose intolerance and milk intake among people in Mgeta and Njombe areas, Tanzania. Master. Norwegian University of Life Sciences (NMBU), Department of Animal and Aquacultural Sciences, Aas, Norway

Mwambene PL et al (2012) Fipa cattle in the southwestern highlands of Tanzania: socio-economic roles, traditional management practices and production constraints. Anim Genet Resour 51:1–14. doi:10.1017/S2078633612000112

NAFIS (2015) Dairy goat production: national farmers information service. https://www.nafis.go.ke/livestock/dairy-goat-production/. Accessed 09 Apr 2015

Narrod C et al (2009) Public-private partnerships and collective action in high value fruit and vegetable supply chains. Food Policy 34(1):8–15

NBS (2012) National Sample Census of Agriculture 2007/08. Large Scale Farms Volume IV. Ministry of Agriculture, F. S. a. C., Ministry of Livestock Development and Fisheries, Ministry of Water and Irrigation, Ministry of Agriculture, Livestock and Environment, Zanzibar, Prime Minister's Office, Regional Administration and Local Governments, Ministry of Industries,

Trade and Marketing, The National Bureau of Statistics, and the Office of the Chief Government Statistician, Zanzibar. Dar es Salaam, National Bureau of Statistics

Njombe AP et al (2011) The Tanzania dairy industry: Status, opportunities and prospects. Paper presented at 7th African dairy conference and exhibition, MovenPick Palm Hotel, Dar es Salaam, 25–27 May 2011

Njombe AP, Msanga YN (2011) Livestock and dairy industry development in Tanzania. In: Title of book. Department of Livestock Production and Marketing Infrastructure Development and the Ministry of Livestock and Fisheries Development. Ministry of Livestock and Fisheries Development, Dar es Salaam, p 17

Njombe AP et al (2012) Efforts to improve the dairy industry in Tanzania. In: The smallholder dairy value chain in Tanzania stakeholder meeting, Ministry of Livestock and Fisheries Development, Morogoro, Date that the Meeting was Held

Nziku ZC et al (2015) Modelling climate change adaptation in vulnerable crop and livestock production systems in Mgeta, Tanzania. AJFAND (in press)

Peacock C (2007) The goats model. A proven approach to reducing poverty among smallholder farmers in Africa by developing profitable goat enterprises and sustainable support services. Working papers, Farm-Africa

Prinsloo J et al (1969) Effect of different sugars on diarrhea of acute kwashiorkor. Arch Dis Child 44(237):593–599

Prosser CG et al (2008) Composition of the non-protein nitrogen fraction of goat whole milk powder and goat milk-based infant and follow-on formulae. Int J Food Sci Nutr 59(2):123–133

Rasic LR, Kurmann JA (1978) Yoghourt. In: Techn. Copenhagen, Technical Dairy Publishing House, 23 COPENHAGEN XP002142806, pp 36

RLDC (2010) Survey on dairy products market in Tanzania. Rural Livelihood Development Company, Dar es Salaam

Ryoba R, Hansen KK (1988) Composition of goat's milk compared to cow's milk. In: Mgheni M, Ekern A (eds) Improved dairy production from cattle and goats in Tanzania. Reports from research projects 1982–1985. Place of publication, Noragric at the Agricultural University of Norway, pp 60–62

Safari J et al (2009) Effects of concentrate supplementation on carcass and meat quality attributes of feedlot finished small East African goats. Livest Sci 125(2):266–274

Sarwar G et al (1998) Free amino acids in milks of human subjects, other primates and non-primates. Br J Nutr 79(02):129–131

Smith KA et al (2012) The role of N_2O derived from crop-based biofuels, and from agriculture in general, in Earth's climate. Philos Trans R Soc B Biol Sci 367(1593):1169–1174

Sturman JA (1993) Taurine in development. Physiol Rev 73(1):119–147

Syrstad O (1993) Evaluation of dual-purpose (milk and meat) animals. World Anim Rev 77(1977):56–59

Tschirley D et al (2014) The rise of a middle class in East and Southern Africa. Wider working paper 2014/119. World Institute for Development Economics Research, Helsinki

URT (2012) National sample census of agriculture 2007/2008 small holder agriculture volume iii: Livestock sector—national report. Ministry of Agriculture, Food Security and Cooperatives, Ministry of Livestock Development and Fisheries, Ministry of Water and Irrigation, Ministry of Agriculture, Livestock and Environment, Zanzibar, Prime Minister's Office, Regional Administration and Local Governments, Ministry of Industries, Trade and Marketing, The National Bureau of Statistics and the Office of the Chief Government Statistician, Zanzibar. Dar es Salaam, National Bureau of Statistics

Wharton B (1968) Difficulties in the initial treatment of kwashiorkor. In: Cance RM, Widdowson E (eds) Calorie deficiencies and protein deficiencies. J&A Churchill Ltd, London, pp 306–321

Zervas G, Tsiplakou E (2013) Goat milk. In: Park YW, Haenlein GFW (eds) Milk and dairy products in human nutrition: production, composition and health. Wiley-Blackwell, London, pp 498–518

Chapter 28
Private Sector Actions to Enable Climate-Smart Agriculture in Small-Scale Farming in Tanzania

Sheryl Quail, Leah Onyango, John Recha and James Kinyangi

The agricultural innovations were invented decades ago. What we need is to distribute them in a form that is useful to people,
Andrew Youn, Founder of One Acre Fund

Abstract The private sector plays the most important role in financing agricultural investments, innovation and information dissemination where constraints on government investment render private sector actions more important. In East Africa, little is known about the participation of small businesses, independent traders, farmer organizations, large-scale wholesalers, marketing boards and cooperatives in climate-smart agriculture (CSA) and their potential role in its diffusion to small-scale farmers. In particular, the informal sector is out of view even though it forms the backbone of rural agrarian economies. This study examines relationships between private sector actors and farmers and examines supply chains of agricultural inputs, as well as agricultural product value chains. The potential for using the Quality Declared Seed (QDS) system to disseminate CSA bean and potato varieties is assessed, as is the commercial maize seed supply chain and its impact on agrobiodiversity. Finally, farmer trust of private sector actors, traders in particular, is evaluated. The data used is from a survey of 100 farmers and semi-structured interviews with traders, local input suppliers, transporters and marketing organizations.

S. Quail (✉)
University of Florida, Gainesvile, FL, USA
e-mail: squail@ufl.edu

J. Recha
International Livestock Research Institute (ILRI), Nairobi, Kenya
e-mail: J.Recha@cgiar.org

L. Onyango
Maseno University, Maseno, Kenya
e-mail: leahonyango@gmail.com

J. Kinyangi
African Development Bank, Nairobi, Kenya
e-mail: J.Kinyangi@afdg.org

R. Lal et al. (eds.), *Climate Change and Multi-Dimensional Sustainability in African Agriculture*, DOI 10.1007/978-3-319-41238-2_28

28.1 Introduction

In East Africa, an urgent and concerted effort by both governments and the private sector is needed to adapt to and mitigate the adverse impacts of climate change. The private sector—from local traders to global agribusinesses with multi-billion-dollar turnover—plays the most important role in financing agricultural investments globally, including in the East African region, where constraints on government investment render private sector actions more important. Agricultural research has shifted from the public to private sector, effectively giving more power to companies whose profit-seeking motive may not always be in the public interest. However, they have the capacity to conduct needed research, to facilitate trade and the distribution of agricultural products, and to disseminate innovations that enable climate-smart agriculture (CSA). This study focuses on private sector actions that enable or hinder CSA in East Africa using a pilot project in Lushoto, Tanzania, as a case study. This study transpires during today's era of public–private partnerships (PPPs) and landscape-level efforts to green global commodity chains, particularly those involved in agriculture (Gyau et al. 2015). The study assessed in part whether certain private sector actors share the goals of small-scale farmers and actually work for the public good (Cayford 2004; The Guardian 2015).

Climate change is projected to disrupt food production in East Africa, in particular for small-scale farmers (Funk et al. 2008) with ripple effects in the informal and small business sectors. Adoption of agricultural innovations is linked to improved food production and food security, and will help farmers adapt to altered weather patterns (Kristjanson 2012). Agricultural innovations are comprised of practices that increase crop productivity, reduce food waste and improve the natural resource base that crops are grown on (Barrett et al. 2002). Access to equipment, farm tools, tree seedlings, seeds, fertilizer and other inputs, storage technology etc. —most of which are channeled through private sector entities—is needed for the realization of those innovations.

Currently, there is very little information on private sector participation in agriculture, particularly climate-smart agriculture meant to help vulnerable small-scale producers. In particular, the informal sector is out of view even though it comprises a large component of local African economies. Yet such information is critical in exploring how best to harness private sector comparative advantages to benefit food security under conditions of climate change in East Africa. This study, therefore, can help fill this research gap. The study examines patterns of private sector activities in CSA as well as impacts of climate change on this sector. It builds on research and development efforts undertaken by Climate Change, Agriculture and Food Security (CCAFS)[1] at one of its pilot projects in Lushoto, Tanzania, a region whose agricultural production is closely linked to demand from neighboring

[1]Led by the International Center for Tropical Agriculture (CIAT), CCAFS is a collaboration among 15 CGIAR research centers with leading scientists in agriculture, climate change, environmental and social sciences to identify and address the most important interactions, synergies

regions, especially Dar es Salaam. Building the capacity of select private sector entities and improving their coordination with the public sector so that smallholders are not excluded may be key to facilitating the scaling up of agricultural innovations that improve food security for smallholders grappling with a changing climate (Jayne et al. 2006). Integration of farmers into agricultural value chains with small- and medium-sized enterprises for niche markets and a growing African middle class is often targeted as an area of development and investment. However, the informal sector that connects small-scale farmers with swelling urban masses will dominate agricultural trade in African cities for years to come.

28.2 Climate-Smart Agriculture

Climate-smart agriculture integrates the three dimensions of sustainable development (economic, social and environment) by jointly addressing food security and climate challenges (Food and Agriculture Organization [FAO] 2010). CSA is composed of three main pillars: to sustainably increase agricultural productivity and incomes, to adapt and build resilience to climate change and to reduce and/or remove greenhouse gas emissions where possible. CSA is an approach to developing technical, policy and investment conditions to achieve sustainable agricultural development for food security under climate change. It is designed to identify and operationalize agricultural development within the explicit parameters of climate change.

CSA requires changes in farming practices. This includes, but is not limited to, taking up *climate-smart crop varieties, changing farming schedules and mitigating while adapting to climate change.* As climate change alters temperature and rainfall patterns, one CSA approach for adapting to new conditions is to switch crops. Climate change is also forcing farmers to change the schedule of their customary farming activities. As the need to mitigate carbon emissions grows, improving the efficiency of fertilizer use, planting trees on farms and improving the management of livestock and rangelands will also be crucial CSA activities (FAO 2010). Farmers usually adopt these actions primarily to enhance and diversify incomes, not because the actions lower emissions. Climate change mitigation is an added benefit (Fig. 28.1).

As the duration of rainy seasons shortens, the introduction and uptake of climate-smart varieties are crucial. Early maturing crop varieties that confer disease resistance while achieving high yields have been shown to be the most efficacious and cost-effective strategy in the CSA toolbox. Such crops are central to the CCAFS approach, and while dissemination of commercial seed varieties is easily established, less is known about non-commercial crop varieties.

(Footnote 1 continued)

and trade-offs between climate change and agriculture. CCAFS carries out research in East and West Africa, Latin America and Southeast and South Asia.

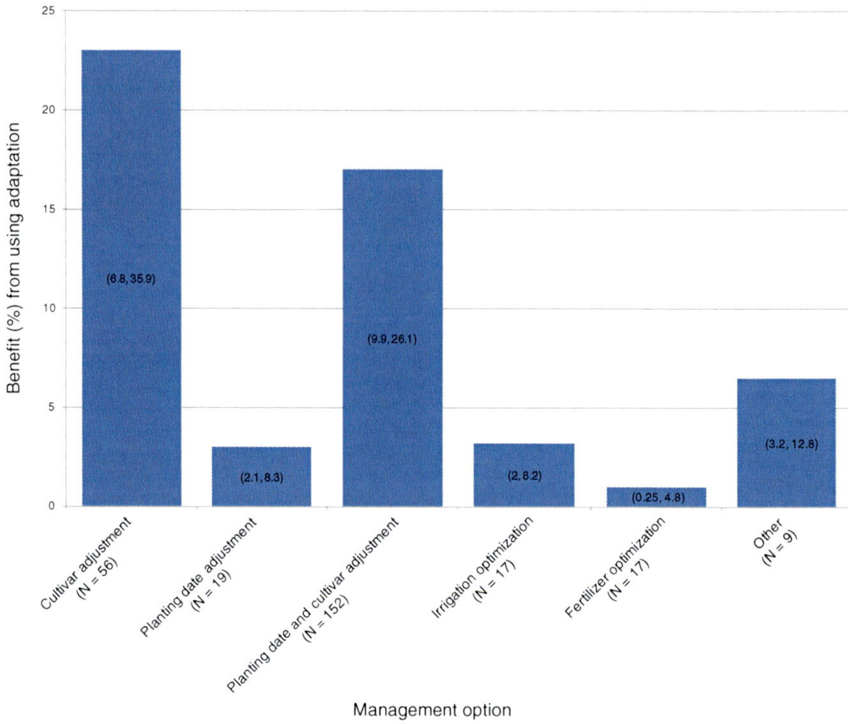

Fig. 28.1 Incremental crop-level adaption. *Source* Challinor et al. (2014); IPCC (2014) WG II, ARS

Integrated pest management (IPM), a process that uses habitat design, biodiversity services, predatory insects, biopesticides and a multifaceted approach to disease containment, is another CSA practice. A review of 85 IPM projects in Asia and Africa found that reduced pesticide use led to increases in crop yields (Pretty and Bharucha 2015). It is commonly believed that not using pesticides would result in crop loss, but those accounts do not include IPM scenarios. Yield increase is one of the three CSA pillars making IPM a necessary component while potentially reducing the cost of production for cash-constrained smallhold farmers (Fig. 28.2).

28.3 Project Description

CCAFS has introduced a suite of CSA practices using participatory action research in collabouration with farmers to document observations about climate and its forecasting and to implement good agricultural practices such as terrace building, tree planting, crop rotation, spacing and appropriate use of inputs. A public–farmer

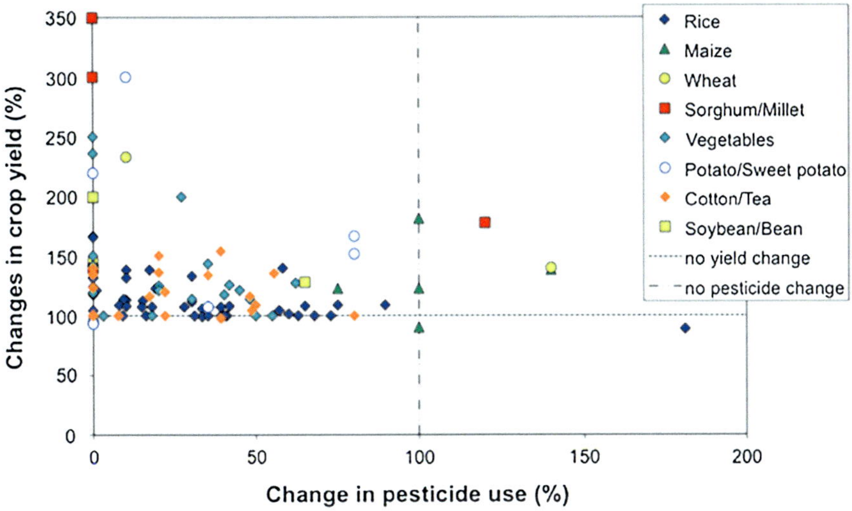

Fig. 28.2 IPM results from 24 African and Asian countries. *Source* Pretty and Bharucha (2015)

partnership was facilitated in which farmers produce bean seed bred by the Selian Agricultural Research Institute (SARI) and potatoes by the International Centre for Potatoes (CIP) using the QDS system to sell climate-smart varieties at the ward level. Tanzania's QDS system is a response to Plant Breeders' Rights of UPOV 91, which provides legal protection of propagating material for commercial seed production. The QDS system was operationalized in 2000 and grants farmers legal protection to produce seed for sale on land not bigger than 5 acres (Ngwediagi 2009). Tanzania is expected to join UPOV 91, and the legal protection of farmer-managed seed systems, which comprise 75 % of the seeds used in Tanzania, is unclear.

CSA crop varieties that are developed by the public sector involve research on several dimensions of crop production, including breeding, crop management, pathology and entomology. Once developed, farmers test the varieties in situ under the supervision of agricultural researchers. A varietal release is then performed by the Tanzania Official Seed Certification Institute (TOSCI), which proceeds with the distinctiveness, uniformity and stability (DUS) test, performance trials and official release. Together with the Agricultural Seed Agency (ASA), TOSCI provides guidelines and training for the newly introduced variety under the QDS system. Bean varieties introduced by the project include Lyamungu 85 and 90 and Selian 97 for improved yields, disease resistance to root rot and angler leaf spot, and drought tolerance. Farmers had previously grown 13 varieties of beans, many of which succumbed to disease. Eleven varieties of potatoes were initially introduced, of which farmers selected six after three growing seasons of experience. Previously introduced varieties such as Kidinya and Obama are susceptible to late blight. They were sourced from Kilimanjaro, Arusha and sometimes Nairobi and presumably brought by traders. The varieties introduced by the project originated in Mbeya.

Fig. 28.3 Diffused light stack house

They are resistant to late blight, high yielding and early maturing. The newly introduced Asante variety matures in three months as opposed to previously planted varieties that matured in four to five months. The Asante variety allows farmers to cultivate potatoes during long and short rainy seasons. Training was given to farmers on how to construct diffused light stacks that improve seed storage. The stacking innovation is made from wood and allows air to flow across layered stacks of potatoes. Early maturing varieties of maize, namely, Panna and Decapu, are produced by commercial seed companies. These varieties were introduced to local input suppliers. Disease-free cassava and bananas were also introduced (Fig. 28.3).

28.4 Lushoto Geography and Agrarian Change

Lushoto lies in the West Usambara Mountains, which are part of the wider Eastern Arc Mountains (ERM). Bathed by the Indian Ocean monsoons, the ERM have maintained a steady climate for millions of years, allowing the range to evolve exceptional biodiversity across narrow elevational gradients (Hall et al. 2009). Historically, the Sambaa people carved complex banana groves out of these forests, similar to Chagga agroforests. Farmers replaced native trees with useful food crops plant by plant, tree by tree until a modified a multi-canopy structure emerged that preserved some degree of hydrologic and ecological function. Useful native species were kept. Beans, pumpkins, tree tomatoes, coco yams, sweet potatoes, cassava, mayombo beans, pigeon pea and sugarcane were grown under banana trees and several native trees deemed useful by farmers. Maize was introduced later and cultivated by women (Johansson 2001).

This region underwent major regime changes over the past century, beginning with German conquest, followed by repressive British soil conservation schemes and the Ujamaa villagization programme that reorganized peasant political structure. It was followed by a declining demand for black wattle, one of Lushoto's first cash crops established under colonial rule. This subsistence-oriented rural economy gave way to small-scale producers of cash crops on increasingly fragmented plots. Many farmed slopes, and a growing segment sold their labour to profitable fertile

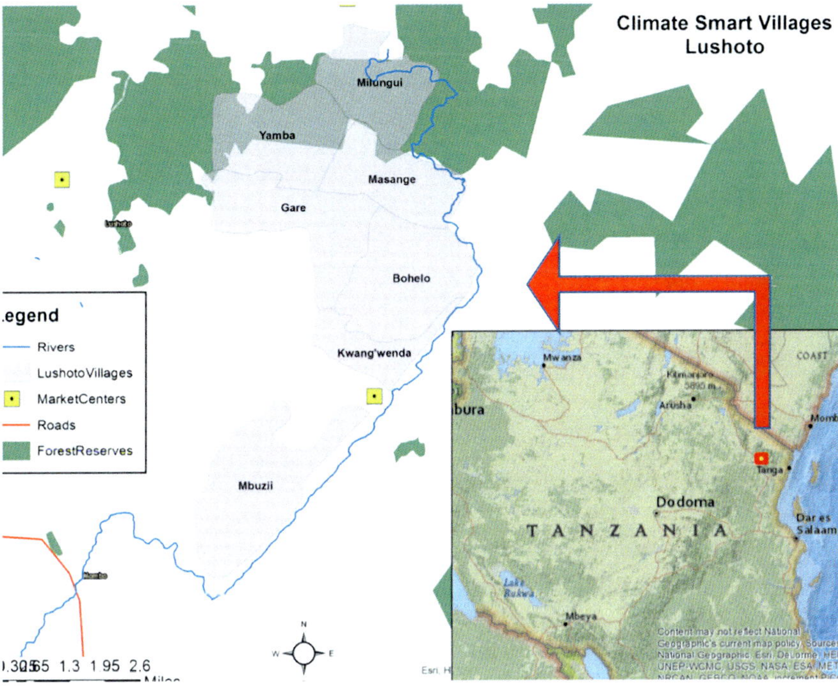

Fig. 28.4 CCAFS project site—Lushoto 1

valley bottom farms (Johansson 2001). Labour once dedicated to traditional banana groves turned to quick profits derived from fertile soils beneath freshly cut forests that produced vegetables for distant urban markets. Over time, a sustainable agroforestry food system slowly lost its diversity, complexity and structure and yielded to an unsustainable slash-and-burn agriculture complemented by over-grazing livestock (Fig. 28.4).

The 1981–2000 Soil Erosion Control and Agroforestry Project (SECAP; Johannson 2001; Namwata et al. 2012) began as a top-down technical package designed to slow deforestation, reforest degraded slopes, prevent soil erosion and improve agriculture. SECAP followed on the heels of government rural development projects characterized by local patronage politics and contention between farmer and extension agent. The project represents Tanzania's longest non-government agricultural initiative, which was sponsored by the German Technical Cooperation Agency GTZ. Following repeated failures and an uninterested farmer base, the project evolved into a more participatory approach. Unrelated to SECAP but occurring during its tenure was the rise of vegetable production that largely replaced the cultivation of maize, which then had to be imported from other regions. Irrigation was a key aspect of this transition in crop production. Irrigated land produced more than three times as much income as non-irrigated plots, and farmers were acutely aware of the role forests played in regulating local atmospheric and hydrologic function that directly impacted the profitability of their crops.

Despite the remarkable rise in the productivity and diversity of the crops grown, Lushoto farmers complained about being worse off. They complained that children were harvesting potatoes instead of attending school; that hard-earned money was spent on school fees, inputs, hired labour and imported maize; and that the vagaries of cyclic drought and floods made agricultural unsustainable (Johannson 2001).

Several other initiatives were begun in Lushoto following SECAP. Despite built capacity in many villages, some fell outside the range of agricultural projects. Even so, many of them host vibrant government-sponsored research particularly for vegetable production. The entrance of CCAFS builds upon earlier initiatives while strengthening existing ones, in particular to build capacity around climate change and adaptation. And the conditions that farmers complained about during the SECAP era continue today.

28.5 The Private Sector in the East African Context

The private sector refers to the part of the economy that is not state controlled and is run by individuals and companies for profit (Gyau et al. 2015). In Tanzania, as in much of Sub-Saharan Africa, the private sector is comprised of formal and informal sectors with the latter absorbing far more economic activity than the former. The Integrated Labour Force Survey estimated that the informal sector represents 60 % of Tanzania's gross domestic product (GDP) and that approximately 75 % of total employment is absorbed by the agriculture sector (ILO 2008). Tanzania's formal sector employs about 10 % of the country's workforce. This distinction between formal and informal sectors is important because much of the literature rationalizing strengthening private sector investment in African agriculture targets the formal sector, both domestic and foreign (FAO 2010; Global Harvest Initiative 2011; Mtengeti et al. 2015), in an era when employment gains in the informal sector are of greater significance.

Private sector actors in this study refer to independent traders (*Madalali*) of varying size, input suppliers, transporters and marketing organizations. The formal sector, which is subject to registration, taxation, labour protection and legal frameworks, differs from the informal one in that the latter asserts its right to self-sufficiency and autonomy often outside legal and policy frameworks.

Investment in agriculture by the formal sector typically occurs with input manufacturing, seed propagation for certain crops, cultivation of high-value and non-traditional crops like cut flowers, fruit, coffee and vegetable exports, and through companies involved in food processing, transport and marketing (FAO 2010). In Tanzania, formal private sector actors include Tanseed, East African Seed, Kibo, Bahkresa, Mohammed Enterprises, Yara, Syngenta, Link Forward, commodity buyers for cotton, tobacco, tea, sugar and coffee and others. Aside from commodity buyers, who may or may not provide extension services and input packages, private companies have limited physical interaction with farmers and are limited to when they organize campaigns to promote specific products during farmer field days and agricultural shows. Input manufacturers often have agents who distribute their products to local agro vet shops, which, in turn, sell their products through local traders.

The informal sector dominates rural agrarian economies but typically is not considered part of the private sector. Thus, it is necessary to describe what exactly the informal sector is if its role in climate-smart agriculture is to be examined. The informal sector is flexible and dynamic in its response to supply and demand and lubricates the exchange of goods originating from the formal sector. It absorbs surplus labour and can provide reasonable wages. This sector is characterized by complex social networks in which both exploitation and fairness co-exist. Low entry costs and small-scale operations allow individuals to create employment for themselves and to alleviate poverty in ways the formal sector cannot (Muller 2004).

The informal sector can be a source of innovation and efficiency. Farm tools are one example of the innovative capacity of this sector. In a study that compared agricultural tools produced in formal and informal sectors, Muller (1980) found that the greatest diversity and innovation in tool making came from blacksmiths, many of whom were farmers themselves. Steel was in short supply in the past, and two large factories produced the majority of hoes consisting of two types. Farmers would buy these and hire blacksmiths to refashion the tools for current and newly introduced crops. Muller noted thousands of tools crafted for different soil types, weather conditions and crops, and the flexibility of blacksmiths to adapt to changing conditions. If adaptive capacity is the hallmark of climate-smart agriculture, investing in small-scale industries may offer greater impact than formal, large-scale industries confined by capital-intensive investments in equipment and machinery. Local blacksmithing survives today in pockets with support from the government parastatal Small Industries Development Organization (SIDO).

The market reforms of the late 1980s and early 1990s scaled back many government interventionist practices and policies in the distribution of agricultural goods (Sitko and Jayne 2014). Restrictions were lifted on the intra-national movement of grains in particular, and private trade was legalized. The vacuum left by retreating governments was filled by small-scale, independent traders who capitalized on low barriers of entry in terms of working capital, cash flow and assets (Jayne and Jones 1997). Although traders in most commodities exist, their presence in grain assembly markets is most evident. In the case of Lushoto, however, traders dominate a profitable vegetable, potato and bean trade far more than grains, which are used for home consumption and are supplemented by imported maize. Small-scale trading is thought to reduce poverty for rural people and equally benefits remote farmers by aggregating their surplus production into marketable quantities, by responding efficiently to market price signals and by absorbing transportation costs. Perceptions by state actors of traders as exploitative, low-balling dealers have rationalized often market-distorting policies through marketing boards that seek to protect farmers from independent traders (Sitko and Jayne 2014). This is not to argue that exploitation in the informal sector does not exist. It does, through the rigging of weights, for example, when buying agricultural products and demanding *lumbesa*, an added amount on top of a sack, occurs. Corrective actions demanded by the government are warranted when this happens.

Trust plays a pivotal role in the functioning of economies. Low levels of trust have been related to chronic underdevelopment and poverty (North 1990) and dampen the ability of people to do business with each other. Personal trust refers to things like neighbors being friendly to each other and the establishment of connections, friends providing informal loans entrusting the money will be returned and other related informal actions. Personal trust is distinguished from impersonal, institutional trust that allows people to do things like deposit money in banks. Impersonal trust is more characteristic of the formal sector. Trust is believed to be strengthened by the rule of law, enforceable contracts and the ability to impose sanctions on cheaters. Trust in Tanzania's agrarian economies can be thought of as a component of "economies of affection," a term coined by Hyden (1980) to describe communities that are united by survival, morality, social maintenance and development. Trust can be a barometer of a community's ability to retain its social values as it straddles subsistence- and market-oriented worlds.

28.6 Study Objectives and Methods

This study is part of a larger, ongoing effort to understand the role of the private sector across the entire food value chain, particularly the local level and through its interactions with the informal sector. This study also attempts to understand better the informal sector's potential role in enabling and scaling up climate-smart agriculture from pilot projects in Tanzania, Kenya and Uganda. This chapter is focused on Lushoto, Tanzania, and draws from data collected in December 2013 and February 2015. Surveys were administered to 100 households in both periods across four villages to determine what and where farmers are selling with special attention to CSA crop varieties. This study also attempted to determine the level of trust farmers have in dealing with traders. Focus group discussions and semi-structured interviews with key informants, including traders or middle (wo)-men, transporters, input providers, a marketing officer and the president of Usambara Lishe Trust, were additional sources of data. In general, this study seeks to understand the role of informal private sector actors, independent traders or middle men and transporters.

28.7 Results

28.7.1 Agricultural Production

Lushoto agricultural production is tightly coupled with demand for crops from regional neighbors, in particular Dar es Salaam. Approximately 10 % of farmers, however, are strictly subsistence oriented and not engaged in any form of trade. The

Table 28.1 Crops grown and livestock raised by farmers

Crop	% Farmers	Crop	% Farmers	Livestock	% Farmers
Maize	96	Wild greens	5	Cattle	83
Beans	95	Lettuce	4	Sheep	46
Potatoes	53	Papaya	3	Goats	40
Avocado	45	Sugarcane	3	Pigs	3
Cassava	39	Spinach	2	Chicken	80
Cabbage	39	Jackfruit	2	Ducks	3
Banana	36	Sunflower	2		
Tomato	35	Mastaferi	2		
Green Pepper	30	Peas	1		
Plums	17	Yam	1		
Passion	14	Chinesc vegetables	1		
Sweet Potato	12	Onion	1		
Peaches	12	Amaranth	1		
Cucumber	11	Lemon	1		
Mango	10	Palm	1		
Coffee	10	Topetope	1		
Carrot	6	Apple	1		

majority grow food crops as cash crops, with the exception of maize and cassava, which are used for home consumption. The December survey found almost all farmers grow maize and beans, followed by potatoes, cassava, bananas, tomatoes and green peppers. The second round of sampling, which was done in different villages, found less production of cassava, bananas and sweet potatoes, in contrast to the first study that counted 92, 97 and 86 %, respectively, of farmers growing these crops (Table 28.1).

Farmers were asked which crops were sold by them and to whom (Table 28.2). This is important, because it is often assumed that farmers grow for themselves first for food security reasons, and only secondarily to sell surplus production for cash. We found that this is not always the logical trajectory of farmer decision-making. Beans comprise 93 % of crops sold to traders and markets to a lesser extent, followed by potatoes (50 %), cabbage (45 %), tomatoes (33 %), bananas (26 %) and green peppers (21 %). Because local markets and traders dominate food crops sales, we report only these because it is assumed that the food sold in local shops is consumed locally (Table 28.2). The exception is Yamba village, which is remote, at high elevation and not near any local markets. In this village, one shop accumulates crops that are then sold to traders or taken to urban areas by the shop owner himself. In addition, although crops taken to local markets are sold for local consumption, a significant portion of them are bought by traders who transport crops out of Lushoto. We also disaggregate data on the percentage of farmers selling to various

Table 28.2 Number of farmers selling crops to traders and markets (n = 100)

Crop	Traders	Local market	Total (%)
Beans	48	45	93
Potato	35	15	50
Cabbage	38	7	45
Tomato	12	21	33
Banana	3	23	26
Green Pepper	8	13	21
Avocado	0	12	12
Maize	9	2	11
Cucumber	1	8	9
Sweet Pot	3	5	8
Lettuce	4	0	4
Mango	1	3	4
Plum	1	3	4
Plum	1	3	4
Papaya	1	2	3
Carrot	1	1	2
Cattle	2	0	2
Sunflower	2	0	2
Coffee	1	0	1
Mnavu	1	0	1
Spinach	1	0	1

outlets by village, because in Mbuzii, in contrast to other villages, all farmers bring their produce to the local market, and only 10 % sell to traders. There is a reason for this. It is also not uncommon for farmers to sell to both traders and markets. However, before proceeding, it is important to expand on who exactly a trader is, because to simply say 'trader' is somewhat vague (Table 28.3).

28.7.2 Independent Traders, Value Chains and Markets

Independent traders are of two types: those involved in the acquisition and aggregation of agricultural output into bulk quantities and those who work for larger enterprises, typically agroinput companies that provide fertilizers, seeds and agrochemicals. Sometimes traders do both. For this study, traders refer to the former, although it is not uncommon for traders to provide inputs in exchange for buying rights to a field of crops. Traders may originate from inside or outside a given locality or community and can be small- or large-scale.

Traders utilize an assortment of tactics to obtain crops from farmers. Some wait along roads to intercept farmers walking to the markets. Others are found in villages where they aggregate crops, while others may come from outlying areas. Larger

Table 28.3 Farmers selling to various outlets

Selling to	Mbuzii (n = 18)		Yamba (n = 30)		Milingui (n = 25)		Kwangwenda (n = 28)		All villages (n = 101)	
	#	%	#	%	#	%	#	%	#	%
Middle man	2	11	24	80	24	96	13	46	63	58
Local market	18	100	21	70	17	68	23	82	79	80
Local shop	0	0	26	87	3	12	4	14	33	28
Transporter	0	0	3	13	6	25	3	11	12	12
Cooperative	0	0	6	20	0	0	0	0	6	5

traders wait at the road readied with a lorry and use trusted village-originated traders to do their buying. Many traders are small-scale and capital-constrained and incur considerable risk. Others are medium- and large-scale. These can better survive risky ventures. For certain crops, traders buy a field of crops weeks in advance of harvest, referred to as pre-harvest buying, when supply is low and prices high. If competition is particularly stiff, a trader has to peruse fields even earlier; although competition can be fierce, traders also report cooperating with each other. Once a deal is struck with a farmer, the pre-harvest buyer assumes all responsibility for a field and associated labour and harvesting costs. Roughly 90 % of traders are from Lushoto, and 10 % are outsiders.

Some smaller traders work seasonally and use hired labour to tend their fields while they buy up crops and revert back to land preparation following the harvest season. Importantly, many Lushoto traders are farmers. Others engage in trading year round and shift to other regions in pursuit of local climates, harvests and lower taxes. Together with their hired lorries and intricate knowledge of markets around the country, the traders swiftly transfer crops to market. They can sell quickly or alternatively hold on to their less perishable (and chemically treated) products and wait for prices to go up. And while the gendered stereotype of a trader is a man, women can also be found engaged in trading albeit at a small scale.

Between 15 and 20 lorries leave Lushoto every day for Dar es Salaam, Tanga, Zanzibar, Moshi, Arusha and, to a lesser extent, Mtwara. Of these, approximately 20 are loaded with potatoes that go to Zanzibar each week, some of which will then be sent to the Seychelles Islands. Lushoto traders compete in Dar es Salaam markets with Iringa, Mbeya and Kilimanjaro traders. Together, they create significant oversupply events in Africa's third-fastest growing city where demand for food is high.

Traders face a number of risks, but their most common complaints are ineffective communication and oversupply events in the markets they deliver to. Post-harvest issues are particularly acute because, with the exception of beans, the majority of crops leaving Lushoto are highly perishable. Traders are very careful to call destination markets in advance to assess the supply of a given crop. If supply is low, they will quickly assemble a load of produce, but if they arrive at a flooded market that was empty two days prior, prices drop considerably, and the traders may not be able to sell. In this scenario, they can wait for inventory to clear knowing that some of the produce will rot. Alternatively, they may take their produce to other markets in search of other buyers. When Dar es Salaam markets are full, it is not uncommon for traders to go to Tanga.

Using their description of market efficiency, traders were asked to rate each market they visited in terms of oversupply events and communication effectiveness (see Fig. 28.5). The oversupply season occurs from August to October following the long growing season and is when traders make most of their money. With the exception of Tandale, Dar es Salaam markets are very inefficient. It's an attractive destination, however, because prices are higher in this market, and inventory moves quickly. Moshi, followed by Tanga, is the most efficient market. No or few oversupply events were reported for Tanga. However, in Moshi, it can take up to two

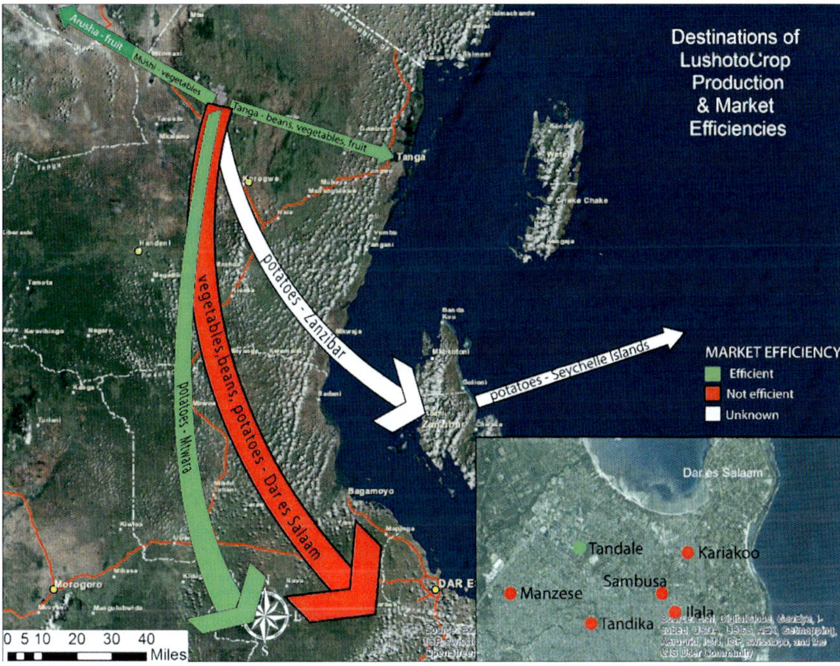

Fig. 28.5 Value chains and efficiency

weeks to sell a truck of produce while it takes only a maximum of one week in Dar es Salaam. Potatoes are pre-ordered by Mtwara buyers who request Lushoto traders to organize transportation, who, in turn, report efficiency in this system. No traders affiliated with the potato trade in Zanzibar could be located so the levels of oversupply and the efficiency of movement of products are unknown. The number of oversupply events experienced by traders in their last ten trips ranged from 1 to 4. Under these circumstances, the price decrease for potatoes is upward of 40 %, 25 % for beans and 50 % for cabbage. Product perishability losses can be as low as 5 % of an entire lorry, although losses of this magnitude are rare.

28.7.3 *Traders and Trust*

Farmers were asked to rate the level of trust they have with traders to whom they sell their produce. Traders were asked to estimate the level of trust they have with buyers in their market destinations. No farmer expressed complete distrust of traders; nor did traders express complete distrust of their buyers. However, a word of caution is in order for this interpretation due to cultural nuances. Tanzanian culture is a polite one and people can verbally soften their true feelings. For example, one trader rated his main market buyer with the highest level of trust while at the same time saying, "What can I do? I have to trust him, but I know he's stealing from me." In a focus group discussion in Mbuzii Village, women said traders in a local market called Soni could not be trusted while the men said the opposite. Of note is the proximity of Mbuzii to the Soni market, which explains why few traders buy in that village. In other words, when given the chance, farmers will eliminate one link in the chain and sell directly to Soni, a popular market and stopping point for produce going to Dar es Salaam (Table 28.4).

All small-scale traders expressed a great deal of frustration with their buyers in Dar es Salaam. The traders related that they had no choice when they arrived at markets other than to accept the prices offered to them. In fact, the small trader feels not only frustrated but also trapped. However, one farmer reported sending tomatoes to the same Dar es Salaam buyer for 30 years and that he and his buyer trust each other completely. That buyer claims to have very cordial relations with many Lushoto farmers and has been buying from the same people for many years. Hence, there is the occasional exception to the rule. Larger traders expressed a great deal of satisfaction with their occupation, stating it pays well, in contrast to smaller traders (Table 28.5).

28.7.4 Pre-harvest Buying

Almost a quarter of traders engage in pre-harvest purchase of crops from farmers, typically two weeks before harvest. It is strictly a function of high market demand, and traders buy an average of three fields. In Lushoto, cabbage, followed by potatoes, carrots and beans are crops prone to this type of trade. As traders compete with one another for these crops, they pay farmers in advance for their harvests. If competition becomes acute, traders extend advanced purchasing to three and four weeks before the harvest. This occurs typically with farmers who irrigate in valley bottoms and are happy to sell before markets become flooded with produce and prices are higher. Profits are quickly plowed back into production and used to hire labour for land preparation as well as to purchase inputs for the next cropping season. Traders absorb all costs from pre-harvest purchase to crop harvest. They often hire the same farmer to harvest crops. This is the riskiest form of trade particularly for fields located in valley bottoms prone to flooding. Traders report an average of two lost crops over the past five years, all from floods, with an average loss of $750. Some traders have informal loss-sharing agreements with farmers in the event of a crop loss to reduce the associated risk. Contrary to the popular stereotype of the exploitive trader preying upon cash-strapped, desperate farmers, farmers report having the upper hand in this relationship (Table 28.6).

Table 28.4 Degree of trust farmers have of traders/buyers

Trader type	Farmer trust of traders/buyers							
	Trust completely		Trust partially		Complete lack of trust		Total # of traders	
	#	%	#	%	#	%	#	%
From village	42	50	16	19	0	0	58	62
From outside Lushoto (usually buyer in urban market)	6	7	1	1	0	0	7	8
Pre-harvest buyer	4	5	13	15	0	0	27	29
Waits on road	1	1	1	1	0	0	1	1

Table 28.5 Types of traders

Trader Type	% of Farmers
Villager	69
Dar origin	8
Pre-harvest buyer	22

Table 28.6 Farmer trust in traders

	Trust completely	Trust partially	No trust at all
Village trader	45	7	0
Trader from outside	16	12	0
Total	61	19	0

28.7.5 Transporters

Only a few transporters were interviewed, one of whom was a hired driver. Trucks are mostly owned by local elites. They hire drivers, who, in turn, take traders and their aggregated agricultural surplus to markets. The transporters incur no risk and report that the two biggest losers are farmers and small-scale traders. Transportation is a profitable business despite the heavy taxation of imported vehicles. It takes three years of use to recover the cost of a truck purchase. The average cost to take a load to Dar es Salaam is $350 in addition to taxes paid to the district upon leaving Lushoto. Lushoto traders like to buy in neighboring districts, Handeni in particular, because of lower taxes.

28.7.6 Other Relevant Actors

Although the bulk of Lushoto's food value chain moves through local entities and the informal sector, two nongovernment organizations (NGOs) and government initiatives exist to facilitate market linkages for small farmers.

The Usambara Lishe Trust (ULT) was initiated by SECAP in 1996 to bypass exploitative traders. ULT works in four villages. Its mission is to provide direct marketing and extension information for approximately 70 high-value niche-market fruits and vegetables, most of which require irrigation. They supply SkyChef, various hotels and supermarkets in the upscale Msasani Peninsula. ULT has grown from an initial base of 60 farmers to 257 across several villages. Before obtaining a refrigerated truck, ULT hired a 3.5 ton lorry for 500,000 tsh ($250) to go to Dar es Salaam—roughly 20 % less than the prices traders pay for the same service. The district recently built a refrigerated packing house for cleaning and storage. The ULT also invested in a refrigerated truck to complete its cold chain mandated by SkyChef. ULT trains its farmers to use IPM practices. It tests for pesticide residues in an effort to meet food safety requirements. Farmers set their own prices based on their production costs. ULT embodies virtually every aspect of sustainable agriculture, including a fair playing field for farmers, reduced pesticide use and reduced post-harvesting losses via a refrigerated cold chain. Yet most farmers decline to join because it would require planting small increments of different crops instead of planting a few crops that can be aggregated for a larger sale that nets them instant cash. Payments by ULT, in contrast, are delayed. Farmers prefer more homogenous fields driven by large-scale demand from urban masses and organized by informal farmer–trader relationships.

The Marketing Infrastructure, Value Addition and Rural Finance Support Programme (MIVARF) began in February 2015. It was designed to add value to fruits and, to a lesser extent, beans. The focal area falls outside CCAFS sites but borders several. Fruits are sold to Dar es Salaam, Arusha, Nairobi, Zanzibar and the Seychelles Islands. Peas have been sold in Arusha, but rejection rates are high even when the peas are transported in a refrigerated truck. None of the CCAFS project villages report growing peas. However, they do grow various fruits, including avocados, mangos, bananas, peaches, plums and apples to a small extent. CCAFS has not linked with the MIVARF programme to connect its farmers for fruit sales.

Two important private sector development organizations should be mentioned, even though no project activities by them were identified in Lushoto. Both work intensely in the agricultural sector with farmer groups and small to medium sized enterprises (SMEs). The Private Sector Programme Support (PASS) was established in 2000 to assist agribusinesses in enterprise development specifically for value-added businesses. PASS deals with private individuals/sole proprietors, farmers groups, companies and, to a very small extent, savings and credit cooperative organizations (SACCOs) and associations. PASS organizes small farmers into groups and links them to markets, input supply chains, credit and advisory services (Temu 2013). Similar to PASS but broader in scope, SIDO is a parastatal that provides SMEs with business development services. It has offices in every region of the country and has incubated several agribusinesses with a focus on value addition. Both organizations are relevant to future private sector development efforts of climate-smart initiatives.

28.7.7 Input Suppliers

Lushoto has four main input suppliers, one of whom was contracted by CCAFS to deliver to its project villages early-maturing/high-yielding maize seed and fertilizers to be used as top dressing. Village-based input shops are now in their infancy in some villages. Those in villages without input supply shops must either travel to Soni or Lushoto or rely on village-based traders to bring inputs. Under these circumstances, those inputs are not easily obtained. Some traders lend farmers inputs in exchange for access to a crop at harvest. Lushoto's input suppliers often extend informal interest-free loans to farmers and traders with repayment periods of one week to a month.

Inputs are sourced from a variety of companies. Shop owners drive to Dar es Salaam and Arusha to stock up on inputs. Some companies deliver to Lushoto. Yara, an input supplier based in Norway, sells a variety of fertilizers that are sourced from Dar es Salaam and Arusha, while urea is imported from China. Maize seeds from Panna, Kibo and East African Seed companies are delivered to the Lushoto market. Maize seed sales have increased dramatically from 1 to 2 tons in 2005 to as many as 130 tons in recent years. Vegetable seeds originate in Arusha. Non-commercial seeds, as in the case of potatoes and beans, move through traders.

Lushoto's intensive horticultural production has been associated with high pesticide and fungicide use, which represents an environmental cost. Pesticide and fungicide sales wax and wane with the season, but the latter is steadily gaining ground indicating that fungal diseases are on the rise particularly during the rainy season. Drought events translate into a drastic decline in agrochemical sales. Mancozeb, a carbamate whose metabolite is listed as a probable carcinogen and as being harmful to fish and wildlife by the U.S. Environmental Protection Agency (EPA), is the highest-selling fungicide. The three principal agrochemical companies are Link Forward (China), Balton (Israel) and Syngenta (Sweden). Link Forward is reported to supply the best-selling products based on price and availability. In contrast to other agrochemical suppliers, Link Forward delivers to Lushoto and, at the time of this study, was in preliminary negotiations with a major input distributer to act as a supply hub for their products. Despite the shortened efficacy of Link Forward's most popular products, farmers are willing to incur the labour costs of more spraying events compared to longer-lasting Syngenta products that cost slightly more—for a savings of $1.50 in the case of Mancozeb. Few farmers use herbicides. Those who do use 2.4 D and Round Up. Balton and government officials provide training about the handling and application of pesticides. Village-based traders, who sell inputs, read package labels for handling instructions.

The use of agrochemicals has been on the increase as has inorganic fertilizers used as a top dressing on crops fertilized primarily by manure (Fig. 28.6). A small minority of CCAFS/Lushoto farmers produce enough manure to sell to others; it is not clear whether the availability of manure is a constraint. In 2012, 48 % of farmers used pesticides; today, 84 % do. The number of farmers using inorganic fertilizers is up from 52 % in 2012 to 87 % today. This combination of manure and inorganic top dressing replaces organic matter and nutrients previously mined from eroded montane soils.

28.7.8 Seeds

Central to the CSA model in Lushoto are early-maturing, disease-resistant and high-yielding crop varieties that were field tested and evaluated by farmers. They

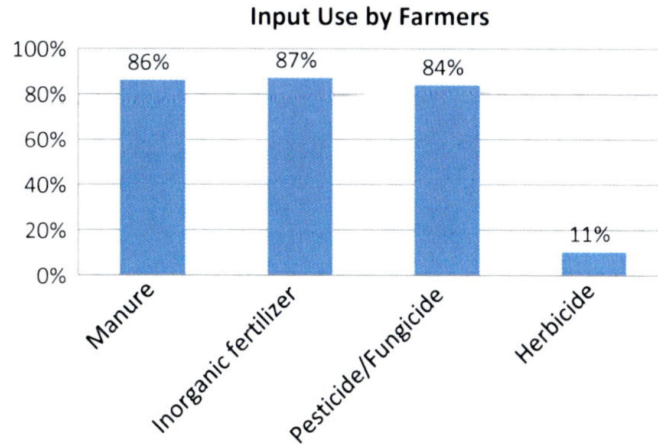

Fig. 28.6 Input use by Lushoto farmers

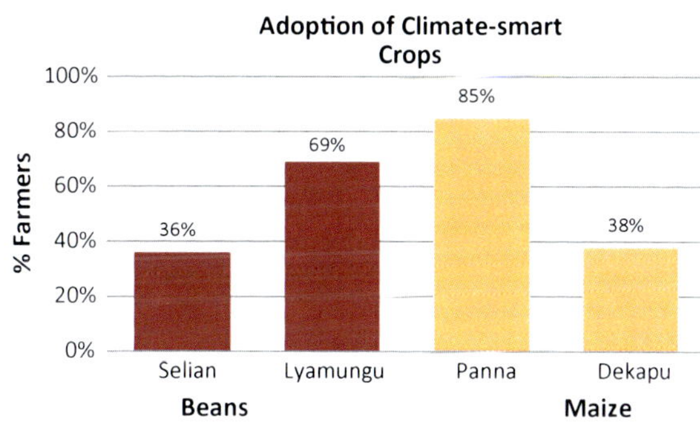

Fig. 28.7 Adoption of CS crops

report high satisfaction with bean performance and profit. Twelve bean varieties were reported as being tested, the most popular of which are Lyamungo (69 %) and Selian (36 %). In one village, predominantly men grow the improved bean varieties as a cash crop. Women complained that growth in its use has caused a price drop for the varieties women grow. Women sell their surplus once home consumption needs have been met first. This price effect was experienced in other villages as well. However, in them, men and women reported growing commercial and subsistence varieties so there was no differential impact by gender. Urban demand for clean, uniform beans is reflected in higher prices than for sacks of mixed beans, which sell for less; the trade-off is reduced agrobiodiversity and possible susceptibility to pests and disease. Farmers also reported high levels of satisfaction with the performance of specific maize varieties. As shown in Fig. 28.7, the Panna and Dekapu varieties are most often grown by farmers. Of the ten maize varieties reported, only two or three are non-commercial varieties. Dissemination of new potato varieties is still in its infancy, and the market prices of these varieties are unknown. The Kidinyo and, to a lesser extent, Obama potato varieties originated in Kilimanjaro and presumably were brought by traders. Kidinyo is the second highest-priced potato on the market, next to the Obama, which is more difficult to

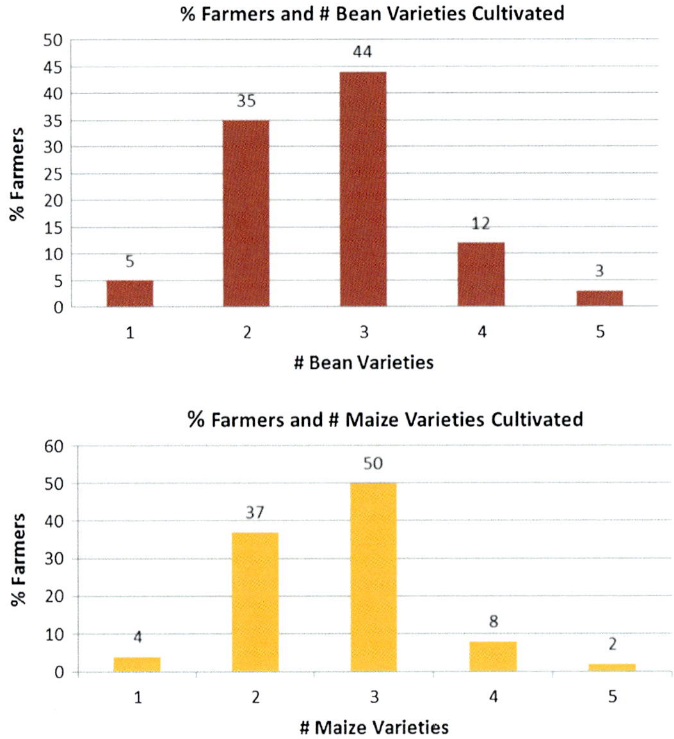

Fig. 28.8 Percentage of farmers and number of maize and bean varieties grown

grow. Farmers report that potatoes from Kenya are of superior quality and outcompete their best grown varieties in larger markets such as Dar es Salaam.

Maintenance of agrobiodiversity is an important component of climate-smart agriculture (FAO 2010). Almost half of farmers grow three varieties of beans, and 50 % grow two types of maize. Farmers in every focus group meeting lamented having lost some crop varieties due to their enthusiasm for climate-smart varieties. A red maize variety was reported to have been wiped out in the last drought. A number of climate-smart traits were reported for some of their long maturing traditional maize varieties such as drought tolerance, longer storage time, pest resistance and taste; yield was not necessarily at the top of their list. Although satisfaction with the new crop varieties was high, many complained that they are more susceptible to insect attacks. A small group of farmers in Yamba Village asked that their local maize variety be tested against the commercial varieties under drought conditions (Fig. 28.8).

Dissemination and pricing of non-commercial bean and potato varieties were assessed. Future plans entail scaling up the sale of farmer-produced seed through village-based SACCOs using the QDS system. Out of 100 farmers, farmer-to-farmer exchanges represent 13 % of bean seed acquisition and 16 % of

Fig. 28.9 Sales of Bean and potato seeds

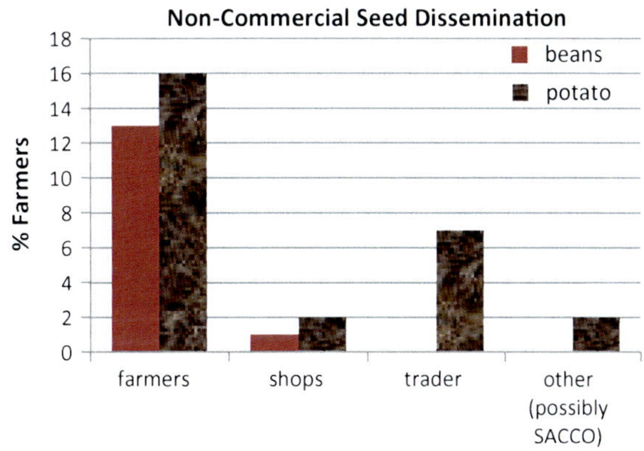

potato[2] seed acquisition (Fig. 28.9). Only a few farmers sell to shops and other outlets. Our data show that seven percent of farmers sell potato and bean seed to traders, which earn 25–50 % more than selling them as food crops.

28.8 Discussion and Recommendations

The private sector in Tanzania is most commonly associated with the informal economy. It mainly consists of individuals who prefer to rely on their own means of subsistence and production independent of a formal government or private sector actor that has failed to provide them with jobs or to feed the larger population. The informal economy with its web of social relationships, trade and the movement of goods is complex and not easily understood.

Traders incur risk in terms of food perishability and price volatility from oversupply of produce and often face high costs of doing business. They provide a needed service by aggregating vegetables sold by farmers to obtain money who then buy inputs and hire labour for other crops. Small and large traders have different levels of access to markets. Smaller traders typically make fewer trips to cities. They also are impacted by economies of scale and are likely to have less secure connections and negotiating power with buyers. Small traders experience more oversupply events and lower prices. Larger traders travel to markets more frequently and are likely to have more secure connections and to be favoured by buyers. As one larger trader commented, "I may lose profit, but I never take a loss." Market buyers incur little or no risk, as is the case for transporters. They may be motivated to artificially create oversupply events in order to drive down prices especially for less connected small-scale traders.

[2]The potato varieties sold are not "climate-smart" varieties.

New ways to reduce risks to traders and improve performance should be prioritized, which may have positive impacts on prices paid to farmers. An example is the introduction of a phone app/Information Communication Technology (ICT) platform for traders to register, track and coordinate trucks going to markets to smooth oversupply events, particularly for Dar es Salaam markets. When oversupply is prevented and trucks are not waiting more than a day to offload perishable products, post-harvest losses are reduced. According to traders, communication is critical, and if markets are full, they leave crops in the ground. Governance problems are reported in government-run markets where entrance fees are paid for the upkeep of the market structure and function. It is unknown whether cold storage is feasible in these areas.

Levels of trust between traders, especially those of Lushoto origin, and farmers are high, which is important to ensure a properly functioning economy. Cultural nuances could explain some of the high level of trust as Lushoto traders are exceptionally friendly and open, in contrast to traders in other parts of the country. Levels of farmer trust in traders tend to decline, however, as the traders grow in size. Input suppliers based in town, and, to a less extent, in villages, extend credit to farmers and traders, which indicates more trust in the system.

Scaling up CSA practices can build on dense social networks that involve traders, particularly for disseminating innovations, and, possibly, non-commercial crop varieties.

Lushoto is known for its efficient input supply chain. Infusing that chain with post-harvest storage and environmentally friendly products should be prioritized. Purdue Improved Crop Storage (PICS) bags, for example, which restrict oxygen in bags, improve the quality of maize and bean storage. Other similar products include biopesticides and bait traps and biological inoculants for beans. NitroSUA, for example, is an inoculant for bean crops developed by Sokoine University of Agriculture and is in the process of being commercially produced. The challenge is for project implementers to be cognizant of innovations and connect them to town input suppliers, who, in turn, supply villages and independent traders, thus lubricating the exchange of goods between the formal and informal sectors.

Efforts to offer index-based insurance for smallholder farmers are currently in process. Offering this service is risky where pre-harvest buying is prevalent. Almost a quarter of the farmers participate in this activity. In the event of crop failure, traders would likely incur the loss if they pre-paid for the crop, and farmers would conceptually receive a double payment for a lost crop from the insurance company and the pre-payment made by the trader. On the other hand, most traders are farmers and farmers who lost a field would be compensated in their role as a trader.

Priority must be given to the use of IPM practices in order to increase yields and reduce pesticide use. Heavy use of pesticide and fungicides is of great and environmental health concern. It also represents increased costs of production. A study of agrochemical applications on cabbage, tomatoes and onions in northern Tanzania indicated usage similar to their application in Lushoto (Ngowi et al. 2007). Pesticides used included pyrethroid derivatives, organophosphates and carbamates similar to those found in Lushoto's input supply shops. Three insecticides in this study were World Health Organization (WHO) Class Ib (highly hazardous), 20 % contained chemicals suspected of endocrine disruption, 24 % were cholinesterase inhibitors and 7 % carcinogens and potential carcinogens. Neither the Ngowi study or this study assessed the presence of fake pesticides. They are estimated to occupy 40 % of the stock in input supply shops. Extension agents are supposed to educate farmers. However, they have little or no training in the use of IPM practices. Farmers need to be trained to do so as should local input suppliers. They should all have the capacity to disseminate IPM practices and knowledge, bait traps, pheromones, biopesticides etc.

Climate-smart bean varieties have been shown to be more popular than traditional varieties. It is likely that similar results will occur once climate-smart varieties of potatoes are more widely disseminated. Lushoto's current potato varieties were most likely brought by traders, and potato seed continues to be sold to traders. Despite the rule that seeds bulked under the QDS system are supposed to be sold within the confines of village and ward boundaries, the chance is high that traders will disseminate these and other non-commercial seed varieties outside ward boundaries. And that is a good thing. A more elabourate business model for seed bulking and other value addition possibilities could tap into PASS and SIDO expertise and resources.

Adaptive seed systems support formal and informal seed systems, each with its own strengths and weaknesses (Westengen et al. 2014). Commercial seed companies have access to larger genepools for breeding. Seventy-five percent of Tanzanian farmers save and trade seed among themselves in farmer-managed seed systems (FMSSs) (Hella 2015). With a more restricted genepool, farmers continue to select seeds adapted for local conditions and trade among themselves, thus reducing costs of production. The QDS system is a hybrid system that assists farmers in bulking publicly bred varieties made from wider germplasm pools. However, the trend is for commercial companies to supplant locally produced varieties with high-yielding, commercially bred hybrids, particularly maize. FMSSs and local varieties are important to Lushoto farmers. However, when treasured varieties are lost due to climate events or market effects, no mechanisms exist to bring banked germplasm back to the field. It is important for those implementing CSA to lend support to the entire seed system and to help farmers maintain agrobiodiversity-associated FMSSs while simultaneously providing them with climate-adapted varieties.

Introducing higher-yielding varieties can involve trade-offs. These varieties can result in increased pesticide use, proportionately lower percentages of consumer food dollar going to farmers and lower prices for farmers as supply increases. Even farmers who fail to adopt the alternative maize and beans seeds are likely to bear some of the costs. As the supply of agricultural produce increases, prices can decrease (Cayford 2004). If a food crop becomes a high-value commodity export, the new price of a once affordable staple can become out of reach for many farmers as was the case of quinoa produced in Bolivia and Peru for American markets. In this case study, we detect a similar effect albeit localized for local bean varieties.

References

Barrett et al (2002) The challenge of stimulating adoption of improved natural resource management practices in African agriculture. In Barrett CB, Place F, Aboud AA (eds) Natural resources management in African agriculture: understanding and improving current practices. Oxon, UK: CAB International Publishing, pp 1–21

Cayford J (2004) Breeding sanity into the GM food debate. Issues Sci Technology 20(2). Avaliable at http://issues.org/20-2/cayford/:

Challinor AJ, Watson J, Lobell DB, Howden SM, Smith DR, Chhetri N (2014) A meta-analysis of crop yield under climate change and adaptation. Nature Clim Change 4:287–291

Food and Agriculture Organization (2010) Private sector agribusiness investment in sub-Saharan Africa. Rural Infrastructure and Agro-Industries Division, Rome

Funk et al (2008) Warming of the Indian Ocean threatens eastern and southern Africa food security but could be mitigated by agricultural development. Proc. Natl. Acad. Sci USA 105 11081-11-86. doi:10.1073/pnas.0708196105

Global Harvest Initiative (2011) Enhancing private sector involvement in agriculture and rural infrastructure development: a policy paper. www.globalharvestinitiative.org. Accessed 20 July 2015

Gyau AM, Freeman OE, Mbow C, de Leeuw J, Catacutan D (2015) Landscape approaches to sustainable supply chain management: the role of agribusinesses. In: Minang PA, van Noordwik M, Freeman OE, Mbow C, de Leeuw J, Catacutan D (eds) Climate-smart landscapes: multifunctionality in practice. World Agroforestry Centre (ICRAF), Nairobi, pp 295–305

Hall J, Burgess N, Lovett J, Boniface M, Gereau R (2009) Conservation implications of deforestation across an elevational gradient. Biol Conserv 142:2510–2521

Hella J (2015) Farmer Managed Seed Systems in Tanzania. A report commissioned by Tanzania Organic Agricultural Movement (TOAM)

Hyden G (1980) Beyond Ujamaa in Tanzania: Underdevelopment and an uncaptured peasantry. Berkeley: University of California, 270 pp

International Labour Organization (2008) Tanzania mainland. Social protection expenditure and performance review and social budget. ILO, Geneva

IPCC (2014) Climate change. In: Field CB, Barros VR, Dokken DJ, Mach KJ, Mastrandrea MD, Bilir TE, Chatterjee M, Ebi KL, Estrada YO, Genova RC, Girma B, Kissel ES, Levy AN, MacCracken S, Mastrandrea PR, White LL (eds) Impacts, adaptation, and vulnerability. Part A: global and sectoral aspects. Contribution of working group II to the fifth assessment report of the Intergovernmental Panel on Climate Change. Cambridge University Press, Cambridge

Jayne TS Jones S (1997) Food marketing and pricing policy in eastern and southern Africa: a survey. World Development 25.9:1505–1527

Jayne TS, Mather D, Mghenyi E (2006) Smallholder farming under increasingly difficult circumstances: policy and public investment priorities for Africa, MSU International Development Working Paper No. 86, 40 pp

Johansson L (2001) Ten million trees later: land use change in the west usamabara mountains. The soil erosion control and agrofrestry project in Lushoto District 1981–2000. Eschborn: Deutsche Gesellscraft fur Technische (GTZ)

Kristjanson P et al (2012) Are food insecure smallholder households making changes in their farming practices? Evidence from East Africa. Food Security 4.3:381–397

Mtengeti EJ, Eva M, Frank B, Lars OE, Ramadhani C (2015) Effect of improved plant nutrition on maize and rice grain chemical composition under smallholder farmingsystems in Tanzania. In: Conference presentation at international conference on climate change and multi-dimensional sustainability in African Agriculture, Morogoro, 3–5 June 2015

Muller J (1980) Liquidation or consolidation of indigenous technology: a study of the changing conditions of production of village blacksmiths in Tanzania. Aalborg: Universitetsforlag

Muller M (2004) The political dynamics of the informal sector in Tanzania, 2nd Module Project. International Development Studies, Roskilde University Center

Namwata BML, Masanyiwa ZS, Mzirai OB (2012) Productivity of the agroforestry systems and its contribution to household income among farmers in Lushoto District, Tanzania. Int J Phys Soc Sci 2(7):369–392

Ngowi AVF, Mbise TJ, Ijani ASM, London L, Ajayi OC (2007) Pesticide use by smallholder farmers in vegetable production in Northern Tanzania. Crop Prot 26(11):1617–1624

Ngwediagi P (2009) Establishment of a plant breeders' rights system in Tanzania: achievements and challenges. In: Institutionalization of intellectual property management: case studies from four agricultural research institutions in developing countries. CAS-IP, Rome, pp 10–19

North Douglas (1990) Institutions, institutional change and economic performance. Cambridge University Press, Cambridge

Pretty J, Bharucha ZP (2015) Integrated pest management for sustainable intensification of agriculture in Asia and Africa. Insects 6:152–182

Sitko NJ, Jayne TS (2014) Exploitative briefcase businessmen, parasites, and other myths and legends: assembly traders and the performance of maize markets in eastern and southern Africa, World Development 54:56–57

Temu X (2013) The characterization of elements that have successfully linked farmers to markets: selected case, Private Agricultural Sector Support (PASS), Tanzania

The Guardian (2015) EU must pull its weight to help create a better global financial system. http://www.theguardian.com/global-development/2015/apr/20/eu-financing-for-development-global-financial-system. Accessed 17 July 2015

Westengen O, Ring K, Berg P, Brysting A (2014) Modern maize varieties going local in the semi-arid zone in Tanzania, BMC Evolutionary Biology, 14.1. 12pp. http://www.biomedcentral.com/1471-2148/14/1

Chapter 29
The Role of Selected Agroforestry Trees in Temperature Adaptation on *Coffea arabica*: A Case Study of the Moshi District, Tanzania

Jacqueline Kajembe, Ignas Lupala, George Kajembe, Wilson Mugasha and Faraji Nuru

Abstract This study was conducted to assess the soil and air temperature levels of dominant tree species in agroforestry systems that influence *Coffea arabica* productivity. A purposive sampling technique was employed to select villages and farms from different agroecological zones. The data collection methods included questionnaires and focus groups with key informants. Soil and air temperatures were measured using temperature sensors. General linear models in SAS were employed to analyze the temperature data and Statistical Package for Social Science for socio economic data. Dominant tree species included *Grevillea robusta*, *Albizia schimperiana* and *Rauvolfia caffra*. There was a significant difference in soil and air temperature regulation among tree species in the midland ($p < 0.05$), with mean temperature differences of 0.5–1.6 and 0.2–0.4 °C for soil and air temperature, respectively. *G. robusta* significantly regulates soil and air temperature in both highland and midland zones ($p < 0.05$) compared to the other studied tree species that had mean temperature differences of 0.2–1.6 and 0.3–0.4 °C for soil and air temperature, respectively. Since moderate temperature favours Coffee productivity, *G. robusta* is recommended in both the midland and highlands; however, it is imperative to investigate how the soil impacted by *G. robusta* affects coffee productivity.

Keywords Agroforestry systems · Temperature · *Coffea arabica* · Agroforestry trees

J. Kajembe (✉) · W. Mugasha · F. Nuru
Tanzania Forestry Research Institute, P.O. Box 1854, Morogoro, Tanzania
e-mail: kajembej@yahoo.com

I. Lupala
Tanzania Forest Services Agency, Morogoro, Tanzania

G. Kajembe
Department of Forest Mensuration and Management,
Sokoine University of Agriculture, Morogoro, Tanzania

R. Lal et al. (eds.), *Climate Change and Multi-Dimensional Sustainability in African Agriculture*, DOI 10.1007/978-3-319-41238-2_29

29.1 Introduction

Climate change is accelerating at a much faster rate than previously thought (IPCC 2014). This is particularly true for small farmers in developing countries whose economic wellbeing and food security hinges on farming. Because of this and its high dependence on climate, agriculture has received a great deal of attention promoting studies and debates over how developing countries might adapt to the impact of microclimate and climate changes (Van Noordwijk et al. 2011). Temperature, as one aspect of the microclimate, is taken as a priority simply because the Intergovernmental Panel on Climate Change (IPCC) in its fifth assessment expects a probable temperature rise in the near future (IPCC 2014). As temperature goes up, its impacts on agricultural crops is expected to be more significant, with implications for millions of smallholder farmers including the increasing burden of agricultural diseases and pests (Doering 2002; Van Noordwijk et al. 2011). Cash crops such as tea and coffee are expected to be the most affected by increasing temperatures (Lott et al. 2009).

Coffee used to be the largest export crop and top source of general export earnings in Tanzania (URT 2002). Despite the prevailing trend of decreasing productivity and declining quality, coffee still makes a significant contribution to smallholder livelihoods (Tanzania Coffee Board 2012). About 95 % of coffee produced in Tanzania is grown by smallholders, while about 5 % is grown on estates. Tanzania produces approximately 67 % of *Coffea arabica* and 33 % of *Coffea robusta*; the former is produced in the Ruvuma, Mbeya, Arusha, and Kilimanjaro regions, and the latter is produced primarily in the Kagera region (Mhando et al. 2013).

Many natural systems are affected by regional climate changes, particularly temperature increases and changes in rainfall patterns. For coffee-producing areas, projections produce alarming scenarios. While the precise impact cannot yet be described, rising temperatures will accelerate the ripening process and thus lead to a decline in product quality (Tanzania Coffee Board 2012). For the case of *Coffea arabica*, the optimal temperature range is 18–21 °C, thus any further increase of temperature would seriously affect the viability of coffee crops (Lin 2007). Temperatures above 24 °C greatly reduce the net photosynthesis; above 23 °C, the development and ripening of the fruit are accelerated, leading to loss of coffee quality (Lott et al. 2009).

Agroforestry (combining trees/shrubs with agricultural crops and/or livestock on the same land) is increasingly recognized as an effective approach for minimizing production risksunder climate variability and change (Verchot et al. 2007; ICRAF 2008). It has attracted considerable attention because of its potential to maintain or increase productivity in agriculture land. People often assume that appropriate agroforestry trees can provide the essential ecological functions that are needed to ensure sustainability and maintain microclimatic conditions (Kidd and Pimentel 1992). Agroforestry provides a financially and ecologically viable way of protecting crops like *Coffea arabica* in areas where microclimatic factors regularly exceed the optimal range (Lin 2007). The trees used in agroforestry are known to bring about changes in microclimatic conditions and other components of the ecosystem through biorecycling of mineral elements and environmental modifications (Ogunkunle and Awotoye 2011).

According to Rao et al. (2007), different trees species in agroforestry systems have different capacities to regulate temperatures and crop production. The importance of trees in microclimate regulation has been studied by Lin et al.(2008), who showed that mixing trees with coffee plants decreases the temperature and humidity fluctuations and reduces the vulnerability to water stresses. Missing from the equation, though, is the identification of which tree species, when integrated with coffee, perform best with regard to the regulation of temperature levels and it its influence on coffee productivity in Tanzania.

The aim of this study is to assess the temperature levels of the dominant tree species in agroforestry systems that could potentially influence the productivity of *Coffea arabica*. The specific objectives of the study were to identify the tree species commonly used in agroforesty systems in different agroecological zones, determine the soil and air temperature regulation ability of dominant tree species under *Coffea arabica* in a given agroecological zone, and assess the ability of a given tree species to regulate air and soils temperature levels in different agroecological zones. Under this study, the following hypotheses were tested:

- *H1*: The soil and air temperature regulation ability of dominant tree species vary significantly in a given agroecological zone.
- *H2*: The soil and air temperature regulation ability of dominant tree species does not vary significantly in a given agroecological zone.

29.2 Materials and Methods

29.2.1 Description of Study Area

The research was conducted in the Moshi District located in the northeastern part of the Kilimanjaro region of Tanzania, as shown in Fig. 29.1. The district lies between 3°03′ and 3° 20′ S and 37°15′ and 37°21′ E and boasts a population of 504,287 (URT 2003). In this district, the temperature and altitude ranges are classified according to agroecological zones, as shown in Table 29.1: highland (where the Chagga home gardens are located), the midlands, and the lowlands. Food and cash crops are produced in all of the agroecological zones. The major food crops include bananas and maize; the main cash crops include coffee and cotton. Coffee is grown in highland and midland zones. Cotton is chiefly produced in the lowland zone, but its production has not been expanding because of uncertainty in the market (Daldo 2011).

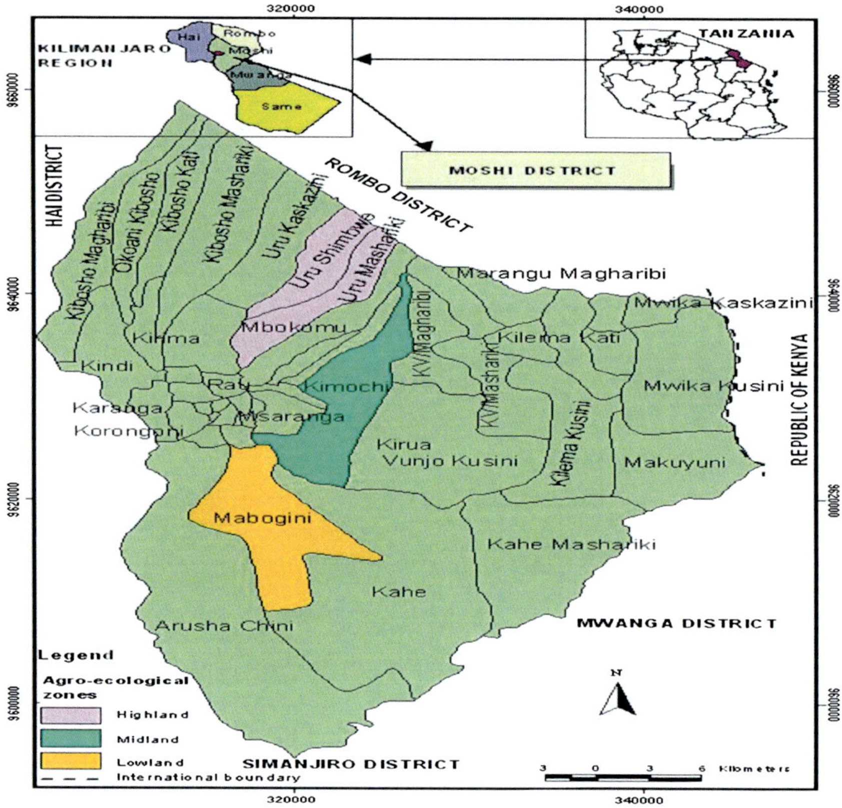

Fig. 29.1 Map of Moshi District with locations of study areas

Table 29.1 Data based on agroecological zones in Moshi District

Agroecological zones	Altitudes (m)	Temperature (°C)	Main crop	Rainfall (mm)
Highland	1500–1800	15–20	Coffee, bananas	1200–2000
Midland	1100–1500	20–30	Maize, beans, coffee	1000–1200
Lowland	900–1100	>30	Maize	400–900

Source Zongolo et al. (2000), United Republic of Tanzania (URT) (2003)

29.2.2 *Study Design and Sampling*

The study area was stratified into three agroecological zones highland, midland, and lowland based on climatic condition because the tree compositions in agroforestry systems vary greatly based on the zones' topography and climatic conditions. A purposive sampling technique was employed to select sample villages and farms based on the availability of smallholder farmers that integrate trees with coffee in their agroforestry systems. The highland and midland zones were selected because the majority of smallholder farmers integrate trees with agricultural crop in their farms. According to Soini (2006), the highland zone features a population density of about 650 persons per sq km versus the lowland zone's 250 persons per sq km. This would most likely make land in the highland zone scarcer; thus, to maximize production, farmers prefer to integrate trees in their farmlands.

For the selected agro-ecological zones highland and midland one village was selected from each zone i.e. Shimbwejuu in the highland and Mwasikusini in the midland. Socioeconomic and biophysical studies were carried out within the selected villages based on the identification and assessment of dominant/preferred agroforestry trees in regulating soil and air temperature. Farmers practicing agroforestry were purposefully selected for the socioeconomic study. In each selected village, three farms were randomly selected from farmers who were involved in the socioeconomic study to participate in the biophysical study.

29.2.3 *Data Collection Methodology*

This study employed a socioeconomic survey and a biophysical study. The methods used in each part were different.

29.2.3.1 Socioeconomic Survey

The socioeconomic study involved qualitative and quantitative data collection. The qualitative data were collected by using the participatory rural appraisal (PRA) approach through field observation, focus group discussions (FGDs), and key informant discussions; the quantitative data were collected using structured questionnaires. A field observation and FGDs were conducted with fifteen key informants from each village by using a checklist of probe questions to guide the identification of commonly used tree species that are integrated with coffee crops in the study areas. According to Kayunze (2003), FGD involves a small number of respondents, usually six to fifteen, under the guidance of a moderator. For the purpose of this study, the key informants included the Ward and Village extension officers, contact farmers, Ward and Village executive officers (WEOs and VEOs), and the Tanzania Coffee Research Institute (TaCRI). The field survey was employed to compare what the farmers had reported with what was observed on the farms. According to (Kothari 1990), field surveys are essential in correlating the community response to the actual phenomena on the ground. A semi-structured questionnaire was also administered to selected farmers to capture information in quantitative forms.

Biophysical Study

The biophysical study focused on assessing the potential of selected dominant/preferred agroforestry tree species in regulating the soil and air temperature. This was accomplished by installing temperature sensors (Plate 29.1) to gather temperature data from a source (e.g., soil or air) and convert it to a form that can be understood by an observer. The temperature sensors used in this study were from Decagon Devices Company. Simultaneous measurements of the soil and air temperatures were carried out in a 1-day campaign at six selected sites, each on a different day as resources allowed only one set of equipment. The soil and air temperature levels were measured under identified dominant tree species integrated with coffee in the study area (Fig. 29.2).

The temperature sensors were placed in the specified plots based on the distance and species canopy of the dominant tree. Soil temperatures were measured at a soil depth of 5 cm, and air temperature was measured from the vantage point of a wooden stick 1.5 m above the ground in a 5 m × 5 m plot. The measurements were taken at two locations (plots) on a farm: one under full shade and the other under no shade as suggested by Wilson et al. (1997). According to Shashua-Bar and

Plate 29.1 Temperature sensors and data logger

Fig. 29.2 Tree area where soil and air temperature measurements were taken

Hoffman (2000), different levels of shading produce different levels of cooling effects. Data were recorded by dedicated logger (Em50 series data loggers) at 1-h intervals for 24 h to observe changes in temperature. To prevent the air temperature sensors from direct solar radiation or other influences, a radiation shield was used to contain the sensors (Plate 29.2).

Plate 29.2 Measurement of soil and air temperature using sensors in Shibwejuu village

29.2.3.2 Data Analysis

Socioeconomic Data

Data collected through FGDs were subjected to content analyses, which are methods for analyzing the symbolic content of communication. The basic idea is to reduce the total content of the communication to categories (Singleton et al. 1993). Through this method, the data collected through verbal discussions with key informants were analyzed in detail, whereby the recorded dialogue was broken down into the smallest meaningful units of information.

Temperature Data

Since the soil and air temperatures were recorded for 24 h, the first step was to calculate the average temperature difference (open area minus shade) at specific times; for each measurement made under a tree shade, there was a corresponding reading in the open area. The difference between the two corresponding measurements taken simultaneously at the two sites was considered to be partly attributable to the shade tree; thus the difference in air and soil temperature values was subjected to a one-way analysis of variance using general linear models (GLM) procedure in the Statistical Analysis System (SAS 1991). Air and soil temperature differences were dependent variables, the location and species were independent variables, and the farms were random variables, as presented in Eq. 29.1 below.

$$Y = R + a \times L + b \times TS + \varepsilon \tag{29.1}$$

where,

Y = Soil or air temperature difference
R = Replication (farms)
L = Location
TS = Tree species
ε = Error.

29.2.4 *Study Findings and Discussion*

29.2.4.1 Preferred Tree Species Used in Agroforesty Systems Under Different Agroecological Zones

Tree species used for shading coffee crops in the Moshi District of Tanzania are shown in Table 29.2. Of the twenty-four species listed, the farmers of these species consider *Grevillea robusta, Albizia schimperiana,* and *Rauvolfia caffra* to be the

Table 29.2 Treespecies used for shading purposes on *Coffea arabica* in Moshi District

Local name	Scientific name	Other uses
Mruka	*Albiziaschimperiana*	Timber, fuelwood, soil fertility improvement
Mwerezi	*Grevillearobusta*	Fodder, fuelwood, timber, poles
Mringaringa	*Cordiaafricana*	Timber, fuelwood, poles
Msesewe	*Rauvolfiacaffra*	Timber, medicinal, brewing
Mwembe	*Mangiferaindica*	Fruit, fuelwood, shades
Mparachichi	*Perseaamericana*	Fruit, fuelwood, fodder
Mfuruanji	*Albiziagummifera*	Timber, poles, fuelwood, soil fertility improvement
Msamana	*Pauridianthaholstii*	Medicinal value, firewood
Mshamana	*Margaritariadiscoidea*	Fodder, soil fertility improvement
Mchio	*Oleacapensis*	Timber, fuelwood, poles
Tundadamu	*Prunusdomestica*	Ediblefruit
Mlimao	*Citrus limon*	Ediblefruit, firewood
Mchungwa	*Citrus sinensis*	Ediblefruit, firewood
Mvule	*Khayaanthotheca*	Timber, poles, fuelwood
Mfurufuru	*Crotonmacrostachys*	Firewood, medicinal, shade
Mfenesi	*Artocarpusheterophullys*	Ediblefruits, firewood
Mdalasini	*Peunuspersica*	Spices, firewood, fodder
Mwarobaini	*Azadirachtaindica*	Medicinal value, firewood
Mlusina	*Leucaenaleucocephala*	Fodder, soil fertility improvement
Mstafeli	*Artocarpusheterophyllus*	Ediblefruit
Mkrismass	*Delonixregia*	Firewood, ornamental
Msederela	*Cedrelaodorata*	Timber, fuelwood
Iber"a	*Psidium guajava*	Ediblefruit, firewood

most dominant tree species used in agroforestry systems in the highland zone and *Grevillearobusta* and *Albiziaschimperiana* to be the most dominating the midland zone.

Both indigenous (*Albizia schimperiana* and *Rauvolfia caffra*) and exotic (*Grevillea robusta*) tree species were identified by the smallholder farmers as the most common tree species integrated with coffee in the study area. The reason behind the selection of those tree species was because they have a more positive impact on surrounding communities' daily life compared to the other species. Akbari and Taha (1992) show the importance of trees incorporated with different crops in the agricultural land as having a hugely positive effect on people's lives, including the amelioration of microclimates. For instance, *Albizia schimperiana* is normally incorporated in agroforestry systems as a shade tree for coffee crops and that this species is in the process of being promoted in agroforestry and plantations as substitute timber. Aiming to protect coffee environmental disasters, Caramori et al. (2004) noted that some of the species, such as *Gravillea robusta*, have been used in different parts of the world; one example is Brazil, where it was proven to have satisfactory results when integrated with coffee on the same land. Also, most of the

Table 29.3 Farmers' criteria to identify preferred trees in highland and midland zones

No.	Tree characteristics	Farmers' preference
1	Organic matter	The one that add a lot of organic matter
2	Crown size	Moderate crown size
3	Shade quality	Moderate light
4	Leaf decomposition	Fast leaf decomposition

tree attributes that are considered favorable as shade trees by farmers in the study area are similar to what was reported in Costa Rica by Albertin and Nair (2004).

Most of the reasons given by these farmers when identifying the most dominant tree species were based on their knowledge, experience, and needs that they think should be achieved on their farms. This indicates that local farmers possess knowledge of the morphological, physiological, and ecological features of shading trees (Soto-Pinto et al. 2007), and thus tree selection is a deliberate act and not just arbitrary shade reduction.

The addition of organic matter to their farms is a major reason given by all interviewed farmers for using trees with coffee (Table 29.3). This is important for the farmers since they have been cultivating the same land for a long time; also, these areas are located on steep slopes that have a high risk of soil erosion and nutrients loss. To have sustainable production, a nutrient exchange between the soil, water, and organic matter is essential to fertility and needs to be maintained.

Farmers preferred moderate tree-shade conditions for coffee growth since photosynthetic rates are generally maximized at this stage (Beer et al. 1998). Albertin and Nair (2004) conducted their study in Costa Rica and supported the use of moderate light since it discourages coffee diseases and it was important for fruit filling and suppressing the wilting of the coffee.

29.2.4.2 Soil and Air Temperature Regulation Ability of Dominant Tree Species

The performance of dominant tree species on regulating thes oil and air temperature levels in a given agroecological zone is presented in Table 29.4. The results indicate that there was no significant difference in air and soil temperatures among the tree species in the highland (p values > 0.05). However, there was significant difference in air and soil temperatures among tree species in the midland zone (p values < 0.05).

The results demonstrate that both *Albizia*and *Grevillea* species perform best in the midland zone versus the highland zone. This can be explained by the lapse rate, which is the change in temperature while moving upwards through the atmosphere. Generally, the temperature declines by 0.6 °C for each 100 m increase in elevation, but it may vary as a function of moisture. It declines by 1 °C per 100 m for dry air and drops about 0.5 °C per 100 m when the air is saturated with moisture (Gommes 2002).

Table 29.4 Performance of dominant tree species on regulation of soil and air temperature levels

Location	Tree species	Soil temperature (°C)	*p* value	Air temperature (°C)	*p* value
Highland zone	*Albizia schimperiana*	0.33[a] (0.9597)*	0.7077	0.0282[a] (0.5737)	0.2204
	Grevillea robusta	0.23[a] (0.9724)		0.3212[a] (1.2704)	
	Rauvolfia caffra	0.23[a] (0.6934)		−0.0164[a] (1.6929)	
Midland zone	*Albizia schimperiana*	0.45[b] (1.0357)	<0.0001	0.1571[b] (0.6457)	0.0180
	Grevillea robusta	1.60[a] (0.9620)		0.3945[a] (0.5393)	

*Standard deviation, mean with the same letter are not significantly different

29.2.4.3 Performance of Given Tree Species on Regulating Air and Soil Temperatures in Agroecological Zones

A comparison of the performance of dominant tree species on the regulation of soil and air temperature levels in different agroecological zone is indicated in Table 29.5. The results show that there was no significant difference of *Albizia schimperiana*on the regulation of soil and air temperature in the highland and midland zones (p values > 0.05). However, there were significant differences in *Grevillea robusta* in the regulation of soil and air temperatures in the highland and midland zones (p values < 0.05).

According to our results, the *Grevillea* species performs best in the highland and midland as compared to the *Albizia* species. It was important to determine the tree species that perform best in both agroecological zones because traditional coffee cultivation is associated with low tree species diversity and simplified forest structure: few stems, low canopy height, and low crown closure. Instead of integrating many trees in the farms, which would result in dense canopy and resource competition, only single or a few trees species can be selected and integrated with coffee to produce the maximum coffee yield and other products.

Table 29.5 Performance comparison of dominant tree species on soil and air temperature levels

Tree species	Soil temperature (°C)		*p* value	Air temperature (°C)		*p* value
	Highland	Midland		Highland	Midland	
Albizia schimperiana	0.3354[a] (0.9597)*	0.4464[a] (1.0357)	0.4280	0.0282[a] (0.5737)	0.1571[a] (0.6457)	0.7901
Grevillea robusta	0.2287[b] (0.9724)	1.6037[a] (0.9620)	<0.0001	0.3212 (1.2701)	0.3945[a] (0.5393)	0.0013

*Standard deviation, mean with the same letter are not significantly different

Shade provision and low access to fertilizers often result in the purposeful integration of canopy trees with coffee crops in smallholder farms. Although indigenous species like *Albiziaschimperiana* could provide shade and add fertility through leaf shading and decomposition, its leaf structure is highly characterized by big pores as compared to *Grevillearobusta,* which allows some light to penetrate under the trees where the crops are located. On the other hand, the growth and development of coffee crops is also dependent on the amount of light to which the coffee is exposed. This argument was supported by Muthuri (2004), who indicated that a key criterion when choosing appropriate tree species is leafing phenology.

29.2.5 Conclusion and Recommendation

Based on the results and discussion, it can be concluded that *Albizia schimperiana, Grevillea robusta,* and *Rauvolfia caffra* are the preferred tree species in the highland and *Albizia schimperiana* and *Grevillea robust a*are most preferred in the midland. The performance of *Albizia schimperiana* and *Grevillea robusta* are significant in the regulation of soil and air temperatures in the midland zone as compared to the highland zone. The performance of *Grevillea robusta* is significant in the regulation of soil and air temperatures in both zones; thus *Grevillea robusta* was found to have the potential ability of regulating temperature for both zones.

Given the fact that the temperature data were recorded within few months, the results can only act as preliminary information. The results are far from being definite, and they can best be considered a baseline from which further research can be extended. The results suggest some directions for further study that could prove fruitful: research support to test soil impacted by *Grevillea robusta* affects coffee productivity and support on implementing the research findings through capacity-building for farmers on nursery establishments and tree planting of the recommended tree species.

References

Akbari H, Taha H (1992) The impact of trees and white surfaces on residential heating and cooling energy use in four Canadiancities. Energy 17(2):141–149

Albertin A, Nair PKR (2004) Farmers perspective on the role of shade trees in coffee production systems: an assessment from the Nicoya Peninsula, Costa Rica. Human Ecology 32:443

Beer J, Muschler R, Kass D, Somarriba E (1998) Shade management in coffee and cacaoplantations. Agrofor Syst 38:139–164

Caramori PH, Leal AC, Morais H (2004) Microclimatic characterization and productivity of coffee plants grown under shade of pigeon pea (*Cajanus cajan*) in Southern Brazil. Agrofor Syst 63:75–82

Daldo (2011) District agricultural and livestock development plans. Moshi District Council, Kilimanjaro Region, Tanzania, 22

Doering OC (2002) Effects of climate change and variability on agricultural production systems. Kluwer, Dordrecht

Gommes R (2002) Environment, climate change and bioenergy division. Food and Agriculture Organization of the United Nations

International Centre for Research in Agroforestry (2008) Transforming lives and landscapes. ICRAF strategy 2008–2015. World Agroforestry Centre, Nairobi, Kenya, 48

IPCC (2014) Climate Change 2014: impacts, adaptation, and vulnerability. Part B: regional aspects. In: Barros VR, Field CB, Dokken DJ, Mastrandrea MD, Mach KJ, Bilir TE, Chatterjee M, Ebi KL, Estrada YO, Genova RC, Girma B, Kissel ES, Levy AN, MacCracken S, Mastrandrea PR, White LL (eds) Contribution of working group II to the fifth assessment report of the intergovernmental panel on climate change. Cambridge University Press, Cambridge, p 688

Kayunze KA (2003) Social science research methods. Sokoine University of Agriculture, Morogoro, p 42

Kidd CV, Pimentel D (1992) Integrated resource management in agroforestry for development. Academic Press, San Diego, p 233

Kothari CR (1990) Research methodology: methods and techniques, 2nd edn. New Age International (P) Limited Publishers, New Delhi

Lin BB (2007) Agroforestry management as an adaptive strategy against potentialmicroclimate extremes in coffee agriculture. Agric For Met 144:85–94

Lin BB, Perfecto I, Vandermeer J (2008) Synergies between agricultural intensification and climate change could create surprising vulnerabilities for crops. Bioscience 58:847–854

Lott JE, Ong CK, Black CR (2009) Understorey microclimate and crop performance in a *Grevillea robusta*-based agroforestry system in semi-arid Kenya. Agric For Meteorol 149:1140–1151

Mhando DG, Haller T, Mbeyale G, Ludi E (2013) Adaptation to changes in the coffee value chain and the price of coffee among coffee producers in two villages in Kilimanjaro, Tanzania. Afr Study Monogr 34(1):27–56

Muthuri CW (2004) Impact of agroforestry on crop performance and water resources in semi-arid central Kenya. Ph.D. thesis, Jomo Kenyatta University of Agriculture and Technology, Kenya, 292

Ogunkunle CO, Awotoye OO (2011) Soil fertility under different tree cropping system in a southwestern zone of Nigeria. Not Sci Biol 3:123–128

Rao KPC, Verchot VL, Laarman J (2007) Adaptation to climate change through sustainable management and development of agroforestry systems. World Agroforestry Center, Nairobi, Kenya. An Open Access Journal published by ICRISAT

Shashua-Bar L, Hoffman ME (2000) Vegetation as a climatic component in the design of an Urban Street: an empirical model for predicting the cooling effect of urban green areas with trees. Energy Build 31(3):221–235

Singleton R, Straits B, Straits M (1993) Approaches to social research. Oxford, New York

Soini E (2006) Bird diversity and land use on the slopes of Mt. Kilimanjaro and the adjacent Plains, Tanzania. ICRAF Working Paper No. 11. World Agroforestry Centre, Nairobi, Kenya

Soto-Pinto L, Villalvazo-López V, Jiménez-Ferrer G, Ramírez-Marcial N, Montoya G (2007) Multistrata coffee systemsin Chiapas, Mexico. Biodivers Conserv 16:419–436

Tanzania Coffee Board (2012) Coffee industry development strategy 2011–2021. [http://www.coffeeboard.or.tz/News_publications/startegy_english.pdf]. Retrieved on 10 February, 2016

United Republic of Tanzania (URT) (2003) The 2002 Population and Housing Census, Central Census Office, National Bureau of Statistics, President's Office, Planning and Privatization, Dar es salaam. [http://www.Tanzania.go.tz/census/regions.tm] site visited on 16/3/2012

URT (2002) Coffee Industry Act. Government Printer, Dar Es Salaam. 2012. Macro economic policy framework for five-year development plan/budget 2012/13 2015/16. Ministry of Finance, Government Printer, Dar Es Salaam

Van Noordwijk M, Hoang MH, Neufeldt H, Öborn I, Yatich T (2011) How trees and people can co-adapt to climate change: reducing vulnerability through multifunctional agroforestry landscapes. World Agroforestry Centre (ICRAF), Nairobi

Verchot LV, Van Noordwijk M, Kandji S, Tomich T, Ong C, Albrecht A, Palm C (2007) Climate change: linking adaptation and mitigation through agroforestry. Mitig Adapt Strateg Glob Change 12(5):901–918

Wilson TD, Brook RM, Tomlinson HF (1997). Interactions between NERE (PARKIA BIGLOBOSA) and under-planted sorghum in a Parkland system In: BURKINA FASO Expl Agric (1998), vol 34, pp 85–99

Zongolo SA, Kiluvia S, Mghase G (2000) UmbweOnana PRA report. Traditional Irrigation and Environmental Development Organization, Moshi

Part VI
Agricultural Risk, Insurance and Policy

Chapter 30
Addressing Climate Change Through Risk Mitigation: Welfare Implications of Index Insurance in Northeastern Tanzania

Jon Einar Flatnes and Michael R. Carter

Abstract While index insurance offers a compelling solution to the problem of covariant risk among smallholder farmers in developing countries, most rainfall-based contracts suffer from poor quality due to a low correlation between the index and farmer losses. Moreover, a lack of historical household-level yield data has made it difficult to quantify the level of basis risk and the impact on farmer welfare. This paper utilizes a unique dataset of plot-level historical rice yields in Northeastern Tanzania to estimate the level of basis risk and the welfare implications of two hypothetical index insurance contracts. One is a standard area-yield contract, while the other uses an index based on publicly available high-resolution satellite data that are mapped to actual yields to minimize basis risk. Our results suggest that the satellite index explains approximately 55 % of the variation in zone-level yields across years. Moreover, despite the presence of large basis risk under both contracts, they are each found to improve the welfare of the average farmer in the sample. Finally, we show that the demand for satellite contracts may be as high as 30 % under reasonable assumptions about loading costs and risk preferences.

Keywords Agriculture · Basis risk · Index insurance · Microinsurance · Remote sensing · Tanzania

JEL Classification O10 · O13 · O16 · Q14

J.E. Flatnes (✉)
Department of Agricultural, Environmental, and Development Economics,
The Ohio State University,
Columbus, OH 43210, USA
e-mail: flatnes.1@osu.edu

M.R. Carter
Department of Agricultural and Resource Economics,
University of California, Davis, One Shields Avenue,
Davis, CA 95616, USA

R. Lal et al. (eds.), *Climate Change and Multi-Dimensional Sustainability in African Agriculture*, DOI 10.1007/978-3-319-41238-2_30

30.1 Introduction

An overwhelming body of literature has left no doubt that risk poses one of the greatest threats to development in low-income economies. Particularly small-scale farmers are plagued by risk, as weather variation is the largest source of risk in agriculture (Cole et al. 2013; Giné and Yang 2009) and because such risk is spatially correlated, making local risk sharing mechanisms ineffective and affecting everyone in the community. While index insurance offers a compelling solution to the problem of covariant risk, these products have generally suffered from low demand. For example, Cole et al. (2013) find that the adoption rate for a rainfall based index insurance product offered to smallholder farmers in two regions in India is close to zero. This somewhat disconcerting observation has prompted several empirical and experimental studies attempting to isolate the determinants of the demand for index insurance. While this evidence suggests that price, liquidity, interlinkage with credit, and trust all have an effect on demand, an issue that has only recently received attention in the development literature is the quality of the contract itself. For example, Clarke (2016) developed a theoretical model demonstrating that basis risk, that is, the risk that a farmer suffers a loss, but the insurance index does not trigger a payout, could make it optimal for a risk-averse farmer to decline index insurance coverage. This result stems from the fact that if a farmer suffers a loss, but the insurance index does not trigger, she is worse off in the bad state of the world with insurance than without insurance due to the payment of the insurance premium.

However, measuring the magnitude and impact of basis risk requires detailed historical data on farmer yields, which are typically not available for smallholder farming in developing countries. While there exist some panel datasets based on longitudinal household studies, these typically have only yield data for a limited number of years and are usually not geographically concentrated enough to allow for analysis within a particular insurance zone. Hence, the limited number of studies that estimate basis risk generally use county-level or district-level data, which do not allow for household-level analysis of basis risk.

In this study, we utilize a unique dataset of historical yields for rice farmers in a concentrated area of Northeastern Tanzania to estimate the magnitude of basis risk under two hypothetical index insurance contracts: an area-yield contract and a satellite-based contract. We then compare the performance of both contracts in terms of their ability to smooth the income of farmers and develop a general approach for evaluating the welfare implications of index insurance contracts, given plot-level yield data. Our results show that both the area-yield contract and the satellite-based contract are welfare-enhancing for the average farmer when offered at actuarially fair prices. Moreover, we demonstrate that while the area-yield contract provides the best coverage for farmers, the satellite-based contract may in fact have a higher demand due to the relatively lower cost of obtaining the index data, which is reflected in a lower premium. Specifically, we show that demand for the satellite-based contract may be as high as 30 % under reasonable assumptions about loading costs and risk preferences. Finally, the satellite-based contract exhibits a lower level of design risk than most weather-based contracts that exist on the market today. In particular, our satellite measure of Gross Primary Production (GPP) explains 55 % of the across-year variance in zone-level yields.

This paper makes several contributions to the literature on index insurance in developing countries. To our knowledge, this is the first study to analyze the welfare implications of a satellite-based index insurance contract for smallholder rice farming in developing countries. Moreover, this paper is also one of the first to analyze the impact of basis risk on the welfare of individual farmers using plot-level panel data. Only one other paper (Jensen et al. 2014) studies basis risk on a household level. In their study, they analyze a particular index-based livestock insurance (IBLI) made available to pastoralists in Northern Kenya and find that IBLI reduces exposure to covariate risk due to high loss events by an average of 63 %.

The use of satellite data to estimate losses has received surprisingly little attention in the index insurance literature. Giné et al. (2010) suggest that such an index is likely to have less basis risk than a rainfall-based index and a few informal studies provide some evidence of this (e.g., Carter and Laajaj 2009). In addition, several studies from the remote sensing literature have found varying levels of correlation between satellite-based indices and observed crop yields. For example, Rosema (1993) developed a sophisticated model of evapotranspiration to simulate crop yields and tested his predictions against several years of recorded biomass observations for 25 sites in Mali. While he finds a relatively low R^2 (.2) using farm-level data, a much stronger correlation (R^2 between .68 and .84) was found when using an improved version of his model to predict maize yields in Zambia and Zimbabwe both at a provincial and at communal level for the years 1994 to 1997 (EARS 2012). Using a satellite-based measure of water use efficiency (WUE) and radiation use efficiency (RUE) to model wheat yield in India, Bhattacharya et al. (2011) find an R^2 of .81 and .64 when comparing predicted yields from the RUE model and the WUE model, respectively, to district-level average yield statistics for three seasons (2002/03–2004/05) in 12 selected wheat-growing districts within four agro-climatic zones in India. Moreover, while many studies have used various satellite-based measures to predict crop yields, most relied on county-level data or experimental plot data over a short time period to validate their models. Hence, by using plot-level data, which cover up to 10 planting seasons, we produce more accurate estimates of the relationship between the satellite-based measures and crop yields.

The remainder of the paper is structured as follows. Section 30.2 discusses how basis risk impacts the quality of an index insurance contract, while Sect. 30.3 describes the data and the study area. In Sect. 30.4, we discuss the methodology used to analyze the performance and the welfare implications of the index insurance contracts and present the results. Section 30.5 concludes.

30.2 The Problem of Basis Risk

Insurance contracts are designed to maximize the welfare of farmers by smoothing income fluctuations at the lowest possible cost. In theory, traditional individual indemnity contracts provide perfect coverage against actual losses. However, they are fraught with problems of moral hazard and adverse selection. Extensive monitoring and verification can reduce the magnitude of these problems, but this is often prohibitively expensive, thus making such contracts infeasible for smallholder agriculture. Index insurance contracts, on the other hand, are based on a verifiable index, which is correlated with, but cannot be influenced by, individual outcomes. Hence, index insurance contracts essentially eliminate moral hazard and adverse selection and typically require no monitoring or verification of individual outcomes, making such contracts affordable even when the insured amount is small.

By construction, index insurance contracts are not intended to cover idiosyncratic losses but rather to protect farmers against covariant yield or price shocks beyond the control of individual producers. Therefore, index insurance contracts are not insurance contracts per se but rather hedging instruments, which may not be effective at smoothing income for individual farmers. In the finance literature, the residual level of risk faced by the farmer is referred to as basis risk, which, if sufficiently high, can significantly reduce the value of index insurance for farmers, even to the point where it is welfare-reducing (Clarke 2016).

To better understand the effect of basis risk, consider the total risk that an individual farmer faces. Following Miranda (1991), we can decompose a farmer's individual yield y_{izt} for farmer i residing in zone z in year t into a covariant component and an idiosyncratic component:

$$y_{izt} = \alpha_{iz} + \beta_{iz}(\bar{y}_z - \bar{y}_{zt}) + \epsilon_{izt} \tag{30.1}$$

Here, $(\bar{y}_z - \bar{y}_{zt})$ is the zone-level deviation from the mean yield in year t and represents covariant losses. β_{iz} measures how well the farmer's yield tracks covariant losses, while ϵ_{izt} is an idiosyncratic shock specific to an individual farmer. As demonstrated by Miranda (1991), index insurance may only be effective if β_{iz} is sufficiently large. For farmers with betas below some critical value, even an index insurance contract that perfectly tracks area yields will not be effective and may actually increase the total risk faced by the farmer. However, most index insurance contracts offered to farmers in developing countries are not based on area yields and thus do not perfectly track covariant losses. This imperfect correlation introduces another source of basis risk, which is referred to as design risk. Design risk arises due to the inability of the index to accurately measure covariant losses within a defined insurance zone and manifests itself as an imperfect correlation between the index and actual average zone-level losses. If we assume that farmers face no price risk and that average yields within a zone can be accurately measured, then an index insurance contract based on area yields will, by default, have no design risk. Hence, an area-yield contract will offer the best protection that an index insurance contract can have within a given zone. However, such contracts are typically infeasible due to the high cost of collecting yield data for each insurance period.

Instead, index insurance projects in developing countries have almost exclusively used precipitation-based indices, which have been shown to carry a relatively high level of design risk, with correlations between precipitation and biomass growth at the weather station ranging from .26 to .70 (Sims and Singh 1978; Price et al. 1998). Even lower correlations have been reported for the link between rainfall indexes and crop production (Martyniak 2007; Staggenborg et al. 2008). Making matters worse, weather stations are typically sparse, and the agro-climatic landscape in Sub-Saharan Africa is generally quite heterogeneous, with the result that basis risk increases sharply with distance from the weather station. Smith and Watts (2009) show that if the spatial correlation between yields at two farms is .5, the correlation between the precipitation index and yields at the farm located some distance away from the weather station would be .35 or lower. Moreover, their simulation results indicate that, in this scenario, there is approximately a 60 % probability that a farmer experiencing a severe yield loss (yields less than 50 % of average) will receive no indemnity payout. Furthermore, in a study of 270 weather-based index insurance products in India over the period 1999–2007, Clarke et al. (2012) show that when there is a 100 % loss at the sub-district level, the average claim payment made was only 12 %. These discouraging results prompt the need for more innovative insurance contracts and further empirical research on the impact of basis risk.

30.3 Background and Data

To create an index which minimizes basis risk and to analyze how the proposed contracts would affect the welfare of individual farmers, we need access to historical plot level yields and satellite index data for the same plots. Ideally, we would rely on individual level yield data from longitudinal household surveys; however, to our knowledge, there are no such surveys that cover more than 3–4 years in East Africa. Moreover, none of these existing household surveys include information on the location of plots, which makes it difficult, if not impossible, to link the high-resolution satellite data to individual farmer yields. Instead, we implemented a retrospective yield survey among smallholder farmers to gather historical data on yields and the locations of the corresponding plots. While this approach might be prone to recall error, we feel confident that the data gathered are quite accurate for various reasons. First, we collected data on rice, which is a major cash crop in the study area, and most farmers we interviewed told us that they kept records of historical yields. Second, the interviewers would refer to major events, such as elections and soccer World Cups, to help farmers remember specific years. Finally, farmers were specifically told that there was no penalty for not remembering data. The interview was intentionally kept short to keep the focus on the yield data, and we gathered data on rice yields, fertilizer use, acreage, planting/harvest times, and the occurrence and severity of extreme weather events for the years 2003–2012. Moreover, we asked farmers to indicate on a detailed satellite map the approximate location of their plot(s).

Our study area consists of four wards located east of the Pare Mountains in the Same district in the Kilimanjaro region of Tanzania (see Fig. 30.1) and covers an area of approximately 20 by 5 km. The main crops grown are paddy (rice) and maize, and rice fields are clustered together with little other vegetation or crops. This makes rice the ideal crop for satellite-based index insurance, since pixels will suffer little due to contamination from non-rice vegetation. The rice clusters are also used as the basis for the insurance zones since the agro-ecological conditions are relatively homogenous within each cluster. In total, we define 10 separate insurance zones (see Fig. 30.1). Given the proximity of the Pare Mountains to the west, most of the water enters the valley through rivers coming down from the mountains. Most of the rice fields are irrigated using canals that link to these rivers. This implies that an index insurance contract based on rainfall from weather stations located in the valley would likely fail to predict drought and flood events. While this area can accommodate up to three growing seasons per year, rice is normally only cultivated once annually, typically between mid-November and mid-March, which covers the short rains period between December and February. While both droughts and floods could impacts crops during any part of the growing season, the rice plants are most vulnerable during the first month and a half following planting, which is also when the rains are the most uncertain. There is also significant variation across zones with respect to drought/flood risk. The three northern zones are generally most prone to drought, as they depend directly on one river coming down from the mountains. The four zones in the Ndungu irrigation scheme face the least risk due to a well-developed canal system. Finally, the three southern zones, which are located in a flood plain downstream from a major lake, face a smaller drought risk but a high risk of flooding.

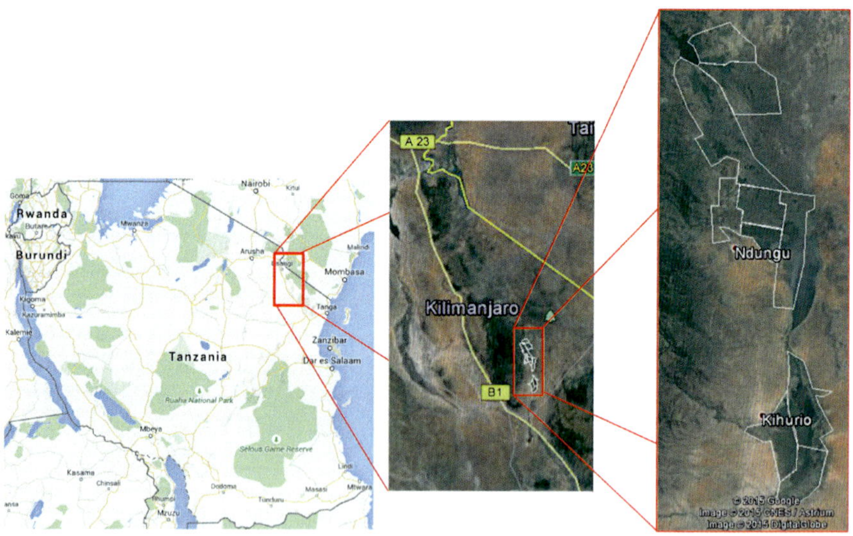

Fig. 30.1 Map of study area

Table 30.1 Summary statistics

Variable	Maore N	Maore SE	Maore SW	Ndungu E	Ndungu N	Ndungu S	Ndungu W	Southern plain N	Southern plain S	Southern plain W	Total
Plot size (acres)	1.25	1.49	1.51	0.92	0.96	0.98	1.44	1.18	1.01	0.86	1.18
Land ownership (%)	97	97	98	92	100	98	100	95	100	91	97
Irrigation use (%)	84	58	87	100	94	97	75	100	100	97	89
Start of season	12-Nov	12-Nov	19-Nov	22-Nov	25-Oct	13-Nov	15-Nov	21-Nov	19-Nov	29-Nov	16-Nov
End of season	20-Mar	27-Mar	1-Apr	17-Apr	26-Mar	15-Mar	15-Apr	14-Apr	9-Apr	7-Apr	3-Mar
Season length (months)	4.3	4.4	4.5	4.9	5.1	4.1	5.1	4.8	4.7	4.3	4.6
Fertilizer use (%)	75	56	57	79	95	88	84	78	60	66	72
Rice yield (kg/acre)	1,653	1,665	1,516	1,721	1,927	1,713	1,630	1,489	1,453	1,545	1,629
Drought index (1–3)	0.81	1.20	0.68	0.35	0.35	0.36	0.67	0.51	0.75	0.59	0.62
Flood index (1–3)	0.43	0.37	0.44	0.24	0.18	0.23	0.25	0.61	0.62	0.68	0.42
Experienced drought (%)	37	56	33	18	18	20	38	27	33	30	30
Experienced drought (%)	19	16	22	12	8	11	17	30	27	31	20
N	37	33	53	12	35	40	4	38	34	32	323

The data presented in this study are from a random sample of 323 farmers from 10 villages (out of 16) chosen at random from local village lists. Farmers were selected proportional to the population in each village and were interviewed in a central location in each sub-village. Table 30.1 contains a set of summary statistics, grouped by each of the 10 insurance zones. We note that average yields vary a fair bit across the zones, with the Ndungu irrigation scheme zones having the highest yields and the southern flood plains having the lowest average yields. Fertilizer use also varies significantly across zones but is quite prevalent in this area. Finally, while there is some variation in planting/harvest dates, most zones tend to plant in mid-November and harvest between mid-March and mid-April.

The satellite measures used to create the primary index are based on high-resolution (between 250 and 1000 m^2 pixels), high-frequency (daily) remote sensing data, which are publicly available, free of charge, from the NASA website. These data are produced by the MODIS satellite, which started recording data in 2001 and is expected to continue to gather data into the foreseeable future. In their raw form, the data cover both the infrared and the visual spectrum and capture data in 36 spectral bands ranging in wavelength from 0.4 to 14.4 μm. Using models that are well-established in the remote sensing literature, these images can be used to create a series of crop health indicators, which have been found to be relatively accurate estimates of the actual measures. In particular, some of the most relevant and commonly used measures of vegetation health include: (1) Normalized Difference Vegetation Index (NDVI), (2) Fraction of Photosynthetic Active Radiation (FPAR), (3) Leaf Area Index (LAI), (4) Evapotranspiration (ET), and (5) Gross Primary Production. While most of these indicators are directly available from the NASA website, converting these data into an index that can be used to predict yields for a particular plot or area requires several steps. In particular, after the index data are either computed using the aforementioned models or downloaded directly from the NASA website, they are adjusted for cloud cover and atmospheric conditions. Then, the daily index data for a season are converted into a single value that can be used to estimate crop yields. This is done by taking the integral of the index data between an assumed planting date and harvest date but adjusting for the background vegetation at planting. Finally, in order to filter out non-rice pixels and to reduce the impact of "bad" pixels, we apply a crop masking model and a spatial smoothing function. Given the massive data processing and complex modeling required to run the data through these steps, we partnered with a private corporation, Vencore Inc., which created a comprehensive model that automatically produces the final index values we need. For a complete description of this work, see Merkovich (2014).

30.4 Methodology and Results

In this section, we analyze a set of hypothetical index insurance contracts in the context of our study area in Northeastern Tanzania. In particular, we study the performance of an area-yield contract based on the historical yield data collected, and a satellite contract that is based on the remote sensing indices described in the previous section. We first define an index insurance contract and then develop and estimate a response function mapping the satellite index values to the yield data. This response function is then used as the basis for the satellite-based index insurance contract. We then analyze and compare the performance of various index insurance contracts and present a method for evaluating the welfare implications of each contract.

30.4.1 Defining an Index Insurance Contract

To create a framework for the remaining analysis, consider an insurance contract C, which makes indemnity payments I_C according to the function $I_C(X_{zt}, p_C)$, where X_{zt} is a zone-level index, and p_C is a set of parameters specifying the relationship between the index and indemnity payments. The actuarially fair premium is equal to $P^{AF}_{C,z} = \int I_C(X_{zt}, p_C)G(X_{zt})dX_{zt}$, where $G(X_{zt})$ is the distribution of the zone-level index X_{zt}. Given sufficient realizations of the index, it is possible to estimate the parameters of $G(X_{zt})$, assuming some underlying distribution, but in practice, $P^{AF}_{C,z}$ may be approximated as $P^{AF}_{C,z} \approx \frac{1}{T}\sum_{t=1}^{T} I_C(X_{zt}, p_C)$, assuming T is sufficiently large.

First, consider an area-yield contract that compensates farmers with an amount equal to actual area-yield shortfalls below the zone mean. Assuming fixed prices, income is equal to yield, which implies that all insurance payouts and incomes can be are denoted in kg/acre. Under this index insurance contract, $I_C(X_{zt}, p_C) = I_C(\bar{y}_{zt}, p_C) = max(\bar{y}_z - \bar{y}_{zt}, 0)$, where $\bar{y}_{zt}$ is the mean yield of all plots in zone z in year t, and $\bar{y}_z$ is the mean of $\bar{y}_{zt}$ over all years in the sample. The blue solid line in Fig. 30.2 plots the normalized indemnity payment $\frac{I_{AY}(\bar{y}_{zt}, p_{AY})}{\bar{y}_z}$ against the normalized index $\frac{\bar{y}_{zt}}{\bar{y}_z}$.

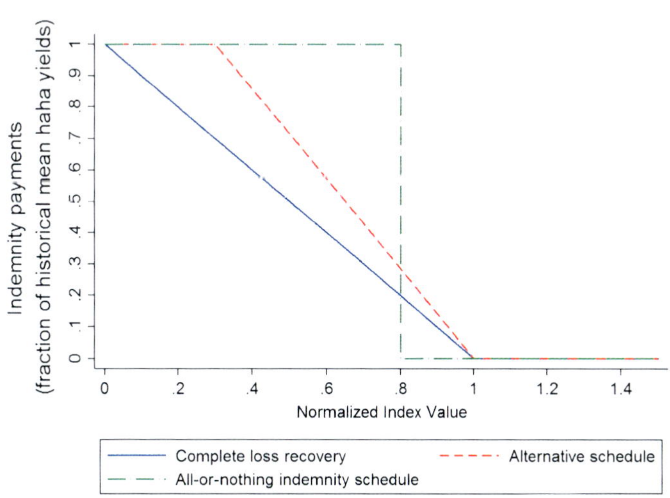

Fig. 30.2 Examples of indemnity schedules

Next, consider a contract for which the index is correlated with, but not equal to, zone-level area yields, such as a rainfall-based or a satellite-based contract. Now, the index is no longer based on $\bar{y}_{zt}$. Instead, define the function $\bar{y}_{zt} = f(X_{zt}) + \epsilon_{zt}$, which maps the realized index values X_{zt} to actual zone-level yields $\bar{y}_{zt}$. Assuming $E[\epsilon_{zt}] = 0$, $f(X_{zt}) = \bar{y}^*_{zt}$ represents predicted yields, which we can use as the index instead of X_{zt}. The process of estimating the function $f(X_{zt})$ is described in Sect. 30.4.2. The magnitude of the design risk under a non-area-yield index insurance contract is now determined by the correlation between $\bar{y}^*_{zt}$ and $\bar{y}_{zt}$, particularly for $\bar{y}_{zt} < \bar{y}_z$. Now, to allow for easier comparison with the area-yield contract, we adjust the indemnity schedule parameters p_{IS}, such that the actuarially fair premiums are the same: $P^{AF}_{IS,z} = P^{AF}_{AY,z}$. The red dashed line in Fig. 30.2 shows an example of how the indemnity schedule for a non-area-yield contract may be adjusted to ensure that the average indemnity payments are the same between the two contracts.

While these index insurance contracts are designed to compensate farmers for any yield shortfalls, they may be less practical to implement in reality, given that payouts depend on losses through a continuous function. Instead, many index insurance contracts use a single (or multiple) strike point(s), below which the insurance pays out a fixed amount. For such a contract, the indemnity payment function may be written as

$$I_C(X_{zt}, p_C) = \begin{pmatrix} \alpha\bar{y}_z & \text{if } \bar{y}_{zt} \leq \hat{y}_{zt} \\ 0 & \text{otherwise} \end{pmatrix} \tag{30.2}$$

where α is a parameter determining the magnitude of the payout, and $\hat{y}_{zt}$ is the strike point. An example of such a contract, with $\alpha = 1$ and $\hat{y}_{zt} = 0.8$, is indicated in Fig. 30.2 as a green dash-dotted line.

30.4.2 Creating a Yield Response Function

In order to minimize the design risk of the satellite-based index insurance contract, we estimate a response function, which maps a satellite index to actual average zone-level data. While a non-parametric method would capture the potential non-linearities in the relationship between the index and yields, we have too few zone-level data points to obtain reliable out-of-sample predictions. Instead, we estimate a simple linear model with zone-level fixed effects:

$$\bar{y}_{zt} = \alpha_z + \beta * Index_{zt} + \epsilon_{zt} \tag{30.3}$$

We estimate Eq. (30.3) for all five indices, both with and without zone-level fixed effects, and as a switching regression with a different intercept and slope for index values below and above the zone mean. Table 30.2 displays the results for a few select regressions with the best fit. Among the five different indices, the GPP

Table 30.2 Regression results for select yield response functions

	[1]	[2]	[3]	[4]
GPP index	0.13*** (0.02)	0.17*** (0.01)	0.10** (0.04)	
GPP < zone mean			−163.6 (109.8)	
GPP < zone mean × GPP index			−0.011 (0.03)	
NDVI index				117.6*** (17.7)
Constant	1564.9*** (54.8)	1490.7*** (7.3)	1594.5*** (69.4)	1253.3*** (48.8)
Zone effects	No	Yes	Yes	Yes
Observations	70	70	70	70
Adjusted R^2	0.284	0.534	0.547	0.468

Standard errors in parenthesis, clustered at the zone level. Dependent variable is zone-level yield
$^{**}p < 0.05$; $^{***}p < 0.01$

index appears to perform the best. This is not unexpected, given that GPP is the most direct measure of produced biomass. Moreover, allowing for a different intercept and slope for index values below the zone mean significantly improves the fit. Finally, the zone-level fixed effects are jointly significant, which is consistent with the idea that the different growing conditions in the different zones affect the relationship between GPP and yield. Comparing the adjusted R^2 of the different models, we find that model Clarke (2016) exhibits the best in-sample fit of the data with an adjusted R^2 of 0.547.

Based on the estimates from this model, we can create predictions of zone-level yields. These predictions are then used as the primary index for the satellite-based insurance contract. Figure 30.3 shows a scatter plot of actual zone-level yields versus predicted yields with a 45 degree line superimposed. It is clear from the plot that the model is effective at distinguishing good years from bad years. In particular, consider all the zone-years for which the actual yields were lower than 1300 kg/acre (corresponding to a 20 % loss relative to the overall mean). All these data points have predicted yields lower than the overall mean of 1600 kg/acre. Hence, an index insurance contract that pays out for any year for which the predicted yield based on the satellite index is lower than the mean would correctly target years when losses have actually occurred. However, the model is less effective at predicting the severity of losses. In particular, for zone years where actual yields were lower than 1300 kg/acre, there is no correlation between actual yields and predicted yields. While we have no good explanation for this result, we believe it might stem from the imperfect biological relationship between produced biomass and yields under extreme conditions. For example, during a drought, the plant will first reduce the production of grains while using the available water to maintain the health of the plant itself, thereby resulting in a yield shortfall while the biomass remains the same.

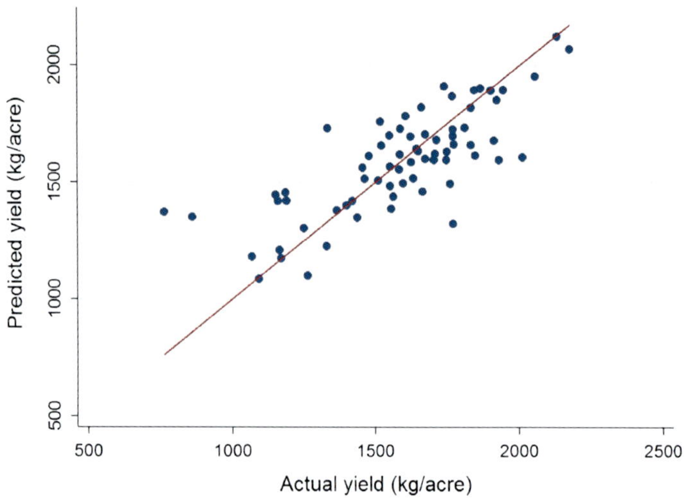

Fig. 30.3 Predicted versus actual area yields in our study area

30.4.3 Relative Contract Performance

When evaluating the quality of index insurance products, it is useful to first consider objective measures of insurance quality such as actual payouts and the distribution of losses under various contracts. The advantage of this approach is that it does not require any assumptions about how farmers value such payouts or the distribution of final incomes with and without insurance. In the next section, we move beyond objective distributions and estimate the welfare impact of different contracts by assuming an underlying utility function.

We begin by studying the design risk of the different contracts described in Sect. 30.4.1. By construction, index insurance cannot protect farmers against idiosyncratic shocks; however, as we show in the subsequent section, the magnitude of idiosyncratic risk relative to covariant shocks still affects the value of index insurance to farmers. To study design risk, we assume that all farmers within a zone are identical and that the only risk facing the farmers is covariant risk. This is equivalent to saying that the farmers only care about zone-level incomes, which would be true under a perfect risk-sharing arrangement within the zone. We then calculate zone-level payouts $I_C(X_{zt}, p_C)$ for both the continuous and the discrete payout function and then compute the resulting incomes π_{zt} for the area-yield contract and the satellite-based contract. This allows us to study insurance payouts as a function of actual normalized losses and the distribution of zone-level incomes after insurance relative to no insurance.

Figure 30.4 shows a plot of zone-level insurance payouts $I_C(X_{zt}, p_C)$ as a function of normalized zone-level yields $\frac{\bar{y}_{zt}}{\bar{y}_z}$ for the satellite-based contract relative to the area-yield contract under the continuous payout function. By construction, the

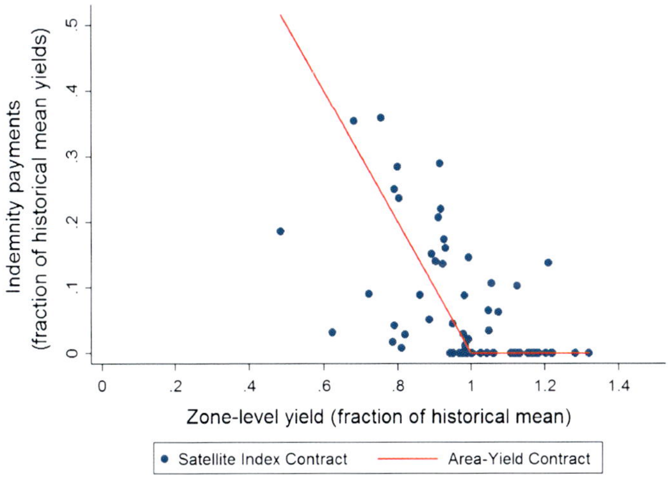

Fig. 30.4 Indemnity payments by zone-level yields for a satellite contract relative to an area-yield contract under a continuous payout schedule

area-yield contract does not carry any design risk; hence, payouts are exactly equal to losses, as indicated by the red solid line. The payouts from the satellite-based contract as a function of normalized zone-level yields are indicated by the blue dots. This plot clearly illustrates the finding from the previous section that while the satellite model is effective at separating good years from bad years, it is less effective at estimating the severity of losses. In particular, 23 out of 25 zone-years with less than 95 % of average yields would receive some kind of payout under the satellite-based insurance contract. However, out of those 25 zone-years, 10 will receive insurance payouts that do not cover at least 95 % of the losses. With regard to false positives, out of 28 zone-years with a higher than 105 % of average yields, only 3 would falsely receive an insurance payout under this contract.

Next, consider the discrete payout schedule. Figure 30.5 plots normalized payouts as a function of normalized zone-level yields for both the area-yield contract and the satellite-based contract. The strike point for each contract is calibrated such that the premiums are equal to that of the area-yield contract with complete loss recovery.[1] Given that the strike point is set lower than the average yield in the zone, the satellite contract now misses several high-loss events, including two of the worst years in the sample. This is again consistent with the idea that the satellite index is unable to pick up the severity of losses. If, instead, we set the strike point equal to 100 % of the average yields, we eliminate the false negatives; however, this also leads to a multiple-times increase in the premium due to false positives.

[1]Only the strike point, $\hat{y}_{zt}$ is adjusted (to $\hat{y}_{zt} = 0.8$ for the area-yield contract and to $\hat{y}_{zt} = 0.87$ for the satellite contract); α is fixed at 1.

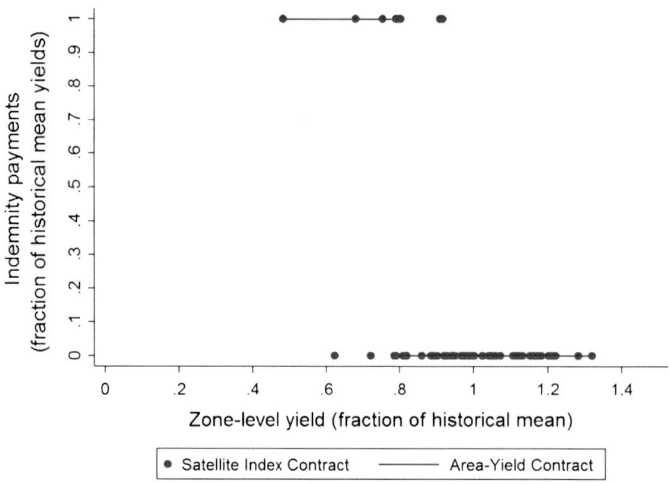

Fig. 30.5 Indemnity payments by zone-level yields for a satellite contract relative to an area-yield contract under a discrete payout schedule

Another, and perhaps more useful, way to view these data is to look at the distributions of final zone-level incomes after indemnity payments are received and premiums are paid. In particular, final incomes, π_{zt}, are calculated as

$$\pi_{zt} = \bar{y}_{zt} + I_C(X_{zt}, p_C) - P_{C,z} \tag{30.4}$$

For ease of comparison, we assume that the premiums are all actuarially fair. Figure 30.6 shows the cumulative distribution function (CDF) of normalized incomes under no insurance, under the area-yield contract, and under the satellite contract, assuming a continuous payout structure. Even after premiums are paid, both contracts offer an improvement over no insurance at the lower end of the tail. In particular, both contracts second-order stochastically dominate the no-insurance option, which implies that if farmers care only about covariant risk (which is a strong assumption), all risk-averse farmers would prefer to buy either of the insurance contracts.

Now, consider the all-or-nothing contracts. Since the final income also includes the premium, the premiums no longer have to be equal to allow fair comparison between the contracts.[2] Figure 30.7 shows the CDF of normalized incomes under the all-or-nothing contracts. As expected, the all-or-nothing payout structure reduces the performance of the area-yield contract, since the continuous area-yield contract is, by construction, optimal. However, the performance of the satellite-based contract improves significantly under the all-or-nothing contract, as

[2]We tried several different combinations and found that the best performance was achieved when $\hat{y}_{zt} = 0.8$ and $\alpha = 0.3$ for the area-yield contract and $\hat{y}_{zt} = 1$ and $\alpha = 0.3$ for the satellite contract.

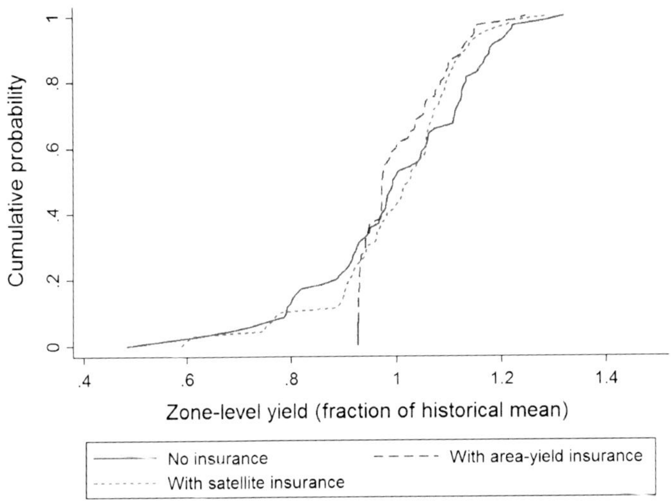

Fig. 30.6 CDF of zone-level yields with and without insurance for contracts with continuous payouts. Distributions assume that premiums are actuarially fair

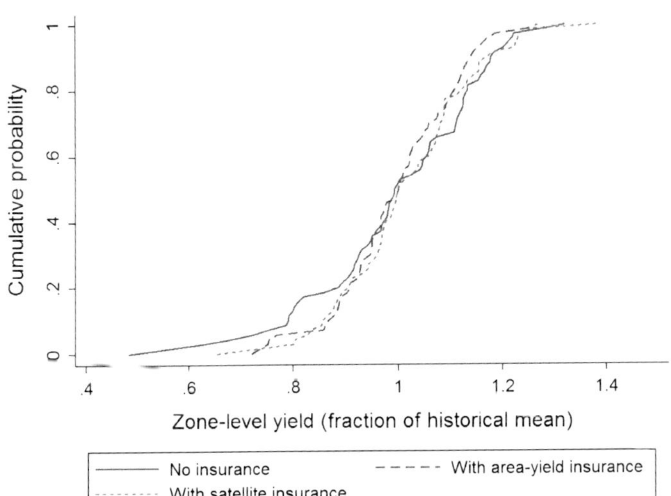

Fig. 30.7 CDF of zone-level yields with and without insurance for all-or-nothing contracts. Distributions assume that premiums are actuarially fair

the lower tail of the distribution moves to the right. This is again a manifestation of the idea that the satellite index does a poor job predicting the severity of losses. Hence, a contract that pays a fixed amount for any predicted losses, regardless of magnitude, will provide better protection for farmers with large yield shortfalls.

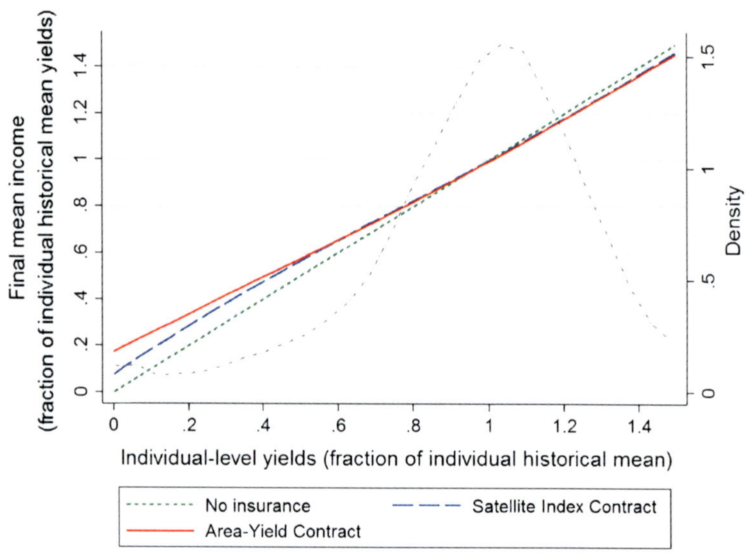

Fig. 30.8 Final mean income by individual-level yields under no insurance, area-yield insurance, and satellite insurance. A kernel density function of individual-level normalized yields is superimposed

Although index insurance cannot address idiosyncratic risk, it is still informative to study how these contracts affect the total risk faced by individual farmers. For a farmer who is not at all exposed to covariant risk, perhaps because she has invested in irrigation, even the "optimal" area-yield contract would not provide any benefit. Conversely, a farmer whose total risk is mostly tied to covariant events would greatly value an area-yield contract and may also benefit from a less optimal contract, provided the design risk is not too large. To analyze the impact of index insurance on individual farmer outcomes, we calculate the final income of each individual in a given year after insurance payouts and premium payments have been made under the different contracts. Figure 30.8 shows the results of fitting a local polynomial regression to these data (expressed as a fraction of individual historical mean yields) as a function of normalized individual yields, assuming the continuous payout contract.

First, note that even the area-yield contract—which, by construction, has no design risk—is not particularly effective at smoothing income for individual farmers, indicating that the majority of the risk faced by the farmers is idiosyncratic. On average, when a farmer suffers a complete loss, the insurance covers only about 20 % of the average yields after the premium has been paid. Under the satellite contract, only 10 % of the loss is covered; however, it does just as well as the area-yield contract when losses are less than 50 %. While these figures seem low, each of these index insurance contracts may still provide valuable benefits to farmers. The next section explores the welfare implications of index insurance and analyzes whether farmers would be better off under each contract than they would without insurance.

30.4.4 Welfare Effects

While the previous analysis allows for an objective comparison of different contracts using actual yield data, it provides no indication of whether any of these contracts would actually enhance the welfare of individual farmers. As shown in Clarke (2016), even an actuarially fair index insurance contract might be welfare-reducing for a risk-averse individual if the basis risk is sufficiently high. Understanding the welfare implications of index insurance is important, because it allows us to determine whether a particular contract should even be marketed to farmers. In particular, if the design risk is sufficiently high, as is the case with many weather-based index insurance contracts, or if the importance of idiosyncratic risk relative to covariant risk is sufficiently large, index insurance may not improve the welfare of farmers even if they face significant production risk.

In this section, we present a simple framework for analyzing the welfare implications of index insurance in general using historical plot-level yield data and hypothetical insurance payouts. Our approach uses expected utility analysis to analyze how a given index insurance contract would affect the welfare of farmers. In particular, we assume a constant relative risk aversion (CRRA) utility function, which is consistent with similar analyses of utility gains from index insurance (Woodard et al. 2012; Clarke 2016). Using the data from our yield study, we first estimate a model of zone-level yields assuming an underlying Weibull distribution. Specifically, if $\bar{y}_{zt}$ is the average yield in zone z in year t, and $\bar{y}_z$ is the average yield in zone z across all years, we estimate the parameters a_1, a_2, b_1, and b_2 of the Weibull probability density function using maximum likelihood estimation:

$$f(\bar{y}_z, a, b) = \frac{a}{b}\left(\frac{\bar{y}_{zt}}{b}\right)^{a-1} e^{-\left(\frac{\bar{y}_{zt}}{b}\right)^a}, \tag{30.5}$$

where

$$a = a_1 + a_2 * \bar{y}_z \tag{30.6}$$

$$b = b_1 + b_2 * \bar{y}_z \tag{30.7}$$

Based on these estimated parameters, we can simulate draws of $\bar{y}^s_{zt}$ for each of the 10 insurance zones. To calculate insurance payouts under the area-yield contract, we compute $I_{AY}(\bar{y}^s_{zt}, p_{AY})$ as defined in Sect. 30.4.1. For the satellite-based contracts, in order to simulate insurance payouts, we first need to create a response function linking $\bar{y}^s_{zt}$ to an index value or predicted area yield, $\bar{y}^{p,s}_{zt}$. To do this, we estimate a switching regression, allowing for a different intercept and slope if average yields are below the zone mean:

$$\bar{y}^p_{zt} = (\gamma_{1zt} + d\gamma_{2zt}) + (\delta_{1zt} + d\delta_{2zt})\bar{y}_{zt} + \varepsilon_{zt}, \tag{30.8}$$

where d is a dummy which equals one if $\bar{y}_{zt} < \bar{y}_z$ and zero otherwise. Now, $\bar{y}^{p,s}_{zt}$ are simulated as random draws from the estimated model, and indemnity payments for the satellite-based contract can be computed as $I_{SI}(\bar{y}^{p,s}_{zt}, p_{SI})$. To calculate actuarially fair insurance premiums for the different contracts, we compute $P^{AF}_{C,z} = \approx \frac{1}{T}\sum_{t=1}^{T} I_C(X_{zt}, p_C)$ for $C \in (AI, SI)$ as defined in Sect. 30.4.1. Also, as described in Sect. 30.4.1, the parameters of the indemnity schedule for the satellite-based contract are adjusted such that their premiums are equal to that of the area-yield contract.

Next, to simulate individual yields, we first estimate the following model based on Miranda (1991) as a random intercepts and coefficients model:

$$y_{izt} = \alpha_{iz} + \beta_{iz}(\bar{y}_z - \bar{y}_{zt}) + \varepsilon_{izt} \tag{30.9}$$

We assume that α_{iz} and β_{iz} are distributed jointly normal and that ε_{izt} are independent and normally distributed. Using the coefficients and the variance-covariance matrix of the random effects parameters from estimating Eq. (30.9), we can simulate individual yields y^{sim}_{izt} by making draws from the resulting distributions.

Using the simulated individual yields and insurance payouts under the three different contracts, we calculate the expected utility of an individual, both with and without insurance:

$$EU_{iz,NI} = \sum_{t=1}^{T} U\left(w + y^{sim}_{izt}\right) \tag{30.10}$$

$$EU_{iz,C} = \sum_{t=1}^{T} U\left(w + y^{sim}_{izt} + I_{C,zt} - P_{C,z}\right) \tag{30.11}$$

where $U(z_{izt}) = \frac{1}{1-\gamma}(z_{izt})^{1-\gamma}$, γ is the coefficient of relative risk aversion, and w is initial wealth, which we assume to be constant across individuals and time and equal to the lowest value such that the final income is still positive for all individuals after insurance premiums are paid. Now, in order to estimate the welfare impact of index insurance, we calculate an individual's willingness to pay (WTP) for each insurance contract under a range of risk preferences. In particular, we solve for an individual's premium $P_{C,iz}$, such that $EU_{iz,NI} = EU_{iz,C}$, and compute the mean WTP across all individuals.

Figure 30.9 shows the average WTP among farmers for the area-yield contract and the satellite-based contract over a reasonable range of risk preferences (0–2.5). As expected, when individuals are risk neutral, the WTP simply equals the

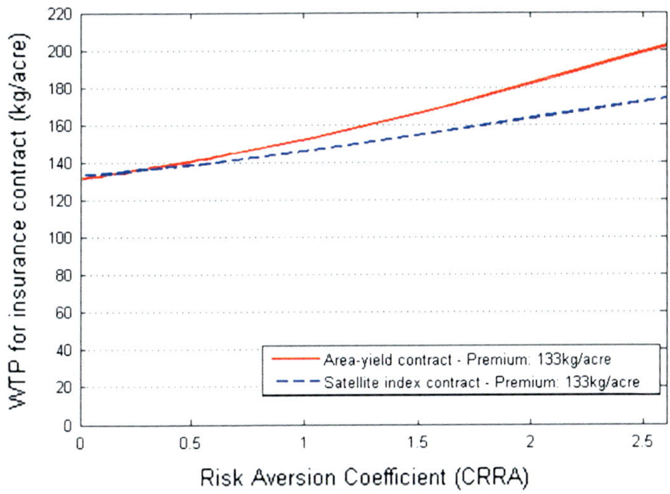

Fig. 30.9 WTP for index insurance contracts

actuarially fair premium of 133 kg/acre. As farmers become more risk averse, the benefit of both insurance contracts increases, which is evidence that if priced at actuarially fair rates, each of the insurance contracts improves the welfare of the average farmer. Moreover, the area-yield contract provides the greatest value to farmers at all levels of risk aversion, which is not surprising given that this contract has no design risk, by construction. In particular, for an individual with a CRRA of 1, the WTP for the area-yield contract is 14 % over the actuarially fair price versus a 9 % premium for the satellite-based contract. Given that the latter contract would likely be significantly cheaper to provide, this might in fact be a more viable contract than an area-yield contract.

The above analysis considers only the average WTP across all farmers. However, given that there is significant heterogeneity between farmers in how closely their yields are correlated with area yields (as measured by β_{iz} in Eq. (30.9)), there will also be heterogeneity in farmers' WTP for index insurance. To estimate the demand for each of the three index insurance contracts, we therefore calculate the percentage of farmers in the simulated population that would benefit from the insurance, given certain assumptions about loading costs. In particular, we assume that the area-yield contract has a premium of 40 % above actuarially fair prices, while the satellite contract has a 20 % premium, reflecting the fact that collecting area-yield data annually for each zone is costly. Using these premiums, we calculate the proportion of farmers benefiting from insurance as

$$\frac{1}{N}\sum_{i=1}^{N} 1\left(EU_{iz,C} > EU_{iz,NI}\right), \tag{30.12}$$

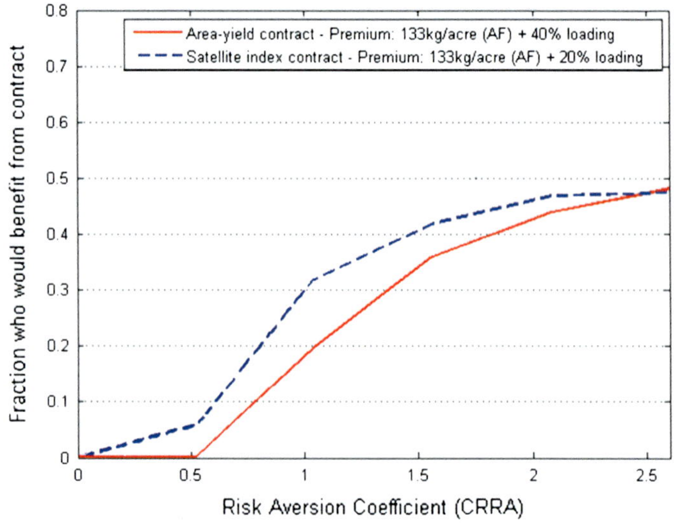

Fig. 30.10 Demand for index insurance contracts

where 1() is an indicator function which equals one if the expression is true and zero otherwise, and N is the total number of simulated individuals.

Figure 30.10 displays the proportion of individuals in the simulated sample who would benefit from each of the index insurance contracts under a range of risk preferences. While an area-yield contract commands the highest WTP, the demand for such a contract would be low due to the high data collection costs. The satellite-based contract, despite providing lesser coverage, would actually have a higher demand, due to its relatively lower premium. In particular, for farmers with a CRRA of 1, demand would be 18 and 32 % under the area-yield contract and the satellite-based contract, respectively. These findings are obviously dependent on the loading cost assumptions; however, these assumptions are consistent with the premiums charged by many microinsurance institutions in developing countries.

30.5 Conclusion and Discussion

While index insurance may in theory offer a promising solution to the problem of covariant risk in smallholder agriculture, its impact has thus far been limited in part due to the poor quality of many index insurance contracts. In particular, the correlation between the index and farmer losses has often been close to zero, implying that farmers are purchasing a lottery ticket rather than actual insurance. While the impact of insurance quality (basis risk) has been studied extensively in the literature on index insurance in a developed country context, very little research has focused on the impact of poor insurance quality in developing countries, and even fewer make use of plot-level panel data to analyze the welfare implications of index insurance on a household level.

We use a unique panel dataset of plot-level rice yields in Northeastern Tanzania to study the welfare implications of two hypothetical index insurance contracts. In particular, using the yield data collected and a rich dataset of plot-level satellite-based crop health indices, we define a standard area-yield contract and a satellite-based contract. We then compare the performance of both hypothetical contracts and use expected utility analysis to estimate farmers' willingness to pay (WTP) and demand for both contracts, under a range of risk preferences. Our results show that while the farmers in our sample are facing substantial idiosyncratic risk, both contracts would increase the welfare of the average farmer when offered at actuarially fair prices. In particular, we find that farmers' WTP for the area-yield contract exceeds the actuarially fair price by approximately 14 % for a farmer with a CRRA of 1, while the same figure is 9 % for the satellite-based contract. Moreover, we estimate that the demand for the satellite contract may exceed 30 % even under unsubsidized commercial rates.

These findings may have important implications for the future of index insurance in developing countries. Most index insurance products in the developing world today are based on precipitation measures, which have been shown to correlate poorly with actual farmer losses, yet little research has been done to understand the welfare implications of such contracts. This paper presents a systematic approach for evaluating the viability of index insurance contracts in general and shows that both an area-yield contract and a satellite-based contract may be welfare-enhancing under certain conditions. However, further research is needed to test the viability of these contracts in a real-world setting, and a next step would be to conduct a small-scale impact evaluation by offering these contracts to farmers though an insurance company.

Finally, it is important to consider some of the limitations of these results. For example, the satellite-based yield response function has been estimated only for rice, which is clustered together with little contamination from other vegetation or non-rice crops. If fields are more scattered, as is the case with maize and sunflowers and several other crops in developing countries, the correlation between the satellite-based measures and yields might be lower. Also, while the satellite based contract may be the most appropriate in data-scarce environments, such as African small-scale agriculture, the area-yield contract would always be a better choice in areas where yield data are already available at a low cost.

Acknowledgments We would like to thank Travis Lybbert, Stephen Boucher, Ghada Elabed, Jean Paul Petraud, Isabel Call, Wenbo Zou, Emilia Tjernstrom, Thomas Barre, Eliana Zeballos, Patrick McLaughlin, Jacob Humber, Asare Twum Barima and Abbie Turianski for their insightful comments and feedback. We are also grateful to VisionFund Tanzania and the enumerators, community leaders, and participants in our survey for their time and service. This research was made possible, in part, through funding provided by the U.S. Agency for International Development through the BASIS Assets and Market Access Collaborative Research Support Program.

References

Bhattacharya BK, Mallick K, Nigam R, Dakore K, Shekh AM (2011) Efficiency based wheat yield prediction in semi-arid climate using surface energy budgeting with satellite observations. Agric For Meteorol 151:1394–1408

Carter MR, Laajaj R (2009) Using satellite imagery as the basis for index insurance contracts in West Africa. http://i4.ucdavis.edu/projects/contracts/files/laajaj-using-satelliteimagery.pdf. Accessed 5 Oct 2012

Clarke DJ (2016) A theory of rational demand for index insurance. Am Econ J Microecon 8 (1):283–306

Clarke DJ, Mahul O, Rao KN, Verma N (2012) Weather based crop insurance in India. Policy research working paper 5985, March

Cole S, Giné X, Tobacman J, Topalova P, Townsend R, Vickery J (2013) Barriers to household risk management: evidence from India. Am Econ J Appl Econ 5(1):104–135

EARS (2012) Environmental analysis and remote sensing, monitoring and early warning validation results. http://www.earlywarning.nl/frames/Frame_val.htm. (crop yield in Africa link). Accessed 15 May 2015

Giné X, Yang D (2009) Insurance, credit, and technology adoption: field experimental evidence from Malawi. J Dev Econ 89(1):1–11

Gine X, Menand L, Townsend R, Vickery J (2010) Microinsurance, a case study of the Indian rainfall index insurance market. World Bank policy research working paper no. 5459

Jensen ND, Barrett CB, Mude AG (2014) Basis risk and the welfare gains from index insurance: evidence from Northern Kenya. Working paper

Martyniak L (2007) Response of spring cereals to a deficit of atmospheric precipitation in the particular stages of plant growth and development. Agric Water Manag 95(3):171–178

Merkovich R (2014) Tanzania crop yield study. Internal report

Miranda MJ (1991) Area-yield crop insurance reconsidered. Am J Agric Econ 73:233–242

Prince SD, Brown de Coltound E, Kravitz LL (1998) Evidence from rain use efficiencies does not indicate extensive Sahelian desertification. Glob Change Biol 4:359–374

Rosema A (1993) Using METEOSAT for operational evapotranspiration and biomass monitoring in the Sahel region. Remote Sens Environ 46:27–44

Sims PL, Singh JS (1978) The structure and function of ten Western North American grasslands: intra-seasonal dynamics and primary producer compartments. J Ecol 66:547–572

Smith V, Watts M (2009) Index based agricultural insurance in developing countries: feasibility, scalability and sustainability. A paper submitted to the bill and Melinda gates foundation. http://www.agecon.ucdavis.edu/research/seminars/files/vsmith-indexinsurance.pdf

Staggenborg SA, Dhuyvettere KC, Gordon WB (2008) Grain sorghum and corn comparisons: yield, economic and environmental responses. Agricultural experiment station working paper, Kansas State University, Manhattan, Kansas

Woodard JD, Pavlista AD, Schnitkey GD, Burgener PA, Ward KA (2012) Government insurance program design, incentive effects, and technology adoption: the case of skip-row crop insurance. Am J Agric Econ 94:823–837

Chapter 31
Assessing the Economic Value of *El Niño*-Based Seasonal Climate Forecasts for Smallholder Farmers in Zimbabwe

Ephias M. Makaudze

Abstract This study demonstrates the potential value of seasonal forecasts to smallholder farmers in Zimbabwe—the majority of whom often suffer severely from drought impacts. Using simulation models to compare crop yield performances of farmers *with* and *without* forecasts, results indicate that: during a "*drought year*", farmers *with forecasts* (WF) record higher yield gains (28 %) compared to those *without forecasts* (WOF); during a "*neutral year*" WF farmers obtain higher yield gains (20 %) than those WOF; however, during a "*good year*", results show no yield gains as WOF farmers perform better. This suggests that during a good year, forecasts may not have a significant impact. Using gross margin analysis, results show WF farmers realizing higher returns (US$0.14/ha) during a drought than WOF farmers who net a negative return (−US$0.15/ha).To conclude, *El Niño*-based seasonal forecasts could play an important role as loss mitigation measures particularly during a drought.

Keywords Seasonal climate forecasts · Smallholder farmers · *El Niño* · Economic value · Drought · With forecasts and without forecasts

31.1 Background

Until recently, drought occurrences in southern Africa were observed to be closely correlated with *El Niño* events occurring in the eastern tropical Pacific (Anyamba and Eastman 1996; Mason and Jury 1997). *El Niño*–Southern Oscillation (ENSO) events impose a strong influence on rainfall patterns and distribution in southern Africa. Its impact is strongest during the peak rainfall months of December–March (Mason and Jury 1997). ENSO events often culminate in severe droughts. For instance, major droughts that affected the region (e.g., 1982–1983, 1986–1987,

E.M. Makaudze (✉)
Department of Economics, University of the Western Cape,
P. Bag X17, Bellville, Cape Town 7580, South Africa
e-mail: emakaudze@uwc.ac.za; emakaudze@gmail.com

R. Lal et al. (eds.), *Climate Change and Multi-Dimensional Sustainability in African Agriculture*, DOI 10.1007/978-3-319-41238-2_31

1991–1992, and 2006–2007) are closely associated with *El Niño* episodes (WMO 1995; Kogan 1998).

In light of the increasing knowledge and understanding of ENSO events, it is now possible to predict major *El Niño* episodes at lead times of about 3–6 months (Mason 2001). Today, many national departments of meteorology in southern Africa, under a collaborative regional climate forum called SARCOF (Southern African Regional Climate Outlook Forum), are collectively involved in monitoring ENSO events with the objective of providing seasonal forecast information to various end-users (water planners, policymakers, farmers, food organizations), especially smallholder farmers. The forecasts denote rainfall probabilities presented in a three-pronged format (normal, above-normal, and below-normal) for any pending season. These forecasts are routinely broadcast via radio, TV, newspapers, farm bulletins, and the internet around early September one month before the seasonal rain starts.

Although seasonal forecasts are broadcast in Zimbabwe and other countries in southern Africa, it is not yet well-established how farmers, particularly smallholders, use forecasts to improve farm management practices or undertake strategic decisions to either avert/mitigate losses or exploit favorable weather conditions to optimize returns. If farmers can use forecasts to improve farm decisions, they not only reduce their vulnerability to *El Niño* but also their dependence on food-aid, which often is subject to political abuse and uncertainty.

Climate variability is the most dominant source of food insecurity in southern Africa. With a majority of smallholder farmers dependent on rain-fed agriculture and vulnerable to climate variability, seasonal forecasts hold promise as a tool for drought mitigation and/or risk management. Essentially, seasonal forecasts offer farmers a realistic opportunity to manage climatic variability. With advance information about the predicted seasonal outlook, smallholder farmers become better equipped to handle climatic anomalies in ways that can reduce vulnerability to climate shocks. Specifically, smallholder farmers would be able to use seasonal forecasts as a tool to avert otherwise costly losses in the form of income, crop, animal, and even human losses.

Useful lessons can be drawn from past extreme drought events. For instance, the 1991–1992 extreme drought event left more than 100 million people (mostly smallholders) within the Southern African Development Community (SADC) region severely affected and more than US$580 million worth of food-aid had to be distributed under emergency measures to avoid massive levels hunger and starvation (SADC 1993). The losses in Zimbabwe were even more catastrophic as the smallholder farm sector lost over two million head of cattle, the main source of draft power and capital wealth; crop yields dropped 54 % below normal and the country had to survive on more than 2 million tons of cereals received as food-aid (Makarau 1992). Climate risks, especially drought, constitute a formidable barrier to investment and the adoption of high yield (but risky) technologies that have the potential to increase production and improve the livelihood of smallholder farmers.

31.1.1 Study Motivation

Southern Africa is one part of the world likely to suffer disproportionately from the negative effects of climate change—in particular, from the associated risks of extreme drought and flood events. The recently observed upsurge in extreme climate events across southern Africa bears testimony to some of these perceived impacts. Lobell et al. (2008) predicts that southern Africa could lose more than 30 % of its staple maize by 2030 due to climate change.

Extreme drought or flood events not only add to stresses on water resources, food insecurity, and human health but are also largely responsible for constraining economic development in many countries across southern Africa. Most countries in this region are regarded as extremely vulnerable to climate change due to their over-dependency on rain-fed agriculture. Vulnerability to catastrophic climate events is worsened by the lack of coherent mitigation policies and concrete plans to deal effectively with climate-change risks. In Zimbabwe, for instance, mitigation efforts have been largely focused on disseminating seasonal climate forecasts to end-users, especially smallholder farmers. However, the impact of this policy has been largely minimal as seasonal forecasts remain widely unembraced by smallholder farmers. Many smallholders continue to face serious household food insecurity problems with a significant proportion being highly dependent on food-aid and/or food-handouts.

A fundamental question regarding this issue is why seasonal forecasts are not adopted by smallholder farmers in Zimbabwe (and equally by other countries in southern Africa), where many continue to face serious food shortages, hunger, malnutrition, and often life-threatening starvation. Plausible arguments to explain this include: skepticism of the forecasts due to past failures; ineffective communication; inappropriate format; and lack of forecast extension skills and forecast education for smallholder farmers. (i) In terms of skepticism, forecasts tend to suffer a credibility problem that arises mainly from the failure of past forecasts (Patt 2001; Patt and Gwata 2002). (ii) Forecasts also are not being effectively communicated to potential beneficiaries, especially smallholder farmers. Although forecasts are disseminated in Zimbabwe, it is the art and skill of their communication that is largely missing—distilling, translating, and transforming information to make it more manageable, user-friendly, understandable, and beneficial to end-users. (iii) Seasonal forecasts are disseminated in probability undertones that may be difficult for a layman farmer to understand. (iv) Even if farmers do understand the probability forecasts, do they know how to apply them to their best advantage? With no forecast extension, farmers may lack the knowledge of how to apply forecasts for maximum benefit. (v) Smallholder farmers are faced with more pressing constraints than the availability or non-availability of seasonal forecasts. As Blench (1999) argues, it is too naïve to expect a farmer to gamble all their resources on a single best-bet strategy. Farm decision-making is a holistic approach hence farmers could be facing more binding constraints than the mere non-availability of seasonal forecasts. Smallholder farmers may be more concerned with issues such as when will the rainfall season start, what will the rainfall seasonal distribution look like, what crop varieties should be grown given the forecasts, etc. These questions go beyond the mere dissemination of forecast information and necessitate the repackaging of seasonal forecasts to meet farmers' needs and expectations.

Considerable effort must be applied in developing forecast extension skills and farmer education to demonstrate the potential benefits to smallholder farmers. The creation of climate education centers in developing countries could mark a turning point in assisting smallholder farmers to cope with climate-related risks and disasters.

31.1.2 Objectives

Seasonal forecasts have had very minimal impact as they remain largely unembraced by smallholder farmers in Zimbabwe. This occurs despite many smallholders facing serious food shortages with a significant proportion perennially dependent on food-handouts. While studies in developed countries reveal that farmers benefit substantially from using seasonal forecasts (Easterling and Mjelde 1987; Solow et al. 1998; Mjelde et al. 2000), the benefits of using seasonal forecasts by smallholder farmers in Zimbabwe (as well as in other countries in southern Africa) is not yet well-established (Vogel 2000; Patt et al. 2005; Makaudze 2009).

This paper focuses on assessing the economic value of seasonal climate forecasts to smallholder farmers in Zimbabwe, which might provide useful insights to policy-makers and facilitate more debate on seasonal forecasts and their potential role as a risk-mitigation tool. Such debates are more pertinent given the perceived climate change and its impact on millions of poor smallholder farmers in Zimbabwe and other countries in southern Africa.

The structure of the remainder of the paper deals with the methodology used for assessing the economic value of seasonal forecasts, a discussion of the data sources, a presentation of the simulation results derived from the Decision Support System for Agro-technology Transfer program (DSSAT v4 2004), and a final conclusion.

31.2 Methodology

The key objective of the study is to assess the economic value of seasonal forecasts to smallholder farmers in Zimbabwe. Forecast information is useful when it helps the decision-maker change course from a less-informed to a more-informed position. In the context of a farmer, a change of course could mean altering management decisions such as changing fertilizer amount, cultivar type, planting date, or plant population, among other things. However to accomplish this, the presumption is that new information incorporated into the agent's decision-making must alter management decisions; in other words, management decisions become altered only as a function of new information received. Thus, a decision process is characterized by the new information about a stochastic event and how this interacts with those variables under the decision-maker's control.

There is a growing body of literature on the economic valuation of seasonal forecasts, starting from theoretical underpinnings of decision-making under uncertainty to the valuation of forecasts on the basis of ex-ante approaches (Mjelde and Dixon 1993; Meza et al. 2003; Hansen et al. 2006). Ex-ante approaches value seasonal forecasts using predictive models that simulate a stochastic (rainfall) event; numerous examples are found in literature (Mjelde et al. 2000; Meza et al. 2003; Reyes et al. 2009; Hansen et al. 2010). Although ex-ante approaches are informative, they can only offer a limited valuation of what the forecast information is plausibly worth because they are largely normative (Msangi et al. 2006). Despite these drawbacks, ex-ante approaches are widely used and this paper follows a similar approach.

To construct the valuation model, one can start by discussing the household decision-making theory using the framework suggested by Rubas et al. (2006), who cast a decision-maker utilizing forecasts to influence farm decisions with the aim of maximizing the underlying utility objective. Decision theory assumes that preferences among risky alternatives can be described by the maximization of a utility function. To present the model mathematically, assume the farmer's problem is to maximize expected utility by choosing from a decision set using only prior knowledge. Mathematically this can be generalized as:

$$u(H) = \max_{D} E_c[u(D, c)h(c)], \tag{31.1}$$

where *max* is maximization operator, $u(H)$ is the maximum expected utility using climatologic information, E_c represents expectation operator for the range of climate conditions of interest, $h(c)$ represents the historical probability density function of climate conditions, u denotes the utility function, and D the decision set. Embedded within this equation are all other aspects that affect the decision process, such as risk aversion, institutional factors, and others.

When climate forecasts (F_i) become available, the probability density function of climate conditions is represented by $g(c|F_i)$. The decision maker's maximization problem becomes:

$$u_i(F_i) = \max_{D} E_{(c|F_i)}[u(D, c)g(c|F_i)], \tag{31.2}$$

where i represents the forecasts and F_i represents one of the many possible forecasts. Expected utility covering the entire forecast system, F, can be written as:

$$u(F) = \max_{D} E[u_i(F_i)Z(F_i)], \tag{31.3}$$

where $Z(F_i)$ is the probability density function associated with the probability of each forecast. Therefore, the value of the forecast system is:

$$V = u(F) - u(H), \tag{31.4}$$

where V represents the difference between the expected utility with the use of seasonal forecasts versus expected utility using only prior knowledge. If V is in utility terms, the difference in utility can be converted into monetary units using certainty equivalence dollars (Mjelde et al. 1993). If risk neutrality is assumed, V could be interpreted in monetary units. With the above approach, one is able to assess the value of using climate forecasts.

The schematic diagram below (Fig. 31.1) illustrates a dateline of activities showing forecast signals and the decision-making processes of a typical smallholder farmer in Zimbabwe. The activity sequence is as follows: (i) The rainfall season starts at the end of October, peaks during the months of January and February, gradually declines during the month of March, and ends in April. (ii) By the end of August, the national department of meteorology broadcasts forecasts for a pending season predicting rainfall outlook. (iii) Upon receiving forecasts, a farmer makes crucial farm decisions, such as the size of land to cultivate, the selection of crop cultivars, the fertilizer quantities to purchase, crop rotation, and so on. (iv) Forecasts are issued covering two growth stages: stage 1 refers to the first three months during the growing season of October–November–December (OND), a period which relates to early germination and initial crop growth; stage 2 refers to the subsequent months of January–February–March (JFM), the most critical phase in crop growth cycle covering crop flowering, pollination, grain-filling, maturation, and the resultant yield. Forecasts offered in two stages (OND and JFM), provides the farmer with the flexibility to modify actions. (v) Lastly, harvest time (April–May) ends with the realization of the final seasonal output.

Seasonal climate forecasts have been broadcast/disseminated in Zimbabwe (and most countries in southern Africa) since the 1997–1998 season. The forecasts are routinely broadcast during the month of August via several communication channels (radio, TV, print media, internet, etc.). As shown in Table 31.1, the forecasts are issued as a three-pronged probability format that underlies the likelihood of the

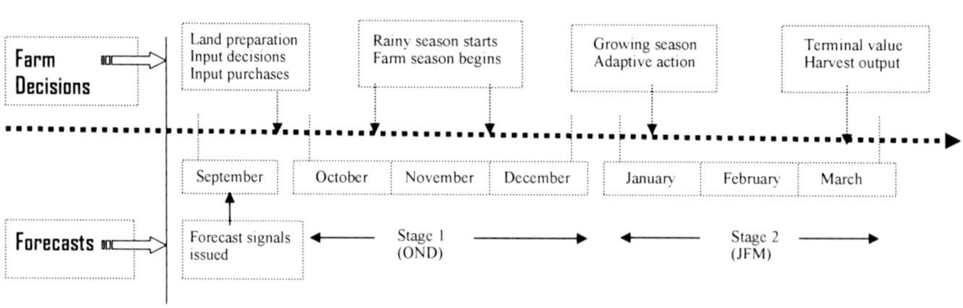

Fig. 31.1 Dateline of forecasts, growing season, and realized output

Table 31.1 Seasonal forecasts issued by Drought Monitoring Centre for growing season 1997–1998 through 2011–2012

Season	ENSO signal	Rainfall forecast probability					
		OND			JFM		
		Above_N	Normal	Below_N	Above_N	Normal	Below_N
1997–1998	El Nino	0.20	0.40	0.40	0.13	0.35	0.53
1998–1999[a]	–	–	–	–	–	–	–
1999–2000	La Nina	0.37	0.40	0.23	0.40	0.37	0.23
2000–2001	La Nina	0.45	0.33	0.25	0.35	0.45	0.20
2001–2002	Neutral	0.35	0.425	0.23	0.30	0.50	0.20
2002–2003	El Nino	0.30	0.40	0.30	0.25	0.40	0.35
2003–2004	Neutral	0.33	0.43	0.25	0.28	0.40	0.33
2004–2005	La Nina	0.30	0.40	0.30	0.25	0.40	0.35
2005–2006	Neutral	0.40	0.35	0.25	0.35	0.40	0.25
2006–2007	El Nino	0.25	0.40	0.35	0.30	0.40	0.30
2007–2008	La Nina	0.35	0.40	0.25	0.38	0.38	0.25
2008–2009	Neutral	0.30	0.40	0.30	0.30	0.40	0.30
2009–2010	El Nino	0.30	0.40	0.30	0.30	0.40	0.30
2010–2011	La Nina	0.35	0.40	0.25	0.35	0.40	0.25
2011–2012	El Nino	0.25	0.40	0.35	0.40	0.35	0.25

Source Adapted from SARCOF statistics
[a]Missing information

season being an *above normal* (good), *below normal* (bad), or *near-normal* (neutral) year. Above-normal rainfall is defined as the wettest 33.3 % of recorded rainfall amounts in each zone, normal is defined as the middle 33.3 % of the amounts, and below-normal rainfall is the driest 33.3 % of recorded rainfall amounts. Figure 31.2a, b illustrates typical announcements of seasonal forecasts pertaining to the 2011–2012 season for the entire SADC region. Figure 31.2a shows the predicted rainfall outlook for the first 3 months (OND) of the rainy season and Fig. 31.2b indicates the second part (JFM).

31.3 Data Sources and Assumptions

The study used two types of data: technical crop input and daily weather data. Technical crop input data (shown in appended supporting Tables 31.5, 31.6 and 31.7) is obtained from two sources: the departments of Research and Specialist Services (R&SS) and Agricultural Technical and Extension Services (AGRITEX). Because of the diverse agro-ecological, soil, and climatic conditions prevailing across Zimbabwe's landscape, the departments of R&SS and AGRITEX have developed detailed crop input management handbooks that provide extensive

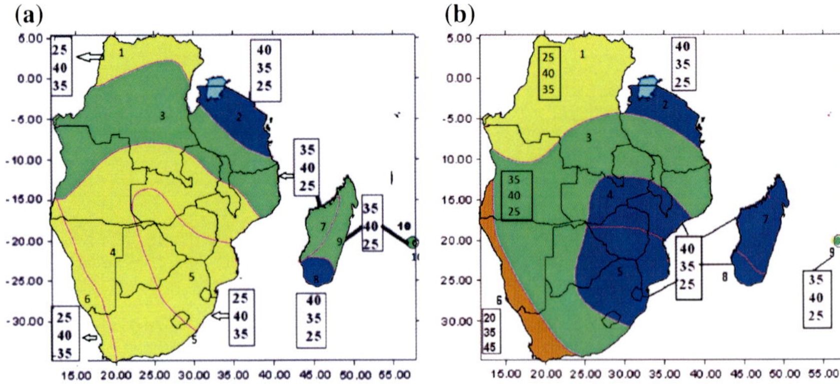

Fig. 31.2 **a** Rainfall forecast for the period OND, 2011. **b** Rainfall forecast for the period JFM, 2012. *Source* SARCOF statement, August 2011

technical details to document suitable cultivars, appropriate fertilizers/herbicides/chemical application rates, ideal planting dates, proper plant space, and plant population (see Appended supporting Tables 31.5, 31.6 and 31.7). The data illustrates the recommended input levels across different agro-ecological regions that allow the attainment of feasible optimal yields under ideal climatic conditions. This is referred to as "traditional farm management" practice throughout this paper.

The second type of data includes the daily weather data on rainfall, temperature (minimum and maximum), solar radiation, and evapo-transpiration obtained from the National Department of Meteorology. The data is obtained for three typical seasons that represent a drought/bad season (*El Niño*), a good season (*La Niña*), and an average season (neutral). Three specific seasons (1991–1992, 2003–2004, and 2004–2005) are selected that exemplify a typical bad, good, and average season, respectively. The data are obtained for four weather stations (Harare, Masvingo, Mutoko, and Bulawayo), each representing agro-ecological region II, III, IV, and V, respectively. (Zimbabwe is divided into five agro-ecological or natural regions (NR) numbered I to V and indicating potential in terms of soil fertility, rainfall, soil-water balance, moisture, and so on, which decreases as one ascends to higher regions from NR I to NR V). Due to data constraints, analysis is limited only to maize, the staple crop.

The DSSAT v4 program is used to run the various maize simulations based on different weather conditions and management practices. A key feature of DSSAT is the "cropping system model" (CSM) that simulates crop growth and development over time for individual crops based on phenology, daily growth, a plant's nitrogen and carbon demand, senescence, etc. DSSAT-CSM requires three key inputs that include weather input, management input, and soil input.

(i) "Weather input" is necessary for generating daily data for weather variables (e.g., maximum and minimum temperatures, solar radiation, precipitation, relative humidity, wind speed, etc.). (ii) "Soil input" consists of three components that include *soil dynamics* (computes soil characteristics), *soil water* (computes soil water processes including infiltration, runoff, and water-table depth), and *soil nitrogen and carbon* (computes soil nitrogen and carbon processes including organic and inorganic fertilizers). (iii) "Management input" characterizes when to plant, harvest, apply inorganic fertilizers, apply crop residue and organic materials, irrigate, and so on.

Seasonal forecasts are incorporated in the DSSAT program (2004) via the "management input" options. Because management input offers a user the flexibility to alter management practices, this provides an ideal option to explore the potential impact of seasonal forecasts on yield outcomes. In this case, three management practices are analyzed based on forecast predictions: *change planting date*, *change crop variety,* and *change fertilizer amounts*. For instance, if forecasts predict a *below-normal* (or bad) season, farmers with forecasts (WF) can alter management practices by planting early, growing short-season and drought-resistant varieties, applying minimal amounts of fertilizers, and so on. In contrast, farmers without forecasts (WOF) rely on traditional management practices and drawing from their own knowledge and experience. Comparing yield performances between WF farmers and WOF obtainable under different management practices that characterize different weather conditions, this paper establishes the potential economic value of seasonal forecasts from the smallholder farmers' perspective.

To keep the analysis tractable, some simplifying assumptions are necessary: (i) WF farmers are utilizing the availed forecasts information for farm management decisions that optimize net returns (yield per hectare, t/ha, or net gross margin, US$/ha); (ii) WOF farmers rely on historical, traditional knowledge and experience to formulate management decisions that maximize net returns (yield per hectare, t/ha, or net gross margin, US$/ha). Traditional management practices include detailed technical information provided by agronomic experts (as discussed earlier). (iii) For both categories (WF and WOF farmers), no herbicides, insecticides, or other chemicals are applied in the production process, (iv) no labor costs are considered, and (v) the farmer's risk behavior is embedded in the decision-making process. (vi) Assuming profit motive, each farmer (whether WF or WOF) pursues an input strategy that seeks to maximize the final outcome, referred to as the *optimal-input* strategy throughout the paper. Alternatively, a farmer can pursue an input strategy where no inputs (fertilizers/chemicals/insecticides, etc.) are applied, referred to as the *zero-input* strategy. (vii) Finally, all simulations are performed under the presumption that forecast information is perfect.

31.4 Main Results

This section presents the main results of the study. The results show how maize yields change in response to varying farm management practices based on WF and WOF assumptions (discussed earlier). The simulations are run based on three farm management practices: change planting dates, change crop cultivars, and change fertilizer application rates. The simulations are repeated for different seasons that exemplify a bad (*El Niño*), good (*La Niña*), and an average (neutral) year.

The first case results illustrate simulated maize yields obtained under weather conditions that characterize a drought year (1991–1992) and based on altering three management practices (change planting dates, cultivar choice, and fertilizer application rates) showcasing both WF and WOF farmers. The second and third cases are replicates of the first performed under weather conditions that underlie a neutral (2003–2004) and good (2004–2005) season, respectively. In each case, observed changes in maize yields are recorded for the selected representative districts (Harare, Masvingo, Mutoko, and Bulawayo) drawn from natural regions (NRs) II, III, IV, and V, respectively. Using this approach, three performance indicators (yield gains/losses, net gross margin, return per dollar invested) that underpin the economic value of seasonal forecasts are derived and compared across regions/districts between WF and WOF farmers.

31.4.1 First Simulation Results Based on a Drought Year (1991–1992)

The results of the first simulation (Table 31.2) show maize yield gains/losses across different agro-ecological regions for a selected typical drought season, 1991–1992. Starting with wet agro-ecological NR II (Harare district), results indicate that under an *optimal-input* management strategy—by planting early—WF farmers realize higher yields of 3.03 and 2.26 t/ha on medium- and long-season maize varieties, respectively, which translates to 0.13 and 0.7 % higher yield performance than WOF farmers. In contrast, if a WF farmer responds by planting late, they realize lower yield levels of 0.94 and 2.05 t/ha on long- and medium-season varieties, respectively, compared to WOF farmers who are realizing higher yield levels (1.33 and 2.69 t/ha) for the same varieties. The results suggest that forecast information yields no additional value if it involves late planting, especially in wet region NR II. The result is sensible given the long- and medium-season varieties would require longer days-to-maturity (145–170 days), which may not be possible given late planting.

Under the *zero-input* management strategy, results show that by planting early, WF farmers do realize higher yield gains on both long- (0.03) and medium-season (0.14) varieties than WOF counterparts. However, similar to the observation above, there are no yield gains if WF farmers plant late compared to WOF farmers.

With respect to semi-arid, agro-ecological NR III (Masvingo district), results indicate that under the *optimal-input* strategy, WF farmers obtain higher yields by either planting early or late compared to WOF counterparts. By planting early, WF farmers obtain higher yields of 13 and 29 % on short- and medium-season varieties, respectively. By planting late, WF farmers obtain higher yield gains of 3 and 34 % on short- and medium-season varieties, respectively. This practice of planting early

Table 31.2 Observed changes in maize yield for farmers WF and WOF under three management strategies during a typical drought season (1991–1992)

Season	District	ΔManagement strategies		ΔPlanting date			Yield gain/loss	
		ΔFertilizer application	ΔMaize variety	With forecasts (t/ha)		Without forecasts (t/ha)	% Δ	
				Early planting	Late planting	Traditional planting	Early	Late
1991–1992	NR II (Harare)	Optimal-input	Long	2.26	0.94	1.33	0.70	−0.29
			Medium	3.03	2.05	2.69	0.13	−0.24
		Zero-input	Long	1.09	0.71	1.06	0.03	−0.33
			Medium	1.32	0.97	1.16	0.14	−0.16
	NR III (Masvingo)	Optimal-input	Medium	1.83	1.90	1.42	0.29	0.34
			Short	1.59	1.41	1.37	0.16	0.03
		Zero-input	Medium	0.74	0.55	0.55	0.35	0.00
			Short	0.45	0.35	0.37	0.22	−0.05
	NR IV (Mutoko)	Optimal-input	Medium	2.53	2.42	2.72	−0.07	−0.11
			Short	1.65	1.62	1.10	0.50	0.47
		Zero-input	Medium	0.90	0.77	0.73	0.23	0.05
			Short	0.52	0.52	0.46	0.13	0.13
	NR V (Bulawayo)	Optimal-input	Short	1.78	1.76	1.59	0.12	0.11
		Zero-input	Short	0.28	0.12	0.15	0.87	−0.20

and/or late can be viewed as a risk-spreading or diversification strategy by WF farmers who may choose to stagger planting dates to overcome early- or mid-season dry-spell risks. It is important to emphasize that for WF farmers, all these decisions are being influenced and guided by forecast signals. Under the *zero-input* strategy, results show yield gains for early planting on both medium- (35 %) and short-season (22 %) varieties. However, there are no yield gains for late planting. Figure 31.3 provides an overview of these results.

The results for driest and most arid agro-ecological regions NR IV and V show WF farmers recording yield gains by either planting early or late and mostly for the short-season varieties. In particular, the highest yield gain (87 %) is recorded in NR V under the *zero-input* strategy. This big difference in yield gains emphasizes the potentially important role forecasts could play as drought mitigation tools,

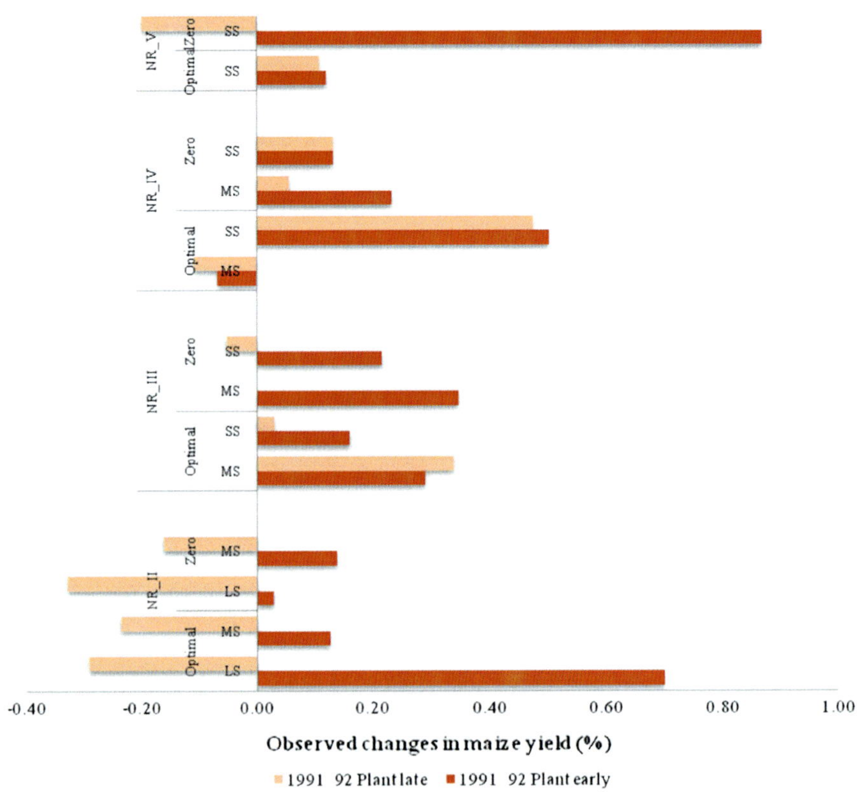

Fig. 31.3 Observed maize yield gains/losses by natural regions (*NR II–V*) for long-season (*LS*), medium-season (*MS*), and short-season (*SS*) varieties under optimal-input and zero-input strategies in 1991–1992 (drought/*El Niño* season)

especially in arid regions such as NR V, where most farmers are not only located but suffer severely due to drought impact.

31.4.2 Second Simulation Results Based on a Neutral (Average) Year (2003–2004)

The results of the second simulation replicate the first case (discussed above), but under different weather conditions that characterize a neutral/average year (2003–2004). Figure 31.4 presents a graphical view of the results. Looking at wet region NR II, WF farmers realize significant yield gains on both long- and medium-season varieties by either planting early or late compared to a WOF counterpart. The highest yield gain (1.15 %) is recorded on long-season varieties when farmers plant late under the *optimal-input* strategy. Under the *zero-input* strategy, yield gains are realized on long-season varieties for both early and late planting, unlike the medium-season varieties.

The results for NR III show no yield gains (under the *optimal-input* strategy) by planting early or late. Modest yield gains are observed under the *zero-input* case during both early and late planting. The results for NR IV show no yield gains accruing for WF farmers either by planting early or late under both *optimal-* and *zero-input* management strategies. In the case of the driest NR V, short-season varieties show positive yield gains mostly under the *zero-input* strategy for both early and late planting.

31.4.3 Third Simulation Results Based on a Good Year (2004–2005)

The results of the third simulation are based on weather conditions that characterize a good rainfall season with results shown in Fig. 31.5. As expected, most regions (except arid NR V) record high yields per hectare due to favorable weather conditions. However, results show WF farmers failing to outperform counterparts WOF as there are no yield gains across most regions. For instance, the long-season varieties show no yield gains by either planting early or late in NR II under the *optimal-input* strategy. It is only the medium-season varieties that record significant yield gains.

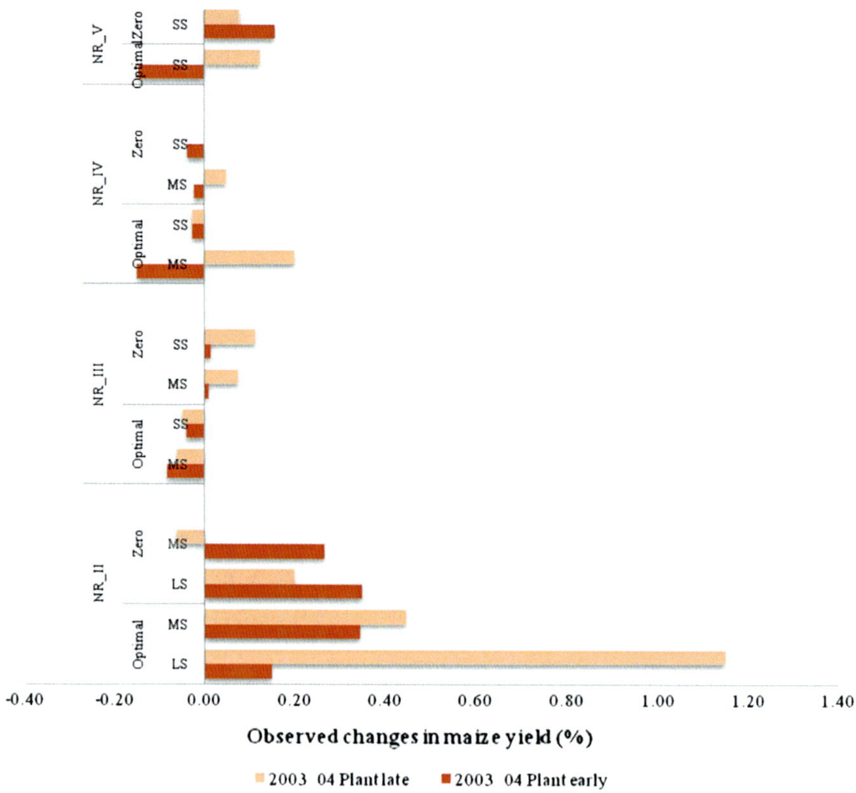

Fig. 31.4 Observed maize yield gains/losses by natural regions (*NR II–V*) for long-season (*LS*), medium-season (*MS*), and short-season (*SS*) varieties under *optimal-input* and *zero-input* strategies in 2003–2004 (average season)

In NR III the short-season variety fails to dominate under both *optimal-* and *zero-input* strategies for WF farmers. This is also the case with NR IV and V which record no significant yield gains by WF farmers during a good year.

The simulation results discussed above indicate the following: (i) For a good rainfall season, regardless of whether pursuing an *optimal-* or *zero-input* strategy, WF farmers across most regions record no significant yield gains compared to those WOF. (ii) The opposite is true during a bad rainfall season as WF farmers obtain higher yield gains, especially by planting early; for a neutral/average season, while most regions record no significant gains, NR II recorded higher yield gains.

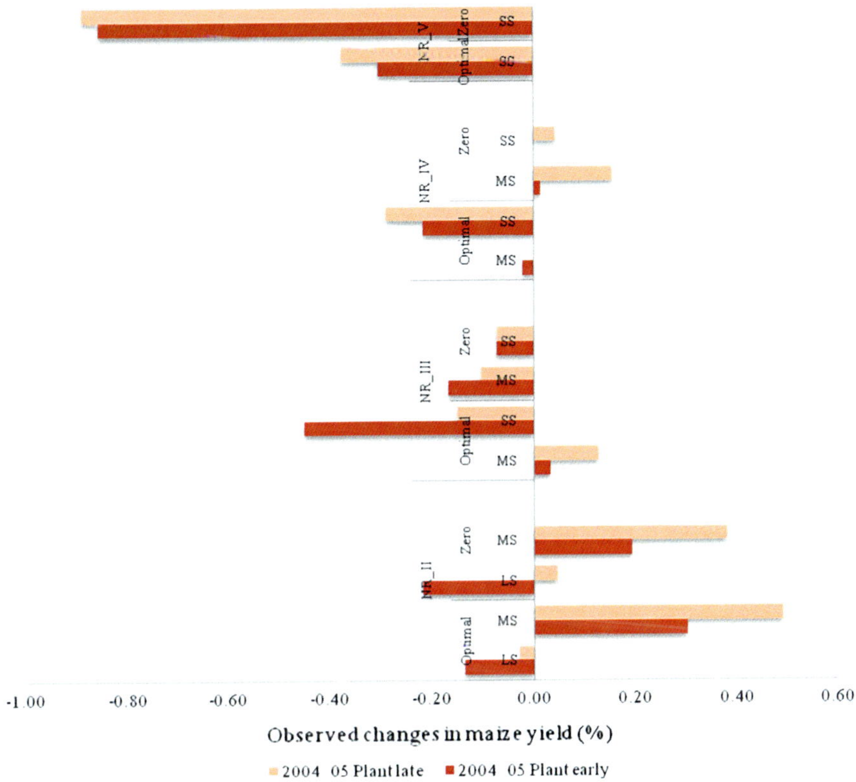

Fig. 31.5 Observed maize yield gains/losses by natural regions (*NR II–V*) for long-season (*LS*), medium- (*MS*), and short-season (SS) varieties under *optimal-input* and *zero-input* strategies in 2004–2005 (*La Niña*/good season)

31.4.4 Value Assessment of Seasonal Forecasts

Based on the simulation results above, the final section presents an economic value assessment of seasonal climate forecasts to smallholder farmers. For this purpose, two indicators are derived and used to gauge the economic value of forecasts: (i) net yield gains/losses based on WF/WOF and (ii) gross margin net return per dollar invested (US$/ha).

31.4.4.1 Value Assessment Using Net Yield Gains/Losses

Using WF/WOF results based on *optimal-* and *zero-input* management strategies discussed above, net yield gains/losses across growing seasons (bad, good, and average) and agro-ecological regions (NR II–V) are computed and summarized in Table 31.3. From the results, the following observations are made: the highest

Table 31.3 WF/WOF proportionate maize yield changes by NRs for the selected seasons

Season	WF yield gains/losses				
	NR	Early planting	Late planting	Net	Overall
1991–1992	II	0.25	−0.26	−0.01	0.28
	III	0.25	0.08	0.33	
	IV	0.20	0.14	0.34	
	V	0.49	−0.05	0.41	
2003–2004	II	0.28	0.43	0.71	0.20
	III	−0.03	0.02	−0.01	
	IV	−0.06	0.05	−0.01	
	V	0.00	0.10	0.10	
2004–2005	II	0.04	0.22	0.26	−0.31
	III	−0.16	−0.05	−0.21	
	IV	−0.06	−0.02	−0.08	
	V	−0.58	−0.63	−1.21	

overall net yield gain (28 %) is recorded during the drought year (1991–1992); this is followed by a yield gain of 20 % during an average/neutral year (2003–2004), with no yield gains (−31 %) recorded during a good year (2004–2005). The following observations can be made: for a *drought year*, NR V records the highest yield gain (49 %), while for a *good year* there are no yield gains observed across most regions except NR II where high yield gains are recorded (26 and 71 %) for both early or late planting.

The results underscore some important implications: (i) Seasonal forecasts are potentially of great value for farmers located in the most arid regions (NR V), particularly during a drought (*El Niño*) year. (ii) Except for NR II, all other regions record negative yield gains during a good year (*La Niña*), implying that forecasts may not make much difference given a good year. It is only in wet NR II that forecasts matter the most during a good year. This is sensible as farmers in better agro-ecological regions would exploit the available forecast information to optimize returns for a predicted good year. (iii) When aggregated across all seasons and regions, WF farmers are overall better-off as they realize a net yield gain of 17.7 % compared to WOF farmers.

31.4.4.2 Value Assessment Using Net Margin Return (US$/ha)

The second approach to valuing seasonal forecasts involves using gross margin analysis. The detailed gross margin values ($/ha) based on simulated maize yields for the three selected seasons (bad, good, neutral) are shown in Table 31.8. A summary of the net return gross margin values are shown in Table 31.4. The results indicate that for a *drought year* (1991–92), WF farmers growing medium-season varieties realize the highest overall net return of $0.52/ha compared

Table 31.4 Comparison of net return values ($/ha) for selected growing seasons between WF/WOF

Season	Variety	Net return ($/ha)								Overall	
		V		IV		III		II			
		WF	WOF	WF	WOF	WF	WOF	WF	WOF	WF	WOF
1991–1992	SS	0.21	0.09	−0.01	−0.08	−0.15	−0.26	–	–	0.04	−0.25
	MS	–	–	0.34	0.43	−0.07	−0.29	0.25	0.12	0.52	0.26
	LS	–	–	–	–	–	–	−0.10	−0.46	−0.10	−0.46
Net										0.14	−0.15
2003–2004	SS	0.14	0.03	0.04	0.07	0.10	0.14	–	–	0.28	0.24
	MS	–	–	0.49	0.27	0.71	0.81	−0.26	−0.48	0.94	0.69
	LS	–	–	–	–	–	–	0.08	−0.47	0.08	−0.47
Net										0.43	0.15
2004–2005	SS	−0.19	0.13	−0.09	0.14	0.16	0.35	–	–	−0.12	0.62
	MS	–	–	0.44	0.46	0.75	0.58	0.80	0.34	1.99	1.38
	LS	–	–	–	–	–	–	0.44	0.55	0.44	0.55
Net										0.77	0.85

to WOF counterparts who realize $0.26/ha. This is not the case with long-season varieties, as both WF and WOF farmers incur losses of $0.10/ha and $0.46/ha, respectively. This result indicates that although both categories of farmers suffer losses due to drought, it is those WOF who suffer most. In particular, WOF farmers located in higher agriculturally potential regions (NR II) growing long-season varieties are the most adversely affected.

For short-season varieties, WF farmers realize modest net returns of $0.04/ha compared to losses by WOF farmers ($0.25/ha). When aggregated across all varieties, results show WF farmers realize overall net returns of $0.14/ha, unlike their WOF counterparts who incur negative returns (−$0.15/ha). A rather important message the results imply is that WF farmers will have the ability to undertake strategic decisions to help avert otherwise severe losses, particularly during extreme drought years.

For a *neutral year* (2003–2004), results showed both WF and WOF farmers recording mostly positive net returns. However, WF farmers predominantly realized higher overall net returns for all varieties. The medium-season varieties in particular record the highest net returns ($0.94/ha), followed by short-season varieties ($0.28/ha), with the long-season varieties recording the lowest net returns ($0.08/ha). Similar to observations above, WOF farmers growing long-season varieties (NR II) experienced the heaviest losses of $0.46/ha. Overall, net results indicate WF farmers realize three times more returns ($0.45/ha) than WOF farmers ($0.15/ha).

Results for the *good year* (2004–2005) differed from the other seasons (*drought/neutral*) discussed above. WOF farmers realized higher returns on all varieties (except medium) and across most regions. Specifically, short- and long-season varieties recorded higher net returns of $0.62/ha and $0.55/ha, respectively. Overall results indicate WOF farmers realized higher net returns of \$0.85/ha compared to their WF counterparts (\$0.75/ha). Because WOF farmers outperform their WF counterparts, the results suggest that forecasts may not make much difference during a good year.

31.5 Conclusion

This study demonstrates the potential value of seasonal forecasts to smallholder farmers in Zimbabwe, a majority who endure heavy losses due to adverse weather, particularly drought. Some important insights can be drawn from the study: (i) If the underlying season is a bad one (implying an *El Niño* year), forecasts play an important role as *loss-mitigation* instruments. As the results indicate, by changing planting dates (early/late), applying appropriate fertilizer rates (optimal/zero), and using suitable maize cultivars (short-, medium-, and long-season varieties), WF farmers are able to reduce and/or minimize yield losses across most regions; in particular, losses could be severe for farmers in better agro-ecological regions such as NR II, who are bound to invest a substantial amount of money in input (seeds, fertilizers, chemicals, etc.) use and purchases. (ii) Forecasts are likely to promote *strategic behavior* that could prove vital for reducing vulnerability of smallholder farmers to catastrophic drought events. As implied by the results, this is particularly true in arid regions NR IV–V, where engaging in a *zero-input* strategy and growing short-season varieties allows WF farmers to realize positive yield gains despite an extreme drought season. (iii) If the underlying season is a good year, no yield gains are observed across most regions (except NR II), suggesting that forecasts may not make much difference.

Climate variability, especially drought, constitutes the most dominant source of food insecurity in Zimbabwe and many countries in southern Africa. With a majority of smallholder farmers practicing dry-land agriculture, seasonal forecasts would play a fundamental risk management role. Wide-scale adoption of forecasts by smallholders would provide farmers with the ability to anticipate variations in crop production early enough to adjust crucial farm decisions and be better prepared to handle climate anomalies in ways that reduce otherwise costly crop, animal, and human losses.

Acknowledgments This paper received support from the Centre for Environmental Economics and Policy in Africa (CEEPA), University of Pretoria, in partnership with the Swedish International Development Authority. The author benefited from various comments provided by CEEPA resource personnel. In addition, the author acknowledges support received from Economic Research Southern Africa (ERSA). Finally, special thanks to Dr. Sebinasi Dzikiti of the Centre for Scientific and Industrial Research (CSIR Stellenbosch, Cape Town), who offered technical expertise on the DSSAT (Decision Support System for Agricultural Technology) program.

Appendix

See supporting supplementary Tables 31.5, 31.6, 31.7 and 31.8.

Table 31.5 Maize recommended optimal fertilizer application rates and planting dates

Crop	NR	Suitable planting date	Maximum expected yield (t/ha)	Fertilizer type Recommendation rate (kg/ha)	
				Compound_D	Ammonium nitrate (AN)
Maize	IIA, IIB	15th Nov	3.0–3.5	300–350	200–300
	III	15th Nov	2.0–2.5	200–300	150–200
	IV	25th Nov–15th Dec	1.5–2.0	200–250	100–150
	V	25th Nov–15th Dec	1.0–1.5	100–200	100

Source FAO

Table 31.6 Maize hybrid variety characteristics and suitable natural regions

Variety	Days to maturity	Suitable NR
Short season	90–135	III–V
Medium season	130–146	II–III
Long season	140–170	I–II

Source FAO

Table 31.7 Maize plant spacing and population recommendations for attainment of optimal yield

Row space (cm)	Within-row space (cm)	Plant population (plants/ha)	NR
90	25	45,000	I and II
90	30	37,000	III and IV
90	45	25,000	IV
150	30	22,000	V

Source FAO

Table 31.8 Gross margins and net return for selected growing seasons (bad, neutral, good)

Item	V		IV				III				II			
Variety	SS		SS		MS		SS		MS		MS		LS	
	WF	WOF	WF	WOF	WF	WOF	WF	WOF	WF	WOF	WF	WOF	WF	WOF
Case (a) Drought year														
Selling price ($/t)[a]	285	285	285	285	285	285	285	285	285	285	285	285	285	285
Simulated yield (t/ha)	1.78	1.59	1.65	1.52	2.53	2.72	1.59	1.37	1.9	1.42	3.03	2.69	2.26	1.33
Gross income ($/ha)	507	453	470	433	721	775	453	390	542	405	864	767	644	379
TVC ($/ha)	418	414	474	471	537	541	530	526	581	571	693	865	719	699
Gross margin, $/ha (loss)	89	39	(3)	(38)	184	235	(77)	(135)	(40)	(166)	171	81	(75)	(320)
Return per $TVC (loss)	0.21	0.09	(0.01)	(0.08)	0.34	0.43	(0.15)	(0.26)	(0.07)	(0.29)	0.25	0.12	(0.10)	(0.46)
Case (b) Neutral year														
Simulated yield (t/ha)	1.67	1.49	1.74	1.79	2.84	2.37	2.09	2.18	3.73	3.98	1.72	1.19	2.77	1.29
Gross income ($/ha)	476	425	496	510	809	675	596	621	1063	1134	490	339	789	368
TVC ($/ha)	416	412	475	477	543	533	541	543	620	625	665	653	730	698
Gross margin, $/ha (loss)	60	13	20	34	266	142	55	78	443	509	(175)	(374)	60	(331)
Return per $TVC (loss)	0.14	0.03	0.04	0.07	0.49	0.27	0.10	0.14	0.71	0.81	(0.26)	(0.48)	0.08	(0.47)
Case (c) Good year														
Simulated yield (t/ha)	1.15	1.65	1.5	1.91	2.74	2.77	2.22	2.61	3.82	3.39	4.58	3.28	3.8	4.14
Gross income ($/ha)	328	470	428	544	781	789	633	744	1089	966	1305	935	1083	1180
TVC ($/ha)	405	415	470	479	541	542	544	552	622	613	725	698	752	759
Gross margin, $/ha (loss)	(77)	55	(43)	65	240	248	89	192	467	353	580	237	331	421
Return per $TVC (loss)	(0.19)	0.13	(0.09)	0.14	0.44	0.46	0.16	0.35	0.75	0.58	0.80	0.34	0.44	0.55

[a]Selling price (t/ha) is presumed the same for all regions
SS short season, *MS* medium season, *LS* long season

References

Anyamba A, Eastman JR (1996) Interannual variability of NDVI over Africa and its relation to EL Niño/Southern oscillation. Int J Remote Sensing 17:2533–2548

Blench R (1999) Seasonal climatic forecasting: who can use it and how should it be disseminated? Overseas Development Institute, London, UK. Natural Resource Perspective 47. Available at www.odi.org.uk/NRP/47.html. Accessed 20 Dec 2014)

Easterling WE, Mjelde JW (1987) The importance of seasonal climate prediction lead time in agricultural decision making. Agric For Meteorol 40:37–50

Hansen JW, Challinor A, Ines A, Wheeler T, Moron V (2006) Translating climate forecasts into agricultural terms: advances and challenges. Clim Res 33:27–41

Hansen JW, Simons JM, Liqiang S, Tall A (2010) Review of seasonal climate forecasting for agriculture in Sub-Saharan Africa. Exp Agric 47(2):205–240

Kogan FN (1998) A typical pattern of vegetation conditions in southern Africa during El-Nino years detected from AVHRR data using three-channel numerical index. Int J Remote Sens 19 (18):3689–3695

Lobell DB, Burke MB, Tebaldi C, Mastrandrea MD (2008) Prioritizing climate change adaptation needs for food security in 2030. Science 319(5863):607–610

Makarau A (1992) National drought and desertification polices. The Zimbabwe situation. SADC regional workshop on climate change, 1992, Windhoek, Namibia

Makaudze E (2009) Do seasonal forecasts and crop insurance really matter for smallholder farmers in Zimbabwe? Using contingency valuation and remote sensing techniques. VCD Verlag Dr. Muller, Weinheim, Germany

Mason SJ, Jury MR (1997) Climatic variability and change over southern Africa: a reflection on the underlying process. Prog Phys Geogr 21:23–50

Mason SJ (2001) El Nino, climate change and southern African climate. Environ-metrics 12:327–345

Meza FJ, Wilks DS, Riha SJ, Stedinger JR (2003) Value of perfect forecasts of the sea surface temperature anomalies for selected rain-fed agricultural locations of Chile. Agric For Meteorol 116(3–4):117–135

Mjelde JW, Dixon BL (1993) Valuing the lead time of periodic forecasts in dynamic production systems. Agric Syst 42(1–2):41–55

Mjelde JW, Peel DS, Sonka ST, Lamb PJ (1993) Characteristics of climate forecast quality: implications for economic value to Midwestern corn producers. J Clim 11:2175–2187

Mjelde JW, Penson JB, Nixon CJ (2000) Dynamic aspects of the impact of the use of perfect climate forecasts in the corn belt region. J Appl Meteorol 39:67–79

Msangi S, Rosegrant MW, You L (2006) Ex-post assessment methods of climate forecast impacts. Clim Res 33:67–79

Patt A (2001) Understanding uncertainty: Forecasting seasonal climate for farmers in Zimbabwe. Risk Decis Policy 6:105–119

Patt A, Gwata C (2002) Effective seasonal climate forecast applications: examining constraints for subsistence farmers in Zimbabwe. Glob Environ Change 12:185–195

Patt A, Suarez P, Gwata C (2005) Effects of seasonal climate forecasts and participatory workshops among subsistence farmers in Zimbabwe. PNAS 102(35):12623–12628

Reyes CM, Gonzales KG, Predo CD, de Guzman RG (2009) Assessing the value of seasonal climate forecasts on farm-level corn production through simulation modeling. Discussion paper series no. 2009–2005, Makati City, Philippines

Rubas DJ, Hill HSJ, Mjelde JW (2006) Economics and climate applications: exploring the frontier. Clim Res 33:43–54

SADC (Southern African Development Community) (1993) Report of the regional drought management workshop, volume 1, workshop proceedings. SADC Drought Management Workshops in Southern Africa, Harare, Zimbabwe

Solow AR, Adams RF, Bryant KJ et al (1998) The value of improved ENSO predictions of U.S. agriculture. Clim Change 39:47–60
Vogel C (2000) Usable science: an assessment of long-term seasonal forecasts amongst farmers in rural areas of South Africa. S Afr Geogr J 82:107–116
WMO (World Meteorological Organization) (1995) The global climate system review: climate system monitoring, June 1991–Nov 1993. WMO, No. 819

Part VII
Agricultural Research for Sustainability

Chapter 32
Kinds of Research: Relationship with Agricultural Research for Sustainability

Isaac J. Minde and Stephen A. Nyaki

Abstract Climate change is now a reality. It is observed in frequent extreme weather events, such as droughts, heat waves, and floods which have significant negative effects on agricultural production. There is, therefore, a need for agriculture to adapt and become more resilient to maintain and even surpass present productivity levels. Consequently, agricultural research also has to change its approaches, methods, and priorities if the negative impacts of climate change are to be addressed. This paper discusses the importance of orienting agricultural research to address the multi-dimensional components of sustainability—environmental, economic, social, and institutional. In this paper, the following approaches were adopted: First, we examined existing agricultural research paradigms in the literature. Then, we assessed the extent to which these paradigms address one or more of the multi-dimensional features of sustainability. Second, we reviewed and modified indicators in the literature which are helpful in ex-ante or post-ante measurements of the degree to which a research study has or will address and contribute to the four sustainability features. Third, we conducted a brief assessment of the kinds of agricultural research conducted by selected institutions in the eastern Africa region (Sokoine University of Agriculture, Makerere University, Association for Strengthening Agricultural Research in Eastern and Central Africa, and the Tanzanian Ministry of Agriculture, Food Security and Cooperatives) to determine the extent to which those research studies explicitly or implicitly address one or more of the sustainability features. Last but not least, we looked at agricultural research for sustainability leadership from the perspectives of gender and local

I.J. Minde (✉)
Department of Agricultural, Food and Resource Economics,
Michigan State University, East Lansing, MI, USA
e-mail: minde@iagri.org

I.J. Minde · S.A. Nyaki
Innovative Agricultural Research Initiative Project,
Sokoine University of Agriculture, Morogoro, Tanzania
e-mail: stephenaniseth@gmail.com

R. Lal et al. (eds.), *Climate Change and Multi-Dimensional Sustainability in African Agriculture*, DOI 10.1007/978-3-319-41238-2_32

institutions. The relationships among the kinds of research (basic/disciplinary/theoretical, applied, or problem-solving, and subject-matter research) and their concern for sustainability were explored. In general, we found that the kind of research with the closest relationship with sustainability was problem-solving research, while the furthest from sustainability concerns was basic or disciplinary or theoretical research. Another finding was that the most popular dimension of sustainability in research is environmental sustainability. Literature on this dimension is abundant, and awareness of it is highest compared to the other three dimensions. However, there is a clear recognition that the rest of the sustainability dimensions are not less important. It is only that awareness of them has not been emphasized as much. This study concludes by urging agricultural researchers and research managers to adopt these indicators of agricultural research for sustainability in ex-ante and post-ante assessments of agricultural research projects. Obviously, it will be difficult for a single researcher to address all four dimensions of sustainability with the same level of rigor. Most likely, a researcher will prioritize one sustainability dimension of the four depending on the objective of the research.

32.1 Introduction

It is now generally agreed that the debate on whether climate change is real is closed. The most obvious signs of climate change include frequent extreme weather events, such as droughts, heat waves, and floods. Climate change has direct, negative impacts on agricultural production. In Africa, approximately 70 % of the population lives in rural areas where agriculture is the main economic activity, contributing on average 25 % of national gross domestic product and 60 % of export earnings and employing 70–90 % of the workforce (UNDP 2014). Due to this overdependence on agriculture, Africa remains potentially very vulnerable to the negative effects of climate change. There is, therefore, a dire need to ensure that our agricultural system functions in a sustainable fashion as it responds to climate change. Simply defined, sustainability refers to making our agriculture economically, socially, and ecologically sound today without compromising the wellbeing of future generations.

Agricultural research has great potential to make positive contributions to agricultural sustainability and, in the process, mitigate the negative effects of climate change. Simple examples of such contributions are identifying what crop varieties can tolerate the rise in temperatures and cope with the moisture stress brought about by frequent droughts. A well-developed and functioning agricultural research system contributes to sustainable food and income security for all agricultural producers and consumers, especially resource-poor households (Maiangwa 2010). Agriculture research provides the public with better knowledge of farming and technology for productivity improvement. A vivid example is given by the

2.8 % average annual growth of Australian agriculture from 1974 to 2004 as a result of multifactor productivity growth (Productivity Commission 2005; Warren 2013). A research approach aimed at ensuring that agricultural sustainability is achieved is called agricultural research for sustainability in this paper.

32.1.1 Objectives of This Paper

This paper seeks to (i) underpin the relationship between kinds of research and their potential contributions to agricultural research for sustainability; (ii) understand and improve the principal elements (indicators) guiding agricultural research for sustainability; (iii) assess the extent to which agricultural research for sustainability is conducted in the eastern African region and by whom; and (iv) recommend ways for improvement.

32.1.2 Approaches

Several approaches were applied in the development of this paper. These were as follows:

(i) Analytical review of literature on research paradigms—agricultural research, agricultural research *and* development, agricultural research *for* development, and agricultural research for sustainability
(ii) Assessment of the extent to which these research paradigms address one or more of the multi-dimensional features of sustainability
(iii) Review and modification of indicators found in the literature which are helpful in in ex-ante or post-ante measurements of the degree to which a research study has or will address and contribute to the four sustainability features
(iv) Assessment of the kinds of agricultural research conducted by selected institutions in the region—Sokoine University of Agriculture (SUA); Tanzanian Ministry of Agriculture, Food Security and Cooperatives (MAFC); Association for Strengthening Agricultural Research in Eastern and Central Africa (ASARECA); and Makerere University—to establish the extent to which these research studies explicitly or implicitly address one or more of the sustainability features
(v) Explore leaders of agricultural research in terms of gender and local leadership.

32.1.3 Definitions of the Kinds of Research

Research is a systematic scientific process of searching for or gaining knowledge and is sometimes done only for knowledge's sake. A research process is scientific if it is undertaken using known procedures that are reliable, measurable, and replicable (Kothari 2004). Johnson (1986) identified three kinds of research: basic or disciplinary, subject-matter, and problem-solving or applied research. As noted by Johnson (1986), this listing is in order of increasing complexity but, unfortunately, decreasing respectability.

Disciplinary research, sometimes called basic or theoretical research, is research undertaken to discover new knowledge or theories describing a particular phenomenon (Sherman 1988; Hansen 2009). It involves the discovery of new technologies that provide basic information on how bio-physical and social systems behave. Subject-matter research is multidisciplinary research on a subject of interest to a set of decision makers facing a set of practical problems (Johnson 1986). "Multidisciplinary" is a key word in this definition because this kind of research draws information from many disciplines. It provides information on how the disciplines can be used to solve occurring or prevent future agro-ecological problems using the findings of basic research (Hadorn et al. 2006). Examples of subject-matter research include understanding watershed dynamics, the impacts of climate change and variability, and the contribution of agroforestry home gardens to household food security and income generation.

Problem-solving research, sometimes called applied research or action research, is research that provides information about key societal problems and solutions to them (Hansen 2009). An example of this research is finding a solution to the problem of maize lethal necrosis disease, which is devastating maize in eastern Africa. According to the United Nations Food and Agriculture Organization (FAO), adaptive research uses other research outputs to improve productivity or solve problems. In this case then, adaptive research is mostly related to problem solving research. It involves the use of the same or similar procedures with the intention of replicating the results without compromising validity or integrity (Sherman 1988; Kothari 2004). The aim is to domesticate earlier findings to a local situation.

32.2 Agricultural Research Paradigms

To date, agricultural research has passed through four broad paradigms (Table 32.1). These paradigms are (i) agricultural research; (ii) agricultural research *and* development; (iii) agricultural research *for* development; and (iv) agricultural research for sustainability.

Table 32.1 Agricultural research paradigms by period and their main characteristics

	Agricultural research	Agricultural research *and* development	Agricultural research *for* development	Agricultural research *for* sustainability
Meaning (key properties)	Mother to all; any research activity aimed at improving the productivity and quality of crops or livestock (Loebenstein and Thottappilly 2007)	Research activities conducted with the intent to make a discovery that could either lead to the development of new products or procedures or improve existing products or procedures (e.g., new drought-tolerant crop varieties, research to improve and upgrade the value chain (Investopedia 2015)	Research activities conducted to gather information and provide solutions to solve global and societal problems (e.g., hunger, malnutrition, and climate change); system that links research with defined developmental outcomes (WLE 2014)	Research that integrates social, economic, institutional, and environmental components and supports projects and perspectives that have positive impacts on future resources, agro-ecosystem health, and human wellbeing (Wies et al. 2011)
			R4D which involves sharing knowledge (between researchers and development practitioners) and learning	
Characteristics	(1) Explores	(1) Discovers new agricultural technology	Solves to key societal problems primarily problem-solving research (WLE 2014)	Preserve welfare of future generations (Wies et al. 2011)
	(2) Solves problems			
	(3) Discovers			
	(4) Informs (Loebenstein and Thottappilly 2007)	(2) Upgrades technologies (Asopa and Beye 1997)		
Period	Locally from 25–255 AD under the Han Dynasty (Jones and Garforth 1997)	Scotland 19th century and in most industrialized countries from the mid-20th century to the present (Asopa and Beye 1997)	Not well known, but ASARECA prioritized AR4D in the 11 member states of Central and Eastern Africa starting in 2005 (Omamo et al. 2006)	Since 1987 after the Brutland Commission of the United Nations (Drexhage and Murphy 2010)
	Scientifically from the 18th and 19th century until the present (Asopa and Beye 1997)			
Drivers	Recognition of the importance of well-coordinated agriculture	Curiosity, famine around the world	Occurring or projected effects to society	Changing climate
				Population growth
				Ecosystem and natural resource mismanagement
	Population growth and periodic threats of famine		Mostly CGIAR in Africa	Food insecurity and changing agricultural systems

Source Compiled by the authors from various sources

32.2.1 Agricultural Research

This paradigm is the mother to all other research paradigms. It encompasses all kinds and types of research undertaken in the field of agriculture with the intention to produce knowledge or solve a particular problem in agriculture. Loebenstein and Thottappilly (2007) describe this field as any research activity aimed at improving the productivity or quality of crops or livestock. Over time, a number of changes on our planet attributed to natural and man-made causes have influenced the way we feed or depend on the environment. To adapt to the changes, such as increased population, better ways of production have been devised through research, for example, fertilizer and machines to increase output. The importance of agriculture research increases day by day amid a myriad of socio-economic (demographic changes, poverty, hunger, malnutrition), political (unrest), and environmental (climatic changes, global warming) factors that threaten and affect present generations and the future.

32.2.2 Agricultural Research and Development

This paradigm contains research that targets short- or long-term results in agricultural development. This term encompasses all activities that lead to the development of new technologies or products. It can also be concerned with improving existing technologies that will help achieve developmental objectives, such as the discovery of new drought-resilient maize varieties (Asopa and Beye 1997; Investopedia 2015). Here, theoretical constructs are tested to observe their projected developmental impacts, and if successful, they will be used for development processes.

According to Alston et al. (2009) and Alston (2010), economists have also examined developmental impact in many agricultural research projects using models that estimate the far-reaching consequences of agricultural research and development. These models take into account production economics, development economics, industrial organization, economic history, welfare economics, political economy, and econometrics.

32.2.3 Agricultural Research for Development (AR4D)

Agriculture research for development is a broad approach that centers on research activities geared toward searching for alternative ways of tackling deep-rooted agricultural problems. These include, among others, climate change adaptation and food security concerns. This research involves the use of developed technology to solve problems (European Union 2008). Here, well-tested theoretical constructs are replicated to bring about desired developmental impacts. This system links research with developmental outcomes and involves learning and knowledge sharing by researchers and development practitioners. The key characteristic of this research paradigm is that the research topics are problems to be solved (WLE 2014). In eastern and central Africa, this paradigm has been given more attention by ASARECA since 2005, when the bloc realized it lacked a common path toward research for development in the region (Omamo et al. 2006).

Most sub-Saharan African countries have long been challenged by the pressures of feeding a growing population, the risks and vulnerability of climate change, along with energy crises, environmental degradation, water scarcity, biodiversity losses, and pandemic diseases. The projected tripling of the world's population by 2050 will require doubling food production in developing countries, creating an urgent need for vibrant, reformed agricultural research, technology, innovation, and knowledge systems (Freibauer et al. 2011).

The difference between agricultural research *and* and *for* development is that, in the former, development involves technological discovery or improvement, while in the latter, development is regarded as the societal impact of using a particular research technology.

According to the European Union (EU), *agricultural research for development* is multi-dimensional research that addresses the agricultural challenges faced by developing and emerging economics. Agricultural research for development provides technological, economic, and institutional knowledge and innovations which contribute to sustainable development (EU 2008). The EU measures the speed of agriculture development with the attainment of the Millennium Development Goals 1 and 7 of eradicating extreme poverty and hunger and ensuring environmental sustainability. In attaining these goals, agricultural research has played a vital role. Research-driven agricultural productivity has had a positive impact on poverty reduction in Africa, Asia, and Latin America, with a high proven rate of return on investment of 100 % per year in some cases (Thirtle et al. 2003; Alston et al. 2004).

32.2.4 Agricultural Research for Sustainability

This research paradigm integrates social, economic, institutional, and environmental components and is aimed at making positive impacts on future resource, agro-ecosystem, health, and human wellbeing (Wies et al. 2011). The intent of this paradigm is to shape agro-ecosystem resource use in a way that benefits present and future generations. This is done by objectively incorporating components of sustainability in agricultural research projects. This paradigm became prominent in 1987 after the United Nations Brutland Commission laid out the idea of sustainability in all research dimensions (Drexhage and Murphy 2010).

In this paradigm, indicators are used as yard sticks for assessing whether agricultural research is oriented toward sustainability. These indicators are specific to each pillar of agricultural sustainability, although some cut across the paradigms of agricultural research.

32.2.4.1 Types of Sustainability

Holmberg and Karlsson (1992) identified the essential aspects of sustainability as economic, social, and environmental components. Institutional sustainability was later introduced to organize the use of common resources and govern the sustainability of systems which depend on social institutions that control access and use of resources and technologies (Lynam and Herdt 1989).

(a) Environmental sustainability

Environmental sustainability is attained if a system maintains a stable resource base and avoids the over-exploitation of renewable resources and depletion of non-renewable resources. Environmental sustainability includes, among other elements, the maintenance of biodiversity, atmospheric stability, and other ecosystem functions not ordinarily classed as economic resources (Saysel et al. 2002; Morelli 2011; Hansmann et al. 2012; Moldan et al. 2011).

(b) Economic sustainability

A system is said to be economically sustainable if it can produce goods and services on a continuous basis, maintain manageable levels of government and external debt, and avoid extreme sectorial imbalances that damage agricultural or industrial production. In the literature, the field of economic sustainability has drawn less attention than environmental sustainability (Holmberg and Karlsson 1992). The majority of sustainability management tools and systems are designed mostly by environmentalists and social scientists. Some refer to economic sustainability but are so unclear that they are inadequate for managing a real business (Doane and MacGillivray 2001).

(c) Social sustainability

A system is classified as fulfilling social sustainability if it strives to achieve gender equity, political accountability and participation, fairness in distribution and opportunity, and the adequate provision of social services, including health and education (Holmberg and Karlsson 1992).

(d) Institutional sustainability

There are two general perspectives on the concept of institutional sustainability. Institutions can be defined, first, as a set of rules (the rules of the game) and, second, as a set of roles for organizations that have attained special status or legitimacy. An example of a rule-oriented institution, according to Brinkerhof and Goldsmith (1990), is a system of land tenure, while a role-oriented institution could be the legal authority established to adjudicate disputes arising out of that land tenure system. Both can be institutionalized, the former as codes of law or custom and the latter as a concrete organization. However, view of institutions has evolved. Currently, institutions defined as organizations cannot perform their activities well and achieve their common predetermined objectives if they do not adhere to principles of institutionalization, that is, to the form of governance, rules, and customs of the organization.

According to Manning et al. (2012), the social, economic, and environmental pillars of sustainability are not mutually exclusive but mutually reinforcing. However, the environmental pillar is pictured to impose boundaries on the other pillars. The economy operates within social relationships, and the whole society is embedded within the natural world (the environment). Conventionally, sustainability or sustainable development is taken as a confluence of the three key pillars (Fig. 32.1a) where the use of resources is said to be efficient if the pillars' common elements of being viable, bearable, and equitable are considered together. Cato (2009) draws these pillars (sustainability circles) as equal sized, indicating equal importance. However, in reality, the economy carries much more sway in decision making, with society bearing the costs and the environment paying the highest price of all. Cato (2009), went on to argue that economists consider what happens to the environment and the people who live in it to happen somewhere else, so these events can be pushed outside the economy and be dealt elsewhere. However, there is no elsewhere, so economic activities and wastes produced should be handled differently. Proper understanding of sustainability proceeds from recognition that society is nested inside the environment, and the economy is part of society (Fig. 32.1b). The inference here is that society and the economy are both dependent on the environment. Therefore, the structure of economic activities that take place within a network of social relationships should be managed in a sustainable manner to benefit the future with fewer or no negative environmental consequences (Cato 2009). I this context, institutions should be introduced to coordinate and manage the interaction of the economy with society and the environment.

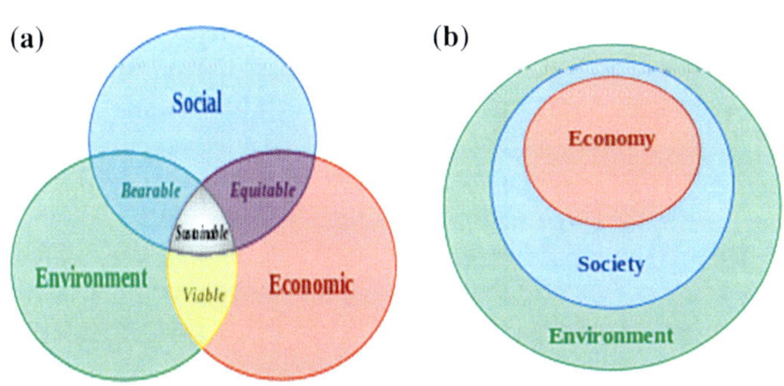

Fig. 32.1 Sustainability and the relationships among its pillars (adapted from Adams 2006; Cato 2009)

32.2.4.2 Kinds of Agricultural Research and Sustainability

Agricultural research falls into two main categories: basic, or applied. However, these categories vary in their focus on sustainability. While improving on the work of Lyson (1998), Goldberger and Buttel (2001) found that agricultural researchers who devoted a large percentage of their time to basic research were less likely to advocate the principles of sustainable agriculture with its economic, environmental, institutional, and social dimensions. This is due to the fact that basic research is curiosity motivated, whereas most sustainability research, like problem-solving (applied) and subject-matter research, is mission oriented.

32.2.4.3 Nature and Scope of Agricultural Research in Africa

Recognizing agriculture as the mainstay of most national economies in Africa, Ocholla and Onyancha (2006) analyzed research nature and trends in the discipline using descriptive infometrics and focusing on seven indicators in 1991–2005 materials from the Agricultural Online Access (AGRICOLA) and Science Citation Index/ISI-E databases. AGRICOLA is bibliographical database of citations for agricultural literature created by the National Agricultural Library (NAL) in South Africa. The database includes journal articles, book chapters, short reports, and reprints. The SCI-E is a multidisciplinary index for scientific journal literature. From the AGRICOLA and SCI-E databases, 2368 and 1254 papers, respectively, were extracted. The distribution of documents by subject, among other factors, was used as an indicator of classification. The dominant research subjects were environmental sciences (274), economics (254), plant production (234), soil cultivation (209), agriculture—multidisciplinary (179), soil fertility (170), agronomy (143), agriculture—general (119), soil science (116), water resources (116), and farm organization and management (114). Animal science research did not feature among the top categories partly because most research in animal science requires longer period of time to observe results.

Akinbamijo (2014) observed that agricultural research plays a key role in fostering innovations and advancing technologies that build the resilience and increase the efficiency, sustainability, and profitability of small-scale farmers in Africa (Box 1). Given the limited resources for conducting agricultural research, proper targeting of potential research areas is of paramount importance. These should be areas that promise to create a multiplier effect in sustaining the current and future generations (Box 1). Small-scale farmers in Africa will be environmentally and economically resilient in the long term if agricultural research concentrates on the following areas of research as outlined by the Forum for Agricultural Research in Africa (FARA).

Box 1: FARA: Priority areas in agricultural research for sustainability

Climate resilience (climate-smart agriculture); development of improved drought- and pest-tolerant crop varieties and drought- and disease-tolerant animal species; improved crop and livestock production techniques (e.g., integrated pest management, mixed crop–livestock production systems); an assessment of the inventory of indigenous knowledge and practices in the context of climate change and resilience; environment and natural resource conservation (e.g., watershed restoration, community forest and range land conservation, enhancement, and protection); biotechnology; eco-tourism and community resource conservation and preservation (Akinbamijo 2014); the role of gender in agricultural development; co-operation between the public and private sectors; and institutional gaps and potentials in resource and organisational control.

32.3 Methodologies for Agricultural Research for Sustainability

Research collaborations that transcend disciplinary and interdisciplinary research approaches on sustainability science require, first, constructive input from various communities or pools of knowledge to ensure that the essential knowledge from all relevant disciplines and actor groups related to the problem at hand are included. Second, research on solution options requires knowledge production beyond problem analysis because goals, norms, and visions should guide transition and intervention strategies. Third, collaborative efforts between researchers and non-academic stakeholders promise to increase legitimacy, ownership, and accountability of the problem and for the solution options. This brings us to the concept of transdisciplinarity, which is a key characteristic that agricultural research for sustainability should pursue (Box 2) (Talwar et al. 2011; Spangenberg 2011).

Box 2: Differences of transdisciplinary research from multidisciplinary and interdisciplinary research

Transdisciplinary research in general and in sustainability science in particular is an interface practice. First, it emerges from socially relevant problems that imply and trigger scientific research questions. Second, it relies on mutual and joint learning processes between science and society embedded in societal and scientific discourses (Lang et al. 2012). It is different from multidisciplinary research, which gathers knowledge from various disciples with the intent of to share knowledge and compare results from the study but not to cross boundaries or generate new integrative knowledge for solving a problem. Transdisciplinary research is also different from interdisciplinary research, which crosses disciplinary boundaries but is more oriented toward problem solving. Transdisciplinary research goes beyond the scope of interdisciplinarity by engaging non-disciplinary frontiers. Transdisciplinary research is encouraged in agricultural research for sustainability because it seeks to transcend disciplinary lines—"redrawing the disciplinary map" into a broader framework that involves practical engagement with local and regional stakeholders and issues of concern. Transdisciplinary research involves not only multiple disciplines but also multiple non-academic participants (Attwater et al. 2005; Stock and Burton 2011; Lang et al. 2012).

According to Lang et al. (2012), the ideal conceptual model of a transdisciplinary research process can be viewed as a sequence of three phases:

(i) Collaboratively framing the problem and building a collaborative research team
(ii) Co-producing solutions-oriented and transferable knowledge through collaborative research
(iii) Re-integrating and applying the produced knowledge in both scientific and societal practice

32.3.1 *Agricultural Research for Sustainability Assessment. a Review*

The assessment of agricultural research for sustainability conducted by Lyson (1998) showed that the research orientation towards sustainability has varied across disciplines. In Lyson's (1998) study, researchers were picked from the fields of agricultural economics, agricultural engineering, agronomy, animal science, biochemistry, chemistry, toxicology, biological science, entomology, food science, forestry, horticulture, natural resources, environmental science, plant pathology, plant science and botany, social science, and veterinary medicine. They were asked to rate on a scale of 1–5 the importance of their research contributions to sustainability. The indicators were research that improves (i) environmental quality; (ii) the profitability of farming; (iii) quality of United States rural life; and (iv) the sustainability of agriculture. For each of these indicators, the environmental aspects of farming were observed to be more clearly aligned with agricultural sustainability, with less attention given to its social-based dimensions.

The multidimensionality of sustainability assessment was well depicted in Lyson's (1998) study because respondents who viewed agricultural sustainability as an important personal goal did not necessarily subscribe to all the four dimensions. Lyson (1998), for instance, observed that, while improving environmental quality ranked highly among scientists in natural resources, agronomy, horticulture and forestry, less than half of veterinary scientists reported it to be an important goal. Improving the quality of rural life was a high priority for the social scientists and economists but much less for researchers in other disciplines. These results show how important disciplinary orientation is for the researcher targeting a particular dimension of sustainability.

32.3.2 Indicators of Agricultural Research for Sustainability

Table 32.2 presents various indicators in the literature that can be used by researchers, research managers, and donors to plan agricultural research for sustainability. These indicators are intended to provide guidance towards sustainability. Understandably, it will not be possible to use all the indicators in one research project, in part due to the disciplinary biases noted. In Table 32.2, indicators are classified according to the four sustainability pillars.

32.4 Experience in Agricultural Research for Sustainability in Selected Research Institutions

In this section, we attempt to assess how the selected research institutes (ASARECA, SUA, MAFC, and Makerere University) oriented their research towards agricultural sustainability from 2000 to 2014. We use the framework and principles of the sustainability pillars and indicators for sustainability presented in Sect. 32.3. The analysis was carried out by reviewing the titles, objectives, methods, and recommendations of 356 study reports: 33 from ASARECA, 173 from MAFC, 22 from Makerere University, and 128 from SUA. Degree of orientation to sustainability was measured on a scale of high, medium, and low. An article had a high orientation toward sustainability if sustainability-component indicators were explicitly documented in the article's title, objectives, or methods (using the list of indicators and principles). A study had a medium orientation towards sustainability if the sustainability-component indicators were included implicitly in the document, including being mentioned in the recommendations. A low orientation to sustainability was noted if the indicators of sustainability were not observed but mentioned in suggestions for future research. The frequency tables used to generate this assessment are found in "Appendix".

Table 32.2 Pillars, criteria, and indicators of agricultural sustainability

Pillars	Criteria	Indicators
Environmental pillar	Ecosystem component	Minimizing emissions of greenhouse gases
	Air: Maintaining or enhancing the supply of quality air function of the agro-ecosystem	Minimizing emissions of pollutants and eco-toxic pollutants
	Soil: Maintaining and enhancing the soil regulation function of the agro-ecosystem	Minimizing soil loss, increasing and maintaining soil chemical quality, increasing or maintaining soil physical quality
	Water: Maintaining and enhancing the water supply function of the agro-ecosystem	Amount of quality surface water and ground water
	Energy: Maintaining and enhancing the supply of energy function of the agro-ecosystem	Amount of energy generated
	Biodiversity: Maintaining and enhancing the supply of biotic resources function of the agro-ecosystem	Increased biodiversity
	Maintaining agro-ecosystem quality and supply of habitat	Improved in functional quality of habitats
	Ecosystem integrity: Maintaining and enhancing ecosystem stability and regulation	Increased resistance and resilience of the ecosystem
Economic pillar	Viability: Maintaining and enhancing the economic function of the agro-ecosystem	Rising farm income
		Minimized dependence on direct and indirect subsidies and optimal dependence on external financing
		Optimizing market activities and farmers' professional training
		Intergenerational continuation of farming activities and land assessment issue
Social pillar	Food security and safety: Maintaining and enhancing the production function of the agro-ecosystem	Balance maintained between production capacity and societal demand for food
		Increased diversity and quality of food and raw materials
		Amount of agricultural land available

(continued)

Table 32.2 (continued)

Pillars	Criteria	Indicators
	Quality of life: Maintaining and enhancing the physical wellbeing of the farming community function of the agro-ecosystem	Optimal labor conditions and health of the farming community
		Increased education of farmers and farm workers
		Greater equality in gender relations
		Increased family access to and use of social infrastructure Increased family integration into society
	Maintaining and enhancing the psychological wellbeing of the farming community function of the agro-ecosystem	Increased feelings of independence among farmers
	Social acceptability: Maintaining and enhancing the wellbeing of society function of the agro-ecosystem	Increased social services
		Reduced pollution levels
		Increased number of acceptable production methods Increased stakeholder' involvement
	Cultural acceptability: Maintaining and enhancing information function of the agro-ecosystem	Increased educational, scientific, cultural and spiritual value features
Institutional pillar	Institutional arrangement, environment, and organizations:	Number of laws and regulations of resource allocation and use formed and enforced prudently
		Property rights in place
		Increased access to information increased for farmers
		Number of cooperatives and associations that are member based
	Maintaining and enhancing institutional arrangements and institutional environments and organizational function of agro-ecosystem	More institutions with increased capacity to balance interests and improve equity

Adapted from Petters et al. (2005), Van-Cauwenbergh et al. (2007), Vandermeulen and Huylenbroeck (2008)

Table 32.3 Degree of sustainability across kinds of agricultural research and research institutions, 2000–2014

Research institution	Kind of research	Degree of sustainability			Total
		High	Medium	Low	
ASARECA	Basic	0	0	0	0
	Subject matter	7	14	4	25
	Problem solving	8	0	0	8
	Total	15	14	4	33
MAFC	Basic	0	0	0	0
	Subject matter	78	25	2	105
	Problem solving	51	16	1	68
	Total	129	41	3	173
Makerere	Basic	0	0	0	0
	Subject matter	5	5	0	10
	Problem solving	8	4	0	12
	Total	13	9	0	22
SUA	Basic	0	0	0	0
	Subject matter	28	25	3	56
	Problem solving	60	11	1	72
	Total	88	36	4	128

In Table 32.3, it is observed that, based on the presented definitions of kinds of research, none of the reviewed research institutions (ASARECA, MAFC, Makerere, and SUA) attempted to conduct basic research. This could be because basic research requires much more time and capital investment, which is not always available. Moreover, basic research involves inventing or developing theories which have no immediate impact on the agro-ecosystem. Conversely, problem-solving and subject-matter research was more common in all institutions with a high orientation toward social, economic, environmental, and institutional sustainability. Unlike problem-solving research, subject-matter research was partly oriented to sustainability because it also seeks to explore the problem at hand rather than theories. In Table 32.4, the environmental sustainability pillar is observed to be the most researched dimension, with more than 36 % of the articles reviewed targeting it. Relatively much less research was conducted on the institutional pillar.

We consider the Consultative Group on International Agricultural Research (CGIAR), through its research centers, to be the leading institution working on sustainable agriculture in the sub-Saharan countries. Through research, CGIAR carries out a mission to promote sustainable agriculture for food security in developing countries. It conducts strategic and applied research, produces international public goods, and focuses its research agenda on problem solving through interdisciplinary programs implemented by one or more of its international centers

Table 32.4 Research institution, kinds of research, and orientation to sustainability pillars, 2000–2014 (frequency)

Research type	Sustainability pillar			
	Environmental	Social	Economic	Institutional
Basic	0	0	0	0
Subject matter	118	104	79	58
Problem solving	114	95	44	30
Total	232	199	123	88

in collaboration with a full range of partners. Those programs concentrate on increasing productivity, protecting the environment, preserving biodiversity, improving policies, and strengthening agricultural research in developing countries (James 1996; CGIAR 2011).

Globally, the CGIAR is recognized as highly active in agricultural research for sustainability. CGIAR has published a 40-year report on the impact of its agricultural research projects since its establishment in 1971. Impacts were produced as a result of international public goods, including improved crop varieties, better farming methods, policy analysis, and associated new knowledge (CGIAR 2011). These research products were made freely available to national partners, who transformed them into locally relevant products responding effectively to the needs of rural households in developing countries. The 40 solid quantitative findings on CGIAR's impact since its founding in 1971 show that its programs and research projects had a mixture of social, economic, environmental, and institutional sustainability orientations even though the idea of sustainability might not have been a key program goal (CGIAR 2011). From 1971 to 2011, CGIAR devoted most of its resources to developing (i) improved crop varieties, generating approximately 7250 new crop varieties; (ii) better farming methods, in which fertilizer tree fallows and a lack of tillage renew soil fertility using on-farm resources; and (iii) policy analysis and associated new knowledge, which improved resource use and the sustainability of the economy, environment, and society in a number of ways (DID 2010; CGIAR 2011).

32.5 Review of Gender and Leadership in Agricultural Research for Sustainability

Gender and leadership status in African agricultural research for sustainability were investigated by reviewing research studies deemed to be very close to addressing sustainability concerns. The objectives of this investigation were (i) to understand the nature of leadership in African agricultural research for sustainability from 2000

Table 32.5 Leadership status of African agricultural research for sustainability

Gender	Origin		Total
	Foreign	African	
Female	10	4	14
Male	43	23	66
Total	53	27	80

to 2015, specifically, the extent to which it has been African led; and (ii) to assess the extent to which this leadership is women managed and lead by women.

Gender and leadership in African agricultural research for sustainability was studied by reviewing 80 online articles focusing on the eastern Africa region. The variables in the review were the gender of the first and second author, year of publication, region studied, type of study, study orientation towards agriculture sustainability, and degree of sustainability. The gender (male or female) and origins (African or foreign) of the first and second authors were checked through an online search. Articles selected from 2000 to 2015 were well checked against the indicators of sustainability to ensure that they reported agricultural research for sustainability by type. The selection was biased toward regional studies, which means that they were conducted in more than one country. Country-specific studies were not included because analyzing them would have required more time than was available. It was noted whether each study was problem–solving, subject–matter, or basic research and how close it was to sustainability given the available indicators.

(i) Leadership

In Table 32.5, 53 of a sample of 80 (66 %) reviewed articles on agricultural research for sustainability articles in Africa were led by foreign (non-African) authors, primarily from the United States (18 %).

(ii) Gender

Regarding the gender context, only 14 of the 80 reviewed articles (17.5 %) were led by women (first author), including 4 female African researchers (Table 32.5).

32.6 Conclusion

Based on the review of literature and analysis of data on the performance agricultural research in the context of the four types of sustainability in the eastern African region, the following concluding points are noted:

(i) The current agricultural research paradigm is agricultural research for sustainability. It was preceded by three main paradigms: agricultural research, agricultural research *and* development, and agricultural research *for* development. We note that there are key drivers responsible for each paradigm. For example, the most recent paradigm is driven by climate change.

(ii) Basic research in the region was observed to be scanty, mainly due to inadequate research funds and a lack of immediate relevance. However, we note that basic research is of critical importance to attaining agricultural sustainability in the future. Problem-solving research was observed to be most common sustainability-linked kind of research across the selected research institutions.

(iii) Transdisciplinarity in agriculture research for sustainability is emphasized as critical because the involvement of the general public in ex-ante and ex-post manners assists in increasing ownership and implementation of research results.

(iv) The environmental pillar is the most popular pillar among researchers targeting sustainability. Institutional sustainability is the least researched pillar.

(v) Indicators of sustainability could help guide future research and be used by researchers, donors, and research managers to properly plan, target, and conduct agriculture research for sustainability.

(vi) Investigation of gender and leadership roles in agricultural research for sustainability revealed that only 10 of 80 research projects sampled were led by women, and among these 10 women were only four Africans. Men were leaders in 66 of the 80 research projects, and approximately 40 % of them were African. These results show that there is a dire need to strengthen research capacity to promote local leadership and increase parity in male and female participation in agricultural research for sustainability.

Appendix

See Table 32.6.

Table 32.6 Frequency distribution of research publications by institution, degree of sustainability, and types of sustainability

Research institution	Degree of sustainability	Environmental sustainability			Social sustainability			Economic sustainability			Institutional sustainability		
		Yes	No	Total	Yes	No	Total	Yes	No	Total	Yes	No	Total
ASARECA[a]	High	8	7	15	13	2	15	10	5	15	2	13	15
	Medium	3	11	14	13	1	14	7	7	14	3	11	14
	Low	0	4	4	0	4	4	0	4	4	0	4	4
	Total	11	22	33	26	7	33	17	16	33	5	28	33
MAFC	High	98	31	129	84	45	129	53	76	129	40	89	129
	Medium	23	18	41	16	25	41	11	30	41	4	37	41
	Low	0	3	3	0	3	3	0	3	3	0	3	3
	Total	121	52	173	100	73	173	64	109	173	44	129	173
MAKERERE[b]	High	9	4	13	11	2	13	11	2	13	4	9	13
	Medium	2	7	9	9	0	9	3	6	9	1	8	9
	Low	0	0	0	0	0	0	0	0	0	0	0	0
	Total	11	11	22	20	2	22	14	8	22	5	17	22
SUA	High	70	18	88	33	55	88	19	69	88	28	60	88
	Medium	20	16	36	17	19	36	9	27	36	5	31	36
	Low	2	2	4	3	1	4	0	4	4	1	3	4
	Total	92	36	128	53	75	128	28	100	128	34	94	128

[a]The numbers in "Appendix" come from a Microsoft Excel sheet. For instance, ASARECA had a total of 15 articles with a high orientation toward agriculture research for sustainability. Among the 15 papers, 8 had an orientation towards environmental sustainability. The balance (7) did not have a high degree of environmental sustainability. As well, 13 papers had a high orientation toward social sustainability, and 2 did not. The meaning and interpretation of the data are the same for other institutions

[b]None of the reviewed articles from Makerere University had a low orientation toward agricultural research for sustainability. This explains the absence of a low row under degree of sustainability

References

Adams WM (2006) The future of sustainability: re-thinking environment and development in the twenty-first century. http://cmsdata.iucn.org/downloads/iucn_future_of_sustanability.pdf. Accessed 12 June 2015

Akinbamijo Y (2014) Priority investments for agricultural research to enhance resilience of poor small-scale farmers in Africa. Building resilience by innovating and investing in agricultural system. Paper presented at 2020 conference in building resilience for food and nutrition security, Addis Ababa, Ethiopia, 15 May 2014

Alston JM (2010) The benefits from agricultural research and development, innovation, and productivity growth. OECD Food, Agriculture and Fisheries Papers, 31. http://www.oecd-ilibrary.org/the-benefits-from-agricultural-research-and-development-innovation-and-productivity-growth_5km91nfsnkwg.pdf?itemId=/content/workingpaper/5km91nfsnkwg-en. Accessed 16 May 2015

Alston JM, Chan-Kang C, Marra MC, Pardey PC, Wyatt TJ (2004) A meta-analysis of rates of return to agricultural research and development. http://ebrary.ifpri.org/cdm/ref/collection/p15738coll2/id/125334. Accessed 14 May 2015

Alston JM, Pardey PG, James JS, Andersen MA (2009) The economics of agricultural R&D. Annu Rev Resour Econ J 1:537–566. doi:10.1146/annurev.resource.050708.144137

Asopa VN, Beye G (1997) Management of agriculture research: a training manual. Module 3: organizational principles and design. http://www.fao.org/docrep/w7503e/w7503e00.htm. Accessed 24 May 2015

Attwater R, Booth S, Guthrie A (2005) The role of contestable concepts in transdisciplinary management of water in the landscape. J Syst Res Behav Sci 22:185–192. doi:10.1002/sres.668

Brinkerhof M, Goldsmith B (eds) (1990) Institutional sustainability in agriculture and rural development: a global perspective. Praeger, New York

Cato MS (2009) Green economics: an introduction to theory, policy and practice. Earthscan, London

Consultative Group on International Agriculture Research (2011) Findings on the impacts of CGIAR research from 1971 to 2011. http://www.cgiar.org/pdf/cgiar_40Findings.pdf. Accessed 10 May 2015

Department for International Development (2010) The impact of research on policy and practice in Indonesia's pulp and paper sector. http://r4d.dfid.gov.uk/Output/186180/Default.aspx. Accessed 28 May 2015

Doane D, MacGillivray A (2001) Economic sustainability: the business of staying in business. SIGMA Project, London

Drexhage J, Murphy D (2010) Sustainable development: from Brutland to Rio 2012. Background paper. United Nations, New York

European Union (2008) Guidelines on agricultural research for development. EU Directorate-General development and relations with African, Caribbean and Pacific states. Brussels. https://ec.europa.eu/europeaid/sites/devco/files/methodology-agricultural-research-for-development-200806_en_2.pdf. Accessed 12 Mar 2015

Freibauer A, Mathijs E, Brunori G, Damianova Z, Faroult E, Gomis JG, O'Brien L (2011) Sustainable food consumption and production in a resource-constrained world. European Commission, Brussels. http://www.scp-knowledge.eu/knowledge/sustainable-food-consumption-and-production-resource-constrained-world-full-report. Accessed 13 Mar 2015

Goldberger JR, Buttel F (2001) Sustainable agriculture research and the land-grant system: a replication. J Sustain Agric 18(2–3):91–104

Hadorn GH, Bradley D, Pohl C, Rist S, Wiesmann U (2006) Implications of transdisciplinarity for sustainability research. J Ecol Econ 60(1):119–128

Hansen NR (2009) Why the distinction between basic (theoretical) and applied (practical) research is important in the politics of science. London School of Economics, London

Hansmann R, Mieg HA, Frischknecht P (2012) Principal sustainability components: empirical analysis of synergies between the three pillars of sustainability. Int J Sustain Dev World Ecol 19(5):451–459

Holmberg J, Karlsson S (1992) On designing socio-ecological indicators. In: Svedin U, Aniansson HB (eds) Society and the environment: a Swedish research perspective. Springer, Dordrecht, pp 89–106

Investopedia (2015) Research and development. http://www.investopedia.com/terms/r/randd.asp. Accessed 30 May 2015

James DW (1996) CGIAR annual report. Consultative Group on International Agriculture Research, Montpellier

Johnson CL (1986) Research methodology for economists: philosophy and practice. McMillan Publishing Company, New York

Jones GE, Garforth C (1997) The history, development, and future of agricultural extension. In: Swanson EB, Bentz RP, Sofranko AJ (eds) Improving agricultural extension. A reference manual. Food and Agriculture Organization, Rome, pp 7–19

Kothari CR (2004) Research methodology: methods and techniques. New Age International, New Delhi

Lang DJ, Wiek A, Bergmann M, Stauffacher M, Martens P, Moll P, Swilling M, Thomas CJ (2012) Transdisciplinary research in sustainability science: practice, principles, and challenges. J Sustain Sci 7(1):25–43

Loebenstein G, Thottappilly G (2007) The mission of agricultural research. In: Loebenstein G, Thottappilly G (eds) The agricultural research management. Springer, Dordrecht, pp 3–7

Lynam JK, Herdt RW (1989) Sense and sustainability: sustainability as an objective in international agricultural research. Agric Econ J 3(4):381–398

Lyson TA (1998) Environmental, economic and social aspects of sustainable agriculture in American land grant universities. J Sustain Agric 12(2–3):119–130

Maiangwa MG (2010) Importance, problems, and reforms of agricultural research in Africa. J Agric For Soc Sci 8(2):127–138

Manning S, Boons F, Von Hagen O, Reinecke J (2012) National contexts matter: the co-evolution of sustainability standards in global value chains. J Ecol Econ 83:197–209. doi:10.1016/j.ecolecon.2011.08.029

Moldan B, Janouskova S, Hak T (2011) How to understand and measure environmental sustainability: indicators and targets. J Ecol Indic 17:4–13. doi:10.1016/j.ecolind.2011.04.033

Morelli J (2011) Environmental sustainability: a definition for environmental professionals. J Environ Sustain 1(1):1–10

Ocholla D, Onyancha OB (2006) The nature and trends of agricultural research development in Africa: an informetric study. S Afr J Libr Inf Sci 72(3):226–235

Omamo SW, Diao X, Wood S, Chamberlin J, You L, Benin S, Wood-Sichra U, Tatwwangire A (2006) Strategic priorities for agricultural development in eastern and central Africa. IFPRI research report 150. http://ebrary.ifpri.org/cdm/ref/collection/p15738coll2/id/125234. Accessed 28 May 2015

Petters A, Sauvenier X, Valckx J, Van-Cauwenbergh N, Wauters E, Bachev H, Biala K, Bielders C, Brouckaert V, Garcia-Cidad V, Goyens S, Hermy M, Mathijs E, Muys B, Vanclooster M (2005) Framework for assessing sustainability levels in Belgian agricultural systems (SAFE). https://www.belspo.be/belspo/organisation/Publ/pub_ostc/CPagr/rappCP28_en.pdf. Accessed 11 May 2015

Productivity Commission (2005) Trends in Australian agriculture. Research paper, Canberra. http://www.pc.gov.au/research/supporting/agriculture. Accessed 20 Mar 2015

Saysel AK, Barlas Y, Yenigun O (2002) Environmental sustainability in an agricultural development project: a system dynamics approach. J Environ Manag 64(3):247–260

Sherman IW (1988) Two sides of a coin: basic and applied research. J Calif Agric 42(3):2

Spangenberg JH (2011) Sustainability science: a review, an analysis and some empirical lessons. J Environ Conserv 38(3):275–287

Stock P, Burton RJF (2011) Defining terms for integrated (multi-inter-trans-disciplinary) sustainability research. Sustain J 3(8):1090–1113

Talwar S, Wiek A, Robinson J (2011) User engagement in sustainability research. J Sci Public Policy 38(5):379–390

Thirtle C, Lin L, Piesse J (2003) The impact of research-led agricultural productivity growth on poverty reduction in Africa, Asia, and Latin America. J World Dev 31(21):1959–1975

United Nations Development Program (2014) African economic outlook: global value chains and Africa's industrialization. doi:10.1787/aeo-2014-en

Van-Cauwenbergh K, Sauvenier X, Valckx J, Wauters E, Bachev H, Biala K, Bielders C, Brouckaert V, Garcia-Cidad V, Goyens S, Hermy M, Mathijs E, Muys B, Vanclooster M, Peeters A (2007) SAFE—a hierarchical framework for assessing the sustainability of agricultural systems. J Agric Ecosyst Environ 120(2–4):229–242

Vandermeulen V, Huylenbroeck G (2008) Designing trans-disciplinary research to support policy formulation for sustainable agricultural development. J Ecol Econ 67(3):352–361

Warren P (2013) Why agriculture research matters: the aid perspective. http://aciarblog.blogspot.com.es/2013/03/why-agricultural-research-matters-aid.html. Accessed 27 Mar 2015

Water, Land and Ecosystems (2014) Research for development. http://wle-mekong.cgiar.org/about/research-for-development/. Accessed 30 Mar 2015

Wies R, Gerlach C, Chaplin T, Koskey M, Schanabel W (2011) Sustainability research definition. University of Atlas, Enschede

Chapter 33
Biological Inoculants for Sustainable Intensification of Agriculture in Sub-Saharan Africa Smallholder Farming Systems

C. Masso, R.W. Mukhongo, M. Thuita, R. Abaidoo, J. Ulzen, G. Kariuki and M. Kalumuna

Abstract Land degradation in the smallholder farming systems in sub-Saharan Africa is mainly related to insufficient adoption of sustainable agriculture technologies. This study was aimed at investigating the potential of biological inoculants to improve crop yields and control plant diseases in a profitable manner. Three *rhizobia* inoculants for soybean or common bean, 2 arbuscular mycorrhizae fungi (AMF) for sweet potato, and 2 *Trichoderma* products for tomato were applied to determine their effect on yields and tomato late blight disease. The study was conducted in Ghana, Kenya, Tanzania, and Uganda, but the treatments varied among the countries. The *Rhizobia* inoculants produced significant soybean or common bean yield increases in Ghana, Kenya, and Tanzania at $p \leq 0.05$ when compared to the untreated control, and an economic analysis of the Ghanaian data found that Legumefix was profitable with a value–cost ratio of >3. There was significant spatial variability in crop yields (coefficients of variation: 37–64 %), indicating a need for further investigation to correct the limiting factors. The sweet

---

C. Masso (✉) · M. Thuita
International Institute of Tropical Agriculture (IITA), C/O ICIPE, Duduville, Kasarani, P.O. Box 30772-00100, Nairobi, Kenya
e-mail: C.Masso@cgiar.org

R.W. Mukhongo
Department of Agricultural Production, School of Agricultural Sciences, College of Agriculture and Environmental Sciences, Makerere University, P.O. Box 7062-00256, Kampala, Uganda

R. Abaidoo · J. Ulzen
Department of Crop and Soil Sciences, Kwame Nkrumah University of Science and Technology, Kumasi, Ghana

G. Kariuki
Department of Biological Sciences, Egerton University, P.O. Box 536-20115, Njoro, Kenya

M. Kalumuna
Agricultural Research Institute, Mlingano, P.O. Box 5088, Tanga, Tanzania

R. Lal et al. (eds.), *Climate Change and Multi-Dimensional Sustainability in African Agriculture*, DOI 10.1007/978-3-319-41238-2_33

potato response to AMF was variable across sites and seasons, and a significant response ($p \leq 0.05$) was shown only under drought conditions in a soil with low organic matter content (1.2 %). The *Trichoderma* inoculants controlled late blight disease in tomatoes significantly better than Ridomil ($p \leq 0.05$), a synthetic fungicide currently used by farmers in Kenya. Biological inoculants can therefore improve the productivity of the sub-Saharan Africa smallholder farming systems, and awareness of them should be created for relevant stakeholders to increase understanding and adoption of technologies for sustainable agricultural intensification.

Keywords Biological inoculants · Smallholder farmers · Land degradation · Technology adoption · Sustainable intensification

33.1 Introduction

In the context of climate change, investment in sustainable agriculture technologies and practices is crucial for adaptation to sustain crop productivity and feed growing populations (Alarcón and Bodouroglou 2011). Land degradation contributes to climate change (World Meteorological Organization 2005). Although fertilizer production and use contribute to greenhouse gas emissions, sustainable fertilizer use based on nutrient stewardship contributes to the mitigation of climate change by reducing deforestation due to improved productivity per unit area (International Fertilizer Industry Association 2009) and sequestering carbon due to increased biomass production (Mujeri et al. 2012). Increasing population pressures in most sub-Saharan Africa countries and decreasing crop yields per unit area have resulted in deforestation and farming on marginal lands, exacerbating the risks of land degradation (Sutton et al. 2013) and contributing to climate change through reduced carbon–sequestration capacity (International Fertilizer Industry Association 2009).

Farmers in sub-Saharan Africa, especially smallholder households, have often pointed several causes of the limited use of fertilizers, including poor accessibility (i.e., unavailability, unaffordability) and the erroneous perception of fertilizers as too risky for soil quality due to insufficient understanding of the inputs (Minde et al. 2008). Alternative solutions, even when partial, seem needed to reverse the low use of inputs and improve smallholder farmers' ability to adopt sustainable agricultural intensification practices. This is especially important because crops with nutrient deficiencies are generally susceptible to plant diseases (Bhaduri et al. 2014). Biological inoculants designed to improve nutrient availability or reduce pest pressure may be worth considering because they are reported to be cost effective, as well as environmentally friendly (Ghosh 2003).

Biological inoculants generally contain natural-occurring microorganisms that are intentionally multiplied and added to a carrier material (Roesti et al. 2006; Ahmad et al. 2013). These inoculants are also known as bio-fertilizers, bio-pesticides, or biological control agents. Selected bio-fertilizers, such as phosphorus solubilizing bacteria and arbuscular mycchoriza fungi (AMF), can increase the availability of given native or applied nutrients, such as phosphorus, potassium, and micronutrients (Wu et al. 2004; Khan et al. 2009). Others, such as *rhizobia*, *azospirillum*, and *azotobacter*, can biologically fix nitrogen (Bhattacharyya and Tandon 2012). Biopesticides, such as *Trichoderma*, have the ability to induce resistance against selected pathogens (Mbarga et al. 2012).

Despite the potential cost effectiveness and environmentally friendliness of biological inoculants, their availability in most sub-Saharan Africa countries remains very limited. Key stakeholders, including policy makers, regulatory bodies, extension services, and farmer organizations, are not sufficiently aware of the benefits of such inputs. This has resulted in low demand for the technologies and poor investment by the private sector in the supply chain (Masso et al. 2015). Insufficient investment in research has also contributed to poor knowledge generation that could influence policy decisions conducive to the adoption of biological inoculants (Masso et al. 2015). This study aimed at investigating the potential of biological inoculants to improve crop productivity in a profitable manner.

33.2 Materials and Methods

33.2.1 Locations of the Study

The data presented herein were collected in the context of a broad 2012–2017 project (COMPRO-II) on the institutionalization of quality-assurance mechanisms and dissemination of top-quality commercial products to increase crop yields and improve the food security of smallholder farmers in sub-Saharan Africa. This project has been implemented in six countries: Ethiopia (West Showa, South West, East and West Arsi, Jima, Ilubabor, and Kemashi agricultural regions); Ghana (Upper West and Northern agricultural regions); Kenya (Western Kenya); Nigeria (Benue, Kaduno, Kano, and Niger states); Tanzania (Mbeya, Ruvuma, Morogoro, and Tanga agricultural regions); and Uganda (Central and Eastern). The data reported herein focus on Ghana, Kenya, Tanzania, and Uganda.

33.2.2 Products and Crops Assessed

Various biological inoculants were assessed, with a particular focus on rhizobia, AMF, and *Trichodema* inoculants. The tested crops included various legumes, maize (*Zea mays*), tomato (*Solanum lycopersicum*), pepper (*Capsicum*), cassava (*Manihot esculenta*), and sweet potato (*Ipomoea batatas*). However, the tested biological inoculants and crops varied among countries, primarily due to local availability. The data reported herein are limited to soybean (*Glycine max*), common bean (*Phaseolus vulgaris*), maize, sweet potato, tomato, and the following biological inoculants: (i) Legumefix, Biofix, and Nitrosua for rhizobia inoculants; (ii) Rhizatech and Symbion Vam plus for AMF inoculants; and (iii) Trichotech and Trianum-P for *Trichoderma* inoculants. For the maize crop, inorganic fertilizers were used because no suitable biological inoculant had been identified at the time of the evaluation.

33.2.3 Treatment Application

33.2.3.1 Assessment of Rhizobia Inoculants in Kenya

The trials in Kenya were conducted in the context of a randomized control trial at 240 farms to assess farmers' ability to learn by doing. The number of plots was kept relatively low to minimize confusion among farmers. Untreated control and treatment with Minjingu and Sympal alone were conducted at all test locations. The sites were also split into two equal numbers of participating farmers. The first group received Biofix or a combination of a P source and Biofix, while the second received Legumefix or a combination of a P source and Legumefix. Minjingu and Sympal were applied at a rate of 30 kg P ha^{-1}. Soil samples from each farm were taken and analyzed. Soil pH in water was determined in a 1:2.5 (w/v) soil:water suspension, and organic C by chromic acid digestion and spectrophotometric analysis (Heanes 1984). Total N was determined from a wet acid digest (Buondonno et al. 1995) and then N analyzed using a colorimeter (Anderson and Ingram 1993). Olsen P was analyzed using the molybdenum blue method, as described by Murphy and Riley (1962).

33.2.3.2 Assessment of Rhizobia Inoculants in Tanzania

Trials were conducted in three locations in Tanga, Tanzania, (Potwe, Ngomeni, and Makolola) using rhizobia inoculants for common bean and soybean with and without a P source. The P sources used were Minjingu hyperphosphate, Minjungu mazao, and diammonium phosphate. The rhizobia inoculants used for common bean were Biofix for bean (MEA Ltd., Kenya) and Nitrosua (Sokoine University of Agriculture, Morogoro, Tanzania), while Biofix for soybean (MEA Ltd.) and Legumefix (Legume Technology Inc., United Kingdom) were used for soybean. The P sources were applied at a rate of 20 kg P ha^{-1}, and the rhizobia inoculants at the rates recommended by the manufacturers on the product labels. Before the experiment setting, soil properties were determined following the methodology described in Sect. 33.2.3.1.

33.2.3.3 Assessment of Rhizobia Inoculants in Ghana

Trials were conducted in 74 farmers' field in the Upper West Region and 40 farmers' field in the Northern Region. Each farm was designated as a block. A block measured 10 m^2, and harvesting was done within the inner 5 m^2. All the experiments were laid out in a randomized complete block design. Maize and soybean were planted in the Upper West Region, while only soybean was planted in the Northern Region. The four treatments for soybean were: (i) uninoculated control; (ii) rhizobia inoculant (5 g kg^{-1} seed); (iii) 30 kg P ha^{-1}; and (iv) 30 kg P ha^{-1} + rhizobia inoculant. Maize crop also received four treatments: (i) control; (ii) Actyva (371 kg ha^{-1}, representing 85 kg N ha^{-1}, 37 kg P ha^{-1}, and 19 kg K ha^{-1}); (iii) Actyva + Sulphan (123 kg ha^{-1}, representing 30 kg N ha^{-1} and 7 kg S ha^{-1}); and (iv) NPK (250 kg ha^{-1}, representing 83 kg N ha^{-1}, 83 kg P ha^{-1}, and 83 kg K ha^{-1}) plus sulphate of ammonia (130 kg ha^{-1}, representing 27 kg N ha^{-1} and 31 kg S ha^{-1}). Soils from randomly selected sites were analyzed following the methodology described in Sect. 33.2.3.1.

33.2.3.4 Assessment of Arbuscular Mycorrhizae Fungi Inoculants in Uganda

Trials were conducted in two agro-ecological zones (AEZs) in Uganda for two seasons: the long rainy season (2014A) and the short rainy season (2014B). In Eastern Uganda, the trial was conducted at the District Agricultural Training Centre (DATIC) in Tororo (Ferralsols), while in Central Uganda, the trial was performed in the Wakiso District at the Makerere University Agricultural Research Institute, Kabanyolo (MUARIK) (Rhodic Nitosols). Soil analysis was performed following the routine procedures outlined by Okalebo (2002). Briefly, soil pH was measured in a soil-water suspension at a ratio of 1:2.5 using a pH meter (Mettler-Toledo, AG 8603). Extractable P was determined using a spectrophotometer (Jenway, 6405 UV/Vis) following Murphy and Riley's (1962) molybdenum blue method. Exchangeable K was assessed from ammonium acetate extracts using a flame-photometer (Jenway, Essex CM6 3LB). Total N and organic carbon were determined using a colorimeter after digestion of the samples.

The tested products were soil applied at the following rates: Rhizatech: 50 g per mound; Symbion Vam plus: 1.3 g per mound; Triple Super Phosphate: 60 kg P ha^{-1}; Urea: 90 kg N ha^{-1}; and muriate of potash: 100 kg K ha^{-1}. Control treatments were included in the study. P fertilizer and a third of the N and K fertilizers were applied at planting, while top-dressing with two-thirds of the N and K fertilizers was done at 2 months after planting (2 MAP). The NASPOT11 sweet potato variety was used. Each treatment was replicated 4 times.

33.2.3.5 Assessment of *Trichoderma* Inoculants in Kenya

Experimental soils were collected from three locations in Kenya: Bungoma (Haplic Acrisol), Egerton (Vitric Andosols), and Chuka (Rhodic Nitisols). The soils were sterilized before tomato planting in greenhouse conditions at Egerton University (Kenya). The technologies tested to control late blight (*P. infestans*) disease in tomato included: Trianum-P (T1), Trichotech (T2), a combination of Trianum-P and Trichotech (T3), and Ridomil (T4) as a positive control and water (T0) as a negative control. The factorial experiment, therefore, consisted of two factors: the three soil and five technology levels. The technologies were applied 2 weeks after tomato transplanting. Disease severity was then monitored at a frequency of 3 weeks for a period of 12 weeks from transplanting.

33.2.4 Data Collection and Analysis

Of the various data collected, soil characteristics, yield, disease severity, input costs, and produce prices are herein reported. Analysis of variance was conducted on selected datasets using SAS Version 9.3 for Windows (SAS Institute). The treatment means were separated using the least significant difference (LSD) at a probability of 5 % when the models were significant. For economic analysis, the value–cost ratio (VCR) was computed based on the input costs and the produce prices, as follows (Eq. 33.1). A VCR threshold value of 3 was used to determine the profitability of a given input combination (Dittoh et al. 2012).

$$\mathrm{VCR} = \frac{\text{Unit price of the produce}\ \left(\$\ \mathrm{kg}^{-1}\right) * \text{Yield gain}\ \left(\mathrm{kg\ ha}^{-1}\right)}{\text{Cost of the inputs}\ \left(\$\ \mathrm{ha}^{-1}\right)} \tag{33.1}$$

where yield gain is the additional yield obtained from the plots receiving the inputs of interest compared to the plots that did not receive the inputs.

33.3 Results

33.3.1 Selected Soil Proprieties of the Various Experimental Sites

In Kenya, half of the soils showed pH levels of less than 5.50, which are generally associated with reduced P availability and increased aluminum toxicity (Havlin et al. 2005). In the majority of farms, soil organic carbon (C_{org}), total N, and available P were lower than the moderate levels recommended by Okalebo (2002) (3, 0.25 %, and 10 mg kg^{-1}, respectively) (Table 33.1). These results confirm the

Table 33.1 Selected soil properties of experimental sites in Kenya, Ghana, Tanzania, and Uganda

Parameter/location	pH (H_2O)	Organic carbon (%)	Total N (%)	Olsen P (mg kg^{-1})	Exchange K (mg kg^{-1})
Kenya (N = 240)					
Mean (±SD[a])	5.53 ± 0.57	1.42 ± 0.48	0.11 ± 0.04	7.00 ± 13.96	166 ± 163
Minimum	4.24	0.55	0.03	0.64	28
Maximum	7.03	2.70	0.19	100.40	846
Ghana (N = 25)					
Mean (±SD)	5.34 ± 0.52	–	0.05 ± 0.03	4.44 ± 1.36	43 ± 16
Minimum	4.47	–	0.03	2.34	23
Maximum	6.56	–	0.13	7.61	78
Tanzania (mean ± standard deviation)					
Potwe	5.62 ± 0.45	1.31 ± 0.37	0.11 ± 0.02	2.94 ± 1.24	46 ± 36
Ngomeni	5.65 ± 0.46	0.95 ± 0.42	0.07 ± 0.02	2.67 ± 0.78	108 ± 53
Makolola	6.70 ± 0.14	1.60 ± 0.22	0.18 ± 0.03	17.00 ± 1.41	577 ± 750
Uganda (mean ± standard deviation)					
DATIC	5.82 ± 0.09	1.25 ± 0.11	0.11 ± 0.02	5.60 ± 2.6	179 ± 14
MUARIK	5.57 ± 0.16	3.50 ± 0.41	0.15 ± 0.01	1.54 ± 0.33	109 ± 2

[a]SD = standard deviation

widespread N and P deficiencies in Western Kenya. The ranges (minimum: maximum) for the soil properties showed significant spatial variability.

The spatial variability of the selected properties was also confirmed in the three other countries (Ghana, Tanzania, Uganda). On average, pH, C_{org}, and nutrient availability were low in most soils. The application of nutrients, such as N and P, is needed to sustain crop productivity in these soils.

33.3.2 Legume Response to Rhizobia Inoculants

Soybean response to treatment in both Kenya and Tanzania was affected not only by the type of inputs but also by the location (Figs. 33.1, 33.2 and 33.3). In Kenya, irrespective of the spatial variability, the use of rhizobia inoculants with or without a source of P (i.e. Minjingu or Sympal) showed high yields compared to the untreated control and fertilizer alone (Fig. 33.1a, b).

The observed yield increase might be related to improved access to nitrogen because the rhizobia inoculants were used to improve the biological nitrogen fixation (BNF) of the legume. The addition of P to the rhizobia inoculants did not always result in a significant yield increase. Irrespective of treatment, spatial variability could be related to differences in the initial soil fertility across sites. In Tanzania, a similar trend was found for both soybean and common bean (Figs. 33.2 and 33.3). Better responses for both common bean and soybean were

(a)
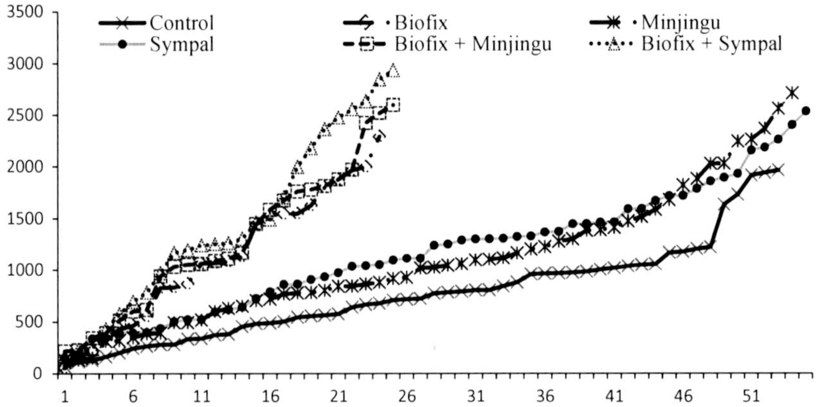

(b)
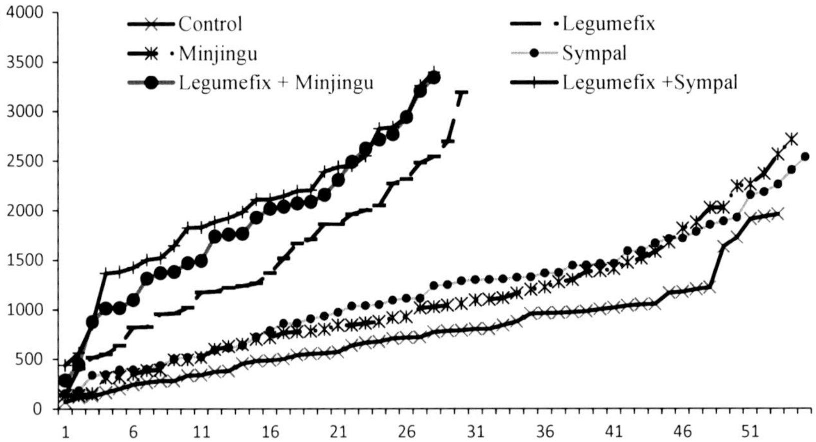

Fig. 33.1 Soybean response to various inputs in Western Kenya as affected by farms' soil fertility typology for the set of 240 smallholder farmers reported in Table 33.1 [**a** Biofix and **b** Legumefix]. The soybean response was not only affected by the inputs, but also the initial soil fertility as shown by the mean values (kg ha^{-1}), the standard deviations (kg ha^{-1}), and the coefficients of variation (CV; %) for the nine treatments [mean ± standard deviation (coefficient of variation) i.e. Control: 791 ± 510 (64); Minjingu: 1098 ± 636 (58); Sympal: 1176 ± 592 (50); Biofix: 1112 ± 650 (58); Minjingu with Biofix: 1272 ± 713 (56); Sympal with Biofix: 1445 ± 854 (59); Legumefix: 1480 ± 754 (51); Minjingu with Legumefix: 1847 ± 786 (43); and Sympal with Legumefix: 2000 ± 730 (37)]

observed in the moderate fertile site of Makolola, while the response was very low at Potwe, which was less fertile. Low fertile soils might require amendment to improve responses to *rhizobia* inoculants. Regression analysis showed that the low pH observed in most locations could not be enough to explain on its own the variability of yield responses to the various treatments (R^2 range: 0.04–0.16). Additional soil properties, as well as rainfall, might have also negatively affected the responses.

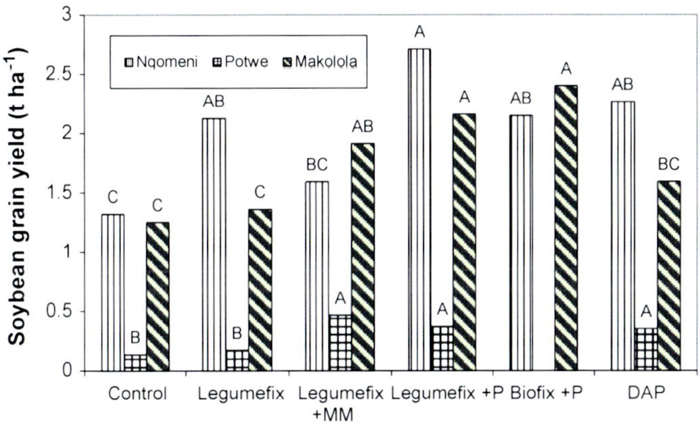

Fig. 33.2 Soybean response to various inputs at three sites in Tanzania

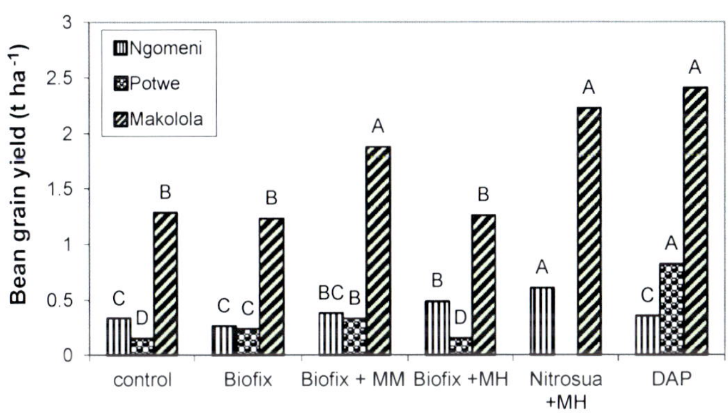

Fig. 33.3 Common bean response to various inputs at three sites in Tanzania

33.3.3 *Sweet Potato Response to Arbuscular Mycorrhizae Fungi Inoculants*

When Rhizatech and Symbion Vam plus with and without N and potassium (K) were applied to sweet potato at two locations in Uganda, no statistically significant difference across treatments was observed during the long rainy season (2014A). A significant difference was observed at the DATIC site only in the short rainy season (2014B), as shown in Table 33.2. When there was no response to the AMF inoculants, there was also no response to the NPK mineral fertilizer treatment. However, the fertility level at both sites was relatively low (Table 33.1), and the lack of response to NPK fertilizers was not expected. This result shows other factors might have affected sweet potato response to the treatments at the sites. Further investigation is necessary to understand the cause of the non-response to the NPK

Table 33.2 Sweet potato response to AMF inoculants at two sites in Uganda for two seasons

Treatment	Season 2014A		Season 2014B	
	DATIC site	MUARIK site	DATIC site[a]	MUARIK site
Yield (metric tons ha^{-1})				
Control	27.6	25.8	7.6b	13.6
Rhizatech	34.3	21.4	11.3ab	12.8
Rhizatech + NK	30.5	27.5	12.8a	13.2
Symbion Vam plus	23.8	28.8	9.9ab	18.0
Symbion Vam plus + NK	27.8	26.3	13.6a	20.1
NPK positive reference	34.5	28.3	12.7a	15.7
LSD ($p \leq 0.05$)	N/A[b]	N/A	4.7	N/A

[a]Means followed by the same letter in the same column are not significantly different based on the LSD test at $p \leq 0.05$
[b]N/A = not applicable; there was no statistically significant difference across treatments

fertilizer. On average, sweet potato did better in 2014A at the DATIC site than the MUARIK site (Table 33.2), which reflects the initial fertility levels at the two sites based on pH, available P, and exchangeable K (Table 33.1). The opposite was observed at the two sites in 2014B, and there was also a treatment effect at the DATIC site. The poor yields at both sites in 2014B relative to 2014A could be explained by the prolonged drought episode during the 2014B season, especially at the DATIC site.

Regarding water-holding capacity, the MUARIK site had a comparative advantage due to high organic matter content (Table 33.1), which could explain its better performance than DATIC during the season of poor rains. The significant difference between the selected treatments and the control at DATIC in 2014B may be related to sweet potato tolerance to moderate drought in the presence of AMF or fertilizers, which could have improved the development of the root system, enabling better water acquisition. For both seasons, there was a significant yield gap, based on the attainable yield of approximately 48 t ha^{-1} for the NASPOT11 (Mwanga et al. 2011). Therefore, there is a need to investigate the limiting factors and poor sweet-potato response to both the NPK fertilizer and the AMF inoculants. Such factors might include soil physics (e.g., compaction), chemistry (e.g., other nutrients, pH, organic matter), biological properties (e.g., microbial activity), and climatic conditions (e.g., rainfall).

33.3.4 Tomato Response to Trichoderma *Inoculants*

The performance of Trichotech, Trianum-P, or a combination of the two at controlling tomato late blight disease in greenhouse conditions was statistically

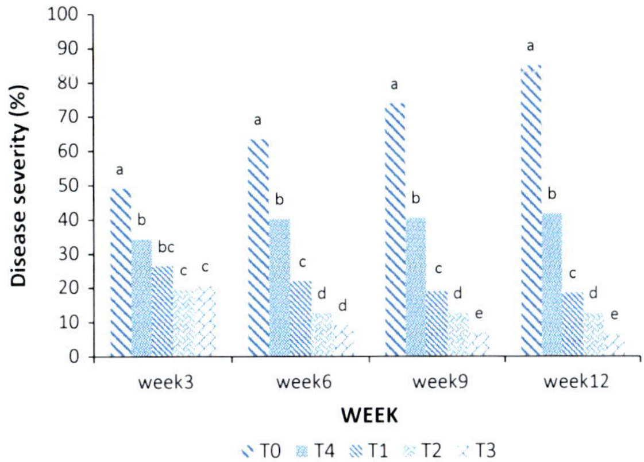

Fig. 33.4 Effect of various technologies on the severity of tomato late blight disease (*P. infestans*) in greenhouse conditions in Kenya across soil types [*T0 = Water (negative control); T1 = Trianum-P; T2 = Trichotech; T3 = Trianum-P plus Trichotech; T4 = Ridomil (positive control)*]

significant compared to the untreated control. In general, the biological inoculants did better than the synthetic fungicide (Ridomil), especially from week 6 after planting. On average, there was a significant benefit from combining Trianum-P and Trichotech, while Trichotech presented a comparative advantage over Trianum-P (Fig. 33.4). Performance was consistent over the 12 weeks of the study. These findings present additional options for farmers who have relied mainly on Ridomil. Given that most smallholder farmers in Africa are resource disadvantaged, they likely would prefer the cost-effective option.

33.3.5 Value–Cost Ratios of Selected Inputs

Based on the VCR threshold value of 3 and data collected in Ghana, the combination of triple superphosphate (TSP) and Legumefix was recommended to smallholder farmers due to profitability (Fig. 33.5). A VCR equal to 3 is generally considered the minimum requirement for a farmer to adopt a given technology when the production or price risk is high (Dittoh et al. 2012). Both risks are expected in the smallholder farming systems in sub-Saharan Africa given the changing climate and the seasonal volatility of both input and output prices at the farm-gate. VCR, therefore, is a simple economic tool used to verify whether it is worth investing in a given technology or practice based on cost recovery and potential profit. When the VCR is equal to 1, the cost is recovered without profit. Smallholder farmers in sub-Saharan Africa, who are generally risk-averse (Kisaka-Lwayo et al. 2005), might be reluctant to adopt the technology in this situation, even though there might still be some agronomic benefits related to the improvement of soil fertility. Hence, the higher the VCR is, the easier it is to convey the extension message promoting novel agricultural technologies or practices in smallholder farming communities given the expected profitability. A VCR equal to 2 is attractive when production and price risks are low, but a high VCR is

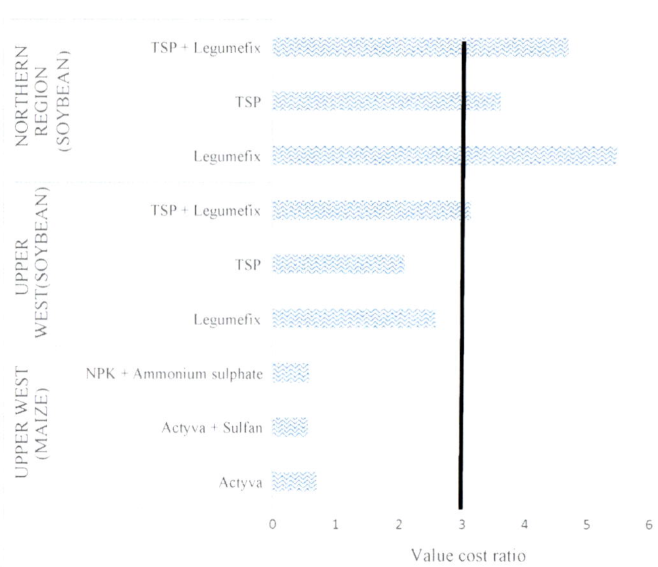

Fig. 33.5 Value to cost ratios (VCR) of selected technologies tested in soybean and maize in two regions of Ghana. A VCR equal to three was considered as the threshold value representing the minimum requirement for a farmer to adopt a given technology when the production or price risk is high, which would be expected in the context of climate change and price volatility

generally recommended when the production and price risks are considered relatively high (Dittoh et al. 2012).

Spatial variability in the VCR (Fig. 33.5) was observed, primarily related to variances in crop responses in Upper West and Northern agricultural regions of Ghana. In both regions, the cost of TSP was \$15.9 ha^{-1}, while the cost of Legumefix was \$0.62 ha^{-1}. Soybean yields were slightly better in the Northern region than the Upper West region (Table 33.3). TSP did not contribute to improvement of VCR in the Northern region for the TSP and Legumefix combination, but it did in the Upper West region. This result shows the need to apply rhizobia inoculants with a P source when the available levels of P are low to moderate.

The application of inorganic fertilizers to maize in the Upper West region in the absence of suitable biological inoculants was found to be not profitable. The relative percentage yield increases compared to the untreated control were higher for maize than soybean (Table 33.3). However, due to input costs and maize prices (data not shown), only soybean was profitable.

Table 33.3 Response of soybean and maize to Rhizobia inoculant and mineral fertilizers in Ghana

Location	Treatment	Grain yield (kg ha^{-1})
Soybean		
Northern Region	Control	830
(N[a] = 40)	Inoculant[b]	1256
	TSP[c]	1312
	TSP + inoculant	1461
	LSD (0.05)	123
Upper West Region	Control	997
(N = 74)	Inoculant	1199
	TSP	1260
	TSP + inoculant	1405
	LSD (0.05)	74
Maize		
Upper West Region	Control	1760
(N = 74)	Actyva	2293
	NPK + sulphate of ammonia	2568
	Actyva + sulphan	2745
	LSD (0.05)	166

[a]N = Number of participating farmers
[b]Inoculant = Legumefix
[c]TSP = Triple superphosphate

33.4 Discussion

33.4.1 Performance of Biological Inoculants

Soybean and common bean responses to rhizobia inoculants showed significant spatial variability in the data obtained in Kenya, Ghana, Tanzania, and other project countries (Ethiopia, Nigeria, Uganda) (data not shown). Crop response to biological inoculants is highly affected by product quality, target crops, crop specificity, management practices, and environmental conditions, including soil fertility and weather, as reported by Woomer et al. (2014). As stated, there was significant spatial variability in soil fertility across the experimental sites in each country, which may explain the inconsistent crop response to the biological inoculants. Although nodulation and soybean are generally sensitive to acidity (Bordeleau and Prévost 1994; Wolf 1999; Havlin et al. 2005), pH alone cannot explain the spatial variability of the crop response. Multivariate analysis is needed to determine the critical factors that affected crop response to the biological inoculants. This knowledge could inform further interventions to improve crop response to inoculation.

Better understanding of initial fertility and, in particular, threshold values to identify needed corrections to given soil parameters and improve the use efficiency of biological inoculants is crucial for wide adoption of this technology in sub-Saharan Africa. To improve inoculation recommendations, the COMPRO-II project is investigating the threshold values of key parameters affecting crop responses to biological inoculants, such as soil organic matter content, soil pH, nutrient content (e.g., N and P levels), and the most probable numbers (MPN) for the target microbial strains.

Sweet potato response to the AMF inoculants was significant only in soil with low organic matter content during the season with drought episodes. AMF, however, are commonly used to improve availability and plant uptake of selected nutrients, such as P and Zn (Harikumar and Potty 2002; van der Heijden et al. 2006; Hu and Rufty 2007; Yaseen et al. 2011). Performance is generally a function of the suitability of the symbiotic relationship between the AMF and the host plant (Öpik and Moora 2012). The low response to both AMF inoculation and NPK fertilization, therefore, was unexpected given the low P levels in the tested soils based on classification of soil fertility developed by Okalebo (2002). While response to P fertilizer in acid soils is generally low due to fixation (Rengel and Marschner 2005), the response to AMF is expected to be better in slightly acid soils (Fattah 2013) due to improved solubilization of P or extended rhizosphere as a result of the fungal hyphae (Lambers et al. 2008; Parewa et al. 2010).

The insignificant response to AMF, at both sites during the long rainy season and at MAURIK during the short rainy season, calls for investigation of better strains that could improve nutrient acquisition in sweet potato. Based on the positive response to AMF inoculation at the site with low organic matter content during the 2014B season with drought episodes and less rain, it is assumed that the main mechanism of action of the AMF product strains could have been improved water-use efficiency, but further investigation is needed to confirm this mechanism of action in sweet potato. Compared to the DATC site, the MUARIK site had an advantage in water retention given the high organic matter content. Porcel and Ruiz-Lozano (2004) reported that AMF can improve moderate-drought tolerance through osmotic adjustment in the roots to maintain a water potential gradient conducive to water entrance into the roots from the soil. The COMPRO-II project is also testing local AMF strains isolated from sweet potato, sorghum, legume, and fallow fields to determine their ability to improve sweet potato production through improved nutrient and water-use efficiency. This is especially important because the best performance observed in this study (34.5 kg ha^{-1}) was still very low compared to the attainable yield (48 t ha^{-1}) reported by Mwanga et al. (2011). Other factors that require further investigation to explain the persistent yield gap in sweet potato production in Uganda include climatic conditions, such as rainfall, and deficiencies in nutrients other than those tested (Pender 2004).

The response to the *Trichoderma* inoculants in greenhouse conditions in Kenya was consistent over time. However, the results were limited to one trial and should be validated in various locations and during different cropping seasons, in addition to performance assessment in field conditions. Trichoterm has a comparative advantage over the other products used to control tomato late blight disease. It is a local product manufactured by Dudutech in Naivasha (Kenya) and could be easily made available to farmers because it is not affected by the tedious procedures for importing biological inoculants.

33.4.2 Importance of the Profitability of Biological Inoculants

If the profitability of the biological inoculants demonstrated in this study was validated on a larger scale, it would represent a great opportunity to improve soil fertility in smallholder farming systems. Involving smallholder farmers in the assessment of the profitability of these biological inoculants is also important to increase adoption of these technologies. In sub-Saharan Africa, crop farming is mainly practiced by smallholder farmers, who are generally resource disadvantaged and have a high level of risk aversion, which discourages them from adopting innovative technologies (United Nations Industrial Development Organization 2014; Kisaka-Lwayo et al. 2005). In a study by Minde et al. (2008), more than 90 % of participating farmers did not use fertilizers, which they considered too expensive. Alternative solutions, even when partial, could play a key role in educating farmer organizations to promote sustainable intensification of the smallholder farming systems.

Using a VCR of 3 as the threshold value, Legumefix was found to be profitable, especially when used in conjunction with TSP. The profitability of the rhizobia inoculant showed spatial variability as a result of changes in yield gains (Fig. 33.5). This finding is also expected for other types of biological inoculants. Therefore, to ensure consistent response across regions and to increase the market share for these technologies, recommendations for use should be tailored to local conditions, particularly crop and biological inoculant combinations, environmental conditions, and adoption of good crop-management practices (Woomer et al. 2014). The current low demand for these technologies, attributable to inconsistent response and insufficient awareness creation, has resulted in minimum investment by the private sector in the products' distribution networks.

In countries, such as Kenya, and for biological control agents, such as *Trichoderma* inoculants, the focus has been on high-value markets intended for export, neglecting the needs of smallholder farmers. Consistent responses are expected to result in a spillover effect that could increase demand for the technologies and, consequently, the public and private sectors' interest in investing in the development of competitive products for smallholder farmers.

33.4.3 Biological Inoculants in the Sustainability of Smallholder Farming Systems

Profitable biological inoculants present a comparative advantage for the sustainability of the sub-Saharan Africa smallholder farming systems because they are significantly more affordable than inorganic fertilizers (Ghosh 2003). The use of inorganic fertilizers is generally constrained by their high cost (Minde et al. 2008) and insufficient economic return (Guo et al. 2009). The integration of legume crops, particularly when inoculated, has been shown not only to improve legume yields but also to increase crop residues, which could improve soil fertility when incorporated into soils (Nezomba et al. 2008; Chianu et al. 2011). Such residues have been reported to contribute to the soil's organic matter pool, which serves as an important indicator of soil fertility (Nezomba et al. 2015). Organic matter also contributes to minimizing land degradation through sustained microbial activities, aggregation stability, water retention, nutrient cycling, and reduced compaction (Robert 2006).

Land degradation in most regions of sub-Saharan Africa has often been associated with low input agriculture. When the production per unit area decreases over time, farmers tend to encroach on forest reserves (deforestation) or marginal lands vulnerable to water erosion (Sutton et al. 2013). Although several initiatives have promoted fertilizer use since the adoption of the Abuja Declaration during the 2006 Africa Fertilizer Summit (Wanzala 2011), average fertilizer use remains very low, less than 10 kg ha^{-1} for all nutrients combined (Dittoh et al. 2012). As few as 1–3 % of farmers could be using fertilizers in selected countries for several reasons, including high costs (Nkonya et al. 2011; Sheahan and Barrett 2014). The adoption of cost-effective products, such as biological inoculants and rhizobia inoculants for biological nitrogen fixation, might contribute to solutions to improve inputs use by resource-disadvantaged farmers seeking to increase soil fertility and generate income (Chianu et al. 2011). The residual N from the legume crops could benefit subsequent crops, as demonstrated by Waddington et al. (2004) in groundnut–maize rotation in Zimbabwe.

Adoption of biological inoculants remains quite low in sub-Saharan Africa despite the potential profitability and benefits. This low uptake has been associated with low awareness, high regulatory barriers, and insufficient understanding of the technologies among regulatory bodies and policy makers (Masso et al. 2015). Consequently, very few such products are in the marketplace, and the quality of selected products was found to be doubtful in a recent study (Jefwa et al. 2014). Innovation platforms and participatory demonstration of the benefits of biological inoculants that include various development partners, such as farmer organizations, extension services, regulatory bodies, policy makers, industry, and national and locally-based international research organizations, would be crucial to improve awareness and uptake of biological inoculants. These efforts could also ensure the enforcement of quality standards for biological inoculants in the marketplace. When profitability is widely demonstrated, increased demand can be expected, leading to increased involvement of public–private partnerships in the development of an efficient distribution network for the technologies. Wide uptake of profitable biological inoculants might result in improved soil and crop productivity and increase the sustainability of the smallholder farming systems, especially when the innovation platforms produce relevant policy decisions and educate farmers to invest their additional income in sustainable agricultural intensification practices.

33.5 Conclusion

Achieving sustainability in the smallholder farming systems demands improved adoption of innovative technologies that enhance soil and crop productivity without significant negative effects on the environment. On average, rhizobia inoculants increased soybean and common bean yields, though there was spatial variability in the responses. *Trichoderma* inoculants could control late blight disease in tomatoes either as well as or better than Ridomil. AMF performance was variable, and the only significant response was observed in drought conditions in soil with low organic matter content. The factors that limited the response to biological inoculants in selected locations should be investigated to address the spatial variability of yields. Elimination of inconsistent responses, combined with the profitability of biological inoculants, particularly rhizobia, could increase the acceptability and the adoption of the technologies by smallholder farmers. The mechanisms of action for AMF inoculants in sweet potato should also be investigated to improve the recommendations for use. Smallholder farmers, extension services, and other relevant stakeholders are not sufficiently aware of the biological inoculant technologies. Therefore, platforms to promote them and ensure consistent quality in the marketplace and field performance, may improve understanding and adoption of the products for sustainable intensification of smallholder farming systems.

Acknowledgments The authors take this opportunity to reiterate their appreciation of the financial support from the Bill & Melinda Gates Foundation through the COMPRO-II project grant to IITA. The authors are also grateful for technical support from the various project partners involved in the data collection. IITA's institutional support and the Innovative Agriculture Research Initiative's (*i*AGRI) facilitation during the development of this article are highly appreciated as well.

References

Ahmad F et al (2013) Phosphorus-microbes interaction on growth, yield and phosphorus-use efficiency of irrigated cotton. Arch Agron Soil Sci 59(3):341–351. doi:10.1080/03650340.2011.646994

Alarcón D, Bodouroglou C (2011) Agricultural innovations for food security and environmental sustainability in the context of the recent economic crisis: why a gender perspective. http://www.un.org. Accessed 30 June 2015

Anderson JM, Ingram JSI (eds) (1993) Tropical soil biology and fertility: a handbook of methods, 2nd edn. CAB International, The Cambrian News, Aberstwyth, UK

Bhaduri D et al (2014) Primary and secondary nutrients—a boon to defense system against plant diseases. Int J Bioresour Stress Manag 5(3):461–466. doi:10.5958/0976-4038.2014.00597.1

Bhattacharyya P, Tandon HLS (2012) Biofertilizer handbook—research, production, application. Fertilizer Development and Consultation Organization, New Delhi, India

Bordeleau LM, Prévost D (1994) Nodulation and nitrogn fixation in extreme conditions. Plant Soil 161:115–125

Buondonno A et al (1995) Comparing tests for soil fertility. II. The hydrogen peroxide/sulfuric acid treatment as an alternative to the copper/selenium catalyzed digestion process for routine determination of soil nitrogen-kjeldahl. Commun Soil Sci Plant Anal 26(9–10):1607–1619. doi:10.1080/00103629509369394

Chianu J et al (2011) Biological nitrogen fixation and socio-economic factors for legume production in sub-Saharan Africa: a review. Agron Sustain Dev 31(1):139–154. doi:10.1051/agro/2010004

Dittoh SA et al (2012) Improving the effectiveness, efficiency and sustainability of fertilizer use in sub-Saharan Africa—policy research paper number 3. http://www.gdn.int. Accessed 30 June 2015

Fattah OA (2013) Effect of mycorrhiza and phosphorus on micronutrients uptake by soybean plant grown in acid soil. Int J Agron Plant Prod 4:429–437

Ghosh N (2003) Promoting bio-fertilizers in Indian agriculture. Institute of Economic Growth, University Enclave. http://www.ipni.net. Accessed 25 Apr 2015

Guo Z, Koo J, Wood S (2009) Fertilizer profitability in East Africa: a spatially explicit policy analysis. Paper presented at the International Association of Agricultural Economics Conference, Beijing, 16–22 Aug 2009

Harikumar VS, Potty VP (2002) Arbuscular mycorrhizal inoculation and phosphorus mobility in phosphorus-fixing sweet potato soils. Malays J Soil Sci 11:45–56

Havlin JL et al (2005) Soil fertility and fertilizers—an introduction to nutrient management, 7th edn. Pearson Prentice Hall, Upper Saddle River, NJ

Heanes DL (1984) Determination of organic C in soils by an improved chromic acid digestion and spectrophotometric procedure. Commun Soil Sci Plant Anal 15:1191–1213. doi:10.1080/00103628409367551

Hu S, Rufty T (2007) Linking arbuscular mycorrhizal fungi with plant health: mechanisms and challenges. Phytopathol 97(7):142

International Fertilizer Industry Association (2009) Fertilizers, climate change and enhancing agricultural productivity sustainably. http://www.fertilizer.org. Accessed 30 June 2015

Jefwa JM et al (2014) Do commercial biological and chemical products increase crop yields and economic returns under smallholder farmer conditions? In: Vanlauwe B, van Asten P, Blomme G (eds) Challenges and opportunities for agricultural intensification of the humid highland systems of sub-Saharan Africa. Springer International Publishing Switzerland, Cham, pp 81–96. doi:10.1007/978-3-319-07662-1_7

Khan AA et al (2009) Phosphorus solubilizing bacteria: occurrence, mechanisms and their role in crop production. J Agric Biol Sci 1(1):48–58

Kisaka-Lwayo M et al (2005) Analysis of risk attitudes of smallholder crop farmers in Umbumbulu district, Kwazulu-Natal, South Africa. In: African Crop Science Society (ed) African crop science conference proceedings, Kampala, 5–9 December 2005, vol 7, pp 909–914

Lambers H et al (2008) Plant nutrient-acquisition strategies change with soil age. Trends Ecol Evol 23:95–103

Masso C et al (2015) Worldwide contrast in application of bio-fertilizers for sustainable agriculture: lessons for sub-Saharan Africa. J Biol Agric Healthc 5(2):34–50

Mbarga JB et al (2012) Mycoparasitic *trichoderma viridae* as a bio-control agent against *fusarium oxysporum* f. sp. *adzuki* and *pythium arrhenomanes* and as a growth promoter of soybean. J Crop Prot 29:1452–1459. doi:10.1016/j.cropro.2010.08.004

Minde I et al (2008) Improving access and utilization of fertilizers by smallholder farmers in the Limpopo province of South Africa. http://oar.icrisat.org/425. Accessed 30 June 2015

Mujeri MK et al (2012) Improving the effectiveness, efficiency and sustainability of fertilizer use in South Asia. Global Development Network (GDN). http://www.gdn.int. Accessed 30 June 2015

Murphy J, Riley JP (1962) A modified single solution method for determination of phosphate in natural waters. Anal Chim Acta 27:31–36. doi:10.1016/S0003-2670(00)88444-5

Mwanga ROM et al (2011) 'NASPOT 11', a sweet potato bred by a participatory plant-breeding approach in Uganda. HortScience 46(2):317–321

Nezomba H et al (2015) Point of no return? Rehabilitating degraded soils for increased crop productivity on smallholder farms in eastern Zimbabwe. Geoderma 239–240:143–155. doi:10.1016/j.geoderma.2014.10.006

Nezomba H et al (2008) Nitrogen fixation and biomass productivity of indigenous legumes for fertility restoration of abandoned soils in smallholder farming systems. S Afr J Plant Soil 25 (3):161–171. doi:10.1080/02571862.2008.10639912

Nkonya E et al (2011) Climate risk management through sustainable land management in sub-Saharan Africa. http://ebrary.ifpri.org. Accessed 30 June 2015

Okalebo JR (2002) Laboratory methods for soil and plant analysis. A working manual, 2nd edn. Tropical Soil Fertility and Biology Program-CIAT and SACRED Africa, Nairobi

Öpik M, Moora M (2012) Missing nodes and links in mycorrhizal networks. New Phytol 194:304–306. doi:10.1111/j.1469-8137.2012.04121.x

Parewa HP et al (2010) Evaluation of maize cultivars for phosphorus use efficiency in an Inceptisol. Int J Agric Environ Biotechnol 3(2):195–198

Pender J (2004) Development pathways and land management in Uganda. World Dev 32:767–792. doi:10.1016/j.worlddev.2003.11.003

Porcel R, Ruiz-Lozano JM (2004) Arbuscular mycorrhizal influence on leaf water potential, solute accumulation, and oxidative stress in soybean plants subjected to drought stress. J Exp Bot 55 (403):1743–1750. doi:10.1093/jxb/erh188

Rengel Z, Marschner P (2005) Nutrient availability and management in the rhizosphere: Exploiting genotypic differences. New Phytol 168:305–312. doi:10.1111/j.1469-8137.2005.01558.x

Roesti D et al (2006) Plant growth stage, fertiliser management and bio-inoculation of arbuscular mycorrhizal fungi and plant growth promoting rhizobacteria affect the rhizobacterial community structure in rain-fed wheat fields. Soil Biol Biochem 38(5):1111–1120. doi:10.1016/j.soilbio.2005.09.010

Robert M (2006) Global change and carbon cycle: the position of soils and agriculture. In: Roose EJ et al (eds) Soil erosion and carbon dynamics—advances in soil science. Taylor & Francis, New York, pp 3–12

Sheahan M, Barrett CB (2014) Understanding the agricultural input landscape in sub-Saharan Africa—recent plot, household, and community-level evidence. World Bank Group. http://elibrary.worldbank.org/doi/pdf/10.1596/1813-9450-7014. Accessed 30 June 2015

Sutton MA et al (2013) Our nutrient world: the challenge to produce more food and energy with less pollution. Centre for Ecology and Hydrology (CEH), Edinburgh

United Nations Industrial Development Organization (2014) Africa industrialization day—agribusinesses' contribution to food security. https://www.unido.org. Accessed 30 June 2015

van der Heijden MGA et al (2006) The mycorrhizal contribution to plant productivity, plant nutrition and soil structure in experimental grassland. New Phytol 172:739–752. doi:10.1111/j.1469-8137.2006.01862.x

Waddington S et al (2004) Progress in lifting soil fertility in Southern Africa. In: New directions for a diverse planet—proceedings of the 4th International Crop Science Congress, Brisbane, Australia, 26 September–1 October 2004

Wanzala M (2011) The Abuja declaration on fertilizers for an African green revolution—status of implementation at regional and national levels June 2011. The New Partnership for Africa's Development (NEPAD). http://www.nepad.org. Accessed 30 June 2015

World Meteorological Organization (2005) Climate and land degradation. https://www.wmo.int. Accessed 30 June 2015

Wolf B (1999) The fertile triangle—the interrelationship of air, water, and nutrients in maximizing soil productivity. Food Products Press, Binghamton, New York

Woomer PL et al (2014) N2Africa final report of the first phase 2009–2013. http://www.N2Africa.org. Accessed 30 June 2015

Wu JL et al (2004) Sorghum diversity evaluated by simple sequence repeat (SSR) markers and phenotypic performance. Plant Prod Sci 7:301–308. doi:10.1626/pps.7.301

Yaseen T et al (2011) Effect of arbuscular mycorrhiza inoculation on nutrient uptake, growth and productivity of cowpea (*vigna unguiculata*) varieties. Afr J Biotechnol 10(43):8593–8598. doi:10.5897/AJB10.1494

Chapter 34
The Economics of Conservation Agriculture in Africa: Implications of Climate Change

Philip Grabowski and Steven Haggblade

Abstract In this chapter, we summarize the available evidence on the agronomic and economic viability of conservation agriculture (CA) in sub-Saharan Africa and assess the likely impact of climate change on the agronomic and economic viability of CA. Using detailed data from Zambia, we compare the net present value of using various CA and conventional practices over a 10-year period and then analyze how those results are likely to change if rainfall becomes more erratic. CA is economically attractive especially when it allows farmers to overcome labor constraints during planting by distributing land preparation labor during the dry season. The results also show that when all three principles of CA are implemented, farmers will likely have more stable and higher yields than conventional tillage methods, although all types of farming will be negatively affected by dry spells. For these benefits to be realized, farmers must retain control of their residues, which, in turn, will require changing community norms about dry season grazing. Furthermore, the development of reliable markets for leguminous crops are necessary for achieving adequate rotations. Given the dynamic nature of smallholder agriculture, driven by climatic as well as socio-economic uncertainty, increasing farmers' capacity to adapt is of utmost importance. Working with farmers to adapt CA to match their specific priorities and constraints provides an opportunity to develop skills for resilient and adaptable farming systems.

Keywords Conservation agriculture · Climate change · Net present value · Crop residues · Legumes

---

P. Grabowski (✉)
Department of Plant Soil and Microbial Sciences, Michigan State University, East Lansing, MI, USA
e-mail: grabow21@msu.edu

S. Haggblade
Agricultural, Food, and Resource Economics, Michigan State University, East Lansing, MI, USA
e-mail: blade@msu.edu

R. Lal et al. (eds.), *Climate Change and Multi-Dimensional Sustainability in African Agriculture*, DOI 10.1007/978-3-319-41238-2_34

34.1 Objectives

Over many centuries, African farmers maintained soil fertility through shifting cultivation and long-term fallow periods. In recent decades, growing demographic pressure on farmland in Africa has resulted in reduced fallow periods that, in turn, lead to growing land degradation, soil erosion and nutrient mining (Stoorvogel and Smaling 1990; Smaling et al. 1997; Dreschel and Gyiele 1999). Although Africa is generally considered land abundant, significant portions of the continent face land constraints, increasing the pressure to develop viable systems of intensification (Crawford et al. 2003; Morris et al. 2007; Jayne et al. 2014). Conservation agriculture represents one among many related efforts to develop sustainable soil fertility management systems (Haggblade et al. 2010).

Conservation agriculture (CA) incorporates three agronomic principles that, together, lay the foundation for development of situation-specific agronomic practices to improve long-term soil fertility management, lower energy costs of tillage and raise farm income. Minimum tillage, usually during the dry season, aims to minimize soil disturbance, thus lowering the energy costs of tillage, reducing erosion and nutrient loss and improving the timeliness of planting and early season weeding by redirecting tillage labor to the dry season. Crop residue retention and mulching aim to suppress weed growth, moderate soil temperature, retain soil moisture and build up soil organic matter (SOM). Legume rotations, the third key principle of CA, aim to reduce disease pressure and improve soil fertility through biological nitrogen fixation.

The specifics of suitable CA agronomic practices necessarily vary by location, crop mixes, soil types and rainfall regimes and across farm households of differing resource endowments. As a result, CA development and adoption have resulted in a wide array of situation-specific management practices and highly variable reported adoption rates (Andersson and D'Souza 2014; Pannell et al. 2014).

With the advent of climate change, weather-induced shifts in cropping conditions will likely amplify the need to further adjust on-farm agronomic practices across a wide array of recommendation domains. Climate change models for southern Africa, for example, anticipate higher temperatures, decreased overall rainfall, as well as increased variability (Boko et al. 2007; Schlenker and Lobell 2010). Many advocates consider CA an important means of helping farmers adapt to climatic variability (Kassam et al. 2009; Milder et al. 2011).

This chapter addresses two key objectives. First, it aims to summarize available evidence on the agronomic and economic viability of CA in sub-Saharan Africa. Second, it assesses the likely impact of climate change on the agronomic and economic viability of CA.

The discussion begins with a review of empirical findings on the agronomic and economic viability of CA across alternative settings in Africa. Next, the chapter provides an analytical assessment of the likely impact of climate change on farming conditions in Africa and on the feasibility and desirability of CA management systems. The empirical analysis that follows builds on a wealth of micro-level data from Zambia to evaluate quantitatively the potential impact of climate change on CA profitability relative to conventional farming practices.

34.2 Factors Affecting the Viability of Conservation Agriculture

34.2.1 Agronomics of Conservation Agriculture

34.2.1.1 CA Agronomic Practices

CA consists of a variety of technologies that can be used to achieve three principles (minimal soil disturbance, mulching with crop residues or green manures and rotations or intercropping with legumes). Each of these three components of CA has its own benefits and challenges as well as complex interactions that result in complementarities when the components are practiced together (Thierfelder et al. 2013a).

Minimum tillage (MT) is often promoted first and emphasized most with the aim of improving soil quality and reducing the labor required for land preparation. MT can be accomplished in many ways, including hand hoe basins, direct seeding with a stick, hoe or jab-planter, and ripping with a tractor or oxen. A key benefit of MT is that it enables early planting through dry season land preparation while conventional tillage in many parts of Africa is not carried out until after the rains have softened the soil (Haggblade and Tembo 2003). One of the major challenges with MT is weed control (Nyamangara et al. 2014), and for this reason, CA is often promoted with herbicides (Andersson and D'Souza 2014), cover crops or the use of cut grasses as mulch to control weeds (Dambiro et al. 2011).

When MT is used, the crop residues are left on the surface of the soil as mulch, instead of being incorporated through tillage. Mulch retention is instrumental for improved water-use efficiency as mulch decreases runoff and evaporation while also minimizing oscillations in soil temperatures (Rockström et al. 2009). In many parts of Africa, farmers find it difficult to retain crop residues as mulch due to free-range grazing of livestock and uncontrolled burning (Baudron et al. 2014).

Rotating or intercropping with legumes every third year allows subsequent crops to benefit from biological nitrogen fixation and aids in pest control. This well-known practice is typically done only on a very small portion of the farm by smallholders due to the lack of attractive markets for the remaining harvest once home consumption is met (Giller et al. 2009). Smallholders also face challenges related to finding improved legume seed and overcoming phosphorous deficiencies in the soil (Mhango et al. 2013).

34.2.1.2 Empirical Evidence of the Agronomic Impact of CA

The yield effect of CA on maize depends on soil characteristics, rainfall patterns and farmers' management practices (weeding, fertilizing, planting date etc.). In general, CA has a positive impact on maize yields in southern Africa when fertilizer is adequate, weeds are controlled and rotations with legumes are used (Thierfelder et al. 2015a, b). CA had a negative yield effect on maize in 20 % of long-term on-farm trials, primarily due to waterlogging, inadequate mulch cover and improper management (Thierfelder et al. 2015a). Negative yield effects on granitic sandy soils that crust have also been documented (Thierfelder et al. 2015b).

Maize yield benefits may be immediate but more often increase gradually over time becoming significant after 3–5 years (Thierfelder et al. 2013b, 2015a, b). These findings are supported by a meta-analysis of long-term maize experiments, which showed generally increasing maize yields over time with MT (Rusinamhodzi et al. 2011). One reason why maize yield benefits may not be immediate is that when soils are nitrogen deprived the biological activity of breaking down cereal residues with a high carbon to nitrogen ratio can lock up soil nitrogen and suppress yields for the first few years (Verhulst et al. 2010; Giller et al. 2009).

There is also significant experimental evidence that CA has the potential to improve maize yields during dry years (Thierfelder et al. 2015b). The combination of minimum tillage with residues increases the infiltration rate of water in most soils (Thierfelder and Wall 2009, 2010) with the exception of certain sandy soils that easily crust, where MT actually reduces infiltration and increases runoff (Baudron et al. 2012). However, the meta-analysis of long-term experiments across the world showed no improvement in yield stability with CA (Rusinamhodzi et al. 2011). Likewise, the results of modeling maize with CA in southern Africa suggest that crop failure is still possible with CA during the lowest rainfall years (Mupangwa et al. 2011).

Maize yield improvements from CA have also been observed outside these carefully controlled experiments (Haggblade and Tembo 2003; Twomlow et al. 2008; Rockström et al. 2009). Ngoma et al. (2015) used household survey data from Zambia to estimate the *ceteris paribus* effect of using minimum tillage land preparation and found significant yield benefits when MT was combined with early land preparation. Work in central Zambia suggests that CA farmers plant, on average, 1–2 weeks earlier than farmers who plow or invert soil by hand hoe and that this contributes about 25 % of observed yield gains under CA (Haggblade et al. 2010).

Legume yield benefits are less pronounced, although dramatic improvements have been observed in drought years in semi-arid environments in Zimbabwe (Thierfelder et al. 2015b).

34.2.1.3 Implications of Climate Change, for CA Agronomics

Precipitation patterns in southern Africa are already highly variable, and the anticipated climatic changes of drier, hotter and less reliable rainfall make the potential benefits of CA more attractive. As rainfall becomes less reliable, it will be increasingly important to find agronomic practices that can improve the infiltration of water, increase soil water-holding capacity through higher SOM levels, reduce soil temperatures and enable earlier planting. With the expected increase in high-precipitation events, it will also be important for farmers to prevent soil erosion. Although CA has potential to provide these benefits, farmers face many challenges in being able to use CA effectively given the constraints in their farming systems. Some of these challenges may also become more difficult to address as the climate changes. For example, weeds may be more problematic, especially where herbicides become less effective (Rodenburg et al. 2011). Likewise, competition for crop residues may increase if farmers increase their herd sizes of small ruminants as an insurance mechanism. Tables 34.1 and 34.2 present a list of the short- and long-term benefits and problems associated with CA, the conditions for where these conditions apply and the expected impact of climate change on these effects of CA.

34.2.2 Economics of Conservation Agriculture

34.2.2.1 Empirical Evidence on the Economic Viability of CA in Africa

CA remains controversial in Africa. Advocates stress the important potential agronomic gains from CA and suggest that early evidence promises accelerated adoption in the future (Friedrich et al. 2012). Skeptics point to generally low adoption levels and highly variable outcomes across locations (Giller et al. 2009; Pannell et al. 2014).

The published work on CA in Africa suggests several generalizable findings. The first concerns the heterogeneity of farming situations, CA packages and outcomes (Food and Agriculture Organization of the United Nations, FAO 2001; Erenstein et al. 2012). Partial and often step-wise adoption of the various components of recommended CA packages results in difficulty accurately measuring adoption rates (Baudron et al. 2012, Mazvimavi and Twomlow 2009; Grabowski et al. 2014).

The majority of published studies nonetheless suggest that CA practices have the potential to generate higher economic returns than conventional alternative management practices (Pannell et al. 2014; Knowler and Bradshaw 2007). Paradoxically, available studies also measure generally low adoption rates (Giller et al. 2009; Pannell et al. 2014). A recent review summarized the situation as follows: "the majority of published economic results for CA are favourable, indicating that farmers would potentially benefit from adoption. However, the contrast between these favourable results and the low adoption of CA in most parts of Africa and South Asia is striking" (Pannell et al. 2014, p. 57). Pannell and colleagues

Table 34.1 Implications of climate change, for the short- and long-term positive effects of conservation agriculture (CA)[a]

	Minimum tillage	Mulch	Conditions for this effect to be meaningful	Hypothesized impact of climate change
Short-term positive effects				
Facilitates early planting	X		Where conventional system does not prepare land until the rains have softened soil	More important if erratic rainfall makes it more difficult to judge when the "real" rainy season has started More important if the duration of the rains is less reliable
Reduced labor for land preparation	X		Where ripping is faster than plowing and where direct seeding is used	More important if yield uncertainty induces diversification of livelihood strategies
Reduced evaporation	X	X	Generally occurs	More important as temperatures rise and rainfall is more erratic
Increased infiltration	X	X	General, dramatic where a hard pan problem is corrected. Not on sandy soils that form a hard crust	More important as rainfall decreases and frequency of dry spells increase
Reduced soil temperature oscillations		X	Generally occurs	More important as temperatures rise and water shortages are more common
Long-term positive effects				
Reduced erosion	X	X	On gently sloping fields or in combination with contour bunds or vegetation strips	More important with increased frequency of heavy downpours and flood years
Increased soil organic matter	X	X	Larger effect on clay soils	Slight reduction possible if increased temperatures accelerate decomposition More important as water shortages increase
Increased soil aggregation	X	X	Generally occurs	More important for increasing infiltration and water holding capacity of the soil
Avoids development of compacted layer from hoeing and plowing			In clay soils	

[a]Adapted from Grabowski and Kerr (2014). *Sources* Baudron et al. (2012), Giller et al. (2009), Haggblade and Tembo (2003), Verhulst et al. (2010)

Table 34.2 Implications of climate change, for the short- and long-term negative effects of conservation agriculture (CA)[a]

Short-term negative effects	Minimum tillage	Mulch	Conditions for this effect to be meaningful	Hypothesized impact of climate change
Reduced Nitrogen availability	X		When soil nitrogen is limited and when residues with high C:N ratio cause immobilization or the lack of tillage significantly reduces mineralization	Leaching will increase during high rainfall events, and higher temperatures may increase volatilization of added nitrogen
Increased weed pressure	X		Under most conditions, magnitude depends on weed seed stock	Potential for increased weed growth with higher temperatures and higher CO_2 concentrations in some contexts Herbicides may become less effective due to decreased persistence in the soil and greater risk of causing crop damage Weed tolerance of herbicides may also increase with higher CO_2 concentrations
Waterlogging	X		Higher rainfall areas or during high rainfall events, especially with heavier soils	More frequent with more erratic rainfall
Stimulation of crop pests and diseases		X	Depending on the conditions for each particular pest and disease	Crops are more susceptible to disease when water stressed
Long-term negative effects				
Increased soil compaction	X		Especially in sandy soils	More problematic because it results in decreased infiltration

[a]Adapted from Grabowski and Kerr (2014). *Sources* Baudron et al. (2012), Giller et al. (2009), Haggblade and Tembo (2003), Verhulst et al. (2010), Rodenburg et al. (2011)

suggested several possible explanations for this seeming contradiction: overly simplistic partial budgeting calculations, over-optimistic assumptions about agronomic impacts from on-station trials, possible publication bias or possibly low net benefits that are not large enough to motivate adoption in the face of learning and transition costs.

34.2.2.2 Key Factors Affecting Economic Viability of CA Relative to Conventional Farming Practices

Four key factors influence the economic profitability of CA relative to conventional farming practices.

First are factor endowments and factor costs. Resource-poor households, without oxen or tractors, are constrained to manage their farms with human labor or hired services. Peak-season labor constraints—particularly at weeding and planting time—have historically limited the farm productivity and output of resource-poor rural households in Africa (Cleave 1974; Ruthenberg 1980; Collinson 1983; Lee et al. 2006). Given the strong seasonality of labor demands in Africa's semi-arid zones, where a single rainy season drives agricultural calendars, labor bottlenecks occur during the first month of the season when the rains arrive and when conventional tillage can begin (Fig. 34.1). As CA involves dry-season (non-peak season) tillage, the viability of CA for any given household depends critically on the opportunity cost of family labor during the off-season. Zambia's cotton farmers, who earn half as much dry-season nonfarm income as other farmers, appear to self-select for CA, in part, because of the low opportunity cost of dry-season household labor (Haggblade et al. 2011). Access to draft oxen or mechanical tillage likewise shapes farm household options. Because farmers who own their own oxen plant first, resource-poor households that rent animal traction services inevitably gain access late and plant late—as much as 4 weeks later in Zambia (Haggblade and Tembo 2003). Given yield losses of 1–2 % per day a farmer delays planting

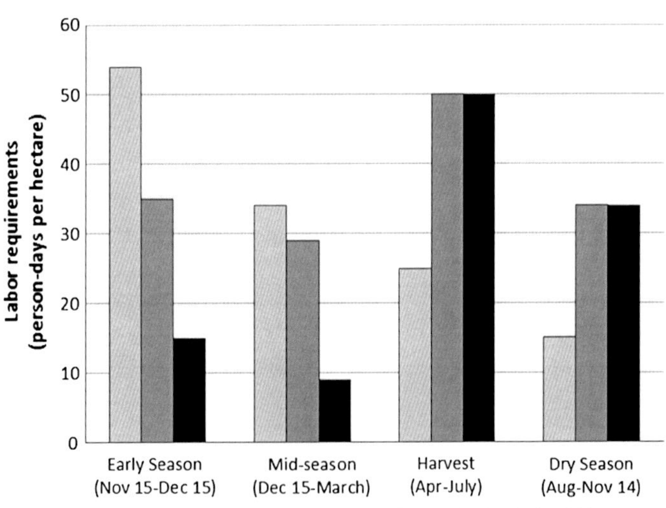

Fig. 34.1 Seasonality of labor requirements in hand hoe maize production in Zambia, under conventional tillage (full soil inversion) and conservation farming (minimum tillage in basins). *Source* Haggblade et al. (2011)

after the first planting rains, this considerable delay severely compromises the viability of any activity relying on borrowed or rented oxen (Howard 1994, Keyser and Mwanza 1996; Nyagumbo 2008; Haggblade et al. 2010).

Second, yield differences relative to conventional practices clearly influence the relative profitability of CA compared to conventional tillage and management. These yield differentials vary quite substantially across locations (with typically higher CA performance in semi-arid and erratic rainfall areas) and across years (with CA yield margins generally higher in sporadic rainfall years and lower in heavy rainfall years; Thierfelder et al. 2015b; Thierfelder and Wall 2009, 2010). Yield also depends critically on planting date, typically favoring CA because dry-season minimum tillage enables early planting with the first rains, as outlined above. Farmer discipline and management skills likewise influence the productivity of any knowledge-intensive management system, including CA (Franzel et al. 2001; Haggblade et al. 2010). Some studies suggest that CA success often depends more on management than on location (Gatere et al. 2013).

Third, input costs influence the relative profitability of CA. Because minimum tillage reduces energy requirements compared to full soil inversion, conservation agriculture yields significant cost savings for land preparation when animals or tractors are used. Zambian commercial farmers first became interested in conservation agriculture because it cut their fuel use dramatically, from 120 to 30 L per hectare (Hudson 1995; The Farmer 1995). On average, other studies suggest a roughly 15 % cost advantage of CA due to reduced fuel costs (Pannell et al. 2014). Herbicide prices tend to work in the opposite direction. Because of increased weed pressure in many settings, at least in the early years of CA adoption, herbicide availability and pricing become an important determinant of the feasibility and viability of CA production. Recent work among cotton farmers in Zambia suggests that herbicide costs account for 10–20 % of purchased input costs under CA (Haggblade et al. 2011). In general, locations with high fuel prices and low herbicide costs tend to favor CA.

Finally, the opportunity cost of fodder matters. Fodder, like grain, has value. In heavy livestock areas, fodder sometimes accounts for as much as 10–20 % of the value of the crop itself (Pannell et al. 2014). Since CA requires crop residue retention, this reduces the value of fodder output for sale or for direct livestock consumption. Because much of the empirical work on CA has omitted fodder valuation, available figures may provide an upward bias in published estimates of CA profitability relative to conventional tillage.

34.2.2.3 Probable Impact of Climate Change, on These Key Factors

Climate change appears likely to influence each of these key parameters (Table 34.3). Most important and well established is the impact of climate change on yield variability. Although climate projections suggest that the volume of rainfall will increase in some areas and decrease in others, the projections generally agree that all areas will see increased variability in the timing and volume of rainfall. To

Table 34.3 Implications of climate change, on the economic viability of conservation agriculture (CA)

Factors affecting the economic viability of CA	Probable impact of climate change
(a) Factor costs (seasonal labor requirements, animal traction, mechanized equipment)	Increased cost of livestock (heat stress, drought, disease, declining pastures)
(b) Input costs (fertilizer, herbicides, fuel)	Increased fuel prices (carbon taxation induce taxing of fossil fuels to reduce CO_2 emissions)
(c) Output value (crop yield, fodder yield, prices of each)	Increased variability of yield

the extent that CA helps farmers ride out increasingly prevalent dry spells (through mulching and increased SOM build-up), CA may help to moderate the yield instability that will increasingly plague conventional farming.

Energy costs may likewise increase. Animal traction, a primary source of farm energy in the semi-arid (non-tsetse fly) zones of Africa, may face increasing production costs given that most climate change scenarios project global warming, increased heat stress on livestock, increasing drought (and flooding), increased animal diseases and pressure on the natural pastures on which most African livestock depend. All of these factors conspire to raise livestock mortality rates, reduce herd size for cattle (though potentially increased for goats where cropping becomes marginal) and increase animal traction costs. This will tend to favor CA as less energy-intensive than conventional tillage with oxen and within CA to favor hand hoe and tractor ripping over ox-ripping.

Related fuel surcharges being considered as part of global carbon taxation will likewise tend to increase petroleum fuel costs. This increase will, in turn, favor CA because of the reduced energy and, thus, the fuel demands of minimum tillage systems. The following section aims to quantify these changes and to estimate roughly how climate change, may alter the relative economic viability of CA compared to conventional management systems.

34.3 Empirical Analysis of the Economics Conservation Agriculture

34.3.1 Data and Methods

Following Pannell et al. (2014), this chapter evaluates the economics of CA relative to conventional tillage using micro-level empirical data to simulate output differentials and input costs over time. Because some benefits (and some costs) of CA emerge slowly over time, we compute the net present value (NPV) of various management packages over a 10-year time horizon.

The analysis compares outcomes for two household groups. First are a large set of resource-poor farm households without a full set of plowing oxen and that depend on household labor or rental markets to cultivate their land. Second are a smaller group of well-off households with access to draft oxen and no cash constraints limiting input access. Within each category, the analysis compares CA and conventional technologies currently used in southern Africa—hoeing, basins, plowing and ripping by oxen or tractor. Given the rapid recent uptake of herbicides (Grabowski and Jayne 2016), the analysis compares the CA technologies with and without herbicides. Table 34.4 summarizes the technologies compared, while a series of annex tables (available on the authors' website) provide the detailed parameters used in the analysis.

Data for this exercise center on semi-arid zones of Zambia, Malawi and Zimbabwe, drawing on a collection of farm-level field investigations and on-farm experimental trials. The base scenario uses data from Zambia, where possible, including maize and groundnut input and output prices from 2010 (Haggblade et al. 2011), milk prices from 2011 (Common Fund for Commodities, CFC 2013) and

Table 34.4 Alternate technology packages compared

Technology packages		Cropped area (ha)		
		Maize	Legume	Total area
Resource-poor households				
Conventional tillage				
1	Hand hoe, full inversion	0.9	0.1	1
2	Ox plowing, rental	0.9	0.1	1
Conservation agriculture				
3a	Basins, hand weeding	0.67	0.33	1
3b	Basins, herbicides	0.67	0.33	1
4a	Ox ripping rental, hand weeding	0.67	0.33	1
4b	Ox ripping rental, herbicides	0.67	0.33	1
5a	Tractor ripping rental, hand weeding	0.67	0.33	1
5b	Tractor ripping rental, herbicides	0.67	0.33	1
Well-equipped households				
Conventional tillage				
6	Ox plowing	2.3	0.2	2.5
7	Tractor plowing, rental	4.8	0.2	5
Conservation agriculture				
8a	Ox ripping, hand weeding	1.67	0.83	2.50
8b	Ox ripping, herbicides	1.67	0.83	2.50
9a	Tractor ripping rental, hand weeding	3.35	1.65	5
9b	Tractor ripping rental, herbicides	3.35	1.65	5

Source Micro-simulation model baseline values, based on current farming practices reported in Haggblade et al. (2011)

rainfall data from Chipata from 1984 to 2002 (Nathan Moore, personal communication 2012). Information that was not available for Zambia or that does not change by country was taken from nearby countries, including the ratio of grain to residues for maize (Gopal Alagarswamy, personal communication about results from the Decision Support System for Agrotechnology Transfer (DSSAT) model 2012) and for groundnuts (Homann-Kee Tui et al. 2015), the on-farm experimental yield effects of CA on maize and legumes (Thierfelder et al. 2015a, b), the yield effect associated with early planting (Haggblade and Tembo 2003), the opportunity cost of residues in terms of livestock production (Homann-Kee Tui et al. 2015; Thorne et al. 2003) and the cost of fencing (Pannell et al. 2014).

34.3.2 Heterogeneity of Outcomes Without Climate Change

Even without climate change, the economic viability of CA varies considerably—by household group (by asset holdings and seasonal opportunity cost of labor), by location (as prices of fodder, the prevalence of foraging cattle, climate and input prices all vary) and over time (as yields and prices of key inputs fluctuate, particularly petro-chemical inputs).

The baseline differentials in Table 34.5 illustrate these differences. Within the large group of resource-poor farm households, conventional tillage with borrowed or rental oxen (T2) emerges as consistently the worst option. Very late land preparation and planting dates reduce yield, while rental charges raise costs. As a result, conventional tillage with rented oxen (T2) returns consistently negative returns (Table 34.5). Conventional hand-hoe production (T1) offers a better option, although returns hover just below zero when the opportunity cost of family labor is valued. The results in Table 34.5 value the opportunity cost of peak-season labor at \$2/day while non-peak season labor is valued at \$0.50/day. These results suggest that conventional hand hoe production generates returns just sufficient to remunerate family labor at these low levels.

The hand hoe CA packages come out best in economic terms, particularly the basins with herbicides to control weeds (T3b). This CA package generates \$275 in net returns, over and above the opportunity cost of family labor, as a result of yield gains over conventional tillage and redeployment of heavy tillage labor to the slack agricultural season (Fig. 34.1). These CA-based basin packages prove particularly viable for households with a low opportunity cost of dry season labor and for households with good access to herbicides. For this reason, private cotton companies in Zambia and tobacco companies in Mozambique and Malawi now make herbicides available to their farmers.

Various CA rental options now available in Zambia involve hiring ripper services from either ox- or tractor-based service providers. Though less viable than CA basins because of the increased land preparation costs, these technologies (T4 and T5) offer returns in excess of either conventional tillage options. Sensitivity

Table 34.5 Net present value (NPV) of alternate management packages, with and without climate change

Technology		Without climate change			With climate change		
		NPV		Probability maize yield < 1 ton/ha	NPV		Probability maize yield < 1 ton/ha
		Year 1	Sum Y1-10		Year 1	Sum Y1-10	
Resource-poor households							
Conventional tillage							
1	Hand hoe, full inversion	−7	−38	0.01	−32	−159	0.10
2	Ox plowing, rental	−35	−178	0.01	−73	−367	0.14
Conservation agriculture							
3a	Basins, hand weeding	35	246	0.00	12	128	0.00
3b	Basins, herbicides	41	275	0.00	18	157	0.00
4a	Ox ripping rental, hand weeding	17	153	0.00	−6	34	0.00
4b	Ox ripping rental, herbicides	23	182	0.00	0	63	0.00
5a	Tractor ripping rental, hand weeding	25	204	0.00	7	111	0.00
5b	Tractor ripping rental, herbicides	31	233	0.00	13	216	0.00
Well-equipped households							
Conventional tillage							
6	Ox plowing	213	1071	0.01	52	656	0.09
7	Tractor plowing, rental	400	2013	0.01	88	1099	0.09
Conservation agriculture							
8a	Ox ripping, hand weeding	237	1384	0.00	103	1098	0.00
8b	Ox ripping, herbicides	252	1456	0.00	109	1170	0.00
9a	Tractor ripping rental, hand weeding	332	2056	0.00	150	1484	0.00
9b	Tractor ripping rental, herbicides	361	2201	0.00	161	1629	0.00

Source Micro simulations over 10-year period

analysis, not reported here, confirms that several key factors influence these rankings, most notably the opportunity cost of dry season labor and fodder values.

Among well-equipped households, the CA-based ripping technologies using animal traction (T8) generally surpass ox-plowing by about 30 % as a result of yield gains from early planting, soil organic matter build-up and improved water infiltration and retention. Herbicides generally prove more profitable than hand weeding (compare T8b with T8a and T9b with T9a), though this result will clearly vary by location and with wage rates for hired labor.

34.3.3 How Does Climate Change Affect the Economics of CA?

Under climate change, the profitability of all farming packages declines as a result of declining average yields. The relative economic attractiveness of various technologies remains unchanged from the baseline scenario discussed above.

The primary difference due to climate change arises from the increased volatility of weather patterns and, thus, of yield outcomes (Fig. 34.2). Conventional packages (T1, T2, T6 and T7) all become significantly riskier under climate change, with the probability of yield falling by over half (to 1 ton/ha) rising from a negligible 1 % to roughly 10 %. Under the various alternative CA management systems, mulching, soil organic matter build-up and more-timely planting combine to reduce riskiness and, thus, lowering the probability of crushing financial losses.

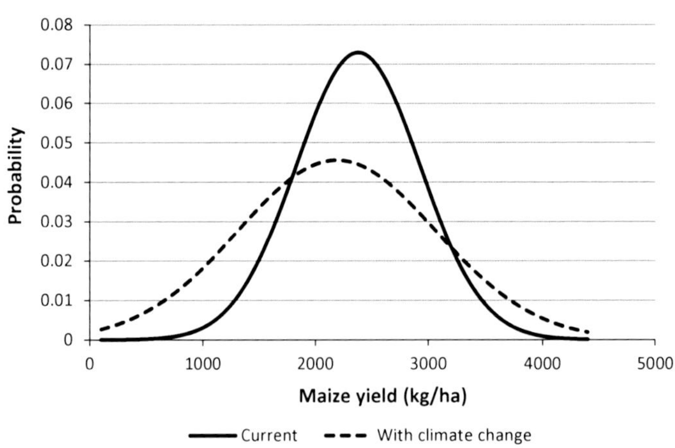

Fig. 34.2 Changes in the distribution of yield outcomes with climate change. *Source* Authors' calculations

34.4 Discussion

The analysis suggests favorable incentives for CA use as it has the potential to reduce yield variability from climate change through gradual improvements in soil fertility, infiltration and water holding capacity. One critical assumption for this analysis is that farmers will be able to find a way to retain control over their crop residues. Where the fodder residue values are average or lower, farmers with CA experience are likely to find it beneficial to leave a significant portion of their residues on their fields as mulch to control weeds and reduce soil surface temperatures. A major constraint to achieving this is the community norm of free-range grazing once the maize harvest is complete that occurs throughout most of southern Africa.

Unless communities decide to change the norms for dry season livestock grazing, CA will either require fencing (which is costly) or will be relegated to remote areas where livestock pressure is lowest. There is evidence that farmers will change these norms when there are adequate financial incentives to do so. For example, livestock grazing norms in southern Malawi changed when pigeon pea (which matures late into the dry season) became the primary cash crop (Zulu et al. 2015). Similar changes occurred in the Kenyan highlands with the rise of feed production (a significant portion of which are legumes) for dairy cattle. This transition occurred more gradually in peri-urban zones along the Kenyan coast (Swallow 2000). Thus, one strategy would be to target CA efforts to places with reasonable prospects for pigeon pea production, intensified dairy production or other financially attractive enterprises that can help motivate changes in dry season grazing norms. Where legumes are part of that strategy, soil fertility benefits would be expected as well.

Legume markets remain an underappreciated but crucial determinant of CA technology viability and adoption rates. In most current systems, the legume rotation component of CA remains the least widely adopted. Because of thin markets, farmers in Zambia prefer to produce only 0.1–0.2 hectares of legumes, just enough for their household needs (Grabowski et al. 2014). Where legume markets remain fragmented and thin, this limits adoption of all three CA components because it dampens incentives to produce legume surpluses for sale. As a result, Zambian cotton companies are beginning to supply soybean seeds and inputs as well as guaranteed soybean purchases for their contract farmers in an effort to stimulate expansion of legume production and improved crop rotation practices (Grabowski et al. 2014). If these pilot efforts prove successful, they will substantially improve incentives for full CA adoption in the region.

Furthermore, farmers who have learned to use CA effectively may be better able to respond to climatic shocks because their experience learning how to dramatically change their agricultural management practices increases their capacity to adapt and thrive in a changing system. CA is knowledge intensive, and effective CA promotion is likely to require building farmers' understanding of agro-ecology and what they can do to be productive despite climatic uncertainty. The application of the ecological term "resilience" to social-ecological systems (Folke 2006) has many applications for smallholder agriculture in Africa. Resilience is characteristic of a system that is able to recover to its original state or transform to a new productive state in response to disturbances. Mitigating the effects of climate change, on smallholder farmers in Africa requires improving the resilience of the production system (such as through the development of production technologies that are less sensitive to variable rainfall) and the broader social-ecological farming system where these production technologies are embedded.

34.5 Conclusions

Farmer agronomic practices clearly govern outcomes, shaping long-term soil quality and short-term farm output. Long before concerns about climate change attained common currency, CA has helped to highlight the importance of farmer management practices in improving outcomes with any given set of available input technologies and production conditions. As demonstrated in our analysis, farmers benefit from CA by overcoming labor constraints at planting time and by following basic agro-ecological principles to improve soil fertility and soil water-holding capacity. These benefits require complex changes in crop management as well as shifts in the community norms for residue management.

With the advent of climate change, weather-induced shifts in cropping conditions will likely amplify the need to further adjust on-farm agronomic practices across a wide array of locations. Given that the specifics of feasible farming packages will continue to change along with a changing climate and resource base, one of CA's cardinal long-term contributions may prove to be its consistent focus on the importance of agronomic practices and experimentation in improving smallholder farming conditions in Africa. The challenges associated with a changing climate highlight the importance of investing in building farmers' capacity to learn and innovate.

Annex: Tables

For the annex tables, please visit www.msu.edu/~grabow21.

References

Andersson JA, D'Souza S (2014) From adoption claims to understanding farmers and contexts: a literature review of conservation agriculture (CA) adoption among smallholder farmers in southern Africa. Agric Ecosyst Environ 187:116–132. doi:10.1016/j.agee.2013.08.008

Baudron F, Jaleta M, Okitoi O, Tegegn A (2014) Conservation agriculture in African mixed crop-livestock systems: expanding the niche. Agric Ecosyst Environ 187:171–182. doi:10.1016/j.agee.2013.08.020

Baudron F, Tittonell P, Corbeels M, Letourmy P, Giller K (2012) Comparative performance of conservation agriculture and current smallholder farming practices in semi-arid Zimbabwe. Field Crops Res 132:117–128

Boko M, Niang I, Nyong A, Vogel C, Githeko A, Medany M, Osman-Elasha B, Tabo R, Yanda P (2007) Africa. Climate Change 2007: Impacts, adaptation and vulnerability. In: Parry M, Canziani O, Palutikof J, van der Linden P, Hanson C (eds) Fourth assessment report of the intergovernmental panel on climate change. Cambridge University Press, Cambridge, pp 433–467

Common Fund for Commodities (2013) The success story of smallholder dairy in Zambia: final report of the CFC/FIGMDP/14 strengthening the productivity and competitiveness of the smallholder dairy sector in Lesotho and Zambia (2007–2011). Common Fund for Commodities, Amsterdam

Cleave J (1974) African farmers: labor use in the development of smallholder agriculture. Praeger Publishers, New York

Collinson M (1983) Farm management in peasant agriculture. Westview Press, Boulder

Crawford E, Kelly V, Jayne T, Howard J (2003) Input use and market development in Sub-Saharan Africa: an overview. Food Policy 28(4):277–292

Dambiro J, Xavier F, Vasco B, Azito M (2011) Introducing conservation agriculture in the Quirimbas National Park of Cabo Delgado, Northern Mozambique. Paper presented at the 5th world congress of conservation agriculture incorporating 3rd farming systems design conference, Brisbane, September 2011

Dreschel P, Gyiele L (1999) The economic assessment of soil nutrient depletion: analytical issues for framework development. International Board for Soil Research and Management, Bangkok

Erenstein O, Sayre K, Wall P, Hellin J, Dixon J (2012) Conservation agriculture in maize-and wheat-based systems in the (sub) tropics: lessons from adaptation initiatives in South Asia, Mexico, and Southern Africa. J Sustain Agric 36:180–206

Folke C (2006) Resilience: the emergence of a perspective for social–ecological systems analyses. Sci Technol 16:253–267

Food and Agriculture Organization of the United Nations (2001) The economics of conservation agriculture. Food and Agriculture Organization, Natural Resources Management and Environment Department. FAO Corporate Document Depository. http://www.fao.org/DOCREP/004/Y2781E/Y2781E00.HTM

Franzel S, Cooper P, Denning GL (2001) Scaling up the benefits of agroforestry research: lessons learned and research challenges. Dev Pract 11(4):524–534

Friedrich T, Derpsch R, Kassam A (2012) Overview of the global spread of conservation agriculture. Field actions science reports. J Field Actions (Special issue 6)

Gatere L, Lehmann J, DeGloria S, Hobbs P, Delve R, Travis A (2013) One size does not fit all: conservation farming success in Africa more dependent on management than on location. Agric Ecosyst Environ 179:200–207. doi:10.1016/j.agee.2013.08.006

Giller K, Witter E, Corbeels M, Tittonell P (2009) Conservation agriculture and smallholder farming in Africa: the heretics' view. Field Crops Res 114:23–34. doi:10.1016/j.fcr.2009.06.017

Grabowski P, Kerr J (2014) Resource constraints and partial adoption of conservation agriculture by hand-hoe farmers in Mozambique. Int J Agric Sustain 12:37–53

Grabowski P, Jayne T (2016) Analyzing trends in herbicide use in sub-Saharan Africa. International development working paper, Michigan State University, East Lansing

Grabowski P, Haggblade S, Kabwe S, Tembo G (2014) Minimum tillage adoption among commercial smallholder cotton farmers in Zambia, 2002 to 2011. Agric Syst 131:34–44

Haggblade S, Kabwe S, Plerhoples C (2011) Productivity impact of conservation farming on smallholder cotton farmers in Zambia. Working paper 47, Food Security Research Project, Michigan State University, Department of Agricultural, Food, and Resource Economics, Lusaka

Haggblade S, Tembo G, Kabore D, Reij C, Ajayi O, Franzel S, Mafongoya P, Place F (2010) Sustainable soil fertility management systems. In: Haggblade S, Hazell P (eds) Successes in

African agriculture. Johns Hopkins University Press and IFPRI, Baltimore and Washington, pp 262–319

Haggblade S, Tembo G (2003) Development, diffusion and impact of conservation farming in Zambia. Working paper 8, Food Security Research Project, Michigan State University, Department of Agricultural, Food, and Resource Economics, Lusaka

Homann-Kee Tui S, Valbuena D, Masikati P, Descheemaeker K, Nyamangara J, Claessens L, Erenstein O, van Rooyen A, Nkomboni D (2015) Economic trade-offs of biomass use in crop-livestock systems: exploring more sustainable options in semi-arid Zimbabwe. Agric Syst 134:48–60. doi:10.1016/j.agsy.2014.06.009

Howard J (1994) The economic impact of improved maize varieties in Zambia. Dissertation, Michigan State University

Hudson J (1995) Conservation tillage. Profit, April

Jayne T, Chamberlin J, Headey D (2014) Land pressures, the evolution of farming systems, and development strategies in Africa: a synthesis. Food Policy 48:1–17. doi:10.1016/j.foodpol. 2014.05.014

Kassam A, Friedrich T, Shaxson F, Pretty J (2009) The spread of conservation agriculture: justification, sustainability and uptake. Int J Agric Sustain 7:292–320

Keyser J, Mwanza M (1996) Conservation tillage. Institute of Africa Studies, University of Zambia, Lusaka

Knowler D, Bradshaw B (2007) Farmers' adoption of conservation agriculture: a review and synthesis of recent research. Food Policy 32:25–48

Lee D, Barrett C, McPeak J (2006) Policy, technology and management strategies for achieving sustainable agricultural intensification. Agric Econ 34:123–127

Mazvimavi K, Twomlow S (2009) Socioeconomic and institutional factors influencing adoption of conservation farming by vulnerable households in Zimbabwe. Agric Syst 101:20–29

Mhango W, Snapp S, Kanyama-Phiri G (2013) Opportunities and constraints to legume diversification for sustainable maize production on smallholder farms in Malawi. Renew Agric Food Syst 28:234–244

Milder J, Majanen T, Scherr S (2011) Performance and potential of conservation agriculture for climate change adaptation and mitigation in Sub-Saharan Africa. Discussion paper 6, Ecoagriculture Partners, Washington, DC

Morris M, Kelly V, Kopicki R, Byerlee D (2007) Fertilizer use in African agriculture: lessons learned and good practice guidelines. World Bank, Washington

Mupangwa W, Dimes J, Walker S, Twomlow S (2011) Measuring and simulating maize (*Zea mays* L.) yield responses to reduced tillage and mulching under semi-arid conditions. Agric Sci 2:167–174. doi:10.4236/as.2011.23023

Ngoma H, Mason N, Sitko N (2015) Does minimum tillage with planting basins or ripping raise maize yields? Meso-panel data evidence from Zambia. Agric Ecosyst Environ 212:21–29. doi:10.1016/j.agee.2015.06.021

Nyamangara J, Mashingaidze N, Masvaya E, Nyengerai K, Kunzekweguta M, Tirivavi R, Mazvimavi K (2014) Weed growth and labor demand under hand-hoe based reduced tillage in smallholder farmers' fields in Zimbabwe. Agric Ecosyst Environ 187:146–154. doi:10.1016/j. agee.2013.10.005

Nyagumbo I (2008) A review of experiences and developments towards conservation agriculture and related systems in Zimbabwe. In: Goddard T, Zoebisch M, Gan Y, Ellis W, Watson A, Sombatpanit S (eds) No-till farming systems. World Association of Soil and Water Conservation, special report 3

Pannell D, Llewellyn R, Corbeels M (2014) The farm-level economics of conservation agriculture for resource-poor farmers. Agric Ecosyst Environ 187:52–64. doi:10.1016/j.agee.2013.10.014

Rockström J, Kaumbutho P, Mwalley J, Nzabi A, Temesgen M, Mawenya L, Barron J, Mutua J, Damgaard-Larsen S (2009) Conservation farming strategies in East and Southern Africa: yields and rain water productivity from on-farm action research. Soil Tillage Res 103:23–32

Rodenburg J, Meinke H, Johnson D (2011) Challenges for weed management in African rice systems in a changing climate. J Agric Sci 149:427–435. doi:10.1017/S0021859611000207

Rusinamhodzi L, Corbeels M, van Wijk M, Rufino M, Nyamangara J, Giller K (2011) A meta-analysis of long-term effects of conservation agriculture on maize grain yield under rain-fed conditions. Agron Sustain Dev 31:657–673. doi:10.1007/s13593-011-0040-2

Ruthenberg H (1980) Farming systems in the tropics, 3rd edn. Oxford University Press, Oxford

Schlenker W, Lobell D (2010) Robust negative impacts of climate change on African agriculture. Environ Res Lett 5:1–8. doi:10.1088/1748-9326/5/1/014010

Smaling E, Nadwa S, Janssen B (1997) Soil fertility in Africa is at stake. In: Buresh R, Sanchez P (eds) Replenishing soil fertility in Africa. Special publication 51, Soil Science Society of America, Madison, pp 47–61

Stoorvogel J, Smaling E (1990) Assessment of soil nutrient depletion in Sub-Saharan Africa: 1982–2000. Report 28, Winand Staring Center, Wageningen

Swallow K (2000) Collective action and the intensification of cattle-feeding techniques: a village case study in Kenya's Coast Province. CAPRi working paper 10, International Food and Policy Research Institute, Washington, DC

The Farmer (1995) Editorial. Conservation tillage helps in a drought year. June 15. Supplement

Thierfelder C, Wall P (2009) Effects of conservation agriculture techniques on infiltration and soil water content in Zambia and Zimbabwe. Soil Tillage Res 105:217–227

Thierfelder C, Wall P (2010) Rotation in conservation agriculture systems of Zambia: effects on soil quality and water relations. Exp Agric 46:309–325

Thierfelder C, Mombeyarara T, Mango N, Rusinamhodzi L (2013a) Integration of conservation agriculture in smallholder farming systems of southern Africa: identification of key entry points. Int J Agric Sustain 11:317–330

Thierfelder C, Mwila M, Rusinamhodzi L (2013b) Conservation agriculture in eastern and southern provinces of Zambia: long-term effects on soil quality and maize productivity. Soil Tillage Res 126:246–258

Thierfelder C, Matemba-Mutasa R, Rusinamhodzi L (2015a) Yield response of maize (*Zea mays* L.) to conservation agriculture cropping system in Southern Africa. Soil Tillage Res 146B:230–242. doi:10.1016/j.still.2014.10.015

Thierfelder C, Rusinamhodzi L, Ngwira A, Mupangwa W, Nyagumbo I, Kassie G, Cairns J (2015b) Conservation agriculture in Southern Africa: advances in knowledge. Renew Agric Food Syst 30(4):328–348

Thorne P, Thornton P, Kruska R, Reynolds L, Waddington S, Rutherford A, Odero A (2003) Maize as food, feed, and fertiliser in intensifying crop-livestock systems in east and southern Africa: an ex ante impact assessment of technology interventions to improve smallholder welfare, ILRI impact assessment series 11. ILRI, Nairobi

Twomlow S, Hove L, Mupangwa W, Masikati P, Mashingaidze N (2008) Precision conservation agriculture for vulnerable farmers in low potential zones. In: Proceedings of the workshop on increasing the productivity and sustainability of rainfed cropping systems of poor, smallholder farmers, Tamale, Ghana, 22–25 September 2008

Verhulst N, Govaerts B, Verachtert E, Castellanos-Navarrete A, Mezzalama M, Wall P, Chocobar A, Deckers J, Sayre K (2010) Conservation agriculture, improving soil quality in sustainable production systems. In: Lal R, Stewart B (eds) Food security and soil quality. CRC Press, Boca Raton

Zulu L, Adams E, Chikowo R, Snapp S (2015) Emergence of community-based livestock management institutions to control herbivory for promoting scaling up of pigeon peas in legumes/maize systems in Central and Southern Malawi. Preliminary report, Michigan State University, East Lansing

Chapter 35
Research and Development Priorities

Rattan Lal, David Kraybill, David O. Hansen, Bal Ram Singh and Lars Olav Eik

The conference was a great success. It addressed an important and a timely theme of 4 dimension of sustainability: environmental, economic, social, and institutional. Participation was very enthusiastic and was representative of all regions of Sub-Saharan Africa (SSA) and the donor community. Discussions were objective and of high quality and all goals of the conference were realized. Above all, there were numerous examples of presentations describing innovative research based on some cross cutting issues and cutting edge science. Among innovative research approaches discussed included: (i) the nexus approach and inter-connectivity, (ii) the societal value of soil and the terrestrial carbon (C) pool, (iii) payments for ecosystem services, (iv) the disease-suppressive soils, (v) the soil-centric green revolution, (vi) focus on nutrition-sensitive agriculture, and (vii) relation between human health and soil quality. Therefore, the objective of this chapter is to summarize the salient points of discussion and identify some key research and developmental priorities.

R. Lal (✉)
The Ohio State University SENR-CMASC, Rm 210 Kottman Hall, Columbus, OH, USA
e-mail: lal.1@osu.edu

D. Kraybill
AEDE/CFAES, The Ohio State University, Columbus, USA
e-mail: kraybill@iagri.org

D.O. Hansen
College of Food, Agricultural, and Environmental Science, The Ohio State University, Columbus, OH, USA
e-mail: hansen.4@osu.edu

B.R. Singh · L.O. Eik
Norwegian University of Life Sciences, Ås, Norway
e-mail: balram.singh@nmbu.no

L.O. Eik
e-mail: lars.eik@nmbu.no

R. Lal et al. (eds.), *Climate Change and Multi-Dimensional Sustainability in African Agriculture*, DOI 10.1007/978-3-319-41238-2_35

35.1 Megatrends

35.1.1 Population in Sub-Saharan Africa

The population of Sub-Saharan Africa is increasing more rapidly than that of any other geographical region and is predicted to grow from the currently estimated 900 million to about 1.4 billion by 2030, and 2 billion by 2050 (United Nations Population Division 2011), while the region's rural population will continue to grow by almost 50 % between 2015 and 2050. It is expected to stabilize at >3 billion by the end of the 21st century. It is the burgeoning population of SSA that has drastically altered the north-south distribution of the world population. By 2020, 6 of the 8 billion people will live in newly emerging economies and developing countries.

The population increase will result in increased *demand for food and this rising demand for food* in SSA region will need to be met by higher crop yields on existing croplands and/or expansion of land under cultivation (Chamberlin et al. 2014). Since 1970s, crop production growth in this region is primarily a result of area expansion, rather than through increased yield per cultivated area, with an estimated 40 % increase in cultivated land area between 1990 and 2012 (Heady 2015). Also there is a big gap in technology transfer from researchers to smallholder beneficiaries in the region due to week infrastructure and poor and ineffective extension advisory services among others.

These are also the regions where the natural resources (e.g., soil, water, vegetation) are in short supply, prone to degradation, and a major challenge to restore. Furthermore, institutional support remains to be weak, and poor infrastructure is a major challenge that hinders sustainable development.

Yet, SSA is the microcosm of developing countries of the world. This is the region where global issues of the 21st century are the most challenging. Important among these, which must be addressed effectively and urgently include: food and nutritional insecurity including hidden hunger and malnutrition and micronutrient deficiencies; climate change and its manifestations of extreme events and uncertainties; water scarcity and deteriorating quality including eutrophication (algal blooms) and contamination; rapid loss of biodiversity, etc. Approximately 1.7 million (2.8 %) of deaths worldwide are attributed to micronutrient deficiencies caused by lower consumption of fruits and vegetables that is regarded as top 10 selected risk factors for global mortality (WHO 2014) and SSA region gets its bigger share. These priority issues require innovative approaches for an effective and targeted intervention.

Almost 1 billion people are undernourished, particularly in Sub-Saharan Africa (239 million) and Asia (578 million). In developing countries, even if agricultural production doubles by 2050, one person in twenty still risks being undernourished —equivalent to 370 million hungry people, most of whom will again be in Africa and Asia (SOLAW FAO 2011). At the same time, the proportion of adults with a body mass index (BMI) of 25 or greater increased from about 29 % in 1980 to about 37 % in 2013 in both developed and developing countries. Even in Africa overweight/obesity increased from about 17 to over 30 % from 1980 to 2008. There have been substantial increases (about 23 %) in prevalence of either overweight or obese in 2013. Because of the established health risks and substantial increases in prevalence, obesity has become a major global health challenge (WHO 2013) For nutrition to improve and for food insecurity and undernourishment/overweight to recede, future agricultural production of nutritious food crops, both for food security and food safety, will have to rise faster than population growth. This will have to occur largely on existing agricultural land, because the scope of area expansion is rather limited in many countries. Improvements will thus have to come from sustainable intensification that makes effective use of land and water resources as well as not causing them harm. Such a trend would imply agriculture remaining an engine of growth, vital to economic development, environmental services and central to rural poverty reduction (SOLAW FAO 2011).

35.1.2 Land Degradation

Africa's soils are fragile to degradation, especially in dryland areas and hence agricultural productivity and food security in the region are equally vulnerable. Soil fertility is steadily declining due to constant nutrient mining on already underperforming soils (Agriculture for Impact 2014), which decreases soil productivity. Besides nutrient mining, soil degradation occurs in various forms of chemical, physical and biological degradation such as loss of topsoil (erosion by wind or water), loss of organic matter, salinization/alkalization, acidification, pollution, compaction/crusting, waterlogging, and others (Diagana 2003). Investments are, therefore needed to restore degraded areas to a productive state in order to break this cycle, enable the renewed use of old agricultural lands for food production and decrease the demand for crop expansion into previously uncultivated areas (FAO 1998). FAO (2012) a estimated that ~240 million people in the region were vulnerable to food insecurity in 2010, and the number is expected to increase by ~17 million in 2020.

Furthermore the concept of land degradation neutrality (LDN), developed by the United Nations Convention to Combat Desertification (UNCCD) Secretariat, needs to be brought into focus. The concept was initially developed as "zero net land degradation" (ZNLD) as a contribution to the discussions on sustainable development goals (SDGs). The concept would be achieved by reducing the rate of land degradation on the one hand, and increasing the rate of restoration of degraded land on the other (Grainger 2014).

Land degradation neutrality would be achieved when "the amount and quality of land resources necessary to support ecosystem functions and services and enhance food security remains stable or increases within specified temporal and spatial scales and ecosystems" (UNCCD). The goal of achieving LDN was cemented into future sustainability and land restoration targets through the development of SDG

15 (Target 15.3) which proposes the "Protection and promotion of sustainable use of terrestrial ecosystems, to halt desertification, land degradation and biodiversity loss."

35.2 Research and Development Priorities

Several ideas discussed at the conference are indeed innovative and may be extremely effective in addressing natural resource management while achieving sustainable development goals (SDGs) of the U.N.

35.2.1 Sustainable Development Goals of the U.N. and Sub-Saharan Africa

"Among numerous SDG (e.g., eliminating hunger and poverty, empowering women), two of the numerous targets are directly relevant to the deliberations of the conference:

(i) SDG target 2.4: By 2030, ensure sustainable food production systems and implement resilient agricultural practices that increase productivity and production, that help maintain ecosystems, that strengthen capacity for adaptation to climate change, extreme weather, drought, flooding and other disasters and that progressively improve land and soil quality."

(ii) SDG Target 15.3: "By 2020, combat desertification, restore degraded land and soil, including land by desertification, drought and floods, and strive to achieve land degradation neutral world."

(iii) SDG#7 of the U.N emphasizes the importance of environmental sustainability. Specifically it has the following components: 7(A) Integrate the principles of sustainable development into country policies and programs and reverse the loss of environmental resources. 7(B) reduce biodiversity loss; and 7(C) have the proportion of population without sustainable access to safe drinking water and basic sanitation.

35.2.2 Nature-Centric Versus Anthro-Centric Approach

The conference specifically addressed the complex global issues (e.g., climate change, food/nutritional insecurity) through landscape management rather than through soil-based approaches. In this context, adopting nature-centric or bio-centric egalitarian approach would lead to preserving habitat for enhancing bio complexity and restoring ecosystem functions and services. Indeed, human well-being depends on learning to live in harmony with nature, while restoring soils using the landscape approach.

35.2.3 Climate Change and Agriculture

Climate change is a serious global issue and will adversely impact SSA through increase in frequency and intensity of drought. Drought is the most severe and widespread problem in SSA which must be addressed systematically. The problem will be especially severe with the projected population of SSA exceeding 3.1 billion by 2100. Greater warming is expected across all seasons in the 21st century ($\sim$3–4 °C rise by 2099), but with more intensity in central southern Africa and the semi-arid tropical margins of the Sahara (Cairns et al. 2013). Under 2 °C warming, the existing variations in water availability across the region could become more pronounced (World Bank 2013). The "4 pour Mille" Initiative proposed by the French government at COP-21 in Paris (30th November to 12 December) recommends carbon sequestration in world soils at the rate of 0.4 % per year to 40 cm depth. A follow up meeting is planned in Marrakech in September 2016 to develop a modus operandi to implement the "4 per Thousand" initiative. Policy makers, scientists, industry, and the general public must help implement this proposal in SSA to mitigate climate change, restore soil quality, and advance food/nutritional security.

Generally, the region is confronted with several climatic risks that could have far-reaching consequences to its agricultural systems in the future. The length of growing period, which indicates the adequacy of moisture availability, temperature, and soil conditions for crop growth, is projected to decrease by up to 20 % for most parts of SSA by 2050 (Sarr 2012). Moreover, various future climate scenarios such as increased occurrences of dry spells during the growing season depict limited diversification options and livelihood transitions for agro-pastoral systems. Owing to the linkages between agriculture, climate change, and food security, SSA is faced with the challenge of advancing agricultural productivity and food security of the growing population, while mitigating the contributions of agriculture to climate change and maintaining soil resources for the sake of posterity. In this context, adoption of climate-smart agriculture, comprising proven practical techniques, such as mulching, intercropping, conservation agriculture (CA), agroforestry, crop rotation, integrated crop and livestock management, improved grazing, soil and water management, weather forecasting, early warning systems, risk insurance, and livelihood diversification, could be a good research and development strategy. The Global Partnership on Forest Landscape Restoration estimates that over 400 million hectare (M ha) of degraded forest landscapes in Africa offer opportunities for restoring, or enhancing the functionality of "mosaic" landscapes that mix forest, agriculture, and other land uses. Restoration of degraded ecosystems can improve human livelihoods, repair ecosystems, and increase the resilience of both people and landscapes to climate change.

35.2.4 Biogeochemical Cycles at Ecosystem and Landscape Scale

Principal biogeochemical cycles affecting climate change and the net primary productivity (NPP) are those of water (H_2O), nitrogen (N), phosphorus (P), and sulfur (S). These cycles are strongly coupled because of a high inter-connectivity among them. Drastic anthropogenic perturbations of the coupled cycling of these elements in SSA and elsewhere have exacerbated the problems of global warming, water scarcity, and contamination, accelerated soil erosion, and aggravated extinction of some soils etc.

Thus, these researchable themes must be addressed at an ecosystem level. Ecosystem is a natural unit consisting of all plants, animals, and micro-organisms (biotic) in an area functioning together with all of the non-living (abiotic) physical factors. Functional ecosystems are characterized by robust biogeochemical coupled cycling of H_2O, C, N, P, and S (Fig. 35.1).

Fig. 35.1 Vist by Minister Le Foll to Ohio State University to discuss '4 pour Mille' initiative

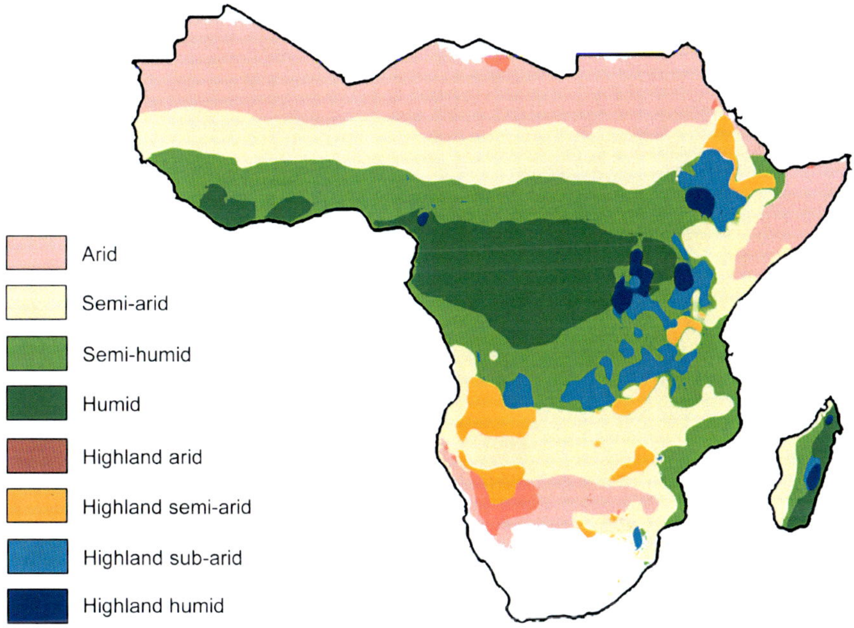

Fig. 35.2 Principle biomes of Sub-Saharan Africa. Redrawn from the data of HarvestChoice (2014)

Ecosystem is a biological community interacting with biotic and abiotic environments through energy transformations and biogeochemical cycling. Biome is similar to an ecosystem. It is a climatically and geographically defined area of ecology with similar climate conditions (e.g., tropical rainforest, savanna, wetlands). Basic studies on biogeochemical cycles etc. must be conducted at ecosystem, biome or landscape level. Principal ecoregions/biome of SSA (Fig. 35.2) includes tropical rainforest, savanna, West African Sahel, East African Highlands, and the desert. These regions must be studied for characterization of their potential and challenges, and systems of sustainable soil use and management identified and implemented. As biological organisms, humans belong to the natural ecosystems. It is the human perception that the natural ecosystem belongs to him/her which has created the problems of "unsustainability".

35.2.5 The Global Commons in Sub-Saharan Africa

Addressing the issue of "global commons" in SSA is important to achieving the SDGs of the U.N. In terms of land tenure, it is important to recognize that "everybody's property is nobody's property" (Gordon 1954). Aristotle stated (Politics, Book II, Chap. 3) "what is common to the greatest number has the least care bestowed upon it. Everyone thinks chiefly of his own, hardly at all of the common interest." Thus, addressing the land tenure issue at village level would be an urgent step forward in SSA.

The problem of "knowledge commons" is another issue that is constraining the scientific progress. Hess and Ostrom (2007) commented on the issue of knowledge commons by using an appropriate analogy, namely, "Two monks were arguing about a flag. One said 'The flag is moving.' The other said 'The wind is moving.' The Sixth Patriarch Zeno happened to be passing by. He told them 'not the wind, not the flag, but the mind is moving'."

35.2.6 Environmental Sustainability and Climate Change

The term ecological implies "interdependence of elements within a system". Thus, ecological sustainability implies "meeting human needs without compromising the health of ecosystems". However, it is important not to forget that "Homo Sapiens" are one of the 8.7 million species on Earth, and wellbeing of other inhabitants of planet Earth must not be compromised.

Environmental sustainability involves the maintenance of natural capital. Therefore, it is closely linked with economic sustainability, institutional sustainability, social sustainability, and agricultural sustainability. These interlinkages are shown in Fig. 35.3.

A strategy to advance sustainable development goals of the U.N. is to address interconnectivity of different manifestations of sustainability. Understanding interconnectivity is essential to achieving SDGs of the U.N. John Muir, a naturalist, stated that "when we try to pick out anything by itself, we find it hitched to everything else in the universe."

Human's inability to work with nature by fitting its activities into nature's cycles perturbs planetary systems and leads to degradation trends and unsustainability of the specific land use or managed systems. To be sustainable, therefore, human must work in harmony with nature and create symbiotic systems. Simply put, sustainable systems are those which: (i) replace whatever (e.g., nutrients) is removed or harvested, (ii) respond prudently to whatever is changed, and (iii) predict what may happen from anthropogenic and natural perturbations and anticipate measures that would enhance resilience. Therefore, the strategy is to strengthen processes which strengthen resilience and restore planetary cycles (e.g., H_2O, C, N, P, and S). The goal is to provide ecosystem services. In this context, sustainable systems have: (i) capacities similar to those of the natural ecosystems to maintain favorable life support systems (e.g., microclimate, water renewability and quality, air quality), (ii) attributes that regenerate renewable resources (e.g., solar energy, water, fish, timber) and (iii) enhance livability, beauty, and aesthetic value of the environment for all biota including human.

Therefore, climate-resilient agroecosystems have similar attributes that enhance resilience such that: (i) productivity is enhanced and sustained even under changing

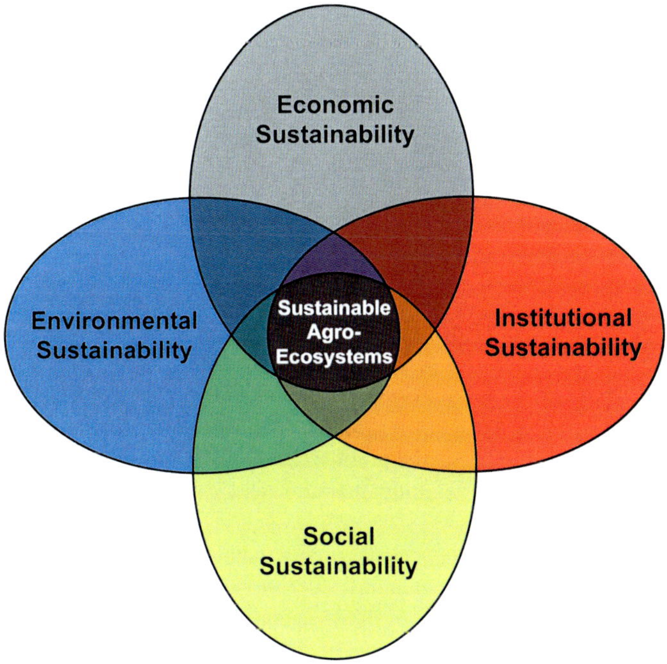

Fig. 35.3 Interlinkages among major dimensions of sustainability

and uncertain climate, (ii) soil quality is restored and soil C budget is positive, and (iii) gaseous emissions (especially of CH_4 and N_2O) are reduced or minimized.

Resilient or adaptable systems must conserve water in the root zone (green water), moderate soil temperature and maintain it within the optimal range (20–30 °C), accentuate root growth and proliferation, improve biotic activity and species diversity, and accentuate agronomic productivity and improve nutritional quality of products for better human and animal health and wellbeing.

35.2.7 Economic Sustainability and Climate Change

Climate change brings about systemic shifts in natural systems, affecting yields, revenues, and costs of food production, and therefore finding and implementing sustainable climate-smart agriculture requires an appreciation for the interrelatedness of the assets under the command of the farm household. Food is produced by millions of smallholders in Africa, each of whom faces his or her own calculus of financial costs and returns of mitigation and adaptation. Farming is hard work and, unless farmers perceive a material gain from new ways of doing things, they are unlikely to adopt them. Technologies and practices to make agriculture sustainable in the face of climate change must reduce farmers' costs per unit of output or, at a minimum, avoid increasing per-unit costs. Otherwise, even the most beautiful technological solutions will not be adopted widely enough to result in a sustainable food economy.

Farmers operate at any given time with a particular set of assets that yield particular livelihood outcomes (food for home consumption, cash income, raw materials for farm-based value-added activities, etc.). The productivity and value of agricultural household assets are influenced by climate change. These assets include land and water resources (natural assets), equipment and materials (physical assets), knowledge and skills (human capital), cash and credit (financial capital), and relationships and networks (social capital). Agricultural households are in danger of exposure to climate shocks so severe that their livelihood assets lose value or are destroyed. If the stock of livelihood assets falls below a critical threshold, the household may become irreversibly trapped in poverty. For example, natural and physical capital can diminish to the point where households cannot afford to invest in education or healthcare, ensuring that the next generation will be as poor or poorer than the parents.

Public and private investments can prevent poverty traps, creating prosperity paths on which households accumulate livelihood assets that lead to improved standards of living and attainment of human capabilities, even in the face of climate change. Both public and private investments are needed because the costs and benefits of mitigation and adaptation for society differ from those of the individual. Where individuals cannot appropriate the format benefit of developing and disseminating new technologies and practices, total economic gain can be increased through public investments. On the other hand, the private sector is generally better at generating innovations and producing goods and services that governments provide. Thus, the greatest good can be achieved by a mix of public and private investment that leads to desired outcomes.

Conservation agriculture (CA) has long been viewed by many scientists as an environmentally sustainable method of farming. CA involves minimum tillage, mulching and persistent groundcover, and crop rotation. But, despite its ecological benefits, CA has been adopted by relatively few farmers in Sub-Saharan Africa, probably because of costs that have been inadequately appreciated by the scientific community. The review paper on CA by Grabowski and Haagblade in this volume concludes that CA is most financially viable in locations and years in which weather conditions are most severe. Additional research is needed to identify how particular weather conditions interact with particular CA regimes to make CA profitable for farmers.

Drip irrigation is a method of reducing water requirements and is particularly relevant as a water source for high-value row crops such as vegetables and fruits. Much of the research on drip irrigation has focused on emitter and storage technologies and costs of both have consequently dropped. In addition, appropriate utilization methods can reduce costs. In this volume, Mahinda, Gachene, and Kilasara examine the financial benefits of alternative water application regimes for farmers using drip irrigation, finding that water application twice per day yields greater benefits, particularly in dry locations, than single applications. More research is needed to find ways of still further reducing the cost of purchasing and maintaining drip irrigation equipment and on alternative management methods for use of the equipment with the aim of making drip irrigation much more widespread than it is in Africa today.

Micro dosing techniques and use of slow release fertilizers for high value crops is another innovative approach to economize fertilizer use on one hand and reduce environmental impacts associated with fertilizer on the other. Furthermore, integrated nutrient management (INM) by effectively using easily available organic residues in combination with chemical fertilizers needs to be adopted. These are areas of research, despite considerable available information, requires that more attention be given to developing and validating the technologies on smallholders farms under changing climatic conditions. Genetic improvements in crops and animals can mitigate some of the causes of climate change and facilitate adaptation to climate change. Genetic improvements require investment by the public and private sectors in gene banks, genomic analysis, and breeding, as well as investments by farmers in improved seeds and animals. In general, the development of new varieties and breeds requires large amounts of financial capital and careful planning of research systems. With planning and capital, the negative consequences of climate change can be reduced. More research is needed on genetic improvements as well as on the economic benefits of targeting particular crops and breeds and on means of disseminating plants and animals that are genetically improved to cope with climate change. Genetic improvements in crop cultivars are also needed to select or breed cultivars with high micronutrient density to enhance the nutritional quality of food and fodder crops for human and animal consumption.

Even the best laid agricultural plans are risky because of weather variability, which is exacerbated by climate change. Seasonal weather forecasts, focused on forecasting weather anomalies, can reduce farmers' downside risk. Makaudze (2009) reports that profitability of farmers using seasonal forecasts in Zimbabwe is greater in bad-weather years compared with that of farmers' not using forecasts. However, it's difficult to know whether the higher profitability is the result of the forecast information leading to improved management or whether farmers who use forecasts are systematically better managers. Research to sort out the possibility of selection bias is needed. Furthermore, improved seasonal forecasting data and methods are needed in Africa.

Risk in agriculture can also be reduced through the use of crop insurance. Creating crop insurance contracts that are sustainable for insurers has proven difficult for several reasons, one of which is the problem of "moral hazard", whereby those who are insured have reduced incentive to exert effort to avoid loss. For this reason, most crop insurance schemes indemnify farmers for differences between predicted and actual values of an index that serves as a loss proxy than cannot be directly influenced by the farmer rather than for the individual farmer's actual losses. A chapter by Flatnes and Miranda in this volume analyzes two index insurance contract schemes in Tanzania. The more innovative of the two contracts uses satellite imagery to lower the cost of data gathering. Based on plot-level data, the authors determined that the index captures 60 % of farmers' actual losses and they estimate that at least 30 % of farmers are likely to sign on to a contract in the absence of subsidies. More research is needed to improve the accuracy of crop indexes, particularly those derived from remote sensing data, and to design contracts that farmers will buy to reduce weather risk.

Another reason why crop insurance has often failed to be sustainable is the high cost of transacting with individual farmers. To lower transactions costs, crop insurance can be linked to agricultural credit in such a way that in bad-weather years, insurers indemnify the creditor rather farmers directly. More research on the theory and actual adoption of interlinked insurance and credit schemes is needed to hone the contracting process so that farmers benefit and so that insurance funds are adequately replenished.

35.2.8 Social Sustainability and Climate Change

As has been amply illustrated in the contributions to this volume, the various dimensions of sustainability discussed are highly interrelated. Unlike the economic and environmental dimensions, social sustainability is less well defined, and indeed has been defined in a number of different ways. Based on the papers which are found in this volume, social sustainability should consider at least three major objectives, the first being meeting basic needs and related equity considerations; the second being the need to change normative patterns of behavior at different levels of society in order to meet bio-physical environmental goals; and the third being how to sustain and work from existing patterns of interaction, prevalent values and traditions (Vallance et al. 2011). Regarding social sustainability, particular attention has been given to the need to consider the social determinants of adaptation to climate change with a focus on the governance, wealth and economic development, technology, information skills infrastructure, institutions and equity (Sen 2000).

Several important topics that merit further research when dealing with social sustainability in the face of climate change were highlighted in this volume.

(a) The first is the critical role that communities and local settings play to ensure social sustainability in attempts to adapt to change at this level. Fortmann (2016) argues that many alternatives proposed by governmental agencies are not sustainable because they do not consider specific needs and circumstances of local communities. She further asserts that research done on how particular communities adapt to specific impacts of climactic change events can prove useful to other communities undergoing similar traumas.
(b) An important equity consideration is the need to study the equity of inter and intra generational equity of resource sharing and how these may be impacted by major climate events. Increased inequality can lead to conflicts and destabilization of social organizations, including families and communities.

(c) There is a need to focus on how conflicts over access to land, water and other natural resources resulting from changing climate events are resolved. This includes increased competition for land between farmers and pastoralists who are increasingly dislocated from traditional range areas due to drought (Little and McPeak 2014).
(d) It is equally important to study the interface between federal and regional governmental and local attempts to deal with impacts of climate change, particularly as they relate to the distribution of scarce natural resources. This includes access to water and the role of regional entities, such as Basin Water Offices in Tanzania and local organizations such as water user associations in determining the conditions under which it occurs (Vedeld et al. 2015).
(e) Gender roles are also impacted by climate change and these changes can profoundly impact the capacity of families and communities to adapt to them. Droughts and floods have resulted in male migration to seek additional off farm income resulting in the need for women to assume additional community decision making roles and to take on farm operation decision-making roles (Nombo et al. 2014; McCornick et al. 2013:24).

35.2.9 Institutional Sustainability and Climate Change

Institutional sustainability refers to the ability of institutions to continue to assist impacted populations to adapt to the impact of climate change events and circumstances over time (Kajembe et al. 2016). Institutions can be those at the local level, such as individual families, local communities, or key community functions, such as local government, executive branches, and religious institutions. They can also be relevant regional or national institutions, such as ministries of water, agriculture and local government. A number of important topics for research dealing with institutional sustainability in the face of climate change are apparent in the papers found in this volume (see especially Kajembe et al. 2016).

(a) Institutional evolution and change are particularly important topics when discussing the role of related national programs that are constantly changing and often times dependent on donor preferences. Of relevance are their normative and operational patterns of interaction the institutions at the community level and how these may change in response to challenges brought on by major climactic events.
(b) Institutional legitimacy refers to how institutions assigned responsibility for overseeing the use and distribution of land, water and other production resources maintain their legislative, executive and judicial authority regarding resources impacted by climate change; Important to consider are the perspectives of individuals directly affected by climate change and the broader societal environment that defines what the institution should be and how it should function.

(c) Bricolage refers to the process of constructing and borrowing different existing institutional elements in order to create different frameworks for decision-making and practices (Cleaver 2002). It is a process that has been followed by different populations in adapting to climate change, both at the community and at the regional and national levels.
(d) Institutional performance refers in a general sense to the extent to which existing institutions succeed in achieving goals at the lowest costs (Ostrom 2003). In regard to climate change impacts, it also relates to how well current institutions address the complex dimensions of access to impacted resources. Several key topics in this regard are management of conflicts arising over the use of scarce natural resources such as water and land; the distribution these resources; related policy formulation; and maintenance of related infrastructure).

35.3 Conclusions

The objective of this conference was to review the state-of-the-science, discuss strategies to implement the proven technological options, identify key knowledge gaps, and prioritize research needs. An important and an urgent issue is to translate knowledge into action at the community level, and identify key policy interventions to promote the adoption of best management practices. The BMPs recommended must ensure agronomic and environmental sustainability. Yet, these goals can only be met if technology is also economically sustainable. A necessary pre- requisite for the latter is institutional and social sustainability.

The Green Revolution of the 1960s bypassed Sub-Saharan Africa. There are numerous reasons for the stagnation or even decline in agricultural production in several regions of Sub-Saharan Africa between 1960 and 2015. Most important among these are weak institutional support, poor infrastructure and the lack of qualified human resources needed for effective transfer of technology. These constraints must be alleviated, especially with regards to the human capital development.

There is also a strong need for a continuous dialogue between the scientific community and the policy makers. In addition to the knowledge per se, political will power is essential to bring about the much needed transformation. No change can happen unless the community to be changed is willing and actively involved in bringing about the much awaited change.

The conference also highlighted several important technical approaches to increase agricultural sustainability in the fact of climate change.

- Genetic improvements in crop cultivars are needed to select or breed cultivars with high density of micronutrients to enhance the nutritional of food and fodder crops for human and animal consumption.

- Micro dosing techniques and use of slow release fertilizers for high value crops can be implemented to economize fertilizer use on one hand and reduce environmental impacts associated with fertilizer on the other.
- Integrated nutrient management (INM) needs to be adopted because it makes effective use of easily available organic residues in combination with chemical fertilizers.

References

Agriculture for Impact (2014) No ordinary matter: conserving, restoring and enhancing Africa's soils. Montpellier panel report, Dec 2014

Cairns JE, Hellin J, Sonder K (2013) Adapting maize production to climate change in Sub-Saharan Africa. Food Secur. doi:10.1007/s12571-013-0256-x

Chamberlin J, Jayne TS, Heady D (2014) Scarcity amidst abundance? Reassessing the potential for cropland expansion in Africa. Food Policy 48:51–65

Cleaver Frances (2002) Reinventing Institutions: bricolage and social embeddedness of natural resources management. Eur J Dev Res 14(2):11–30

Diagana B (2003) Land degradation in sub-Saharan Africa: What explains the widespread adoption of unsustainable farming practices? Draft working paper, Department of Agricultural Economics and Economics, Montana State University, Bozeman, MT

FAO (1998) Soil fertility initiative for sub-Saharan Africa. In: Proceedings of the SFI/FAO consultation. World soil resources report no. 85, Rome

FAO (2012) The state of food security in the world 2012. FAO, Rome

Fortmann L (2016) Climate change and social sustainability: a case for polycentric sustainabilities. Paper presented at the climate change and multi-dimensional sustainability in African agriculture conference. Hilux Hotel, Morogoro, Tanzania, 3–5 June 2014

Grainger A (2014) Is land degradation neutrality feasible in dry areas? J Arid Environ 112(Part A):14–24

Gordon HR (1954) The economic theory of a common property resource: the fishery. J Polit Econ 62:124–142

HarvestChoice (2014) AEZ (16-class, 2009). International Food Policy Research Institute, Washington, DC, and University of Minnesota, St. Paul, MN. http://harvestchoice.org/data/aez16_code

Heady DD (2015) The evolution of global farming land: facts and interpretations. The Poverty, Health and Nutrition Division. The International Food Policy Researdh Institute (IFPRI), Washington, DC. http://fsg.afre.msu.edu/fsp/Headey_ICAE_Land_Paper_v3.pdf. Accessed 29 Sept 2015

Hess C, Ostrom E (2007) Understanding knowledge as a commons: from theory to practice. MIT Press, Cambridge

Kajembe, Pal Vedeld GC, Babili IH, Silayo DS, Mosha DB (2016) Institutional sustainability in the face of climate change: empirical insights from irrigation Institutions in Iringa Rural District, Tanzania. Paper presented at the climate change and multi-dimensional sustainability in African agriculture conference. Hilux Hotel, Morogoro, Tanzania, June 3–5

Little ZD, McPeak JG (2014) Resilience and Pastoralism in Africa South of the Sahara, with a Particular Focus on the Horn of Africa and the Sahel, West Africa. Paper presented at the 2020 Conference Paper 9, Washington, USA

Makaudze E (2009) Do Seasonal Forecasts and Crop Insurance Really Matter for Smallholder farmers in Zimbabwe? Using Contingency Valuation and Remote Sensing Techniques. VCD Verlag Dr. Muller, Germany

McCornick P, Smakhtin V, Bharati L, Johnston R, McCartney M, Sugden F, Clement F, McIntyre B (2013) Tackling change: future-proofing water, agriculture and food security in an era of climate uncertainty. International Water Management Institute, Colombo
Nombo CI, Kabote SJ, Mamiro DP, Synnevag G, Mattee AZ, Urassa JK, Matata L (2014) Adaption to climate change: changing gender relations in the Meatu and Iramba districts in Tanzania. In: Sustainable intensification to advance food security and enhance climate resilience in Africa. Springer, Heidelberg, pp 587–599
Ostrom Elinor (2003) How types of goods and property rights jointly affect collective action. J Theor Polit 15(3):239–270
Sarr B (2012) Present and future climate change in the semi-arid region of West Africa: a crucial input for practical adaptation in agriculture. Atmos Sci Lett 13:108–112. doi:10.1002/asl.368
Sen A (2000) Development as freedom. Oxford University Press, Oxford
SOLAW FAO (2011) The state of the world's land and water resources for food and agriculture. Summary report. FAO, Rome
Vallance S, Perkins HG, Dixon J (2011) What is social sustainability? A clarification of concepts. Geoforum 42:342–348
Vedeld P, Liheluka E, Mbeyale GE (2015) Institutional water resources management and livelihood adaptation: a case from Kilombero rural areas, Tanzania. Paper presented at the climate change and multi-dimensional sustainability in African agriculture conference, Hilux Hotel, Morogoro, Tanzania, June 3–5
United Nations Population Division (2011) World population prospects: the 2010 revision. UN Population Division; United Nations Population Division, New York
WHO (2013) Global, regional and national prevalence of overweight and obesity in children and adults 1980–2013: a systematic analysis. Lancet 384(9945):766–781. http://www.ncbi.nlm.nih.gov/entrez/eutils/elink.fcgi?dbfrom=pubmed&retmode=ref&cmd=prlinks&id=24880830
WHO (2014) Global strategy on diet, physical activity and health. World Health Organization report. Geneva, Switzerland, pp 14–18
World Bank (2013) Turn down the heart: climate extremes, regional impacts and the case for resilience. World Bank, Washington

International Conference on Climate Change and Multi-dimensional Sustainability in African Agriculture: June 3–5, 2015

Summary of Discussions

Session 1: Multi-dimensional Sustainability

Safeguarding environmental, economic, social and institutional sustainability: A balancing act

Environmental and Economical Sustainability

- SSA: One billion people in 2015, 25 % of people food insecure
- Population increase: 2.1 billion in 2050 and 3.8 billion in 2100
- Human dignity: Meeting the needs of current and future generations
- Sustainable intensification: "Producing more from less" while enhancing resilience of managed ecosystems
- 300–400 % yield increase needed to meet the demand of the projected future population
- Many pathways to sustainable agricultural intensification (conservation agriculture, biochar, agroforestry, desert control, pasture management, water harvesting)
- Important to monitor environmental effects of agricultural activities and farming systems
- "Soil stewardship" vital to maintain soil's carbon content and productivity
- Climate change, a market failure. Important to include non-economic variables in economic planning.
- Natural capital generates ecosystem services, which are vital to sustainability of all other systems, including the economy; therefore, the economy should be viewed as a subsystem of natural systems
- Emerging paradigm of economic sustainability: Economic activities of today should be undertaken so as to preserve, replace, or expand capital, including natural capital, to maintain at least the same level of utility or income tomorrow as today
- Accounting for and accommodating the interconnectedness of environmental, economic, and social systems enhances the resilience of each of these systems.

R. Lal et al. (eds.), *Climate Change and Multi-Dimensional Sustainability in African Agriculture*, DOI 10.1007/978-3-319-41238-2

Institutional and Social Sustanibilities

- Important to include both formal and informal structures and their rules in planning, implementing, and evaluating resource-based activities
- Institutions are changing rapidly in Tanzania. Changes may take place from both top and bottom
- Importance of legitimacy and trust in institutions if they are going to work
- Social sustanibilities (plural) should be emphasized and not just one solution brought in from the outside
- Importance of bottom-up approaches in which communities take command over their own destiny
- Consumption in the rich world is not sustainable, still poor communities get the blame for environmental degradation and are also the most vulnerable to the effects of climate change.

Session 2A: Soil Management–Soil Fertility

Major Highlights

Land degradation is happening in Tz

- Need more monitoring

Increasing yield can spare land

- 5 t/ha maize = no expansion

Improved agronomic practices

- Increased yield 49–163 % over farmer practice
- Increased profit by $397 per ha?

Reduce environmental impacts

- Timing and application method.

Common Themes

We *must* increase yields on existing farmland
Location, location, location

- Site specificity is key.

Major Takeaways

Smallholder farmers as entrepreneurs

- Put it in $ (or Tsh) terms

Who should pay for it?

- Consumer?
- Education is needed

Soil analytical capacity and nutrient recommendations

- HUGE need!
- Farmers want to know their soil status
- Detailed nutrient recommendations.

Session 2B: Integrated Systems-Value Chains

Summary of key issues.

Sustainable Seed Delivery Systems—By R.B. Jones

- Consequences of CC on which requires development of crop varieties that are tolerant to effects of CC
- Currently existing seed systems: Formal and Informal
- Four stages in seed production: No breeding; original breeding; strong breeding systems and robust breeding
- AGRA paradigm in seed systems is demand driven
- Seed value chain: Education, breeding, production and delivery
  - Challenge: dysfunction link between public seed systems and dissemination/distribution
- For success in production, balanced crop nutrition is required
- Seed policies and regulations should be right; and
- Seed delivery to be done by the private sector but alignment should be there with public seed systems.

Pro-poor Value Chains in Zanzibar—By Rashid et al.

- Existing value chains in Zanzibar are long and complex: clogged with middle men and exclusion of farmers

- Zanzibar agriculture challenged with drought, low input and traditional production technologies
- Zanzibar livestock sector challenged with lack of animal health service delivery, unavailability of feeds and poor breeding techniques
- Production and market potential exist but market requires delivery reliability and sustainable production and supply of goods/services
- Proposed solution is to develop specific value chains and tap the market (e.g., the Tourism sector)
- Approach: identify stakeholders, train them in value chain issues from production to post harvest and link them to hotels
- Value adding activities focused on quality control, sustainable production, strong communication channels and timely delivery
- Value chain support given-supplementary feed, slaughter and storage facility and transport systems
- Expected output: Emerson hotel in Zanzibar given samples for testing in order to provide feedback to chain developed.

Electronic Smart Subsidies in Tanzanian Agriculture—By Gabagambi

- Controversial subject for discussion as policy makers are hesitant to pilot the program
- Modern input drive production but small holder farmers still do not use them; Why? Expensive!
- Subsidy introduced through NAIVS but still success rate is low as yields are not increasing. Why?
- Inputs did not reach the beneficiaries; smuggling; delayed delivery, and adulteration
- NAIVS challenges-long chains; many hands; temptations; leakage and quality reliability
- Hence develop innovative input distribution system ESSA
- Advantages of ESSA-Use ICT, reduce paper work, link farmers direct to agro-dealers, driven by agro-dealers
- Requirements; GoT to establish input account with BoT, and farmers and agro-dealers MUST be registered
- Potential challenges: New generation of problems; status quo from profiteers of the current system, status of agro dealers, political resistance and GoT reluctance even to pilot it.

Key Questions/Discussion

Sustainable seed system:

- Reliable seed systems under PPP arrangements is necessary for production and distribution of seed and the CC effects.

Pro-poor value chain in Zanzibar:

- Specific value chain development is a potential market for small holder farmers in Zanzibar.

Electronic smart subsidies in Tanzania:

- How is the government ready to accept ESSA and the influence of political capital in input supply system in Tanzania?

Section 3

Session 3A: Genetic Resources

Paper 1: Institutional Aspects of Genetic Resources in Respect of Climate Change in Sub-Saharan Africa

- Genetic resources are central to the strengthening food security and building of a more resilient agricultural system in the face of climate change
- Climate change poses new risk to crop productivity in SSA
- GR have great potential in developing well adapted varieties to the new unstable environment in the face of climate change—this means more demand for GS
- Climate change will also accelerate genetic erosion
- Therefore there is a critical need to collect and conserve endangered species
- No country in SSA is sufficient in Genetic resources
- Most countries in SSA do not have the necessary resources and Institutional Capacity to embark on conservation of GS and maintenance of Gene banks
- Therefore concerted efforts should be put on collection, conservation, sharing and use of Genetic Resources to mitigate the effects of Climate Change.

Paper 2: Adaptation to Climate Changes in SSA: The Role of Genetic Resources and Seed System

- Crop adaptation plays a key role in enabling farmers to adapt to the impacts of Climate Change
- Cultivar adjustment is the most effective on farm strategy
- Formal versus informal seed system s-case of sorghum and maize in Tanzania
- 24 % of seed is from formal seed source and 76 % is from informal seed source
- OPV are recycled for more than 10 years and started to differentiate and undergo adaptation
- Ensure global accessibility of GR=public good
- Introduction of new GR requires evolving seed systems that allows interaction between formal and informal seed system approaches to cultivar development, release and distribution
- Strike a balance between variety protection and farmers' right-integrated approach.

Paper 3: Economic Aspects of Genetic Resources in Addressing Agricultural Productivity in the Context of Climate Change

- Climate change affects both genetic resources and agricultural productivity
- Interaction of CC and non-climate factors will amplify the vulnerability of agricultural system
- Strengthening current conservation of GR is critical for future adaptation of agricultural system to CC
- In order to guide adaptation, planning and investment estimates of costs and benefits of conservation of improved GR should be done
- Access to relevant information will create awareness among the farmers and other stakeholders
- This will lead to right to ownership and access among farmers esp women
- Diverse Genetic resources are important for adaptation efforts in order to enhance Agricultural Productivity
- Concerted efforts at national, sub-regional, regional and International levels to conserving, improving and utilizing Genetic Resources should be stepped-up
- Use of Genetic Resources technologies in addressing Agricultural Productivity should be integrated in breeding, adaptation and development efforts.

Session 3B: Summary of Proceedings

Major Highlights of Session Presentations

1 paper on Organic Farming
2 papers on Conservation Agriculture
1 paper on Crop/Livestock intensification

Major Highlights

Organic Farming is expanding (area and farmers); incorporates all 4 aspects of sustainability (environmental, economic, social and institutional) and seeks to promote human health but has not been researched widely and stimulate further growth.

Conservation Agriculture

Based on herbicides, residue retention, min. tillage and crop rotation and return of OM to the soil is critical for sustainability. However, there are issues to consider—infestation by termites and free grazing of livestock.

SOC stability (improved soil structure, soil hydrology and aeration water retention) and increases root biomass and root volume but results are site-specific.

Sustainable Intensification—Africa Rising

Action research and development (in TZ) using the *Research to Impact Pathway* led to conclusion that greatest impact is recorded by adopting 'blind' testing by farmers themselves followed by selection of what works for them.

Common Themes or Ideas Found in the Presentations

1. All systems (OF, CA, SI) emphasize that return of OM to the soil is critical for soil health and sustainable production
2. All systems have capability to enhance agronomic viability for land use and crop production
3. Economics of both all systems need to be expounded to generate the relevant figures as supporting evidence.

Take Away Messages

1. More research in OF required to generate data to convince farmers/consumers to convert to OF and SUA was challenged to play a leading role in OF research in TZ
2. The contribution of CA to improved productivity should consistently focus on improving agronomic management

3. Research to explore options for reducing the change should involve stakeholders themselves to enhance adoption
4. Biochar has its greatest potential in drought sensitive light soils (sandy) for increasing water retention.

Session 4A: Water Management

The Social Dimension of Water Management and its Consequences in an Era of Declining Water Supply: A Synthesis of Past Research and Future Directions (Mvena)

- **There is a problem of water scarcity** in Tanzania caused by climate change, population increase, unsustainable water consumption patterns—However there is lack of attentions
- Again there is uneven water distribution due to geographical, class (rich use more water in 5 minute shower than the whole day consumptions of poor in rural) and by sector (industry,irrigation, domestic, etc.) use
- Agriculture in general is the main user of water in Tanzania (Lack of conservation people cultivate even in the stream bed, lake Haubi has disappeared), **export of water virtue water**
- **The consequences to humanity**; Less water means: less crops, change of crops, collapse of industries (increased use while water is unavailable), destruction of ecosystems, disease, poverty, water conflicts, dietary shifts (forced to eat sorghum instead of maize, since cattle consume more water then opt for goat or chickens; eat edible insects by FAO etc.).

Hydrological Monitoring of Headwater Catchment for Climate Change Adaptation Studies

- GCMs predict Temperature to increase by at least 2 % in Tanzania by 2050 (worst case); this is predicted to cause production decline of Maize by 13 %, sorghum 8.8 % and rice by 7.6 %
- To move forward a more resilient and sustainable agriculture is needed; our interaction with small holder farmers should focus on dealing with inter annual (natural) climate variability vs focus on climate change
- Understanding the reality of climate variability is essential for resilience and use it to inform management strategies
- Understanding the hydrology of the headwater catchment is essential to define the water resources.

Agricultural Research for Sustainability

- Crop diversity offers many ecosystem and human services
  - climate adaptation, nutrition and food security, IPM, carbon
- Fertilizer trees reduce fertilizer use by 75 %.
- To incentive tree planting, given a choice. Farmers overwhelmingly chose fertilizer subsidy.
- Takes three years to get N results from trees. Big investment up front but pays off later.
- Comment: AGRA—seed voucher program different than fertilizer voucher program that suffers from lack of private sector competition.

Kinds of Research and their Potential Contribution to Sustainability

- Developed indicators of sustainability based on four pillars: environment, economic, social and institutions to compare four universities from 2000 to 2014 —ASARECA, MAFC, Makerere, and SUA
- MAFC scored highest, SUA second
- Areas of research
  - Climate change resilience, drought and pest (disease) tolerant crop and animal species, IK for climate change and resilience, natural resource conservation, biotech, ecotourism

Potential Role of Biological Innoculants in Sustainable Food Production in SSA

- Research on innoculants for fertilization and some root borne disease
- Fertilizer too expensive for 91 % of farmers
- Some successes, some failures
- Soil microbial health is associated with pH and SOC
- One important product—NitroSUA, support from iAgri. Performed better than NPK and ammonium sulphate and others
- Lower cost
- Low quality control, need enabling policy environment
- SUA is discussions with ministry regarding regulatory process
- How to work with agrodealers.

AIP for Agricultural Sustainability: The iAgri Case

- AIP—agricultural innovation platform: forum designed to convene key stakeholders and value chain actors
- G-SOKO—tracks movement of farm produce from warehouse on behalf of smallholders in EAC. Farm to buyers
- Supports relationships between academia and industry. Attract co-investment for PPP
- Support business growth among TNZ agribusinesses
- Micro-irrigation systems
- Rainwater harvesting
- IK foods for nutrition and income, restore agr biodiversity
- Malted and fermented flours from sorghum and millet
- Broiler production.

Session 5A: Soil Management—Land Rehabilitation

Institutional Aspects of Land Degradation and Rehabilitation in Africa

Drivers of land degradation

- Poor management, population pressure, insecure land tenure, poor infrastructure, access to market and impact of climate change

Impact of land degradation

- Extreme poverty, food insecurity, etc
- Degraded land should be called **underperforming land** so that it is not abandoned for **degradation of another land**
- Return on invest on land management is **35 times** profitable as compared to other land use.

What is needed

- Creating enabling environment
- Strengthen leadership and political support
- Map soil fertility and vulnerability to degradation and potential for rehabilitation
- Use farmers for transformation
- Land tenure
- Capitalize local knowledge
- Simplify knowledge sharing

- Improve science extension interface
- Priotize adaptation to cc.

Towards restoring more than we degrade

- If 12 % of global degraded land is restored, it could feed 200 million people and rise 40b per year.

Extent of Salt Affected Soils and Their Effects in Irrigated and Lowland Rainfed Rice Growing Areas of South Western Tanzania

Extent of the problem

- Salinity problem is 100 % in all schemes in Tanzania. This has resulted to Land abandonment (loss) of between 5 and 25 % in the scheme and crop loss by 5–100 %.

Causes

- Poor drainage system
- Poor constructions.

Solutions

- Need for provision of drainage system, Rehabilitation of the structure and use of salt tolerant cultivars while embarking on a salt soil management
- Salinity is a problem and there is a need to create an awareness to government officials (responsible)on proper construction and management of irrigation schemes.

Soil and Nutrient Losses and the Role of Gender in Land Degradation in Southwestern Uganda

- Soil erosion is the major cause of declining crop production in Southwest Uganda
- Increase in production has been due to extensification instead of intensification.
- There is decreased of extensification but its due to diminished land
- Production increment are due to largely increase in conversion of wetlands and forests into small scale agriculture
- Indigenous knowledge has shown promising and interesting observations
- Due to change of cover/ degradation now people are highly adopting planting of trees
- Increasing production per unit area should be of more importance

- Improving indigenous knowledge in addressing degradation is more appropriate approach
- Most of soil conservation were reported to be too laborious and costly to women while they are responsible for contributing to food security by 70 %.
- While men own most of land shares, their contribution to land conservation/management is very limited resulting in escalation of the land degradation problem in Southwestern Uganda.

Session 6: International Year of Soils (IYS) 2015

Highlights of Session Presentations

- Introduction to the IYS2015 by Chairperson, Stefan Schlingloff FAO, Technical Officer, Land and Water Division.
- Environmental issues of measuring agricultural sustainability—by PK Nair, UF USA
  - Work in multidisciplinary team to achieve sustainable agriculture
  - Agroforestry as an opportunity for sustainable agriculture—to bridge the divide btn ecology and economy—science base technology.
- Soil—a limited resource under changing climate—by Bam Ram Singh NMBU Norwegian Society of Soil Science
  - All life depends on soil—Soil is crucial in global problem—threatened by climate change—Soil multi-functionality—Yet, soil is a limited resource and takes long time to restore.
- Global Issues and IUSS
  - Introducing to the establishment of the IYS2015 and the World Soil day celebrated on 5th Dec every year
  - The history of soil management-from the beginning of Earth to impact on human development (Anthropocene); presentation of IUSS structure and functions in support of Soil science with multiple role (from technical to outreach to influence policy)
  - Soil science in the future—think beyond the traditional practice (eg. Peri-urban agriculture, soil-less production).
- Soils and Climate Change (CSA) by Stefan Schlingloff FAO, Technical Officer, Land and Water Division
  - Soil is the origin and solution to Climate change issues; Sustainable land management practices are instrumental for adaptation of agriculture, Soils are important sink for C, and Climate Smart Agriculture (CSA) addresses both, Adaptation and Mitigation; Information available at FAO on CSA, Projects and database.

Common Themes or Ideas from Session 6

- Soils—central in sustainable development in the face of global challenges—Climate
- Change—CSA/Climate-resilient sustainable agriculture (CRSA)
- Land degradation—SSA is the most vulnerable
- Multi-functionality of Soil—food and nutrition, water, air, energy, ecosystem services
- Soils should receive enough attention it deserve—protection, capacity building
- Soils—be considered in multi-disciplinary/sector approach to address global challenges.

Takeaway Message

Soils Sustain Life

Awareness Creation

- Soil is a scarce valuable resource
- Soil should not be taken for granted—Take care of the soil for it to take care of you
- Soil should be part and parcel in personal, local, national and international development agenda
- Soil scientists and partners—Advance Soil science as discipline of study and be integrated in addressing global challenges such as Climate Change and food security.

Private Sector Solutions for Sustainability

Understanding the Role of Informal Sector in Climate Smart Agriculture (CSA)

Highlights

- CSA hinged on 3 pillars: Sustainability, Adaptation and Reduce greenhouse emissions
- Approach: Climate smart varieties
  - Change in farming schedules
  - Mitigation where possible
  - Climate information services

- Actors: Independent traders [middle (wo)man] and transporters Market and value chain organizations.

Common themes

- Private Sector introduces synergistic effects to climate change adaptation, making it more holistic
- Sustainable intensification: increased production; less area cultivated; less degradation
- Thrust: Productivity, Safety and the Environment
- Need for strong regulatory structures to ensure safety and fair play.

Major takeaways

- Harness the power of communication for pricing and risk avoidance
- Smallholder farming tend to "mine" the soil
- In case of forest replanting, more focus on varieties than numbers
- Climate change presents an opportunity for wealth creation
- There is a need for a vehicle to propagate skills to the farmers, PPP model bridges this gap
- Chemical fertilizers and organic manure to play complementing roles
- Private sector imparts a pushing effect to increased production
- Private sector to supply inputs and markets
- Conducive environment necessary for private sector to take root.

Session 8A: Extension Systems for Agricultural Sustainability

Major Highlights of the Session

- The need for sustainable agriculture calls for a vibrant extension system that will address the climatically changed scenario to take care of the rural poverty, poor nutrition and low productivity.
- Extension framework is very complex and needs to take into consideration policy environment, the farming system, the market and the communities themselves. In other words there is a need of looking at the entire agricultural innovation system.
- There is thus a need to reposition the extension systems to meet the current challenges which has to do with the feeding of increasing population and curb hunger and malnutrition. In order to achieve that we have to go for pluralistic extension provision both from the public and private sectors.
- Private sector extension is an investment in the business to achieve new markets, greater efficiency and high profitability. Actors in the food value chain need to invest in extension in order to get quality materials and at the same time keep or maintain the customer.

Common Themes/Ideas

- Few technological advances have been adopted and unfortunately this is normally leveled against extension despite the fact that other institutions control the resources and also make various decisions including policy ones.
- We need to improve on policy, technology, market accesses and organization of farmers. For this to be achievable and **sustained** we need combined efforts both in the public institutions as well as the private sector. In short pluralistic extension services are needed to address the concerns of a diverse clientele.
- Opportunities for extension services are increasing because of the growing food markets due to population growth as well as increased urbanization.
- There has been a serious concern with the implementation of the extension programs since they do not necessary start with the farmers needs and rarely address the technological and social capacity requirements.

Major Takeaways

- Communicate for behavioral change and allow full participation of the beneficiaries
- Venture into new opportunities of extension provision by use of ICT such as mobile phones
- Combine efforts by public and private extension service providers
- Extension agents need to be educated in new techniques and new frontiers such as biotechnology
- Improve on the policies especially that which address technology and market access
- Collaborations between all actors at various levels is important to realize effective extension service.

Session 9A: Landscape Approaches

Presentation

- Managing Landscape for Environmental Sustainability
  - Landscape is being degraded

    anthropogenic and natural processes
  - Need for strategies for ecosystem and landscape restoration

    Conserving biodiversity and integrating multiple functions
    Conserving and managing water resources (maximizing green water)
    Improving ecosystem functions and services
    Promoting ecological restoration

  - Africa needs well thought solutions as "Soil problems in Africa aren't temporary, the fixes can't be either"

- Integrated Management in Africa: Synthesis of Findings from a Continental Review:
- Characteristics of Integrated Landscape Management
  - multiple landscape benefits
  - Land management practices should lead to multiple landscape objectives
  - addressing synergies and trade-offs among land use objectives
  - markets, policies and programs should achieve multiple objectives
  - inclusive governance through collaborative processes
- Need for
  - Adaptive collaborative management
  - Long-term research partnership between landscape initiatives and Universities on systems, impacts and governance aspects of landscape
- Upgrading Legacy Soil Maps for Climate Resilient Agriculture: A Case of Kilombero Valley in Tanzania
  - Need to update the 1959 legacy map
  - Combined field and laboratory work to develop a detailed digital map to provide spatial soil information crucial for decisions needed to adapt agriculture for climate change
- Transboundary Agro-ecosystem Management Programme for the Kagera River Basin: Implementing SLM best practices with catchment approach
  - Increasing pressure on Rresource base and Ecosystems
  - Introduced a number of success SLM techniques such as bench terraces, water retention pits, contour ditches
  - Lessons for successful adoption and scaling up

    Participatory approach involving all stakeholders
    Commitment by stakeholders to changing behaviour
    techniques resulting in improving livelihoods
    Support and Incentives
    Cooperation between Government and partners.

Session 9B: International, National and Local Policies for Agricultural Sustainability

Major Highlights

Regional and National Policies on Sustainable Agriculture

- Member states different
- No regional policy on agriculture
- Countries to not have climate change policies, they are at different stages of development
- COMESA is harmonizing policies in the region

Exploring Greenhouse Gases and Transformation in Changing Climate

- Agric. Intensification increase GHG emission
- Transformation (practical, political and personal) is necessary to reduce impact of changing climate

The Role of Agro-ecology in Climate Change and Mitigation

- Agro-ecology can mitigate CC
- Agro-ecology practice s are advantageous but no policy support
- Challenges of scaling out agro-ecology practices include policies which do recognize agro-ecology.

Common Themes

- Agricultural intensification and climate change
- Climate change/CA policies.

Major Takeaways

- We can reduce the impact of climate change if we change the way we do things
- Working together is more beneficial then working individually. We need to formulate strategies engaging stakeholders to collaborate
- Let us push for policies that will promote practices that reduce impact of CC.

Conference is a Success

- Objectives achieved
- High scientific quality

- Thematically relevant presentations
- Good participation
- Objective discussions
- Consensus on several issues.

Examples of Research Innovation

- The Nexus Approach: Inter-connectivity,
- The Societal Value of Soil and Terrestrial Carbon Pools,
- Payments for Ecosystem Services (e.g., carbon sequestration, water quality, biodiversity),
- Disease-Suppressive soils,
- Soil-Centric Green Revolution,
- Nutrition-sensitive agriculture,
- Improving human health by enhancing soil quality, and
- Promote urban and peri-urban agriculture.

Some Interesting Ideas

- Kilimo Kwanza (Uttam Kheti)
- Sustainable transformation
- Beating hunger through business and science
- Creating wealth through climate change
- Assumption that farmers do not know?

Anthro-centric Versus Nature-centric

- Global issues (e.g., climate change, biodiversity, water quality) can be addressed through landscape restoration via ecological approaches
- Adopting nature-centric or bio-centric egalitarian approach would lead to preserving habitat for enhancing bio-complexity and restoring ecosystem functions and services
- Human wellbeing depends on learning to live in harmony with nature, and restoring **soils** using the landscape approach.

Everybody's Property is Nobody's Property (Gordon, 1954)

"What is commons to the greatest number has the least care bestowed upon it. Everyone thinks chiefly of his own, hardly at all of the common interest."

Aristotle (Politics, Book II, Chap. 3)

"The Knowledge Commons"

Two monks were arguing about a flag. One said, "the flag is moving." The other said, "The wind is moving." The Sixth patriarch, Zeno, happened to be passing by. He told them, "Not the wind, not the flag; mind is moving."

Charlotte Hess and Elinor Ostrom (2007)

Index

A
Abiotic stresses , 9, 440
Adaptation, 24–27, 31, 33, 37, 38, 65, 80–82, 84, 89, 91, 242, 243, 247–249, 251, 253–255, 412, 422, 426
Adaptive capacity, 308
Adoption, 378, 379, 386, 388–390
Afforestation, 160, 163
Agricultural expansion, 75, 78
Agricultural intensification, 488, 499
Agricultural model inter-comparison and improvement project (AgMIP), 262, 277
Agricultural productivity, 173, 175, 178–181
Agriculture, 242–245, 253, 254, 468–473, 475, 478, 479
Agroforestry, 155, 157, 159–162, 165
Agroforestry systems, 555, 557, 561
Agroforestry trees, 554, 557
Attributes of environmentally sustainable systems, 7

B
Balanced nutrition, 310
Basis risk, 569–573, 585, 588
Biodiversity, 367, 371, 375, 376
Biological inoculants, 640, 641, 650, 651, 653
Bulk density, 188, 190, 191, 196, 198

C
Carbon footprint, 132, 134, 135
Carbon sequestration, 370, 376, 383, 385
Catchment approach, 397, 399, 402, 409
Catena, 190
C3 crops, 260
Climate change, 172, 174–176, 180, 182, 242–250, 252–255, 412, 413, 417, 422, 425, 427, 428, 433–436, 438–441, 444, 449, 468, 469, 475, 476, 478, 479, 660, 662, 664, 665, 667, 668, 670–672, 674
Climate change and agriculture, 19, 434, 683
Climate resilience, 436
Climate-smart, 508, 513, 521
Climate-smart agriculture (CSA), 434–436, 445, 447, 449, 450
Community based, 47, 160, 397, 401
Coffea arabica, 554, 555
Conditions for institutional sustainability, 38
Conservation agriculture, 660, 661, 663–665, 667
Coupled human and natural systems, 17, 18, 21
Crop adaptation, 328, 331, 332, 335, 337–339
Crop residues, 661, 663, 673
Crop varieties, 331, 339

D
Deforestation, 150, 155, 158, 162, 163
Desertification, 62–67, 82–84
Developing countries, 317, 319
Digital soil mapping, 347, 348, 351, 355
Drip irrigation, 228, 230, 231, 233–235
Drought, 4, 10, 591–593, 597, 599, 600, 602, 606–608
Drought resistant crops, 208
DSSAT, 262

E
Ecological economics, 369, 370
Ecology–economy conflict, 365
Economic aspects, 172, 175, 178, 180, 181
Economic returns, 228, 229, 230, 231, 233, 235
Economic sustainability, 13, 14, 364, 621, 622, 634, 686, 687, 694
Economic value, 594, 599, 600, 605
Ecosystem services, 151, 369, 371, 372, 376, 378, 382, 383, 390
El Niño, 591, 592, 598, 600, 602, 608
Environmental footprint

R. Lal et al. (eds.), *Climate Change and Multi-Dimensional Sustainability in African Agriculture*, DOI 10.1007/978-3-319-41238-2

Environmental sustainability, 433
Extension, 468–470, 472–479

F

Farmer field schools, 394, 476
Farmer options, 247, 253
Food insecurity, 64, 67, 70, 72, 74
Food security, 308–310, 312, 316, 323, 434–438, 448

G

Genetic resources, 172, 174–177, 179–181, 328, 330, 333, 335, 340, 341
Global commons, 685
Goat's milk, 507–510, 512–521
Good governance, 33, 34, 38

H

Human nutrition, 511
Hydraulic conductivity, 188, 190, 196, 198

I

Incentives, 407
Income security, 162
Index insurance, 569–575, 577–580, 585–588
Innovative solutions, 425
Inorganic fertilizers, 488
Institutional bricolage, 31, 36, 38
Institutional legitimacy, 32, 36, 39
Institutional performance, 28, 32, 33, 37
Institutions, 24, 25, 26–32, 34–39
Integrated assessment models, 19, 21

K

Kilombero valley, 348, 361

L

Land degradation, 62–67, 69, 70, 79, 86–88, 150, 151, 155, 158, 159, 161, 163–165, 640, 654, 681, 682
Land degradation neutrality (LDN), 65, 66, 69, 70, 74, 75, 78, 79, 81–84, 86, 87, 89
Land management, 187, 189, 193, 197
Land restoration, 64, 66, 69, 70, 72, 73, 75, 78, 79, 82, 84–86, 88
Landscape management, 83, 91, 682, 710
Legacy soil maps, 347, 361
Legumes, 661, 670, 673
Livelihoods, 163, 164

M

Machine learning, 353, 356
Macronutrients, 490–494, 499
Microinsurance, 588
Micronutrients, 486, 488, 490, 493, 494, 497–499
Mineral fertilizer, 132, 135
Mitigation, 24, 25, 33
Multiple institutional logics of action (MILA), 28, 39

N

Natural capital, 13, 14, 17, 369, 686, 694
Net present value (NPV), 668, 671
Nitrogen-fixing trees (NFT), 372, 384
Nutrient mining, 136, 140

P

Policy, 366, 378, 379, 381, 386, 389, 390
Policy interventions, 10, 692
Population trends, 88, 681
Poverty, 63–65, 67, 70, 73–75, 78, 80, 88
Processes and factors of environmental sustainability, 4
Public-private partnership (PPP), 140, 508

R

Reforestation, 161, 163, 165
Remote sensing, 5, 87, 447, 571, 576, 690
Representative concentration pathway (RCP), 263–277
Resource economics, 162

S

Salinity, 101, 103, 105, 106, 108–111, 113, 115, 116
Salt-affected soils, 99, 101–104, 107, 108, 113–115, 117
Salt tolerance, 115, 116
Seasonal climate forecasts, 593, 594, 596, 605
Seed systems, 328, 332, 334, 335, 337, 339–341
Semi-arid areas (ASALs), 228, 229, 235
Shambani graduates Ltd., 506
Smallholder farmers, 592–594, 605, 608, 640, 641, 646, 649, 653
Social, 242–247, 252–254
Social values, 366
Sodicity, 101, 105, 106, 108, 110, 111, 113
Soil erosion, 186, 187
Soil fertility, 136, 146, 147, 196
Soil improvement, 372, 374, 384
Soil quality and health, 7, 8, 713
Sorghum production, 229, 230, 233, 235
Southern corridor, 99, 100, 104, 106, 107, 113–115
Sub-saharan africa(SSA), 432, 434, 435, 436, 444, 447, 449

Sustainable agriculture, 308, 316
Sustainable development goals (SDG), 65, 73, 88, 681, 682, 685, 686
Sustainable intensification, 5, 8, 9, 11, 128, 139, 413, 424, 428, 653, 655
Sustainable land management, 396, 397, 399, 402, 405, 407, 409
Sustainable land use, 164
Sustainable livelihoods framework, 19, 20
Sustainable transformation, 469, 470, 472, 473, 478, 479

T
Tanzania, 260–263, 274, 396, 398, 399, 468, 469, 471–475, 477–479, 507–511, 513, 516, 517, 520, 521
Technology adoption, 386–390, 476, 640
Temperature, 554, 555, 557, 558, 560, 562–564
Transboundary agroecosystems management, 396

V
Varieties, 308, 312, 313, 315, 318–320
Vulnerability, 242, 243, 245, 247, 249, 252, 253

W
Water governance, 243–245, 245–250, 252, 254, 255
Water harvesting, 209
Watering regimes, 231, 235
Water management, 202, 209
Water scarcity, 202, 205, 207, 208
Water use efficiency, 201, 223, 260, 273, 424, 571, 652, 661
Water withdrawals, 203, 205
With forecasts and without forecasts, 591